Cannabis
A Complete Guide

Cannabis
A Complete Guide

Ernest Small

Agriculture and Agri-Food
Ottawa, Ontario, Canada

CRC Press
Taylor & Francis Group
Boca Raton London New York

CRC Press is an imprint of the
Taylor & Francis Group, an **informa** business

Her Majesty the Queen in Right of Canada, as represented by the Minister of Agriculture and Agri-Food Canada

First published 2017 by CRC Press

Published 2019 by CRC Press
Taylor & Francis Group
6000 Broken Sound Parkway NW, Suite 300
Boca Raton, FL 33487-2742

First issued in paperback 2021

ISBN 13: 978-1-03-209740-4 (pbk)
ISBN 13: 978-1-4987-6163-5 (hbk)

Publisher's Note
The publisher has gone to great lengths to ensure the quality of this reprint but points out that some imperfections in the original copies may be apparent.

Library of Congress Cataloging-in-Publication Data

Names: Small, Ernest, 1940- author.
Title: Cannabis : a complete guide / Ernest Small.
Description: Boca Raton : Taylor & Francis, 2017. | Includes bibliographical references and index.
Identifiers: LCCN 2016009739 | ISBN 9781498761635
Subjects: LCSH: Cannabis. | Marijuana.
Classification: LCC SB295.C35 S63 2017 | DDC 633.7/9--dc23
LC record available at http://lccn.loc.gov/2016009739

Visit the Taylor & Francis Web site at
http://www.taylorandfrancis.com

and the CRC Press Web site at
http://www.crcpress.com

Contents

Preface

William Shakespeare (in *Julius Caesar*, Act 4, scene 3) wrote:

There is a tide in the affairs of men
Which, taken at the flood, leads on to fortune;
Omitted, all the voyage of their life
Is bound in shallows and in miseries.
On such a full sea are we now afloat,
And we must take the current when it serves,
Or lose our ventures.

Indeed, society is currently facing a flood of new information and developments about cannabis, the subject of this book, and rather momentous decisions must be made for the future.

"Cannabis" in its broad sense refers to the cannabis plant (*Cannabis sativa*), especially its fiber products (such as textiles, plastics, and dozens of construction materials), edible seed products (now in over a hundred processed foods), psychoactive chemicals (employed both as illicit and medicinal drugs), and all associated considerations. In short, cannabis is a generic term referring to all aspects of the plant, especially its products and how they are used. Concern over illegal drug usage of cannabis has distorted evaluation of all other issues, with polarized camps arguing the merits of their perspective and the inferiority of their opponent's viewpoints. Indeed, it is very difficult to find evaluations of cannabis that are not free of bias, distortion, emotion, and selective consideration of evidence. The literature and especially the Internet are crowded with opposing claims. On the one hand, some proponents insist that cannabis provides the perfect path to economic, ecological, and medical salvation for the world. Others, however, contend that it represents a fraudulent hoax that will deteriorate financial, moral, physical, and societal health.

Science is a search for truth and provides indispensable guidance to society for the creation and adoption of new technologies. Regrettably, scientific research on virtually all aspects of the plant species *C. sativa* has been suppressed for most of the twentieth century, a victim of the sometimes observed tendency to avoid examination of sensitive or sinister subjects. Ignorance, however, generally exacerbates problems and has likely contributed to worsening the substantial harm that has become associated with cannabis. Currently, there is an explosion of interest in marijuana forms of cannabis, in part because of developing medical applications, but also because of increasing tolerance of recreational usage. Nevertheless, governments have long maintained a costly war against the consumption of cannabis, and there is official reluctance to alter the status quo. Although so-called narcotics are widely viewed as intrinsically evil, the leading controlled drug plants have some legitimate, useful applications. Much of the world is now insisting on a reappraisal of both the industrial (nonintoxicating) and drug aspects of cannabis, and indeed, there are promising new applications that deserve to be assessed and, in some cases, adopted. Because *Cannabis* is first and foremost a plant, evaluation of its potential for harm and benefit needs to take account of its botanical nature, about which much remains to be explored. This book is an analysis of the various economic potentials of *Cannabis*, based particularly on its extraordinary biological properties.

Up to a century ago, most people lived on farms and were well acquainted with how critically important crop plants are for human survival. Today, the plants that sustain us are of limited concern for the majority, except for occasional grumbling when shortages result in dramatic increases in the price of particular foods. Nevertheless, cultivated plants are critical for human welfare, and *C. sativa*, once considered a foundational contributor to civilization, has the potential of once again becoming a major contributor to society.

This volume does not examine to any appreciable extent the sociological and political aspects associated with cannabis but does present sufficient historical, cultural, medical, and legal background to provide context for scientific and economic issues. The conflicting claims for medicinal virtues and toxicological vices are examined based mainly on the most recent authoritative scientific reviews. The attempt is made consistently to reflect majority scientific opinion, but the reader will be aware that many aspects of cannabis are controversial. Some of the presentation may shock those who have been conditioned by the past century of negative information and fearmongering about every aspect of cannabis. Some may be offended by the inclusion of graphic details about marijuana usage, but this information is already widely available and is necessary to understand the subject. This book attempts to document both the virtues and vices of what is surely one of the world's most important species.

Aside from the relevance to specialists, the general public should find the presentation attractive because of the huge interest today in marijuana. Unfortunately, society has become so specialized and compartmentalized that most people have limited appreciation of the importance of science to their lives, except when a topic like marijuana becomes sensationalized. This review of cannabis can serve as a vehicle for public education in the realm of science and technology. Indeed, toward the goal of disseminating the important information in this book to a wide audience, the presentation is user-friendly, concise, and well illustrated, in the hope that nonspecialists will find the topics both informative and entertaining.

Acknowledgments

Brenda Brookes (Figure 0.1) has provided invaluable assistance to my research for many years, coauthoring dozens of publications, including three books. Her excellent work is evident throughout this volume in the numerous illustrations she prepared.

My deepest gratitude to the following for careful, insightful, and very knowledgeable reviews of the chapters indicated: Dr. Jace C. Callaway (Chapter 8), Dr. Axel Diederichsen (Chapter 17), Dave Marcus (Chapter 15), Dr. John M. McPartland (Chapters 11, 12, and 13), Steve G.U. Naraine (Chapters 12 and 14), and Dr. Ethan B. Russo (Chapters 11, 12, and 13). Of course, all misinterpretations, errors, and inadequacies are the responsibility of the author.

Sources of illustrations are provided in their captions, except as noted in the following. All scanning electron microphotographs are by Ernest Small and Tanya Antle, unless otherwise acknowledged. Similarly, unless otherwise acknowledged in the figure captions, photos are by Ernest Small. A few drawings without acknowledgments were prepared collectively by the graphics personnel of Agriculture and Agri-Food Canada. Brenda Brookes and Barry Flahey prepared many illustrations, and Ms. Brookes skillfully assembled and enhanced all of the illustrations for publication.

As acknowledged in the captions where they are shown, some figures are reproduced consistent with the following Creative Commons Licenses—CC BY 2.0 (Attribution 2.0 Generic): http://creativecommons.org/licenses/by/2.0/; CC BY 3.0 (Attribution 3.0 Unported): http://creativecommons.org/licenses/by/3.0/; CC BY 4.0 (Creative Commons Attribution 4.0 International license): https://creativecommons.org/licenses/by/4.0/deed.en; CC BY ND 2.0 (Attribution NoDerivs 2.0 Generic): https://creativecommons.org/licenses/by-nd/2.0/; CC BY SA 2.0 (Attribution ShareAlike 2.0 Generic): https://creativecommons.org/licenses/by-sa/2.0/; CC BY SA 2.5 (Attribution ShareAlike 2.5 Generic): https://creativecommons.org/licenses/by-sa/2.5/; CC BY SA 3.0 (Attribution ShareAlike 3.0 Unported): http://creativecommons.org/licenses/by-sa/3.0/; CC BY SA 4.0 (Attribution ShareAlike 4.0 International): https://creativecommons.org/licenses/by-sa/4.0/deed.en; CC0 1.0 (Universal Public Domain Dedication): https://creativecommons.org/publicdomain/zero/1.0/deed.en.

FIGURE 0.1 Brenda Brookes, contributor to this book, in an experimental plantation of *Cannabis sativa.*

Author

Dr. Ernest Small (Figure 0.2) received a doctorate in plant evolution from the University of California at Los Angeles in 1969 and has since been employed with the Research Branch of Agriculture and Agri-Food Canada, where he presently holds the status of Principal Research Scientist. He is the author of 14 previous books, 6 of which received or were nominated for major awards. He has also authored over 350 scientific publications on plants. Dr. Small's career has included dozens of appearances as an expert botanical witness in court cases, acting as an adviser to national governments, presenting numerous invited university and professional association lectures, supervising postgraduate students at various universities, participating in international societies and committees, journal editing, and media interviews. He has been an adjunct professor at several universities and currently has this status at Ryerson University in Toronto. Dr. Small is widely known for his work on hemp and marijuana, which has included development of a standard strain that has been the basis of all licensed medicinal marijuana in Canada for more than a decade, supplying over 100,000 patients. Dr. Small published two previous books and more than 40 research papers on *Cannabis*. He has received several professional honors, including the following: election as a Fellow of the Linnean Society of London; the G.M. Cooley Prize of the American Association of Plant Taxonomists for work on the marijuana plant; the Agcellence Award for distinguished contributions to agriculture; the Queen Elizabeth Diamond Jubilee medal for contributions to science; the George Lawson Medal, the most prestigious award of the Canadian Botanical Association, for lifetime contributions to botany; and the Lane Anderson Award, a $10,000 prize for science popularization, received for "Top 100 Food Plants." His most recent books, *Top 100 Exotic Food Plants* and *North American Cornucopia: Top 100 Indigenous Food Plants*, were finalists for the Botanical and Horticultural Libraries' award.

FIGURE 0.2 The author in his plantation of *Cannabis* in Ottawa, Canada, in 1971.

Executive Summary

In past centuries, the cannabis plant (*Cannabis sativa*) was one of the world's most admired crops, furnishing a range of products often considered indispensable. In recent decades, however, exaggerated fear of the abuse potential of cannabis has resulted in extremely punitive and counterproductive suppression, not just of marijuana consumption but also of the useful industrial (nondrug) values of the cannabis plant and its many products. Worst of all, the search for unbiased scientific knowledge has been drastically curtailed, an egregious example of how political agendas remove some subjects from objective examination. Additionally, human prejudice against recreational drugs in general and relentless condemnation of marijuana in particular have made cannabis a taboo subject. However, over the last half-century, society has become increasingly aware that the evil nature of cannabis has been overstated and that there are potentially invaluable benefits deserving exploration. Very recently, the constraints limiting cannabis research have been loosened, and a tidal wave of research and development has been unleashed. Certainly, there are pros and cons related to the many ways that cannabis can be employed, and intelligent risk/benefit analysis requires knowledge. That knowledge is rapidly becoming overwhelming. This volume attempts to bring together very old and very new information that needs to be considered to best guide the development of cannabis.

Cannabis sativa is best known as the plant source of marijuana, the world's most widely consumed illicit recreational drug. However, it is also extremely useful as a source of stem fiber, edible seed oil, and medicinal compounds, all of which are undergoing extraordinarily promising research, technological applications, and business investment. Indeed, despite its capacity for harm as a recreational drug, cannabis has phenomenal potential for providing new products to benefit society and for generating extensive employment and huge profits. Misguided policies until recently have prevented legitimate research of the beneficial properties of cannabis, but there is now an explosion of societal, scientific, and political support to reappraise, indeed remove, some of the barriers to usage. Unfortunately, there is also a corresponding dearth of objective analysis. Toward redressing the limitation of information, this book is a comprehensive reference summarizing botanical, business, chemical, ecological, genetic, historical, horticultural, legal, and medical considerations that are critical for the wise advancement and management of cannabis.

Cautions

This work presents extensive information gathered from the literature, and some error, omission, and misinterpretation are inevitably incorporated into compilations of this type. Moreover, scientific knowledge concerning the material is rapidly evolving. Liability arising directly or indirectly from the use of any of the information is specifically disclaimed.

The medicinal and nutritional information provided is not intended to replace the medical advice of trained healthcare professionals.

The economic information provided is not intended as investment advice.

Most aspects of cannabis, the subject of this book, are criminalized, albeit there is a general trend to reduce criminal penalties, and the world's prison systems could not possibly house the huge numbers who have consumed marijuana illicitly. Nevertheless, depending on jurisdiction, those who deliberately or even inadvertently transgress the laws and regulations governing cannabis risk arrest, imprisonment, loss of employment, loss of property, and loss of standing in the community. The information in this book is not intended to be used in any way that contravenes the legal system of any jurisdiction. Indeed, given the enormous personal costs associated with becoming a victim of criminal prosecution, it is necessary to exercise extreme caution in any kind of association with cannabis.

Value judgments and opinions regarding several of the topics discussed in this book are currently the subject of contentious debate and disagreement. Indeed, intellectual freedom to analyze and express points of view is of greatest value precisely when issues are so controversial. While the book attempts to maintain a dispassionate, objective perspective, it is not possible to represent every viewpoint in a manner that will satisfy all, or even a majority, of interested parties. In particular, the author's assessments regarding medical, social, ethical, commercial, and criminal aspects should not be regarded as necessarily reflecting the views of any government, publisher, agency, or individual contributing in any way to this work. Indeed, it is safe to say that no one will agree with everything in this book, and everyone will have different ideas about what is correct.

1 Introduction

THE CANNABIS PLANT

Cannabis sativa, best known as the source of marijuana, is the world's most recognizable, notorious, and controversial plant. As befits a species that has captured the world's attention, it is impressive in appearance (Figure 1.1). While the structure of plants may seem much simpler than that of animals, the architectural adaptations of *C. sativa* are very complex and are cleverly designed to carry out a wide variety of functions (Chapter 6). Cannabis plants vary enormously in height depending on environment and whether selected for stem fiber (the tallest kind), but are typically 1–5 m tall. Simmonds (1976) stated that hemp has been known to grow to 12 m in height, but it should be kept in mind that, as discussed later, other plants called "hemp" sometimes grow to such heights and are often confused with *C. sativa*. The main stalk is erect, furrowed (especially when large), with a somewhat woody interior, and it may be hollow in the internodes (portions of the stem between the bases of the leaf stalks). Although the stem is more or less woody, the species is frequently referred to as an herb or forb (an herbaceous flowering plant that is not grass-like, i.e., not like grasses, sedges, or rushes). Both herbs and forbs are defined as lacking significant woody tissues, so these terms are not really accurate. As discussed in this book, in many respects, deciding on appropriate terminology for cannabis is contentious.

"CANNABIS"—A COMPREHENSIVE TERM

"Cannabis" in its broad sense refers to the cannabis plant, especially its psychoactive chemicals (employed particularly as illicit and medicinal drugs), fiber products (such as textiles, plastics, and dozens of construction materials), edible seed products (now in over a hundred processed foods), and all associated considerations. In short, cannabis is a generic term referring to all aspects of the plant, especially its products and how they are used.

Biologists and editors conventionally italicize scientific names, such as *Homo sapiens*. Italicized, *Cannabis* refers to the biological name of the plant (only one species of this genus is commonly recognized, *C. sativa* L.). Nonitalicized, "cannabis" is a generic abstraction, widely used as a noun and adjective and commonly (often loosely) used both for cannabis plants and/or any or all of the intoxicant preparations made from them. In this book, "cannabis" is employed in its broadest sense, as explained in the previous paragraph.

THE WIDESPREAD MISUNDERSTANDING THAT MARIJUANA IS "FLOWERS" OF *CANNABIS SATIVA*

"Herbal marijuana" is the most frequently consumed form of cannabis, both for medical and nonmedical purposes. Herbal marijuana is obviously plant material from *C. sativa*, but from precisely what botanical organs does it originate? As pointed out in Chapter 12, in the past, low-grade marijuana (sometimes derisively termed "ditchweed," although this term more narrowly refers to wild-growing low-tetrahydrocannabinol [low-THC] weedy plants) often was made up of a combination of foliage, twigs, "seeds" (technically one-seeded fruits called achenes), and material from the flowering section of the plant. Today, only "sinsemilla" (material from the flowering part of the unfertilized female plant) is commonly harvested.

Most plants have numerous flowers, and botanists employ technical terms to describe the ways that flowers are arranged on branches or branch systems. The term "inflorescence" refers to

1

FIGURE 1.1   *Cannabis sativa.* Photo by Barbetorte (CC BY 3.0).

(1) a group or cluster of flowers on an ultimate branch and/or (2) the entire branching system bearing flowers. When the flowers are fertilized and develop fruits, the branching systems are termed "infructescences." In many *Cannabis* strains, the ultimate branches bearing flowers have been selected to develop very congested, short branching systems bearing many flowers. These are the so-called "buds" of marijuana—desired because they are extremely rich in THC. "Buds" are technically "inflorescences"—a combination of the flowers and the ultimate small twigs of the branching system subtending the flowers. In the standard terminology of horticulture, "buds" are meristems (growing points or locations where cells divide) of stems or flowers or are embryonic stems, leaves, or flowers that will develop and enlarge with time. Like a number of other standard terms, the marijuana trade has adopted and converted the word "bud" to mean something different from its conventional meaning.

Marijuana is frequently referred to as the "flowers" of *C. sativa.* Indeed, in pre-Second-World-War drug literature, herbal marijuana was often known by the now largely antiquated pharmacological phrase "Cannabis Flos" (literally, Latin for "cannabis flowers"). As shown in Figure 1.2, the term is still occasionally encountered. In common language, a flower may be broadly understood to be "something that grows in a garden," but in technical botany, a flower is usually defined as a reproductive structure composed of one or more of sepals, petals, stamens, and pistils. (This is a narrow sense botanical definition; there are broader definitions available.) Female flowers of *C. sativa* lack sepals and stamens and (as explained in Chapter 6) lack typical petals. A female flower, illustrated in Figure 1.3b, is virtually devoid of THC, so defining or characterizing marijuana as the flowers of the plant (which in fact are present) is technically erroneous. (Parenthetically, jurisdictions that define illicit marijuana as the flowers of the plant are subject to legal challenges, since the material so defined is harmless from an abuse potential perspective.)

FIGURE 1.2   Medical marijuana preparation entitled "Cannabis flos," from the Netherlands firm Bedrocan, illustrating the use of this obsolescent phrase to denote material manufactured from the flowering parts of the plant. Photo by "Medische-wiet The Dutch Patient" (CC BY SA 3.0).

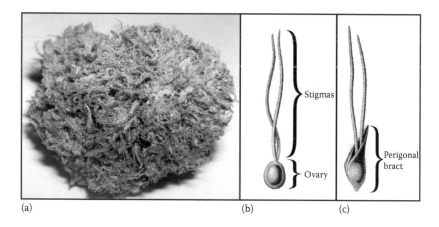

(a)   (b)   (c)

FIGURE 1.3   Figures presented to illustrate that marijuana is not the "flowers" of *C. sativa* since they are devoid of THC. (a) A "bud" of the marijuana strain Bubba Kush. Most of the visible green material is made up of tiny leaves, which are moderately rich in THC. Photo by Coaster420, released into the public domain. (b) A female flower. This is virtually devoid of THC. (c) A female flower inside a surrounding perigonal bract. The perigonal bracts contain the majority of the bud's THC but are not visible in (a) as they are nestled deeply amidst the tiny leaves. The reddish-brown threads in (a) are dried, overmature stigmas, shown in the fresh, green stage in (b). b and c are extracted from Figure 1.5.

"Bracts" are the key component of marijuana that contributes to drug potential. Botanically, a "bract" is a modified or specialized leaf, especially one associated with flowers. The structures termed bracts in *C. sativa* are quite small, resembling miniature unifoliolate leaves (i.e., leaves with just one leaflet), and they are indeed associated with the flowers. As presented in Chapter 11, a "perigonal bract" (illustrated in Figure 1.3c) covers in a cup-like fashion each female flower, and enlarges somewhat, becoming densely covered with tiny secretory glands that produce the bulk of the THC that the plant produces. (The terms "bracteole" and "perigonium" are sometimes encountered as synonyms of "perigonal bract" as the phrase is applied to *Cannabis* but are also used in different senses when applied to other plants.) In sinsemilla marijuana, which is produced by protecting the female flowers from being pollinated, the bracts remain quite small and are very densely covered with secretory glands. By contrast, pollinated flowers develop into "seeds" (achenes) and the perigonal bract becomes much larger and the density of secretory glands is lessened considerably. In

C. sativa, in addition to the tiny perigonal bracts, the flowering axis produces tiny leaves that are unifoliolate (with just one leaflet; "unifoliate," descriptive of plants with just one leaf, is incorrect) that are scarcely different from the perigonal bracts, and as one proceeds down from the tip toward the base of the branch bearing flowers (the axis of the bud), there are increasingly larger bracts that transition into small leaves with more than one leaflet. In the bud illustrated in Figure 1.3a, the green material that is visible constitutes both perigonal bracts and tiny young leaves. The smaller tiny leaves, like the perigonal bracts, are richly covered with tiny secretory glands, while the larger leaves within the bud have a lesser density of glands and so less THC on a relative concentration basis. As explained in Chapter 13, the larger leaves within buds are often trimmed away to make the THC concentration of the buds larger. To emphasize the key point in this paragraph, strictly speaking, marijuana (sinsemilla) is not literally "flowers," although a small amount, perhaps about 2%, is made up of female flowers virtually lacking THC; rather, it is THC-rich material (bracts, tiny leaves) *associated* with the flowers. The distinction made here is academic, admittedly, and is unlikely to change the widespread practice of referring to marijuana as flowering material. As pointed out by Small and Naraine (2016a), although the stigmas of the female flowers are originally devoid of THC, they are sticky, and gland heads rich in THC tend to fall away from the bud but are trapped on the stigmas, so in fact, the flowers secondarily acquire appreciable THC.

WHY CANNABIS IS CONTROVERSIAL

It hardly needs to be pointed out that cannabis is immensely controversial, accused of both deadly sins and marvelous virtues. It is famous (or infamous) because its chemicals have been considered to be the cause of considerable evil and harm by some, but of pleasure and cures by others (Figure 1.4). Indeed, cannabis is reminiscent of the malevolent Dr. Hyde and the saintly Mr. Jekyll—split personalities epitomizing good and evil within an individual (Small and Catling 2009). Democratic societies are currently struggling to evaluate just how bad and how good cannabis is. This book is intended to address these issues in sufficient but not overwhelming detail for the consideration of an informed public as well as decision makers.

FIGURE 1.4 The alleged good and evil sides of cannabis. Prepared by B. Brookes.

SEXUAL REPRODUCTION IN *CANNABIS*

We humans are preoccupied with sex, which also happens to be a subject of immense importance for cannabis. Most animals are divided into males and females (so male and female reproductive cells are produced on different individuals), although some are hermaphrodites. By contrast, most plants produce male reproductive elements (pollen) and female cells (eggs) on the same individual. *Cannabis sativa* is among the small minority of plants following the animal rather than the plant reproductive pattern. Most populations are divided into plants bearing only female flowers or only male flowers (Figure 1.5). Male plants are termed "staminate," so-named because the essential male floral organs are stamens, while female plants are termed "pistillate," so-named because the essential female floral organs are pistils, the egg-containing organs. Male plants die after shedding pollen, whereas female plants persist after their flowers are pollinated, maturing and shedding seeds until killed by frost. Female plants grown in a greenhouse, or in climates lacking a cold winter, can

FIGURE 1.5   (a) Flowering female plants of *C. sativa*. (b) Flowering male plants. (c) Painting of *C. sativa* from Köhler, F.E., *Medizinal-Pflanzen, Volume 2*, Gera-Untermhaus, Berlin, Germany, 1887. Left side shows female organs, right side shows male organs. (A) Flowering male branch. (B) Fruiting female branch. (C) Cluster of male flowers. (D) Fruit (achene) surrounded by perigonal bract. (E) View of wide (flat) side of "seed" (achene). (F) View of narrow side of seed. (G) Female flower, showing ovary and two stigmatic branches. (H) Female flower surrounded by young perigonal bract.

remain alive for years, although declining steadily in vigor. This potential longevity has led some to term the plants "annual or perennial depending on climate," but it is clear that the species is normally an annual. Sex expression, examined in Chapter 4, has been remarkably manipulated in domesticated plants, generally to the extreme detriment of males. Indeed, as detailed in Chapter 4, femaleness has become very highly valued in cannabis plants, while males are now regarded as the decidedly inferior sex. Another curiosity is that, unlike most animals, sexual expression in *C. sativa* is modifiable by a variety of stresses, and it is even possible (as discussed in Chapter 4) to induce females to become males, and vice versa.

Cannabis sativa produces leafy branches in the early part of its seasonal life cycle, and during the early growth period, male and female plants are virtually indistinguishable. Most populations are induced to flower by shortening days in late season, the timing of floral induction being one of many adaptive features of the plant and a critical consideration in maximizing the productivity of the plant for the various purposes for which it is grown (Chapter 5).

HOW THE FOUR KINDS OF HORSES ARE SIMILAR TO THE FOUR KINDS OF *CANNABIS SATIVA*

Cannabis plants are extremely diverse, and this has generated extraordinary widespread misunderstanding concerning the classes or categories deserving recognition, not just by the general public but also among professionals in numerous disciplines. It is no exaggeration to say that both the popular literature (notably as reflected by information on the Web) and the professional literature (particularly scientific publications) present highly confused and confusing interpretations of how variation among cannabis plants is structured and what terminology is appropriate. The root of misunderstanding of variability in *Cannabis* is that humans, not nature, have generated the most conspicuous differences. The topic is examined in depth in Chapter 18, but before considering the rather voluminous information in this book, it is important for clarity of understanding to appreciate the four principal kinds of plant that are significant to human welfare. These are (1) "wild" weedy plants, (2) plants selected for valuable fiber in the stems, (3) plants selected for edible oil–containing seeds, and (4) plants selected for intoxicating and medicinal drugs. The variation pattern exhibited by domesticated kinds of *Cannabis* is paralleled by many other examples of how humans have enslaved wild species, domesticating them (changing them genetically) into different utilitarian classes with characteristics uniquely suited to different purposes. As noted in the following discussion, just as *Cannabis* is composed of four basic economic classes, similarly, there are four functional groupings of horses.

Although there are numerous horse breeds today, three main types have been recognized, differing in ways that humans have chosen: heavy horses, light horses, and ponies. Heavy horses (also termed draft, draught, and dray horses) have strong bodies, broad backs, rounded withers (between the shoulder blades), and short thick legs—all features maximizing ability to pull large wagons (Figure 1.6a) and plow fields. Light horses (riding horses) have long bodies with backs that are narrow enough for humans to ride comfortably and long legs that stride easily. This class includes quarter horses, thoroughbreds (Figure 1.6b), and miniature horses. Ponies (often confused with miniature horses) usually have notably thick manes, tails, and coats; relatively small heads; thick necks; and short legs. As exemplified by the Shetland pony, ponies are commonly used as pets, riding horses for children, and horses for small wagons (Figure 1.6c). Numerous domesticated plants and animals have been so drastically altered by selection that they cannot survive without the assistance of humans. Domesticated kinds of *C. sativa* and domesticated horses, however, are frequently very hardy, and when they escape to the wild, they are often capable of living on their own (Figure 1.6d). Just as horses can be divided into the four classes discussed in this paragraph, *C. sativa* can be similarly divided into three domesticated and one wild class, as discussed next.

Chapter 18 provides an extensive analysis of the theory and practice of classification of living things with particular reference to *C. sativa*. As discussed there, biological classification of

FIGURE 1.6 Four basic kinds of horses. (a) A team of Clydesdales, representative of the heavy horse class. Photo (public domain) by Matthew Varga, U.S. Air Force. (b) Race horses, representative of the light horse class. Photo by John Picken (CC BY 2.0). (c) A team of ponies, representative of the pony class. Photo by Quistnix (CC BY 2.5). (d) Feral horses in Nevada. Photo by Del Brown (CC BY SA 2.0).

exclusively wild plants and animals is based only on natural genetic relationships. However, classification of living things that have been substantially altered by humans is often also based on utilitarian considerations, particularly the ways that they have been genetically modified for particular purposes. Just as domesticated horses exhibit three discernibly different kinds selected for different purposes, as well as wild ("ruderal") free-living populations, an analogous pattern is found among cannabis plants. The many different kinds of plant in *C. sativa* can be grouped into four basic categories, the first three of which include cultivated plants that have been selected for one of three economic products:

1. Fiber from the main stalk (employed for textiles, cordage, and numerous recent applications).
2. Oilseed (oil-rich seed employed for human food, livestock feed, nutritional supplements, industrial oils, and occasionally as a biofuel).
3. Psychoactive drugs from the flowering parts (used mostly illicitly for recreation and more recently legally as medicinals).
4. "Wild" (weedy) plants that have escaped from cultivation and grow independently in nature.

WILD PLANTS

Cannabis sativa is very widely encountered outside of cultivation growing as a weed (Figure 1.7; Chapter 3), and the existence of such "wild" plants is the basis for considerable arguments concerning appropriate classification (Chapter 18). The word "wild" can refer in a general way to plants or animals reproducing in nature without human care. However, the term is used in several different precise senses, as detailed in Chapter 18, and it is important to understand the sense in which some cannabis plants are "wild." Wolves and feral dogs can both be termed "wild," but wolves represent the ancestors of dogs, while feral dogs are merely escapes, more or less identical to pet dogs, although often extensively hybridized. The Australian dingo, however, represents an escaped dog that has reevolved adaptations to living in the wild. Accordingly, "wild" can mean (1) groups (like wolves) never altered by humans, (2) groups altered by humans that have

FIGURE 1.7   Wild *C. sativa* growing as a weed along a roadside at the edge of an upland deciduous forest, near Saratov city, Russia. Photo by Le.Loup.Gris (CC BY 3.0).

merely escaped (like feral dogs), and (3) groups altered by humans that have escaped and reevolved characteristics more suited to wild existence. Occasionally, one also encounters (4) "wolfdogs"—hybrids between wolves and dogs, which sometimes transfer genes between the two. As discussed in Chapter 18, "wild" cannabis plants appear to belong to groups 2, 3, and 4, but there do not seem to be genuinely wild plants that have not been changed genetically by humans. The world's so-called wild cannabis plants are likely extensively interbred with cultivated plants, and it appears that the ancient wild ancestor of *C. sativa* that existed in pre-Neolithic times (i.e., prior to 10,000 BC) is no longer extant (see Chapter 18).

As detailed in Chapter 3, plants of *C. sativa* growing outside of cultivation have distinctive adaptations, which are not present in domesticated plants. As discussed in Chapter 17, the genes that adapt wild plants to the stress of wild existence are very valuable for improving cultivated forms of *C. sativa*. Unfortunately, for decades, there have been enthusiastic, expensive, and short-sighted efforts to eradicate wild plants in North America, although their potential to be used as illicit drugs is insignificant.

Law enforcement personnel in the United States commonly use the phrase "ditch weed" for wild-growing *C. sativa*. Because almost all wild-growing plants in North America cannot produce intoxication, all poorly intoxicating plants are now often referred to as ditch weed. The slang term "weed" is the most popular of the dozens of terms used informally to refer to one or both of marijuana and marijuana plants. In the Netherlands, one encounters the term "Nederweed" ("Netherweed"), and in Europe, one finds "Euroweed."

FIBER PLANTS

"Hemp" usually refers to *C. sativa* plants used for fiber and also is the term employed for the fiber obtained from the stalk (i.e., the main stem). (As discussed next, when hemp is grown for oilseed, it is distinguished as "oilseed hemp" or "hempseed.") In past centuries, hemp was a staple resource for both civilian and military purposes, used mostly for textiles and cordage. The shipping industry for many centuries relied on hemp products (Figure 1.8). Based mostly on fiber, hemp was once touted, rather unrealistically, as "the new billion dollar crop" (Popular Mechanics 1938), with the claim that it "can be used to produce more than 25,000 products, ranging from dynamite to Cellophane." Nevertheless, *C. sativa* cultivation for fiber almost ceased in Western countries after the Second World War. However, as noted later, in the last several decades, there has been a resurgence of interest of fiber applications, mostly for nontraditional uses.

At present, there are only small, niche markets for the production of hemp fiber for various purposes. Traditional usage of the fiber for clothing, cordage, and paper continues, but these products are very expensive and appeal to a very small clientele. However, the hemp industry has been reinvigorated by new fiber-based products (Roulac 1997; Bouloc et al. 2013). Both the outer (bark, phloem) long fibers and the short internal (hurds, wood) fibers are now being employed in specialty pulp products and composites. These usages include fiberboard, insulation, pressed fiber products, masonry products (concrete, stucco, plaster, and tiles), carpets, straw-bale construction materials, livestock bedding, and a very wide range of plastics, as detailed in Chapter 7. The automotive industry has particularly pioneered the development of pressed fiber and molded plastic components. The considerable rot-resistance of the fiber is being exploited in geotextile products, such as landscaping fabric. The usage of hemp for these new fiber applications has been primarily in Europe, and subsidization was important in establishing the new hemp-related industries. Chapter 7 presents an extensive discussion of fiber aspects related to *C. sativa*.

OTHER SO-CALLED "HEMPS"

The name "hemp" can be confusing. It usually refers to *C. sativa*, but the term has been applied to dozens of other species representing at least 22 genera other than *Cannabis*, often prominent fiber

FIGURE 1.8   Hemp was indispensable for sails and rigging for navies during the "Age of Sail" (from the sixteenth to the mid-nineteenth century). (a) Traditional seeding, harvesting, and processing hemp in nineteenth century Netherlands. Painting dated 1873. (Public domain, Website Geheugen van Nederland/ Koninklijke Bibliotheek.) (b) Large-diameter (17 cm) hemp rope. Hemp anchor cables could exceed 60 cm in diameter. Photo by Ji-Elle (CC BY 3.0). (c) This 1832 painting by Pierre-Julien Gilbert (1783–1860) shows the inconclusive combat between the British HMS Tremendous (in foreground) and HMS Hindostan (left) against the French frigate La Canonnière (right), on April 21, 1806. (Public domain.)

crops. Montgomery (1954) listed over 30 "hemp names." So-called hemps include ambari hemp (deccan hemp, best known as kenaf, *Hibiscus cannabinus* L.), Manila hemp (abaca, *Musa textilis* Née), Mauritius hemp (*Furcraea foetida* (L.) Haw.), roselle hemp (*Hibiscus sabdariffa* L.), New Zealand hemp (*Phormium tenax* J.R. Forst. & G. Forst.), sisal hemp (*Agave sisalina* Perrine), and sunn hemp (*Crotolaria juncea* L.).

"INDIAN HEMP"

Especially confusing is the phrase "Indian hemp," which has been used both for intoxicating Asian drug varieties of *C. sativa* (so-called *C. indica* Lamarck of India), for jute (*Corchorus capsularis* L., also called Bengal hemp, Calcutta hemp, and Madras Hemp; see Ash 1948), and for *Apocynum cannabinum* L. (also known as American hemp as well as by other names), which was used by North American Indians as a fiber plant (see Figure 2.3).

OILSEED PLANTS

Cannabis sativa is employed as a source of a multipurpose fixed (i.e., nonvolatile) vegetable oil, obtained from the seeds (technically fruits called "achenes"; Figure 1.9, left). As documented in

FIGURE 1.9   Hempseed, the most economically promising nondrug product of *C. sativa*. Left: Seeds. Photo by Jorge Barrios (released into the public domain). Right: A display of commercial consumer products made with hempseed or hempseed oil. Photo by Dave O (CC BY SA 2.0).

Chapter 8, the seeds of *C. sativa* in recent decades have become an important source of edible oil. The seeds have traditionally been called "hempseed," and this expression has been used also for varieties of *C. sativa* grown especially for the oilseed. Although oilseed use was relatively unimportant historically compared to fiber applications, the commercial products made from hempseed have much greater significance and potential today than the fiber usages. The seeds of *C. sativa* are increasingly being recognized as a legitimate source for medicinals, nutraceuticals (nutritional extractives), and functional (i.e., nutritionally fortified) foods (Figure 1.9, right). Indeed, while "medical marijuana" is widely (with justification) held to have impressive therapeutic potential, as discussed in Chapter 8, "medical hempseed" also has remarkable therapeutic capacities.

INTOXICATING DRUG PLANTS

Forms of *C. sativa* producing elevated amounts of intoxicating chemicals were selected, particularly over the last thousand years in Asia (Figure 1.10), where the consumption of inebriating drug preparations (such as marijuana and hashish) have been consumed for ritualistic, religious, and hedonistic purposes. During the last century, the usage of marijuana has increased to the point that cannabis has become the world's leading illegal recreational drug. The chemistry and variation patterns of the cannabinoids (particularly the chief intoxicant THC) are examined in detail in Chapter 11, and nonmedical drug usage is documented in Chapter 12. The latter chapter provides extensive information on how recreational cannabis drugs are prepared and used, as well as the resulting physiological and psychological effects. While this information may disturb those unacquainted with cannabis drugs, it is widely available and mostly familiar to a substantial proportion of people, especially the young, and is needed to understand the possible associated harms, which are examined extensively in Chapter 12.

GENETIC RELATIONSHIPS OF DIFFERENT KINDS OF CANNABIS PLANTS

As explained in Chapter 18, when essentially all individuals of a group *can* interbreed freely (as within *C. sativa* and within *Homo sapiens*), it is of interest to determine whether, despite this ability to

FIGURE 1.10   An American soldier beside intoxicating marijuana plants in Kandahar, Afghanistan. Public domain photo by U.S. Army.

combine genes freely, there are nevertheless genetically distinctive subgroups. The existence of subgroups can be of practical interest. For example, it is well known that certain racial or ethnic groups of human beings tend to suffer from certain inherited diseases. For economic plants like *Cannabis*, the genes present in different subgroups can be invaluable for breeding improved crop cultivars (Chapter 17). A chief reason why subgroups develop in groups within which interbreeding can occur freely is geographical separation: when subgroups are too distant from each other to interbreed, and especially when they are in places with different climates and other stresses, they are often free to diverge genetically. Indeed, both in cannabis plants and in humans, it is obvious that subgroups developed in historical times. The genetic subgroupings of *C. sativa* that recent research suggests deserve recognition are discussed in detail in Chapter 18. Just how these genetic subgroupings are related to the utilitarian groups discussed earlier (fiber plants, oilseed plants, marijuana plants, and weeds) and what classification and nomenclature are appropriate are complicated issues, summarized in Figure 1.11 and explained in detail in Chapter 18.

THE CRIMINALIZATION AND SUPPRESSION OF CANNABIS

Cannabis sativa is infamous as the world's most widely utilized illicit plant. Because cannabis has been considered to be a leading drug of abuse, it has been seriously criminalized (Figure 1.12) since the Second World War, and almost all research and economic development—both drug and nondrug aspects—were suppressed for most of the twentieth century. After the Second World War, *C. sativa* became the leading illicitly cultivated black market crop in the Western World, with law enforcement dedicating huge efforts to eradicating the plants wherever they were discovered (Figure 1.13). Most scientific investigations authorized in Western countries were either forensic studies to aid law enforcement or medical and social research specifically intended to document and reduce harmful effects. As presented in Chapter 15, criminalization of cannabis has been associated with enormous law enforcement costs and social upheaval.

There have been many casualties of the "war on drugs" that has been waged with unusual ferocity against marijuana for decades. Science itself has been a principal casualty. For most of the last

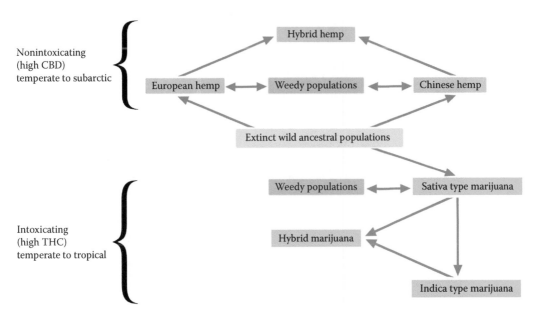

FIGURE 1.11 Evolutionary relationships and gene flow patterns among the different genetically distinctive kinds of hemp (nonintoxicating *C. sativa*), the different genetically distinctive kinds of marijuana, related weeds, and presumed wild ancestral populations. The connections illustrated in this simplified flowchart are examined in detail in Chapter 18.

FIGURE 1.12 Symbolic representations of the illegality of cannabis. Prepared by B. Brookes.

century, the stigma attached to "narcotics" was so severely associated with marijuana that scientists risked their careers attempting to initiate studies of virtually any aspect of cannabis, except its harmfulness. It is well to recall the martyrdom of the illustrious Russian crop geneticist N.I. Vavilov (1887–1943), who carried out scientific studies and made invaluable seed collections of *C. sativa*. As noted in Chapter 17, attempting to present scientific truth to power (the dictator Joseph Stalin) resulted in his imprisonment (Figure 1.14) and death. In democracies, it is not unreasonable for the elected governance to restrict or even prevent taxpayer-funded scientific research. More controversial, but at least debatable, is prohibition of research that bears on ethical issues (such as human reproduction). What must not be prevented is the search for truth that contradicts ignorant dogma, as the sorry history of cannabis demonstrates.

FIGURE 1.13   Eradicating a marijuana plantation.

FIGURE 1.14   Mug shots of the Russian geneticist and agronomist Nikolai Ivanovich Vavilov (1887–1943), distinguished student of *C. sativa*, and martyr for the cause of scientific truth. (Public domain photos.)

One of the tragic but much less obvious consequences of the criminalization of cannabis has been the short-sighted destruction of seed collections of *C. sativa* acquired by agriculture departments (mostly in North America) and the directives to refuse additional collection. As noted in Chapter 17, "seed banks" are collections of seeds, particularly of crop plants, and represent value considerably exceeding all of the monetary holdings of conventional banks. It is of the highest priority that the "germplasm resources" of *C. sativa* be conserved for the future welfare of the world, in the same way that other valuable crops like wheat, barley, corn, and potato are maintained.

THE RELEGITIMIZATION OF CANNABIS

By the last decade of the twentieth century, several developments contributed to a surge of scientific and technological development of *C. sativa*. First, in many countries (with the conspicuous exception of the United States), after a half century of prohibition of cultivation, there was a resurrection of production of the plant for nondrug purposes (Chapters 7 through 10). Second, nondrug hemp has acquired a reputation for being phenomenally beneficial for the environment and has become a leading symbol of sustainable agriculture (Chapter 16). Third, in much of Western society, there has been a growing tolerance of the extremely widespread recreational use of marijuana, as reflected by a romantic, idealized image in the media, less enthusiastic law enforcement, and even de facto decriminalization in some jurisdictions (Chapter 12). Fourth, there has been a substantial and increasing usage of marijuana prescribed for medical purposes (Chapter 13).

As noted in the following, decriminalization of cannabis for (1) industrial hemp, (2) medical marijuana, and (3) recreational marijuana has occurred or is occurring in many jurisdictions, the result of sociological, philosophical, political, and legal developments—complex and very contentious subjects which are examined only briefly in this book.

THE DECRIMINALIZATION AND RESURRECTION OF INDUSTRIAL (NONINTOXICATING) CANNABIS

By the middle of the twentieth century, the nondrug fiber and oilseed uses of *C. sativa* were widely viewed as obsolete, with no reasonable potential for legitimate development. Moreover, the issues of recreational and medical uses of *Cannabis* made it very difficult to rationally consider the redevelopment of industrial hemp for purposes that everyone should agree are not harmful. The extent of hostility to even harmless forms of *C. sativa* was indicated by remarkable limitations of human rights in democratic countries. For example, in Queensland, Australia, it was illegal to simply publish or possess information on growing industrial hemp before amendments to the Drugs Misuse Act 1986 were proclaimed on September 27, 2002, by the Queensland Parliament (Olsen 2004).

By the beginning of the twenty-first century, most Western countries witnessed the reintroduction of cultivation of nonintoxicating hemp, after at least a half century of total prohibition. The cultivars authorized are considered to be safe enough to be grown (generally under license) for the production of fiber and oilseed products. The delay in reauthorizing hemp cultivation was caused by fear that acceptance of the new crop would (1) be interpreted by the public as de facto acceptance of the legitimacy of all aspects of the species *C. sativa*, (2) act as a stepping stone to the legalization of marijuana, (3) impede the war on drugs, and (4) require costly monitoring to ensure that licensed crops are treated according to regulations.

When it became evident during the last decade of the twentieth century that hemp cultivation was being promoted, several economic analyses were conducted in various countries (Riddlestone et al. 1994; Gehl 1995; McNulty 1995; Ehrensing 1998; Kraenzel et al. 1998; Marcus 1998; Pinfold Consulting 1998; Thompson et al. 1998; Johnson 1999). These analyses are more or less obsolescent, since hemp cultivation has actually been revived in many countries, and the resulting hemp products have been tested in the marketplace for the last two decades. Fortenbery and Bennett (2004) is a more recent analysis of fiber potential but is rather discouraging of future development. Robbins et al. (2013) examines the potential in Kentucky, and Johnson (2015) analyzes the potential in the United States; these reviews are much more optimistic. Hemp is now grown commercially in about three dozen countries, with the notable exception of the United States (although this seems about to change). Earlier economic analyses generally failed to foresee that hempseed rather than fiber applications would become the most promising aspect of industrial hemp development. A variety of imaginative, innovative hemp fiber and hempseed products have appeared in the marketplace in the last two decades and have provided considerable impetus to increasingly promising industries (Small and Marcus 2002). Chapters 7 to 10 examine the extraordinary qualities, applications, and potentials of industrial hemp.

"HEMP" VS. "MARIJUANA"

As has been noted, *C. sativa* has been selected primarily for three different purposes: fiber (from the bark of the stem), edible seeds and seed oil, and intoxicating preparations (mostly from the flowering parts of the female plants). The common names "hemp" and "marijuana" (much less commonly spelled marihuana) have been applied loosely to all three classes, although historically, "hemp" has been used primarily for the fiber kind of plant as well as for its harvested fiber, and "marijuana" for the drug kind as well as for drug preparations made from it. The industries concerned with the nonintoxicating uses for fiber and oilseed have been at pains to distance themselves from the marijuana uses of *C. sativa* because of the stigma long attached to illicit drugs. Great efforts are made to point out that "hemp is not marijuana." The key phrase that has been used to distinguish plants authorized for noneuphoric drug uses (both fiber and oilseed) is "industrial hemp." Industrial hemp is now commonly employed to designate fiber and oilseed cultivars of *C. sativa* with very limited content of the intoxicating chemical THC.

THE DECRIMINALIZATION AND RESURRECTION OF MEDICAL CANNABIS

As noted in Chapter 13, cannabis has been extensively employed medically since ancient times. The illegality of cannabis during most of the twentieth century tragically retarded research and development of therapies. Ironically, black market marijuana used by thousands of people suffering from various conditions made it apparent that cannabis can alleviate symptoms and led to heroic efforts by patients, doctors, social activists, and lawyers to make medical marijuana available. Currently, medical marijuana has been authorized in several jurisdictions, and its use is rapidly expanding in Western countries. In the last several decades, there have been great advances in the scientific understanding of how cannabis affects human physiology, and new therapeutic products and technologies are either under development, being tested, or in some cases already accepted as useful. The literature on medical aspects has become extremely voluminous, and by no means is there agreement on the value of cannabis for treating particular conditions. Indeed, there is quite ferocious debate about the wisdom of employing medical marijuana for most medical issues. Chapter 13, by far the largest chapter in this book, reviews in detail the most recently available evidence bearing on the wisdom of employing cannabis for over two dozen illnesses.

Regardless of majority evaluation by the medical profession, in some jurisdictions medical marijuana has been remarkably commercialized, with the development of so-called medical dispensaries that are more reminiscent of supermarkets than clinics (Figure 1.15) and some physicians supplying medical marijuana in a manner reminiscent of street drug dealers. These unsavory developments—reflecting failures of adequate regulatory planning—are discussed in Chapter 15.

THE DECRIMINALIZATION AND RESURRECTION OF RECREATIONAL CANNABIS

Cannabis generally remains highly criminalized, particularly in some Asian countries, where it can result in capital punishment. Most of the Western World prohibits the recreational use of marijuana, but legalization has occurred in Uruguay and several U.S. states and is expected in other areas, particularly in the Americas. De facto legality of recreational marijuana has been the case in the Netherlands for decades (Figure 1.16), although not officially accepted. In democratic countries, there has been a general softening of penalties, or at least of prosecution, coinciding with increasing public tolerance of illicit usage. However, there remains considerable opposition and uncertainty about whether and how to modify current restrictions regarding recreational marijuana. Complicating the issue, investment in a recreational marijuana industry is widely viewed as potentially immensely profitable and business forces are driving developments. These issues and factors bearing on wise regulatory policies are examined at length in Chapter 15.

FIGURE 1.15 A conception of a large dispensary distributing medical marijuana products. Prepared by B. Flahey.

BENEFIT/HARM ANALYSIS OF THE MANY WAYS THAT CANNABIS IS USED

Cannabis sativa is remarkable—indeed incredible—with respect to the range of useful products it produces and the myriad ways these commodities can be employed (Figures 1.17 and 1.18). However, because it has been possible to develop various industries and products only for a limited period, their potential remains to be explored and evaluated. This book is concerned with examining in detail the current state of knowledge of the comparative merits and disadvantages of cannabis. Several of the chapters present a benefit/harm analysis of the individual ways that cannabis is employed—for fiber-based goods (Chapter 7), oilseed products (Chapter 8), essential oil (Chapter 9), minor items (Chapter 10), nonmedical drug uses (Chapter 12), medicinal applications (Chapter 13), and environmental enhancement (Chapter 16).

THE INTERFACE OF CANNABIS SCIENCE AND PUBLIC POLICY

If it were not for the notorious recreational drug usage of marijuana, there would be very limited interest in regulation of the plant and its products, and its potentials would have been well explored by now. But there is much that has been evaluated only recently, and even more that has not been examined at all. This volume presents an up-to-date evaluation of the "facts" about cannabis, particularly the pros and cons of employing it for various industrial and medical purposes. The following

FIGURE 1.16 Scenes in Amsterdam exemplifying the city's extensive commercialization of recreational marijuana. (a) A coffee shop (claimed to be the oldest coffee shop in the city), one of the Bulldog chain, typical of shops where marijuana is purchased and consumed on site. Photo by Daniel Farrell (CC BY SA 2.0; original photo trimmed). (b) Products in a store window. Photo by Nickolette from Bulgaria (CC BY 2.0). (c) Display of seeds and other products in a store. Photo by Jo Guldi (CC BY 3.0).

chapters summarize scientific knowledge about various aspects of the subject. However, it is well to keep in mind that human beings consider not only scientific knowledge but also whether the probable results of applying that knowledge are, on balance, favorable. Put simply, will the inevitable harm from reducing current restrictions on cannabis outweigh the potential good?

Automobile speed limits provide an instructive parallel (World Health Organization 2014). An increase in average speed of 1 km/h increases fatalities by 4% to 5%. A person hit by a car traveling at 50 km/h has an 85% chance of dying, but if struck by that same car at 30 km/h, the risk is only 5%. In high-income countries, speed contributes to about 30% of deaths on the road. In recognition of the dangers of automobile speed, there are many mechanisms that are employed—education, police traffic enforcement, speed zones, speed traps, speed bumps, and road designs that separate cars from pedestrians and bicyclists. Inevitably, grandstanding politicians succeed in lowering traffic to a crawl at intersections where a child has been killed by an irresponsible driver, and certainly, protection of vulnerable neighborhoods must be of special concern. But in the final analysis, it is human psychology on a much broader scale that is most determinative of how fast drivers go and indeed why people obey any law. In general, people obey laws for two reasons: (1) to avoid legal

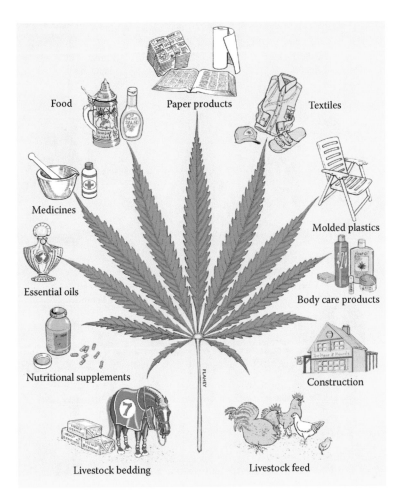

Food Paper products Textiles

Medicines

Molded plastics

Essential oils

Body care products

Nutritional supplements Construction

Livestock bedding Livestock feed

FIGURE 1.17 Major legitimate uses of *C. sativa*. Drawn by B. Flahey.

consequences and sanctions and (2) because the laws are perceived to possess legitimacy (Tyler 1990). In the case of marijuana, current laws are widely disrespected and disobeyed, with enormous financial and personal costs to society. Just as driving speed limits need to be determined with a view to what the majority will voluntarily honor, even in the face of some inevitable harm, so public policy regarding cannabis needs to be revised in recognition not only of "the facts" but also evolving public opinion.

CURIOSITIES OF SCIENCE, TECHNOLOGY, AND HUMAN BEHAVIOR

- Tracing the origins of the names for plants that have been used since prehistory is difficult, and this is especially so for cannabis. The earliest name for the cannabis plant could be the Sanscrit *sana*, meaning a hollow reed-like plant or cane, the name applied perhaps because the stem is often hollow. Corresponding to this is the Persian *canna* and *kannap*, hence the Arabian *cannab*, a small reed or cane; the Greek *kanna* and *kannabis*, a reed and anything made from it; and the Latin *cannabis*, from *canna*, a reed or cane, which led to the genus name *Cannabis*.
- The origin of the English word "hemp" is obscure. It appears to have arisen from the old Latin *hanapus* and the Old High German *hanaf*, referring to a bowl or basket. This

FIGURE 1.18   The four principal uses of *C. sativa* prepared by B. Brookes.

corresponds with the English "hamper," meaning a hemp bag or wicker basket. It has been contended that both the genus name *Cannabis* and the word hemp are based on a language of Central Asia or the Near East (see previous discussion). While "hemp" seems to be quite unrelated to "cannabis," it may have resulted from a process called the Gothonic sound-shift, whereby there is a substitution of *h* for *k* and of *f* or *p* for *b* in Teutonic languages.

- It is commonly assumed that the term "marijuana" is derived from the (Mexican) Spanish *mariguangot* (perhaps related to the Portuguese *maran guango*), meaning "intoxicant," or from the Mexican/Spanish slang *Maria y Juana*, for "Mary Jane" (Piper 2005). The Spanish for cannabis is cañamo, which seems close enough to the English marijuana, but it has also been contended that the derivation of "marijuana" is obscure. "Marijuana" was mentioned in ballads sung by the prominent Mexican revolutionary general Pancho Villa and his men in the 1890s, popularizing the term. The American newspaper tycoon William Randolph Hearst (1863–1951) was known for his dislike of Mexicans, and he further popularized the word in his newspaper chain in the 1930s as a means of criticizing them. According to one conspiracy theory, Hearst exercised his influence against hemp cultivation because he owned vast timber holdings, which fed the paper industry, and he was concerned that should hemp be used to produce paper, he would lose financially. Marijuana was corrupted to "MaryJane," a name that still is occasionally mentioned.

(a)   (b)   (c)

FIGURE 1.19 Mimics of *C. sativa* (public domain paintings). (a) *Hibiscus cannabinus* L. (family: Malvaceae). (Courtesy of Moninckx, J., *Moninckx atlas, Vol. 2.* Amsterdam, the Netherlands, 1682–1709.) (b) *Datisca cannabina* L. (family: Datiscaceae). (Courtesy of Sibthrop, J., Smith, J.E., *Flora Graeca. Vol. 10*, Taylor, London, 1840.) (c) *Eupatorium cannabinum* L. (family: Asteraceae). (Courtesy of Zorn, J., Oskamp, D.L., *Afbeeldingen der artseny-gewassen met derzelver Nederduitsche en Latynsche beschryvingen. Vol. 1*, J.C. Seep en Zoon, Amsterdam, the Netherlands, 1796.)

- Marijuana was sometimes called Mezzrole, after Milton "Mezz" Mezzrow, a musician. He moved to Harlem in 1929, where he sold marijuana cigarettes, as recorded in his autobiography *Really the Blues*.
- "Pot," slang for marijuana, is occasionally used as a name for the marijuana plant. One possible derivation is that "pot" is the shortened form of the Mexican/Spanish *potiguaya*, meaning marijuana.
- "Hashish," also spelled hasheesh or haschisch, is a concentrated, highly intoxicating form of the resin of *Cannabis*. The word seems to be derived from Arabic, meaning "herbage."
- "Neosemanticisms" are old words given new meanings. Examples are dope, grass, herb, skunk, tea, and weed, all of which have been applied to marijuana.
- In Mandarin Chinese, the way that a word is pronounced (inflected) gives it different meanings. For example, the word *ma* can mean mother, scold, horse, or hemp. This observation led to a team of University College London researchers, headed by Sophie Scott, to overturn the long-held theory that language is handled only in the left temporal lobe of the brain, as has long been known for the brains of native English speakers. Using magnetic resonance imaging, which indicates brain cell activity, it was found that Mandarin speakers do use the left temporal lobe, but also use the right temporal lobe, normally associated with music.
- There are dozens of species with an epithet (the second word in scientific names) like *cannabinus*, indicative of similarity with *C. sativa*, but the resemblance is always superficial (note Figure 1.19). Many plants have leaves with an odd number of leaflets with sawtooth edges, the leaflets palmately arranged (arising independently from the top of the leaf stalk), and these tend to mislead many into thinking that they are viewing a marijuana plant.

2 Prehuman and Early History of *Cannabis sativa*

THE FAMILY TREE AND PREHUMAN ANTIQUITY OF *CANNABIS SATIVA*

Cannabis sativa is an angiosperm—a member of the flowering plants that dominate terrestrial parts of Earth. Although there is some evidence of an older origin, most fossil evidence demonstrates that flowering plants appeared at least by the Lower Cretaceous geological period, about 125 million years ago, and were diversifying into modern plant families by the Middle Cretaceous, 100 million years ago. The Cannabaceae family traditionally has been defined as comprised of two genera, *Cannabis* and *Humulus* (Small 1978a). Grudzinskaya (1988) added the fossil genus *Humulopsis* to the Cannabaceae and split *Humulus* into two genera (although only *Humulus* is currently accepted). *Humulus* species are vines and easily distinguished from *Cannabis*. However, the fruits (achenes) are very similar and could be confused with each other. Older texts commonly use the obsolete orthography Cannabinaceae and Cannabiaceae for the family (Miller 1970). Recent molecular evidence indicates that the family is best considered as composed of about 10 genera (Sytsma et al. 2002; Yang et al. 2013; Figure 2.1). McPartland and Guy (2004a), on the basis of parasite relationships of *Cannabis* and related families, suggested that the Cannabaceae lineage evolved no earlier than 34 million years ago. Except for pollen grains, fossils tracing back millions of years when *C. sativa* first evolved are lacking, and its age of origin has not been determined with accuracy.

THE INCREDIBLY PARALLEL HISTORIES OF *CANNABIS* AND ITS CLOSEST RELATIVE, *HUMULUS* (HOP)

The common hop *Humulus lupulus* (Figure 2.2), the closest relative of *Cannabis*, is a remarkable plant with numerous uses (DeLyser and Kasper 1994). (Note that "hop" refers to the plant, while "hops" refers to its fruits [cones] employed to flavor beer.) The divergence of the genus *Cannabis* from its sister genus *Humulus* has been estimated on the basis of molecular data to have occurred approximately 21 million years ago (Yang et al. 2013; Divashuk et al. 2014). Because of the genetic closeness, hop provides a standard of comparison. The relationship of the common hop to humans is astonishingly parallel to the relationship of *C. sativa* to humans. As detailed in the following, both have numerous qualities preadapting and predisposing them to being developed by people for a wide range of similar products and purposes. This parallelism is no accident; it is deterministic, reflecting how close genetic relationship predisposes related plant species to being domesticated in similar ways—the "homologous series" of Vavilov (Kupzow 1975).

- Both *C. sativa* and *H. lupulus* are commonly found beside streams and rivers, their seeds seemingly distributed by water movement.
- Both *C. sativa* and *H. lupulus* have numerous small secretory glands producing a resin, but while the aliphatic acids of the hop plant resin provide flavor for legal intoxicants (beer and ale), the intoxicant THC of the marijuana plant is mostly illegal for recreational intoxication.
- Like hemp, hop stems contain considerable fiber and have been used in making paper and twine. The stems were also once used in basketry and wickerwork.
- For intoxicant purposes, both hops and marijuana are usually grown in the absence of males. Seedless hop cones are the counterpart of marijuana "buds." Highest quality buds and hops are both grown vegetatively (as clones).

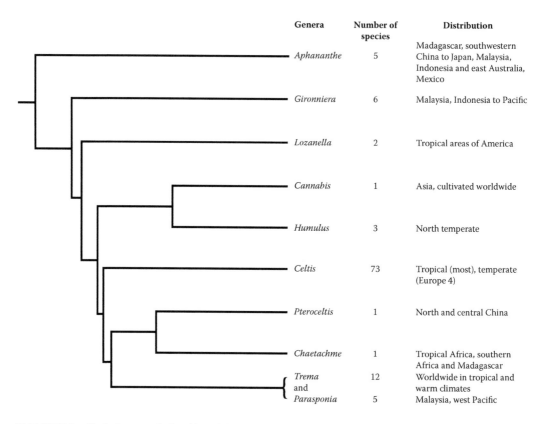

Genera	Number of species	Distribution
Aphananthe	5	Madagascar, southwestern China to Japan, Malaysia, Indonesia and east Australia, Mexico
Gironniera	6	Malaysia, Indonesia to Pacific
Lozanella	2	Tropical areas of America
Cannabis	1	Asia, cultivated worldwide
Humulus	3	North temperate
Celtis	73	Tropical (most), temperate (Europe 4)
Pteroceltis	1	North and central China
Chaetachme	1	Tropical Africa, southern Africa and Madagascar
Trema and Parasponia	12	Worldwide in tropical and warm climates
	5	Malaysia, west Pacific

FIGURE 2.1 Evolutionary relationships of the genera of the Cannabaceae, based on chloroplast DNA. Figure constructed from information in Yang et al. (2013), in which it is noted that *Trema* and *Parasponia* appear to warrant amalgamation into a single genus. Prepared by B. Brookes.

- The tiny secretory glands of *C. sativa* that synthesize the intoxicant chemicals of the plant are harvested (as "trichomes," "pollen," "crystal," or "hashish powder" as discussed in Chapter 12) and used as an extremely concentrated form of cannabis. Similarly, the tiny secretory glands of hop are harvested, often using similar techniques (Bishop 1966; Rigby 2000), and marketed commercially as a medicinal product.
- The early stages of biosynthesis of the flavoring chemicals of *H. lupulus* bear similarities to those that produce the intoxicant chemicals of *C. sativa* (Raharjo et al. 2004).
- *Cannabis* and *Humulus* can be grafted to each other (Crombie and Crombie 1975).
- Like marijuana, the consumption of hops was once illegal: In England about 1500, after learning of how well hops preserved beer in continental Europe, British brewers started adding hops to ale (sweet beer made without hops), turning it into bitter beer. Henry VIII (1491–1547), responding to a petition to ban hop, described as "a wicked weed that would endanger the people," outlawed the use of hop by brewers. His son, Edward VI (1537–1553), rescinded the ban in 1552.
- Like cannabis, in the nineteenth century, hops were an ingredient in many patent medicines. One of these was "hop bitters," composed of hops in 30% alcohol. Its advertising slogan was "Take hop bitters three times a day, and you will have no doctor bills to pay."
- Like some cannabinoids of cannabis that under some condition produce sedation (discussed in Chapter 13), hops have been found to have sedative chemicals including the volatile alcohol dimethylvinyl carbinol (Small 2016). The terpene myrcene is prominent in both *Cannabis*

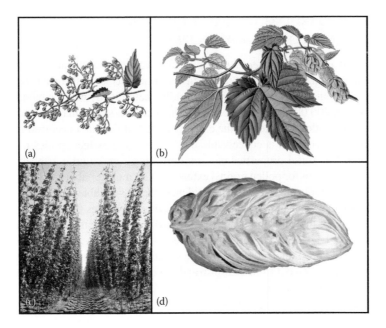

FIGURE 2.2 Common hop (*Humulus lupulus* L.), a vine closely related to *C. sativa*. (a) Male branch. (b) Female (fruiting) branch. (a and b: From Köhler, F.E., *Medizinal-Pflanzen, Volume 2*, Gera-Untermhaus, Berlin, Germany, 1887.) (c) Hop plantation, the vines trained to grow upwards on wires. Photo by Goliath (CC BY 3.0). (d) Cone (homologous to a "bud" of *Cannabis*), long-sectioned, showing yellow lupulin glands (homologous to the stalked glandular hairs of *Cannabis*). Photo by David Gent, U.S. Department of Agriculture, online at Bugwood.org (CC BY 3.0).

and *Humulus*, and as noted in Chapter 9, myrcene-rich marijuana is thought to be exceptionally sedative. There is a long tradition of using hops as a sedative, including putting the cones in pillows and planting hops beside bedrooms to encourage sleep. The sedative value of hops led to it being used as a cure for "uncontrolled sexual desires and a quarrelsome nature." In present-day Germany, hops are part of sleep preparations (Wichtl 2004).

• Like many chemicals in *Cannabis*, hops are rich in antibiotic compounds. Several hop constituents are very effective against gram-positive bacteria, which may explain why hop essences were once commonly used to treat tuberculosis patients.

• Hop cones and hemp seeds are employed in food and both contain compounds with considerable antioxidant effects. Antioxidants protect against substances called free radicals, generated during metabolism, which are thought to worsen a range of diseases, including heart disease, stroke, and certain cancers. In the case of cannabis, several of the cannabinoids are antioxidants, as well as such components as vitamin E in the seeds. In hops, the health-promoting antioxidants include prenylated flavonoids, which are accordingly found in beer. Unfortunately, to maximize health benefits, one would have to drink 450 L (about 1000 American pints) of beer daily. No one has calculated how much cannabis consumption is needed to realize benefits from its antioxidants.

• It has often been suggested that hops are intoxicating, like marijuana. This may be partly due to the peddling by British merchants in the nineteenth century of a substance they called hopeine, alleged to be a narcotic derived from the finest wild American hops. In fact, it was a mixture of an aromatic oil and morphine. Many have been tempted to smoke hop leaves, although they are devoid of the mood-altering chemicals of marijuana.

UNCERTAINTY REGARDING THE EARLY HISTORICAL ASSOCIATION OF *CANNABIS* AND PEOPLE

Extinct genera (such as *Australopithecus* and *Paranthropus*) connect the human genus *Homo* to the other great apes (chimpanzees, gorillas, orangutans). Extinct species of the genus *Homo* are thought to have evolved less than three million years ago, while anatomically modern humans (*Homo sapiens*) may be less than 200,000 years of age. Given the recentness of human existence, it is obvious that *C. sativa* is much older, and clearly, the early evolution of the plant occurred in the absence of selective pressures by people. It is conceivable that some of the extinct species related to modern humans discovered the useful properties of *Cannabis*. However, cultural artefacts and fossilized remains of *C. sativa* (except for pollen grains) extend back at most for about 10,000 years in Eurasia (Fleming and Clarke 1998), and the older the evidence, the less reliable it is.

Agriculture, which began as long ago as 11,000 BC in some places (Hancock 2012), is the foundation of civilization. Of the thousands of plant species that humans have used for various purposes, only a few dozen have been critical to the advancement of civilization, and *C. sativa* is one of these. Indeed, it is one of the most ancient of crops. The earliest archaeological evidence for human use of the plant has been speculated to be hemp strands in clay pots from tombs as old as 10,000 BC (Kung 1959; Chang 1968), although this interpretation is doubtful. *Cannabis* may have been harvested by the Chinese 8500 years ago (Schultes and Hofmann 1980), but it should be kept in mind that harvesting could have been from wild-growing plants. *Cannabis* has certainly been deliberately grown for at least 6000 years (Fleming and Clarke 1998). As with many major crops that trace to very early times, the early history of *C. sativa* is poorly known because it was cultivated and used well before the appearance of writing.

LOCATION OF THE PREHUMAN DISTRIBUTION RANGE

Cannabis sativa is widely regarded as indigenous to temperate, western or central Asia, but perhaps as far east as eastern Asia (Li 1974). However, no precise area has been identified where the species occurred before it began its association with humans. De Candolle (1885), the first authoritative student of the biogeography of crop plants, speculated that the ancestral area was the southern Caspian region. Other authors (e.g., Walter 1938; Sharma 1979) have suggested that the plant is native to Siberia, China, or the Himalayas. Piomelli and Russo (2016) stated, "*Cannabis* originated in Central Asia and perhaps the Himalayan foothills." Certainly, the plant is of Old World origin and in prehistorical times could have naturally occupied many areas across the breadth of Asia, as evidenced by the present success of wild-growing (ruderal) plants, which are widespread in Asia.

Fossilized pollen grains of *C. sativa* that are preserved in sediments of lakes and bogs have some potential for discerning ancient distribution areas of the species. However, the grains of *C. sativa* and *H. lupulus* are quite difficult to distinguish (Fleming and Clarke 1998), and wild populations of both species frequently occur near streams and rivers, making it difficult to identify which species left pollen deposits in wetlands such as lakes and bogs where pollen is often preserved.

As discussed later, there are discernible areas in Eurasia where *C. sativa* has been selected for fiber or marijuana (Figure 2.7), but it is well known from the study of other crops that such areas may represent secondary centers of selection—i.e., they were transported from an original, often quite distant indigenous area (Harlan 1951).

The chief reason that there is uncertainty regarding the primeval location of *C. sativa* is that for at least the last 6000 years, it has been transported widely, providing extensive opportunities for establishment outside of its original range (Abel 1980; Clarke and Merlin 2013). Since the present distribution of wild-growing plants in Asia is entirely or substantially the result of distribution by humans, it is not a reliable guide to the original indigenous area. Because the species has been spread and modified by humans for millennia, there does not seem to be a reliable means of accurately determining its original geographical range, or even whether a plant collected in nature

FIGURE 2.3 Left: Indian hemp (*Apocynum cannabinum*), a traditional stem fiber source of Native North Americans, which has been confused with *C. sativa*. Photo by Steve Dewey, Utah State University, Bugwood.org (CC BY 3.0). Right: Native American family (Chief Sevara, Utes tribe), photographed about 1899, showing costume partly prepared from native North American fiber plants. Public domain restored photo, Library of Congress.

represents a primeval wild type or has been modified by domestication (Schultes 1970). The seeds of some wild-growing populations in India are remarkably small, unlike those collected from any other area of the Old World, but whether this is indicative of a distinctive ancient wild form is unclear. As discussed in Chapter 17, whatever ecological constraints once limited *C. sativa* to its ancestral home range, over the millennia, it has become adapted to grow in much of the world.

The hop genus *Humulus* is the closest relative to *Cannabis*, and indeed, the two genera were once considered to be the only ones in the Cannabaceae family. Given that they likely once had a common ancestor, the geographical distribution of *Humulus* might be informative of where *Cannabis* was once native. *H. lupulus* L. is native to virtually the entire northern hemisphere; *H. japonicus* Siebold & Zucc. ("*H. scandens*") is indigenous to temperate Asia; and *H. yunnanensis* Hu is confined to Yunnan province in China (Small 1978a). The common area of the three species is China, and indeed, some have speculated that the country could be the original home of *Cannabis* (Boutain 2014).

There has been speculation that *C. sativa* occurred in North America in pre-Columbian times, but this is almost certainly the result of confusion with "Indian hemp," *Apocynum cannabinum* L. (Figure 2.3), a native of North America, the source of a stem fiber employed by native North Americans for clothing, hunting nets, fishing lines, and twine.

THE HABITAT OF ANCIENT WILD *CANNABIS SATIVA*

The circumstances and adaptations of extant wild-growing populations of *C. sativa* provide a basis for judging its ecology before human influence. The species thrives in mammalian-manured, continuously moist but well-drained soil, in open areas with limited competition from other plants (Figure 2.4). This suggests that ancestral *C. sativa* grew on the alluvial soils near streams and other water bodies, and depended on herds of wild, large, mammalian grazers to deposit excrement (Figure 2.5).

THE "CAMP-FOLLOWER" MODEL OF EARLY DOMESTICATION OF *CANNABIS SATIVA*

Cannabis sativa is the most widely cited example of a crop that is postulated to have evolved initially as a "camp follower" (Anderson 1954; Schultes 1970). Humans at the hunter-gatherer stage are thought to have been nomadic, often traveling among temporary camps and creating trails among

FIGURE 2.4 Ruderal (weedy) hemp near Ottawa, Canada. This photo shows several characteristic habitat features of *C. sativa*: (1) The plants are in an open, sunny location. (2) They are growing near a manure shed, in nitrogen-rich soil. (3) A stream is nearby, maintaining a moist substrate. (4) The soil near the stream is alluvial (sandy and well drained). (5) Competition from other plants is limited.

FIGURE 2.5 An interpretation of the prehuman ecology of *C. sativa*. The habitat requirements of modern ruderal hemp (natural adaptation to well-manured, moist but well-drained soils and open sunny locations, as shown in Figure 2.4) suggest that the ancestral plants thrived near streams frequented by mammalian herds. Przewalski's horse, native to the steppes of central Asia, is illustrated. Drawn by B. Flahey.

these. Abandoned campsites and paths would tend to be open (unshaded), located frequently near lakes or streams, and the soils would be enriched by deposition of organic materials (excrement and unused remains of harvested animals and plants). Seeds and roots from gathered plants that humans would have selected for their usefulness would also be deposited in these open, fertilized areas. This amounts to selective planting of desirable plants in protected situations where they will receive excellent light and soil—a precursor of cultivation. Inevitably, people would have noticed and eagerly harvested materials from the plants that were growing along their routes and former homesteads,

FIGURE 2.6   An interpretation of the early domestication of *C. sativa* in accord with the "camp-follower" and "dump-heap" hypotheses of crop origin. The plant would have been collected from the wild as a source of stem fiber, edible seeds, and inebriating resin. Seeds discarded on refuse dumps near temporary camps would have found ideal conditions (manured soil, an open sunny location, probably proximity to a water supply, and limited competition), and consequently would have become desirable companions for mankind. The pipe-smoking shown represents artistic license, as ancient methods of smoke inhalation in the Old World are controversial (Clarke and Merlin 2013). Drawn by B. Flahey.

especially in garbage dumps, and such plants would have been among the first that would have been considered for deliberate planting. As described by Anderson (1954), this explanation is variously known as the "rubbish heap" or "dump heap" hypothesis (in archaeology, rubbish heaps are referred to as "kitchen middens"). It is interesting that, in parallel, some monkeys have been shown to create "monkey gardens"—concentrations of preferred food plants in areas where they have discarded seeds (Rindos 1984). Uncultivated, colonizing plants that grow vigorously in human-cleared areas are known as weeds. It is no accident that many, probably the majority, of the world's major domesticated crops are related to, or are known to have originated from, such plants. The ability to be weedy clearly preadapts plants to being domesticated. *Cannabis sativa* is superbly adapted for the role of camp follower. It is very weedy by nature. It is also a nitrophile (nitrogen-loving species) and would have grown exceptionally well in the nitrogen-rich manured soils around early settlements. Its propagules are thought to be distributed by streams, which, as noted previously, are often near campsites, as well as by people and animals, including domesticates. Because *Cannabis* has products (stem fiber, edible seeds, and intoxicating tissues) that could have been easily harvested and utilized by prehistoric peoples, it was almost certainly associated with humans in very early times (Figure 2.6).

HOW ADAPTATION TO STREAMSIDE SOILS LED
TO WATER-BASED FIBER EXTRACTION

Alluvial soils near streams and rivers are typically sandy or silty and well-drained—exactly the kind of soil conditions in which *C. sativa* grows best. It would seem likely, therefore, that humans commonly encountered the plant near water. People would have noticed that the stems of plants that fell into the water disintegrated in several weeks because of rotting but left behind the valuable fibers (see "water retting" in Chapter 7). Subsequently, they likely deliberately steeped plants in the water to obtain the fiber. When agriculture was adopted, plants were likely grown near streams and rivers to take advantage of the suitable soils and conveniently close water that could be used to extract the fiber.

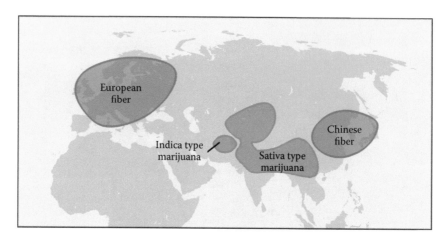

FIGURE 2.7 Approximate postulated geographical locations of the two groups of fiber plants and the two groups of marijuana plants domesticated more than a millennium ago and subsequently transported to other parts of the world. Detailed information concerning the evolution, classification, and nomenclature of the groups is presented in Chapter 18. Prepared by B. Brookes.

EARLIEST GEOGRAPHICAL CENTERS OF THE FOUR KINDS OF DOMESTICATED *CANNABIS SATIVA*

For most of its historical association with humans, *C. sativa* was primarily valued as a source of stem fiber, considerably less so as an intoxicant, and only to a very limited extent as an oilseed crop. Fiber aspects of the species are examined in Chapter 7. "Hemp" (*C. sativa* grown for fiber) is one of the oldest textile fiber crops, with extant remains of hempen cloth trailing back six millennia. *Cannabis sativa* grown for fiber was introduced to western Asia and Egypt and, subsequently, to Europe somewhere between 1000 and 2000 BC. As illustrated by the two green distribution areas in Figure 2.7, there is evidence that hemp domestication occurred particularly in China in very early times (millennia ago) and later in Europe (where hemp became widespread after 500 AD). Detailed information on intoxicant races of *C. sativa* is provided in Chapters 11 to 13. Intoxicating races were domesticated in southern Asia. In most of the range, the so-called "sativa type" dominated, whereas in a much smaller area in southwestern Asia, the so-called "indica type" was selected. Detailed information concerning the evolution, classification, and nomenclature of the four domesticated groups is presented in Chapter 18.

EARLY OLD WORLD GEOGRAPHY OF FIBER AND MARIJUANA CLASSES OF *CANNABIS SATIVA*

As illustrated in Figure 2.8, dating back at least a millennium in the Old World, there developed a remarkable north–south separation of *C. sativa* selections grown mostly for fiber and those grown particularly for intoxicating drug preparations. In Europe and northern Asia, *C. sativa* was grown virtually exclusively for fiber, occasionally for its edible seeds (also useful for lubricating and illumination oil). In southern Asia and Africa, the nonintoxicant uses of the stem fiber and oilseed were sometimes exploited, but the plants were particularly employed for drugs for recreational, cultural, and spiritual purposes. Clearly (as discussed in Chapter 18), strong selection for fiber in the north led to the evolution of races of *C. sativa* with characteristics maximizing fiber production. Conversely, strong selection in the south led to the evolution of races of *C. sativa* with characteristics maximizing the production of inebriating drug content. A side effect of the north–south split is adaptation to the different length of daylight encountered in the two areas, as discussed in Chapter 5 dealing with photoperiodism. Northern fiber-type races are particularly adapted to relatively early

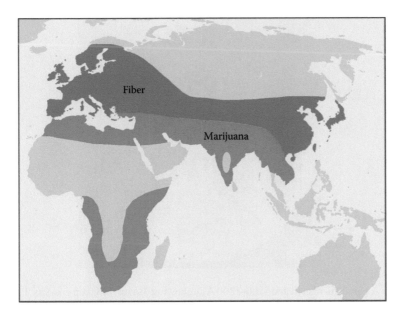

FIGURE 2.8 Approximate pre-Columbian distribution of fiber *C. sativa* (in green) and marijuana *C. sativa* (in red). Prepared by B. Brookes.

flowering to survive in the shorter growing seasons of the north. At least the sativa type marijuana class of plant is adapted to flowering quite late, taking advantage of the longer growing seasons of the south (the indica type was selected for a relatively short growing season).

EARLY MIGRATIONS OF FIBER AND MARIJUANA KINDS OF DOMESTICATED *CANNABIS SATIVA*

Of the four kinds of domesticated *C. sativa* noted in Figure 2.7, the European fiber kind and the sativa type marijuana kind became predominant in being cultivated outside of Eurasia. Major transportation routes of the European fiber (hemp) kind of plant and of the sativa type marijuana kind of plant to the New World in post-Columbian times up to the mid-nineteenth century are shown in Figure 2.9.

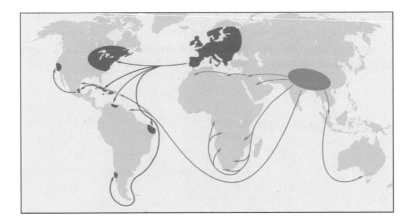

FIGURE 2.9 Principal distribution routes of *C. sativa*. In green: transport of European fiber hemp to the Americas, from the late sixteenth to the mid-nineteenth centuries. In red: transport of sativa type marijuana, from ca. 1000 AD to the early twentieth century. Prepared by B. Brookes.

FIGURE 2.10   A porcelain dish (dated 1700–1800 AD) showing Magu, the hemp goddess. Photo by Daderot, showing an exhibit in the Asian Art Museum, San Francisco (CC0 1.0).

During this period, the Americas served as important areas of cultivation of fiber hemp, while cultivation of marijuana races became widespread in much of coastal Africa. An excellent documentation of historical usage and cultural diffusion of *Cannabis* is provided by Clarke and Merlin (2013).

CURIOSITIES OF SCIENCE, TECHNOLOGY, AND HUMAN BEHAVIOR

- Words for cannabis (the plant or its preparations) are apparently absent from the Old Testament (but see Bennett 2010, who provides evidence to the contrary). As pointed out by Mechoulam et al. (1991), this is odd because the Assyrians who interacted considerably with the Old Testament Jews (Assyria thrived from about 2500 BC to 605 BC) extensively employed cannabis as an inebriant and medicament. Mechoulam et al. suggest that the writers of the Old Testament wished to suppress pagan influences and so censored all mention of intoxicating cannabis (and opium, as well).
- Ancient China was the first site of extensive domestication and usage of *Cannabis*, and one of the earliest reflections of its value is the mythological Chinese "hemp maiden" or "hemp goddess," Magu (Figure 2.10), described as a beautiful young woman with long birdlike fingers. Tracing to Taoist legends, she is also known as the Goddess of Longevity and the symbolic protector of females.

3 The Ecology of Wild *Cannabis sativa*

THE NATURE OF "WILDNESS"

When plants grow exclusively in nature, the term "wild" accurately describes them. However, many plants have been removed from their natural area and changed genetically to suit the needs or whims of humans, and sometimes, such genetically changed plants escape from cultivation and reestablish in nature. Sometimes, hybridization between genetically changed plants and their wild ancestors occurs, changing the latter. In either of these latter two circumstances, the term "wild" can be misleading because humans have in fact removed some of the genes that contributed to the wildness of the plant. The word "wild" has been used in a broad sense to include all populations growing outside of cultivation and, in a narrow sense, to refer to populations of a species that are uninfluenced genetically by domestication. As discussed in Chapter 18, wild-growing plants of *Cannabis sativa*, insofar as has been determined, are either escapes from domesticated forms or the results of thousands of years of widespread genetic exchange with domesticated plants, making it virtually impossible to determine if unaltered primeval or ancestral populations still exist. Moreover, because the species has been spread and modified by humans for millennia, there does not seem to be a reliable means of accurately determining its original geographical range or even whether a plant collected in nature represents a primeval wild type or has been influenced by domestication.

Plants of *C. sativa* growing outside of cultivation exhibit various degrees of "wildness," as noted in this chapter. Several terms are used to denote plants of different extents of "wildness," and it is critical to be aware of the ambiguity of these terms in discussing *C. sativa*. Plants that develop as a result of seeds unintentionally scattered from cultivated plants are said to be "volunteers," a label used in agriculture. Such plants are not really wild, although they may occur in uncultivated places. For the most part, volunteers appear in or very near the field where the maternal plants were grown. The word "spontaneous" is used in floristics (the study of plant geography or distribution) to denote plants that appear locally as a result of human activities, but do not spread. Such plants can be domesticates (e.g., tomatoes growing only on refuse heaps where tomato seeds were discarded; cereals growing only near mills where the seeds were processed) but may also be wild somewhere in the world but growing in foreign locations that are not particularly hospitable and so incapable of spreading. This localized appearance of foreign plants often occurred when their seeds were unintentionally transported and discarded in disposable ship ballast, subsequently growing only where the ballast was discarded. The term "ruderal" (applied both to plants and their habitats) means growing in waste places or rubbish and is descriptive of the habitat of perhaps the majority of weeds. One also encounters "feral," applied to wild *C. sativa* (and other weeds), although mostly, the word is used for escaped domesticated animals (such as dogs and horses) that are living outside of human control. Both the words feral and ruderal are ambiguous, since they are applied to (a) those escaped domesticates that basically retain all of their domesticated characteristics but nevertheless establish and spread vigorously outside of cultivation and (b) types of plants that differ dramatically from domesticates, with adaptations specifically suited to wild existence.

The term "naturalized" is employed to denote plants that have managed to colonize and grow successfully in areas where they were once absent. The term "weedy" applies to plants that similarly grow successfully in areas where they were once absent but often do so particularly aggressively and successfully (extremely aggressive weeds are often called "invasives"). As discussed in the next section, populations of *C. sativa* that can be described as naturalized and/or weedy have evolved adaptations (particularly seed characteristics) suiting them to wild existence. In this book, the terms "weedy" and "ruderal" should be understood to refer to "wild *C. sativa*" with specific adaptations for living in nature.

EVALUATING WILDNESS IN *CANNABIS SATIVA* POPULATIONS

As noted later in the section on seed ecology in this chapter, plants domesticated either for marijuana or industrial (nonintoxicating) usage have seeds that differ in several characteristics from plants that are adapted to growing in the wild. Syndromes of contrasting features of the seeds clearly differentiate plants that have been domesticated (for fiber, oilseed, or drug use) from plants that have lived outside of cultivation for many generations. The seed characteristics found in domesticated plants adapt them to cultivation, and in contrast, the seed characteristics found in plants growing outside of cultivation adapt them to life outside of cultivation. Indeed, intermediate development of these seed characteristics indicates the extent to which a given population has been domesticated or, alternatively, is re-adapting to living in the wild.

Named hemp cultivars and marijuana strains of *C. sativa* almost always exhibit the domesticated syndrome of seed characteristics, as indeed do almost all plants of the species that are cultivated. This domesticated syndrome is quite deleterious for free-living plants in nature, so plants growing outside of cultivation that have the domesticated syndrome of seed characteristics almost certainly have been planted (usually illegally) or have recently escaped from cultivation.

Cannabis sativa growing outside of cultivation is basically a weed, growing mostly in habitats created or modified by humans (Figure 3.1).

GEOGRAPHY OF WILD *CANNABIS SATIVA*

In Eurasia, weedy hemp is particularly widespread in southeast and central Asia and widespread in many European countries. To a lesser extent, *C. sativa* is found outside of cultivation in South America, Australia, and Africa (Davidyan 1972). According to Haney and Kutscheid (1975), *C. sativa* seldom becomes naturalized as a result of escapes from cultivated hemp in subtropical and tropical areas, perhaps indicative of the species being naturally adapted to a north-temperate climate. Wild plants in the Old World have adapted to various habitats for thousands (probably millions) of years, while those in North America have a history of only a few hundred years. Not surprisingly, in Eurasia, the species grows wild over an enormous range of climates and altitudes, much greater than in North America. Vavilov (1926b) observed vast stands of wild hemp in Eurasia. In the Himalayas, *C. sativa* occurs at altitudes of thousands of meters, whereas in North America, uncultivated plants are uncommon at sites over a few hundred meters.

In North America, *C. sativa* has been collected in natural areas from British Columbia to New Brunswick in Canada (Small 1972b, 1997; Small et al. 2003), and from the 48 conterminous states and Washington, DC (least collected in Mississippi and Idaho). However, many plants apparently growing outside of cultivation are in fact illicitly grown plants or are recent ("spontaneous") escapes from cultivation (in either case, the seeds reveal that they are domesticated races). In North America, weedy populations (possessing the characteristics of plants adapted to wild existence) are best established in the American Midwest and Northeast, and in southern Ontario and southern Quebec (especially along the St. Lawrence and lower Great Lakes), all areas where hemp cultivation was concentrated historically. Many wild North American populations may have been derived from escapes during the resurgence in cultivation in both Canada and the United States during World War II. Naturalized hemp is uncommon in the western United States, rare in the United States south of 37°N latitude, and very rare in Mexico (Haney and Kutscheid 1975).

The rate of spread of wild hemp was studied by Haney and Bazzaz (1970), using herbarium collection data for the United States from 1870 to 1970. Over the century, they observed a steady increase in the number of counties of the United States in which wild hemp was collected, from less

FIGURE 3.1 Ruderal (weedy) *C. sativa* in Turkey (above) and in eastern Canada (below).

than 50 in 1870 to more than 400 in 1970. Lush, extensive stands are developed in Iowa, Illinois, and Missouri, and more limited stands with smaller plants in most of the northeastern United States and Canada. In the western United States, where wild hemp is least successful, only small scattered colonies of hemp growing outside of cultivation are generally found, but these are often near cities, suggesting that they may be recent escapes.

Were it not for long-continued efforts by law-enforcement personnel to eradicate plants in North America, *C. sativa* would be more commonly encountered growing in natural areas. Given the resurgence of licensed cultivation of industrial hemp in Canada and the possibility that the United States will also permit such growth, escapes from cultivation will probably result in wild-growing plants becoming more common (perhaps even invigorated) in the future in North America.

HABITATS AND PLANT COMMUNITIES OCCUPIED

Vavilov (1926b) noted that in Eurasia, "Wild and weed hemp grow chiefly on the borders of the fields, in ravines, hollows, on rubbish heaps near settlements or habitations in general, near sown plots on soils that are rich and not sod-bound." As a weed in North America, *C. sativa* occurs in farmyards, waste places, vacant lots, in disturbed areas of pastures, occasionally in fallow fields, and along or beside roadsides, railways, ditches, creeks, fence rows, borders of cultivated fields, bridge embankments, lowland drainage tributaries, and open woods (Haney and Bazzaz 1970; Haney and Kutscheid 1975). The species seems very poorly adapted to penetrating into established stands of perennial vegetation and generally invades such areas only after the soil is freshly disturbed. As a colonizer, weedy hemp spreads slowly, except in drainage channels (the appellation "ditch weed" reflects its preadaptation to moist depressions). Weedy hemp in the United States is often found in alluvial sites disturbed by flooding (Haney and Bazzaz 1970). Haney and Kutscheid (1975) found that disturbance of the soil was extremely important in determining whether or not *C. sativa* would establish in Illinois, and moreover, the less fertile the site, the more hemp was dependent upon soil disturbance. Haney and Kutscheid (1975) also observed that wild hemp populations in Illinois do not persist at a given site, unless continuously disturbed, but may remain in a given area provided nearby disturbed sites are regularly generated. These authors noted that nearly all the populations they studied were on sites repeatedly disturbed by cultural activities, such as mowing, cultivation of soil, stream channel improvements, and refuse disposal. They also concluded that ruderal hemp is progressively more restricted to optimum and disturbed sites toward the ecological limits of its range. A survey of 1774 wild hemp stands in Kansas by Eaton et al. (1972) showed that 94% of the stands were associated with such past soil disturbances as construction, land clearing, tillage, livestock trampling, or flooding. Wild hemp was found in borders of cultivated fields (55%; primarily fence rows), on nonagricultural land (24%; primarily right-of-ways and bridge embankments), native pastures (12%), abandoned fields (4%), cultivated pastures (3%), and in cultivated fields (2%). Wherever it is found, ruderal *C. sativa* usually associates with other weeds, as one would expect from the habitats and sites described here.

CLIMATE LIMITATIONS

Reflective of its extensive geographical distribution, ruderal *C. sativa* occurs in a wide range of climates. Most accounts of the meteorological limits of *C. sativa* refer to the domesticated plant, which has narrower tolerances than the wild-growing counterpart.

TEMPERATURE

Cannabis sativa does not tolerate cold temperatures well, but once again, the weedy forms are more stress-tolerant than domesticated plants. In northern areas, the seeds germinate at lower temperatures and the seedlings and young plants survive frost better than do cultivars (Haney and Kutscheid 1975). Haney and Kutscheid (1975) observed that ruderal plants in Illinois germinated in early April, at lower temperatures than most other weedy annuals, and moreover that frost-damaged seedlings were never observed. Ruderal plants may grow more rapidly in cold temperatures than cultivars grown locally, but it should be remembered that wild populations have been selected for optimal growth under local environments for many generations while cultivars have often been selected in foreign environments. One may presume that wild plants that have become adjusted to near-tropical conditions near the equator have different physiological adaptations from wild plants adapted to temperate and subarctic areas.

WATER

Both domesticated and wild plants of *C. sativa* are tolerant of hot, arid conditions provided that the roots are adequately supplied with water, but ruderal plants in Europe have been observed to be much more drought resistant than cultivars (Janischevsky 1924). In North America, Haney and Bazzaz (1970) noted that wild hemp in sandy soils in Illinois survives dry conditions in deep, loose-textured soils by virtue of the roots growing to gain access to deep water sources. The general absence of weedy hemp in western North America has been explained in part on the basis of the relative dryness of the West (Haney and Bazzaz 1970). While *C. sativa* prefers moist soils, it does not tolerate waterlogging. As noted in Chapter 7 regarding hemp cultivars and in Chapter 12 regarding marijuana strains, some cultivated biotypes are susceptible to fungi in humid climates, but presumably, wild plants that have been present in an area for many generations have become adapted to the local humidity.

LIGHT INTENSITY

Both domesticated and wild plants of *C. sativa* develop best in full sun, and weedy plants thrive in open areas. However, some wild plants have been observed growing moderately well in shaded habitats in Europe (Janischevsky 1924) and Canada (Small et al. 2003). Janischevsky (1924) concluded that wild plants were more shade-tolerant than cultivated plants. In general, however, *C. sativa* usually develops comparatively poorly in the shade.

PHOTOPERIOD

As detailed in Chapter 5, induction of flowering in most populations of *C. sativa* (except those adapted to the northernmost and tropical areas) is controlled by relative length of light from day to day, the plants normally flowering in the shortening days of late summer, although drought often speeds up maturation. Ruderal populations that have become adapted to their local region flower early enough to allow seeds to mature before a killing frost. Because of the briefer season, wild plants tend to be shorter at more northern latitudes and higher altitudes, although more suitable cultural conditions promote larger plants everywhere.

EDAPHIC (SOIL) LIMITATIONS

Cannabis sativa grows best on sandy loams. Because it is very intolerant of waterlogged soils, weedy hemp seldom survives on soils high in clay that retain water. Haney and Kutscheid (1975) found that weedy hemp in Illinois appeared to tolerate many soils (but not those with more than 40% clay), provided that the roots receive adequate aeration. Well-drained bottom land is particularly attractive to ruderal hemp, and it is often found near streams and creeks.

Nitrogen is the element that most plants in nature find to be so deficient in availability that growth is limited. As expected, therefore, most of the world's plant species are adapted to substrates in which nitrogen is in short supply, and very few are nitrophiles (nitrogen-loving). Plants classified as nitrophiles are dependent on high levels of available nitrogen in the substrate and strip soils readily of nitrogen. Wild *C. sativa* is a nitrophile, growing lushly in nitrogen-rich habitats such as well-manured sites, but growing very poorly when the element is deficient (Figure 3.2). Indeed, one of the dependable indicators of the possible occurrence of weedy *C. sativa* is the presence of nearby manure or livestock. Vavilov (1926b) noted that wild hemp in Russia thrives in low places and ravines into which wild animal excrement is washed and on soils fertilized by grazing cattle. Manure not only supplies nutrients, but also the humus is important in retaining moisture that hemp demands (Dewey 1914). Weedy hemp is often found on the margins of crop fields, where the plants can take advantage of fertilizer run-off.

Weedy hemp in the United States has been collected on sandy soils very low in nitrogen, but the plants are dwarfed (Haney and Bazzaz 1970). Like many weeds, *C. sativa* is very flexible, able to survive as a dwarf in infertile ground but responding with dramatically increased growth to a good supply of soil nutrients (Janischevsky 1924).

It is well known that *C. sativa* requires ample calcium to grow well. This is consistent with the species' natural adaptation to circumneutral soils that are not notably acidic. It also seems to reflect

FIGURE 3.2 *Cannabis sativa* plants subjected to deficiencies of (left to right) nitrogen, phosphorus, and potassium and (at right) a control that received complete nutrition. These phenotypic responses indicate considerable flexibility with respect to soil infertility, allowing the species to produce at least some seeds under extreme conditions. Small et al. (1975) reported that the dwarfed plants were more or less comparable to the control in concentration of the intoxicating ingredient THC.

the large amounts of calcium that the plant accumulates as calcium carbonate at the base of the very numerous cystolithic hairs that cover the above-ground shoot, and as numerous calcium oxalate crystals within the leaves and stems (Bergfjord and Holst 2010).

SEED ECOLOGY

Cannabis sativa reproduces by small, slightly flattened one-seeded fruits (classified as achenes), usually referred to as "seeds." Haney and Kutscheid (1975) stated, "Seed production data probably provide the greatest insight to the ecology of hemp." Wild (ruderal or weedy) hemp, almost everywhere it occurs, was derived from plants that escaped from cultivation in the past, and re-evolved characteristics suited to wild existence. As detailed in the following, the seeds of wild *C. sativa* are remarkably suited to survival in nature by means of a number of adaptations. Wild populations that have been free from the selective influences of cultivation for many generations manifest extreme development of these adaptations, and conversely, populations only recently escaped from cultivation show only weak or no development of these features.

APPEARANCE

The seeds (achenes) of wild plants have morphological features that easily distinguish them from the seeds of plants that are cultivated, either for fiber, oilseed, or illicit drugs (Small 1975a; Figures 3.3 through 3.5). These features, discussed in the following, are clearly adaptive.

Size

In nature, all plants arrive at a compromise between the number of seeds produced and their size. Like humans, some plants put considerable energy into generating a small number of progeny, which they very carefully nurture. Very large, one-seeded fruits like coconut, which provide a huge

FIGURE 3.3 Achenes ("seeds") of *C. sativa* (areas of attachment to the plant are uppermost). Left side shows two achenes of a domesticated plant, right side shows two achenes of a ruderal plant. The domesticated fruits are larger, lack a camouflagic persistent covering layer derived from the perianth, and lack an elongated attachment base that facilitates disarticulation in the wild form.

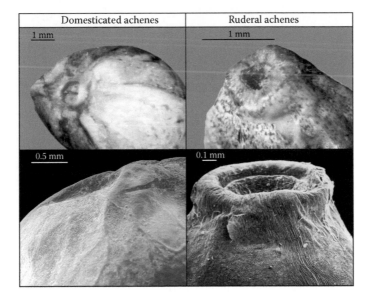

FIGURE 3.4 Comparison by light microscope (above) and scanning microscope (below) of attachment area of domesticated achenes (left) and wild (ruderal) achenes (right). In the wild fruits, a well-developed abscission area is present, and a basal "neck" that facilitates disarticulation is evident.

amount of "milk" for the embro, exemplify this. More typically, plants produce a very large number of quite small seeds in the expectation that only a few will find suitable habitats and escape destruction. Dandelions exemplify the extreme of producing a huge number of tiny seeds. *Cannabis sativa* is intermediate in producing a medium number of moderately sized seeds, by plant standards. The species has adopted a strategy that allows small plants to produce at least some seeds, and larger plants to produce many seeds.

The seeds of wild plants of *C. sativa* are smaller than those of domesticated plants (Figure 3.3). Seed weight in *C. sativa* varies enormously, from more than 1000 seeds to the gram in some wild Asian plants to less than 15 seeds to the gram for some cultivated plants (Vavilov 1926b; Watson and Clarke 1997). Usually, the seeds of wild strains are smaller than 3.8 mm in length, in contrast to the larger seeds of domesticated selections. The ecology of the species may differ considerably according to the size of the seeds, and this remains to be studied. The smallest seeds occur in ecotypes

FIGURE 3.5   (a) Light microscope view of perianth (petal tissue) covering seed (achene) of a wild form of *C. sativa*. Notice that pigmentation (dark areas) is associated principally with vascular (fluid-conducting) tissue. (b) Scanning electron micrograph of perianth-derived achene covering layer; note areas of straight, aligned cells making up vascular tissue, and areas of nonpigmented areas (regions of cells with wavy margins). (c) Scanning electron micrograph of exocarp (fruit wall) of an achene of a wild form, showing portion covered by persistent perianth and areas where this covering has sloughed off. Photos from Small, E., *Can. J. Bot.*, 53, 978–987, 1975.

native to alpine foothills of the Himalayas, possibly an adaptation to surviving at an ecological limit that is so stressful that it permits only a minimal amount of seed reserves to be accumulated before the plants expire. Large size of seeds in domesticated plants is usually the result of selection for a desired product in the seeds (frequently for food), but also, larger seeds provide a greater store of food reserves for successful germination and establishment. As discussed in Chapter 8 dealing with oilseed cannabis, larger seed size of domesticated seeds could be partly due to the practice of deep planting, requiring large vigorous seedlings to be able to grow out of the deep soil.

Seed Shedding

In nature, plants reproduce mainly by distributing propagules, mostly seeds and fruits (occasionally vegetative tissues), commonly by wind, water, gravity, and cooperating wild animals. Humans have domesticated many wild plants, frequently specifically to harvest the seeds or fruits. Most wild plants cast off their seeds or fruits as soon as they mature, by various mechanisms. This has two undesirable consequences from an agricultural perspective: when a seed or fruit drops away, it is more difficult to collect; and when seeds or fruits do not remain attached to the plant at maturity, it necessitates repeated collection of propagules from each plant over the weeks that they sequentially mature. Selecting mutations that inactivate the separation mechanism (abscission, i.e., breaking away of fruits at their base so they fall away) or the dehiscence mode (i.e., opening of fruits to release seeds) with the result that the mature seeds or fruits remain on the plant greatly facilitates harvest. This reduction of "shattering" (natural shedding of seeds at maturity) is the most important way that humans have domesticated the majority of crops (Harlan 1995; Fuller and Allaby 2009). Cereals currently supply more than half of the calories consumed by humans (Small 2009), and in all of them, a "domesticated syndrome" of characteristics is recognizable whereby the "grains" (fruits technically termed caryopses) have lost the features in their wild ancestors that cause them to detach and scatter away (see, for example, Sakuma et al. 2011).

Although the precise anatomical and morphological features that facilitate release of cereal grains and the seeds of *C. sativa* differ, one can recognize a syndrome of seed characteristics that differentiate wild *C. sativa* from domesticated kinds of fiber hemp, oilseed hemp, and marijuana plants. The key feature enabling rapid release of seeds at maturity is the presence of an "abscission zone" (area of weak cells that weaken at maturity) at the base of the seed. At the base of the cannabis seed, one often views a circular scar (sometimes referred to as a "horseshoe" in European literature), the remnants of a plane of weakened tissue that served to disarticulate the seed from the stem to which it was attached. In conjunction with this feature, the base of the seed is often attenuated (elongated and narrowing), a characteristic that assists the seed to fall away easily from surrounding tissues of the infructescence (fruit-bearing portion of the plant). The combination of an attenuated base and a well-developed abscission zone in wild seeds (illustrated in Figures 3.3 and 3.4) facilitates disarticulation as soon as the fruits are ripe, and this is essential given the considerable predation by birds on seeds that remain attached to the plant.

In addition to the features described here, humans have selected seed-bearing crops that mature most of their seeds more or less simultaneously, to minimize harvest loss. As well, a highly congested fruit axis (adjacent seeds very close together along with bracts and young leaves) makes it very difficult for seeds to fall away from the plant and facilitates harvest of the seeds. A corn cob illustrates just how close together seeds can be packed and how most of them mature more or less simultaneously. These features are evident in plants selected for yield of seeds (as described in Chapter 8). By contrast, wild *C. sativa* plants mature seeds sequentially over a long season, and the seeds are relatively well separated so that they do not interfere with each other's ability to fall off the plant.

Protective Thick Shell

In contrast to domesticated seeds, wild seeds have a comparatively thick fruit wall or "hull." This (1) protects the seeds against mechanical abrasion; (2) makes it more difficult for herbivores to break open the seeds and consume the embryos; (3) makes it more difficult for water to penetrate, so that the seeds won't germinate and will remain dormant, providing for germination over several years; and (4) keeps oxygen from entering and degrading the storage oil. Possibly, the thick shell also serves to protect the seeds from being digested while they are in the gut of some animals, so that they will be deposited and germinate in the animal's excrement, an environment to which cannabis is adapted.

Camouflage

A camouflagic mottled layer covers the seeds of wild *C. sativa*, providing some protection for the fallen seeds against mammalian and insect herbivores. The layer is developmentally homologous with the perianth—the petals and sepals of many flowers. The dark appearance of wild seeds also contributes to their being inconspicuous. By contrast, as discussed in Chapter 8 dealing with oilseed cannabis, light-colored seeds have often been selected in cultivated plants.

GERMINATION BEHAVIOR

Dormancy

Unlike the seeds of cultivated varieties of *C. sativa*, wild seeds of the species are generally at least somewhat dormant and germinate irregularly, features that obviously adapt the plants to the environmental fluctuations typical of wild habitats. The dormancy requirement is typically satisfied by a period of cold treatment, aging, or some other natural condition furnished by nature. Haney and Kutscheid (1975) found that some fully mature Kansas wild hemp seeds could be germinated in the laboratory within three weeks of maturity in the fall, while seeds stored at room temperature reached maximum germinability in three months, but seeds stored at 5°C reached maximum germinability only after five months (immature seeds, if viable, required a considerable period before they could be germinated). Janischevsky's (1924) study of Russian wild hemp seeds showed that fewer than 10% could be germinated immediately after maturation, but that repeated watering and drying cycles increased germination (water-soluble germination inhibitors may be present in wild hemp). Vavilov (1926b) and Scholz (1957) noted that the germination of Russian wild hemp proceeded very slowly and intermittently, the seeds often remaining dormant for weeks or even months, with typically only 10% germinating promptly.

Longevity

Dormancy is a natural adaptation delaying germination, but in *C. sativa* the delay is not much longer than a few years, since the seeds are not naturally long-lived. Haney and Kutscheid (1975) reported that seeds from ruderal Kansas populations declined in viability from 70% to 4.4% in 15 months of soil burial, an observation suggesting that seeds do not persist in a viable state in the soil for more than two or three years. Indeed, recommendations to control weedy volunteer hemp often mention that the site should be viewed for possible reappearance of the plant for two or three years (e.g., Illinois Bureau of Investigation, undated; Eaton 1972). Small et al. (2003) observed that volunteer plants appeared for up to four years following experimental cultivation of wild plants in Ottawa, Canada. Goss (1924) tested germination of hemp seeds that had been buried in soil for 24 years and observed no germination in tests conducted over several years. Under artificial controlled conditions, the seeds of *C. sativa* can be stored for decades, as described in Chapters 7 and 17.

Environmental Factors Controlling Germination

Although seeds of *C. sativa* will germinate in the light or in darkness, B.J. Eaton (unpublished typescript circulated 1972) noted that light partially inhibited the germination of wild hemp of Kansas. This may represent a natural adaptation for seeds to await the darkness of being buried by soil before germination. Wild (feral) seeds germinate better after a period of storage at cold temperatures, while domesticated seeds germinate quickly on being watered (Small et al. 2003), reflecting adaptation to remaining dormant overwinter by wild seeds. Clearly, wild seeds benefit from a period of cold stratification to overcome germination inhibitors (Small and Brookes 2012). As noted in Chapter 7, wild seeds in northern areas germinate at lower temperatures in the spring than the seeds of domesticated plants, indicating adaptation to the cold spring temperatures of north temperature climates.

SEEDLING DEVELOPMENT

Seedlings normally appear in early spring and grow rapidly in suitable habitats. Once germination begins, a radicle (root of the embryo) emerges and grows into the soil (Figure 3.6a). The next significant event is the emergence from the fruit of the two cotyledons (primary leaflets). One of the two cotyledons of the seedlings is usually larger than the other (Figures 3.6b,c,d and 3.7),

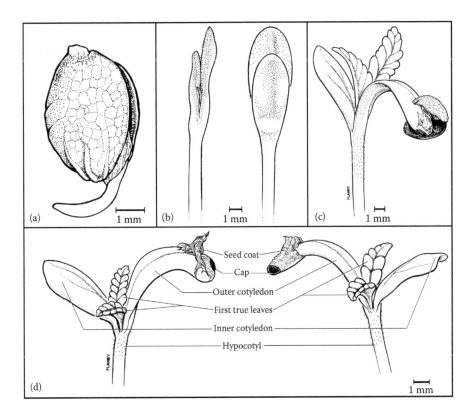

FIGURE 3.6 Seedling development of *C. sativa*. (a) Germinating achene ("seed"). (b) Cotyledons recently emerged from achene. Left, edge view of cotyledons. Right, lamina view of cotyledons. Note that one cotyledon (the "inner cotyledon") is characteristically larger than the other (the "outer cotyledon"). (c) Pericarp (fruit coat) persisting on the "outer cotyledon" of a seedling. (d) Complementary views (rotated 180°) of a seedling in which a portion of the seed coat is persistent as a "cap" on the outer cotyledon. Compare anatomical details in Figure 8.3. Drawings by B. Flahey.

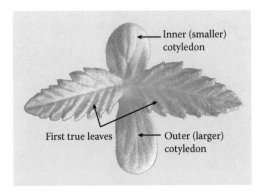

FIGURE 3.7 Top view of young seedling. Photo by Avriette (CC BY 3.0).

a phenomenon termed "anisocotyledony." This has been recorded for *C. sativa* in the literature (e.g., Walter 1938; Berkman 1939; Clarke 1981), although it is not well known. The first paired true leaves are simple (each with just one leaflet). The second and/or third paired true leaves are compound (each with more than one leaflet), with serrate (saw-edged) margins, and the leaf arrangement during early development is decussate (opposite, each succeeding pair rotated 180°), a combination of features that distinguishes *C. sativa* seedlings from those of most other weedy species.

Small and Antle (2007) discovered that in about 10% of seedlings, the clamshell-like fruit coat (pericarp) persists on the larger ("outer") cotyledon (Figure 3.6c) but falls off by the time the seedling becomes several days old. A dark, somewhat thickened area of the seed coat (the "cap") also persists on the tips of the outer cotyledons in about 25% of seedlings (Figure 3.6d). Seedlings of most plants are very difficult to identify when they have just developed cotyledons, but the features noted in this paragraph can be very helpful in identifying them as *C. sativa* (Small and Antle 2007).

SEED DISTRIBUTION AGENTS

A variety of animals could serve to distribute the seeds of *Cannabis* in nature (Figure 3.8). Humans, other mammals, birds, and insects have been proposed as biotic disseminating vectors for wild hemp, and water movement may also be important. Since the plant often grows in hilly and submontane habitats, gravity could also play a role. Because wild *C. sativa* is dioecious, the most effective dispersal agents should distribute at least a seed of each sex to a given site, although pollen is distributed so widely that even isolated plants may participate in reproduction.

FIGURE 3.8 The range of animals potentially significant in dispersing seeds of *C. sativa*. Prepared by B. Brookes.

Birds

Since birds are strongly attracted to the seeds, Haney and Bazzaz (1970) suggested that they are likely the most important wild animals distributing them in North America. Janischevsky (1924), working on the ecology of ruderal Russian hemp, noted that birds are usually observed consuming seeds on cannabis plants but are much less frequently seen on the ground in pursuit of fallen seeds. Darwin (1859) observed that "Some seeds of... hemp... germinated after having been from twelve to twenty-one hours in the stomachs of different birds of prey." Presumably the only hemp seeds that could germinate after passing through the digestive tract of birds were not macerated by the beak or gizzard. B.J. Eaton (unpublished typescript circulated 1972) fed wild hemp seeds to upland game birds, noting that quail passed approximately one viable seed/700 seeds consumed, and doves passed one viable seed/12,400 seeds consumed. It is also possible that some seeds are transmitted by adhesion to claws or bills (Merlin 1972).

Hemp seed is a superb bird feed (Chapter 8), and not surprisingly, wild bird feed has frequently been the cause of plants becoming established in a locality (Potter 2004). In recent years, most jurisdictions in Western countries have insisted that hemp seeds be steam sterilized. However, sometimes, a small percentage (much less than 1%) of such sterilized hemp seeds in commercial bird feed can germinate (Small, unpublished), indicating that hemp seed in current bird feed may still be a continuing source of spontaneous plants.

Water

Weedy hemp in North America is often found in alluvial sites disturbed by flooding, and flood waters may serve to distribute the seeds by water transport (Haney and Bazzaz 1970). Hemp seed has been widely used as fish bait in fresh waters (Potter 2004; see Chapter 10), and this too could result in distribution and establishment. Fish are known to distribute some seeds (Anderson et al. 2009), but whether this is the case for *Cannabis* has not been investigated.

Mammals

Ruderal hemp clearly depends heavily on human activities for dispersal. As noted in Chapter 2, large wild herds of mammalian grazers probably were important to providing manured habitats for *Cannabis*, and the species characteristically grows in moist areas, so the mammals may have distributed seeds caught up in mud on their hooves. In more recent times, domestic livestock may similarly serve as distribution vectors. Small rodents are attracted to the seeds of *C. sativa* and may play a role in their dispersal. The extent to which the seeds of *Cannabis* can survive a journey through the digestive system of mammals remains to be examined, but since some seeds will survive the digestive tract of birds, the same is likely true for mammals.

Insects

As illustrated in Figure 3.9d, insects, especially ants, commonly drag seeds of certain plants to their nests in order to consume a fleshy edible portion that these plants supply on the outer part of the seed. Plants often produce such an "elaiosome," i.e., a fleshy edible appendage on seeds or fruits serving to attract dispersal vectors, constituting an adaptation for distribution of the seeds. Janischevsky (1924) alleged that he had discovered such a symbiotic relationship between wild hemp and the red bug *Pyrrhocoris apterus* L. He observed it apparently feeding on the attenuated base (the attachment area) of the achene, and concluded that the base was an elaiosome. *Pyrrhocoris apterus* is widely distributed in Europe and Asia, and Janischevsky speculated that the wide distribution of ruderal hemp was in part due to the insect. However, the insect is a generalized feeder that has no fidelity to *Cannabis* (Kristenová et al. 2011), and the bases of wild achenes of *Cannabis* and its relative *Humulus* do not develop genuine elaiosomes, although the detachment zone of the achenes is a weak area of the protective pericarp and might offer some limited nutrition to insects. Whether insects are significant dispersers of the seeds of *Cannabis* is unclear, but it is possible.

FIGURE 3.9 Comparison of (a) achene of a wild form of *C. sativa*, (b) seed of castor bean (*Ricinus communis* L.), and (c) achene of Japanese hop (*Humulus japonicus* Siebold & Zucc.). In castor bean, the basal (attachment) area is a genuine elaiosome (an edible tissue attracting insects, particularly ants, for seed dispersal), as shown in (d). In *C. sativa* and its close relative *Humulus*, including Japanese hop, the basal area probably does not function as a classical elaiosome. Photograph b (public domain) is by S. Hurst, U.S. Department of Agriculture; c (public domain) is by C. Ritchie, U.S. Department of Agriculture; d is by B. Brookes (from Small, E., *Biodiversity*, 12, 186–195, 2011).

ALLELOPATHY

Allelopathy refers to the chemical influence (generally inhibitory) of one species by another. Most often in plants, chemicals are released into the environment, where they affect the development and growth of neighboring plants. Allelopathic chemicals can originate from any part of the plant but frequently come from the roots and enter the soil surrounding the plant. Such "chemical warfare" tends to suppress competing nearby plants. *Cannabis sativa* has minor allelopathic properties (Inam et al. 1989; McPartland 1997b; McPartland et al. 2000; Pudełko et al. 2014), the compounds that it releases into the soil possibly assisting it to restrain at least to a small extent some competing plants. The cannabinoids and terpenes that *C. sativa* produces are located in the aerial parts, not the roots, but portions of the plant that fall to the ground because of age or wind could perhaps leach chemicals into the soil and tend to be protective. The fallen plants in the autumn may provide a mulch rich in allelopathic chemicals that protect seeds that remain in the location and germinate the next spring. The seeds themselves are sometimes "contaminated" with resin from their surrounding perigonal bracts, and the terpenes and cannabinoids in the resin could be protective against a variety of organisms. *Cannabis* seeds have been shown to be antibiotic (Ferenczy 1956), but whether this is due to resin contaminating the seed surface or to internal compounds is not clear. For additional discussion of allelopathy in *C. sativa*, see the section "Adaptive Purpose of the Cannabinoids" in Chapter 11.

ROOT ECOLOGY

Root ecology is a key to the success of wild hemp, as revealed by Haney and Kutscheid's (1975) study of wild Illinois hemp. Root development is conditioned by soil texture, compaction, and

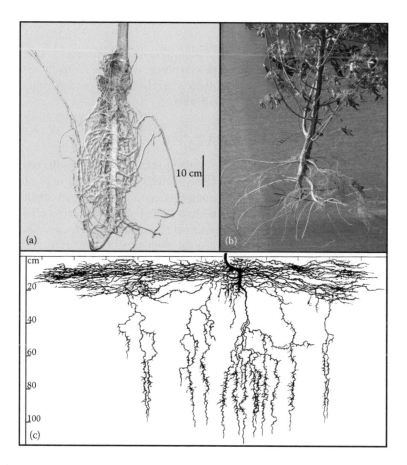

FIGURE 3.10 Root systems of *C. sativa*. (a) A deep tap root developed in friable soil with a low water table. (b) A shallow root system with well-developed lateral fibrous roots, exemplifying roots produced in substrates with high water tables or compacted soil. The 10-cm scale applies to both (a) and (b). (c) Vertical profile of root system in a sandy-loamy soil in a hemp field in Klagenfurt, Austria. Notice the two distinctive root development zones, one near the surface, with some roots growing quite deeply. From Kuchera, L., *Root Atlas of Central European Arable Weeds and Crops*, DLG Verl.-Ges, Frankfurt am Main, Germany, 1960 (in German) (CC BY SA 2.5).

moisture content. In coarse textured, well-drained, deep soils, a main taproot develops, with slim lateral branches (Figure 3.10a). The taproot can extend more than 2 m down, allowing access to a low water table. In medium-textured, moderately water-retentive soils, the primary root develops to a depth of about 1 m, with extensive laterals concentrated in two locations: near the surface and at about 1 m (Figure 3.10c), a bet-hedging strategy enabling acquisition of both surface and moderately deep water. If the water table is near the surface (generally undesirable for good growth of *C. sativa*), the root system is shallow (Figure 3.10b). Particularly in very moist soils, considerable lateral branches develop near the top, so much so that the root system may appear to be fibrous rather than tap-rooted. Similarly, when grown in pots, the ability to develop a long tap root is constrained, and the root system becomes highly branched.

Symbiotic relationships of the roots and fungi are often beneficial for plants. The vesicular-arbuscular mycorrhizal fungus *Glomus mosseae* (Nic. & Gerd.) Gerd. & Trappe has been recorded on *Cannabis* (McPartland and Cubeta 1997). The ecology of symbiotic microorganisms associated with *Cannabis* has scarcely been examined. (See entries regarding symbiotic bacteria associated with the roots of *Cannabis*, in "Curiosities of Science, Technology, and Human Behavior" in this chapter and in Chapter 18.)

STEM ECOLOGY

Shoot and foliage architectural adaptations are discussed in detail in Chapter 6. Domesticated fiber plants are drastically altered by comparison with wild plants, which are relatively wind resistant due to low stature and woodier, flexible stems. Domesticated oilseed plants and marijuana strains more closely resemble wild plants in the nature of their stems.

PHENOTYPIC PLASTICITY: A KEY TO SUCCESS

Phenotypic plasticity is "the ability of individual genotypes to alter their growth and development in response to changes in environmental factors" (Barrett 1982). It is flexibility of response and allows a population to survive in a broad range of environments, especially marginal conditions. It is a key component of the genetic system of weeds (Bradshaw 1965; Baker 1974) and is often critical to the ability of species to diversify and adapt in response to natural and human-caused selection (West-Eberhard 2003). By virtues of several of its ecological adaptations, *Cannabis* has exceptional adaptive phenotypic plasticity. These adaptations include: ability to grow large and produce many seeds in ideal environments or to survive as a dwarf and produce at least a few seeds in very inhospitable circumstances; roots develop as a widespreading fibrous system near the surface, where deep soils are too waterlogged, or grow very deep as a taproot to find water in dry soils (Figure 3.10); damaged main stalks regrow abundantly from lower branches (see Chapter 6); and tolerance of shade, despite being primarily a full-sun plant.

BIOTIC COMPETITORS

Most observations of pests and diseases of hemp are based on domesticated plants, and the applicability of these observations to ruderal plants is unclear. Weedy *C. sativa* differs from its cultivated relative in several respects that bear on susceptibility to diseases and pests. Wild plants are much less homogeneous than most cultivars and usually exist in much smaller stands, rarely growing as densely as hemp monocultures. Wild plants of *C. sativa* probably have the advantage of natural biological control (i.e., a wide variety of predators and parasites that regulate the pests) and are toughened by limitations of nutrients, unlike the luxury consumption that results in relatively delicate cultivated plants. All these factors suggest that while weedy hemp is susceptible to most of the same pests and diseases as are its pampered cultivated cousins, the degree of damage is probably much less.

Human Eradication Efforts

In some areas, wild-growing *Cannabis* is listed in regulations calling for eradication of noxious weeds—objectionable not because of its intoxicating qualities but simply because weeds are considered undesirable. In most of the world, weedy *Cannabis* is of limited concern, although there have been long-continued efforts by law enforcement to eradicate ruderal plants in North America. In contrast to the huge social costs, the deleterious effects of *Cannabis* as a weed in North America are relatively minor. Discovery of extensive growth of ruderal hemp on a farm often invites unwelcome attention, from the legal authorities as well as from delinquents who mistakenly believe that ruderal hemp in North America is as intoxicating as high-quality marijuana. As an agricultural weed, however, ruderal hemp is of limited importance (Small et al. 2003). Decades of eradication have exterminated many of the naturalized populations in North America. Mechanical eradication or tilling for two or three years is effective at destroying populations, and young plants are easily eliminated by herbicide applications. Liquid-propane flaming is also occasionally used to control *C. sativa*. All of these techniques are effective while the plants are still reasonably immature, but once seeds are produced, it is generally necessary to postpone additional controls until early in the next season. In Kansas, most wild hemp stands were found to be unsuitable for tilling (Eaton et al. 1972), and disturbing the soil frequently assisted wild hemp to invade and establish. Planting a competitive perennial grass such as fescue or smooth bromegrass has been recommended to eliminate the recurrence of plants in non-crop areas (Weber 1978).

PESTS AND DISEASES

As detailed in Chapter 7, there are numerous animal, fungal, bacterial, and viral species that affect domesticated *C. sativa*, although generally the species is resistant to all biotic attacks. The extent to which the same problems affect wild plants has been inadequately documented, but it may be presumed that the wild plants are generally less affected than their cultivated counterparts because they are tougher by nature.

CURIOSITIES OF SCIENCE, TECHNOLOGY, AND HUMAN BEHAVIOR

- Birds are well known to occasionally become drunk by feeding on fermented berries, and there may be a parallel situation with respect to hempseeds. As noted in Chapter 11, while the seeds (achenes) of *C. sativa* do not contain intoxicating constituents, resin from the surrounding bracts can coat the seeds, and thus marijuana varieties could produce seeds with appreciable intoxicant ability. As pointed out in Chapter 8, cannabis seeds with adherent resin have been shown to be capable of making birds giddy. Because birds love the seeds passionately, it is possible that some birds have learned to appreciate the intoxicating qualities of marijuana strains.
- Eliminating wild hemp, the present governmental policy throughout North America, is a very bad idea for many reasons. As discussed in Chapter 17: the plants are too low in THC to have abuse potential; although they are weeds their damage potential to agriculture is insignificant; and they harbor valuable genes for breeding improved cultivars of industrial hemp and medical marijuana. Afzal et al. (2015) found another potential usage of wild hemp: as a source of friendly bacteria that can improve the root ecology of crops. The bacteria in question are not the well-known nitrogen-fixing "rhizobacteria" that form symbiotic relationships with plants but are "endophytic bacteria," which enter plant roots, thrive within internal tissues, and provide growth benefits by producing plant hormones which increase nutrient availability. Afzal et al. (2015) concluded that "Wild *C. sativa* is a good source of agriculturally beneficial endophytic bacteria."
- Wild cannabis grows so abundantly in Kyrgyzstan that the country has facetiously been termed "Marijuanastan." There have been extensive efforts to eradicate the plants, which allegedly have relatively high levels of THC. Given that marijuana strains are commonly grown in the area, it is likely that wild plants there have higher levels of THC than in North America.

4 Sex Expression

WHAT IS "SEX" IN FLOWERING PLANTS?

Male animals produce sperm, females produce ova (eggs), and the combination of one sperm and one egg produces a new individual. Some animals are hermaphrodites, producing both sperm and eggs, but generally, identifying male and female animals is straightforward. Higher (flowering) plants are different, and introductory botany courses provide details on the varied patterns that one encounters. An essential difference between animals and flowering plants is that adult plants do not directly produce either sperm or pollen; rather, tiny multicellular female plants are produced in the ovary of the flowers (these are parasitic, relying for sustenance on nutrients absorbed from the flower), and even tinier multicellular male plants known as pollen grains are also produced inside the flowers (in the anthers). The tiny females are stationary, awaiting visits by the mobile tiny males. The reason that flowering plants have adopted this system is that they lack a penis to introduce sperm to the eggs, or they don't grow in a habitat where water is available in which the sperm can swim to the eggs in the ovaries. Rather, flowering plants rely on pollinators or the wind to make the transfer, and when the tiny male plants arrive at the pistil (female part of the flower), they produce a "pollen tube" (analogous to a penis), which introduces the sperms to the eggs. Only pedantic botanists concerned with technical details insist on referring to the tiny male and female plants (these are termed "gametophytes") as the only plants that truly possess sex and insist on interpreting the large flower-producing plant as sexless (these are termed "sporophytes").

Most flowers produce both sperm-producing and egg-producing gametophytes; i.e., the eggs and sperm emanate ultimately from the same flowers, so the plants bearing such flowers cannot reasonably be identified either as male or female (they are functionally hermaphrodites). But, as noted in the following discussion, in a minority of species (like *Cannabis sativa*; Figure 4.1), there are some plants just with flowers that produce only pollen and some plants just with flowers that produce only ovaries (and eggs), and it is common practice for the former to be identified as "males" (even though strictly they just produce tiny male plants, which produce sperm; such plants are technically termed "staminate"), and it is also common practice for the latter to be identified as "females" (even though, strictly, they just produce tiny female plants, which produce eggs; such plants are technically termed "pistillate").

MALENESS AND FEMALENESS IN *CANNABIS* IN COMPARISON TO OTHER FLOWERING PLANTS

The sixth century AD *Qi Min Yao Shu* ("Essential Arts for the People"), which has been characterized as the first Chinese scientific treatise, stated (Mignoni 1997–1998): "If we pull out the male hemp before it scatters pollen, the female plant cannot make seed. Otherwise, the female plant's seed production will be influenced by the male plant's scattering pollen during this period of time. The fiber of the male plant is the best." This passage appears to be the first recorded comprehension of sexual differentiation in *C. sativa*, preceding the recognition by about a thousand years, by Western botanists, that sex exists in at least some plants. Textbooks commonly identify European botanists of the seventeenth and eighteenth centuries as the discoverers of sex in plants (*Nature* 1933), but they appear to have rediscovered what was known long ago in China. Certainly, it was not until the eighteenth century that the true nature of sex in flowering plants became widely appreciated (Anonymous 1933).

FIGURE 4.1   Flowering plants of *C. sativa*: male at left, female at right. Note the comparatively frail appearance of the male.

The wild plants of *C. sativa* are among the small minority (4% according to Yampolsky and Yampolsky [1922], 6% according to Renner and Ricklefs [1995], 7% in publications cited by Divashuk et al. [2014], or some undetermined higher figure according to Bawa [1980]) of flowering plants with male reproductive organs (stamens) and female reproductive organs (carpels) confined to separate plants (i.e., the populations are "dioecious," with unisexual flowers, those on a given plant either entirely male or entirely female). Familiar food plants with separate male and female plants include asparagus (*Asparagus officinalis* L.), date (*Phoenix dactylifera* L.), hop (*Humulus lupulus* L.), kiwi (*Actinida deliciosa* (A. Chev.) C.F. Liang & A.R. Ferguson), papaya (*Carica papaya* L.), and spinach (*Spinacia oleracea* L.). Staminate plants, with male flowers only, are routinely called males, and pistillate (carpellate) plants, with female flowers only, are called females, and this standard terminology is followed here.

DIFFERENCES BETWEEN MALE AND FEMALE CANNABIS PLANTS

After seeds germinate, *C. sativa* normally grows vegetatively (i.e., without producing flowers) for several months. Following the summer solstice, shortening days (actually nights that become longer) induce flowering in most populations, as discussed in the next chapter. Floral buds ("primordia") are normally initiated in midsummer, growing from axils of leaves (the crotches between the leafstalks and the adjacent stem), at first along the top of the main stalk, later along branches. Once flowering commences, production of fully developed foliage ("fan leaves") slows down, with newer leaves being relatively small and bearing fewer leaflets. Flowers develop proceeding from the base of flowering branches upward to the top of the inflorescence (the branching system bearing the flowers). The flowers of *Cannabis* are small but very numerous.

Before flowers are developed, male and female plants of *C. sativa* cannot be reliably distinguished by appearance. Nevertheless, the sexes are different, not just with respect to reproductive organs: male plants at maturity tend to be 10%–15% taller, although less robust than the female plants, with slimmer stems, less branching, smaller leaves, and a more delicate appearance (Figure 6.1). Minor morphological differences (such as density of stomates and glandular trichomes) have also been alleged to be present between preflowering males and females (Truță et al. 2007). Before sex in plants was widely understood, eighteenth century European botanists (males at the time, reflecting their perception of masculine superiority) often referred to the vigorous females as males and the wimpy males as females (Bouquet 1950; Stearn 1974).

Male plants die after shedding their pollen. Female plants protected from frost can remain alive for years (gradually losing vitality), although the species is normally an annual. However, cloned biotypes of female marijuana plants are often regenerated for many years by repeated cuttings, which do produce new vigorous plants, while the older plants show their age.

MALE PLANTS AND THEIR FLOWERS

Male plant morphology is specialized for pollen dispersal by wind; males are taller than females, with large, showy, highly branched inflorescences that provide large exposure of the exposed stamens to the wind (Figure 4.2). The male inflorescences are loose, axillary, and relatively diffuse (technically they are cymose panicles or thyrses), as shown in Figure 4.3. Male flowers are pedicellate (i.e., with individual stalks), with five greenish or whitish tepals (distinct petals and sepals are not present) and five stamens with flaccid filaments (stalks) opposite the tepals. The male flowers fall away after anthesis (pollen shedding).

Male flowers at anthesis are extremely attractive to bees, including bumble bees and honey bees, which collect substantial amounts of pollen (which can therefore be present in commercial honey). Pollen-collecting flies are also often present. However, these insects do not visit the female flowers and so do not play a role in pollination.

FIGURE 4.2 A well-developed male plant. A plant of this size produces enough pollen to fertilize every plant in even the largest hemp plantations, and this phenomenal capacity means that very few, if any, males are needed in most cultivated fields.

FIGURE 4.3 Flowering branch with male flowers shedding yellow pollen. Photo by Erik Fendersson, released into the public domain.

POLLEN DISPERSAL

The sole purpose of the males is to produce pollen (Figure 4.4a and b), and they excel at this task: a single flower can produce about 350,000 pollen grains (Faegri et al. 1989), and there are hundreds of flowers on larger plants. *Cannabis* pollen rapidly loses viability after three days, but some grains can live for more than a week (Choudhary et al. 2014). Pollen from wind-pollinated plants tends to fall to the ground at distances approximating the inverse square law (i.e., proportionally to the inverse of the square of the distance from the source), although wind direction can considerably modify how far the pollen is transported (note Figure 4.4c and d). While the inverse square law dictates that most wind-distributed pollen is deposited close to the source, some can be distributed for very long distances (Proctor et al. 1996), so it is likely that there is occasional genetic interchange among remote populations. It has been claimed that *Cannabis* pollen has been carried by wind for over 300 km (Clarke 1977). Cabezudo et al. (1997) noted that *C. sativa* pollen, apparently from marijuana cultivated in North Africa where it is common (Aboulaich et al. 2013), was transported by wind currents to southwestern Europe. Hemp pollen is a significant allergen for some people (Chapter 13), so its presence is often monitored. Stokes et al. (2000) recorded that in August, in the midwestern United States (where cultivation of hemp is not permitted, but weedy hemp is common), hemp pollen represented up to 36% of total airborne pollen counts! However, Lewis et al. (1991) found smaller amounts in Texas (mostly around 1% of *Cannabis* + *Humulus*, the two kinds of pollen taken together because they are very difficult to separate). Because the pollen of *Cannabis* spreads remarkably, an isolation distance of about 5 km is usually recommended for generating pure-bred seed, exceeding the distance for virtually every other crop (Small and Antle 2003). Because of widespread clandestine cultivation, the pollen can be found, at least in small concentrations, over much of the planet.

FEMALE PLANTS AND THE POLLINATION OF THEIR FLOWERS

Female flowers occur in axillary inflorescences (technically termed "spicate cymes") that are smaller and relatively obscure compared to male flowers. They are also much more congested than the diffusely distributed male flowers. The tiny female flowers consist of a unilocular (one-celled) ovary and a short apical style with two long filiform (thread-like) stigmatic branches (Figure 4.5b–d). Very small papillae (thread-like surface extensions; Figure 4.5a) on the stigmas are the landing platforms for pollen grains and represent the pollen-receptive part of the stigmatic branches. The stigmas are densely covered with receptive papillae to receive pollen (Figure 4.6).

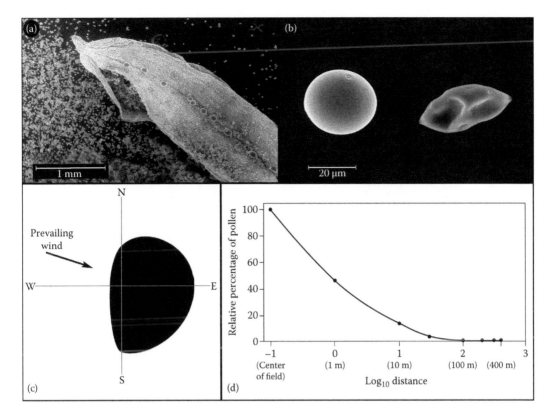

FIGURE 4.4   Pollen production and distribution in *C. sativa*. (a) Scanning electron microscope photo showing copious release of pollen from an anther (dehiscence is longitudinal, i.e., along the long axis). (b) Scanning electron microscope photo showing hydrated pollen grain (left) and dehydrated grain (right). (c) Geographical pattern of pollen dispersion in relation to prevailing wind direction, based on a hemp plantation study site in Ottawa, Canada. The pollen source is located at the intersection of the axes. (d) Mean relative percentage of pollen collected from a hemp study site plotted against $\log_{10}$ of the distance from the edge of the hemp field, illustrating inverse square law pattern of decreasing spread of pollen from the source. All figures are from Small, E., Antle, T., *J. Ind. Hemp*, 8(2): 37–50, 2003.

The pollen-receptive part of the stigmatic branches may extend all the way to the ovary (Figure 4.5c) or only part way (Figure 4.5d). Unlike the male flowers, the female flowers are essentially sessile (lacking stalks). A perigonal bract (sometimes called a floral bract) subtends each female flower (Figure 4.5b) and grows to envelop the fruit (this is important in intoxicating resin production, and additional detail is given in Chapter 11). In contrast to the male flowers, the female perianth (sepals and/or petals) is not at all recognizable as conventional floral bloom material, consisting of a thin undivided layer adhering to the ovary (this unusual anatomical feature is very important ecologically as discussed in Chapter 3, and for classification purposes, as discussed in Chapter 18).

REMARKABLE DEVELOPMENT OF UNPOLLINATED FEMALE FLOWERS

Because *C. sativa* produces enormous amounts of pollen, the stigmas of female flowers normally are pollinated, and this stops them from growing longer than about 3 mm. By contrast, illicit and medicinal marijuana are usually produced in the absence of pollen. The style-stigma parts of such virgin pistils expand notably in length (averaging over 8 mm), so that the female flowers have extremely prominent stigmas (Figure 4.7). Frequently, the stigmas also increase greatly in diameter

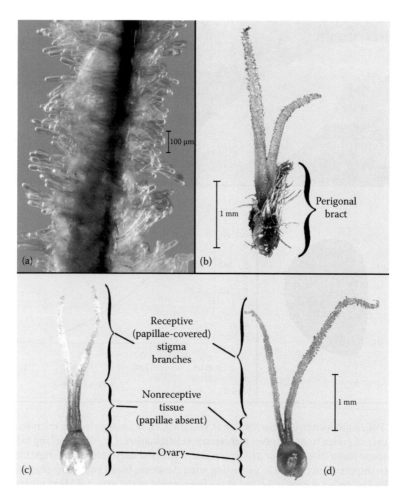

FIGURE 4.5 Female flowers of *C. sativa*. (a) Close-up of a stigma showing the pollen-receptive papillae. (b, c) Young pistils with two stigma branches. (b) Cultivar Yvonne, showing the two stigma branches emerging from a young, protective perigonal bract, which is covered with glandular and cystolithic trichomes. (c) Cultivar Canma, showing a pistil with the two pollen-receptive (covered with papillae) stigma branches emerging from a common, nonreceptive base ("style") attached to the ovary. (d) Cultivar Canda, structurally like b, but the two style/stigma branches emerge independently from the ovary.

and occasionally develop more than the normal two branches (Figure 4.8). From an evolutionary viewpoint, this expansion of pollen-receptive tissue is an apparent adaptation for increasing the probability that pollen will be encountered, resulting in fertilizing the females when males are extremely scarce (Small and Naraine 2016a).

As discussed by Small and Naraine (2016a), the phenomenon of stigmas continuing to grow when they remain unfertilized is rare. Close to 90% of flowering plants have animal-pollinated flowers, and these (including their stigmas) necessarily have relatively fixed proportions, since their architecture must be compatible with the corresponding dimensions of their pollinators. By contrast, wind-pollinated species are not so constrained, and the flexibility of stigma size response to lack of fertilization, described here for *C. sativa*, has been documented for several other wind-pollinated plants, including spinach (*Spinacia oleracea* L.) and a walnut species (*Juglans mandshurica* Maxim.). The female flowers of maize (*Zea mays* L.) feature a pair of thread-like styles each terminated by a stigma (in maize, a style + stigma is termed a "silk"). The styles of unfertilized maize flowers continue to elongate for 10 or more days, and indeed, excessively elongated silks are widely employed

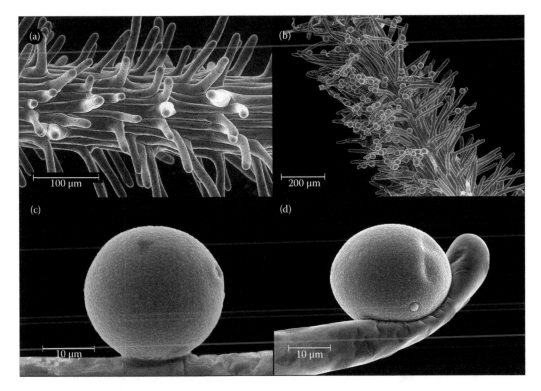

FIGURE 4.6 Pollen collection on the stigma of *C. sativa*. (a) Unpollinated papillae of stigma. (b) Numerous pollen grains trapped on papillae of stigma. (c, d) Pollen grains on stigmatic papillae. Scanning electron microscope photos by E. Small and T. Antle.

FIGURE 4.7 "Buds" (unfertilized, congested female inflorescences) of marijuana showing extremely prominent stigmas. Left: Terminal branch of a marijuana strain (Purple Erkle) with clustered female flowers, the whitish appearance of the stigmas indicating that many are receptive to being fertilized by pollen. Right: Terminal branch of a marijuana strain (Bullrider) with clustered female flowers, the brownish appearance of the stigmas indicating that the stigmas are overmature for pollen reception. The transition from white to brownish or orange stigmas is commonly used as an index of the ideal stage of maturity for harvesting the female inflorescences for marijuana (buds). Photos by Psychonaught, released into the public domain.

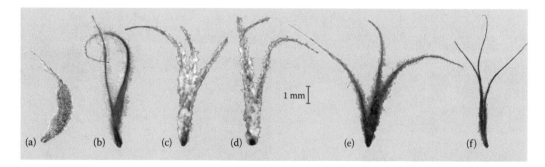

FIGURE 4.8 Old unfertilized stigmas that have expanded and/or divided into extra branches. (a) Stigma of an unfertilized wild plant showing a massively developed basal portion. (b) A two-branched stigma. (c–f) Three-branched stigmas. (b–f) Based on a dried "bud" of the medicinal marijuana strain Dinachem. From Small, E., Naraine, S.G.U., *Genet. Resour. Crop Evol.*, 63, 339–348, 2016.

FIGURE 4.9 Stigmas from a "bud" of the medicinal marijuana strain Dinachem, showing how THC-rich secretory gland heads accumulate on the papillae of the stigmas. From Small, E., Naraine, S.G.U., *Genet. Resour. Crop Evol.*, 63, 339–348, 2016.

by corn growers as a sign that the crop has been inadequately fertilized. Whether the stigmas or the styles or both (as in *C. sativa*) enlarge, the effect is the same: making it more likely that the stigmas will encounter pollen and so the flowers will be fertilized.

The expanded amount of stigmas in marijuana "buds" has important consequences. First, the high-THC secretory gland heads of *Cannabis*, where most THC is located, tend to fall away from dried marijuana, significantly decreasing pharmacological potency, but many gland heads become stuck to the receptive papillae of the stigmas, reducing the loss (Figure 4.9). Second, although stigmas constitute a small proportion of marijuana—about 0.5% according to Small and Naraine (2016a)—their distinctive chemistry could have health effects. This has not been examined in *Cannabis*, but it is known that in some plants the stigmas can be toxic (Small and Naraine 2016a).

MONOECIOUS ("HERMAPHRODITIC") KINDS OF *CANNABIS SATIVA*

Many modern cultivars, especially those selected for stem fiber production, are monoecious (with both male flowers and female flowers on many of the plants in a population). In nature, monoecious cannabis plants are rarely encountered. Monoecious hemp plants were first deliberately developed in 1935 (Bócsa 1998). In monoecious forms, staminate flowers are normally located on the upper part of flower-bearing stems and are produced before the female flowers on the lower parts of stems; male flowers are also produced before transitional hermaphroditic flowers (some of which are sometimes sterile), which are also often encountered. The expression of sex in monoecious strains is quite variable, with numerous occurrences of intergradation between maleness and femaleness (Finta-Korpel'ová and Berenji 2007; Faux and Bertin 2014). In a monoecious population, individual plants may be entirely female, entirely male, predominantly of one sex (i.e., most flowers are of one sex), or more or less intermediate. Some flowers may be hermaphroditic (with both a pistil and stamens), and the inflorescence on a plant may develop both male and female flowers. Populations may exhibit various proportions of the different kinds of plant, including 100% male, 100% female, and a spectrum of plants with intermediate sexuality (a population structure that has been termed "subdioecy"). Sometimes, plants have the appearance (habit) of females but bear only male flowers (such plants have been called "feminized males"), and the opposite of this (plants with female flowers but looking like males) have been called "masculinized females." Environmental stresses during development can cause a reversal in the kinds of flowers that continue to develop. So-called "sexual chimeras" are plants that develop secondary branches with flowers of the opposite sex compared to the rest of the plant.

While male plants almost always die after shedding pollen, the presence of even a few female flowers on hermaphroditic plants seems to protect them against dying before seed set (personal observation). However, in a plantation setting, there is a much reduced need for the prodigious pollen production that is normal in the wild plants, so hermaphroditic plants tend to be bred that are predominantly female.

In dioecious hemp cultivars (with male flowers only on some individuals and female flowers only on others), rare monoecious individuals have been observed to occur spontaneously. It has been estimated that this happens with a frequency of 0.1% (Finta-Korpel'ová and Berenji 2007). Recently escaped plants are occasionally monoecious, but monoecy is associated with inbreeding depression and is therefore very rare in wild *C. sativa*, which is naturally strongly outcrossing (Heslop-Harrison and Heslop-Harrison 1969). Monoecy typically reduces plant vigor and productivity. Monoecy is also associated with smaller, less vigorous pollen grains. The pollen grains of *C. sativa* are usually spherical and three pored, with the pores most often arranged along the equator of the grain at equal intervals. Migalj (1969) found that the acetolyzed pollen grains of dioecious strains tended to have a diameter averaging about 33 μm, while the grains of monoecious strains were smaller, with a diameter averaging about 27 μm, and the pollen of dioecious plants was also more uniform, while the grains of monoecious plants were more variable in size and in number of pores. Zhatov (1983) reported that hemp pollen viability in dioecious strains may reach 90% under favorable weather conditions but usually ranges from 60% to 80% and that after three days of storage pollen viability was 50%, dropping to about 16% after seven days; pollen viability in monoecious strains tends to be lower than in dioecious strains.

Clearly, monoecious cultivars would not be as popular as they are if their advantages did not outweigh their disadvantages. The predominant advantage for fiber production is uniformity of plant height and habit, in contrast to dioecious cultivars, in which the males and females mature at different times (although as pointed out by Bócsa 1998, forms in which males and females mature simultaneously were first described in 1935). The widespread use of monoecious cultivars promotes predictability of time of maturation and uniformity of fiber. In the case of oilseed production, all plants of monoecious cultivars are seed-producing, so space in the field is not occupied by unproductive male plants. For both fiber and oilseed cultivars, mechanized harvesting is facilitated by the uniformity of plants produced by monoecious cultivars. Although monoecious cultivars tend to have an unstable sexual expression, monoecious cultivars facilitate the harvest of both stems for fiber and seeds for oil by reducing crop heterogeneity (Faux et al. 2013).

CROSS-FERTILIZATION AND SELF-FERTILIZATION

Self-fertilization is possible in *C. sativa*: pollen from the male plants of a population can fertilize the females of the same population, and in monoecious plants, it is even possible for a plant to fertilize itself. However, inbreeding depression is pronounced. Van Lai (1985) observed that inbreeding resulted in plants with smaller and less fertile pollen, poorer fiber, and (especially) lower seed production, all of these effects being more evident in monoecious than in dioecious experimental lines. In contrast to the deleterious effects of inbreeding, hybrid vigor is frequently observed. (As noted by Stepanov 1974, 1976, and others, whether or not hybrid vigor results from a particular cross depends on genetic factors in the parents.) To promote outcrossing, male plants of a given population tend to come into flower one to three weeks before female plants have receptive stigmas. Dioecious plants are thought to be mostly or entirely cross-pollinated, while monoecious plants are selfing to some degree: 20%–25% according to Finta-Korpel'ová and Berenji (2007; cf. Horkay 1986).

GENETIC DETERMINATION OF SEX

Inheritance of sexual expression in *Cannabis* has been studied extensively but is not completely understood (Hoffmann 1970; Truţă et al. 2007). Sexual differentiation in dioecious strains is based on a pair of sex chromosomes, the male being heterogametic, i.e., XY (Hirata 1924, 1927, 1929), the Y chromosome frequently described as larger (Yamada 1943; Sakamoto et al. 1998; unlike mammals, but like some other plants), producing an approximately 50:50 sex ratio. Bócsa and Karus (1998) pointed out that females tend to be slightly more frequent in most hemp crops, with a ratio of 107–113 females to every 100 males. Menzel (1964) observed a heteromorphic (i.e., differently sized) pair of chromosomes in males of dioecious strains, but not in monoecious plants (consistent with the presence of an XX pair in the latter). Simmonds (1976) noted that "Genetic studies of monoecious lines have not clearly established whether such lines are all XX or are XX, XY, and YY, so the control of sex expression is not yet fully understood." Faux et al. (2014) found that the five monoecious strains they examined all were of the XX type. Ainsworth (2000) concluded that sex expression appears to be somewhat determined autosomally, with an X/autosome dosage type chromosome system. Flachowsky et al. (2001) reviewed this topic and concluded that monoecism is determined by two autosomal genes.

SEX-ASSOCIATED DNA MARKERS

As noted previously, there are often differences in certain sex-determining chromosomes between males and females of *C. sativa*, but these are often unreliable for purposes of identifying whether a plant is male or female. However, the DNA of the chromosomes is composed of sequences of nucleotides, and differences in sequences (polymorphisms) among individuals are now commonly employed to analyze relationships and for identification of cultivars. DNA sequences in *C. sativa* can also be used to identify sex. Sequences characteristic of male plants have been identified (Mandolino and Ranalli 1998; Mandolino et al. 1999; Törjék et al. 2002), and the same is true for females (Hong et al. 2003). Immature plants, even seedlings, can be sexually distinguished by DNA sequence, as demonstrated by Techen et al. (2010). Other studies of male-associated and female-associated DNA markers include Sakamoto et al. (1995, 2000, 2005), Flachowsky et al. (2001), Mandolino et al. (2002), Peil et al. (2003), Shao et al. (2003), Cristiana Moliterni et al. (2004), and Rode et al. (2005).

ENVIRONMENTAL DETERMINATION OF SEX

Sex development in *C. sativa* is labile, modifiable by a wide range of environmental factors, including nutrient excess or deficiency, temperature, mutilation, and light regime (Schaffner 1919, 1921;

Cheuvart 1954; Arnoux 1963). The proportion of female plants has been reported to be increased after exposure of seeds to ultraviolet light and by irradiation with gamma rays (Nigam et al. 1981) and decreased by shorter day-length during the growing season (Schaffner 1923, 1931) and higher nitrogen concentrations in the soil (Arnoux 1966a, 1966b). Such factors can result in sex reversal, and indeed, the aberrant production of plants with male, female, and intergradient flowers. In a survey of over 1400 U.S. herbarium specimens, 55% were male, but only 41% of the plants collected along streets and highways were male; Haney and Bazzaz (1970) speculated that this could be due to the higher carbon monoxide levels near roadways. This is supported by the finding that carbon monoxide has been shown to favor the development of female flowers (Heslop-Harrison and Heslop-Harrison 1957).

MODIFICATION OF SEX BY HORMONAL TREATMENT

Sex development can be influenced by hormonal treatments (Heslop-Harrison and Heslop-Harrison 1969). The application of auxins and ethylene feminize *Cannabis* (Heslop-Harrison 1956; Mohan Ram and Jaiswal 1970), whereas gibberellins are masculinizing (Atal 1959; Mohan Ram and Jaiswal 1972; Galosh 1978; Chailakhan 1979). Cytokinins, which naturally occur in roots of *C. sativa*, may also promote feminization, as evidenced by the fact that cutting away some of the roots tends to masculinize the plants (Hall et al. 2012). Mohan Ram and Sett (1982b) and others have shown that male flowers can be induced on female plants by chemical treatments. In particular, Mohan Ram and Jaiswal (1970) and Mohan Ram and Sett (1982a) demonstrated that ethrel (also called etephon and other names) favors the development of female flowers on male plants. Ethrel, very widely used on crops, works by being metabolically converted to ethylene (C_2H_4), a gas that functions as a hormone, regulating many aspects of growth and development of plants. Chemicals that inhibit the biosynthesis or the activity of ethylene (including aminotoxyvinylglycine, silver thiosulphate, and silver nitrate) induce the formation of male flowers, and by contrast, the precursors or activators of the biosynthesis of ethylene (like etephon) induce the formation of female flowers (Mohan Ram and Sett 1982a,b).

Cristiana Moliterni et al. (2004) concluded that the ability of some dioecious *Cannabis* strains to undergo sexual reversion has a genetic basis. Some biotypes such as the cultivar Carmagnola are very resistant to treatments that induce sex reversions, while plants of the cultivar Fibranova are much more easily induced to develop flowers of the opposite sex.

REDUCTION AND ELIMINATION OF MALES FOR HUMAN PURPOSES

CULLING OF MOST MALE PLANTS WHEN SEEDS ARE PRODUCED FOR INDUSTRIAL HEMP

Before monoecious cultivars were available, farmers realized that very few male plants were required for producing seed. Hemp grown for seed to reproduce the next year's fiber crop in the early decades of the twentieth century in the United States required amazingly low numbers of male plants; most were removed except for one or two per 25 square meters as soon as they could be identified (Dempsey 1975). Eliminating most of the males could have altered the gene frequencies of the material being grown, for better or worse. Hemp growers in the nineteenth century complained that their material was becoming less productive after several years, requiring importation of new seed stocks, and it is conceivable that eliminating males in fields used to produce seeds could have contributed to the problem.

CULLING OF ALL MALE PLANTS FOR THE PURPOSE OF PRODUCING MARIJUANA

For production of marijuana, male plants are usually eliminated before they can shed pollen to fertilize the females, as unfertilized female inflorescences ("buds") are highly valued (Figure 4.7;

see Chapter 12). Female marijuana plants have as much as 20 times the concentration of THC as corresponding males (Clarke and Merlin 2013). By contrast, male fiber plants, although also less productive than corresponding females, produce a higher quality of fiber and before the twentieth century were often so highly valued that they were harvested separately by hand, when labor was cheap. Today, males are considered undesirable for fiber because they senesce earlier and degenerate, thus decreasing the overall quality of fiber when harvesting is carried out just once. In former, labor-intensive times, when the male and female plants were hand-harvested separately, selection pressures were probably more or less equal for the sexes, or perhaps there was some preference for male plants. Monoecious varieties are commonly utilized today for fiber, so that all plants mature simultaneously and their quality is uniform. For production of oilseed, dioecious varieties are still often employed, although at present there are very few varieties exclusively used for oilseed production. Several "dual-purpose" varieties are grown for simultaneous production of fiber and oilseed, and these may be monoecious or dioecious. Because female plants are more valued for oilseed and drugs, selection has been much more directed to the females than the males.

SELECTION OF FEMALE-PREDOMINANT MONOECIOUS STRAINS

As noted above, monoecious strains can differ in proportion of female flowers on given plants, and proportion of mostly female plants in a population. Monoecious strains have chiefly been selected to generate fields of uniform plants for harvest of fiber prior to sexual reproduction, but of course seeds are needed to produce the next season's crop. For this purpose, relatively few male flowers are needed (unlike wild plants, which disperse huge amounts of pollen for long distances to find scattered plants, in a plantation a relatively small amount of pollen suffices). Energy that the plants use to produce male flowers is largely wasted from the farmer's perspective. Accordingly, the proportion of male flowers has tended to be decreased by selection.

GENERATION OF SEEDS PRODUCING "ALL-FEMALE" PLANTS

McPhee (1925) observed that selfing a normally female (pistillate) plant (using pollen from occasional spontaneous male flowers appearing on the female plant or from male flowers induced to develop by manipulating the environment) gave rise to seeds producing only female plants. In fact, the knowledge of this phenomenon is much older. Ascherson and Graebner (1908–1913) commented that older authors had observed isolated female plants bearing viable seeds that produced exclusively female plants. The first registered "unisexual hemp," the cultivar Uniko-B, which produces almost exclusively female plants, has been marketed since 1965. (Strictly, the F_2 [produced by open pollination of the F_1] is marketed as the cultivar, as the F_1 seed requires considerable manual labor to produce.)

   Using female plants, selfed with pollen that they are forced to produce by chemical or environmental sex-reversal techniques, is now a common way to produce "feminized seed," which is often marketed by the marijuana seed supply industry. Of course, this produces an extremely inbred plant, but once a female plant has been discovered with exceptionally attractive psychotropic properties, it is desirable to propagate its characters. As noted in the following, this is ideally done with 100% fidelity simply by cloning the plant; however, while the seeds produced from self-pollination will not reproduce the maternal plant exactly, the resulting plants will be closer in nature than had the seeds been generated by cross-pollination. As noted previously, only female plants are normally used for drug production, and the use of feminized seeds makes it unnecessary to remove male plants that are produced.

   Industrial hemp seed in sufficient quantity to plant entire fields can't be economically produced by the above technique, but an analogous hybridization trick is being employed. Special hybrid seeds are obtained by pollinating females of dioecious lines with pollen from monoecious plants, and these are predominantly female (so-called "all-female," these generally also produce some hermaphrodites and occasional males). All-female lines are productive for some purposes (e.g., they are very uniform, and with very few males to take up space they can produce considerable grain), but the hybrid seed is expensive to produce. As pointed out by Bócsa (1998), field populations of monoecious hemp left to reproduce naturally will revert to dioecious hemp over time, and so monoecious hemp strains can only be maintained by constant regeneration by humans.

USE OF ENVIRONMENT AND/OR HORMONES BY MARIJUANA GROWERS TO INCREASE FEMALES

As noted previously, environment and hormonal treatments can influence sex ratio or result in flowers of a different sex developing on plants that started off as the opposite sex. Clandestine marijuana growers have used these techniques to increase the proportion of females produced by a given lot of seeds (hardly necessary with today's knowledge of cloning female plants or using feminized seeds). In particular, marijuana growers have sprayed the growth regulator ethrel on young plants to feminize (one might say emasculate) the plants as much as possible.

CLONING OF FEMALE PLANTS

Humans propagate many crops vegetatively (e.g., apples, potatoes, and strawberries) as clones, a tactic to avoid the variability produced by sexual reproduction, in order to maintain a uniform genotype that is especially desirable. This is the method increasingly being used to propagate (female) strains of high-THC *Cannabis*, particularly the most desirable biotypes (Chandra et al. 2010b). Genetically identical plantlets can be generated rapidly using modern biotechnology (Lata et al. 2009, 2010a,b). In perhaps an ultimate departure from normal plant sexual reproduction, propagules of marijuana strains, generated by tissue culture, have been encapsulated to form "artificial" or "synthetic" seeds, sometimes called "synseeds" (Chandra et al. 2010a, 2013; Lata et al. 2011), a technique that is often employed in horticulture (Ravi and Anand 2012).

SUMMARY OF THE DEMISE OF MALE *CANNABIS* UNDER DOMESTICATION

Sexual selection is often recognized as a special kind of natural selection (Darwin 1859). It involves competition within a gender for the opposite sex and is important in evolution. In nature, males often are especially important in sexual selection. Human selection of the sexual characteristics of domesticated species is also a powerful evolutionary force, but by contrast, the males of domesticates have lost much of their importance. Farmers often favor females of livestock. Bulls are much harder to manage than cows, do not produce milk or calves, and humans only require a limited number to reproduce their captive herds, a situation reminiscent of the plight of the male cannabis plant. As detailed previously (and summarized in Figure 4.10), male *Cannabis* plants have also suffered significantly under domestication: (a) humans have created many cultivars that are monoecious (the plants bearing both male and female flowers), but a preponderance of female flowers has been favored; (b) cultivars have been created by hybridization that are entirely female; (c) seeds that produced female plants only have been created by selfing of female plants that have been tricked into producing a few pollen grains; (d) for marijuana production, male plants are usually eliminated; (e) for seed production, male plants have sometimes been reduced; and (f) clones maintained for drug production are female. A curious aspect of the sexual evolution of *Cannabis* under domestication has to do with the fact that humans have turned a normally dioecious species into forms that are monoecious. This constitutes reversing the normal pattern thought to exist in nature—that dioecious species have evolved from monoecious ones (Lewis 1942).

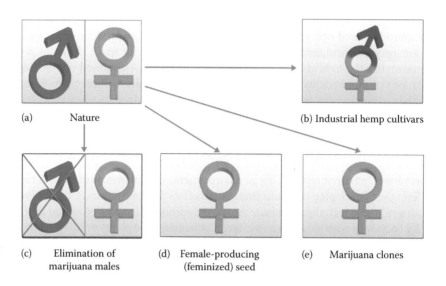

(a)   Nature   (b) Industrial hemp cultivars

(c)   Elimination of   (d)   Female-producing   (e)   Marijuana clones
       marijuana males   (feminized) seed

FIGURE 4.10   Reduction of frequency of male function in domesticated *C. sativa*. (a) Approximately equivalent occurrence of male and female plants in nature. (b) Generation of hermaphroditic plants (with reduced numbers of male flowers) in many industrial hemp cultivars. (c) Removal of male plants before they shed pollen, to generate seedless ("sinsemilla") marijuana buds. (d) Generation of "all-female" seed, which produces predominantly female plants, for production of female marijuana buds, and "female-predominant" industrial hemp hybrid cultivars. (e) Vegetative (clonal) reproduction of female plants for production of marijuana.

CURIOSITIES OF SCIENCE, TECHNOLOGY, AND HUMAN BEHAVIOR

- Shen-Nung was a legendary mythical emperor of China reputed to have lived about the twenty-eighth century BC. He is said to have learned to talk by the age of three days and to walk after one week. At the age of three years, he made the first plow and invented agriculture. He is reputed to have personally tested hundreds of herbs for their poisonous qualities, in order to find remedies, until one that was quite poisonous killed him. Despite this rather spectacular lapse of judgment for a genius, Shen-Nung is considered the Father of Medicine in China. Indeed, not long ago, it was common practice in China for drugstores to offer a discount on medicines on the first and fifteenth of each month in honor of this legendary patron of the healing arts. In China, *yin* symbolizes the weak, passive, negative feminine influence in nature, whereas *yang* represents the strong, active, positive, masculine force, and in most drug plants one finds them in combination. But *Cannabis* poses a problem for medicinal use according to Chinese traditions, because there are separate female and male plants. Shen-Nung advised that only the female plants should be cultivated for medicine. Thus, drugs from female *Cannabis* were used to treat cases involving a loss of yin from the body, such as are said to occur in menstruation, gout, malaria, beriberi, rheumatism, constipation, and absentmindedness.
- The heights of the king and queen of France were used to predict the comparative height of male and female plants. It was thought that if the king was taller, the male plants would grow taller than the females, and vice versa. If male plants were taller, the fiber from the plants would be better. (In fact, male plants are virtually always taller than the females and do produce superior fiber in comparison to the females.)
- Male plants of *C. sativa* shed pollen grain prolifically. Honey bees love the pollen and collect it avidly when available. Because *C. sativa* is a wind-pollinated plant, the flowers do not offer nectar to bees, which is the principal material that bees can use to make honey.

However, honey typically contains pollen grains that bees have collected, and so wherever hemp is grown outdoors and honey bee colonies are nearby, the resulting honey is very likely to have at least some content of *Cannabis*. The male flowers do secrete some resin, which can coat the pollen grains, so that one might suspect the bees could become intoxicated. However, as noted in Chapter 13, insects lack the cell receptors necessary for THC, the intoxicant chemical of marijuana, to be effective. In theory, humans who consume honey containing pollen grains from drug varieties of *Cannabis* could become intoxicated. In practice, this is a far-fetched scenario.

- As noted in the preceding text, male plants of *C. sativa* are relatively delicate in appearance and die at a much younger age. Confirming the weakness of the male sex, Haney and Kutscheid (1975) observed that two-thirds of many Illinois wild hemp populations they studied were female, and since the sex ratio in theory is 50:50, there is obviously a much larger mortality of male plants.

5 Photoperiodism

INTRODUCTION TO REPRODUCTIVE PHOTOPERIODISM IN *CANNABIS SATIVA*

Photoperiodism is a physiological reaction of organisms to the relative duration of daily illumination. More precisely, "Photoperiodism is the ability of organisms to assess and use the day length as an anticipatory cue to time seasonal events in their life histories" (Bradshaw and Holzapfel 2007). In effect, the sun is acting as a master timer, controlling the initiation of biological developments (Figure 5.1). In the following discussion, the term "photoperiodism" is used with specific reference to induction of flowering. Although Garner and Allard (1920) are given considerable credit for clarifying the influence of photoperiod on plant development, Tournois (1912) is thought to have first discovered photoperiodism in plants (see Jarillo et al. 2008 for a review of the subject). Based on studies of hemp and its relative Japanese hop, Tournois observed that flowering was promoted by short duration of daylight over consecutive day/night intervals (giving rise to the expression "short-day plants") and delayed by days with long periods of light. As with most other plants in which flowering is induced by a requirement for a given duration of darkness, a single exposure of a few minutes of light during the dark period can disrupt flowering and delay maturation (moonlight and lightning have no effect). "Long-night" plants would have been a better label than "short-day" plants, since the length of darkness, not the length of light, is the stimulus. In the northern hemisphere, nights (i.e., length of dark periods) increase after the summer solstice (about June 21), providing the flowering stimulus. Although not yet clarified for *Cannabis*, it appears that many flowering plants use a photoreceptor protein such as phytochrome or cryptochrome in the foliage to sense seasonal changes in day length and chemically transmit a signal to the plant's buds to flower.

Cannabis has been evaluated to be a quantitative (facultative) short-day plant—that is, flowering is normally induced by a required sequence of days each with a minimum uninterrupted period of darkness ("critical photoperiod"), but there is some flexibility in that at least some normally photoperiodic plants will flower eventually regardless of photoperiod. However, as detailed in this chapter, the photoperiodic nature of different populations in *Cannabis* differs notably, and some populations are not photoperiodic. For those plants that do respond to photoperiod, the photoperiod of *Cannabis sativa* has been reported to range from nine hours (Heslop-Harrison and Heslop-Harrison 1969) to 14 hours (Borthwick and Scully 1954). The critical photoperiod required to induce flowers is 10–12 hours of light for most hemp cultivars, often 13–14 hours for marijuana strains.

Cannabis may produce flower buds or at least undifferentiated primordial buds under continuous illumination (Borthwick and Scully 1954; Heslop-Harrison and Heslop-Harrison 1969); however, before these open, some cultivars require short days, while others will flower in continuous light, but only after a long period of growth (Schaffner 1926; Borthwick and Scully 1954; Heslop-Harrison and Heslop-Harrison 1969). The critical daylength may be longer for male plants than for female plants in a given population (e.g., 13 hours for males, 12.5 hours for females), which is consistent with the fact that males normally come into flower faster in response to shortening days (Borthwick and Scully 1954). Potter (2009) found evidence that "suggests that there is more than one critical daylength in *Cannabis*, one of which induces flowering and a shorter daylength at which vegetative growth is hormonally inhibited."

Flowering is induced in *Cannabis* mainly by lengthening daily hours of darkness following the summer solstice, but also to some extent by intrinsic, genetic factors. However, environmental stresses also have some effect on flowering time. Weather, site conditions, and management practices can modify the timing of floral initiation (Lisson and Mendham 2000b; Lisson et al. 2000a,

FIGURE 5.1   In photoperiodism, the sun operates as a master timer, the daily balance of light and dark stimulating key developments in organisms. Prepared by B. Brookes.

2000b,c). High temperatures accelerate flowering (e.g., van der Werf et al. 1994a), and drought is an especially important factor in speeding up maturation.

Marijuana breeders or generators of hybrid seed often need to coordinate the flowering times of male and female plants. Clones of both sexes can be maintained in a vegetative state under long days and brought into flower by short day treatment. However, males tend to be harder to control, and once they commence flowering, it is very difficult to have them revert to a vegetative state.

AUTOFLOWERING (DAY-NEUTRAL) PLANTS

As documented in this section, flowering may occur regardless of light regime in some populations. Plants at the extremes of the geographical range of *C. sativa*, either at their northernmost locations of survival or near the equator, appear to often come into flower because of intrinsic developmental reasons rather than photoperiodic regime (Figure 5.2). These plants that are indifferent to photoperiod are technically termed day-neutral, more commonly "autoflowering" in the marijuana literature. At or near the equator, seasonal photoperiodic cycles are insignificant, and indeed, the seasons are often longer than required for full development. In the northernmost locations, the season is so short that the plants would not have time to mature seeds if they waited until days began to shorten. Clearly, therefore, there are adaptive (survival) reasons why day-neutral races would have evolved in the northernmost regions where *C. sativa* occurs. Near the tropics, however, the day/night difference is small and does not provide much of a stimulus to retain photoperiodic capacity, and in any event, the plants have such a long season that they can grow almost indefinitely. Regardless of whether a population is day-neutral or photoperiodic, time of flowering is critical to the survival of populations of *C. sativa*, so that both wild and domesticated plants that have grown for many years in a region may be expected to have been selected for an appropriate flowering time.

Autoflowering strains can be very useful for growing marijuana indoors, since they can be grown under continuous light, which induces faster, greater growth (Potter 2014), and early-maturing strains may be especially suitable for production schedules. The autoflowering hemp cultivar FINOLA is ideal for growing in northern Scandinavia, where it can take advantage of the very long summer daylight.

Hybridization can play a sometimes unpredictable role in determining flowering time. Autoflowering marijuana strains that flower fairly early regardless of photoperiod have been described in the underground literature as having been generated by hybridization of short-season and long-season plants. It does seem that hybridization can produce odd effects on photoperiodic response;

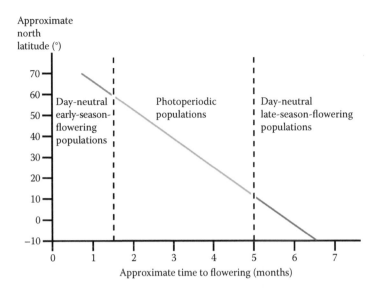

FIGURE 5.2 Conceptual diagram showing early-flowering day-neutral populations of *C. sativa* adapted to and occurring in the northernmost areas and late-flowering day-neutral populations adapted to and occurring in near-equatorial regions. Plants adapted to the intervening latitudes are induced to flower by shortening days, a relatively larger number of short days required for plants nearer the equator. The exact relationships between latitude and indifference or susceptibility to photoperiodism have not been determined, and because the climate and growing season at a given latitude vary according to longitude, a given latitude is unlikely to delimit the exact range of day-neutral plants. Prepared by B. Brookes.

I have observed hybrid-generated seedlings come into flower in less than two weeks, at a height of only 5 cm!

LATITUDINAL PHOTOPERIODIC ADAPTATION

A worldwide, north–south pattern of clinal (geographically graduated and genetically fixed) photoperiodic adaptation correlated with stature has evolved in *Cannabis* (Figure 5.3). "Bergmann's Rule" states that within a taxonomic group of animals, individuals are larger in colder environments (an ecogeographic generalization with mixed validity). For plants, the reverse is often the case: the shorter, colder season at higher latitudes (or altitudes) limits growth and accordingly stature. Annual plants like *Cannabis* are designed by nature to maximize propagule production, achieved in part by growing as large as possible within the limitations of the length of their season and the cultural conditions of their growth sites. It seems clear that the historical migration of *Cannabis* throughout much of the world for purposes of cultivation was accompanied by strong selection for local photoperiodic regime. During domestication, some populations could have been selected for photoperiodic insensitivity (like some cultivars of strawberry and other crops), but this has not been important for *Cannabis* (at least until recently). Wild plants and cultivars are photoperiodically adapted to their local climate; plants adapted to growth in northern areas tend to come into flower readily with shortening days, allowing time for seeds to mature before a killing frost; and conversely, plants adapted to areas closer to the equator tend to come into flower slowly with shortening days, in order to grow for a longer period in the milder environment. (This has been recorded numerous times in the non-English, twentieth century European literature; for examples, see Ranalli 1998.) Russian (U.S.S.R.) agronomists classified hemp into four ecogeographical maturation groups, respectively adapted to a longer season: Northern, Middle-Russian, Southern, and Far Eastern (Serebriakova and Sizov 1940; Davidyan 1972), so that races of *Cannabis* are available to meet the local photoperiodic requirements of most regions of the country.

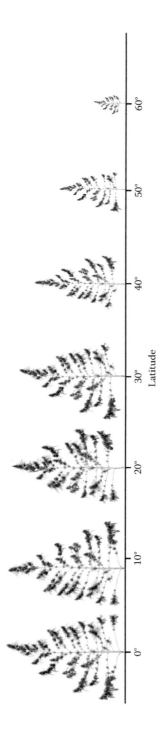

FIGURE 5.3 North–south gradient of stature of *C. sativa* reflecting adaptation to growing season, largely controlled by photoperiodic control of flowering time. Toward the equator, the season is longer than needed, and the plants tend to come into flower after they have reached a given stature. Toward the north of their range limit (about 66°N), the increasingly shorter season only allows progressively smaller plants to develop. Compare Figure 5.2. Prepared by B. Brookes.

Time of flowering is a critical determinant of yield in *C. sativa*. In most biotypes, induction of flowering substantially stops vegetative plant growth, so stem fiber and biomass are affected. As discussed in the chapters dealing with particular products for which the plant is grown (fiber, oilseed, marijuana, biomass, essential oil, or any other purpose), quality is often also affected. Problems that can arise from growing a strain in an inappropriate photoperiodic regime are examined in the following.

As noted previously, most hemp cultivars (mostly originating from temperate areas of the world) have critical photoperiods ranging from 10 to 12 hours. Most plants used for pharmacological or recreational marijuana originate from south-temperate and subtropical areas and so are adapted to being induced to flower by critical photoperiods that are closer to 12 hours than to 9 hours. Growers of indoor marijuana wish to provide as much light as possible to maximize growth but need to shorten the lighted period to induce flowering. Not surprisingly, in view of the fact that 12 hours is usually minimal to ensure flowering of marijuana strains, reducing light availability to a 12-hour period daily is almost universally recommended to induce flowering of plants grown for marijuana. Potter (2014) reported that about 90% of 200 marijuana strains examined flowered seven to nine weeks after the initiation of short-day treatment.

THE INTERACTION OF PHOTOPERIODISM, KILLING FROSTS, AND SEED PRODUCTION

The purpose of chickens is to make eggs, and similarly, the purpose of plants is to make seeds—as many as possible given a particular species' natural reproductive abilities and the permissiveness of its environment. In most of its range, *Cannabis* is capable of continuing to mature seeds until degraded and killed by frost. Like numerous other plants with many flowers, frequently, many flowers are fertilized, and seeds mature sequentially over weeks or even months, if seasonal conditions permit. If a region experiences an early frost, fewer viable seeds are produced, but if an unusually long season ("Indian summer") occurs, more seeds are produced. Accordingly, while the particular inherited timing of flowering is a natural limitation of how many seeds can be produced, there is some variation from year to year depending on climate variation.

The information presented in the previous paragraph needs to be distinguished from another concept—determinate vs. indeterminate growth. Tomato plants exemplify this point (they are perennials but are grown as annuals). Determinate tomato plants (usually small herbaceous shrubs) stop growing in late season, and there is a limited number of fruits (not necessarily a small number) that can be harvested. Indeterminate tomato plants (usually vines) keep on growing, keep on producing flowers that are fertilized, and keep on maturing new fruits. *Cannabis* normally exhibits determinate growth: once flowering is induced, vegetative growth (production of new branches and leaves) substantially slows and ceases while flowers develop, get pollinated, and transform into seeds. In *Cannabis*, as in most short-day plants, short days cause the plant to stop producing buds that produce leaves and branches and to start producing buds that produce flowers. Both vegetative growth and reproductive growth are more or less determinate (the species is, after all, an annual), but sometimes, under greenhouse conditions (which are hardly natural), female plants may continue growth.

PHOTOPERIODIC PROBLEMS FOR PLANTS OF NORTHERN ORIGIN GROWN CLOSE TO THE EQUATOR

There have been extensive studies of flowering time of European fiber hemp cultivars (e.g., Amaducci et al. 2008a,b,c). (By comparison, relatively limited research has been conducted on flowering photoperiodism of *C. sativa* adapted to semitropical and tropical regions.) For the northern hemisphere, higher latitude of origin is associated with earlier flowering and seed maturation (e.g., De Meijer and Keizer 1996a). Growing hemp cultivars adapted to the relatively short season

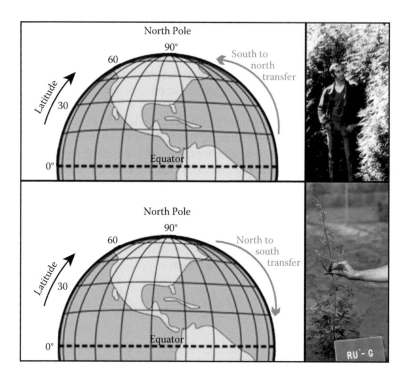

FIGURE 5.4 Examples of adaptation to local photoperiod that result in inadaptive induction of flowering when grown in an unsuitable latitude. Top: A race of *C. sativa* photoperiodically adapted to a long southern season but coming into flower very late when grown in a short northern season. These plants from southern Europe have remained vegetative when grown in Ottawa, Canada (latitude 45.4°N), shortly before a killing autumn frost. Bottom: A race of *C. sativa* photoperiodically adapted to a short northern season but coming into flower very early when grown in a long southern season. This male plant from Kazakhstan (ca. 50°N) is mature, about two weeks from death, although grown in Oxford, Mississippi (latitude 34.4°N) in the southern United States, where the growth season extends for several more months, and male plants adapted to the climate would continue to grow and reach several meters in height.

of more northern latitudes, in longer-season regions (closer to the equator), results in earlier than desirable flowering (Figure 5.4, bottom), a phenomenon that has sometimes been described as "pre-flowering" (Amaducci et al. 2008a).

European fiber varieties (representing most of the world's certified cultivars of *C. sativa*) were developed for regions north of the 45th parallel (Watson and Clarke 1997). These are mostly adapted to long summer daylengths, flowering in the autumn in response to shortening days. In equatorial regions, daylength is always short enough for these northern varieties to initiate flowering, and northern cultivars flower prematurely when 1–2 m high, limiting yield. This occurs, for example, in Australia (Ditchfield et al. 1997; Hall et al. 2014). Indeed, moving northern European cultivars as little as 10°–15° closer to the equator reduces performance substantially (Dippenaar et al. 1996).

PHOTOPERIODIC PROBLEMS FOR PLANTS OF SEMITROPICAL ORIGIN GROWN IN NORTHERN LATITUDES

The previous discussion points out the deleterious effect of moving plants, adapted to a northern photoperiod regime, closer to the equator. Conversely, when plants adapted to the photoperiod of semitropical climates are grown in north-temperate climates, they may mature so late that they succumb to cold weather before they can produce seeds (Heslop-Harrison and Heslop-Harrison

1969; Figure 5.4, top). For fiber production in relatively high-latitude and high-altitude locations, it is sometimes desirable to grow cultivars that mature very late (even too late to produce seeds before winter) in order to continue stalk production for as long as possible. (Depending on variety, however, growing for an excessively long period may lower fiber quality.) Of course, seeds for such cultivars would have to be produced in areas with longer growing seasons, in locations closer to the equator.

Photoperiodic differences are particularly apparent when *Cannabis* populations obtained from different latitudes are grown together in experimental plantations. In Ottawa, Canada, where I have grown over 1000 accessions outdoors, those from the northernmost locations (Siberia) sometimes produced seeds in less than a month after planting, while some from near-equatorial locations (India, Africa) sometimes remained vegetative after five months (and were killed by frost). When hemp cultivation was authorized in Canada in 1998 (after more than a half century of prohibition), the only source of cultivars with reliably low tetrahydrocannabinol (a requirement) was the European Union; embarrassingly, most of the cultivars were so late-maturing that they were unsuitable for Canadian locations. (It is possible to harvest vegetative plants of hemp for fiber, but Canadian plants are chiefly grown for oilseed.)

PHOTOPERIODIC PROBLEMS FOR SEMIEQUATORIAL DRUG STRAINS IN NORTHERN CLIMATES

Most drug forms have historically been cultivated in areas south of the north-temperate zone, sometimes close to the equator, where they may be photoperiodically adapted to near-12-hour days and an associated long season. In Chapters 12 and 13, two groups of marijuana plants are discussed; many strains of the less common "indica type" are able to mature in relatively northern locations. Although indica type strains originate from relatively southern areas of the Northern Hemisphere, they seem to mature earlier than sativa type strains because of adaptation to a shorter season due to drought. By contrast, non-marijuana plants (both wild and legally cultivated) are mostly found in north-temperate climates and are photoperiodically adapted to mature by the fall season in such locations. When drug strains are grown in north-temperate climates, maturation is much delayed until late autumn, or the plants succumb to cold weather before they are able to produce seeds. Before illicit marijuana growers became acquainted with the fact that most marijuana strains are very late-maturing, they often found that their clandestine outdoor plants remained vegetative, not producing the congested flowering tops (buds) that are most valued. Particularly in California, hybridization and selection produced marijuana strains that are capable of flowering outdoors (Clarke and Merlin 2013). Of course, photoperiod can easily be controlled indoors by varying light (or dark) period, which is one of the reasons why marijuana is commonly grown in buildings.

UNSUITABLE PHOTOPERIOD MAY NOT BE THE ONLY CAUSE OF POOR PERFORMANCE

In addition to photoperiodic adaptation, climate adaptation determines the success of *Cannabis* crops selected in one part of the world but grown in a quite foreign location. The point is that lack of adaptation to the stresses in a new location, caused by local climate and/or local biotic agents, can worsen the deleterious effects of unsuitable photoperiod. Most hemp cultivars (mostly fiber strains) were developed for relatively cool northern regions and do not perform well when moved closer to the hotter equator (Watson and Clarke 1997). This has occurred, for example, when European cultivars were planted in South Africa: flower initiation was induced in the five varieties grown only 34 days after sowing, extreme heat hastening maturation (Dippenaar et al. 1996). In addition, when cultivars are grown in foreign locations, adaptation to pests and diseases can be a problem. Northern fiber cultivars, when grown near the equator, have proven to be very susceptible to a wide range of biotic agents for which they have no resistance (Watson and Clarke 1997).

CURIOSITIES OF SCIENCE, TECHNOLOGY, AND HUMAN BEHAVIOR

- The floral industry relies on photoperiodism to bring some ornamental plants into flower to meet seasonal demand, especially for major holidays. The most important of these is poinsettia (*Euphorbia pulcherrima* Willd. ex Klotzsch; Figure 5.5). For proper floral development, the plants need to be kept in complete darkness (uninterrupted by occasional light) for a period of 12 to 14 hours daily for 8–10 weeks. Commercial greenhouse growers generally accomplish this by the use of shade curtains.
- Plants are not the only species to employ daylength to initiate important processes in their life cycles—so do many animals, especially mammals (Bradshaw and Holzapfel 2007). A familiar example to dog owners is the seasonal shedding of fur. Another example is the singing of male canaries, which is much more frequent in the spring because the photoperiod causes their testes to grow, and they want to attract females. Humans exhibit a wide variety of seasonal behaviors, but it is uncertain whether any of these are related to photoperiodic regime (Weir 2001; Bronson 2004).
- Certain plant species have flowers that open and close at approximately fixed times during the day, regardless of weather or season. For example, some species of morning glory do indeed open their flowers in the morning, while four-O-clock opens its flowers in late afternoon. The eighteenth century Swedish botanist Carl Linnaeus (the Father of Binomial Nomenclature, for which additional information is given in Chapter 18) designed (but may not have ever actually planted) a 12-hour daylight floral clock, composed of flowering species arranged in a circle, the times of opening or closing of their flowers roughly indicating the time of day. Botanists have replicated Linnaeus' clock in different places, only to often find that the clock did not work because the plants that flowered at the anticipated times in Sweden would not do so at different latitudes because of their photoperiodic requirements. Moreover, the circadian (daily) rhythms of the flowers are rather variable because of climate. For those wishing to test out Linnaeus' clock at their home location, see http://www.nytimes.com/2015/01/29/garden/planting-a-clock-that-tracks-hours-by-flowers.html?_r=0.
- Perennial ornamental plants can be arranged in a garden, sequentially according to their flowering times, to form a "bloom calendar." However, because flowering time of many species is determined by photoperiodism, and often influenced by environment, the suitability of given species will differ according to latitude and climate.

FIGURE 5.5   Poinsettia, second only to marijuana in frequency of being induced to flower indoors by control of photoperiod. Photo by Petr Kratochvil, released into the public domain.

6 Shoot and Foliage Architecture

This chapter is concerned with how human selection of *Cannabis sativa* for different purposes (fiber from the stem, drugs from the inflorescence, and oil from the edible seeds) has altered the external appearance of the "shoot" (above-ground part of the plant) by comparison with that of wild-growing plants. Stem features that are of particular adaptive importance to *C. sativa* include its main stem ("stalk") and patterns of branching with respect to the disposition of the foliage and reproductive organs. Leaf features that are of importance include size and width of the leaflets. Because male plants are less robust than females, die after flowering, and have not been nearly as subject to strong selection by humans as the females, features discussed in this chapter pertain particularly to female plants.

ANATOMY AND MORPHOLOGY

This chapter has considerable information on the anatomy (internal structure) and morphology (external structure) of *C. sativa*. Additional anatomical and developmental features are described in several of the chapters in this book where they are particularly relevant. Chapter 3, dealing with ecology, contrasts adaptive features of wild and domesticated plants. Chapter 4 describes male and female plants in detail. Chapter 7, on fiber content, deals with stem features, which are critical to fiber characteristics. Chapter 8, on oilseed, deals with structure of the seeds. Chapter 11, on chemical aspects, describes the drug-secreting glands of the plant. Mediavilla et al. (1998) present a detailed analysis of the morphological life cycle of the plant. Chapter 8 in Hayward (1938) provides a detailed analysis of the developmental anatomy of *C. sativa*. The composite plate shown in Figure 6.1, perhaps the best scientific drawing of the hemp plant ever prepared, illustrates the most important parts of the cannabis plant.

BASIC LEAF BOTANY OF *CANNABIS*

The leaf of *Cannabis* is probably more widely recognized than the foliage of any other plant. The smallest leaves are located at the branch apices and are "unifoliolate" (bearing one leaflet; "unifoliate" is incorrect, since it means "one-leaved"). The larger leaves are multifoliolate (with several leaflets). The multifoliolate leaves tend to be decussate on the lower stem (with opposite pairs, the succeeding pairs turned 180°), usually alternate near the stem apex, petiolate (with a leaf stalk), palmately compound (the bases originating from a common point) with an odd number (3–13) of coarsely serrate, lanceolate leaflets, with the proximal pair (i.e., closest to the stem) notably small. The foliage and stems of some populations are sometimes anthocyanin-streaked, and frost often causes plants to similarly become suffused with purple; as discussed in Chapter 12, this represents one of the kinds of coloration that has been preferentially selected in some marijuana strains.

SHOOT ARCHITECTURE OF WILD PLANTS

Figure 6.2 shows the appearance of well-developed wild plants, which characteristically produce a main stem from which many branches arise. As with numerous annual herbaceous plants, ultimate size depends on the availability of nutrients, water, and light; and crowding from competition tends to suppress lower branching and promote vertical growth. In a given wild population, one may find plants that are less than 30 cm in height, and some exceeding 2 m. The widespread assertion on the Internet that there is a unique wild species, "*Cannabis ruderalis*," that is quite short,

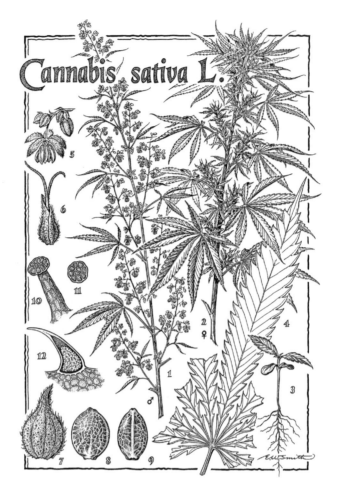

FIGURE 6.1 Illustration of the organs of *C. sativa*. 1. Flowering branch of male plant. 2. Flowering branch of female plant. 3. Seedling. 4. Leaflet. 5. Cluster of male flowers. 6. Female flower, enclosed by perigonal bract. 7. Mature fruit enclosed in perigonal bract. 8. Seed (achene), showing wide face. 9. Seed, showing narrow face. 10. Stalked secretory gland. 11. Top of sessile secretory gland. 12. Long section of cystolith hair (note calcium carbonate particles at base). Reproduced with the permission of Economic Botany Library of Oakes Ames, Harvard University, Cambridge, MA, drawn by E. Smith.

is rubbish—very short plants growing outside of cultivation have simply developed in a stressful environment or are adapted to short seasons and so do not have time to become large. Janischevsky (1924), the author of the name *C. ruderalis*, noted that when his alleged species grew in well-manured sites, the plants grew very large.

SHOOT ARCHITECTURE OF PLANTS GROWN IN DIFFERENT FIELD CONFIGURATIONS

The height and branching pattern of *C. sativa* have been altered in domesticated plants in ways that maximize production of the desired product (stem fiber, drugs from the inflorescence, or seeds). These differences have become genetically fixed by selection but are accentuated by density of planting because, just as with trees in forests, crowding suppresses branching and promotes vertical growth. The various agricultural field configuration patterns that are commonly encountered are shown in Figure 6.3 and are discussed in the following paragraphs.

FIGURE 6.2 Strong branching patterns typical of well-developed, open-grown, wild (ruderal) female plants of *C. sativa*. Left: collected from a weedy site near Ottawa, Canada. Right: cultivated near Toronto, Canada, from seeds from Georgia (Eurasia).

Figure 6.3a and b shows shoot configurations typical of marijuana strains of *C. sativa*. As discussed in Chapters 12, 13, and 18, there are two basic classes of marijuana plants, "indica type" (shorter ones; Figure 6.3a) and "sativa type" (taller ones; Figure 6.3b). All of these plants are naturally (genetically) very well branched (like wild plants). Moreover, they are traditionally planted (outdoors at least) at low density, leaving room for the branches to develop well. Maximizing branch production is desirable to produce many flowers, as the perigonal bracts (discussed in Chapter 11) around the female flowers produce most of the intoxicating chemicals that are desired. As discussed in Chapter 4, male plants are removed to prevent production of seeds, which are not the desired product.

As discussed in Chapter 9, a very recent market has developed for the production of essential (volatile) oil, a product substantially from the perigonal bracts, exactly the same source of chemicals for marijuana. Plants of the same architecture as sativa type marijuana (Figure 6.3b) have been used as sources of essential oil (indeed, such strains are very suitable for the purpose, although they pose security problems).

As discussed in Chapter 8, there has been comparatively limited selection of strains of *C. sativa* in historical times specifically for oilseed production. Since plants that are big and well branched produce many flowers (such as those shown in Figure 6.3b), when allowed to produce seeds, they do so very well. Such plants have been commonly used as sources of oilseeds. In more recent times, as discussed in Chapter 8, short plants with numerous flowers (and hence seeds) congested on short branches (as shown in Figure 6.3e) have been grown at moderate densities to produce oilseeds, a strategy that reduces the production of stem tissue in a given area while maximizing the production of seeds on a given acreage. Plants with limited (or at least short) branching are naturally superior than irregularly branching plants for the purpose of fully and uniformly occupying a field and maximally utilizing solar irradiation.

As noted previously, different densities of planting are used to increase or suppress branching, but the different kinds of *C. sativa* have been genetically selected—some to grow well at high concentrations,

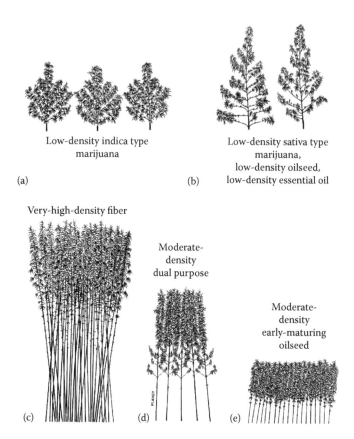

Low-density indica type
marijuana

(a)

Low-density sativa type
marijuana,
low-density oilseed,
low-density essential oil

(b)

Very-high-density fiber

Moderate-
density
dual purpose

Moderate-
density
early-maturing
oilseed

(c)		(d)		(e)

FIGURE 6.3   Common categories of shoot architecture of cultivated *C. sativa* in field configurations maximizing production of the desired harvest product. (a) Short, conical, well-branched, female, marijuana plants of indica type (discussed in Chapters 12, 13, and 18) are grown well spaced to maximize development of both foliage and flowers containing cannabinoids. (b) Tall, well-branched plants are grown well spaced to maximize production of flowers (for harvest of either marijuana or essential oil) or seeds (for planting). (c) Fiber cultivars are grown at very high density to produce unbranched, tall plants that maximize quantity and quality of stem fiber. (d) "Dual-purpose" cultivars are grown at moderate density, which tend to be somewhat branched and of medium to tall height, a compromise strategy for production of both stem fiber and oilseeds. (e) Some modern oilseed cultivars are grown as short, relatively unbranched plants to maximize production of seeds while minimizing production of stem tissues, and to facilitate machine harvesting. Drawn by B. Flahey.

others at low concentrations. Unlike fiber races that have been selected to grow well at extremely high densities, drug strains tend to be less tolerant of high population densities (De Meijer 1994a).

Susceptibility to pests and diseases differs according to density of cultivation, and likely, there has been selection for tolerance to the usual cultivation density for particular kinds of *C. sativa*. The very dense plantations in which fiber crops are grown raises the humidity around the stalks and increases infections by fungal diseases. On the other hand, the dense canopy may be protective against many insects. By contrast, both drug and many oilseed crops are grown in open rows, and the increased sunlight is attractive to flea beetles and birds.

SHOOT ARCHITECTURE OF FIBER CROPS

Because fiber is produced mostly in the main stem (stalk) of *C. sativa*, and the longest fiber bundles are produced in the main stem, tall plants are best for harvest of fiber, and humans have been unconsciously selecting tall plants for thousands of years. There has also sometimes been deliberate selection for height of fiber plants. Forster (1996) noted that to obtain seeds for planting fiber plants in

FIGURE 6.4 Densely grown hemp, illustrating the development of tall, slim stalks and suppression of branching. Photo by Adrian Cable (CC BY 2.0).

FIGURE 6.5 Comparison of densely grown and open-grown fiber hemp. Left: the tall hemp cultivar Petera, illustrating suppression of branching from main stalk under dense growth. Photo courtesy of Anndrea Hermann. Right: an uncharacterized fiber strain from Germany, open-grown in Canada, illustrating more evident branching from the main stalk.

FIGURE 6.6   A Chinese fiber land race (note map in Figure 2.7 and discussion in Chapter 18). This plant is 2.5 m in height. It was grown at a spacing of 1 m from other plants and is essentially open-grown. The strong lateral branching is typical of many Chinese land races (European fiber cultivars tend to be less branched). The appearance of this Chinese fiber plant is indistinguishable from most sativa type marijuana plants (see Chapter 18).

Chile, "During seed production, preferred males of ideal stature are selected and allowed to fertilize the females." Such a procedure would strongly favor the development of tall plants. As discussed in Chapter 7 dealing with fiber production, branching is suppressed by growing the plants at extremely high densities (Figures 6.4 and 6.5). As detailed in Chapter 7, the slim, unbranched plants produced at high planting densities maximize quantity and quality of fiber. Woody tissues in the stem of fiber selections have been suppressed so that the stems are much hollower than in any other category of *C. sativa*. This makes the stems weaker and less flexible, but the high density of planting protects the plants from being lodged (blown over) by wind. Because of the limited branching, seed production is much more limited than in strains used for oilseed. However, sometimes, "dual-purpose" cultivars are grown (Figure 6.3d) with intermediate characteristics between the fiber (Figure 6.3c) and oilseed (Figure 6.3e) kinds, so that both products can be harvested, albeit in relatively modest amounts.

As discussed in Chapter 7, there are two basic kinds of fiber plants, European and Chinese. The latter, although much older, tends to be less selected specifically for stem fiber characteristics, as evidenced by the development of more branches, which produce more seed (Figure 6.6).

REDUCING THE HEIGHT OF MARIJUANA PLANTS

As discussed in Chapter 12, in Asia, one method of preparing hashish once often involved using hands or leather to collect (by adherence) sticky resin from the inflorescences at the top of the plants (alternatively and more conventionally today, hashish is prepared by filtering techniques, described in Chapter 12). Accordingly, strains suitable for hashish collection based on stickiness should not be too tall. As Bouquet (1950) recorded: "The cultivators, dressed in leather, moved about through the plantations. The resin sticks to their clothes, which are scraped from time to time with a blunt

FIGURE 6.7   An indica type marijuana strain from Afghanistan (this class of drug plants is mapped in Figure 2.7 and discussed in Chapters 12 and 18). Note the limited stature, short internodes producing a compact appearance and very wide leaflets. Photographed at the U.S. National Institutes of Health, University of Mississippi (Oxford) marijuana plantation site.

FIGURE 6.8   Marijuana (*C. sativa*) growing in a mine shaft tunnel for the Canadian medical marijuana program. The limited stature of this strain was necessary to accommodate the limited height available. Similarly, growers of illicit marijuana have selected shorter strains that can be grown under artificial light in rooms with low ceilings.

curved knife. This method of collection shows clearly that in those regions the plant does not grow to any great height." In a similar vein today, dwarf varieties of tree fruits have been bred. An added benefit of low stature is greater wind resistance. The indica type group of marijuana strains is naturally much shorter (Figures 6.3a and 6.7) than the higher-stature sativa type of marijuana plant (Figure 6.3b), but as discussed in Chapter 12, the tall sativa type plants are more popular.

As discussed in Chapter 12, for the last half-century, marijuana plants have frequently been grown clandestinely indoors to avoid detection by law enforcement, a situation in which tall plants are frequently too large (especially when overhead lighting and ventilation are installed in a room). Legitimate, authorized medicinal marijuana growers also often find tall plants to be too awkward to raise in greenhouses and specially fitted secure rooms. It is possible to adjust height by controlling the photoperiod, but plants that are naturally shorter are often grown where limited height is necessary (note Figure 6.8). "Breeders continue to develop early-maturing and high-yielding varieties that are short and compact for indoor grow room use and to avoid detection outdoors" (Clarke and Merlin 2013).

Indoor growers sometimes resort to removing the tops (Figure 6.9), pinching stem buds to promote branching, trellising, and other techniques to limit the height of plants (Clarke 1981). Potter (2009) observed that the height of indoor plants can be shortened by growing them under continuous light or by brushing the plant in early development (like plants buffeted by wind, the stems become thicker and shorter to resist movement).

"SEA OF GREEN" AND "SCREEN OF GREEN" CULTIVATION METHODS

"Sea of Green" (SOG) and "Screen of Green" (or "SCOG" [Screen of Green Growing]) refer to indoor cultivation methods of inducing low continuous indoor marijuana canopies in order to maximize use of limited space.

FIGURE 6.9 A marijuana strain of *C. sativa* that has been topped in early growth (the terminal meristem of the stem leader destroyed), causing extensive branching and the development of numerous "buds" (congested flowering branches). Photo by Chrisgedwards (CC BY 3.0).

Sea of Green is a method developed in the Netherlands that grows marijuana so that the buds mature and can be harvested while the plants are young, short, and well branched. Because illumination is intercepted mostly by the upper part of a marijuana plant, and the lower parts receive relatively limited amounts, the technique is a way of shortening the light-intercepting canopy so that most of it is well lit and growth is optimized. Effectively, this is a way of eliminating much of the lower leaves, which are somewhat parasitic on the plant when the plant is tall. Much more intense light is available outdoors, so the technique is intended for indoor growth. The method is based on growing a higher density of plants (for example, one plant per 30 cm^2) than usually established but harvesting them faster. Productivity can be considerable (Knight et al. 2010). The tendency of strains to grow fast, tall, and/or branch and the maturation time will determine planting density. A wire or cord trellis may be spread horizontally over the canopy to support the heavy maturing branches, using twist ties to attach the branches to the trellis framework. Tying down the main stalk (leader) at a low level stimulates the growth of lower branches, promoting even filling in of the canopy. Some trimming may be necessary to encourage uniform development of the canopy, as well as openness below the canopy to allow ventilation. Lower, senescing leaves should be removed to increase air circulation. The trellis should be positioned about 20 cm above the top of the growth medium.

SCOG is essentially the same as the SOG method but specifically employs a horizontal trellis ("screen") through which plants grow. Training of plants tends to be more severe than in the SOG method, keeping the plant canopy short. (The vertical growth in SCOG is more limited than in SOG, the canopy of SOG described as a "forest" and that of SCOG as a "field.") Chicken wire with 5-cm openings is commonly employed as the screen, but nylon netting can also be used (although it is less sturdy).

These methods are labor intensive, more suitable for small spaces of limited height than for large grow rooms and for circumstances where maximum yield from limited lighting is desired. Not surprisingly, these techniques have been particularly employed for clandestine growth, where maximizing production in very limited indoor spaces is an imperative.

REDUCING THE HEIGHT OF FIBER PLANTS

Ranalli and Venturi (2004) and Amaducci (2005) describe a practice sometimes employed in Italy that dwarfs fiber hemp (producing so-called "baby hemp") so that it can be processed by the same equipment used for the much shorter stems of flax. Seeds are sown at high density (100–120 kg/ha), and when the hemp stems are about 1.2 m tall, they are sprayed with desiccants (generally the herbicide glyphosate) to stop their growth. However, the process is considerably less efficient than normal hemp technology. Westerhuis et al. (2009) examined the possibility of producing dwarfed hemp simply by postponing the sowing date or of planting two successive short-season crops, and while the small plants so produced were quite usable for hemp fiber, the feasibility and economics of producing dwarf fiber plants have not seemed worthwhile.

Another curious technique to control height at harvest time was employed in the middle of the twentieth century in Russia (Kirby 1963). Because of very high demand for labor in the autumn, three different varieties that respectively matured very early, in mid-season, and in the fall were planted, thus distributing availability of workers and machinery throughout the growing season. An added advantage was that retting (described in Chapter 7) of summer-harvested stems proceeded more quickly, required less space, and produced higher-quality fiber in the warmer weather.

RESISTANCE TO CATASTROPHIC STEM DAMAGE

Cannabis sativa normally has a dominant leader stem that produces a central stalk. As discussed in this section, the species has an amazing capacity to recover from catastrophic damage to the main stem.

The European corn borer (*Ostrinia nubilalis* = ECB; Figure 6.10a) is a major Lepidopteran pest of *C. sativa*. Young larvae eat the leaves until the insects are half-grown and then bore holes into the stems. A typical entrance hole resulting from an attack on the main stem is shown in Figure 6.10b. The insect is indigenous to the Old World, where it apparently once reproduced mainly in association with *Cannabis* and its close relative *Humulus* (although also attacking many other plant species). It was not exposed to corn (i.e., maize, *Zea mays*), which is indigenous to the Americas, until post-Columbian times ("European hemp borer" would have been a better choice of name). In a study of infestation of a large experimental field, Small et al. (2007) discovered that ECB damage to *C. sativa increased* the shoot weight of the plant by 20%, concomitantly enlarging seed production, suggesting that *C. sativa* is adapted to the insect. The expanded productivity observed was due to branch proliferation at the site of the attack (see Figure 6.10c and d). Figure 6.10e shows silhouettes of a normal and an ECB-damaged plant, and it is evident that the increased number of branches resulting from the damage has produced more biomass and more seeds. (The insect preferred larger stems but was unaffected by tetrahydrocannabinol content, which varied considerably in the experimental field.)

There is controversy whether insect damage may, at least in a limited sense, be good for plant productivity. McNaughton's (1983) classic paper in this regard proposed that in some circumstances, plants can respond to herbivory by just growing faster ("compensation" or "overcompensation"). Verkaar (1986) surveyed papers purporting to support the hypothesis that grazing can have positive effects on plant growth and fitness and concluded that "the hypothesis may only be tenable under very particular circumstances."

Horticulturally, it is well known that destroying leader buds to induce proliferation of flowers or fruits in a range of plants can increase productivity, so it is logical that insects that carry out this activity might also be beneficial to crop production. Moreover, as noted previously, humans have engaged in the practice of damaging stems to increase the productivity of *C. sativa*. Pate (1998b) observed that when growing hemp for seed, the number of flowers per plant and the number of seeds produced can be increased by topping the plants when 30 to 50 cm high. Dewey (1902) recorded that hemp grown in North America at the turn of the century was sometimes topped to make it spread and produce more seed. Moes (1998) found that following severe hail damage to a hemp plot in

FIGURE 6.10 Response of *C. sativa* to the European corn borer (ECB) (*Ostrinia nubilalis*). (a) Left: female adult; photo by Frank Peairs, Colorado State University, Bugwood.org (CC BY 3.0). Right: larva; photo (public domain) by Keith Weller, U.S. Agricultural Research Service. (b) Photograph of an ECB infestation site on a *C. sativa* stem. Note frass around entrance. (c, d) Photographs of site of branch proliferation caused by ECB damage. (e) Silhouettes of normal plant (left) and plant developed after ECB damaged the leader. Figures b–e based on Small, E., Marcus, D., McElroy, A., Butler, G., *J. Ind. Hemp*, 12, 15–26, 2007.

Manitoba, axillary branches developed at nodes below damaged stems, produced inflorescences, and provided a substantial (albeit reduced) seed yield.

FOLIAGE ARCHITECTURE

There is great variation of the size of leaves and the shape of leaflets in *C. sativa*. The evolution and ecology of leaf size are complex subjects and are related to the total number of leaves and their turnover rate (e.g., Whitman and Aarssen 2010). In numerous plants, the leaves of domesticated forms are larger than is the case in related wild species. This is likely due to the greater photosynthetic capacity of large leaves, the result of selection by humans to be more productive in a given limited area. This pattern seems to be true for the three classes of domesticated *Cannabis* (fiber, oilseed, and marijuana), all of which tend to have larger leaves than do wild *Cannabis* plants. In *Cannabis*, compared to the foliage of wild plants, the photosynthetic area of individual leaves is often larger in domesticated plants by virtue of (1) having more leaflets and (2) having leaflets that are larger, especially wider (compare the wild leaves of Figure 6.11a and b with the domesticated leaf in Figure 6.11c). This pattern of larger leaves with wider leaflets in domesticates compared to wild relatives is frequently encountered in other crops with compound leaves, for example, in carrot, *Daucus carota* L. (Small 1978b), and in alfalfa, *Medicago sativa* L. (Small 2011b). On the basis of modeling considerations for tomato leaves, Sarlikioti et al. (2011) concluded that for a given leaf area, bigger but fewer leaflets were better at intercepting light than more but smaller leaflets.

FIGURE 6.11   (a, b) Examples of wild races of *C. sativa* with very narrow leaflets. (a) is the type specimen of *C. ruderalis* Janischevsky (which is correctly classified as *C. sativa* var. *spontanea*). (b) is the type specimen of *C. sativa* var. *spontanea* Vavilov. (c) A large leaf with wide leaflets of a fiber cultivar (*C. sativa* var. *sativa*).

Environment can modify leaf size. The leaves of many plant species growing in the wild are often small simply because of environmental modification—from the more stressful conditions encountered in nature. In *Cannabis*, however, the leaflets of wild plants are typically small even in excellent growth conditions. When fiber cultivars are grown closely together, as done conventionally, branching is suppressed and they lose most of their lower leaves. The fewer leaves that survive near the top of the plants are larger, partly as a matter of physiological compensation, but also as a genetically controlled tendency to produce larger leaves.

Some kinds of Chinese fiber land races and southern Asian (indica type) marijuana races are noted for their large leaves with wide leaflets—a clear reflection that they are the products of considerable domestication. As discussed in Clarke and Merlin (2013), these groups are ancient and have undergone millennia of selection.

Larger leaves (and larger leaflets) in domesticated *Cannabis* may be the result of greater photosynthetic demand, but why should leaflets be narrower in related wild plants? Brown et al. (1991) examined the hypothesis that the feeding efficiency of leaf-eating insects is lowered on leaves that are small, dissected, or needle-like—patterns that make the insect work harder to reach the edible lamina. It seems possible that the smaller, narrower leaflets in wild plants of *C. sativa* are adaptive in making their foliage less accessible to herbivores, and the reduced need for such protection in domesticated plants has allowed them to develop bigger, wider leaflets. It is also possible that smaller and narrower leaflets are more resistant to wind damage, another advantage in wild plants.

As pointed out in Chapter 12, the two groups of marijuana plants differ in leaflet width, with sativa type plants having narrower leaflets than indica type plants. The underground marijuana literature sometimes also contends that the leaves of indica type plants tend to have fewer leaflets than those of sativa type. Coincidentally, indica type plants have much shorter internodes, resulting in pronounced crowding of the foliage and darker green foliage. These variables seem to be correlated in the same ways that shade leaves differ from sun leaves. Many plants develop smaller, lighter-green leaves in the sun and larger, darker-green leaves in the shade (e.g., Nobel 1976; Givnish 1988), and the crowded (therefore shaded) leaves of indica type plants seem to be consistent with this observation.

PROTECTIVE UNICELLULAR HAIRS

Most of the above-ground plant surfaces of *C. sativa*, especially of the foliage, develop stiff, pointed hairs, technically termed "trichomes." Glandular trichomes, which are multicellular, are discussed in Chapter 11. This chapter addresses two kinds of "simple" unicellular trichomes. Although cannabis material is almost always identified chemically for court purposes, the hairs have also been employed as identification features, as they frequently differ from hairs in other plants (Thornton and Nakamura 1972). In particular, the presence of rigid cystolithic hairs predominantly on the top of the leaves and relatively flexible simple hairs on the bottom is suggestive that herbal material is marijuana. The two kinds of hairs are described next.

CYSTOLITHIC TRICHOMES

"Cystolithic" trichomes (Figure 6.12) are a mechanical defense against herbivores. By definition, such unicellular hairs have small particles of calcium carbonate embedded in the base (as shown in Figure 6.12a; "cystolith" is derived from the Greek *kustis* and *lithos*, meaning "bag of stones"). In *Cannabis*, cystolithic hairs are predominantly present on the adaxial ("upper") surfaces of leaves. This feature tends to make the plant unpleasant to chew and less palatable to herbivores, protecting them from being eaten. Reinforcing this unpalatability is the presence of calcium oxalate crystals in many cells of the leaves (Figure 6.12b). Both calcium carbonate cystolithic hairs and calcium oxalate crystals are antiherbivore features which many other species of plants also possess.

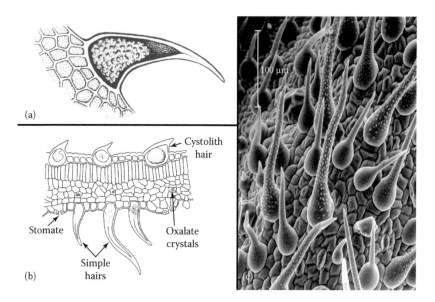

FIGURE 6.12 Cystolith hairs (unicellular structures with calcium carbonate particles embedded in the base). (a) Diagram of a single cystolithic hair (from Figure 6.1). (b) Section of leaf showing cystolith hairs on adaxial ("upper") surface and simple hairs on abaxial ("lower") surface. From Meijer, J.G., Structuur van de Inflorescentieschubben van *C. sativa*. http://www.geheugenvannederland.nl/?/en/items/SAE01:2074, 1904 (public domain). (c) Scanning electron micrograph of adaxial ("upper") surface of a young leaf of *C. sativa*, showing protective cystolithic trichomes arrayed in one direction. Photo by E. Small and T. Antle.

The hairs are also quite stiff, acting like miniature thorns. They tend to point toward the tip of leaves (Figure 6.12c), and because they are mostly orientated in the same direction, they provide a roughness to the surface of foliage, which probably discourages larger animals. This roughness is responsible for causing skin irritation and dermatitis in people who handle *C. sativa* extensively without suitable protective clothing.

Simple Trichomes

Also common on much of the plant are additional unicellular hairs that lack the basal stony concretions and tend to be slimmer. These are silicified, tending to increase the unpalatability of the foliage. In *Cannabis*, simple hairs are present on both surfaces of leaves, but in young leaves, they can be dominant on the abaxial ("lower") surface. These hairs also are protective as a physical barrier, insulating the foliage from insects, as illustrated in Figure 6.13. Insects tend to prefer the lower surface of leaves, where they are relatively hidden from predators and where it is cooler, so protection of the lower leaf areas is particularly important.

CURIOSITIES OF SCIENCE, TECHNOLOGY, AND HUMAN BEHAVIOR

- According to a traditional Japanese children's story, Ninja warriors planted a batch of hemp when they began training, with the intent to leap over it every day. By the end of the season, the warriors were expected to jump over the 3–4 m (10–13 feet) high mature hemp.
- In the French Ardennes, it was thought to be absolutely essential that the women become drunk on the night of the first Sunday in Lent, in order that the hemp would grow tall.
- In medieval times in southwestern Germany, men and women leaped hand-in-hand over a bonfire, while calling for the hemp to grow. It was believed that those who jumped highest would cause the hemp on their farms to grow tallest.

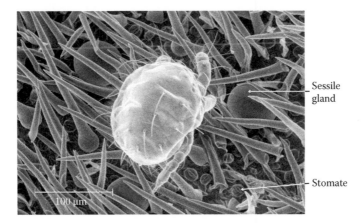

FIGURE 6.13 Scanning electron micrograph of abaxial ("lower") surface of a young leaf of *C. sativa*, showing a mite attempting to penetrate a protective forest of simple hairs and young sessile resin glands (the spherical structures). Note the presence of stomates, which in most plants tend to predominate on the lower shaded surfaces, where these pores will lose less water by evaporation and so lessen the danger of the plant drying out. Prepared by E. Small and T. Antle.

- When sowing hemp seeds in old France, farmers would pull up their pants as far as possible, believing that the hemp would grow only as tall as their pants.
- Still another European practice to make hemp grow tall was to sow hemp seeds on days dedicated to saints believed to be tall.
- Hemp has been observed to grow 15 cm per day.

7 Fiber

INTRODUCTION TO FIBER

Hemp is a natural fiber, and to appreciate its current importance, it is desirable to have some background into the nature of fiber and the world market for it. "Fiber" has several meanings, but for purposes of this chapter, it refers to thread-like material, either obtained from natural sources or human-made, and used in various forms (especially woven into fabrics, matted as in paper, or glued together as in fiberboard). Wood fiber provides over three-quarters of all fiber produced and also dominates the number of species that are available to produce fiber (wood fiber can be obtained from over 10,000 species of trees and over 1000 species of bamboo). Except for the category "manmade cellulosics," wood fiber is excluded from the following analysis (see Figure 7.1). "Mineral fibers" (mostly made of glass, steel, asbestos, or carbon) are also excluded from this discussion. There are two basic classes of fiber: natural and synthetic. The world's *natural* fiber market includes fibers extracted directly from plant and animal species. Cotton (from *Gossypium* species) and wool are the leading natural fibers. Other plant species significant in world trade because they are deliberately cultivated for fiber include jute (*Corchorus* species), kenaf (*Hibiscus cannabinus* L.), roselle (*Hibiscus sabdariffa* L.), sunn hemp (*Crotalaria juncea* L.), flax (*Linum usitatissimum* L.), ramie (*Boehmeria nivea* (L.) Gaudich.), abaca (*Musa textilis* Née; = Manilla hemp, Manila hemp), sisal (*Agave sisalina* Perrine), and henequen (*Agave fourcroydes* Lem.). As illustrated in Figure 7.2, important vegetable fibers originate from different parts of plants (leaves, stems, seeds, fruits), depending on the species. Plant fibers are also obtained as by-products from crops, notably from straw left after cereals (particularly wheat, sorghum, and rice) are harvested, and from bagasse (the fibrous residue remaining after sugar from sugar cane is extracted). In contrast to natural fibers, *synthetic* fibers are prepared from fossil fuels. Examples include polyester, polypropylene, and nylon. *Manmade cellulosics* is an intermediate category (sometimes included in synthetics and sometimes termed "regenerated fibers"); high-cellulose material, primarily salvaged from timber processing and crop residues (especially cotton), are chemically processed and converted to produce manufactured fibers. Rayon and acetate are examples. The world's fiber market today is dominated by synthetic fibers, especially polyester, which is made mostly from ethylene derived from coal. Polyester constitutes three-quarters of all synthetic fibers. The world's *textile* market generally uses fiber for fabrics, particularly for clothing. Cotton currently accounts for almost 40% of the total textile fiber market (and 85% of the natural fiber textile market), but polyester is more important, accounting for over 50% of the total textile fiber market. For years, polyester has been gaining market share while cotton has been losing ground. Animal fibers such as wool and silk, which are protein based, have also been losing popularity. Today, hemp constitutes only about 0.3% (on a tonnage basis) of the world's natural fiber production (excluding wood fiber).

HISTORY

ANCIENT HEMP HISTORY

For most of recorded history, *Cannabis sativa* was primarily valued as a fiber source, considerably less so as an intoxicant, and only to a limited extent as an oilseed crop. Hemp is one of the oldest sources of textile fiber, with extant remains of hempen cloth trailing back at least six millennia. For thousands of years, hemp has been most valued for rope because of its strength, durability, and water resistance (Bócsa and Karus 1998).

Estimates of the time that hemp was harvested by the Chinese range from 6000 years (Li 1974) to 8500 years (Schultes 1970; Schultes and Hofmann 1980) or even 10,000 years (Allegret 2013).

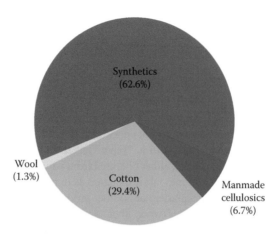

FIGURE 7.1   Relative composition of the global fiber market (in 2014), excluding wood fiber except for man-made cellulosics. Synthetic fibers are oil based, the remaining categories originate from harvested plants and livestock. (Compare Small 2013b.)

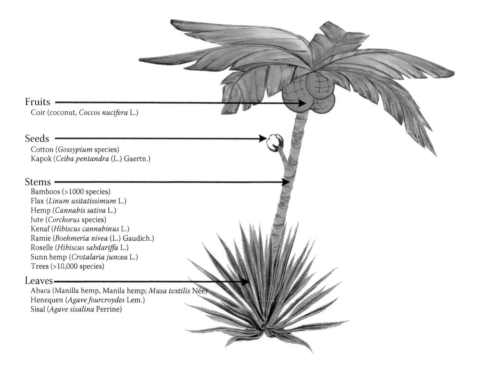

FIGURE 7.2   A synthetic plant showing parts of the world's major species utilized for vegetable fibers.

For millennia, hemp has been a respected crop in China (Touw 1981; Clarke and Merlin 2013), where it became a very important fiber for clothing (note Figure 7.3). The Chinese also manufactured strings, ropes, textiles, and paper from hemp (Li 1974). To this day, China remains the world's chief producer of hemp fiber.

Hemp grown for fiber was introduced to western Asia and Egypt and subsequently to Europe between 1000 and 2000 BC. Cultivation in Europe became widespread after 500 AD. The crop was first brought to South America in 1545, in Chile, and to North America in 1606, in Port Royal, Acadia (Small 1979b).

FIGURE 7.3 Traditional ethnic Chinese hemp dresses, exhibited in the Yunnan Nationalities Museum, Kunming, Yunnan, China. Photos by Daderot (released into the public domain).

THE GOLDEN AGE OF HEMP IN THE WESTERN WORLD

Hemp was one of the leading fiber crops of temperate regions from the sixteenth through the eighteenth centuries. It was an important European crop until the middle of the nineteenth century. Hemp was widely used for rot-resistant, coarse fabrics as well as for paper, and was the world's leading cordage fiber (used for rope, twine, and similar purposes) until the beginning of the nineteenth century. Until the middle of the nineteenth century, hemp rivaled flax as the chief textile fiber of vegetable origin and was described as "the king of fiber-bearing plants—the standard by which all other fibers are measured" (Boyce 1900).

Ships once used enormous amounts of hemp (Figure 7.4). Hemp was the raw material for sails because the fabric is very water and rot resistant. Most ship sailcloths, rigging, and nets up to the

FIGURE 7.4 Major maritime uses of hemp in the past. Hemp was used for (a) sails, ropes, and rigging, (b) nets, and (c) oakum for caulking seams in wooden ships, and was indispensable for navies. Prepared by B. Flahey.

late nineteenth century were made from hemp fiber. The "age of sailing ships" is usually defined for Western countries as lasting from the sixteenth century to the mid-nineteenth century (peaking in importance in the nineteenth century, the "Golden Age of Sailing"). During the age of sailing ships, *Cannabis* was considered to provide the very best canvas, and indeed this word, as well as the genus name *Cannabis*, is derived from the same ancient language words for hemp. A single ship could require as much as 60 tonnes of hemp rope—30 km for rigging alone; an anchor cable could exceed 60 cm in diameter. To replace worn-out hemp materials, larger ships could require 100 tonnes of hemp fiber every year (Bócsa and Karus 1998).

THE LABOR-INTENSIVE PRE-TWENTIETH CENTURY PERIOD

Traditional planting, harvesting, and processing of hemp were astonishingly labor-intensive, as indicated in Figure 7.5. In Europe, prisoners were often conscripted to process hemp. As shown in Figure 7.6, beating hemp stems to extract the fiber was a common form of hard labor in Victorian prisons. As shown in Figure 7.7a, "picking oakum" was another especially laborious task, needed to prepare the considerable amounts necessary to make ships water-tight (Figure 7.7b). Convicts were provided with worn-out ropes and were required to untwist them into corkscrew strands, unroll, and de-tar these into a fluffy mass of fibers, then roll the fibers from thigh to knee to produce a loosely twisted continuous strand. Only occupants of jails and poorhouses picked oakum, and the phrase "picking oakum" came to mean "getting into trouble." When slavery and poorhouses were abolished, labor for preparing oakum became so limited that it contributed to the demise of hemp.

The hemp industry flourished in Kentucky, Missouri, and Illinois between 1840 and 1860 because of the strong demand for sailcloth and cordage (Ehrensing 1998). From the end of the Civil War until 1912, virtually all hemp in the United States was produced in Kentucky (note Figure 7.8), and although simple machinery was adopted to facilitate collection and processing, a large labor force was still required. As documented by Hopkins (1951), slaves conducted the heavy labor prior to the American Civil War, and the diminishing production afterward was carried out by subsistence farmers and low-wage hired hands.

THE TWENTIETH CENTURY GREAT DEMISE OF HEMP IN THE WESTERN WORLD

Several developments, listed in decreasing order of importance in the following, drastically curtailed the importance of hemp fiber outside of Asia. (1) The use of steam- and petroleum-powered motorized ships greatly reduced the need for hemp fiber for naval purposes. (2) Hemp rope tends to hold water in the interior, and to prevent internal rotting, the ropes were tarred, a laborious process that was made unnecessary when abaca was substituted. Abaca rope proved preferable for marine use because it was lighter, could float, and had greater resistance to salt water corrosion. (3) The Industrial Revolution (approximately 1760–1840 in Britain) initiated sustained economic growth and living standards in the Western world but also accentuated differences for the cost of fiber production between rich temperate regions and poor tropical and semitropical regions. As a fiber crop, hemp (like flax) is best adapted to temperate areas, in contrast to other leading fiber crops such as cotton, jute, and sisal. Outside of Asia, production costs (largely determined by labor) in recent centuries have been much cheaper for tropical and semitropical fiber crops, and this contributed to making hemp much less competitive. (4) Hemp fiber was once important for production of coarse but durable clothing fabric. In the nineteenth century, softer fabrics took over the clothing market. As the world has judged, cotton is a remarkably more attractive choice for apparel. The invention of the modern cotton gin by Eli Whitney in 1793 enormously increased the efficiency of cotton production and has been claimed to have contributed to the demise of hemp fiber, which is relatively difficult to separate cleanly from other parts of the plant. Increasing limitation of cheap labor for traditional production in Europe and the New World led to the creation of some mechanical inventions for preparing hemp fiber, but too late to counter growing interest in competitive

FIGURE 7.5 Traditional nineteenth century European hemp extraction and processing technologies. (a) Cutting down plants at base. (b) "Water retting," the process of immersing stems for a week or more, so decay microorganisms loosen attachment of fiber to other tissues. (c) An alternative to retting: hand-stripping the fiber-bearing "bark." (d) Using a "hand break" to crudely separate fiber from retted stems. (e) Left: Beating dried hemp stalks with a hand tool to crudely separate fiber from retted stems. Right: Hackling (drawing partially cleaned hemp bark fiber through a bed of nails to clean off remaining undesired tissues). (f) Spinning fiber into thread. (a–d) From Lallemand, M.G., Levy, M. 1860. L'illustration Journal Universel 926, 1860. (e and f) From Anonymous, *Galerie industrielle*, Eymery, Paris, France, 1822 (in French).

FIGURE 7.6 A prison for prostitutes in London, showing them using mallets to clean off debris from harvested hemp. One of a series of six paintings (1731) converted to engravings (1732), called The Harlot's Progress, prepared by the English artist William Hogarth. (Public domain illustration.)

(a)   (b)

FIGURE 7.7 Activities related to oakum. (a) Oakum preparation in prison. From Mahew, H., Binny, J., *The Criminal Prisons of London, and Scenes of Prison Life*, Griffin, Bohn, and Company, London, U.K., 1864. (b) Caulking of a ship's hull with oakum. From Von Henk, L.F.W., Nieth, E., von Werner, A., *Zur See*, 3rd ed., Verlagsanstalt und Druckerei, Hamburg, Germany, 1895 (in German).

crops. (5) Human-made fibers began influencing the marketplace with the development of rayon from wood cellulose in the 1890s. Largely during the twentieth century, commercial synthetic fiber technology increasingly became dominant (acetate in 1924, nylon in 1936, acrylic in 1944, polyester in the 1950s), providing competition for all natural fibers, not just hemp. (6) Hemp rag had been much used for paper, but the nineteenth century introduction of the chemical woodpulping process considerably lowered demand for hemp. (7) A variety of other, minor usages of hemp became obsolete. For example, the use of hemp as a waterproof packing (oakum), once desirable because

FIGURE 7.8   Early twentieth century postcards showing hemp harvesting and extraction scenes in Kentucky.
(a, b) Machine-assisted harvesting. (c) Harvesting by hand. (d) Bundling cut stems into "shocks" for field retting. The shocks shed water like pup-tents, promoting even drying. (e) A field with hemp bundled into shocks.
(f) Using a "hand break" to crudely separate fiber from stems. (Public domain illustrations.)

of resistance to water and decay, became antiquated. (8) The growing use of the cannabis plant as
a source of marijuana drugs in the Western world in the early twentieth century gave hemp a very
bad image and led to legislation prohibiting cultivation of hemp (note Figure 7.9). During the two
World Wars, there were brief revivals of hemp cultivation by both the allies and Germany because
of difficulties importing tropical fibers. In particular, abaca and sisal fiber from the Philippines
and Netherlands Indies were cut off in late 1941, and there was a concerted effort to reestablish the
industry in the United States (Wilsie et al. 1942, 1944; Hackleman and Domingo 1943; note Figure
7.10). In 1952, the U.S. Department of Agriculture issued a revision of Robinson's (1935) guide
to cultivating hemp in the United States but lost interest in the crop subsequently. After the war,
however, hemp cultivation essentially ceased in most of Western Europe, all of North America, and
indeed in most non-Asian countries, although production continued at a diminished level in Asia,
eastern Europe, and the Soviet Union.

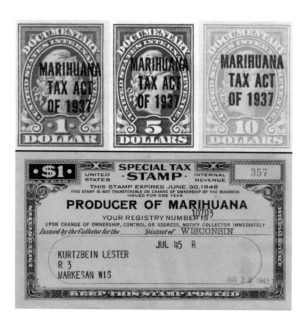

FIGURE 7.9   Stamps required to satisfy the regulations of the U.S. Marihuana Tax Act of 1937. The act governed the importation, cultivation, possession, and/or distribution of marijuana. Medical usage required a levy of a dollar an ounce (28 g), any other usage was taxed at $100.00 an ounce. Importers were required to register and pay an annual tax of $24. Marihuana Tax Act stamps were required to be affixed to order and export forms. Violations of the act resulted in a fine of up to $2000 and/or imprisonment for up to five years. The Marihuana Tax Act of 1937 stopped not just the use of the plant as a recreational drug but also commercial production and trade in industrial hemp. Top: stamps issued in 1937 (public domain photo by U.S. Customs and Border Protection). Bottom: stamp issued in 1945 (public domain photo). Stamps such as these have become collector's items, selling for over $1000.00 apiece (Wirtshafter and Krawitz 2005).

FIGURE 7.10   A poster (public domain) that was widely circulated to promote the production of hemp for the U.S. war effort. During World War II, the Japanese cut off supplies of abaca from the Philippines. In response, the U.S. government contracted the building of over 40 hemp fiber processing mills throughout the Midwest to produce cordage for the navy.

THE RECENT RENAISSANCE

In Asia (particularly in China), in most of the Soviet Union, and in most of Eastern Europe, hemp cultivation was not prohibited as it was in most of the remaining world during the twentieth century. In these areas, hemp production continued to a lesser or greater degree depending on local markets (Ceapoiu 1958; De Meijer et al. 1995). A surge of interest in reestablishing the hemp industry in Western countries began in the 1990s, particularly in Europe and the British Commonwealth. At the time, governments generally were hostile to growing any form of *C. sativa* for fear that this was a subterfuge for making marijuana more acceptable. Throughout Western nations in the 1990s, interest in reviving traditional nondrug uses of *C. sativa*, as well as developing new uses, has had to contend with the dominating image of the plant as a source of marijuana. Nevertheless, cultivation resumed in the temperate-climate regions of many Western countries. Examples of first resumption of cultivation include the following: Australia (Tasmania), 1990; Austria, 1995; Canada, 1998; England, 1993; Germany, 1995; the Netherlands, 1994; Spain, 1986; Sweden, 1995. Some Western European countries, such as France and Spain, never prohibited hemp cultivation and also participated in the 1990s in the revival of hemp cultivation. The impetus for growing hemp in the West was economic, motivated by the general need to find new profitable crops. Critical to the successful initiation of hemp cultivation in most countries was persuasion of governments that the hemp industry would not compromise programs controlling forms of *Cannabis* that could be used to produce marijuana. About three dozen countries currently grow significant commercial hemp crops. As of 2016, the United States has been the only notable Western nation to persist in prohibiting hemp cultivation, although, the majority of U.S. states have enacted resolutions or legislation favoring the resumption of hemp cultivation, and cultivation has been initiated in some states. However, federal U.S. laws have precedence. The reluctance to authorize hemp cultivation has been particularly related to continuing suspicion that cultivating hemp would facilitate and promote "narcotic" usage of the species (Figure 7.11). There has been a widespread perception in the United States that those promoting industrial hemp are pro-marijuana interest groups. Indeed, "Groups such as the National Organization for the Reform of Marijuana Laws (NORML) have picked up the hemp crusade in order to claim the benefits of industrial hemp as an advantage of marijuana legalization" (Caulkins et al. 2012).

FIGURE 7.11 The view that cultivating hemp (the outer perimeter of the plantation) is a stalking horse or subterfuge for recreational marijuana usage. Prepared by B. Brookes.

FIGURE 7.12   The two basic classes of fiber in the stem. The top of this hemp stem was subjected to ret-ting (decomposition of the softer tissues), separating the phloem (bast) fiber from the woody core, which is composed of xylem fiber. The phloem fibers are amalgamated into bundles; note how these bundles intersect to form a net-like supportive girdle. As discussed in the text, the woody core provides vertical strength, while the phloem fiber network provides flexibility, preventing excessive bending and torsion. Compare Figure 7.18. Photo by Natrij, released into the public domain.

THE TWO BASIC CLASSES OF STEM FIBER: PHLOEM (BAST) AND XYLEM (WOOD)

Two basic classes of fiber occur in the stems of *C. sativa*: phloem (bast) fiber and xylem (wood) fiber, illustrated in Figure 7.12. These are associated with the two vascular (fluid transportation) systems of plants: xylem tissue, which functions to transport water and solutes from the roots to other parts of the plant, and phloem tissue, which transports photosynthetic metabolites from the foliage to nourish other parts of the plant. Additional details on these fibers are presented in the following discussion.

ANATOMY OF THE STEM

THE ECONOMICALLY IMPORTANT STEM TISSUES

Three kinds of fiber cells in the stems of *C. sativa* are of economic value: primary phloem, second-ary phloem, and xylem. The fiber cells of hemp are alive initially but die at maturity as their cell walls become blocked by deposit of lignin. The very valuable primary phloem (bast) fibers are initi-ated in the apical meristem at the tip of the growing main stem (Figure 7.13, left) and subsequently elongate (much more so in the internodes, i.e., between the nodes where the leaves arise). The pri-mary phloem fibers are slightly separated from the epidermis of the stem by several layers of cells making up the cortex (Figures 7.13 through 7.15). Upon the completion of internode elongation, a cambium (thin cylinder of meristematic tissue running the length of the stem), located internally to the primary fibers, produces (a) secondary phloem fibers toward the outside of the stem (but inside the primary phloem fibers) and (b) xylem (woody hurds tissue) toward the center of the stem.

As a result of the growth processes described previously, the mature hemp stem consists of several concentric cylinders of tissue (see Figures 7.13 through 7.15). The multicellular cortex is

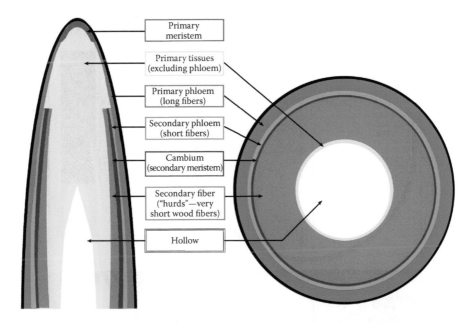

Primary
meristem

Primary tissues
(excluding phloem)

Primary phloem
(long fibers)

Secondary phloem
(short fibers)

Cambium
(secondary meristem)

Secondary fiber
("hurds"—very
short wood fibers)

Hollow

FIGURE 7.13 Simplified diagrams (left: long section, right: cross-section) of a hemp stem showing locations of principal tissues. Particularly note the primary phloem, which contains the high-quality long fibers. The short fibers in the secondary phloem are of lesser value but can contribute to various modern applications. As noted in the text, the very short wood fibers in the hurds have become a valuable commodity for many lower-priced products. Prepared by B. Brookes.

found immediately internal to the unicellular epidermis; as with other stem fiber crops, removal of the cortex (not to be confused with "decortication" described later) by "retting" (also described later) is a key initial step in fiber extraction (a partly retted hemp stem is shown in Figure 7.12). Internal to the cortex is the primary phloem fiber tissue, in which the principal fiber of interest is found; the primary fibers are amalgamated into rope-liked "bundles." Internal to the primary fiber bundles (toward the center of the stem) are the secondary phloem fibers, which are also amalgamated into bundles. The secondary phloem fiber is of considerably lower quality than the primary fiber. Because secondary fiber bundles are more common in the lower third of the mature stem, the upper third of the stem produces higher-quality fiber than the lower third. The next significant concentric layer is the cambium, a meristematic tissue that generates the secondary fiber toward the outside and the wood (xylem) tissue toward the stem center. A pith made up of undifferentiated cells initially occupies the center of the very young stems. As the cambium produces the woody xylem toward the center, the pith is crushed and degenerated. The center of the pith becomes hollow, but only to a limited extent at the internodes (where the leaves arise), and even less so toward the base of the stalk. The woody tissue and the remnants of the pith constitute the "hurds."

Ambiguities Regarding Stem Tissue Anatomical Terms

The pith remnants constitute less than 1% of the hurds, but in some descriptions, the entire hurds are mistakenly called pith. The phrase "woody core" is often applied to all tissues internal to the cambium, and the phrases "woody fibers" and "wood fibers" pertain to the hurd fibers. "Shives" rather than "hurds" is more often used for flax than for hemp, and "core" is more frequently applied to kenaf and jute. The term "bark" is often used to indicate all stem tissues external to the cambium, so that "bast fibers" is synonymous with "bark fibers" (De Meijer 1994a; De Meijer and Keizer 1996a).

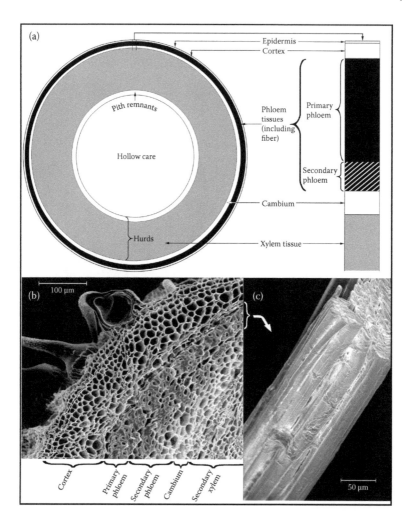

FIGURE 7.14   Structure of a hemp stem with special reference to fiber. (a) Scaled diagram of a cross-section of a mature hemp stem, showing detail (at right) of the outer portion. The relative proportions of primary and secondary fiber, hurds, and hollow core vary with maturity, position in stem, conditions of cultivation, and genetic background. In fiber extraction, the epidermis and cortex are removed, and the valuable phloem fibers are separated at the cambium from the woody central core (hurds). (b) Scanning electron micrograph of a portion of a hemp stem cross-section. (c) Scanning electron micrograph of a bundle of primary fibers. (a) Prepared by B. Brookes. (b and c) Prepared by E. Small and T. Antle.

HOW THE THREE KINDS OF FIBER CONTRIBUTE TO STEM STRENGTH

Tall stems such as occur in *C. sativa* require structural support, and the three kinds of fibers are important in this respect.

XYLEM FIBER

In trees, wood tissues (which include very tough tracheid cells) are responsible for the vertical strength of the stem, and indeed, in *C. sativa*, wood tissue (the "hurds" described previously) occupying the central part of the stem is also the main source of strength. The wood fibers of *C. sativa* are notably shorter than the phloem fibers and are significantly stiffer and less flexible. Like the trunk of a tree, the stalk becomes thicker (and woodier) toward the base, for support. The stalk may exceed 50 mm

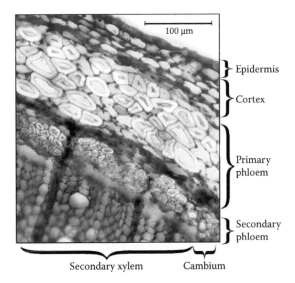

FIGURE 7.15 Cross-section of the outer part of a hemp stem, the tissues biologically stained to contrast their differences. Photo courtesy of Charlene Forrest and Jane P. Young (additional information is available in Forrest and Young 2006).

in diameter in uncrowded plants, but in thick stands, the stalk usually varies from 6 to 20 mm. The progressive thickness toward the base is due mostly to more hurd tissue being formed.

PRIMARY PHLOEM FIBER

As noted previously, primary phloem fibers glued together in fiber bundles occur in the outermost part of the stem. The phloem fibers (both primary and secondary) serve only to a limited extent to supplement vertical stem strength; their principal function is to provide flexibility so that the stem can bend but not break in response to wind and other environmental forces. The very high tensile strength of the phloem fibers limits bending and turning (torque). The long axis of the phloem fibers is more or less parallel to the stem axis—an arrangement that naturally keeps the stem upright and resists stem bending.

SECONDARY PHLOEM FIBER

As noted previously, secondary phloem fibers glued together in fiber bundles develop between the outer primary fiber bundles and the internal xylem. The secondary fibers are much shorter than the primary fibers, and the secondary fiber bundles are smaller in diameter than the primary fiber bundles. Because the primary phloem fibers stop being formed when the stem is young and slim, the secondary fibers take on the task of adding to the growing need for strength of the stem as it enlarges. As noted previously, the progressive thickness of the stem toward the base is due mostly to more hurd tissue being formed, but this woody tissue is supplemented toward the base of the stalk by secondary fibers. Because females live longer than males, they develop thicker, stronger phloem fiber cells and more secondary fiber bundles than do the males. Proportionately, males have a greater amount of primary fiber in the stems, resulting in higher quality.

LIGNIFICATION

Cell wall lignification deposits lignins (ill-defined phenolic polymers) in the extracellular polysaccharidic matrix (i.e., between cells, which are accordingly more strongly glued together), adding strength to fiber cell bundles as they age (Ros Barceló 1997). This has mixed benefits: it decreases overall fiber quality, increasing the tensile strength of the fibers but reducing break and torque resistance and elasticity (Bócsa and Karus 1998).

FIBER QUALITIES

The primary bast fibers (Figure 7.16) are the most valuable product of the stems. Measuring 3–55 μm long (Van der Werf 1994a) or up to 100 μm according to Crônier et al. (2005), they are among the longest cells found in plants. They are 7 to 50 μm in diameter and are amalgamated in fiber bundles (do not confuse these bundles of cells with the fibril bundles or with the microfibril bundles noted next), which can be 1–5 m long, although averaging only about 30 μm in diameter (Shahzad 2012). The fibers in the bundles are cemented together by a complex mixture of pectins, hemicelluloses, and smaller amounts of lignin. Hemp fiber cells are made up of "microfibrils"—bundles of cellulose molecule chains; the microfibrils are themselves organized in bundles ("fibrils") oriented in a helical pattern along the length of the fiber. The fibers are notable for their high tensile strength (Fan 2010), due substantially to their thick, cellulosic secondary wall and their low microfibril angle (Mohanty et al. 2000). ("Microfibril angle" is the angle between the direction of the helical windings of cellulose microfibrils in the secondary cell wall of fibers and tracheids and the long axis of the cell. A low angle, i.e., the microfibril orientation is close to the fiber axis, increases strength.) As noted earlier, the primary phloem fibers arise as cells near the shoot apex and continue to elongate, growing intrusively through other stem tissues that have stopped growth. Thickening of the secondary wall continues until the internal part of the cell (lumen) is almost entirely filled and the cell dies. The degree of elongation, thickness of the cells wall, and hence strength of the cells are modifiable somewhat by environmental conditions (Schäfer and Honermeier 2006). Van der Werf (1993), reporting on Ukrainian hemp, stated, "Many of the cultivars which have a high fiber content in the stem have been found to have larger fiber bundles than the older cultivars which have a low fiber content. Large fiber bundles are undesirable as they decrease fiber fineness, one of the most important quality parameters for cordage and textile purposes."

As the stem matures, the cambium produces additional (secondary) bast fibers, which are short (about 2 mm long), about 25 μm wide, and more lignified. The woody core fibers of the hurds are even shorter, 0.5–0.6 mm long, and like hardwood fibers are cemented together with considerable lignin. The secondary bast fibers are of notably smaller value than the primary bast fibers, and in

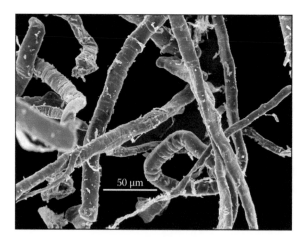

FIGURE 7.16 Scanning electron microphotograph of individual primary hemp fibers. Photo by Setral Chemie GmbH (CC BY 3.0).

turn, the woody core is of still lesser value. The secondary bast fiber bundles tend to adhere to the woody core, and so some of this fraction is difficult to separate.

Various authors differ in their reports on the composition of the fiber and hurds. Gümüşkaya et al. (2007) reported that the bast fibers are 57%–77% cellulose, 9%–14% hemicellulose, and 5%–9% lignin, while the hurd fibers are 40%–48% cellulose, 18%–24% hemicellulose, and 21%–24% lignin. The European Food Safety Authority (EFSA) Panel on Additives and Products or Substances Used in Animal Feed (2011) reported that the fiber contains 80%–83% cellulose (presumably including hemicellulose) and 17%–20% lignin, while the hurds contain 35% cellulose, 18% hemicellulose, and 21% lignin. De Groot et al. (1994) reported that the hurds contain 33%–37% cellulose, 16%–20% hemicelluloses, and 17%–22% lignin. Van der Werf et al. (1994a) stated that the lignin content in hemp varies from 4% in the bast fraction to 21% in the hurds fraction.

Both genetic and environmental conditions determine stem dimensions, as well as size and chemical characteristics of the bast fibers of economic interest for fiber hemp. These aspects of the stem are also determined to some extent by growth conditions. Diameter of stem, size of bast fibers, and chemical content of fiber are determined to some extent by growth conditions, particularly density of planting (for literature, see Bócsa and Karus 1998; Amaducci et al. 2005; and Khan et al. 2011). The stems of plants grown very closely (12 cm apart) may be only 4 mm in diameter, while the stems of the same variety grown well spaced (45 cm apart) may be as thick as 26 mm (Bócsa and Karus 1998). Physical and mechanical properties of the fibers differ among the bottom, middle, and top of the stems (Li et al. 2013). Breeding is important to create cultivars with improved fiber fineness, tensile strength, and other qualities, but production and processing technologies are thought to be critical in making best use of the natural qualities of hemp fiber (Müssig 2003; Finta-Korpel'ová and Berenji 2007).

Time of harvest affects fiber quality. Centuries of experience have resulted in hemp being harvested for fiber while it is in early flower. This was confirmed by Liu et al. (2015), who found that "Fibers harvested at the beginning of flowering exhibited high tensile strength and strain, which decreased with plant maturity. Reduction in strength was related to the increase in proportion of secondary fibers and decrease in cellulose deposition leading to inferior properties of fibers."

Hemp long fiber is one of the strongest and most durable of natural fibers, with high tensile strength, wet strength, resistance to decay, and other characteristics that make it technically suited for various industrial products (Hemptech 1995; Karus and Leson 1996). Hemp fiber dyes well, resists mildew, blocks ultraviolet light, conducts heat, absorbs water well, and has natural antibacterial properties. Both the long, lignin-poor bast fibers and the hurds have considerable applications, as detailed later. As will be noted, depending on technology used, the hurds are not completely separated from the bast during retting. Frequently, bast fibers remaining attached to the hurds increase the quality of the hurds, and conversely, hurds imperfectly separated from the bast decrease the quality of the latter.

There is increasing demand for lightweight, biodegradable, sustainably produced, recyclable materials that can be used alone or in composites, and hemp fiber seems to have excellent properties in these respects (Small 2014b). Shahzad (2012) reviewed the industrial qualities of hemp fiber and came to the following conclusions: (1) The natural strength and stiffness of hemp fiber are very desirable for reinforcement of composite materials. (2) The mechanical properties of hemp fibers are comparable to those of glass fibers, but their biggest disadvantage is variability of properties. (3) Composites made of hemp fibers with thermoplastic, thermoset, and biodegradable matrices have exhibited good mechanical properties. (4) Several surface treatments applied to improve fiber/matrix interfacial bonding have resulted in considerable improvements in the mechanical properties of composites.

FIBER EXTRACTION TECHNOLOGIES

As with other bast fiber crops, the most desirable ("long") fibers are found in the phloem-associated tissues. The traditional and still major first step in fiber extraction is to ret ("rot") away the softer parts of the plant, by exposing the harvested stems to microbial decay in the field ("dew retting," shown in Figure 7.17) or submerging the stems in water ("water retting," shown in Figure 7.5b).

FIGURE 7.17 Traditional dew retting. (a) Windrowed fiber hemp in process of dew retting. Such harvesting of the stems lays the stalks in swathes on the ground, where dew and rain showers stimulate decay. Photograph taken in 1930 on the Central Experimental Farm, Ottawa, Canada. (b) Shocked fiber hemp in process of dew retting. Photograph taken in 1931, near Ottawa, Canada. Photographers unknown.

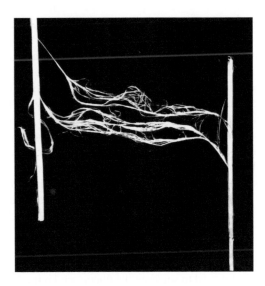

FIGURE 7.18   A hemp stem, bent sharply after retting, breaking the woody central portion (hurds), leaving the bark fibers unbroken. The two portions of stem are separated in this photograph and are joined by the tough, long bark fiber bundles (cf. Figure 7.12).

During retting, certain bacteria or fungi selectively remove pectic substances that bind the fibers to less desirable parts of the stem (the cellulose, which makes up most of the fiber cells, is very resistant to decay). The result is to slough off the outer parts of the stem and to loosen the inner woody core (the "hurds") from the phloem fiber bundles (Figure 7.18). Water retting, which produces higher-quality fiber than dew retting, has been largely abandoned in countries where labor is expensive or environmental regulations exist. Water retting, typically by soaking the stalks in ditches, can lead to a high level of pollution and bad odor of the wastewater because of organic fermentation. Most hemp fiber used in textiles today is water retted in China (Zhang et al. 2008). Retting in tanks rather than in open bodies of water helps to control the effluents. Unlike flax, hemp long fiber requires water retting for preparation of high-quality spinnable fibers for production of very fine textiles.

Occasionally, hemp is "stand retted"—the standing crop is dehydrated by the application of a desiccant herbicide and retting occurs while the crop is erect (and dead). Rarely, hemp is frost retted—the stems are allowed to ret outdoors overwinter. A variety of experimental retting techniques have also been attempted, such as retting in plastic bags (Li et al. 2009) and ensilage (Idler and Pecenka 2007; Idler et al. 2011). Avoiding retting entirely by processing freshly harvested stems directly into products has been proposed (Idler et al. 2011) and, at least in theory, is an attractive possibility.

As with other bast fiber crops, hemp phloem fibers are arranged in bundles parallel to the stem axis and are embedded in a pectic polysaccharide network. The pectin network cementing the fibers together is the major obstacle to obtaining high-quality fiber. Traditional water retting is effective because bacteria that are present secrete pectinolytic enzymes; filamentous fungi producing pectinase are more important in dew retting. A commonly used technique to improve fiber separation is chemical processing with sodium hydroxide or diluted sulphuric acid. Steam explosion is a potential technology that has been experimentally applied to hemp (Garcia-Jaldon et al. 1998). Material separated into crude phloem fiber bundles is the raw material, and this is subjected to steam under pressure and increased temperature, which "explodes" (separates) the fibers so that one has a more refined (thinner) hemp fiber that currently is only available from water retting. The high temperature and moisture soften the fibers and hydrolyze the pectins and hemicellulose, so that there are both mechanical and chemical changes associated with steam explosion. Still additional potential methods that have been considered to augment or replace traditional retting include ultrasonic techniques, enzymatic retting (Pakarinen et al. 2012), and the use of selected or improved decay microorganisms (Tamburini et al. 2004; Thygesen et al. 2005; Valladares Juárez et al. 2009; Di Candilo et al. 2010).

As mentioned previously, removal of the cortex (the tissue external to the desired phloem fiber) by retting is a key initial step in fiber extraction. "Decortication" (the term not referring to the cortex tissue) refers to a mechanical force (bending, shearing, impacting, or kinking) that separates the high-quality, flexible, outer phloem fibers from the low-quality, stiff, and more brittle woody fibers (hurds) in the internal part of the stem. In traditional hemp processing, the long fiber was separated from the internal woody hurds in two steps, *breaking* (stalks were crushed to loosen the adherence between the internal hurds and the external long fibers; this was accomplished in the past with mechanical tools but in more recent times by using rollers that broke the woody core into short pieces) and *scutching* (mechanical beating to remove the hurds). These mechanical forces can separate not just most of the hurds but also the short (secondary phloem) fibers ("tow") and long (primary phloem) fibers ("line fiber," "long-line fiber"). A single, relatively expensive "decorticator machine" can carry out all of these processes.

For highest-quality fiber (generally for clothing), as shown in Figure 7.19, hemp needs not only to be scutched (removal of most of the hurds from the phloem fiber) but also hackled ("hackles" are steel "brushes" traditionally used to separate the fibers; see Figure 7.20). Hackling additionally combs the fibers, making them align in parallel, and removes remaining pieces of stalks, broken fibers, and extraneous material. However, mostly for nontextile applications, cruder alternatives may be employed to produce a less pure grade of fiber. This involves production of "whole fibers" (i.e., harvesting both the long fibers from the cortex and the shorter fibers from throughout the stem) and technologies that utilize shortened hemp fibers. The approach is currently dominant in Western Europe and Canada and commences with field dew retting (typically two to three weeks). A principal limitation is climate—the local environment should be suitably but not excessively moist at the close of the harvest season. Once stalks are retted, dried, and baled, they are processed to extract the fiber. In general, in the European Union (EU) and Canada, fibers are not separated into tow and line fibers but are left as "whole fiber." Based on experience in the EU, where the "fiber line" product may contain 2%–25% hurds that were not separated, the production of 1.0 kg of fiber product produces 1.7 kg of separated hurds as a by-product (Carus et al. 2013). In the EU, the bast fiber sells for about twice the value of the hurds (Carus et al. 2013). In the EU, highly purified fiber (with only 2%–3% hurds) is employed for automotive applications (described later), while the less purified fiber (with 25% hurds) is used in cigarette paper. In Western Europe, the fiber is often "cottonized," i.e., chopped into short segments the size of cotton and flax fiber, so that the fibers can be processed on flax processing machinery, which is very much better developed than such machinery is for hemp.

FIGURE 7.19 Raw cleaned hemp fiber (the thread-like materials are bundles of primary hemp fibers; cf. Figure 7.14c). Photo by Rasbak (CC BY 3.0).

FIGURE 7.20   A hackle employed for aligning and cleaning raw hemp fiber. In general, hackles are combs or boards with long steel teeth for dressing fiber plants. Photo by CVB (CC BY SA 3.0).

HOW DOMESTICATION HAS ALTERED *CANNABIS SATIVA* FOR FIBER PRODUCTION

Fiber hemp plants, in contrast with *C. sativa* plants grown for marijuana or oilseed, and also in contrast with wild plants, have been selected for features maximizing stem fiber production. Selection for fiber has resulted in strains that have much more primary phloem fiber and much less woody core than encountered in marijuana strains, oilseed cultivars, and wild plants. Fiber varieties may have less than half of the stem made up of woody core, while in nonfiber strains, more than three-quarters of the stem can be woody core (De Meijer 1994a). Moreover, in fiber plants, more than half of the stem exclusive of the woody core can be fiber, while nonfiber plants rarely have as much as 15% fiber in the corresponding tissues. Also important is the fact that in fiber selections, most of the fiber can be the particularly desirable long primary fibers (De Meijer 1995a).

Since the stem nodes tend to disrupt the length of the fiber bundles, thereby limiting quality, tall, relatively unbranched plants with long internodes have been selected. Another strategy has been to select stems that are especially hollow at the internodes (Figure 7.21, right), with limited hurds, since this maximizes the production of long phloem fiber (although the decrease in woody tissues makes the stems less resistant to lodging by wind). Similarly, limited seed productivity concentrates the plant's energy into production of fiber, and fiber cultivars often have low genetic propensity for seed output. Selecting monoecious strains overcomes the problem of differential maturation times and quality of male and female plants (males mature one to three weeks earlier). Male plants in general are taller, albeit slimmer, less robust, and less productive (although they tend to have superior fiber). Except for the troublesome characteristic of dying

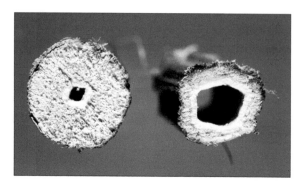

FIGURE 7.21   Cross-sections of stems at internodes of a fiber plant (right) and of a marijuana plant (left). Fiber cultivars have stems that are more hollow at the internodes, i.e., with less woody tissues, since this allows more energy to be directed into the production of phloem fiber.

after anthesis, male traits are favored for fiber production. In former, labor-intensive times, the male plants were harvested earlier than the females, to produce the best fiber. Fiber strains have been selected to grow well at extremely high densities, which increases the length of the internodes (contributing to fiber length) and increases the length of the main stem (contributing to fiber bundle length) while limiting branching (making harvesting easier). The high density of stems also contributes resistance to lodging, desirable because woody supporting hurd tissue has been decreased by selection. The limited branching of fiber cultivars is often compensated for by possession of large leaves with wide leaflets, which increase the photosynthetic ability of the plants.

Since fiber plants have not generally been selected for drug purposes, the level of intoxicating constituents is often limited, usually much less than 1%. However, some hemp strains grown in subtropical Asia (where fiber hemp is a very minor crop and the strains are mostly unimproved land races with fiber content below 20%) are of variable tetrahydrocannabinol (THC) content and may have a content of THC as high as 3%.

FIGURE 7.22   Artist's concept of resurrected traditional paper and clothing uses of hemp fiber. Drawn by B. Flahey.

ECONOMIC PRODUCTS

Clothing and paper are the traditional hemp consumer products, and these are mostly obsolescent, although there are attempts to resurrect these items for niche and boutique markets (Figure 7.22). However, as presented in the following, a variety of quite new applications have revived the hemp fiber industry.

TEXTILES, FABRICS, AND CLOTHING

"Textiles," "fabrics," and "clothing" refer to sheets of fiber networks prepared by weaving, pressing (to make felts), or bonding; none of these ambiguous terms necessarily distinguishes woven from nonwoven products. Although quite expensive, hemp clothing and other woven products (Figures 7.23 and 7.24) have a natural appeal to a sector of the population that considers them desirable simply because hemp seems to be an especially sustainable crop or because purchasing hemp goods

FIGURE 7.23   Hemp clothing and other woven textile products.

FIGURE 7.24   Elegant hemp shoe. Photo by *Cannabis Culture* (magazine) (CC BY 2.0).

is viewed as a way of expressing approval of marijuana. Hemp clothes are resistant to abrasion but are typically abrasive because of relatively coarse and nonhomogeneous fiber bundles and so have been used historically for outer garments and work apparel (Bócsa and Karus 1998) and, sadly, for prisoners and slaves. However, appropriate processing and blending with other natural fibers can significantly improve the "feel" of hemp, and sources in China supply hemp textile blends indistinguishable from fine linens in texture.

Hemp fiber sufficiently cleaned of hurds to be suitable for textiles ("textile grade fiber") is produced primarily in China and to a smaller extent in France and several east European countries. Weaving of hemp fiber into textiles and apparel is primarily done in China. Production of fine hemp fabrics is insignificant in the EU and North America. Outside of China, processing costs are higher than for other fibers, because hemp fibers vary from the standard specifications for length and diameter established for the equipment used in most textile and apparel factories, necessitating the use of specialty machines. The hemp apparel industry outside of China is based on fiber, yarn, and fabrics imported from eastern Europe and China. China's established extraction technology and spinning facilities, to say nothing of much lower labor costs, make it very difficult to develop a hemp textile industry in Europe and North America. The fact that spinning facilities for natural fibers are so concentrated in China makes it almost impossible to competitively produce hemp fabrics elsewhere. In theory, a domestic hemp textile industry could be established outside of China, by developing specialized harvesting, processing, spinning, and weaving equipment and perhaps new technologies. In practice, China controls the hemp textile market and probably will remain dominant for the foreseeable future.

In addition to textiles used in clothing, coarser woven cloth (canvas) made with hemp is used for upholstery, bags, sacks, and tarpaulins. Such products can be manufactured from a relatively crude grade of hemp fiber without weaving. In both the EU and North America, there is production of nonwoven, relatively coarse fabric-like materials (e.g., Figure 7.25) using hemp fiber. Needle punch carpets (made by using barbed needles to assemble a web of compacted, interlocked fibers) are usually constructed today with synthetic fibers, but there is interest in using hemp fiber. Alternatively, the fibers in nonwoven carpets can be held together by thermosetting methods. Composites using hemp in combination with other natural fibers, postindustrial plastics, or other types of resins are being used to produce nonwoven matting for padding (for example, for mattresses and futons), backing for woven carpets, sound insulation, lining of cribs and pens for young livestock, and other applications.

Since European cultivation of hemp is substantial, there is fairly extensive production of such nonwoven products in the EU. However, hemp is grown mostly for oilseed in Canada (hence in North America), so a nonwoven hemp industry is much less developed. Stemergy (originally Hempline), in Ontario, Canada, the first firm to grow hemp for commercial purposes in North America since the

FIGURE 7.25 Multipurpose, nonwoven matting, fabricated from hemp. (Photo by E. Small, sample provided by Kenex Ltd., Pain Court, Ontario.)

Second World War (starting with experimental cultivation in 1994), specializes in the production of hemp fiber for upholstery and carpeting, and several other Canadian enterprises are also interested in hemp primarily as a fiber rather than as an oilseed.

SUBSTRATES FOR PLANT GROWTH

Horticultural Planting Media

Plants are often grown initially from seeds germinated in porous, water-retaining material used as substitutes for soil (fertilization is of course necessary) and then transplanted to regular soil. Some crops, such as cress, are raised to maturity in soilless materials. Peat is the most widely used organic substrate. Plant wastes, rockwool, and synthetic materials are also used. In recent years, hemp fiber has been added to this market niche.

Biodegradable Mulch

"Mulch" refers to material layered onto a soil surface to improve growth of plants (by conserving moisture, either lowering soil temperature or retaining heat, improving soil fertility and health, and preventing weed growth) or simply to enhance visual appearance. Common friable or granular mulch materials include compost, bark chips, wood chips, and gravel. Pulverized hemp hurds are sometimes used as a mulch but are not competitive for this purpose.

Unlike the materials discussed in the previous paragraph, mulches can also be continuous, and plastic sheeting is common for this purpose. At present, the main ground-covering, sheet-like mulches are polymeric (polythene, spun-blown polypropylene), but some are made of glass fiber or natural fibers. Both woven and nonwoven fabrics can be used; woven and knitted materials are stronger and the open structure may be advantageous (e.g., in allowing plants to grow through), but nonwovens are cheaper and better at suppressing weeds. Sheet-like mulches made from fibers, whether woven or not, are widely used for crops and are sometimes referred to as "agricultural textiles" and "geotextiles" (although "textiles" are usually understood to be woven). "Mulch fleece" (pressed, nonwoven fiber blanket) is sometimes made of hemp fiber and is popular in the EU marketplace (Carus et al. 2013).

Flax and hemp fibers exposed to water and soil have been claimed to disintegrate rapidly over the course of a few months, which would make them unacceptable for products that need to have long-term stability when exposed to the elements. Coco (coir; *Cocos nucifera* L.) fiber has been said to be much more suitable, due to higher lignin content (40%–50%, compared to 2%–5% in some bast fibers). Coir is much cheaper than flax and hemp fibers. However, this evaluation does not do justice to the developing hemp mulch market. The degree of rot resistance of hemp fiber makes its use in ground matting desirable for certain applications because the ability to last outdoors for many years is frequently undesirable. For example, the widespread current use of plastic netting to reinforce grass sod is quite objectionable, the plastic persisting for many years and interfering with lawn care.

CORDAGE PRODUCTS

String and rope manufactured from hemp (Figure 7.26) are available from specialty outlets, but except for certain applications, hemp cordage is largely obsolete. Hemp ropework is used to make hemp hammocks, which are popular, and take advantage of the natural strength, water resistance, and decay resistance of hemp fiber. Hemp fiber is sometimes employed to make biodegradable twine to replace plastic ties used to attach plants to supporting poles.

PRESSED AND MOLDED FIBER PRODUCTS

Molded and pressed (nonwoven) natural fiber products, with or without the addition of bonding media, are extensively used for a wide range of applications (Figure 7.27). While many different fibers can be used, hemp fiber (both phloem and hurd) has become a significantly basic resource in Europe.

FIGURE 7.26   Hemp cordage.

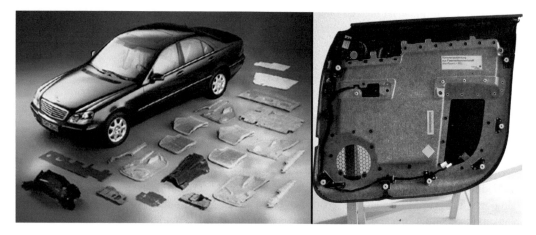

FIGURE 7.27   Molded and pressed fiber products. Left: C-class Mercedes-Benz automobiles have more than 30 parts made of natural fibers, including hemp. (Courtesy of Daimler.) Right: Interior carpeting of a car door made with a biocomposite of hemp fiber and polyethylene. Photo by Christian Gahle, Nova-Institut GmbH (CC BY SA 3.0).

Paper and Specialty Pulp Products

The oldest surviving paper is over 2000 years of age, comes from China, and was made from hemp fiber (Fleming and Clarke 1998). Egyptian papyrus sheets might be thought to be an older form of paper but are not "paper" as this term is understood by experts because the fiber strands are woven, not "wet-laid" (Van Roekel 1994; Van Roekel et al. 1995). Until the early nineteenth century, hemp and flax were the chief paper-making materials. Wood-based paper came into use when mechanical and chemical pulping were developed in the mid-1800s in Germany and England. Today, at least 95% of paper is made from wood pulp. Before then, paper was made from rags, most commonly hemp rag. Using hemp directly for paper was considered too expensive, and in any event, the demand for paper was far more limited than today. In 1769, an early North American newspaper, the *Boston Newsletter*, encouraged people to recycle their rags for the manufacture of paper with this poem:

> Rags are as beauties, which concealed lie,
> But when in paper how it charms the eye,
> Pray save your rags, new beauties to discover,
> For paper, truly, everyone's a lover.
> By the pen and the press such knowledge is displayed,
> As wouldn't exist if paper was not made.

FIGURE 7.28   Left: Hemp paper products (writing paper, notebook, envelopes). Right: Hemp cigarette paper, the most profitable paper product currently manufactured from hemp.

"Specialty pulp" is the most important component of the hemp industry of the EU (Karus and Vogt 2004) and is expected to remain important. In France, a large market for high-quality paper, predominantly cigarette paper (Figure 7.28, right), has developed; such paper is completely free of the intoxicating resin, so there is no possibility of inhaling cannabinoids. Other specialty pulp products made from hemp are bank notes, technical filters, hygiene products, bible paper, art papers, and tea bags. Several of these applications take advantage of hemp's high tear and wet strength. In Europe, decortication/refining machines are available that can produce 10 tons/hour of hemp fiber suitable for such pulp use. Capacity for hemp pulp production and value-added processing is limited outside of Europe. Although specialty pulp has a large market in the EU, Carus et al. (2013) caution that no economic expansion is expected and the market is risky because wood pulp with specific additives could be substituted.

Hemp paper is useful for specialty applications such as currency and cigarette papers, where strength is needed. The bast fiber is of greatest interest to the pulp and paper industry because of its superior strength properties compared to wood. Hemp core fibers are generally considered too short for high-grade paper applications (a length of 3 mm is considered ideal), and too much lignin is present. While the long bast fibers have been used to make paper almost for two millennia, the woody core fibers have rarely been so used. However, the short, bulky fibers found in the inner part of the plant (hurds) could probably be used to make cheaper grades of paper, provided that appropriate technology was developed (De Groot et al. 1998).

The pulp and paper industry based on wood has examined the possible wider use of hemp for pulp, but only on an experimental basis. The long phloem fibers of hemp do have technical properties that make them desirable for strengthening paper (Correia et al. 2003). The possibility of growing hemp specifically for a resurrected hemp-based paper industry has been examined (De Meijer 1993; Capelle 1996), and there have been attempts to clarify the genetics of characteristics that would improve the use of hemp for paper (Hennink 1994). Hemp's long fibers could make paper more recyclable. Since virgin pulp is required for added strength in the recycling of paper, hemp pulp would allow for at least twice as many cycles as wood pulp. However, various analyses have concluded that the use of hemp for conventional paper pulp is not profitable (Lewis et al. 1948; Fertig 1996; Selkirk and Spencer 1999; Lisson and Mendham 2000a). Hemp is not competitive for newsprint, books, writing papers, and general paper (grocery bags, coffee cups, napkins), although there is a specialty or novelty market for those specifically wishing to support the hemp industry by purchasing hemp writing or printing paper despite the premium price (Figure 7.28, left). In Europe, hemp pulp is about five times as expensive as wood pulp (Carus et al. 2013).

Hemp paper is high-priced for several reasons. Economies of scale are such that the supply of hemp is minute compared to the supply of wood fiber. Hemp processing requires non-wood-based processing facilities. Hemp paper is typically made only from bast fibers, which require separation from the hurds, thereby increasing costs. This represents less than 50% of the possible fiber yield of the plant, and future technologies that pulp the whole stalks could decrease costs substantially.

Hemp is harvested once a year, so that it needs to be stored to feed mills throughout the year. Hemp stalks are very bulky, requiring much handling and storage.

Transportation costs are also very much higher for hemp stalks than for wood chips. Waste straw is widely available from cereals and other crops and, although generally not nearly as desirable as hemp, can produce bulk pulp far more cheaply than can be made from hemp. In addition to agricultural wastes, there are vast quantities of scrub trees that can supply large amounts of low-quality wood fiber extremely cheaply. Moreover, in many areas unsuitable for agriculture, fast-growing trees can be grown, and such agro-forestry can be very productive and environmentally benign. And, directly or indirectly, the lumber/paper industry is politically powerful, able to stimulate substantial subsidies and/or supports, which is most unlikely for hemp.

A chief argument that has been advanced in favor of developing hemp as a paper and pulp source has been that as a nonwood or tree-free fiber source, it can reduce harvesting of primary forests and the threat to associated biodiversity. It has been claimed that hemp produces three to four times as much useable fiber per hectare per annum as forests. However, Wong (1998) notes evidence that in the southern United States, hemp would produce only twice as much pulp as does a pine plantation. It remains true, however, that hemp and other annual bast crops can be a potential lumber substitute in areas lacking trees (cf. Chapter 16).

BUILDING CONSTRUCTION PRODUCTS

The classic fable of the *Three Little Pigs* is based on houses that they built, respectively, of straw, sticks, and bricks. A big bad wolf was able to blow down the houses made of straw and sticks but not the brick house. While the fairy tale is intended to instruct young children about the value of hard work to produce worthwhile creations, it does not do justice to the construction value of "wood" and "concrete" manufactured from hemp straw (Figure 7.29). As discussed next, hemp straw is invaluable for producing excellent construction materials for buildings.

FIGURE 7.29 Humorous representation of the strength and durability of hemp-based construction. Prepared by B. Brookes.

Construction Wood Products (Pressboard, Moldings)

In North America, particleboards and fiberboards, which generally contain less than 10% adhesive or matrix, are sometimes referred to as composites. Wood is the principal material used in sheet composite board (particleboard, pressboard, and oriented strand board such as Aspenite or Sterling board), but nonwood fibers are also employed. Flax, jute, kenaf, hemp, and wheat straw are used for this purpose. Wheat straw is the dominant nonwood fiber in such applications. Although it might seem that long hemp phloem fibers are desirable in composite wood products because of their length and strength, in fact, the short fibers of the hurds have been found to produce a superior product of appreciable strength (Nikvash et al. 2010; Figure 7.30). However, hemp fiberboard is more expensive than wood particleboard and so is questionably competitive. Hautala et al. (2004) fabricated a plywood-like composite from hemp fiber strips and epoxy resin, and this equalled plywood in strength, but once again, plywood made with wood is cheaper. In China and the EU, small amounts of particleboard based on hemp hurds are marketed.

Concretized and Masonry Construction Products (Stucco, Building Blocks, Tiles)

Utilizing the ancient technique of strengthening clay with straw to produce reinforced bricks for constructing domiciles, plant fibers have found a number of comparable uses in modern times. Today, polypropylene or glass fiber is often employed to reinforce cement and plaster. Similarly, hemp fibers added to concrete increase tensile strength while reducing shrinkage and cracking. Fiber-reinforced cement boards and fiber-reinforced plaster are other occasionally produced experimental products. Whole houses have been made based on hemp fiber. Hemp bast fibers are produced at much more cost than wood chips and straw from many other crops, so the use of cheaper hemp hurds is appropriate.

The uses noted in the previous paragraph are based on hemp simply as a mechanical strengthener dispersed within a matrix of a material. Much more significantly, hemp can be chemically combined with materials. For example, hemp with gypsum and binding agents may produce light panels that might compete with drywall. Hemp and lime mixtures make a high-quality plaster. Hemp hurds are rich in silica (which occurs naturally in sand and flint), and the hurds mixed with lime undergo mineralization ("petrification") to produce a stone-like material. Using the bast fibers in addition to the hurds does not seem to increase the strength of hemp–lime concrete, which in any event is relatively weak; hemp–lime formulations typically have about 5% of the compressive strength of residential grade concrete and requires load-bearing supplements (De Bruijn et al. 2009); however, mechanical properties increase with the mortar density (Elfordy et al. 2008) and hemp–concrete blocks can be self-supporting. Hemp–lime concrete weighs only about 15% as much as concrete

FIGURE 7.30 Fiberboard sample made with hemp. Photo by E. Small, sample supplied by K. Domier, University of Alberta, Edmonton.

(it can float on water), is easier to handle, lacks the brittleness of concrete, and does not need expansion joints. Table 7.1 indicates some advantages of hemp–lime technology in comparison with conventional construction of buildings. A big advantage is that producing hemp–lime concrete requires about half the energy needed for equivalent quantities of cement and masonry (Balatinecz and Sain 2007). Hundreds of houses have been built in Europe, Asia, and North America using hemp–lime construction (note Figure 7.31). Several guides to building with hemp have been published (Bevan and Woolley 2008; Benhaim et al. 2011; Allin 2012; Stanwix and Sparrow 2014). Hemp–lime construction has economic and environmental benefits (Bevan and Woolley 2008; Ip and Miller 2012; Duffy et al. 2013). The mineralized material can be blown or poured into the cavities of walls and in attics as insulation. The foundations, walls, floors, and ceilings of houses have been made using hemp hurds mixed with natural lime and water. Sometimes, plaster of Paris (pure gypsum), cement, or sand is added. The resulting material can be poured like concrete but has a texture vaguely reminiscent of cork—much lighter than cement and with better heat and sound-insulating properties. Hemp–lime stucco can be sprayed onto surfaces (Figure 7.32). Experimental ceramic tiles made of hemp have also been produced (Figure 7.33), which are noticeably warmer to the touch than conventional tiles and so are ideal for bare feet on bathroom floors.

TABLE 7.1
Pros and Cons of Hemp–Lime Construction

Advantages	Disadvantages
Construction	
Simpler construction than traditional timber frame	Short construction season
Homogenous material facilitates airtightness and provides an ideal surface to plaster	Long drying time
Low risk of thermal bridging	Labor-intensive construction
No risk of insulation slumping within the wall leaving uninsulated air voids	Inexperience of contractor can cause complications; some care and training required
Low-skilled construction method	Limited best practice protocols for inexperienced users
Performance	
Good thermal performance	
Damping of temperature fluctuations	
Breathable wall contributes to humidity regulation and passive control of internal environment	
Reasonable acoustical performance	
Excellent fire resistance	
Structural	
Additional stiffness provided to timber frame construction; alkaline environment protects against wood rotting	Not load supporting
Lighter construction makes foundation of building less extensive and thus more ecological	Long carbonation time to reach full strength
Environmental	
Low embodied energy	Storage and transport of high volumes of materials required
Excellent recyclability of waste materials as well as end of life building recyclability	

Source: After Duffy, E., Lawrence, M., Walker, P., *Civil Environ. Res.*, 4, 16–21, 2013.

FIGURE 7.31 Hemp–lime blocks (made with a mixture of woody hemp core [hurds], lime binders, and water) in building construction. (a) Hemp–lime block. Photo by Scott Lewis (CC BY 2.0). (b) A building in the Philippines with walls constructed of prefabricated hemp–lime blocks (public domain photo). (c) Hemp–lime blocks utilized as insulation on the exterior of standard wood-frame construction. Photo by Olivier DuPort (CC BY 3.0).

FIGURE 7.32 Spray application of hemp–lime composite to a building. Photo courtesy of Steve Alin.

FIGURE 7.33 Hemp "ceramic tile." Photo by E. Small, tile furnished by Kenex Ltd., Pain Court, Ontario.

ANIMAL BEDDING BASED ON HURDS

The woody core (i.e., the hurds or shives) of hemp is highly absorbent and spongy and so makes remarkably good animal bedding, and indeed, such usage traces back more than a century (Dewey 1916; Dewey and Merrill 1916). The hurds appear to be unsurpassed for horse bedding and also make an excellent litter for cats and other pets (Figure 7.34). It has been claimed that the Queen of England's pampered horses sleep on hemp. The hurds can absorb up to five times their weight in moisture (typically 50% higher than wood shavings), do not produce dust (following initial dust removal), and are easily composted. Hemp bedding is especially suited to horses allergic to straw. In Europe, the animal bedding market accounts for about half of all sales of hurds (Carus et al. 2013). Because hemp hurds are a costly product, and animal bedding is in very high demand, this will likely remain the most important application of the hurds. The high absorbency of hemp hurds has led to their occasional use as an absorbent for oil and waste spill cleanup. Hemp as an industrial absorbent has generated some interest in Alberta, for use in land reclamation in the oil and gas industry.

FIGURE 7.34 Animal litter/bedding made from hemp hurds (woody stem tissues of hemp). Photo by Salix (CC BY 3.0).

Plastic Biocomposites

With respect to fiber, a "composite" is often defined as a material consisting of 30%–70% fiber and 70%–30% matrix (Bolton 1995). After sisal, hemp is the most widely used natural fiber to reinforce composites (Shahzad 2012). This paragraph addresses plastic-type composites ("fiber-reinforced plastic" or "fiber-reinforced polymers"). Fibers are introduced into plastics to improve physical properties such as stiffness, impact resistance, and bending and tensile strength. Manmade fibers of glass, Kevlar, and carbon are most commonly used today, but plant fibers offer considerable cost savings along with comparable strength properties. "Natural fiber polymer composites" contain much more polymer ("plastic") than traditional wood-based fiberboard do, often 50%, and although more expensive, they are also more versatile (Balatinecz and Sain 2007). Although hemp composite plastics are considered to be innovative today, in fact, until the 1930s, hemp-based cellophane, celluloids, and other products were common.

There is a substantial market for hemp fiber plastic composites in the EU, where automobiles account for about 15% of the hemp fiber produced (Carus et al. 2013). The market for hemp fiber in North America is small, but much of the limited crop is used for plastics in automobiles. Natural fibers in automobile composites are used primarily in press-molded parts ("compression molding" is accomplished by applying pressure and usually heat to material in a confined cavity). There are two widespread technologies. In thermoplastic production, natural fibers are blended with fibers of material like polypropylene and polyethylene. Most commonly, this mixture is formed into a (nonwoven) mat, which is pressed under heat into the desired three-dimensional form. In thermoset production, the natural fibers are soaked with binders such as epoxy resin, polyester resin, or polyurethane, placed in the desired form, and (along with heat and pressure) allowed to harden through polymerization. Thermoplastics are easier to recycle (Bourmaud and Baley 2007) and are often used where limited structural strength is required. A wide range of thermoset resins are under development that are compatible with hemp fiber, and because some of these are plant based (derived from soy, canola, or corn), it will likely be possible to produce hemp-based biocomposites that are made 100% from plants. Resin-transfer molding, a process to produce thermoset-resin-based reinforced composites of all shapes and sizes (Sèbe et al. 2000), is often employed for high-stress products such as furniture and boats. Hemp has also been used in other types of thermoplastic applications, including injection molding (material in a fluid state is forced under pressure into the cavity of a closed mold). Injection molding is particularly useful for creating a variety of hemp plastic products. Extrusion molding (in which heated or unheated plastic is forced through a shaping die to produce a continuous form such as a film, sheet, rod, or tube) is an additional possibility, but not yet common. The characteristics of hemp fibers have proven to be superior for production of molded composites (Lu and Korman 2012). In European manufacturing of cars and trucks, natural fibers are used in plastic door panels, dashboards, instrument panels, seat backs, package trays, arm rests, sun visors, passenger rear decks, trunk liners, and window pillars. It has been estimated that 5–10 kg of natural fibers can be used in the molded portions of an average automobile (excluding upholstery). The demand for automobile applications of hemp is expected to increase, depending on the development of new technologies (Carus et al. 2013). At present, in the EU automobile biocomposite industry, hemp has about a 15% market share, in competition with flax, jute, kenaf, and sisal fibers (Carus et al. 2013).

American industrialist Henry Ford (1863–1947), in advance of today's automobile manufacturers, constructed a car with body components made of resin stiffened with hemp and flax fiber. The car was able to withstand 10 times the impact of an equivalent metal panel but never entered general production (Shahzad 2012). Ford's use of hemp has been emulated by other car companies (Figure 7.35). Rather ironically in view of today's parallel situation, Henry Ford's hemp innovations in the 1920s occurred at a time of crisis for American farms, later to intensify with the depression. The need to produce new industrial markets for farm products led to a broad movement for scientific research in agriculture that came to be labeled "Farm Chemurgy" (Hale 1934) that today is embodied in chemical applications of crop constituents.

FIGURE 7.35   Lotus Eco Elise, body constructed of hemp plastic composite, introduced in 2008. Car photo by Stuart Chapman. (CC BY 2.0; hemp background added by B. Brookes.)

Of course, all other types of transportation vehicles from bicycles to airplanes can use the hemp plastic technology pioneered by the automobile industry. Natural fibers have considerable advantages for use in conveyance (Carus et al. 2013) because of low density and weight reduction, favorable mechanical, acoustical, and processing properties (including low wear on tools), no splintering in accidents, good energy absorption, occupational health benefits (compared to glass fibers), no off-gassing of toxic compounds, and price advantages.

There is also considerable potential for a variety of industries producing consumer goods to adopt the use of hemp-based plastics in the manner that the automobile industry has demonstrated is feasible. Hemp plastic products can be manufactured to meet needs for hardness, density, heat resistance, and biodegradability, and abilities to be ground, milled, planed, and drilled. Goods that have been generated to date include furniture (especially seats and the backs of chairs), molded basins, recreational products, dishware, jewelry, sporting goods (notably surfboards, skateboards, and snowboards), musical instruments, and luggage.

COMPRESSED CELLULOSE PLASTICS

Most hemp plastics are composites, combining hemp fiber and synthetic thermoplastics. However, it is possible to manufacture plastics using only cellulose. Cellulose makes up about a third of all vegetable material, and indeed, this natural polymer is a principal component of most plastics. In the absence of resin adhesive material, mechanical pressure is important so that the material in 100% cellulose plastic is self-binding. The cellulose content of hemp fiber is quite high (about 70%), making it a suitable starting material. Plastics made of pure hemp cellulose are relatively expensive and are currently used for high-end items (see Figure 7.36).

AGRONOMY

SECURITY REQUIREMENTS

Security regulations for cultivating hemp in most Western countries are usually stringent and represent a significant cost. Depending on jurisdiction, such requirements may involve the use of

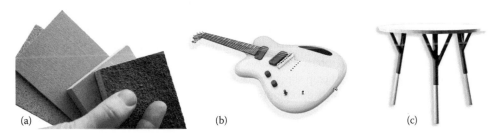

FIGURE 7.36 Products manufactured from compressed hemp cellulose. (a) Hemp board samples. (b) Guitar (hemp cellulose body by AWS Zelfo, body form by Drum Param). (c) Table (design: Elise Gabriel). Photos provided by Richard Hurding, Zelfo Technology.

approved cultivars obtained from authorized sources, secure fencing and storage facilities, careful maintenance of records, governmental inspections, sampling to ensure material has insignificant levels of THC, and personnel free of recent criminal records. The legislative burden that accompanies the cultivation of industrial hemp puts the crop at a unique disadvantage, and it is a tribute to its value that it can be profitably grown. Likely as it becomes more evident that current regulations are needlessly demanding, the regulatory framework will become more tolerant.

GENERAL GROWTH REQUIREMENTS

In most respects, domesticated forms of *C. sativa* have narrower physiological tolerances to stresses than their wild-growing counterparts do (as discussed in Chapter 3). Nevertheless, the considerable plasticity of wild *C. sativa* that allows it to survive in inhospitable environments is also evident in domesticated plants of the species. Cultivated forms of *C. sativa* can be grown over a wide range of agro-ecological conditions.

The physiology of fiber *Cannabis* has been extensively studied (see Van der Werf [1994a] and Van der Werf et al. [1996] for extensive analyses). Both wild and cultivated plants that grow for many generations in a particular location tend to evolve adaptations to their local climates, and these adaptations may make a given biotype quite unsuitable for a foreign location.

As detailed in Chapter 5, induction of flowering in most populations of *C. sativa* is controlled by relative length of light from day to day, and it is essential to employ planting stock that will mature in time for harvest according to the local light/dark regime.

The soil and climate requirements of fiber hemp have been said to be comparable to those of wheat (Ranalli 1998). Generally, temperate, mild, relatively cool conditions are best, and the crop is usually grown in areas with humid atmospheres and a rainfall exceeding 65 cm, with rainfall especially available in the early growing season. Given that marijuana strains are often grown in semitropical and/or very dry regions, and indeed that such strains have been historically occasionally used for multiple purposes, including fiber and oilseed, it is clear that *C. sativa* has the potential to be grown for fiber in areas beyond its traditional production regions. The following discussion is based mostly on cultivars suitable for temperate areas.

ROOT GROWTH

Bócsa and Karus (1998) summarized the growth characteristics of domesticated hemp (which are much like those of wild hemp, described in Chapter 3) in response to soil compaction and moisture. The primary root can grow down to depths of 3.0 m, and from it secondary roots can extend as much as 80 cm. Porous, dry soils encourage downward penetration of the primary root, while compacted and/or wet soils discourage deep growth. Amaducci et al. (2008d) studied root performance of two hemp cultivars. They found that the highest root density occurred in the top 10 cm of soil, but some roots penetrated as deeply as 2 m. The ratio between aboveground and belowground biomass was 5.46 (in contrast, Bócsa and Karus 1998 reported that the root mass of fiber hemp contributes a much smaller proportion: only 8%–9% of the plant's entire biomass). Male plants have a weaker root systems, in parallel to their less robust shoots. According to Bócsa and Karus (1998), in contrast to most crops, in early growth fiber hemp shoots grow so fast that they place stronger demands for water and nutrients on their more slowly developing roots, explaining the considerable need for high levels of water and nutrient availability.

Soil and Fertilization

Ideal soils for hemp are fertile, friable, noncompacted, medium-heavy loams, including silty loam, clay loam, and silty clay (Ranalli 1998). Higher clay content tends to reduce yields. However, sandy soils provide limited fertility and are drought-prone, and the needed fertilization and irrigation may make production uneconomical. Soils should be neutral to slightly alkaline, the pH between 5.8 and 7.5 (5.8 to 7.0 is ideal according to Bócsa and Karus 1998). Abundant organic matter, particularly livestock manure, is particularly beneficial (however, soils high in peat are unsuitable). Although manure is not much used today in growing hemp, Maiden (1894) wrote that "The quantity of manure necessary will depend on the richness and warmth of the soil and upon the climate. In England 10–25 tons rotten dung to the acre [22–56 tonnes/ha] is not considered too much."

Cannabis requires a rich supply of nutrients for good growth, and fertilizer is generally needed. Fertilization rates have spanned wide ranges in the past, probably mostly based on intuition rather than determination of need and response. Ehrensing (1998) compiled the following ranges based on a variety of reports throughout the world (kg/ha)—nitrogen (N): 40–200; phosphate (P_2O_5): 30–120; and potash (K_2O): 0–200. Soil analysis prior to sowing is recommended to assist in determining rate of fertilizer application. Analysis of nutrient concentration in leaves of fiber hemp has been employed to assist in assessing nutrient requirements (Iványi and Izsáki 2009; Iványi 2011). Up to 110 kg/ha of nitrogen, up to 80 kg/ha phosphate, and up to 90 kg/ha of potash may be appropriate. It has been pointed out that soils that grow corn (maize) well have appropriate fertility for hemp, although hemp reportedly requires less fertility, needing approximately the same fertility as a high-yielding crop of wheat.

Most annual domesticated crops have been bred to be nitrophiles, with the capacity to utilize large amounts of nitrogen for productive growth (Emerich and Krishnan 2009). Modern agriculture in fact is, to a considerable degree, based on the creation of crops that can utilize nitrogen fertilizers. The "Green Revolution" of the middle of the last century greatly increased agriculture production, especially in the developing world, by selecting new cultivars that are especially capable and efficient at using nitrogen fertilizers (Borlaug 2000). As discussed in Chapter 3, wild *C. sativa* is a natural nitrophile, thriving in well-manured substrates and stripping soils readily of nitrogen. The nitrogen requirements of fiber hemp have been extensively studied (Van der Werf et al. 1995c; Struik et al. 2000; Amaducci et al. 2002a; Vera et al. 2010; Prade et al. 2011). Cultivars are typically fertilized with nitrogen at a rate of 100 kg/ha/season (Bócsa and Karus 1998), which is higher than the recommended rates for some modern high-yielding field crops. Uniformity of stems is desirable for harvesting and processing, and excessively high levels of nitrogen fertilization (200 kg/ha) have been observed to increase variability of height and weight of hemp stems compared to moderate (80 kg/ha) nitrogen rates (Van der Werf et al. 1995c). Another undesirable consequence of too high nitrogen fertilization is decreased content of bark fiber (Van der Werf et al. 1995b).

Bócsa and Karus (1998) stated that fiber hemp has a substantial need for phosphorus throughout the season, in part to effectively utilize nitrogen, but also to contribute to the elasticity and tensile strength of the fiber cells and bundles.

Bócsa and Karus (1998) also pointed out that fiber hemp has a substantial need for potassium, especially in midseason, when the fibers are rapidly developing. Potassium fertilization of fiber hemp was studied by Finnan and Burke (2013), who found that frequently recommended rates are between 140 and 230 kg K/ha depending on soil type and soil potassium level. However, these authors concluded that potassium requirements are lower than for many other crops and suggested that an optimal potassium fertilization strategy for soils with moderate to high levels of potassium (>70 mg/L) is to simply restore, after harvest, the preharvest level.

WATER RELATIONS

It has long been known that cultivars of *C. sativa* can tolerate hot, arid conditions, provided that the roots are adequately supplied with water (Dewey 1914). Regular rainfall, especially during the first six weeks after seeding, is extremely beneficial. Soils should be well drained but capable of providing sufficient capillarity movement of water from lower depths to the surface. Wet, flooded, or waterlogged soils are very poorly tolerated (young plants up to three weeks or so are very sensitive to wet soils and flooding). Large-scale cultivation of hemp has generally been conducted in areas with sufficient rainfall that irrigation is unnecessary. Soil should be moist when the seeds are planted, and irrigation may be required until germination has occurred (usually within three days after sowing). Industrial hemp is sensitive to drought and needs ample moisture throughout the growing season, especially during the first six weeks of its growth, and it may be necessary to irrigate to keep the soil moist. European experience indicates that 50–70 cm of rainfall is needed during the season, although Lisson and Mendham (1998) stated that the literature suggests a range of 25 to 60 cm annually, depending on climate. Although well-rooted plants endure drought relatively successfully, severe droughts dwarf the crop and hasten flower and seed production. Lisson and Mendham (1998), based on studies in Tasmania, found that water extraction by the roots was mostly confined to the upper 80–90 cm of the soil but occasionally went down to at least 140 cm.

In Chapters 12, 13, and 18, the indica type marijuana group is discussed (which is not suitable for fiber production). This is established in the arid area of Afghanistan and western Turkmenistan, and when strains from this region are grown in high-humidity climates, their dense flowering tops retain moisture and succumb to "bud mold" caused by *Botrytis cinerea* and *Trichothecium roseum* (McPartland et al. 2000). There are land races employed for fiber production in the same areas from which indica type strains originated, and these are likely also unsuitable for fiber production in areas of high humidity.

TEMPERATURES

Cannabis is well suited to warm temperate regions, although it can grow in a range of temperatures. The species tolerates heat well but not cold. Cultivated plants grow best between 14°C and 27°C but tolerate both colder and warmer conditions. The warmer range promotes growth (Van der Werf et al. 1995b). Optimal temperatures for photosynthesis (which is not necessarily reflective of best temperatures for growth) vary from 25°C to 30°C, depending on variety. (Chandra et al. 2011b did not find evidence of differences between drug and fiber varieties.) Optimal temperature for hemp germination has been reported as 24°C, and this rather elevated temperature reflects a crop that is generally not planted early in the season because the seedling develops slowly at low temperatures. Seeds of cultivars tend not to germinate well until the ground is warm (preferably at least 10°C). However, the seeds will germinate at temperatures as low as 0°C and as high as 45°C, and often, fiber hemp has been planted as soon as the danger of a hard frost has passed. Seedlings and mature plants are resistant to light frosts of short duration. Seedlings can sometimes survive short exposures to −8°C to −10°C (Bócsa and Karus 1998), but the plant is best adapted to an extended frost-free season. Mature plants endure light frosts (as cold as −5°C to −6°C), but not a hard frost or long-continuing temperatures around freezing. Van der Werf (1993), summarizing Russian literature on temperature tolerance, noted that: sowing is recommended when soil temperatures are 8°C–10°C; hemp seedlings have been observed to survive short frosts as low as −10°C; before flowering, hemp has been observed to survive frosts as low as −6°C; and a frost of −1°C at flowering decreases seed yield and seed quality.

There is little information available on comparative cold adaptation of cultivars from different latitudes, but possibly those that have been grown for many years in northern locations are relatively tolerant. Grigoryev (1988) discussed the ecology of hemp cultivated under the short-season, cold conditions of central Russia, where the crop has been raised near the northern limit of agriculture (about 66° N latitude): from Arkhangelsk to Mezen to the Pechora River (Ust-Tsilma) to Tobolsk Province, Surgut (61° 17'N), and on the Kolyma River (Verhnekolymsk, at 65° N, 153° E). The cold tolerance of germinating seedlings of Russian varieties described by Grigoryev is noteworthy: some are capable of germinating at 1°C, withstanding frosts of –1°C for 2 weeks and up to –15°C for 24 hours without damage (much more extreme tolerance than recorded elsewhere). For many Russian fiber cultivars, frosts of about –7°C are tolerated, even when the soil is quite moist, but once the stage of flowering is reached, low extended temperature delays growth and frost can destroy the plant. Grigoryev suggested that hemp could be grown as a winter crop, although it is universally sown in the spring, not the fall.

Seed Germination

Germination of domesticated seeds typically takes three to seven days, and is more or less simultaneous, since dormancy has been selected against. However, old seed can germinate somewhat irregularly, and seeds planted at different depths may also tend to produce seedlings at different times. Hemp seeds scattered in the field during harvesting sometimes produce volunteer seedlings in the fall, which are killed by cold (seedlings may also be produced in the spring, and those can survive to maturity). This reflects the lack of dormancy in domesticated *C. sativa*. Janischevsky (1924), who extensively studied wild hemp in southern Russia, never observed fall-produced seedlings. Temperatures for germination were discussed earlier in this chapter. As noted in Chapter 3, light has been reported to be a partial inhibitor of germination. Seeds germinate readily in the dark (Haney and Kutscheid 1975), and commercial seed is generally planted 1–2 cm (sometimes to 4 cm) deep, sufficient for considerable shading, but also useful for moisture availability. After imbibition of water, the radicle (primary root) expands, the hypocotyl (base of the shoot) emerges, and the cotyledons (primary leaflets) unfold above soil level (Mediavilla et al. 1998). The initial growth of the radicle can be quite rapid—up to 10 cm in 48 hours.

Maintaining Seed Viability in Storage

For commercial hemp production, the longevity of fresh seeds decreases fairly rapidly to about 70%–80% after two years of storage in a sheltered but otherwise uncontrolled climate, and if not planted at least by then, it is often recommended that they be discarded. In commercial practice, a minimum germination percentage of 85% to 90% has been recommended (Bócsa and Karus 1998).

There has been considerable study of the storage factors influencing the viability of seeds of domesticated hemp. As reviewed in the following, low temperatures (just above 0°C) and low humidity both prolong seed longevity. Kondo et al. (1950) found that hemp seeds stored for 19 years with the desiccant calcium chloride germinated, indicating the importance of controlling humidity. Crocioni (1950) studied seed moisture, temperature, air, and light in relation to the longevity of hemp seeds of the Carmagnola variety stored at Bologna, Italy, for 3.5 years. He found that seed moisture and temperature were the most important factors, while air (oxygen availability) and diffused light instead of darkness were of uncertain influence. A temperature a few degrees above 0°C and a low seed-moisture content, up to 8.6%, were judged best for preserving germination capacity. Seeds of 10%–15% moisture content kept at high temperatures rapidly lost their germinating capacity. Seeds stored at low temperature retained satisfactory germination for 16 months, whatever the moisture content, and so did seeds of about 8% moisture content stored at or below 27°C. Seeds stored longer than 16 months retained viability better at a seed-moisture content of about 6% than did seeds with a higher moisture content stored at a low temperature. Toole et al. (1960) studied germination of

Kentucky hempseed (much of which was of Italian origin). They reported that seeds stored at 5.7% moisture and 21°C did not decrease in germination after six years; seeds at 9.5% moisture maintained full viability for 5.5 years at both 0°C and −10°C. Lemeshev et al. (1995) outlined plans for storage of *Cannabis* seeds at the Vavilov Research Institute Gene Bank in St. Petersburg, Russia, the largest germplasm collection of *Cannabis* seeds in the world. For their active collection (referred to as a working collection), used for short-term purposes of reproducing and distributing seed, the policy was to store the seeds at about 15°C and 10% moisture content. For "medium-term storage" (up to 10 years), collections were stored at 4°C–6°C and 7% moisture content. For long-term storage (10–20 years) the storage temperature was −20°C, and the moisture content was 6%.

Additional but less detailed reports on temperature and moisture requirements for conservation of germination capacity of hemp seeds are Laskos (1970), Demkin and Romanenko (1978), and Parihar et al. (2014).

Small and Brookes (2012) summarized practical aspects of hemp seed viability as follows.

* To maintain the germinability of seeds of *C. sativa*, they should not be stored under the ambient conditions usually encountered in the high-humidity areas where hemp is typically produced.
* Maintaining *C. sativa* seed at a moisture content of 5% to 8% is sufficient to retain substantial germinabilty for at least six years. A lower moisture content does not seem to improve germinability over this time interval.
* Maintaining *C. sativa* seed at a temperature of 5°C is sufficient to retain substantial germinability for at least six years. A temperature of −20°C can improve germinability, but not greatly.
* For commercial storage of *C. sativa* seed in a viable state for one to several seasons, the most economical investment would usually be drying the seed, at least to 8% but preferably to 6% moisture content and maintaining this level. Where this is difficult, seed should be kept refrigerated, at least at 5°C for periods of up to several years, but at lower temperatures for periods of the order of 10 years.
* For long-term scientific or germplasm banking of *C. sativa* seed in a viable state for up to a decade, a moisture content of 6% coupled with a storage temperature of −20°C is sufficient. Whether more extreme conditions would be of benefit for a longer period has not been determined.
* Although the presence of oxygen hastens seed deterioration of some species, which can therefore profit from storage in an atmosphere such as nitrogen gas (thus excluding oxygen), *C. sativa* seeds do not seem to benefit from such treatment.

Planting Requirements

A well-tilled, fine, level, firm seedbed promotes uniform germination. The soil bed should be free of weeds and debris. Seeds are usually sown at a depth of 10 to 30 mm, the deeper figure appropriate for drier soils (as noted in Van der Werf 1993, in dry seedbeds, hempseed has been planted up to 5 cm deep). However, shallower depths promote soil warming, and more rapid emergence and establishment. Soils with a large clay content are inappropriate, but if nevertheless used, shallower planting may be desirable if rain is expected that could crust the upper layer, making seedling emergence more difficult. Larger seeds can probably withstand deeper depths better that small seeds. Use of a roller at sowing may assist germination by facilitating good contact between seed and soil. Seeds are normally sown in rows spaced from 7 to 20 cm apart, using seed drills. The optimal rate of sowing at any one location depends on variety and local environmental conditions. Fiber strains are typically sown at a minimum rate of 250 seeds/m^2 (approximately 45 kg/ha), and up to three times that density is sometimes recommended. For fiber production, sowing rates vary from 40 to 150 kg seed/ha; in Western Europe, seeding rates are generally 50 to 70 kg/ha.

TABLE 7.2
Seeding Rate for Industrial Based on Seed Size and Density

Weight (g) of 1000 Seeds	Seeding Rate (kg/ha) to Get 100 Seeds/m²	Seeding Rate (kg/ha) to Get 150 Seeds/m²	Seeding Rate (kg/ha) to Get 200 Seeds/m²	Seeding Rate (kg/ha) to Get 250 Seeds/m²
10	10	15	20	25
12	12	18	24	30
14	14	21	28	35
16	16	24	32	40
18	18	27	36	45
20	20	30	40	50
22	22	33	44	55
24	24	36	48	60
26	26	39	52	65

Source: After Baxter, W.J., Scheifele, G. 2009. *Growing Industrial Hemp in Ontario*, Ontario Ministry of Agriculture and Food, Toronto, ON, Canada, 2009. http://www.omafra.gov.on.ca/english/crops/facts/00-067.htm.

The planting rate should be higher if germination is low (although seed less than 85% viable should be discarded) or on poor soils (which should be avoided), and large seeds may require a heavier planting rate (see Table 7.2). In any event, seed density recommendations from the supplier should be followed.

Extensive studies have been conducted on the effects of planting density. The seeding rate strongly influences the yield and quality of fiber produced by the crop. Van der Werf et al. (1995a) found that the proportion of stem in the total dry matter increases with increasing plant density, which is desirable. Another desirable result of higher density is that stands produce thinner stems with a higher percentage of bast fiber and less woody core, i.e., a higher bark-to-core ratio (Van der Werf 1991). Moreover, the fineness of the bast fiber increases at higher densities, which contributes to fiber quality for spinning purposes (Jakobey 1965). However, excessively high densities waste seeds (Amaducci et al. 2002b) and can be counterproductive. Fiber plantations typically have 200–250 plants/m², but plant densities between 80 and 400 plants per square meter have been found to have little effect on either total biomass or stem dry matter yield, and excessive seeding rates result in self-thinning, i.e., high mortality of smaller plants (Van der Werf et al. 1995a,c). Mechanical harvesting and industrial processing of fiber hemp usually benefit from uniform size and diameter of stems, and high densities can increase the variability of stem sizes, which is undesirable (Van der Werf and van den Berg 1995).

In Europe, dual-purpose crops of hemp, grown for both fiber and oilseed, are planted at row spacings from 20 to 40 cm, 20 seeds/m within the row, considered sufficient to suppress weeds. Higher plant densities are more effective at weed suppression (Hall et al. 2014; Jankauskienė et al. 2014), but planting density is designed mainly to maximize yield, not minimize weeds.

ROTATIONS

Hemp has been and continues to be grown sometimes on the same land for several years in succession. However, most agricultural species benefit by alternating with several different crops in sequence over a period of years (usually three to five), and the same is true for *C. sativa*. Rotation tends to promote soil fertility and reduce diseases. Often, some crops complement each other in rotations, and frequently a particular sequence is best. Hemp is typically rotated with cereals and major vegetables. It particularly benefits from following a nitrogen-enriching legume crop, especially alfalfa or clover (soybean is less beneficial).

WIND, RAIN, AND HAIL DAMAGE

Wind, rain, and hail can damage tall crops, especially if their stems are weak. The very tall, slim stems of fiber hemp are subject to lodging from wind, but are grown at such high densities that surrounding plants provide support, usually preventing significant damage. Advanced oilseed cultivars are usually relatively short (so less subject to wind damage) and grown at lower densities (so more susceptible to wind), but lodging is usually not a problem, and in any event, plants that are bent over can often still be productive. Heavy rain and hail can damage foliage of fiber cultivars, but rarely the stems.

PESTS

All plants are susceptible to a wide range of "herbivores" or "pests" (collectively including animals, microorganism pathogens, and weeds), and crops grown as monocultures are usually subject to extensive damage from certain species. For several major crops, losses due to pests can exceed 50% (Oerke 2006). *Cannabis sativa* is attacked by a wide diversity of insects and diseases. McPartland (1997a,b, 1998b) and McPartland et al. (2000) provide authoritative reviews of hemp diseases and pests, and the most significant problems are mentioned here. McPartland (1998a) pointed out that the frequent claim that hemp is "pest-free" is inaccurate; rather, hemp is "pest-tolerant." As a crop, *C. sativa* is remarkably resistant to many pests. Indeed, Kok et al. (1994) demonstrated that fiber hemp suppresses three major soil pathogens: the fungus *Verticillium dahlia* and the root-knot nematodes *Meloidogyne hapla* and *M. chitwoodi*. Significant weed, insect, and disease problems are rare for fiber hemp, and it is generally grown without the use of postplanted (after-seeding) herbicides, insecticides, and above-ground fungicides. There are a few problems with pest mammals and birds, which as discussed later, can be serious predators of the seeds.

Susceptibility of *C. sativa* to pests and diseases differs according to circumstances of cultivation. For example, fiber crops are grown in very dense plantations and the raised humidity around the stalks increases infections of fungal diseases; as well, the dense canopy is protective of many insects. By contrast, both drug and oilseed crops are grown in open rows, and the increased sunlight is attractive to flea beetles and birds. Presumably, susceptibility of wild populations also changes with plant density. Susceptibility also differs according to genetic background of the plant, for example, between fiber and drug cultivars.

Weeds

Weed management prior to planting is required as hemp does not establish well in competition with other plants. Planting hemp too early, when weeds can outcompete it under cold conditions, is undesirable. Similarly, an unexpected cold spell can give weeds the opportunity to establish in a hemp stand. For fiber hemp, which is always densely grown, the developing canopy closes in five to seven weeks, eliminating the need for subsequent weed elimination.

Higher Plant Parasites

Two vascular plant parasitic genera occur on *Cannabis*: broomrape (*Orobanche*) on the roots and dodder (*Cuscuta*) on the stalks and branches. Branched broomrape (*O. ramosa* L., also known as hemp broomrape) is the most significant vascular plant parasite of hemp, while *O. aegypticaca* Persoon and *O. cernua* Loefling have also been recorded on hemp (McPartland et al. 2000). Broomrape is a formidable pest in Europe. It spends most of its life cycle hidden underground, and each plant can produce up to 500,000 dust-like seeds that are sticky and adhere well to the seeds of *C. sativa* (McPartland et al. 2000). The Canadian Seed Growers Association regulations concerning hemp state that the presence of broomrape in industrial hemp crops is cause for refusing pedigreed status and forbid the cultivation of pedigreed hemp seed on land that has grown tobacco during the last three years (tobacco is an alternate host of broomrape). Several species of dodder have also been found on hemp, with *Cuscuta campestris* most frequently recorded on cultivated and wild *C. sativa* in North America (McPartland et al. 2000).

Mammalian Pests

Domesticated mammalian grazers, especially cattle, horses, and goats, have been observed to nibble small amounts of cultivated hemp, but they do not cause significant damage (McPartland et al. 2000). Wild mammals, including deer, rabbits, raccoons, rats, field voles, mice, and groundhogs (wood-chucks) have caused significant feeding damage to hemp (McPartland et al. 2000). Woodchucks especially are capable of causing great destruction to young plantations of hemp (McPartland et al. 2000).

Birds

Walter (1938) recorded that members of the crow family, such as magpies and jackdaws, are particularly attracted to hemp in Europe. The extinct passenger pigeon (*Ectopistes migratorius*) was once a major pest of Kentucky hempseed fields (Allen 1980). Haney and Bazzaz (1970) noted that mourning doves (*Zenaidura macroura*) frequented hemp in Illinois and in southwestern Iowa, and hemp was found to be the most important food of mourning doves in Iowa by McClure (1943). Mourning doves are also very frequent visitors to hemp fields in southern Ontario. Bobwhite quail (*Colinus virginianus*) and ringtail pheasant (*Phasianus colchicus*) have also been observed to feed heavily on hemp seeds in the American Midwest (Robel 1969; McPartland et al. 2000). Hemp seeds improperly prepared without removal of the resin-rich perigonal bract can make birds giddy (Matsunaga et al. 1998).

Insects

Nearly 300 insect species have been found to attack *Cannabis*, but few cause significant damage (McPartland 1996b, 1998a). Stems are the organs of value in fiber hemp, and stem borers are troublesome. The most damaging are the caterpillars of the European corn borer (*Ostrinia nubilalis*; see Chapter 6) and hemp borers (*Grapholita dilineana* and *G. tristrigana*), which eat much of the plant including the seeds. The corn borer is a specialist of *Cannabis* and the closely related genus, *Humulus*, and turned to corn only after maize was introduced to Europe (McPartland 1998a). Beetle larvae, particularly of the hemp flea beetle (*Psylliodes attenuata*), eat roots while the adults eat leaves and inflorescences. Adult weevils and curculios chew into leaves while the grubs feed on the pith of stems and roots. The worst is the cabbage or hemp curculio, *Ceutorhynchus rapae*. Flower beetle grubs (*Mordellistena micans*, *M. parvula*) feed on various parts of hemp, and grubs of a variety of scarab beetles (e.g., the European chafer, *Melolontha hippocastani*) are also a problem. Numerous other damaging species occur in North Temperate regions, and still more species are important in semitropical countries. European cultivars grown in nonnative environments have been observed to be attacked by a wide range of pests and diseases for which they have no resistance (Watson and Clarke 1997). Lago and Stanford (1989) catalogued phytophagous insects on a large plantation of high-THC *Cannabis* in Mississippi. McPartland (1998b) discusses insects to be expected on cultivated hemp in Canada.

Other Invertebrates

Other animals that are significant herbivores of hemp include nematodes and slugs. Of these, nematodes are capable of the greatest damage. Only about a half dozen nematodes have been recorded as causing significant damage on hemp, and for the most part, the greatest damage has been recorded in hot climates (McPartland et al. 2000).

Fungi

Fungi are responsible for most diseases of *Cannabis*, and the long stalks of fiber cultivars are especially susceptible to stalk-canker fungi (McPartland 1996a, 1998b; McPartland and Cubeta 1997).

McPartland (1998a) recorded 88 species of fungi (represented in the literature by over 400 names) and concluded that only a few cause economic losses. Of these, the worst is gray mold, *Botrytis cinerea*, which thrives in cool, humid conditions and can cause considerable damage in wet years (Van der Werf et al. 1996). Hemp canker (of the stem) is commonly due to *Sclerotinia sclerotiorum* and is usually viewed as the second most important disease of hemp (*Fusarium* species and many other genera also cause stem cankers). Root rot of *C. sativa* is often due to *Fusarium solani*. Wilts are caused by *Fusarium* and *Verticillium*, and blights, leaf spot, and mildews of *Cannabis* are due to several fungi. Damping off of seedlings is also caused by a variety of fungi, but most by the oömycetes (not true fungi) of the genus *Pythium*. Seeds are sometimes treated with fungicides to reduce soil diseases.

Bacteria

Four bacteria are significant pathogens of *C. sativa* (McPartland et al. 2000), with one of these split into four "pathovarieties" (designated by pv.): *Pseudomonas syringae* pv. *cannabina* (causes bacterial leaf spot and bacterial blight), *P. syringae* pv. *mori* (causes "striatura ulcerosa"), *P. syringae* pv. *tabaci* (causes wildfire), *P. syringae* pv. *mellea* (causes Wisconsin leaf spot), *Xanthomonas campestris* pv. *cannabis* (cause xanthomonas blight), *Erwinia tracheiphila* (causes bacterial wilt), and *Agrobacterium tumefaciens* (causes crown gall).

Phytoplasma (mycoplasma-like organisms or MLOs) are specialized bacterial parasites of plant phloem tissues and transmitting insects (vectors), and a species of this has been found in *C. sativa* (Phatak et al. 1975).

Viruses

Five viruses regularly infect European cultivated *C. sativa* (McPartland et al. 2000): hemp streak virus (HSV), alfalfa mosaic virus (AMV), cucumber mosaic virus (CMV), arabis mosaic virus (ArMV), and hemp mosaic virus (HMV). Aphids are the most important transmitters of viruses in *C. sativa* (McPartland et al. 2000).

Harvest

For a fiber crop, extensive experience has indicated that hemp is best cut in the early flowering stage or while pollen is being shed, well before seeds are set. However, later harvesting, when pollination is finished and the first seeds begin to ripen, has been practiced in some areas. Mediavilla al. (2001) documented that fiber yield reaches a maximum at the time of flowering of the male plants. Keller et al. (2001) experimentally demonstrated that a harvest time at the beginning of seed maturity facilitates bast fiber extraction without reducing tensile strength, while harvesting after the flowering of the male plants results in reductions of fiber quantity and quality. Rather ignorantly, until 2001, EU regulations specified that, for purposes of subsidization, the crop could not be cut until the seed was 50% mature in form and size. The authoritative guide by Bócsa and Karus (1998) recommends harvesting at full flowering of male plants and at first appearance of flowers for female plants, as subsequent lignification reduces fiber quality. General advice regarding the best time to harvest monoecious fiber varieties has been contradictory, and it is advisable to follow the recommendations accompanying particular cultivars.

Putting tall whole plants through a conventional combine results in the straw winding around moving parts and the fibers working into bearings, causing breakdown, fires, high maintenance, and frustration. Especially on small acreages, crops are harvested with sickle-bar mowers and hay swathers, but plugging of equipment is a constant problem. Slower operation of conventional combines has been recommended (0.6–2 ha/hour). Large crops are ideally harvested with specialized equipment (e.g., Figure 7.37).

Hemp has often been grown as a dual-purpose crop, i.e., for both fiber and oilseed. In France, the principal grower of dual-purpose varieties, the grain is taken off the field first with a combine, the cutting blade raised to leave most of the stalks for later harvest, and the seeds threshed.

FIGURE 7.37   Specialized equipment required for harvesting tall *C. sativa* grown for stem fiber in France.
Photo by Aleks (CC BY 3.0).

Growing hemp to the stage that mature seeds are present compromises the quality of the fiber, because of lignification. As well, the hurds become more difficult to separate. The lower-quality fiber, however, is quite utilizable for pulp and nonwoven usages. In the EU, hurds are usually not completely separated from the phloem fiber (2% to 25% hurds are retained in the so-called "total fiber line," depending on the application).

Yields

Reports of fiber yields in the literature often do not make clear whether the information is based on wet or dry weight (only the latter is appropriate), whether based on entire plant or stalk only, or whether the report deals with primary phloem (line, long) fiber and tow secondary phloem (short) fiber collectively or just the line fiber. About two-thirds of the weight of a dried plant (stalk + foliage) is stalk. About one-fifth of dry, retted, defoliated stems is made up of phloem fiber (primary + secondary collectively). In Europe, dry matter yields have ranged from 5800 to 19,500 kg/ha, and dry stem yields from 5000 to 13,700 kg/ha. In North America, dry stem yields have varied between 2500 and 28,000 kg/ha (all of the preceding statistics based on Fortenbery and Bennett 2004). Struik et al. (2000) reported that hemp potentially can yield 25,000 kg/ha above-ground dry matter, 20,000 kg/ha stem dry matter, and 12,000 kg/ha cellulose. Salentijn et al. (2015) noted that yield of dry bast fiber varies from 1200 to 3000 kg/ha. Cellulose yield is usually 7000–10,000 kg/ha (Zatta et al. 2012). In modern European cultivars, at harvest time (when most of the lower foliage has been dropped), the stems make up about 85% of the above-ground dry weight of the plant (Van der Werf et al. 1998). Figure 7.38 illustrates a typical breakdown of the composition of weight of a hectare of fiber hemp based on a fresh (green) weight of 40,000 kg and a dry stem weight yield of 10,500 kg. The hurds constitute about 70% of the stalk dry matter (Dang and Nguyen 2006), but generally, the hurd yield can be two to three times as much as the yield of bast fiber (five times as much, according to Dewey 1916).

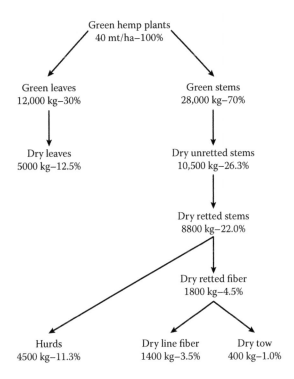

FIGURE 7.38 Estimated fresh and air-dry weight of a hectare of fiber hemp, relative to living hydrated plant, showing relative composition of foliage, unretted stems, retted stems, line fiber (primary phloem fiber), and tow (secondary phloem fiber). After Dempsey, J.M., *Fiber Crops* University of Florida, Gainesville, FL, 1975, except that the yield of hurds has been added.

STORAGE

Harvested, dried, retted stalks are appropriately baled when dried to less than 15% moisture, and drying should continue during indoor storage until moisture content is about 10%. In the Old World, retted stalks have often been stored outdoors. However, moisture can be wicked up from bare ground (possibly even from gravel floors), affecting quality. Buyers generally demand bales of given sizes and may require avoiding polyester and plastic twines since these will contaminate manufactured products.

BREEDING

"Breeding" as a term may simply mean sexual reproduction, but here it refers to creating (usually by selection) new variations of a plant that increase its usefulness for humans. Plant breeding is carried out by several techniques, outlined in the following. Modern plant breeding in Europe has produced several dozen hemp fiber strains, although by comparison with other fiber crops, there are relatively few described varieties of hemp. Since World War II in Europe, breeding has been concerned most particularly with the development of monoecious varieties. Breeding for fiber qualities of *C. sativa* has been reviewed by Salentijn et al. (2015).

"Mass selection" is probably the oldest and most widely practiced method of selection for out-breeding crops like *C. sativa*. The seeds of what appear to be relatively desirable plants are used to propagate the next season's crop, with the result that over several years, the characteristics judged to have merit tend to become uniform. Landraces have evolved in this manner. The process tends to be slow. Because such selection tends to maintain the status quo, it is employed to maintain the purity of registered cultivars.

"Individual selection" concentrates on evaluating the progeny from elite plants (those that appear to be champions with respect to desired traits). Once the superior progeny have been identified, they can be used as breeding stock (involving controlled pollination) to generate a supply of plants with superior traits. Because most of the characteristics of value in *C. sativa* can only be evaluated after flowering has occurred, careful records need to be kept so that the breeder will know which seeds to keep and which to discard. However, the German fiber hemp breeder G. Bredeman pioneered a technique ("Bredeman's method") that allowed breeders to evaluate the fiber merits of (at least male) plants before they flowered, so they could be used directly to generate improved strains. He cleverly slit the vertical stems of male plants, removing half the stem for fiber analysis, and allowing the other half (supported by a stake) to develop pollen. Only the male plants with the highest fiber content are permitted to pollinate the female plants. The seed from all the females are harvested, but only seed from the females with the highest fiber are used for additional breeding. Using this methodology, great gains in the productivity of fiber hemp resulted in Europe.

"Marker-assisted selection" is based on selecting a "marker" that is linked to a trait that a breeder wishes to select. The markers may be morphological or chemical (especially protein) but are commonly DNA tags identifying particular locations in the genome. This indirect selection process can dramatically improve the efficiency of selecting plants with desirable gene combinations. Markers can be employed to transfer single genes or to follow the inheritance of many genes. Marker-assisted selection is considered to have excellent potential for *C. sativa* (Mandolino et al. 1996; Mandolino and Ranalli 1998; Mandolino and Ranalli 2002; Mandolino and Carboni 2004). Genetic markers in *C. sativa* have been found for femaleness (Shao et al. 2003) and maleness (Mandolino et al. 1999, 2002; Törjék et al. 2002; see Chapter 4).

Hybridization followed by selection has long been the major tool of plant breeding to increase yield, vigor, and uniformity. So-called combination or pedigree breeding emphasizes the creation of cultivars by transfer of single genes or gene combinations, achieved by hybridization, backcrossing, and selection.

"Heterosis breeding" is a combination of hybridization and selection that takes advantage of particularly desirable genetic partners. Hybridizing distantly related organisms tends to produce heterosis (hybrid vigor, sometimes related to preventing harmful recessive genes [alleles] from combining), the opposite of inbreeding depression (often related to harmful recessive genes being combined from both parents). In heterosis, an F_1 hybrid tends to have an increase in some desirable characteristics (such as height, productivity, or disease resistance) compared to the mean of the same traits of its parents. Most European cultivars appear to share so much of a common genetic heritage that when they are mated heterosis does not develop, but when hybridized to East Asian cultivars, the hybrids do exhibit hybrid vigor (De Meijer 1998). Lyster H. Dewey of the U.S. Department of Agriculture (see his publications in the Literature Cited), who has been referred to as "the first hemp breeder" (Bócsa 1998), hybridized European and Chinese landraces to produce vigorous cultivars in the early twentieth century.

"Hybrid cultivars" are not merely cultivars that are the result of past hybridization between different biotypes (probably all cultivars of *C. sativa* hold this status); as explained in the following, they are the result of a specific marriage of parental strains or cultivars and must be generated anew each season. They are reminiscent of mules, the progeny of horses and donkeys, renowned for their endurance that exceeds that of their parents but themselves virtually unable to produce offspring. In modern crop hybrid cultivar breeding, the parents can be deliberately inbred, so that when they are mated, their genetic interaction in the F_1 hybrid is especially desirable. Compatible combinations generally need to be determined by trial and error. Dewey (1927) may have created the first documented intervarietal hybrid cultivars. When the plants within an F_1 hybrid population are allowed to interbreed, genes determining their superior characteristics segregate and recombine and are so variable that they are of limited advantage to farmers. Accordingly, farmers must purchase certified hybrid seed every growing season from the breeder, who produces the seed by repeatedly crossing a given set of parents. Actually, "pseudohybrid" cultivars, many of them produced in France, are F_2 hybrid populations.

Several popular hybrid cultivars are mostly monoecious. These are produced by crossing female dioecious hemp (males are rogued out of the field) pollinated by monoecious hemp.

Cultivars of many crops today are transgenic, the result of DNA recombination. In transgenic plants (more generally, genetically modified organisms), a desired "transgene" (coding for a useful trait) has been identified, isolated, cloned, and inserted. Feeney and Punja (2003) demonstrated the feasibility of transgenic hemp by transferring a gene encoding the enzyme phosphomannose isomerase into hemp.

Mutation breeding is based on exposing seeds to chemicals or radiation to generate mutants with desirable traits. If a mutated plant is produced, it can be hybridized with normal plants to transfer the altered gene. Tens of thousands of cultivars of various plants have been created. The technique does not seem to have been used to generate any commercial variety of *C. sativa*.

Doubling the number of chromosomes of a crop plant has sometimes been found to make it more productive. *Cannabis sativa* normally has a somatic (diploid) number of 20 chromosomes (expressed as $2n = 20$) (Small 1972a). Doubling of chromosome number is most frequently achieved by exposing meristems (growing points) to a mitotic toxin. (Mitosis is the normal way that cells divide in two, each new cell receiving a full complement of chromosomes. The toxin, usually colchicine, prevents normal division, so that the new cell receives all of the new chromosomes that normally would have been divided between two cells.) Tetraploid selections of *C. sativa* (i.e., with $2n = 40$) have been generated frequently (e.g., Warmke and Blakeslee 1939; Zhatov et al. 1969; Sidorenko 1978). They proved to be fertile, with larger seeds and seed yields, but since their fibers were coarse and not as useful, they were abandoned (Bócsa 1998; Ranalli 2004).

CULTIVARS

Several dozen European fiber hemp cultivars make up the bulk of modern registered *Cannabis* cultivars (see Chapter 17). Most of these originate from European land races (De Meijer 1995a, 1995b, 1998). These were selected mostly for the temperate regions of Europe, and their photoperiod adaptations are often unsuitable for more northern or southern regions. European cultivars appear to represent a relatively limited range of genetic variation.

ECONOMIC STATUS

More than 30 countries produce hemp crops. China, Canada, and France (in decreasing order) are the leading producers. Although China has virtually always been the largest producer, and hemp can be grown in most places in the country, the crop has nevertheless been minor in Chinese agriculture (Wang and Shi 1999). In Europe, the main producers are France, the United Kingdom, and the Netherlands. Recently, cultivation in Germany has decreased greatly. Italy had an outstanding reputation for high-quality hemp, but productivity has waned for the last several decades. Minor production of hemp for fiber still occurs in Austria, Russia, the Ukraine, Poland, Hungary, the countries of the former Yugoslavia, Romania, North Korea, Chile, and Peru. There has been recent interest in Australia and South Africa in cultivating hemp.

Figure 7.39 shows twenty-first century global hemp yield by year, categorized by seed and fiber production. These data are based on Food and Agriculture Organization (FAO) statistics and are known to be incomplete (some countries have not reported information, and the reliability of some data is undetermined). FAO data for fiber production show a steady decline in worldwide production of hemp from 1966 to 1990, with subsequent stabilization at a relatively low level compared to the past (Figure 7.40).

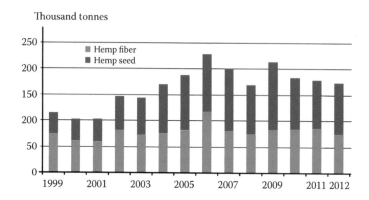

FIGURE 7.39   Yearly world production of hemp fiber and hemp seed, from 1999 to 2012. Based on FAOSTAT data (http://faostat.fao.org/DesktopModules/Admin/Logon.aspx?tabID=0), which are incomplete.

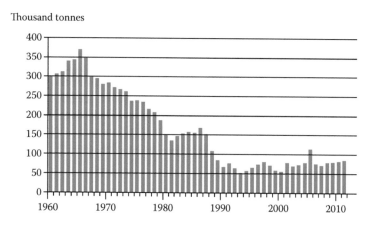

FIGURE 7.40   Yearly world production of hemp fiber from 1961 to 2012 (FAOSTAT data).

Figure 7.41 reports acreage by year for the last two decades, for Canada, China, and the EU, the three largest areas of production (statistics for the latter part of the twentieth century are provided by Sponner et al. 2005). (The reliability of the Chinese data is undetermined. Since China is reportedly the world's largest producer, the acreage shown in Figure 7.41 seems low.) Note that China and Europe have traditionally grown hemp primarily for fiber, whereas Canada grows hemp primarily for oilseed. "The increasing demand for fibers, seeds and cannabinoids in the medical field in recent years has led to a new record acreage of 22,000 ha cultivation area in 2015" (Michael Karus, regarding EU acreage, in *Nova Institute Newsletter*, June 2015).

As noted earlier, hemp constitutes only about 0.3% of the world's natural fiber production, excluding wood fiber. As shown in Table 7.3, the world value of hemp fiber is about 6% that of flax (the most comparable bast fiber), and about 0.05% that of cotton, the leading natural annual fiber crop. Curiously, all three of these crops are also important oilseeds.

MARKET DEVELOPMENT AND FUTURE NEEDS

China has dominated fiber hemp production for millennia, largely for textile applications, mostly for clothing and other woven applications. Specialized harvesting, processing, spinning, and weaving equipment are required for preparing fine hemp textiles. The refinement of equipment and new technologies are viewed as offering the possibility of making fine textile production practical in

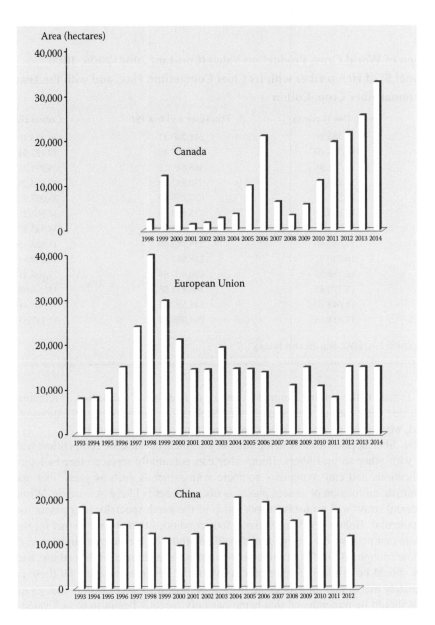

FIGURE 7.41 Acreage for the last two decades for China (based on FAOSTAT data), the EU (source: European Commission), and Canada (source: Health Canada, Office of Controlled Substances; the Canadian data refer to licensed acreage, not cultivated acreage, which is unavailable).

Western Europe and North America, but at present, China controls this market and probably will remain dominant for the foreseeable future. However, the future of fiber hemp development is likely based on alternative usages.

Since the early 1980s, the EU provided considerable subsidization for the development of new flax and hemp harvesting and fiber processing technologies (because of the similarities of flax and hemp, the technologies developed for one usually are adaptable to the other). In addition, various European nations and private firms contributed to the development of hemp technologies. Accordingly, Europe is far more advanced in hemp development with respect to all fiber-based applications than other parts of the world, and harvesting and processing machinery for fiber hemp are highly advanced

TABLE 7.3

Comparison of World Gross Production Value (Constant 2004–2006, 1000 International $) of Hemp Fiber with Its Chief Competitor, Flax, and with the Leading Natural Annual Fiber Crop, Cotton

Year	Hemp Tow Waste ($)	Flax Fiber and Tow ($)	Cotton Lint ($)
2000	13,268.80	248,240.11	26,453,159.46
2001	12,778.49	293,647.48	30,122,510.10
2002	17,771.98	368,689.12	26,991,342.79
2003	15,886.71	370,852.88	27,823,265.26
2004	16,562.23	477,793.49	35,059,563.58
2005	17,898.03	475,695.28	34,981,211.11
2006	25,771.76	312,188.00	34,943,970.60
2007	17,537.40	255,183.66	35,825,750.15
2008	16,230.72	248,941.39	32,138,649.77
2009	18,138.52	180,409.89	29,868,117.11
2010	18,440.45	148,833.05	33,896,488.68
2011	18,648.86	148,597.71	37,306,419.46
Mean	17,411.16	294,089.34	32,117,537.33

Source: Data from FAOSTAT, http://faostat.fao.org.

in Europe. France is the leading European country in fiber hemp cultivation. It remains to be seen whether Europe will continue to dominate in the application of nonwoven applications of hemp fiber and, indeed, whether hemp fiber will become more competitive in the future.

Currently, fiber applications of hemp are very limited because of competition with synthetic fibers and with other natural fibers. Hemp fiber can potentially replace other biological fibers in many applications and can sometimes compete with minerals such as glass fiber and steel. As forests diminish, cultivation of annual plants as fiber sources is likely to increase. While crop residues like cereal straw will probably supply much of the need, specialty fiber plants such as hemp also have potential. Bolton (1995) formulated four conditions that should be met for hemp fiber to become more competitive: (1) the material should be produced at a large enough scale, (2) the price should be low enough, (3) the fiber characteristics should be adequate for the end use, and (4) proven technology should be available for the processing of the new raw material. Of these criteria only (3) is adequately met at this time for hemp fiber to become truly successful in today's marketplace. However, it should be remembered that hemp has only recently begun to be cultivated for fiber in most Western countries after an absence of many years, and it is premature to conclude that its future is limited.

A principal roadblock to hemp fiber development is extraction technology. Hemp fibers are extracted by retting—by subjecting the stems to rotting, mostly either by exposure to humidity in the field or being submerged in water. The latter method produces superior fiber, but the water becomes chemically polluted. Such pollution is prohibited in most developed countries, but environmental regulations are lax in some places (notably China and some eastern European countries), where most of the world's hemp fiber is produced. Since the 1990s, European countries have attempted to produce nonpolluting fiber-extraction technologies, but these are not yet competitive, and hemp fiber has been successful in Europe in part because of considerable state subsidies. Bócsa and Karus (1998) observed that, "By and large, only EU subsidies make hemp cultivation a profitable venture under current economic conditions." Horn (2010) noted that, "At present in the EU registered hemp fiber processors can receive a subsidy of approximately £60 t[-1] of extracted fiber under the flax and hemp regime to provide additional income for their activities and support production."

Several conclusions can be drawn regarding the economic status of hemp, at least tentatively:

1. Hemp is a niche crop, of relatively minor importance today, but one that stimulates considerable investment interest for diversification and product development. Hemp is likely to continue experiencing the risks inherent in a niche market for some time.
2. Hemp is benefitting from the general reconsideration of the relative merits of synthetic fiber, wood fiber, and natural fibers that is occurring, generally favoring a greater usage of natural fibers for economic, product suitability, and environmental reasons.
3. Since fiber hemp is basically a temperate-region crop, it competes less directly with tropical bast crops such as sunn hemp, ramie, and jute and the semitropical kenaf, in comparison to the considerably more competitive flax, which is also an annual-temperate-region bast crop. Lloyd and Seber (1996) compared the advantages and disadvantages of hemp, flax, and kenaf, and their analysis suggests that, overall, none is necessarily superior, and each crop is preferable depending on circumstances.
4. *Cannabis sativa* has considerable potential as an oilseed crop. Canada has become the leading country of hempseed production since industrial hemp was reintroduced in 1998 (Blade 1998), and the EU is also growing more of this oilseed. Dual-purpose cultivars (harvested for both oilseed and fiber) compromise the two crops, and it remains to be determined which will become more profitable in the future.
5. China, the world's leading producer of hemp, has an established tradition of producing high-quality textile-grade fiber and textile products. China's hemp fiber industry has the advantages of cheap labor and tolerance of water retting, which is environmentally unacceptable in most of the West. New technology is being developed in Western nations to compete in this niche, but it is likely that China will remain dominant in the foreseeable future.
6. The economic resurgence of hemp fiber in the marketplace is based on nontraditional usages, particularly in the production of a very wide range of pressed fiber and insulation products and plastics. The greatest success of hemp products has been in the automobile, construction, and agriculture industries. Market penetration has been prominent in the EU, in part because of subsidization (related to "green support" policies), but mainly because of superior characteristics of hemp fiber for particular applications.
7. It is probable that industrial hemp cultivation will be resumed in the United States, although this remains a contentious issue. If this occurs, it is likely to result in a surge of production, applications, and development of fiber hemp and a substantial impetus to all aspects of the hemp industry.
8. Research bearing on the agricultural production and technological exploitation of hemp fiber has been very active during the last decade and is critical to the growing success of hemp industries.

CURIOSITIES OF SCIENCE, TECHNOLOGY, AND HUMAN BEHAVIOR

- The oldest surviving paper is more than 2000 years of age, comes from China, and was made from hemp fiber. Egyptian papyrus sheets that might be thought to be an older form of paper are not "paper" as this term is understood by experts because the fiber strands are woven, not "wet-laid." Until the early nineteenth century, hemp and flax were the chief paper-making materials. Wood-based paper came into use when mechanical and chemical pulping was developed in the mid-1800s in Germany and England. Today, at least 95% of paper is made from wood pulp.
- The Hmong are one of China's largest minority ethnic groups (they are known as the Miao in China) and are also found in Thailand, Laos, and Vietnam. By tradition, they are laid to rest in hemp garments (a common practice in China). Hemp must be used for funeral

dress or the ancestors will refuse the dead person's soul in the afterworld. Each son and daughter must give their departed parent hemp trousers or a hemp skirt to be worn in the coffin. Depending on the number of children, the deceased may be buried in as many as a dozen sets of clothing.

- In ancient Japan, hemp was important in symbolic rites at Shinto shrines and Buddhist temples. Objects that had to be made of hemp included bell ropes as thick as legs and the *noren*—a short curtain that hangs over the doorways and brushes the top of the head as one enters the room, in order to cause evil spirits to flee out of the body (both of there are illustrated in Figure 12.11c). Priests generally dressed in hemp robes. Shinto priests and the faithful also used hemp in ceremonies. One such use was the waving of a *gohei*, a short stick with hemp fibers attached to the end. Shaking these sticks above someone's head drove the evil spirits away.
- Japanese martial artists, including Samurai warriors, dressed in hemp. In prebout ceremonies, the reigning Sumo wrestling grand champion carried a giant hemp rope around his ample girth to purify the ring and drive away evil spirits.
- An old Japanese legend explains why the earthworm has rings around its body. There once were two women hemp weavers, who competed to make the best dress for an upcoming holiday. The fast worker had her dress ready on time, but the slow, careful one had only completed the neck of her dress, which she had decorated with whitish rings. The slow worker persuaded her husband to carry her in a large jar on his back, so that only her neck with the completed top of the dress was visible. When the two women met in a public market, the slow worker peeked out the top of the jar and mocked the quality of the dress of her competitor. A shrill argument resulted, and the agitated husband accidentally dropped the jar, revealing his almost naked wife. Ashamed, she buried herself in the earth so she would not be seen and turned into an earthworm, with rings around its body.
- Hemp garments were more or less exclusively worn by wealthy Japanese more than a thousand years ago. *Yukatabira*—absorbent hemp bathrobes—were put on after soaking in hot springs. The cotton kimono was the common person's version of these expensive bathrobes.
- In an old Japanese religious tradition, rooms of worship were purified by burning hemp leaves by the entrance. This would invite the spirits of the departed, purify the room, and encourage people to dance.
- Gigantic phalli representing deities were stationed beside roadsides in old Japan, especially at crossroads, to bar passage of malignant beings. Wandering pilgrims and travelers prayed at the foot of these monuments and were expected to leave small offerings of hemp leaves and rice to each one that they passed.
- In Japanese traditional marriages, hemp was a symbolic gift of acceptance and obedience from the groom's family to the bride.
- In 1948, in occupied Japan, American General Douglas MacArthur (1880–1964) and his colleagues rewrote the Japanese constitution. They included a Hemp Control Act forbidding cultivation, totally wiping away several thousand years of hemp culture.
- In Shinto belief, hemp symbolizes purity. Accordingly, when a new emperor ascends the throne of Japan, he is bound by tradition to wear hemp garments, and there must be a roll of hemp at the foot of the royal throne. Anticipating the ascent of a new emperor, a group of Shinto farmers planted an illegal hemp crop, and so when Emperor Hirohito died in 1989, they had material to make a new hemp robe for the new emperor.
- Reminiscent of the Chinese tradition of using hemp as burial garments, in Norwegian folklore, hemp cloth symbolized the beginning and end of life and was used for clothing for both birth and burial.
- In the Norwegian valley of Gausdal, people approaching a hemp field would respectfully lift their hats as a greeting to the vette, a nature spirit that lived there.

- In eighteenth century Europe, the ends of slim hemp sticks were dipped in sulfur and used as matches.
- The first and second drafts of the American Declaration of Independence were written on hemp paper. The final version was copied onto animal parchment and signed on August 2, 1776. The Magna Carta and the *King James Bible* were also written on hemp paper.
- Hemp was so important in England in the sixteenth century that King Henry VIII (1491–1547) passed an act in 1533 that fined farmers who failed to grow at least a quarter acre of hemp for every 60 acres of arable land that they owned.
- In colonial America, citizens of several colonies were required by law to grow hemp.
- In early America, hemp ropes were commonly used by hangmen to execute the condemned (Figure 7.42), and gallows humor often concerned hemp. A "hempen collar" was a hangman's noose. A "hempen widow" was the wife of an executed man. "To die of hempen fever" was a way of saying that a man had been hanged. In the Wild West, vigilantes were sometimes called "hemp committees," and "sowing hemp" was a way of saying that someone was on his way to being hanged.
- The 1892 World's Fair in Chicago featured hundreds of architectural "marble" columns that were actually made up of hemp and plaster of Paris. Today, similar cement-like materials made of hemp and plaster of Paris are being used in house construction.
- It has been claimed that hemp was the first "war crop." In ancient China, land barons waged war against each other. Chinese archers made their bowstrings from bamboo fibers, until hemp's greater strength and durability were discovered. Archers equipped with bowstrings made of hemp were at a great advantage, so Chinese monarchs set aside large portions of land exclusively for hemp for weapons. Ironically, as noted in this chapter, hemp was also a major war crop during the Second World War, when foreign supplies of fiber were cut off.

FIGURE 7.42 The Great Hanging at Gainesville, Texas. During the American Civil War, in 1862, 41 suspected Unionists were hanged by Confederates, the largest mass hanging in the history of the United States. This illustration appeared in *Frank Leslie's Illustrated Newspaper*, February 20, 1864.

- Millions of tobacco smokers have been unaware that they have been smoking the cannabis plant. Cigarette papers have widely employed some hemp fiber, its strength and wet-resistance making it ideal for the purpose.
- It is frequently claimed that the earliest Levi jeans were made of hemp. The Levi Strauss Company has denied this, but the composition of early fabrics is uncertain, and the company's records in its San Francisco headquarters burned down during the earthquake fires in 1906.
- Designer Giorgio Armani created an all-hemp tuxedo for actor Woody Harrelson for his attendance at the 1997 Oscars (he was nominated for his performance in *The People vs. Larry Flynt*).

8 Oilseed

INTRODUCTION TO EDIBLE FIXED OIL

In the context of cannabis, "oil" could refer to (1) "fixed" (vegetable) oil from the seed, (2) essential oil from the glandular secretory trichomes, or (3) "hashish oil," i.e., solvent extracts rich in cannabinoids, particularly tetrahydrocannabinol (THC) ("liquid hemp" is a recent expression referring particularly to cannabidiol [CBD]-rich concentrates, especially for vaping; this is sometimes inappropriately called "hemp oil"). This chapter deals with the fixed oil from the seeds of *Cannabis sativa*; the essential oil from the flowering part of the species is discussed in Chapter 9, and hashish oil is discussed in Chapter 12. To avoid ambiguity, the oilseed industry often prefers the phrase "hempseed oil" rather than "hemp oil." Although "hemp seed" and "hemp-seed" are often encountered, hempseed has become sufficiently familiar that its designation as one word is justified. Moreover, the term hempseed is parallel with other oilseed crops such as linseed and rapeseed.

Fixed oils are basically triglycerides, which happen to be not only the chief constituents of vegetable fats but also the main constituents of the body fat of animals, including humans. They are nonvolatile at room temperature (they do evaporate, but very slowly) and are usually obtained from seeds by extraction processes. "Vegetable oils" are fixed oils and include hempseed oil, the subject of this chapter. By contrast, components of essential oils (volatile oils, ethereal oils, terpenes) evaporate rapidly, typically producing aroma. Essential oils are frequently obtained by distillation and are mostly used to add scent or flavor to prepared materials, but are sometimes employed for medicinal purposes.

Vegetable oils are usually distinguished from vegetable fats by the former being liquid at room temperature, while the latter are solid. Vegetable oils are of course obtained from plants, mostly from "seeds" or "grain" in commercial practice. Vegetable oils are used for food, fuel ("biofuel"), medicines, a wide variety of consumer products such as cosmetics, and industrial processing applications, such as paints and other technical coatings. Some vegetable oils are used for multiple purposes, while others are primarily employed for one product or purpose.

All of the world's major vegetable oils are edible and are produced primarily for human food, although most of this oil is used for other purposes, such as lubricants and soaps. The leading plant species responsible for the production of commercial edible oil are palm, soybean, Canola, and sunflower (Figure 8.1). Together, they account for about 90% of global edible oil production. (Soybean, Canola/rapeseed, cotton, and sunflower are the world's leading "oilseeds," valued not just for edible oil but also for edible seeds and industrial oil.) Other edible oils with high economic value are produced from the seeds of coconut, corn, cotton, mustard, peanut ("groundnut"), safflower, and sesame, or the fruits of olive.

The "smoke point" of an oil or fat is the temperature at which smoke becomes evident during cooking, often indicating that compounds in the oil are breaking down. (More dangerous is the "flash point," usually a considerably higher temperature at which vapors from the oil can ignite.) Cooking oils break down after repeated use, and some should not be used for deep frying more than twice. The smoke point of a particular vegetable oil species varies, sometimes substantially, and mostly depends on the percentage of an oil's unsaturation, where saturated oils have a higher smoke point. Most oils used for frying and cooking have smoke point values above 200°C. The smoke point of hempseed oil is approximately 165°C, which is too low for higher-temperature cooking and frying. For culinary purposes, as discussed later, hempseed oil is best used fresh and uncooked, as a salad oil for example, but can also be incorporated into a wide range of processed commercial prepared foods, such as baked breads.

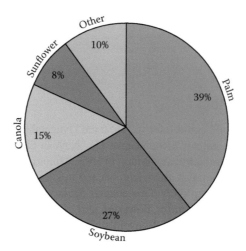

FIGURE 8.1   Crops responsible for the world's production of edible oil (2012 data, based on http://trumax .ca/2012/marketing/2012-crop-recommending-sales-positions/).

INTRODUCTION TO HEMPSEED AND HEMP OIL

Cannabis sativa reproduces by small, slightly flattened, lenticular (biconvex) one-seeded fruits (Figure 8.2b). As noted frequently in this book, the fruits of *Cannabis* are botanically termed "achenes," although they are usually referred to as "seeds." Depending on the cultivar, these usually vary from 2.5 to 4 mm in diameter and 3 to 6 mm in length (as noted in Chapter 3, wild seeds are notably smaller than those of domesticated plants). Thousand-seed-weight of cultivated plants generally ranges from 8 to 27 g (European cultivar thousand-seed-weight is mostly 17 to 25 g). However, thousand-seed-weights of up to 67 g have been recorded (Watson and Clarke 1997). Seeds of monoecious cultivars are

FIGURE 8.2   Hempseed and its raw products. (a) A terminal inflorescence of a domesticated strain, with numerous mature "seeds" (achenes). (b) Hemp seeds (a match is shown for scale). (c) Hempseed oil.

usually smaller than those of dioecious cultivars. The seeds are produced in large numbers, clustered against the stems, on the upper end of the main stalk (Figure 8.2a) and on the ends of branches. As discussed in this chapter, edible oil is extracted from the seeds (Figure 8.2c).

The true "seed" portion is enclosed within the fruit wall (pericarp), which forms the protective "hull" or "shell" (Figure 8.3). Most of the seed is filled by an embryo, principally the two cotyledons (embryonic leaves), which are rich in oils, proteins, and carbohydrates, upon which the germinating seedling relies for nourishment. A rudimentary nutritive tissue (endosperm, rich in aleuron bodies, which are protein storage organelles) is also present (Figure 8.3; Theimer and Mölleken 1995).

The taste of hempseed oil is pleasantly nutty, sometimes with a touch of bitter aftertaste. Unsaturated fatty acids make up over three-quarters of the fatty acids in hempseed oil. This high degree of unsaturation is responsible for extreme sensitivity to oxidative rancidity. The oil has a relatively short shelf life. It should be extracted rapidly under nitrogen (to prevent oxidation), stored in dark containers that protect against light damage, and refrigerated to avoid degradation caused by heat (curiously, all conditions that also promote the shelf life of cannabis drugs). Addition of antioxidants may prolong the longevity of the oil. Steam sterilization of the seeds, often required by law, allows oxygen to penetrate the protective shell and hasten rancidity. Accordingly, sterilized or roasted hemp seeds, and products made from hemp seed that has been subjected to heating, should be avoided for human foods.

Hempseed oil varies in color from off-yellow to dark green (Figure 8.2c). The green color is due to the presence of chlorophyll, which is extracted along with the oil (Matthäus and Brühl 2008). Hempseed oil has often been merchandized in clear glass in order to attract the consumer's interest, but this practice shortens shelf life. Chlorophyll is a photosensitizer, which increases the susceptibility of the oil to oxidation, and this increases the need for protection from light by using bottles made of dark glass. (The presence of some chlorophyll in the achenes, should they be exposed to light, could result in some "auto-oxidation," but this has not been studied. The presence of antioxidants in

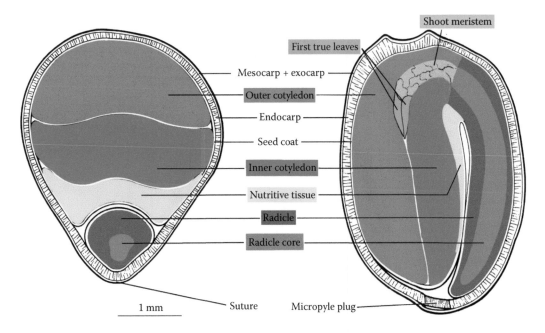

FIGURE 8.3 Diagrams of sectioned achenes of *C. sativa*. Left: Cross-section. Right: Longitudinal section through widest dimension. Most of the seed is made up of the oil-rich cotyledons. The nutritive tissue (endosperm) feeds the embryo during early development and persists as a layer around it (Ram 1960). Diagrams based on Small, E., Antle, T., *J. Ind. Hemp*, 12, 3–14, 2007 (drawn by B. Flahey). Compare anatomical details in the developing seedling shown in Figure 3.6.

the achenes is an obvious adaptation to reduce such oxidation.) The number of green seeds present should be minimized after harvesting, to reduce the presence of chlorophyll in the oil. Insufficient protection of the oil from light can be indicated by a change in the color of the oil over time from green to yellow.

HISTORY

Cannabis seeds were discovered in Chinese tombs over 4500 years of age (Jiang et al. 2006) and have been employed for at least 3000 years as food for both humans and livestock (Schultes 1973). Indeed, hempseed was one of the "five grains" of ancient China, along with foxtail millet, broomcorn millet, rice, and barley or wheat (Huang 2000; Figure 8.4) and remained a staple until the tenth century, when other grains became more important (Cheatham et al. 2009). Archaeological evidence for the food use of hempseed in ancient times in Europe is very limited, but given the existence of traditional European hempseed recipes, it seems that for at least centuries, the seeds were employed for food to a minor extent (Leson 2013). In the past, hemp seed has generally been a food of the impoverished social classes, or a famine food. Often, the whole seed, including the hull, was eaten. Crushed peanut-butter-type preparations have been produced from hempseed in Europe for centuries but were rather gritty since technology for removing the hulls was rudimentary when it existed, and interest in producing commercial hulled hempseed for culinary purposes dates back only to about 1990. In some areas of Southeast Asia, the hull was removed by filtration, after grinding the seed in water. The resulting hempseed "milk" was then heated until the proteins solidified into curds, which were subsequently pressed to form a solid mass, much like tofu from soy, but without the need for chemical precipitants.

FIGURE 8.4   Chinese emperor of the five grains (also known as Zao Jun, the kitchen god, who allegedly lived 2737–2699 BC), to whom the grains, presumably including hempseed, were ritually offered. Illustration from Werner, E.T.C., *Myths and Legends of China*, George G. Harrap, London, U.K., 1922.

As will be discussed, in very recent times, hemp seeds have been "discovered" to have nutritional-therapeutic, medicinal properties. In fact, hempseed has been employed in the treatment of various health disorders for millennia in traditional eastern medicine (Callaway 2004). Historical accounts indicate that "hemp seeds" were used for many medical purposes: as an analgesic, for sores and skin diseases, and for coughs, jaundice, and colic. However, it is unclear whether hemp seeds alone were employed or also the fruit bracts, which would have added cannabinoids (discussed in Chapter 11). In ancient China, various parts of the plant were used medicinally, including the foliage and roots (Wang and Wei 2012). In recent times in China, hempseed has been used to treat blood problems and constipation (Wang and Wei 2012). A traditional Chinese medicine called "hemp seed pill" (made in part with hempseed) has been demonstrated to be safe and effective for alleviating constipation (Cheng et al. 2011). Maltos-Cannabis, a drink formulated with hempseed, was popular in Scandinavia in the early twentieth century as "a health medicine that has been employed with great success against pulmonary diseases, anemia, gastric catarrh, scrofula, neurasthenia, asthenia and emaciation" (Dahl and Frank 2011; Figure 8.5).

The cultivation of hemp as an oilseed crop reached a zenith in nineteenth and early twentieth century Russia, when, in addition to the edible uses, the seed oil was employed for making soap, paints, and varnishes. Until about 1800, hempseed oil was one of the more popular lighting oils, being cheaper than whale oil, but kerosene subsequently replaced both for this purpose. However, for most of history, the seeds were of very minor economic importance, and by the middle of the twentieth century, commercial use was negligible, and cultivated plant selections suitable for dedicated oilseed production were virtually unavailable until the 1990s. For most of the latter part of the twentieth century, the seeds were usually employed as wild bird and poultry feed, although occasionally also as human food. World hemp seed production (mostly in China) fell from about 70,000 tonnes in the early 1960s to about 34,000 tonnes at the beginning of the twenty-first century (Figure 8.6).

At the close of the twentieth century, reminiscent of how new hemp fiber applications resurrected the fiber crop mostly in Europe (as discussed in Chapter 7), a similar development of oilseed products, particularly in Canada, witnessed the founding of an expanding hempseed industry by the year 2000 and a consequent increase in world production as a food crop (Figure 8.6). *Cannabis sativa* is now being grown as a major new source of edible and industrial oilseed products. With the growing recognition of the health benefits from the dietary use of hempseed oil, discussed in the section "Nutritional Qualities of Hemp Seed and Oil," hempseed production has been increasing. Indeed, the economic prospects for continued development as an oilseed crop are considerably better than for continued development as a fiber crop.

FIGURE 8.5 An advertisement (public domain) for Maltos-Cannabis, a beverage based on hemp seed and malt sugar. It was often supplied to young children to protect them against harm (note the mother holding a child with one hand and with the other sending the Grim Reaper away).

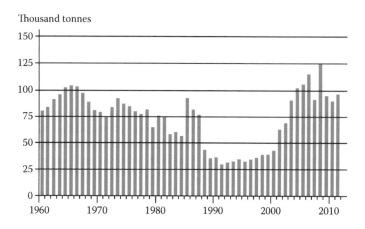

Thousand tonnes

FIGURE 8.6 Yearly world production of hempseed from 1961 to 2012 (FAOSTAT data).

OIL EXTRACTION AND PROCESSING TECHNOLOGY

PRESSURE EXTRACTION

Edible vegetable oils are frequently extracted mechanically from seeds and fruits by the use of oil presses. Depending on design, these may be true "mills" or "oil seed mills" (which crush and macerate the material), true "presses" (which compress), or a combination ("expeller presses," which squeeze continuously fed material, providing separated streams of oil and solid residue). The design of oil presses has evolved from ancient times, when just grinding stones were commonly employed (Figure 8.7, left), to complex, huge commercial machinery (Figure 8.7, right). Several physical characteristics of hempseed (dimensions, hull proportion, hardness, and fracture characteristics) are very suitable for oil pressing (Berenji et al. 2005). Unlike fiber processing technology, which today is economical only on a large-scale basis, it is possible to carry out hempseed oil extraction on a cottage industry scale, using small-capacity screw presses or hydraulic presses, which can extract 60%–80% of the oil (Matthäus and Brühl 2008), although large-scale processing is much more efficient. The raw seed is first cleaned, then crushed or macerated, and finally squeezed under high pressure, extruding most of the oil and leaving behind a

FIGURE 8.7 Examples of simple and complex edible vegetable oil extraction systems. Left: A simple, grinding wheel oil press. Photo by Avishai Teicher (CC BY 2.5). Right: A complex vegetable oil extraction factory. Photo by Bracodbk (CC BY 3.0).

pasty material ("pomace"), which is occasionally used as fertilizer but is usually compressed into "oil cake" ("de-oiled cake" would be a more accurate phrase) or "seed cake," with about 10% oil remaining in the solid material. The oil cake is usually fed to livestock, where regulations allow, so it is a valuable secondary product, although sometimes mistermed as a waste product. Seed particles are removed from the oil by filtration, sedimentation, and/or centrifugation (these operations may be considered part of "refining" the oil, which may also include deodorizing and/or bleaching with diatomaceous earth). The friction from pressing produces heat, which can degrade flavor, aroma, and nutritional value, so the temperature is kept to about 40°C by cooling apparatus to produce "cold-pressed oil." The phrase "cold-pressed oil" is regulated in some jurisdictions, but not in others, where it is sometimes used simply to convey the impression that a given oil is of particularly high quality. However, it is true that genuine cold-pressed oil can preserve bioactive compounds such as essential fatty acids (EFAs), phenolics, flavonoids, and tocopherol (Teh and Birch 2013). The desirable properties of these compounds are generally preserved at temperatures less than 50°C, with only brief exposure to temperatures this high. For edible purposes, the quality of hempseed oil is improved by using only the first pressing, as a virgin oil. However, the key to high quality is to use high-quality grain. Carrying out extraction in an environment free of light and oxygen can assist in minimizing degradation. Hempseed oil is mostly employed as an edible oil, and so it is standardly cold-pressed. The relatively primitive way that hempseed oil was produced in the past amounted to cold-pressing, although this was not necessary for such nonedible uses as the manufacture of soaps, paints, varnishes, and lamp fuel oil. Should nonedible uses of hempseed oil become popular in the future, different extraction systems would likely be employed.

Solvent Extraction

Some vegetable oils are extracted using heat and/or solvents. These methods are faster, extract more oil, and are much more economical, but have been considered to be unsuitable for edible hempseed oil because of degradative changes that they cause (Dimić 2005). According to Callaway and Pate (2009), "Solvent extraction is used for the inexpensive industrial processing of many vegetable oils, although it is not suitable for the production of human food or animal feed because residual solvents (typically hexanes) contaminate the final product." Nevertheless, solvent extraction systems are under investigation for hempseed oil.

When hempseed oil is extracted by solvents, this is most often conducted with n-hexane (Oomah et al. 2002; Kostić et al. 2013). Solvent extraction using n-hexane is relatively inexpensive and efficient, but the residual solvent in the oil is potentially hazardous for food purposes, and extraction time is longer compared to cold pressing. Lin et al. (2012) examined the effects of using ultrasonic sound waves to enhance traditional solvent extraction of hempseed oil and claimed that there are benefits.

"Supercritical fluids" are liquefied gases that evaporate rapidly and completely at atmospheric pressure. Supercritical carbon dioxide is CO_2 that is transformed from a gas to a liquid state by keeping it above the "critical" temperature and pressure point. Supercritical carbon dioxide fluid extraction is a new commercial and industrial technique that uses liquid CO_2 as an extractive solvent under pressure. The technique, like the use of conventional chemical solvents, extracts a higher percentage of oil than possible with presses, but unlike most solvents, CO_2 is nontoxic and nonflammable. This technology is now used in the food industry to extract heat-sensitive, easily oxidized compounds such as polyunsaturated fatty acids (PUFAs) and has been experimentally employed to extract the seed oil of hempseed (Da Porto et al. 2012a,b). The resulting seed residue is virtually devoid of residual oil, and its protein content can be processed into various food products with longer shelf lives. The main disadvantages of this technique, according to Kostić et al. (2014), are high costs, high expenditure of time compared to cold pressing, and relatively low throughput when compared to industrial-scale processes.

NUTRITIONAL QUALITIES OF HEMP SEED AND OIL

According to an ancient legend (Abel 1980), Buddha (Prince Siddhartha Guatama, died 480 BC; Figure 8.8), the founder of Buddhism, survived a six-year interval of asceticism by eating nothing but one hemp seed daily. This apocryphal story holds a germ of truth—hemp seed is quite nutritional, primarily because of the very high content of unsaturated fatty acids (of the order of 80% of the fatty acids) and digestible protein (ca. 25%). In addition, it has been suggested that other components, including trace amounts of terpenes and cannabinoids, could have health benefits (Leizer et al. 2000). Good general accounts of dietary aspects of hempseed oil are Jones (1995), Conrad (1997), Pate (1998b), Leson et al. (1999), Callaway (2002, 2004), Leson and Pless (2002), Oomah et al. (2002), Small (2007), Matthäus and Brühl (2008), and Bureau (2010). The value of hempseed oil from the point of view of its primary components is discussed in the remainder of this section.

Fatty Acids

Almost all reputable medical organizations have advised that saturated fat is a significant risk factor for cardiovascular disease, although this has been disputed by some recent studies, most notably that of Chowdhury et al. (2014). The following discussion is based in part on the conventional view that unsaturated fats are healthier than saturated fats.

Just as with exaggerated claims for the virtues of fiber hemp and medical marijuana, there have also been hyperbolic declarations about the marvels of hempseed oil. Nevertheless, hempseed oil is remarkably nutritious. Udo Erasmus' (1993) book *Fats That Heal, Fats That Kill* pronounced hemp to be "the most perfectly balanced, natural essential fatty acid-rich oil available" and "nature's most perfectly balanced oil." In fact, the composition of omega-6 and omega-3 EFAs in hempseed oil is quite optimal for human metabolism. Hempseed oil contains significant amounts of omega-6

FIGURE 8.8 Statue of Buddha at a temple in South Korea. The ancient usage of *Cannabis* for oilseed and fiber coincided to a considerable degree with the development of Buddhism in Asia. "Big Buddha Seeds" is one of the larger online suppliers of marijuana strains, several with "Buddha" in their names. However, Buddhism is opposed to the use of any inebriant, as reflected by the "Fifth Precept": "abstain from wines, liquors and intoxicants that cause heedlessness." Photo by Steve46814 at en.wikipedia (CC BY 3.0).

gamma-linolenic acid (GLA) and omega-3 stearidonic acid (SDA), which are also produced in human bodies from the EFAs. It seems that these "super unsaturated" fatty acids are responsible for at least some of the numerous health claims that are currently ascribed to hempseed oil. In any event, the best way to achieve a balanced diet is to consume a wide variety of healthy foods, and this applies to sources of fats and oils. Given the health-promoting qualities that hempseed possesses, as described in the following discussion, hempseed deserves to be added to the human diet.

The quality of an oil or fat is most importantly determined by its fatty acid composition. Polyunsaturated fatty acids (PUFAs) are fatty acids that contain more than one double bond in the backbone of their molecule. By contrast, saturated fatty acid molecules have no double bonds. Saturated fat molecules can pack together more easily than unsaturated fats, which is why the former are solids at room temperatures while unsaturated fats are liquids. Hempseed oil is of high nutritional quality because it contains high amounts (generally over 80%) of PUFAs (Figure 8.9), mostly the EFAs linoleic acid (18:2, 50%–60% content in the achenes, depending on strain) and alpha-linolenic acid (18:3, 20%–25%). GLA (18:3, 1%–6%) and SDA (18:4, 0%–3%) are also metabolically important fatty acids, which are present in hempseed oil. (The C:D ratio notation is a shorthand way to describe these molecular species, where C specifies the number of carbon atoms in the molecule and D gives the number of double bonds.) Additionally present is the monounsaturated fatty acid (MUFA) oleic acid (18:1, 10%–16%), also considered healthy, although oleic acid and other MUFAs are not essential for health. Numerous studies have demonstrated that fatty acids are important for human health (e.g., Connor et al. 1993; Holub and Holub 2004; Fedor and Kelley 2009; Panza et al. 2009; Wendel and Heller 2009). Animal experimentation using hempseed as a dietary source has shown that it has significant cardioprotective effects (Prociuk et al. 2006; Al-Khalifa et al. 2007), preventing cholesterol-induced stimulation of platelet aggregation (Prociuk et al. 2008). Indeed, studies of hempseed intake in both humans and laboratory animals have reported that dietary hempseed can induce significant improvements in serum lipid profiles in humans and other animals (Callaway et al. 2005; Karimi and Hayatghaibi 2006; Schwab et al. 2006; Kang and Park 2007).

GLA is a widely consumed supplement known to affect vital metabolic roles in humans, ranging from control of inflammation and vascular tone to initiation of contractions during childbirth (Dimić et al. 2009). GLA has been found to alleviate psoriasis, atopic eczema, and mastalgia and may also benefit rheumatoid arthritis and cardiovascular, psychiatric, and immunological disorders (De Luca et al. 1995; Clarke 1996; Deferne and Pate 1996; Fan and Chapkin 1998; Yu et al. 2005). Aging, diet, and pathology (diabetes, hypertension, alcoholism, etc.) may impair GLA metabolism, making supplementation desirable. As much as 15% of the human population may benefit from addition of GLA to their diet

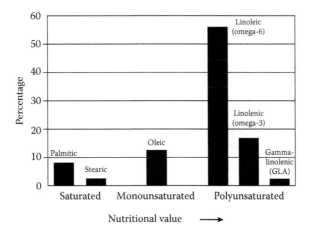

FIGURE 8.9 Mean content of principal fatty acids in hempseed oil. Based on 62 different accessions grown in southern Ontario and reported in Small and Marcus (2000).

(*The Furrow* 1990). At present, GLA is available in health food shops and pharmacies primarily as soft gelatin capsules of borage (*Borago officinalis* L.) or evening primrose (*Oenothera biennis* L.) oil, but hempseed is almost certainly a much more economic source of GLA and EFAs. Although the content of GLA in the seeds is lower, hemp is far easier to cultivate and higher-yielding.

Essential Fatty Acids

Linoleic acid and alpha-linolenic acid are the only two fatty acids that must be obtained from the diet to maintain human health and, for this reason, are considered essential to human welfare (Callaway 2004). Several other fatty acids are sometimes termed "conditionally essential" because they can become essential under some physiological or disease conditions (GLA and SDA, for example, discussed in this and the next section). Linoleic acid and alpha-linolenic acid were once designated "vitamin F," a terminology that has been abandoned. The relative amounts of the EFAs vary among different strains (Small and Marcus 2000; Kriese et al. 2004) and with seed maturity (Peiretti 2009). In contrast to shorter-chain and more saturated fatty acids, the EFAs usually do not serve primarily as energy sources. Instead, the bulk of dietary EFAs serve as raw materials for the production of cell membranes and as precursors for the biosynthesis of many of the body's regulatory biochemicals in the course of normal metabolism (Spielmann et al. 1988). The omega-3 fatty acids are available in other oils, particularly fish and flaxseed, but these tend to have unpleasant flavors compared to the mellow, slightly nutty flavor of hempseed oil.

While the value of unsaturated fats is generally appreciated, it is much less well known that many dieticians consider the Western diet to be nutritionally unbalanced by an excess of linoleic (the omega-6 EFA, unsaturated with two double bonds, named "linoleic" because it was first identified from linseed oil), over alpha-linonenic acid (the omega-3 EFA, unsaturated with three double bonds) (Simopoulos 2002; Figure 8.10). GLA is also an omega-6 fatty acid, and its omega-3 analogue is SDA. Both are present in nutritionally significant amounts in hempseed oil, and at the same ratio as their EFA precursors. A century ago, the typical North American diet ratio of omega-6 to omega-3 fatty acids was about 1–3:1; today, it is about 10–14:1. In hempseed, linoleic and alpha-linolenic occur in a ratio between 2.1 and 3:1. This ratio is considered optimal in healthy human adipose tissue and is apparently unique among common plant oils (Deferne and Pate 1996). Rapeseed oil has a similar ratio of omega-6 to omega-3, but the overall amounts of EFAs are lower, and both GLA and SDA are lacking. (Note, however, that some researchers, such as Harris 2006, dispute the medical significance of the ratio of omega-6 to omega-3, although this opinion is clearly in the minority.) The primary reasons why an optimal ratio has been proposed are the observed negative effect from diets that are too heavy in omega-6 and observations of the

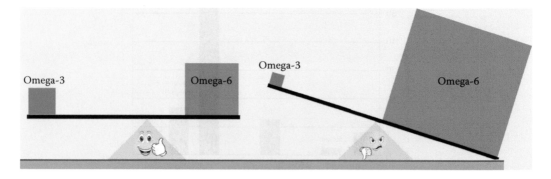

FIGURE 8.10 Health significance of relative balance of omega-6 to omega-3 fatty acids. Left: An ideal dietary balance requires two to three times as much omega-6 as omega-3 fatty acids. Right: The seriously unbalanced diet typical of much of society today has far too much omega-6 compared to omega-3 fatty acids. Prepared by B. Brookes.

enzyme kinetics for delta-6-desaturase, the enzyme that metabolizes both EFAs to GLA and SDA. Too much omega-3 from flaxseed oil, for example, can actually suppress the normal production of GLA in humans (Schwab et al. 2006). Long-chain omega-3 fatty acids from fish, such as EPA and docosahexanoic acid (DHA), seem to reduce inflammation, prevent heart arrhythmias, dilate bloods vessels, and counter blood clotting. By contrast, excessive dietary omega-6 fatty acids promote an inflammatory response and encourage clotting. When insufficient omega-3 is provided (relative to omega-6), there seems to be an increased incidence of common diseases, including heart disease, Crohn's disease, asthma, Alzheimer's, and some kidney diseases. Rodriguez-Leyva and Pierce (2010) pointed out that hempseed oil could be useful for reduction of the symptoms of chronic illnesses like coronary artery disease, hypertension, diabetes, arthritis, osteoporosis, inflammatory and autoimmune disorders, and cancer, although human research is required to validate these claims specifically for hempseed oil.

The essential omega-6 fatty acid linoleic acid, which is present in elevated amounts in hempseed oil, is of particular significance to cardiovascular health. Linoleic acid is metabolized to produce arachidonic acid, which begins a vast and important biochemical metabolic pathway known as the arachidonic acid cascade. From this, important classes of chemicals, including eicosanoids, prostaglandins, leukotriens, and even endocannabinoids (see Chapter 13), are produced to regulate many biological processes (Simopoulos 2008). Hempseed is thought to have potential for reducing platelet aggregation and, therefore, blood clots, which initiate heart attacks and stroke; however, animal studies and limited human studies to date have not yet clarified the relative value of hempseed oil consumption (Rodriguez-Leyva and Pierce 2010). Hempseed protein contains a significant amount of the amino acid arginine, which is an important chemical in the normal regulation of blood pressure.

Seo et al. (2012) noted, "Cardiovascular disease is a principal cause of mortality in many countries, and accounts for up to 16.7 million deaths annually. Cardiovascular disease is primarily caused by atherosclerosis, a chronic inflammatory disease of the arteries that generally clinically manifests as thrombosis. In the past, atherosclerosis was thought to be due to a passive accumulation of cholesterol in the blood vessel wall. However, current data corroborate the hypothesis that atherosclerosis involves chronic inflammatory features… Hempseed might prove to be a promising…anti-atherosclerotic… Although several previous studies have attempted to characterize the effects of dietary hempseed intake, its effects on atherosclerotic heart disease have yet to be thoroughly elucidated."

Minor Fatty Acids

There are other fatty acids in small concentrations in hemp seed that have some dietary significance, including SDA (18:4, up to 2.5% in the achenes), an omega-3 fatty acid (Callaway et al. 1996), and eicosenoic acid (20:1, about 1%), an omega-9 fatty acid (Mölleken and Theimer 1997). Nutritional supplements featuring SDA are currently made from black currant (*Ribes* species) seed, but some hemp cultivars are potential alternative sources. SDA is apparently not an important human dietary supplement because we normally make it ourselves from dietary alpha-linolenic acid. However, SDA supplementation may be helpful for infants and the elderly, who tend to be deficient in delta-6-desaturase activity (Pate 1998b). Eicosenoic acid is important in the production of cerebrosides, which are components of nerve membranes and the "white matter" of the brain.

Tocopherols

Tocopherols are major dietary antioxidants in human serum (Mölleken et al. 2001). Alpha-, beta-, gamma-, and delta-tocopherol represent the vitamin E group. They can comprise up to 0.1% of hempseed oil. These fat-soluble vitamins are essential for human nutrition. About 80% of the tocopherols in hempseed oil are in the gamma form. The vitamin E content of hempseed is comparatively high relative to other dietary oils. Natural antioxidants in hempseed oil, such as alpha-tocopherol, stabilize the highly polyunsaturated oil, thus keeping it from going rancid, at least within the intact seed (Kamal-Eldin and Appelqvist 1996; Sapino et al. 2005; Yu et al. 2005).

STEROLS

Up to about 0.7% of hempseed oil may be phytosterols, such as sitosterol and campesterol (Matthäus and Brühl 2008). Sterols in hemp seeds, like the tocopherols, probably serve the same antioxidant function of protecting the seed oil and are also desirable from a human health viewpoint. Phytosterols are membrane constituents in all plants. Medically, they are known to reduce total blood cholesterol and low-density lipoprotein (LDL) cholesterol levels in serum and so seem to be therapeutic for atherosclerosis (Malini and Vanithakumari 1990; Mölleken et al. 2001; Miettinen and Gylling 2004; Patch et al. 2006).

PROTEIN

Hempseed protein has recently become very popular as a nutritional supplement, although evidence for its health value is relatively limited. Hemp seeds contain 25%–30% protein, with all eight of the amino acids that are essential in the human diet and a reasonably complete amino acid spectrum, although lysine is relatively low as in most vegetable protein (Tang et al. 2006); however, arginine is relatively high (and is discussed in the next section). About two-thirds of hempseed protein is of the edestin (legumin) type, which is easily digestible. Hemp protein also includes vegetable albumin, which is also easily digested and has a reasonable balance of amino acids, comparable to that of soybean or egg white (Callaway 2004). Although the total protein content in hemp seed is smaller than that of soybean, hemp protein is much higher than in grains like wheat, rye, maize, oat, and barley (House et al. 2010), and it is well suited to human consumption (Wang et al. 2008b). The oilcake remaining after oil is expressed from the seeds is employed as a very nutritious feed supplement for livestock, but it also has potential for production of a high-protein flour for humans.

Proteins are potential allergens, but human allergies to hemp protein have rarely been reported (Stadtmauer et al. 2003; Gamboa et al. 2007; Nayak et al. 2013). Fractions of hemp protein have been experimentally demonstrated to reduce hypertension in rats (Girgih et al. 2011). Hempseed lacks gluten proteins and is suitable for those with gluten intolerance.

Arginine

Rodriguez-Leyva and Pierce (2010) noted that hempseed contains "surprisingly high levels of the amino acid arginine, a metabolic precursor for the production of nitric oxide, a molecule now recognized as a pivotal signaling messenger in the cardiovascular system that participates in the control of hemostasis, fibrinolysis, platelet and leukocyte interactions with the arterial wall, regulation of vascular tone, proliferation of vascular smooth muscle cells, and homeostasis of blood pressure... individuals may be able to decrease their risk for cardiovascular disease by following a diet that is high in arginine-rich foods."

PHENOLICS

Phenolics (phenols) include chemical compounds with a hydroxyl group (-OH) bonded to an aromatic hydrocarbon group (with one or more six-carbon rings; sterols, noted previously, are structurally similar, but the hydroxyl group bond is different). The largest group of phenolics is the flavonoids. The cannabinoids are also phenols. CBD, with two phenolic groups, is a polyphenol. Phenolic compounds are effective antioxidants and often have additional physiologically beneficial effects such as cardioprotection and anti-inflammation. Hemp seed is rich in phenols and polyphenols (Vonapartis et al. 2015).

OTHER SIGNIFICANT DIETARY CONSTITUENTS

Hempseed is considered to be a good source of carbohydrates (20%–30%), dietary fiber (10%–15%; about 20% of the fiber is digestible), and minerals (4%–6%), particularly calcium, iron, magnesium, phosphorus, potassium, sulfur, and zinc (Theimer and Mölleken 1995; Deferne and Pate 1996; Pate 1998b).

ENVIRONMENTAL CONTROL OF THE DEVELOPMENT OF FATTY ACIDS

In Chapter 11, it is noted that the concentration of THC, the principal intoxicant of *C. sativa*, depends to an extent on the environment in which the plant develops, the genetics of the seed, and if the female flower has been pollinated. Similarly, the relative percentage of oil in the seeds has been observed to vary with variety, year of cultivation, climatic conditions, and location (Matthäus and Brühl 2008; Grigor'ev et al. 2010). Environment can also alter the fatty acid *quality* (profile) of *Cannabis* seed. This was demonstrated by Przybylski et al. (1997), who compared oilseed quality of hemp grown in Canada (under colder conditions) with the same varieties grown in Europe (under warmer conditions). The Canadian-grown seed oil was about 15% higher in unsaturated fatty acids, with about 10% more of alpha-linolenic acid and GLAs. Theimer and Mölleken (1995) similarly noted that hemp grown in warmer areas tends to produce oil with more saturated fatty acids, and indeed, this pattern has been recorded for other oilseed plants. It appears that a cooler climate may be preferable for development of the unsaturated fatty acids, but if the growing season is too short, grain productivity can be low and the fatty acid profile may be inferior. The more unsaturated a fatty acid is, the lower is its freezing point (i.e., changing from a liquid to a solid form). Callaway and Pate (2009) suggested that this protects the seeds from low winter temperatures in colder regions. Ross et al. (1996) noted that, with increasing maturation of the seeds, consistent with observations of many other oilseed plants, the percentage of unsaturated fatty acids increases and the percentage of saturated fatty acids decreases in the mature seed.

HOW DOMESTICATION HAS ALTERED *CANNABIS SATIVA* FOR OILSEED PRODUCTION

REVIEW OF KEY INFORMATION PRESENTED IN PREVIOUS CHAPTERS

In Chapter 3, it was noted that "seeds" (achenes) in domesticated plants of *C. sativa* differ in several respects from those of wild plants. This information is not repeated here, but it should be noted that oilseed hemp seeds show all of the features characteristic of domesticated plant seeds.

Chapter 5, dealing with photoperiodism, pointed out that plants of *C. sativa* are locally adapted to increasingly shorter seasons of northern latitudes by becoming smaller, and this pattern applies to plants grown for oilseed, as well as those cultivated for fiber and intoxicating chemicals. The very popular, northern, early-flowering oilseed cultivar FINOLA is autoflowering (day-neutral); the flowering date is not affected by day length, although the flowering time may be shortened by other environmental variables, such as drought.

Chapter 6, which discusses the evolution of shoot architecture and planting density patterns in the various kinds of domesticated plants, supplements the information presented here on oilseed varieties. It was pointed out that plants grown either for marijuana or seeds are spaced sufficiently apart to provide for branches (hence flowers, seeds, and THC content) to develop well, and likely, farmers learned the appropriate planting density required to maximize seed production (while keeping in mind that weeds are a common problem when planting density is low).

Chapter 7, which discusses fiber production, noted that there are currently very few cultivars bred specifically for oilseed production, and indeed, most hemp seed is currently obtained from so-called "dual usage" plants (employed for harvest of both fiber and seeds), which are not capable of producing as much seed as oilseed varieties. As noted in Chapter 7, growing hemp to the stage that mature seeds are present compromises the quality of the fiber, because of lignification in the stem. As well, the woody hurds that are useful as a secondary product become more difficult to separate. The lower-quality fiber, however, is quite utilizable for pulp and other nonwoven usages. It seems that generally in the past, as in the present, when seeds were harvested from cultivated *C. sativa*, they came from plants that were usually grown additionally for other purposes, either for fiber or marijuana. Of the dual-usage cultivars, the European cultivars Uniko B and Fasamo and the Canadian cultivar Anka are particularly suited to being grown as a source of oilseed. FINOLA, an oilseed cultivar, can also be grown for dual usage.

THE TRADITIONAL USE OF TALL, HIGHLY BRANCHED PLANTS TO PRODUCE HEMPSEED

From time immemorial, China has been the world's major producer of hempseed. Small and Marcus (2000) examined the growth of Chinese hemp land races, which were quite branched (Figure 8.11) and so capable of very high production of seeds. Chinese hemp grown for dual usage or just for seeds are typically planted at lower densities than hemp grown only for fiber, as this promotes branching, although it should be understood that the genetic propensity for branching of cultivars has been selected. It appears clear that considerable branching is a characteristic that farmers have stressed in order to maximize seed production.

Dewey (1914) noted that a Turkish type of land race called Smyrna was commonly used in the early twentieth century in the United States to produce birdseed because (like most marijuana types

FIGURE 8.11 A highly branched Chinese land race capable of considerable seed production.

of *Cannabis* and unlike fiber types) it is quite branched, producing many flowers, hence seeds. Indeed, Dewey's description of Smyrna is reminiscent of the well-branched kind of Chinese land race shown in Figure 8.11.

Based on the preceding discussion, it seems that large, well-branched plants have been the standard form of plant used particularly for seed production in the past. However, such plants have been grown not just for seeds but also either for fiber or for marijuana, and so they are "general purpose" kinds.

The Use of Short, Compact Plants to Produce Hempseed

Until very recent times, the widespread cultivation of hemp *primarily* as an oilseed was largely unknown, except in pre–World War II Russia. It is uncertain whether the kind of Russian land races once grown as oilseeds are still extant. It is difficult to reconstruct the type of hemp plant that was grown in Russia as an oilseed crop because (1) such cultivation has essentially been abandoned and (2) land race germplasm in the Vavilov Research Institute (St. Petersburg) seed bank, the world's largest public cannabis seed collection, has been extensively hybridized (Small and Marcus 2003; Hillig 2004b) due to inadequate maintenance. A land race certainly was grown in Russia specifically for seeds, and Dewey (1914) gave the following information about it: "The short oil-seed hemp with slender stems, about 30 inches high, bearing compact clusters of seeds and maturing in 60 to 90 days, is of little value for fiber production, but the experimental plants, grown from seed imported from Russia, indicate that it may be valuable as an oil-seed crop to be harvested and threshed in the same manner as oil-seed flax."

While oilseed land races in northern Russia would have been short, early-maturing plants in view of the short growing season, in more southern areas, oilseed landraces would be expected to have had moderate height and to be spaced more widely to allow abundant branching and seed production to develop. Curiously, land races from Russia have been observed to be shorter as sowing latitude decreases, not taller (J. Callaway, personal communication).

There are only a few cultivated varieties currently available that have been bred specifically for the production of grain. Some very recently bred oilseed cultivars are short, compact, and ideal for high-density planting. These include FINOLA, formerly known as FIN-314 (Figure 8.12), which is relatively short and little-branched, matures early in north-temperate regions, and is ideal for high-density planting and harvest with conventional equipment. It appears that modern hempseed breeders have intuitively or intentionally reconstructed the kind of plant that used to be grown in Russia for oilseed. Low stature is desirable in oilseed selections to avoid channeling the plants' energy into stem tissue, in contrast to fiber cultivars, for which a very tall main stalk is desired. Compact clustering of seeds also decreases stem tissue, promotes retention of seeds, and facilitates collection. The efficiency of grain production by crops is often measured by "harvest index," the ratio of harvested grain to above-ground dry matter, and crop breeders are strongly motivated to increase the harvest index by maximizing grain yield while minimizing other plant tissues. Modern selection is also occurring with regard to mechanized harvesting, particularly the ability to grow in high density as single-headed stalks with very short branches bearing considerable seed, an architecture that not only maximizes harvest index but also facilitates machine harvesting.

Plants with limited (or at least short) branching are naturally superior to irregularly branching plants for the purpose of fully and uniformly occupying a field and maximally utilizing solar irradiation. This is obviously desirable for optimizing production, a very important goal in modern plant breeding that will become increasingly significant. As a matter of interest, short plants with compact branching is an architecture that has been selected not just for oilseed production but also for producing marijuana (see the discussion of the "indica type" of marijuana plant in Chapters 12, 13, and 17).

FIGURE 8.12 Fields of *C. sativa* FINOLA, the first hemp cultivar developed exclusively for grain. The low stature facilitates machine harvest and the limited branching minimizes production of stem tissue while allowing a substantial number of plants to be grown in a given area, maximizing production on an acreage basis. Top: The breeder, Dr. J.C. Callaway, is shown (photo by Anita Hemmilä, Finola Inc.). Bottom: A harvester gathering the crop (photo by J.C. Callaway, Finola Inc.).

SELECTION FOR SEED CHARACTERISTICS

Seed Whiteness

This section examines relative seed whiteness (lightness of hue), which appears to have been selected in domesticated *Cannabis* as a result of human preference.

Seeds that are edible and therefore attractive to various herbivores need to be inconspicuous, and Chapter 3 discussed how a camouflagic mottled layer based on perianth (petal tissue) covering the achenes of wild *C. sativa* serves to hide them from herbivores. It was also pointed out that this layer tends to be sloughed off in domesticated strains because it is no longer needed since humans protect the plants against herbivores. The exposed outer layer of the seed (actually fruit) has been observed to differ in shading—either light or dark. Figure 8.13a contrasts the quite dark seeds of a domesticated marijuana strain (typical of the "seeds" of numerous criminal confiscations observed in Canada) and the much lighter seeds of a fiber variety (indeed, most European cultivars have seeds that tend to be lighter shades of brown or gray). In these samples, the camouflagic perianth layer is absent and the color pigmentation resides in the "pericarp" (achene wall, surrounding the true seed). (It should be noted that achenes exposed to sunlight for long periods may become bleached and that light coloring is also characteristic of immature seed.)

Larger achenes are appropriately planted deeper in the soil, and this may be related to their color. Kluyver et al. (2013) proposed that ancient agricultural practices buried seeds quite deeply, leading to an increase in seed size under domestication so that seedlings would have the energy to grow out of the soil. Deeply buried seeds are probably more protected against herbivores and may therefore be more tolerant of light coloration which would tend to attract herbivores. However, darkness of the pericarp of domesticated achenes does not seem to be correlated with their size. Differences in darkness of pericarps among domesticated strains of *C. sativa* may be the result of random fixation, but they may also reflect a frequently observed preference for light-colored achenes in other species, as exemplified in Figure 8.13b and c (for additional examples of similar color selection of fruits and seeds, see Heiser 1988; Small 2013a).

The presence of lighter-colored achenes in European fiber hemp cultivars has been recorded by Vavilov (1931) and Serebriakova (1940). Lighter-colored achenes also are present in Chinese fiber strains, and indeed, Clarke and Merlin (2013) hypothesized that Chinese fiber strains imported into Europe in the nineteenth century contributed genes to European land races and were responsible for

FIGURE 8.13 Selection for whitish achenes ("seeds") under domestication. (a) Left: dark domesticated achenes (lacking a perianth layer) of a marijuana selection of *C. sativa*. Right: whitish domesticated achenes (also lacking a perianth layer) of a fiber cultivar. (b) Left: normal brown achenes of coast tarweed (*Madia sativa* Mol.). Right: white achenes of a cultivar. (c) Left: normal brownish achenes of golden chia (*Salvia columbariae* Benth.). Right: white achenes of a cultivar. Photos (public domain) for b and c by Steve Hurst, U.S. Department of Agriculture.

the origin of lighter-colored achenes in European cultivars. However, human preference for lighter-colored propagules seems to be so universal that probably such selection occurred independently in Europe and China. It is possible that lighter-colored achenes arose in *Cannabis* not because of a human preference for lighter color but because lighter color is associated with some other aspect of the achenes that is of value. Diederichsen and Raney (2006) found that in a large collection of oilseed flax (*Linum usitatissimum* L.) lighter-colored (yellow) seeds were heavier and had a higher oil content than darker-colored (brown) seeds, and it seems possible that the lighter color of the flax seeds is the result of correlation with selection for larger, more nutritious seeds.

Seed Size vs. Seed Quantity

Although some forms of *C. sativa* have quite large seeds, until recently, oilseed forms appear to have been selected mainly for a heavy yield of seeds. In Europe, most cultivars have been selected for fiber yield, and these do not differ much in oilseed potential (Mölleken and Theimer 1997). By contrast, some drug strains (which have been selected for prodigious production of flowers), when left to go to seed, can yield a kilogram of seeds on a single plant (Clarke and Merlin 2013). Piluzza et al. (2013) reported that the seeds of fiber cultivars are larger than those of drug strains, which is consistent with fiber plants having a more extensive historical food usage for seeds than those of drug forms.

Fatty Acid Quality

Percentage and quality of oil in the seeds do not appear to have been important in the past, in part because techniques for analyzing the nutritional chemicals were simply not available until fairly recently. Theimer and Mölleken (1995) concluded that breeding to obtain hemp varieties producing oils with specifically desired fatty acids had not taken place, although selection for oil quality is now being conducted.

Hull Thickness

As noted in Chapter 3, domesticated achenes are thinner-walled than wild achenes, and thinness of pericarp (hull) is an important criterion for modern hemp oil seed breeders since the pericarp is a waste product. Small and Marcus (2000) surveyed 62 accessions and found the hull varied from about 30% to 42% of the weight of the seed.

HEMP SEED AND OIL CULINARY LIMITATIONS

Hemp seeds and hempseed oil are tasty, with an attractive nutty taste when fresh. Untreated hempseed is normally not susceptible to oxidative deterioration during storage because the protective seed hull prevents oxygen from reaching the fatty acids. However, because of fear of viable seeds being used to produce plants, in some jurisdictions, only sterilized seeds or hulled seed is commercially available. Because of the very high content of unsaturated fatty acids (of the order of 80%), heat-treated hempseed and its extracted oil are easily oxidized. Hulled hempseed is normally marketed as a hand food, in closed containers, and once opened, it should be consumed promptly.

There are two consequences of the susceptibility to oxidation of hempseed oil. (1) Heat produces oxidation much more rapidly for hemp than for most of the common edible oils, so hempseed oil deteriorates so rapidly when heated that it is unsuitable for frying. As noted in the early part of this chapter, the smoke point, i.e., the temperature when smoke is produced, is low for hempseed oil, adding to its poor performance as a frying oil. (2) The storage life is also limited, unless the oil is frozen and protected from light.

Callaway and Pate (2009) made the following recommendations regarding cooking temperatures: "In general, the use of hempseed oil in any type of cooking should be limited to the temperature of boiling water. Interestingly, the internal temperature of baking bread does not surpass this threshold. At most, the temperature of hempseed oil should not exceed about 120°C, which is approximately the temperature found in pressure-cooking, and then only for relatively short periods of time."

Hempseed oil is best stored in glass, ceramic, or glazed-metal containers. The oil should be consumed within a short time after a bottle is opened because of degradation due to oxygen in air. Cold storage of the oil, bottled under nitrogen, in sealed, light-proof containers, and possibly with the addition of antioxidants, will prolong shelf life. Small bottles that can be used up in a short time are recommended. Matthäus and Brühl (2008) noted that once bottles are opened, within two months a fishy or varnish-like smell reminiscent of linoleum, paint, or putty can result, and such oil should not be used for human consumption. Because of its susceptibility to oxidation, hempseed oil is mostly used fresh as a salad oil or in processed foods with short shelf lives.

PROCESSED FOOD PRODUCTS FOR HUMANS

For human consumption, the achene is hulled (=dehulled); that is, the inedible shell covering is removed. The remaining edible portion is the embryo. Hulled hemp seed (Figure 8.14, right) is a very recent phenomenon, first produced in quantity in Germany. The quality of modern hemp seed for human consumption far exceeds anything produced historically. Commercial hemp seed is now often found canned or vacuum-packed for the human food markets (Figure 8.14, left). "Hemp nut" is hulled hemp seed. ("HempNut" was a trademark that lapsed into the public domain. "Hemp Hearts," also trademarked, is used in North America.) Modern seed hulling uses mechanical separation and cleaning to produce a smooth, white, gritless hempseed embryo that needs no additional treatment before it is consumed. This product should be distinguished from the protein-rich, oil-poor seed cake that remains after oil has been expressed from the seed. Ordinarily, seed cake from other oilseeds is used for livestock feed. Instead, a protein-rich powder is sometimes sieved from hempseed cake for human food markets.

Hemp seeds and hempseed oil are incorporated into many food preparations, sometimes mimicking familiar foods (Figure 8.15). These include nutritional (granola-type) or snack bars, "nut butters" and other spreads, bread, pretzels, cookies, yogurts, pancakes, muffins, porridge, fruit crumble, frozen desserts ("ice cream"), pasta, burgers, pizza, salt substitute, salad dressings, mayonnaise, "cheese," and beverages ("lemonade," "beer," "wine," and "coffee nog"). One of these products, "hemp milk," comprising a filtered watery mixture of crushed seed, is new in Western countries but is a traditional drink in southern China (Tang et al. 2009). Alcoholic beverages made with hemp utilize hemp essential oil as a flavorant. Hemp food products currently have a niche market, based particularly on natural food and specialty food outlets. As the production area increases and prices decrease, foods from hemp seed will become more common in the Western diet because the taste and nutritional profile are exceptional.

FIGURE 8.14 Hulled (shelled) hempseed. Left: A can of hulled hemp seed. Right: A serving of hulled hempseed. Photo by G. McKeith (CC BY 3.0).

FIGURE 8.15 Some packaged, processed edible food products made with hemp seed and/or hemp seed oil. Bottom photo by Dave O (CC BY SA 2.0).

RAW HEMPSEED SPROUTS FOR HUMANS

Sprouts (recently germinated, live, edible seedlings) of various species (e.g., mung, alfalfa, mustard, cress) are widely marketed and consumed as an especially healthy form of vegetable, since the germination process often makes some nutritious constituents more available, and sometimes detoxifies constituents that protect seeds from being consumed by wild herbivores. Improperly produced sprouts can be contaminated with microorganisms, and periodic outbreaks of disease occur because of inadequate commercial production practices. Nevertheless, in a world dominated by nutritionally unhealthy processed foods, sprouts represent a marvellously healthy contribution to our diet. Hempseed sprouts are just germinated hemp seeds, and while information on their nutritional qualities has not been carefully researched, almost certainly, they are at least as nutritious as commonly sold seed sprouts. Young hemp seedlings have hardly begun to produce secretory trichomes or cannabinoids, so the production of cannabinoids should not be an issue. However, sprouts of *C. sativa* are living plants of *C. sativa* and therefore may be illicit in many countries. Sprouts can be used to produce mature plants and so are of concern to law enforcement. Of course, viable seeds of *C. sativa* can also be used to produce mature plants, and so these may also be forbidden to the general public (and unfortunately, as noted in this chapter, sterilizing hemp seed greatly hastens rancidity). Hempseed sprouts could be marketed frozen (or otherwise preserved), or in a processed state, although sprouts are almost universally marketed fresh. At present, the prospects of developing a hempseed sprout industry seem dismal (at least in North America, which, unlike much of the world, controls commerce in viable seeds). However, there may well be a biotechnological way of generating sprouts that satisfy regulations. As with most consumer products produced from *C. sativa*, many would find "hemp sprouts" worth buying, while others would find reasons to condemn "marijuana sprouts." This issue requires education, as hempseed sprouts do not contain significant amounts of cannabinoids, and hemp is not marijuana.

RECIPE SOURCES

Hempseed recipes date back many years, as evidenced by the following from a fifteenth century Italian cookery book (after Van Winter 1981): "To make a hemp dish for twelve guests, boil one pound of shelled hemp seeds until they burst. When they are ready, add a pound of white well ground almonds, grind the mixture in a mortar with bread crumbs, then bind it with a clear meat or poultry broth and rub it through a horse-hair sieve into a saucepan. Place in on a stove, well away from the flame, and stir it frequently. When it is almost ready, add half a pound of sugar, half an ounce of ginger, a little saffron and rose water. When it is cooked and has been put into bowls, sprinkle it with sweet spices."

With the growing popularity of hempseed, recipe books for hemp have become common. Unfortunately, it is sometimes difficult to distinguish from the title whether these books are dedicated to hempseed or to the preparation of edible marijuana foods. Examples of recipe books dedicated to hempseed include Benhaim (2000), Cicero (2001), Dalotto (1999), Hiener and Mack (1999), Krieger and Krieger (2000), Leson et al. (1999), Rose et al. (2004), Suzanne (2009), and Woodland Publishing (2005).

ANIMAL FEED

BENEFICIAL ASPECTS

Hemp seeds have traditionally been employed as bird and poultry rations, but feeding the entire seeds to livestock has been considered to be a poor investment because of the current high cost involved (subsidization in Europe has allowed such usage, especially in France, although most of the hempseed in Europe is imported from China, as birdfeed). Higher yield and better harvesting practices may make whole hempseed an economical livestock feed in the future as the cultivation area continues to increase.

There have been many studies of how much hempseed or hempseed by-products can be used as livestock feed. Gibb et al. (2005) found that including 14% hempseed in the finishing period of steers had no detrimental effect on their growth or feed efficiency and resulted in brisket tissues that were lower in saturated fat and higher in unsaturated fat (and presumably healthier for human consumption). The seed cake (press-cake, seed meal) left after expressing the oil has proven to be excellent for cows, sheep, and lambs (Mustafa et al. 1999), and Hessle et al. (2008) also found that hempseed cake is excellent feed for cattle. Karlsson et al. (2010) observed beneficial effects from hempseed cake on milk quality of dairy cows (but beyond an ideal proportion of the diet, increasing hempseed cake decreased quality of milk). Reminiscent of the study of Gibb et al. (2005), Silversides and LeFrançois (2005) found that feeding hempseed cake to hens resulted in no significant differences for egg production, feed consumption, feed efficiency body weight change, or egg qualities (except that the eggs had lower saturated fats and higher unsaturated fats and were therefore healthier for human consumption). Khan et al. (2010) fed powdered whole seeds up to 20% of the diet of broilers and concluded that the feed was beneficial and that the eggs had sufficient amounts of alpha-linolenic acid to be marketed as omega-3 eggs. Eriksson and Wall (2012) found that inclusion of hempseed cake did not affect the productivity of organically produced broilers. Hempseed oil has been used up to 12% in laying hen diets, and hempseed up to 20%, without exerting adverse effects on growth performance (Gakhar et al. 2012). Goldberg et al. (2012) reported similar results for hempseed oil. Hullar et al. (1999) found that hempseed cake was beneficial for pigeons. Konca and Beyzi (2012) found that hempseed cake may constitute up to 10% of the diet of Japanese quail. Webster et al. (2000) observed that hempseed cake was a good feed for sunshine bass (*Morone chrysops* × *M. saxatilis*), and it is used to feed carp in France. Pedrosa (2008) and Rema et al. (2010) found that hempcake and hempseed oil were useful feeds for juvenile turbot (*Scophthalmus maximus*). Lee et al. (2010) found that hempseed speeded up the growth and resulted in larger size of *Drosophila melanogaster* flies (not to suggest that these insects are suitable as human food).

The European Food Safety Authority (EFSA) Panel on Additives and Products or Substances Used in Animal Feed (2011), a comprehensive committee-based review of hempseed as animal feed, came to the following evaluation: Three classes of feed materials may be derived from hempseed: whole seed (26% to 37.5% lipids, 25% crude protein, and 28% fiber), hemp seed meal/cake (about 11% lipids, 33% crude protein, and 43% fiber), and hemp seed oil (about 56% linoleic and 22% alpha-linolenic acid). Hemp protein isolate from the seeds could also be used as feed, if economics allowed. Hempseed and hempseed cake can be used as feed materials for most animal species. Felines, however, do not metabolize the EFAs very effectively and should not be fed a diet that is rich in vegetable fatty acids. Suggested maximum incorporation rates of hempseed in the complete feed could be 3% in poultry for fattening, 5%–7% in laying poultry and 2%–5% in pigs for both hempseed and hempseed cake, 5% in ruminants for hemp seed cake, and 5% in fish for hemp seed. With fish, however, the limiting factor is the amount of vegetable fiber, which increases intestinal motility while decreasing nutritional absorption and increasing pollution levels of farmed fish. These suggested amounts should be considered conservative (i.e., experience may show that higher amounts are acceptable).

Callaway and Pate (2009) noted that the tough shell of whole hempseed can limit digestibility if it is part of a meal or if whole seeds are fed to animals. However, hempseed hull contains phytosterols and other nutritional components, so it is not without value. Availability of sandy gravel to poultry and other birds (so they can utilize it in their crops for grinding) can facilitate avian consumption of whole hempseed.

Utilization of hemp stalk, leaves, and other portions of the plant (which may contain some seeds) as silage for livestock is discussed in Chapter 10.

Harmful Aspects

Hemp seeds improperly prepared without removal of the resin-rich perigonal bracts can make birds giddy (Matsunaga et al. 1998). Yousofi et al. (2011) found that female rats fed only a 100% hempseed diet from premating to lactation produced fewer pups and less milk (and they recommended against such a severe diet for pregnant humans!). Stadtmauer et al. (2003) recorded a case of human anaphylaxis related to ingestion of hempseed in a restaurant setting, yet no other reports have been published to support this observation.

Russo and Reggiani (2013) examined oilseed cake from Italian and French cultivars for the presence of several antinutritive properties (phytic acid, condensed tannins, trypsin inhibitors, cyanogenic glycosides, and saponins). They expressed concern about the levels of phytic acid, which they cautioned could lead to mineral deficiencies if fed to excess. So far, there are no reports to suggest that this actually happens, and there is no evidence to suggest that this may occur to a greater extent than can be expected from other vegetable food that contain similar amounts of phytic acid.

The negative effects of feeding marijuana and other nonseed drug preparations to livestock and companion animals are discussed in Chapters 12 and 13.

NUTRACEUTICAL EXTRACTS (DIETARY SUPPLEMENTS), FUNCTIONAL FOODS, AND FORTIFIED FOODS

As explained previously, both hempseed and hempseed oil have nutritional and therapeutic properties. The hempseed industry is now producing a range of extracts to maintain or increase health (e.g., Figure 8.16) and is promoting the nutritional advantages of incorporating hempseed and hempseed oil in foods. Nutritive therapeutic seed products from *C. sativa* are based mostly on the seed oil and particularly on the exceptional fatty acid profile. Quite unlike the highly regulated and legislated prescription environment in which cannabis is employed as a drug, the medicinal-nutritional use of seed products is part of the free marketplace, particularly in the health food industry.

FIGURE 8.16   Hemp oil in capsule form sold as a dietary supplement.

With the expanding market for herbal components in health foods, marketing terms have developed, such as "medical foods," "pharma foods," and "phytofoods." Two terms, "nutraceuticals" and "functional foods," have become widespread, but their meanings have varied somewhat (Small and Catling 1999), particularly in North America. The term "nutraceutical" (sometimes spelled "nutriceutical" in the past) was coined by Dr. Stephen DeFelice of the Foundation for Innovative Medicine, a New Jersey–based industry group. His definition was "a food derived from naturally occurring substances which can and should be consumed as part of the daily diet, and which serves to regulate or otherwise affect a particular body process when ingested." However, the term "nutraceutical" is now commonly applied to an extremely wide variety of preparations with perceived medicinal value, but not necessarily with apparent food value. The phrase "functional food" is also not without controversy, especially in North America. Some have contended that fruits and vegetables should be included in "functional foods" because they are so nutrient packed, while others reserve this term for foods that have a measurable effect on a medical condition. For example, in a randomized controlled double-blind clinical trial, Callaway et al. (2005) showed that dietary hempseed oil relieved symptoms of atopic dermatitis (eczema). The same research group also demonstrated that dietary hempseed oil reduced the ratio of LDL cholesterol to high-density lipoprotein cholesterol in normal volunteers (Schwab et al. 2006). Both of the clinical trials demonstrated various aspects of hempseed oil as a functional food in humans, while the comparative oils did not. In contrast, examples of fortified food would be niacin (vitamin B6) added to cereal grains, sodium chloride (table salt) fortified with iodine in the 1920s to prevent the development of goiter, and cow's milk fortified with vitamin D in the early 1930s to aid in absorption of calcium and phosphorus, thus preventing rickets. The most useful distinction between nutraceuticals and functional foods is simply whether the health-promoting constituents are consumed separately as supplements (nutraceuticals) or as food (functional foods). Given the nutritional qualities of hempseed's constituents, products incorporating hempseed can also legitimately be termed functional foods, although in a broad sense they have also been "fortified." Whether consumed directly or added to food, either naturally derived or produced as a cheaper synthetic version of the active natural ingredient, the phrase "nutritional supplements" highlights a very important trend that has developed for many plants considered to be medicinal. Unlike pharmaceuticals, which are usually potentially toxic medications that should only be prescribed by a medical professional, nutritional supplements for the most part can be purchased from a health food store, herbal practitioner, or independent distributor or they can simply be consumed in commercial fortified foods. Because they are much less expensive than drugs, herbal preparations or extracts, as additions to diet, have been advanced as a new, cost-effective health care system.

COSMECEUTICAL PRODUCTS (NUTRITIONAL COSMETICS)

Since the 1990s, hempseed oil has become very significant as a "cosmeceutical" (cosmetic–nutraceutical), i.e., a body care preparation that promotes the health of skin and allied parts of the body because of the topical absorption of biochemicals. These products include bubble baths, creams, lip balms, lotions, moisturizers, perfumes, shampoos, and soaps (Figure 8.17). One of the most significant developments for the hempseed industry was investment in hemp products by Anita and Gordon Rodderick, founders of The Body Shop, a well-known international chain of hair and body care retailers. This was a rather courageous and principled move that required overcoming legal obstacles related to trace amounts of THC in these products. The Body Shop marketed an impressive array of hemp nutraceutical cosmetics in the 1990s (Figure 8.17, top), and this gave the emerging hemp industry considerable credibility.

Skin readily absorbs essential fatty acids (EFAs), so that lotions rich in these substances can replenish cells damaged by sun and dry air (Wirtshafter 1995). However, EFAs are present in the oil as triglycerides, which are poorly absorbed through the skin. On the other hand, it is important to keep the dead top layers of skin moist, as this protects the lower layers of dermal tissue from

FIGURE 8.17   Body care products made with hempseed oil.

the environment. Also, if skin is compromised, and blood vessels are exposed or near the surface, then these triglycerides will enter the circulation at these sites, where they are also metabolized locally to promote healing. Linoleic acid, alpha-linolenic acid, gamma-linolenic acid, and stearidonic acid specifically have several functions related to skin care, once absorbed: they influence cell membrane functions, including fluidity, transport of electrolytes, and activity of hormones, and they also stimulate cell immunology. These fatty acids applied topically are considered to have potential for treating atopic dermatitis (neurodermatosis) and psoriasis (Vogl et al. 2004). Dietary hempseed oil, consumed as a functional food, has been shown to improve clinical symptoms in patients with atopic dermatitis (Callaway et al. 2005).

INDUSTRIAL (NONEDIBLE) PRODUCTS

The vegetable oils have been classified by "iodine value" as drying (120–200), semidrying (100–120), and nondrying (80–100) oils, determined by the degree of saturation of the fatty acids present (Raie et al. 1995). The suitability of coating materials prepared from vegetable oil depends on the nature and number of double bonds present in the oil's fatty acids. Linseed oil, a very good drying oil, has a very high percentage of linolenic acid. Hempseed oil has been classified as a semidrying oil, like soybean oil, and may be more suited for edible rather than industrial oil purposes. Nevertheless, hempseed oil has found applications in the past in oil paints, varnishes, sealants (wood preservatives), lubricants for machinery, and printing inks (Roulac 1997), although petrochemical extracts have made these uses obsolescent and resurrection of such industrial end uses is challenging because hempseed oil is currently expensive (De Guzman 2001). Occasionally, such products are still found in consumer goods. Larger production volumes and lower prices may be possible, in which case hempseed oil may again find industrial uses similar to those of linseed (flax), soybean, and sunflower oils, which are presently used in paints, inks, solvents, binders, and polymer plastics. Hempseed shows a remarkable range of variation in oil constituents, and selection for oilseed cultivars with high content of PUFAs and other valued industrial constituents is in progress.

In Germany, a laundry detergent manufactured entirely from hempseed oil has been marketed. Soap made with hempseed oil is shown in Figure 8.18. Callaway and Pate (2009) noted, "The idea of using hempseed oil in a soap or shampoo does have appeal, but without adequate precautions taken in the formulation and packaging, these highly unsaturated hydrocarbons will oxidize faster than vegetable soaps made from palm or olive oils. Such an oxidized product, containing oil polymers, may leave a residual greasy feeling on the skin, which can be difficult to completely rinse away."

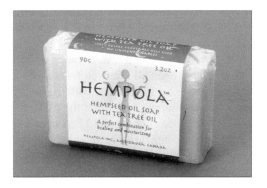

FIGURE 8.18   Soap made with hemp oil.

FIGURE 8.19   An automobile modified to run on vegetable oil and biodiesel. This "hemp car" was a popular attraction at hemp fairs in Canada for many years.

BIODIESEL

Biodiesel, defined as a fuel at least partly made of vegetable oil (Pinto et al. 2005), has been touted as an environmentally friendly alternative to fossil fuels. Hempseed oil can be chemically converted to function as a biodiesel by transesterification with methanol (Li et al. 2010; Ahmad et al. 2011; Ragit et al. 2012; Rehman et al. 2013; Su et al. 2013). Because hempseed oil is highly unsaturated, its freezing point is lower than that of saturated oils, for example, palm oil, and this is a valuable advantage for biodiesel engines operating in very cold environments (Callaway and Pate 2009). In modern times, hempseed oil has been used experimentally as diesel fuel (Figure 8.19), but far cheaper vegetable oils are available, and the cost would have to be reduced considerably to make hempseed biodiesel economically competitive as a fuel. Moreover, using the world's diminishing croplands to grow fuels instead of vital food resources is a contentious ethical issue.

For information on the use of nonseed hemp carbohydrates as a feedstock to produce biofuels, see Chapter 15.

THC CONTAMINATION OF OILSEED PRODUCTS

Hemp seeds contain virtually no THC (Mölleken and Husmann 1997), the chief intoxicant chemical of marijuana. However, THC contamination can result from contact of the seeds with the resin secreted by the epidermal glands on the leaves and floral parts, and also by the failure to sift away all of the perigonal bracts (which have the highest concentration of THC of any parts of the plant) that cover the seeds (Ross et al. 2000). Seed oil prepared from seeds coated with resin may have low levels of THC, and the same is true for foods made with the seeds. A study of THC levels on hempseed produced in recent years in Europe found that the maximum contamination level was 12 mg THC/kg or 12 ppm (EFSA Panel on Additives and Products or Substances Used in Animal Feed 2011), but levels rarely exceeded 5 mg THC/kg in either seed or oil. The presence of cannabinoids is disadvantageous from a regulatory point of view, despite the lack of any scientifically demonstrated THC toxicity at even higher levels. The presence of even trace amounts of THC (and indeed other cannabinoids) in extracted oilseed products intended for human food has been of considerable concern to some and indeed has been a roadblock to much wider development of foods based on cannabis oilseed.

Although much of the Western hemp-growing world uses 0.3% THC in the plant as a maximum concentration for authorized cultivation (0.2% in Europe since 2000), regulations in various countries tolerate only a much lower level of THC in human food products manufactured from the seeds. Permitted levels in hempseed products in different countries range from 10 ppm down to 0.005 ppm. Limits have been set in part because of concerns about possible toxicity, where THC "toxicity" is assumed from THC's transient psychoactivity at sufficient dosage (<10 mg). An extensive analysis of literature dealing with the assumed toxicity of hemp is in Orr and Starodub (1999) (see Geiwitz 2001 for a critique).

The EFSA Panel on Additives and Products or Substances Used in Animal Feed (2011) examined the occurrence of THC accumulation in meat, milk, and eggs resulting from livestock consuming hemp (oilseed or oilseed derivatives or the plants used as fodder). At the time of their review, 95% of hempseed was used to feed livestock (quite unlike North America, where hempseed is extensively used in edible products and cosmetics). Regardless of whether hempseed is consumed directly by humans (in processed products) or indirectly as meat, milk, or eggs from livestock fed on hempseed, the critical consideration is how much THC is absorbed by humans. The Federal Institute for Risk Assessment in Germany estimated a provisional tolerable adult THC intake of 1–2 µg/kg/day as a food contaminant, and from this estimation, a precautionary guidance value for THC in hempseed oil of 5000 µg/kg (5 ppm) was set in 2000, but only in Germany (Matthäus and Brühl 2008).

There has also been some concern about potential interference with drug tests after consuming hempseed foods (Grotenhermen et al. 1998, 2003). Cannabinoids are very lipid soluble and accumulate in fatty tissue throughout the body. They are released very slowly and can remain in the body for up to a month after the consumption of marijuana. Over a period of a week or so, cannabinoid metabolites from hempseed oil can show up in urine (Callaway et al. 1997; Lehmann et al. 1997). The Drug Enforcement Agency and the Office of National Drug Control Policy of the United States raised concerns over tests conducted from 1995 to 1997 that showed that consumption of hempseed products available during that period led to interference with drug-testing programs for marijuana use. However, at that time, most of the hempseed used for North American food products was imported from China, where the typical THC value of field hemp is at least 1%. Federal U.S. programs utilize a THC metabolite level of about 50 parts per billion in urine. Leson (2001) and Leson et al. (2001) reported that this urinary level was not exceeded by consuming modern hempseed products, provided that THC levels are maintained below 5 ppm in hempseed oil and below 2 ppm in hulled seeds. The availability of hempseed oil in the 1990s was due to the use of hemp varieties with high levels of THC and to collection techniques that allowed THC to accumulate on the seeds (Leson et al. 2001). Bosy and Cole (2000) found that by 2000, THC levels in hempseed products in North America had been reduced considerably by the industry. Lachenmeier and Walch (2005) similarly found that by the early years of the twenty-first century, THC levels in hempseed food products in Europe had been strongly lowered, by using varieties with low THC in field tests (<0.3%) and by employing more efficient seed cleaning technologies. However, Petrović et al. (2015) found THC levels varying from 3 to 70 ppm in hempseed oils in Croatia and implied that, in some cases, the oil possibly originated from marijuana strains. Clearly, the use of authorized cultivars is required to avoid excessive THC in hempseed products.

AGRONOMY

See Chapter 7 dealing with fiber hemp for information that is pertinent to aspects of the cultivation of oilseed hemp. The following presentation provides details only when the agronomy of oilseed hemp is significantly different. Detailed information on agronomics of the popular cultivar FINOLA is available at http://www.finola/fi/.

SOIL AND FERTILIZATION

Vera et al. (2004) observed gradual increases in seed yield of the cultivars Fasamo and FINOLA as total nitrogen increased from 40 kg nitrogen (N) ha^{-1} to 160 kg ha^{-1}. Vera et al. (2010) observed a strong increase in seed yield of the oilseed cultivar FINOLA and the dual-purpose cultivar Crag, in response to nitrogen fertilization. The test soils employed were not deficient in phosphorus (P), but the plants benefitted slightly from P application to the seeds. The test soils employed were deficient in sulfur (S), relative to most grain crop standards, but the plants did not benefit from added S. Iványi and Izsáki (2010) examined oil content and protein content of seeds of a dual-purpose variety and found maximum seed yield developed at 80 kg N/ha, while additional nitrogen did not increase these parameters. It appears that, like fiber hemp, oilseed hemp benefits dramatically by supplemental nitrogen, especially in the early development of vegetative growth, and much less dramatic effects were observed from the addition of other elements. However, unlike fiber hemp, oilseed hemp requires mineral nutrition specifically for flowering and seed production. Berger (1969) stated, "During flowering and fruit formation, the demand for potassium and in particular for phosphorus is very high."

TEMPERATURES

Hemp seed production for a given variety requires a warmer climate and a longer season (5 to 6 weeks) than the corresponding fiber crop, to allow time for seed maturation (Bócsa and Karus 1998).

PLANTING

Chapter 13, dealing in part with the cultivation of medical marijuana, points out that a choice can be made to maximize floral production (hence high-grade marijuana) on a given plant by giving it lots of growing room so that it will be very large, or maximizing production per unit area by growing many small plants close together. The same considerations apply for seed production. A very large plant can produce more than 1 kg of seeds, but growing many small plants closely together can produce more seed on a given area.

For seed (grain) production, sowing rates are often 15 to 25 kg/ha, but vary widely, from 1 to 45 kg/ha, depending on the cultivar and conditions of cultivation. (In the absence of contradictory information, the seed supplier's recommendations regarding planting conditions should be followed.) For a cultivar with a 1000-seed-weight of 16 g, planted at 23 kg/ha, the resulting density will be 100–150 plants/m^2 if the germination is high (>80%) (typically about half of the sown seed is lost during emergence and subsequent thinning). Townshend et al. (2012) found that the dual-purpose cultivar Fasamo produced the highest oilseed quantities at 150–225 plants/m^2 but that oil quality did not appear to vary with different plant densities. Based on the cultivars Fasamo and FINOLA (a dedicated oilseed variety), Vera et al. (2006) found that increasing seeding rate (using rates of 20, 40, 60, and 80 kg/ha, corresponding approximately to 49, 78, 105, and 133 plants/m^2, respectively) decreased both the density and size of weeds (as much as one-third) but increased seed yield (also as much as one-third).

Like fiber hemp, seeds may be planted in rows 15–18 cm apart, but such close planting limits interrow removal (by harrowing) of weeds that develop. Nevertheless, Vera et al. (2006) found that a row spacing of 36 cm did not change weed problems more than a row spacing of 18 cm. In Europe, dual-purpose crops of hemp, grown for both fiber and oilseed, are planted at row spacings from 20 to 40 cm, 20 seeds/m within the row, considered sufficient to suppress weeds.

PESTS

Birds, as discussed in Chapter 3, are the most serious pest problem for production of seeds. Hemp grain growers should be aware that flocks of voracious migratory birds are a considerable source

FIGURE 8.20   Birds and hempseed. Left: American goldfinch (September plumage) in a hemp plantation near Toronto, Canada. Right: Flocking migratory birds over a FINOLA hempseed field in Finland (photo courtesy of J. Callaway, Finola Inc.).

of damage to mature hempseed, particularly in small plantations (Figure 8.20). Callaway and Pate (2009) noted that "it is not uncommon to observe geese trampling hemp crops at high northern latitudes, to gain access to the nutrient-rich seed." Not only do birds consume the seed, but they also spread diseases, such as dry mold, as they hop from one seed head to another. Stored grain must be protected from birds, as they will spend considerable time pecking at seed bags to obtain hempseed, in preference to other stored seed.

For weed control with oilseed hemp, which is less densely grown than fiber hemp, mechanical removal may be necessary. There are no available herbicides that are effective for weeds in hemp that do not also eliminate hemp. Except for possible preplanting applications of herbicides to clear fields several weeks before planting, use of any pesticide is strongly discouraged and considered unnecessary by the oilseed industry, which is currently concerned with organic production for food and pharmaceutical use. Hemp borers (*Grapholita dilineana* and *G. tristrigana*) can eat much of hemp plants, including the seeds, but in practice this has not been observed to be a significant problem.

Harvest

A given seed of *C. sativa* is ripe three to eight weeks following pollination. However, as discussed in Chapter 5, photoperiod usually determines flowering time and hence is a natural limitation of when and how many seeds are produced at a given latitude. The cultivar FINOLA is autoflowering (indifferent to local photoperiod). While long days (daily light longer than 15 hours) inhibit flower initiation in most kinds of *Cannabis*, FINOLA flowers happily in Finland about latitude 60°N, where most cultivars will not set seed. Most cultivars of *C. sativa*, grown in suitable climates, continue to develop and mature seed over an extend period of time, and plants have both immature and ripe seeds when grain harvest occurs. All grain plants have undergone strong selection for simultaneous ripening and retention of seeds, to maximize harvest. Oilseed hemp, especially the dual-use varieties (employed for both stem fiber as well as seed production), requires breeding to improve these features. Retention of seeds on plants is promoted by selection against the basal abscission zone of the achenes (so they can't fall away), and by selection for congestion of the branch-bearing achenes (the concentrated bracts and smallest leaves in the infructescences tend to block the seeds from falling away).

Industrial hempseed is best harvested when most of the seed is mature. Harvesting is commonly done when 70% of the seed is ripe and the seed has a moisture content of about 20% (sometimes as much as 30%). Shattering (seed falling off the plants) and bird predation increase if the seed is left in the field for a longer period. Swathing (as done for field retting of fiber hemp; Figure 7.17a) is not recommended for seed production because drying increases the toughness of the stems, making combining more difficult; the seeds do not dry well in the swath; and if rained upon, the seeds may start to sprout. Some exceptions are found in very dry cultivation areas, such as Alberta, where hemp is grown for seed under irrigation. Swathing also poses a significant risk of contaminating the seed with bacteria in the soil, and this is especially true for organic cultivation. Straight combining is preferable, and in western Canada, this is normally carried out when the grain has 12%–20% moisture, depending on cultivar. Waiting for mature fruiting heads to dry down more increases the risks of fibers wrapping around moving combine parts, increasing the risk of fire. Additionally, drier seeds are susceptible to cracking, resulting in oxidation and rancidity. However, after harvest, Callaway and Pate (2009) recommend that for storage, hempseed be air-dried at low temperatures (less than 45°C), to a moisture content of about 9%, employing a moisture meter calibrated for hempseed. A variety of harvesting equipment has been employed for hempseed, often using combines modified for the purpose for tall cultivars (Figure 8.21). FINOLA, as a grain crop, does not require any combine modifications.

Hemp has often been grown as a dual-purpose crop, i.e., for both fiber and oilseed. In France, the principal grower of dual-purpose varieties, the grain is taken off the field first, leaving most of

FIGURE 8.21 Harvesting hempseed. (a) Use of an adapted combine in France. Photo by Aleks (CC BY SA 3.0). (b and c) Harvesting the FINOLA variety in Finland (photos courtesy of J. Callaway, Finola Inc.).

the stalks for later harvest. Growing short varieties dedicated to grain production eliminates the harvesting problems associated with tall hemp varieties (see Chapter 7).

YIELDS

Ideally, hemp seed yield should be based on air-dry weight, with no more than 10% moisture. Hemp yield reports are sometimes difficult to interpret and could be exaggerated by as much as 50% when moist weights are reported. Most reports of seed yield have not been based on varieties selected for oilseed production. Conventional yields in Europe (where hemp has generally not been grown just for seed) are as low as 400 kg/ha, and a yield of 1 tonne/ha is considered good for the standard fiber varieties. In Canada, good yields are of the order of 1 tonne/ha. Yields for FINOLA in Alberta are typically over 2000 kg/ha, with a reported record yield of just over 2000 kg/ha. Maximal yields in China are reportedly as high as 1800 kg/ha (Fortenbery and Bennett 2004).

STORAGE

Grains can be stored temporarily at 12% but should be maintained at 8%–10% for long-term storage. Callaway and Pate (2009) recommend that seeds be dried to just below 10% moisture for storage and pressing, taking care that mold growth does not develop.

BREEDING

Almost no modern breeding of *Cannabis* for oilseed characteristics had taken place by 1995 (Theimer and Mölleken 1995). One of the earliest efforts to increase oil content was by Bócsa et al. (2005) in Hungary. There has been increasing interest in selection for oilseed characteristics since then. The first and foremost breeding goal is to decrease the price of hempseed by creating more productive cultivars. While the breeding of hemp fiber cultivars has proceeded to the point that only slight improvements can be expected in productivity in the future, the genetic potential of hemp as an oilseed crop has scarcely been addressed.

Canadian experience with growing hemp commercially for seed has convinced many growers that it is better to use a single-purpose cultivar than a dual-purpose cultivar. The recent focus of Canadian hemp breeders has been to develop cultivars that are similar to FINOLA, with high seed yields, low stature (to avoid diverting the plants' energy into stalk, as is the case in fiber cultivars), early maturation (for the short growing seasons of Canada), and a desirable fatty acid spectrum (especially higher levels of SDA and GLA).

Because hempseed food products are considered to have great economic potential, there is considerable pressure on the hemp industry in North America to keep THC levels below 0.3%.

CULTIVARS

To date, Canada has been much more concerned with the oilseed hemp industry than anywhere else in the world. However, after industrial hemp was first licensed in 1998, the focus was on fiber hemp varieties (almost the only kind available), which proved uneconomical. Today, the industrial hemp industry in Canada is almost completely concerned with oilseed production, using cultivars dedicated for the purpose or dual-purpose varieties. Commonly grown cultivars in Canada are Alyssa, Anka, CRS-1, CFX-1, CFX-2, Delores, and (especially) FINOLA. Currently, there are relatively few cultivars dedicated to oilseed production, although with increasing world interest in developing oilseed hemp, analyses of oil characteristics of cultivars grown for fiber or for both fiber and oilseed have been conducted, and these studies provide information on germplasm suitable for development of additional dedicated oilseed cultivars. Vogl et al. (2004) surveyed oilseed qualities of some European cultivars (mostly grown for fiber). Small and Marcus (2000) examined oilseed characteristics of 62 accessions. Chen et al. (2010) surveyed the qualitative and quantitative oilseed aspects of eight hemp cultivars grown in various regions of China. Anwar et al. (2006) surveyed oil characteristics of indigenous hemp found in various regions of Pakistan. Table 17.3 provides information on hemp cultivars, including those used primarily for producing oilseed.

THE POTENTIAL USE OF OILSEED CULTIVARS TO PRODUCE CBD AND "HIGHLESS MARIJUANA"

As documented in Chapters 11 and 13, the resin of *C. sativa* is usually dominated by CBD, THC, or both. While most contemporary medicinal and recreational strains are predominantly dedicated to THC, CBD has considerable medicinal potential, indeed exceeding the likely value of THC. Since the resin of hemp cultivars is dominated by CBD relative to THC, it is obvious that these varieties are a potential source of this invaluable cannabinoid. Because oilseed cultivars produce more flowers than fiber cultivars, and the principal source of resin is from the flowering tops of *C. sativa*, oilseed cultivars are a much more promising source of CBD. Just as oilseed cultivars that are short and compact represent the most efficient architectural strategy for maximizing seed production, this is also precisely the way to maximize floral production and resin on a given acreage. However, industrial cultivars of *C. sativa* have not been selected for high resin production. Indica type marijuana, discussed in Chapter 12, which has been selected for high resin production in individual plants and on a given acreage, usually produces considerable amounts of CBD but also produces considerable THC, which is still a barrier at present to most authorized industrial hemp production.

To produce CBD as an extract, entire plants can be employed, including the foliage. For medicinal purposes, there is currently somewhat of a demand for so-called "highless marijuana"—a nonpsychotropic product that is high in CBD but low in THC, which can be smoked. (See the discussion of CBD in Chapter 13 for additional information on highless marijuana.) As discussed in Chapter 13, for biomedical safety reasons, it is far more preferable to imbibe vaporized cannabis (supplied as solvent extracts) than to smoke marijuana (as an herbal) and still more preferable to ingest preparations orally. Nevertheless, to meet a demand for highless marijuana in the form of "buds" (congested flowering heads without seeds; see Chapter 12), plants of the architecture of some modern oilseed cultivars are ideal.

The use of industrial hemp for the production of CBD has not received much consideration yet, although there is commercial interest in using field-grown plants for this purpose. Since the hemp industry is already growing plants outdoors under license, with relatively small security requirements, why shouldn't it also be permitted to harvest the CBD or even produce "highless marijuana"? The industry is not allowed to do so at the moment in most countries, and no doubt, the very concept of highless marijuana will be so enigmatic to many that authorizing such production is likely to meet opposition, or at least slow acceptance.

ECONOMIC STATUS

Hempseed development has been retarded because of its image problem as a relative of marijuana, coupled with trace amounts of THC in the oil, but these concerns have now been mostly overcome. The dietary advantages and attractive taste of hempseed and hempseed oil have led to their widespread use in processed food products, although hempseed is not yet competitive as a commodity with the major edible vegetable oils. There has also been minor success in developing the use of hempseed oil in nutritional cosmetics and hempseed extracts as dietary supplements. Hempseed oil is unsuitable as a frying oil but has good potential for penetrating the salad oil market. Hempseed also has potential as animal feed (especially for both livestock and companion animals). The traditional use of hempseed oil to manufacture industrial products such as paints and lubricants is obsolete but could be resurrected if economic factors allow and breeding produces cultivars that produce high levels of desired oilseed components.

The European Union has historically concentrated on developing *Cannabis* as a fiber crop, much more than as an oilseed crop, but many in Europe are realizing that the success of fiber development has been due primarily to subsidization. Hempseed development in the EU has lagged behind the progress in Canada, which has not provided subsidies to hemp. Canada has concentrated on the development of the hempseed industry, with little attention to hemp fiber. This is consistent with Canada's specialization on production and processing of the major temperate region oilseeds, such as Canola (rapeseed), soybean, sunflower, mustard, safflower, and flaxseed, while Canada lacks a fiber crop industry. Because of the extensive development of oilseeds in Canada, there is extensive capacity to produce high-quality cold-pressed hemp oil. Canada has made great advances in the growing, harvesting, and processing of hempseed during the last decade and has become the world leader in providing raw materials and products for the natural foods, nutraceuticals, and cosmetics industries. Whereas China used to supply most of the hempseed used for food in North America, Canadian-grown seeds have taken over most of this market. The United States will likely follow Canada's dedication to oilseed hemp as industrial hemp becomes legalized there.

China, the world leader in production of hempseed, can produce it cheaply, but imported seed must be sterilized, thus creating delays, adding costs, and lowering grain quality. Seed that has been sterilized tends to go rancid quickly, and so it is imperative that fresh seed be available, at least for human foods. Accordingly, domestic production is a great advantage. Another extremely significant advantage that domestic producers have over foreign sources is certified organic production, which is in demand.

MARKET DEVELOPMENT AND FUTURE NEEDS

The economic prospects for continued development of *C. sativa* as an oilseed crop are much better than for its continued development as a fiber crop, at least in industrialized countries. As noted in Chapter 14, there are considerable prospects for greatly expanded production of medicinal marijuana, and very likely, this will be a stimulus to the industrial hemp industry, as the stigma of marijuana continues to decrease with education and changing demographics.

Creation and marketing of new hempseed products have been increasing for the last two decades. The key need is the development of high-yielding cultivars to increase the competiveness of hempseed in relation to other oilseeds.

The cultivation of dual-purpose crops (fiber and oilseed) is problematical for oilseed production. The profitability of hemp straw is limited at present, while that of the seeds is much higher, so it seems preferable to concentrate on oilseed production. Because fiber hemp and oilseed are best produced as dedicated crops, there is a measure of competition between them. China's supremacy in the production of high-quality hemp textiles at low prices will be very difficult to match in the Western world, although as noted in Chapter 7, Europe has pioneered in the development of hemp applications that do not require the traditional production of very high-quality fiber and has created new technologies and machinery for this purpose. Nevertheless, the future of hemp fiber remains challenging. Europe has shown only limited interest until very recently in developing oilseed hemp, but with the growing realization that fiber hemp has limited growth potential, this is changing. A tradition of concentrating on profitable oilseed products is already well established in Canada, in part because domestic production of oilseeds can be carried out using oilseed production and processing technology that is already available.

The present productivity of oilseed hemp—about 1 tonne/ha under good conditions, occasionally 1.5 to 2 tonnes/ha, is not yet sufficient for the crop to become competitive with major oilseeds. An average productivity of at least 2 tonnes/ha will be necessary to transform hempseed into a major oilseed, a breeding goal that is achievable. At present, losses of 30% of the seed yields are not uncommon, so that improvements in harvesting technology should also contribute to higher yields. Hemp food products cannot escape their niche market status until the price of hempseed rivals that of other oilseeds, particularly rapeseed, flax, and sunflower (Marcus and Small 2002).

After increasing yield per hectare, a second breeding goal is for larger seeds, as these are more easily shelled (hulled). A third goal could be breeding for both large seeds and thin, easily removed shells, so that the fresh seeds could be consumed directly as a snack food. Fourth is breeding for specific seed components. Notable objectives are increasing the health-promoting GLA, improving the amino acid spectrum of the protein, and increasing the antioxidant level, which would not only have health benefits but could increase the shelf life of hemp oil and foods.

It may be noted that while breeding for large seeds could be beneficial for market and processing goals, it would add to the cost of planting seeds, since a heavier weight of seeds would be required to achieve a given plant density (although larger seeds are beneficial in tending to survive better and to grow faster initially). While larger seeds may be beneficial for oilseed cultivars, smaller seeds may save money at planting time for fiber hemp farmers.

Watson and Clarke (1997) suggested that breeding for cultivars with very small seeds could result in such seeds being useful for baking. This would seem analogous with the very small seeds of opium poppy (*Papaver somniferum* L.), widely employed in cuisine. However, the industry can simply shred large hemp seeds to produce small particles the size of poppy seeds.

The stringent security regulations for cultivating hemp in most Western countries significantly increases the cost of production and puts the crop at a unique disadvantage. In particular, the federal government of the United States, which still bans commercial production of hemp, is in the curious position of allowing importation of large amounts of hempseed products while prohibiting its own farmers from growing the crop. Given the enormous ingenuity and capacities of the American agricultural sector, one can predict that oilseed hemp would experience a considerable leap in popularity should the United States join the fraternity of nations now producing the crop.

CURIOSITIES OF SCIENCE, TECHNOLOGY, AND HUMAN BEHAVIOR

- The paintings of Rembrandt (1606–1669), Vincent Van Gogh (1853–1890), and Thomas Gainsborough (1727–1788) were prepared primarily on hemp canvas, often with paints based on hempseed oil.
- Reminiscent of the practice of throwing rice at the wedding couple (symbolic of encouraging fertility), the wedding rituals of south Slavic regions included showering the bride with hempseed when she entered her new home.
- In the Middle Ages of Europe, until the twentieth century, women in England tried to visualize their future husbands by scattering hemp seeds in a garden or churchyard on Midsummer's Eve, while chanting:

  Hempseed I set,
  Hempseed I sow
  The man that is my true love
  Come after me now.
- Benet (1975) provided the following account of how girls in Ukraine wishing to advance the date of their marriage carried hemp seeds in their belts, mounted a heap of seeds, and recited:

  Andrei Andrei,
  I plant the hemp seed on you.
  Will God let me know
  With whom I will sleep?

  The girls subsequently removed their shirts and spit water on the seeds to protect them from being eaten by birds. Finally, they ran around the house naked three times.
- In Ireland, young ladies sowed hemp seed during Halloween in the belief that if they looked behind them while sowing, they would see an image of their future husbands.
- In Poland and Lithuania, a soup made from hemp seeds called *semieniatka* was eaten ritually on Christmas Eve. It was believed that hemp soup offerings should be left for the

departed, as dead relatives would visit every Christmas Eve. A similar meal was eaten in Latvia and Ukraine in the celebration of Three Kings' Day (Benet 1975).

- In Latvia, hemp seed is traditionally included in festival foods eaten during St. John's Day, June 21.
- According to an old European belief, if chickens are fed hemp seeds on Christmas Eve, they will lay all year round.
- The first diesel engine was designed to run on vegetable oils, one of which was hemp oil.
- In 1941, Henry Ford (1863–1947) built a car body from a mixture of resins and plants including hemp and demonstrated that it could run on fuel made from plants, including hemp biodiesel.
- Hemp seed is legally available to most consumers in North America only after it has been steam-sterilized, to prevent the seeds being used to grow plants. Unfortunately, this denatures the protective antioxidants and results in the oil in the seeds quickly going rancid. Robert F. Stroud, the "Birdman of Alcatraz" (played by Burt Lancaster in the film of the same name), became an expert on birds during the long years he avoided execution on California's death row. (Stroud, shown in Figure 8.22, a notoriously dangerous psychopath who had murdered two men, died in prison after spending 54 years in solitary confinement.) His authoritative 1939 book *Diseases of Birds* expressed his frustration with the fact that the seeds he had to use had been sterilized:

  "I want to make it perfectly clear right now that anything said in these pages about the virtues of hemp seed applies to fresh, unsterilized hempseed—most assuredly not to the rancid trash now on the market... Because the seed is rich in the reproductive vitamin, an unlimited supply of it should be kept before the hens making eggs to insure a high percentage of hatchability... The oil of hemp seed becomes rancid very quickly and what was once a valuable food becomes deadly poison. For this reason, hemp seed must always be used with care."

- There are trace amounts of THC, the intoxicating component of marijuana, in hempseed oil and sometimes in the meal obtained from hemp seeds, and these may appear in food products made from them. Some American lawyers have argued, on behalf of clients who had positive drug test results for marijuana, that this was due to having consumed hempseed products. Accordingly, U.S. military forces have been forbidden from eating such

FIGURE 8.22 Robert Stroud, the "Birdman of Alcatraz," who extolled the virtues of hempseed as a birdfeed. (U.S. government, public domain photo.)

products. In fact, modern methods of preparing hempseed products now ensures that only insignificant levels of THC are present, which are too small to produce a positive test with contemporary drug tests.

• Movie actor Woody Harrelson, a supporter of all things related to cannabis, prepared a documentary titled *Go Further*, which chronicled his adventures along the West Coast aboard a bus fuelled by hempseed oil.

• Anka, Carmen, and Deni were the first registered varieties of hemp bred in North America. They were created by the late Canadian breeder Peter Dragla. Anka was named in honor of his wife, Carmen for his daughter, and Deni after the daughter of Woody Harrelson, who supported Dragla's research.

9 Essential Oil

INTRODUCTION TO ESSENTIAL OIL

Essential oil should not be confused with vegetable oil. "Essential oils" (also known as volatile oils and ethereal oils) are said to be "nonfixed" (meaning that they can evaporate quickly), while vegetable oils are "fixed" (meaning that they are relatively stable). The edible vegetable oil of *Cannabis sativa*, i.e., "hempseed oil" from the seeds, was discussed in Chapter 8. The essential oil of *C. sativa* should also not be confused with "hashish oil," i.e., solvent extracts rich in cannabinoids, employed as a highly concentrated form of marijuana, discussed in Chapter 12. The phrase "hemp oil" should be avoided because it could refer either to vegetable oil or essential oil. The phrase "cannabis oil" could refer to either of these or to hash oil and so is particularly ambiguous.

Essential oil is an indistinct category of compounds synthesized primarily as secondary metabolites in plants. Essential oil is responsible for scent in numerous plants (Baser and Buchbauer 2010; Figure 9.1). Approximately 3000 essential oils have been described, of which about 10% have commercial importance in cosmetics, food, and pharmaceuticals (FAO 1995). The composition of essential oils can vary considerably within a given species, and as noted in this chapter, this is particularly true for *C. sativa*.

Essential oils are complex mixtures of organic (hydrocarbon) chemicals and particularly include terpenes and oxygenated compounds such as alcohols, esters, ethers, aldehydes, ketones, lactones, phenols, and phenol ethers (Guenther 1972). Terpenes typically dominate essential oils. Over 30,000 have been chemically characterized from the plant world, a larger number than for any other class of natural plant products (Dewick 2002). Terpenes are made up of units of isoprene: $CH_2=C(-CH_3)-CH=CH_2$. Monoterpenes consist of two isoprene units, sesquiterpenes consist of three. "Terpenoids" are related compounds, although the term is often used as a synonym of terpenes. Many terpenes are extremely odoriferous, detectable by smell at very low concentrations. According to Casano et al. (2011), "terpenes are strongly inherited and little influenced by environmental factors."

THE NATURE OF CANNABIS ESSENTIAL OIL

The characteristic odors of *Cannabis* plants are due to their essential oil. *Cannabis* essential oil is a mixture of volatile compounds, including monoterpenes, sesquiterpenes, and other terpenoid-like compounds. About 140 terpenoids are known in *C. sativa* (Giese et al. 2015), although none is unique to just this species. The two most common terpenes in the plant world are alpha-pinene and limonene, respectively (Gardner 2011), and both are present in the essential oil of *C. sativa*. Other common terpenes of *Cannabis* include myrcene, linalool, beta-caryophyllene, caryophyllene oxide, nerolidol, and phytol. Some terpenes in *Cannabis* are quite pleasant in odor: limonene is fruity (lemons are rich in this chemical), linalool has a rather sweet smell. Mediavilla and Steinemann (1997) found that *Cannabis* essential oils with high sesquiterpene concentrations smelled bad, while oils with high monoterpene percentages (but a low alpha-humulene or caryophyllene oxide concentration) had pleasant smells. Depending on biotype, monoterpenes represent 48%–92% of the volatile terpenes and sesquiterpenes represent 5%–49% (Mediavilla and Steinemann 1997). Monoterpenoids usually make up most of the essential oil of *Cannabis* (Hendriks et al. 1975; Lemberkovics et al. 1981). The aroma of *C. sativa* is particularly due to the monoterpenes pinene and limonene, which frequently comprise over 75% of the volatiles (Hood et al. 1973) and often dominate "headspace" odor near the plant. However, the monoterpenes

FIGURE 9.1 Essential oils from flowering plants such as these are responsible for an incredible range of pleasant odors. This public domain painting, entitled "Smell," was prepared about 1617 by Jan Brueghel the Elder and is housed in the Museo del Prado. Photo credit: The Yorck Project.

evaporate relatively faster than other components, so the composition of essential oil actually in the harvested plant (and capable of being extracted) may differ from the volatiles released around the fresh plant. Consequently, the odor of the living plant is not necessarily indicative of the relative composition of the plant's essential oil (Ross and ElSohly 1996) or of the odor of the dried plant.

The composition of essential oils has been found to vary considerably among strains and cultivars of *C. sativa* (Fournier and Paris 1978; Osman et al. 1985; Mediavilla and Steinemann 1997; Novak and Franz 2003; Hillig 2004a; Bertoli et al. 2010; Fischedick et al. 2010; Casano et al. 2011). Elzinga et al. (2015) were unable to detect clear differences in terpene composition between sativa type and indica type marijuana strains. However, as noted later, marijuana strains tend to have pleasanter odors than hemp cultivars.

The terpenes of *Cannabis* are manufactured in the same epidermal glands (secretory glandular trichomes) in which the cannabinoids of *Cannabis* (discussed in Chapter 11) are produced (Malingré et al. 1975; Fournier and Paris 1978; Meier and Mediavilla 1998). The cannabinoids and terpenoids make up the resinous secretion of the glands. Indeed, the cannabinoids and terpenoids have a parental biosynthetic precursor in common (pyrophosphate). Unlike the terpenes, the cannabinoids are odorless (Clarke and Watson 2002). Terpenes may account for about 1% of marijuana but can comprise as much as 10% of the secretory glands (Gardner 2011).

Rothschild et al. (2005) identified volatile terpenes from *Cannabis* pollen. Pollen does not have secretory glands. It is possible that the cells of pollen grains can synthesize terpenes, but the report could be based on contamination from the anther glands. (Similarly, Paris et al. 1975 identified cannabinoids in pollen, which could also have resulted from contamination.)

As discussed in Chapter 11, there is controversial evidence for trace amounts of cannabinoid production outside of the epidermal glands (in laticifers and tissue cultures), and whether the same is true for terpenes remains to be clarified.

POSSIBLE PLANT-PROTECTIVE FUNCTIONS
OF THE ESSENTIAL OIL OF *CANNABIS*

A considerable variety of chemical constituents of *Cannabis*, including the cannabinoids, are known to be toxic to many organisms (e.g., Radwan et al. 2008, 2009). The essential oil also appears to contain compounds toxic to a wide variety of microbial and animal species that attack plants. What adaptive significance the essential oil has for wild plants of *C. sativa* is unclear, although terpenes such as those in *Cannabis* are repellent to some insects (Thomas et al. 2000), are antimicrobial (Fournier et al. 1978; Novak et al. 2001; Nissen et al. 2009), and are antifungal, antiviral, and antiparasitic (Paduch et al. 2007). The terpenes of *C. sativa* might also contribute to its capacity to repel competing plants (see discussion of allelopathy in Chapter 3).

It has been suggested that the more fragrant terpenes like limonene and pinene function particularly well against insects preying on the floral parts, while the relatively bitter sesquiterpenes in the foliage act against grazing animals (Langenheim 1994). Beta-caryophyllene is usually the predominant sesquiterpenoid of *C. sativa* (Mediavilla and Steinemann 1997), and in addition to repelling insects, it attracts predatory green lacewings, reinforcing protection against insect herbivores (Russo 2011a). As pointed out by Potter (2009), "The very different balance of monoterpenes in the sessile trichomes on the foliage and the predominantly capitate stalked trichomes on floral tissues is supporting evidence that these trichomes have different functions. Both types contain bitter sesquiterpenes which can act as anti-feedant repellents. The increased monoterpene content of capitate stalked trichomes would be expected to lower the viscosity of the contents, thereby making it more able for them to ensnare insects... The monoterpenes are more volatile, and being hydrophobic they are highly persistent in the atmosphere. Insect olfactory systems are devoid of the mucous membranes found in mammals, and they are especially sensitive to such lypophylic chemicals. Monoterpenes are thereby detected by insects at considerable distances from the plant. In many cases these monoterpenes are repellent to insects (e.g., α-pinene and ants) those insects apparently misidentifying the monoterpene as an alarm pheromone."

VALUE OF CANNABIS TERPENOIDS AS COMMERCIAL REPELLENTS

Nerio et al. (2010) reviewed the repellent value of essential oils in general to control insects and other invertebrates harmful as pathogens affecting human and livestock health, and damaging materials of value to humans, especially stored food. They noted that synthetic chemicals currently dominate the repellent market, but natural plant extracts have potential to provide repellents that are safer for humans and the environment. For additional observations, see the discussion of natural pesticides in Chapter 10.

MEDICAL SIGNIFICANCE OF TERPENES IN *CANNABIS*

The common terpenes in *Cannabis* are present in many other plants. These terpenes are known to have a variety of medicinal effects (Table 9.1): several are anti-inflammatory or psychologically soothing, and some have specific therapeutic applications for human illnesses and disorders (Buchbauer 2010).

As discussed in Chapter 12, the sedative terpene myrcene may be responsible for "couchlock," a state of extreme lethargy reputedly associated with indica type marijuana strains. Beta-myrcene is the predominant monoterpene in *C. sativa* (in both marijuana strains and industrial cultivars). Piomelli and Russo (2016) stated: "The sedation of the so-called indica strains is falsely attributed to CBD content when, in fact, CBD is stimulating in low and moderate doses! Rather, sedation in most common *Cannabis* strains is attributable to their myrcene content, a monoterpene with a strongly sedative couch-lock effect that resembles a narcotic. In contrast, a high limonene content (common to citrus peels) will be uplifting on mood, while the presence of the relatively rare terpene

TABLE 9.1

Medicinal Properties of Common Terpenes of *Cannabis sativa*

Terpene	Pharmacological Activities
Alpha-pinene	Anti-inflammatory, antibacterial, bronchodilatory
Beta-caryophyllene	Anti-inflammatory, protects lining of digestive tract, antimalarial
Beta-myrcene	Analgesic, anti-inflammatory, sedative, muscle relaxant
Caryopyhllene oxide	Anti-fungal, deceases platelet aggregation, treats nail infections
Limonene	Antidepressant, immunostimulant, antibacterial
Linalool	Antianxiety, sedative, local anesthetic, anticonvulsant
Nerolidol	Anti-malarial, sedative
Phytol	Sedative, prevents certain congenital malformations

Source: After Russo, E.B., *Br. J. Pharmacol.*, 163, 1344–1364, 2011.

in *Cannabis*, alpha-pinene, can effectively reduce or eliminate the short-term memory impairment classically induced by THC."

Ross and ElSohly (1996) found that when plant material was stored in closed paper bags for three months, about half of the beta-myrcene evaporated away, indicating that the psychological properties of marijuana as influenced by terpenes may change with storage conditions and time.

Of particular interest is evidence of interaction of the natural terpenoids and cannabinoids in marijuana, which may be therapeutic (McPartland and Russo 2001; Russo 2011a). Races of *C. sativa* have been selected which are high in particular terpenes (Gardner 2011), and these might be of value should particular cannabinoid-terpene combinations prove to be of significant medical application. Whether simply combining cannabinoid and terpenoid extracts would be more efficient than breeding biotypes that naturally combine certain cannabinoids and terpenes will need to be considered, since far cheaper sources of terpenes are available than *C. sativa*.

WHY ELIMINATING TERPENES BY IRRADIATING MEDICAL MARIJUANA MAY BE HARMFUL

Until recently, there has been only one official national supplier of medical marijuana in Canada, and as of the writing of this book, the federal government of the United States and the government of the Netherlands also have just one national supplier. The herbal marijuana supplied has been subjected to gamma radiation before being released to patients. The objective is to ensure microbiological safety by sterilizing any potentially toxic microorganisms that are present, which are of concern because many patients are immune-compromised and especially susceptible to infection (Chapter 13). However, an unintended result is the reduction of terpenes, especially monoterpenes, which are especially sensitive to irradiation (Fan and Gates 2001). Just as irradiated food sometimes has a flat or cooked aroma that consumers find objectionable, so patients who are familiar with marijuana sometimes find the dearomatized nature of irradiated medical marijuana so objectionable that they turn to the black market. Moreover, as pointed out previously, there is reason to believe that the terpenes normally in marijuana contribute to its medical efficacy, so that irradiated marijuana may in fact be less healthy. See Chapter 14 for additional information on the wisdom of irradiating medical marijuana.

EVOLUTION OF ESSENTIAL OIL IN DOMESTICATED *CANNABIS*

Some wild populations of *C. sativa* produce very nauseous smells (especially noticeable in closed spaces), raising the possibility that humans have selected plants with relatively pleasant odor due to particular essential oil components. The odor of fiber strains is quite divorced from the quality and quantity of fiber in the stem (the fiber has no particular smell), but the odor of harvested marijuana is unavoidable and so has been more susceptible to selection. It is plausible that in the past, marijuana land races with pleasant odors were selected more often than was the case for fiber strains. Marijuana strains tend to be more attractive in odor than fiber strains, and many of the popularly marketed marijuana strains have relatively pleasing odors. Humans appreciate fragrance for aesthetic reasons, and indeed, in former times, fragrant resins were often employed in religious sacraments. An attractive fragrance would have reinforced the spiritual value of intoxicating forms of cannabis and so likely would have been selected. Clarke and Merlin (2013) observed that "Pioneering marijuana breeders continued selecting primarily for strong potency (high Δ^9-THC content), followed by more aesthetic considerations of flavor, aroma, and color. Modifying adjectives such as 'minty,' 'floral,' 'spicy,' 'fruity,' 'sweet,' 'purple,' 'golden,' or 'red' were often associated with selected varieties." Upton et al. (2013) noted in regard to marijuana that "the aromas as described in modern advertising include: peculiar, narcotic, strong, sweet to sour, fruity to pungent, agreeable, aromatic, fresh and sweet, euphoric, spicy, citrusy, musty, skunky, acrid, juniper, floral, sour, diesel, vanilla, complex, blueberry, pineapple, perfumed, piney, sandalwood, mango, skunky-cheese, and more." General adjectives such as complex, fresh, perfumed, and pungent that are applied to modern strains of marijuana can be difficult to interpret, in contrast to more specific comparative terms such as blueberry, cheesy, citrus, juniper, mango, pineapple, piney, sandalwood, and vanilla.

An unpleasant odor often does not disqualify material from being consumed by humans. The odor of some common marijuana strains is quite objectionable (note the popular strain Skunk). Moreover, it is apparent that some strains with a foul smell are appreciated by many. Some currently popular strains are described by terms such as acrid, diesel, musty, peculiar, and sour. Bouquet (1950) noted, "Ganja [marijuana] has a pronounced fetid smell, much appreciated by addicts." It should be noted that judgment of the relative attractiveness of odor depends appreciably on one's upbringing. The smells of ethnic foods that have become familiar in childhood may seem to be very attractive to those habituated to them, but quite unattractive to others.

The possibility that the terpenes present in the essential oil modify the physiological effects of the cannabinoids is discussed in Chapter 13. If so, it is possible that unconscious selection for medically effective terpene profiles has been significant in the biochemical evolution of *Cannabis* marijuana strains.

ODOR AND OTHER ORGANOLEPTIC QUALITIES OF MARIJUANA

Industries that offer products that are consumed by mouth, like marijuana, are very concerned about organoleptic preferences (taste, odor, and texture) of their offerings since these are critical criteria by which consumers judge acceptability. For marijuana, probably odor (which is interconnected with taste) is the only organoleptic factor of interest, although the abrasiveness of the foliage, caused by the presence of cystolith hairs, may also be significant since there has been some consumption by mouth.

Odor or "aroma" (a combination of smell and taste) is quite subjective, and characterizing *C. sativa* in these respects is problematical. The odor of *C. sativa* has been described by a variety of divergent terms in the literature, such as agreeable, disagreeable, strong, aromatic, heavy, and odd. Intoxicating plants have been alleged to have smells that are "narcotic" and "euphoric," although clearly this requires imagination.

The taste of marijuana is unpleasant—typically bitter, acrid, and resinous—and the mouth feel is also unpleasant—crunchy and sticky. In southern Asia, "bhang" is a low-intoxicant preparation of *Cannabis* leaves, typically combined with milk products (THC is soluble in fat) and sometimes eaten by lower classes, but the taste (as in all marijuana edibles) is masked by more pleasant ingredients. In the illicit drug counterculture/underground trade, hundreds of strains of *C. sativa* are offered, and many of these have names suggesting either or both odor and taste (e.g., Lemon-Lime

Kush, California Orange Bud, and Fruity Juice). However, these differ primarily in olfactory, not taste qualities, likely mostly because of different profiles of the terpenes that are present. Although the terpenes are volatile, some remain in the resin glands unless they are crushed.

ODOR HAS BECOME A KEY SELLING CRITERION FOR MARIJUANA

Plants are capable of producing an enormous range of odors due to variation of essential oil content. The "scented-leaved" or "fragrant-leaved" geraniums (species of *Pelargonium*) are a mixed group of odoriferous species, including several dozen used for culinary purposes. The species smell like fruits (apricot, apple, peach, and strawberry), citrus (lemon, lime, and orange), "nuts" (coconut and filbert), spices (allspice, cinnamon, ginger, and nutmeg), mint, pine, camphor, and especially roses (Small 2006). It should be noted that the perception of smell is at least somewhat unique to individuals (as noted previously, early experiences often determine lifelong preferences), and one person's apricot-scented plant may be another's sour milk.

Cannabis also presents a wide range of odors. The hundreds of marijuana strains currently marketed differ in THC content, the chief determinant of psychological effect, so this is the main basis on which the product is purchased. However, many consumers have acquired the conviction that strong smell is an indicator of potency, and it is now common practice among marijuana users to judge the quality of dried cannabis or hashish at least partly by intensity of smell. Accordingly, the terpene content of marijuana has become a critical quality concern for both illicit dealers and authorized sellers.

As noted previously, strains of *Cannabis* may have been selected with particular terpene profiles in combination with particular cannabinoids, for medical value, but the possibility exists of simply combining cannabinoids with given terpenes from commercial (non-*Cannabis*) sources. Indeed, in theory, a variety of chemicals with desired odors or medicinal properties could simply be added to marijuana. Regulations in particular jurisdictions may prevent adding materials to medicinal marijuana, but in the long-term, it would seem that if genuine medical benefit is gained from adding given terpenes to marijuana, there should not be an objection. If it also is established that no harm results from adding terpenes to marijuana simply for attractive aroma (in the same manner that lemon and pine aromas are widely added to detergents), then one can anticipate that "flavored" or "aromatized" marijuana will become popular. Such a practice is reminiscent of the tobacco industry's offerings of scented and flavored products.

THE IMPORTANCE OF TERPENE ODOR FOR OILSEED *CANNABIS*

"Fresh, cold-pressed hempseed oil from good quality seed typically offers a delicious combination of citrus, mint and pepper flavors from the oil" (Callaway and Pate 2009). Volatile terpenes are primarily responsible for the delicate flavors of the oil (Mediavilla and Steinemann 1997), and it is necessary to control temperature and drying regime carefully to preserve the aroma (Callaway and Pate 2009).

ODOR DETECTION AND LAW ENFORCEMENT

Facetiously, it may be pointed out that the odor of marijuana has affected human evolution, since the distinct smell has widely attracted law enforcement officers, resulting in the incarceration of millions, reducing their Darwinian fitness (potential for leaving progeny). Police have become quite expert in recognizing the odor of marijuana and have even used devices to aid in their evaluation of the strength of the odor and to identify faint odors (note the "smelloscope" or "smellometer" shown in Figure 9.2, right). Caryophyllene oxide, a minor odoriferous component of marijuana, has been used to train police "sniffer dogs" (Figures 9.2, left, and 9.3) to detect marijuana (Martin et al. 1961; Nigam et al. 1965). Caryophyllene oxide is present in the volatile oil of other plants, such as

FIGURE 9.2 Marijuana odor detection aids employed by the police. Left: German police dog. Dogs can be trained to identify the presence of marijuana by the presence of volatile components. Photo by Brian Snelson/ exfordy (CC BY 3.0). Right: "Nasal Ranger" field olfactometer, which has been used to quantify the strength of odors from marijuana grow-houses. Photo courtesy of St. Croix Sensory, Stillwater, Minnesota.

FIGURE 9.3 Sniffer dogs easily detect contraband marijuana by the odor of terpenes. Prepared by B. Brookes.

mugwort (*Artemisia vulgaris* L.) and clove (*Syzygium aromaticum* (L.) Merr. & L.M. Perry), but almost always the dogs alert to marijuana and not to flavoring herbs and spices.

In many jurisdictions, detection of the smell of marijuana is sufficient justification to permit police to inspect a person, vehicle, or premise. In the United States, the Fourth Amendment of the U.S. Constitution protects citizens from unreasonable search and seizure, so the accuracy of judgment of odor as a guide to the presence of marijuana has been extensively examined in courts, but less so by scientists. A rather preliminary review by Doty et al. (2004) suggested that many common plants and some animals have scents identical to marijuana, so that claims by officers of smelling marijuana are uncertain. The fact is, however, that in Western countries, relatively few people transporting highly odoriferous material are likely to be carrying anything but marijuana. In a study of Alaskan marijuana police cases involving possible marijuana in buildings, Myrstol and Brandeis (2012) concluded, "Detection of marijuana odors was not found to be a good predictor of whether or not a search would result in the discovery of less than one ounce of marijuana… Detection of marijuana odors was found to be significantly associated with the discovery of relatively 'large' amounts of marijuana—that is, quantities of four ounces or more, as well as 25 or more plants." As one would expect given the olfactory skills of dogs, canine detection is more accurate than human detection but is not infallible. "Limited work has been published on canine and human detection of marijuana odor, yielding mixed results and high variability" (Rice and Koziel 2015). It has been shown that handler belief that marijuana is present increases the likelihood that dogs will detect it (Lit et al. 2011).

AGRONOMIC PRODUCTION

Commercial preparations of the essential oil, often called "*Cannabis* flower essential oil" and "hemp essential oil," have been prepared from the female or monoecious inflorescences and/or the younger foliage. Mediavilla and Steinemann (1997) observed very considerable variation of pleasantness of both fiber cultivars and marijuana strains. However, marijuana strains tend to be more attractive in odor than fiber hemp, although the essential oil of most hemp cultivars tends to have considerable myrcene, which is pleasant in odor and is often used in the perfume industry. Marijuana strains produce much higher numbers of flowers than fiber strains, and because the (female) floral parts provide most of the essential oil, marijuana strains are naturally adapted to essential oil production. Switzerland has permitted strains with higher THC content to be grown than is allowed in most other countries, giving the nation an advantage with respect to the essential oil market. Accordingly, Switzerland has been a center for the production of essential oil of *C. sativa* for the commercial market. Nevertheless, essential oil in the marketplace has been produced from low-THC *Cannabis*, for example, in France and Canada.

Parameters affecting the production of essential oil have not been carefully examined. Extraction of the essential oil by steam distillation of fresh plants has a yield of only 0.05%–0.29%, depending on growth, harvest, and drying conditions (Upton et al. 2013). Essential oil is maximized under the same conditions that THC is maximized, since both are produced primarily by glandular hairs. However, growing exclusively female plants to prevent seed formation has apparently not been practiced for essential oil production, although pollination prevention in greenhouse experiments showed significantly higher essential oil yield (Mediavilla 1998). Mediavilla (1998) reported that "Highest yield of essential oils was gained when about 50% of the seeds had reached maturity. The ideal harvest time for best quality (scent scores) was... between female flowering and seed maturity. Unfortunately the yield and the quality never [are both] highest at the same time. Therefore the optimal harvest time depends on whether the farmer or his customer is more interested in yield or quality. The time interval when both yield and quality are high is rather small." Because essential oil is water soluble, rain and high humidity can deteriorate production (Mediavilla 1998). This is a common problem for numerous essential oil crops but is avoided by growing in dry environments. Growing *C. sativa* in a greenhouse can greatly increase yield; Potter (2009) found a yield of 7.6 mL/m^2, equivalent to 77 L/ha^2. However, the costs involved in producing essential oil indoors would be prohibitive.

As noted in Chapter 12, production of high-THC marijuana "buds" leaves "waste" (stems and foliage), which can serve as material from which intoxicating THC can be extracted as a secondary or salvage product. Similarly, as noted in Chapter 8, production of oilseed also leaves residue (floral bracts, foliage), which can serve as material from which nonintoxicating, medicinal CBD can be extracted as a secondary or salvage product. In principle, waste material from oilseed production could also be used as a source of terpenes.

COMMERCIAL PRODUCTS

The essential oil of *C. sativa* has some limited commercial value (examples of products are shown in Figure 9.4). Yields are very small—about 10 L/ha (Mediavilla and Steinemann 1997), so the

FIGURE 9.4   Products made with *C. sativa* essential oil. Left: Two bottles of hemp essential oil. Right: Pastilles flavored with hemp essential oil. The essential oil is largely a volatile product distilled substantially from the floral bracts.

essential oil is expensive (Weightman and Kindred 2005 cited a retail price of £1000/L [=ca $1700/L]). Essential oil of different strains varies considerably in odor, and this may have economic importance in imparting a scent to cosmetics, shampoos, soaps, creams, oils, perfumes, and candles and a flavor to foodstuffs (particularly candy and beverages) or medicines. Alcoholic beverages made with hemp utilize hemp essential oil as a flavorant.

Aromatherapy—the therapeutic use of volatile oils—has become popular, and it is possible that cannabis volatile oils could achieve considerable market penetration. There is no evidence at present that cannabis essential oil is as effective as presently utilized aromatherapy oils. Nevertheless, there is a large market for cannabis products of whatever nature merely because *C. sativa* is notorious, and it would not be surprising if cannabis essential oils marketed for aromatherapy achieved market success.

REGULATORY RESTRICTIONS

Terpenes are natural flavor ingredients of numerous plants, and as long as they are in plant tissues, they are usually not of concern for safety considerations. Crude and purified terpene extracts are concentrated chemicals that often do have potential toxicity, and their use is frequently subject to safety regulations. The common terpenes from *Cannabis* are available from sources other than *Cannabis*. The principal terpenoid extract of *Cannabis* has simply been the volatile oil, and the chief regulatory concern associated with it is THC contamination. *Cannabis* essential oil is not an authorized product in many jurisdictions because of concern about THC content. THC has relatively low volatility and water solubility (Malingré et al. 1975) and so is not expected to be present in appreciable concentrations in extracts prepared by steam distillation. It is possible to produce *Cannabis* essential oil capable of satisfying the regulatory needs for very low THC levels in food and other commercial goods.

ECONOMIC PROSPECTS

Market Potential for Cannabis Essential Oil

Today, extracted cannabis essential oil is simply a novelty. The world market for hemp essential oil for flavoring or adding aroma to products is very limited at present and probably has limited growth possibilities, although as noted previously, the aromatherapy market may have some potential.

BREEDING CULTIVARS FOR ESSENTIAL OIL HARVEST

As with oilseed cultivars, tall cultivars are disadvantageous in directing much of their energy into the production of stem tissue rather than reproductive tissue. Short plants with condensed flowering axes are rich in flowers; hence, resin-synthesizing trichomes are a much better investment from the point of view of essential oil production on an area basis (exactly as for oilseed cultivars). As with marijuana strains and oilseed cultivars, male plants are of little or no use, and male expression could be reduced considerably. There are no cultivars selected mainly for essential oil, and given the limited market potential, breeding new varities just for terpenes seems unlikely at present. In Britain, the short-statured oilseed cultivar FINOLA has been employed for essential oil production (Weightman and Kindred 2005).

BREEDING MARIJUANA STRAINS FOR ATTRACTIVE TERPENE PROFILE

As discussed in this chapter, the terpenes of *Cannabis* may have medicinal value as natural components of medical marijuana, and certainly, the aroma associated with the terpenes has become a key criterion judged by consumers. At present, medicinal marijuana is substantially sold as a natural ("organic") herbal, without additives, and if this situation persists, strains with attractive terpene profiles have potentially immense value. However, there seems no necessary reason why terpenes cheaply extracted from other plants can't simply be added to marijuana lacking these attractive terpenes, so that breeding plants for terpene profile seems like a debatable investment.

CURIOSITIES OF SCIENCE, TECHNOLOGY, AND HUMAN BEHAVIOR

- Although terpenes are effective repellents against many insects, some insects synthesize terpenes (often from precursors obtained by consuming plants) and use them as a defense against predators (Eisner 1970). Insects using terpenes as defensive chemicals include caterpillars of swallowtail butterflies, some ants, and some termites. Some Silphidae beetles secrete foul-smelling terpene alcohols from a rectal gland. Like a skunk, one of these species, *Necrodes surinamensis*, ejects this secretion and can rotate the end of its abdomen to spray in all directions (Roach et al. 1990).
- In 1981, U.S. President Ronald Reagan stated, "Trees cause more pollution than automobiles do." The specious rationale for this controversial statement was that "photochemical smog" (also known as ground level ozone pollution), which is created when automobile emissions are broken down into ozone and other chemicals by strong sunlight, is amplified by the presence of various volatile organic compounds, particularly terpenes, which are released to the atmosphere in large amounts by coniferous trees in hot weather. While Reagan's claim was ill-considered, the possibility that terpenes released by trees may have had a significant role in historical climate change (for example, reactive terpenes could have weakened the ozone layer protecting the planet) has been taken seriously in some scientific publications (e.g., Hari and Kulmala 2008).
- Humans can distinguish several million different colors and almost half a million different sounds, but our sense of smell is remarkably more acute, capable of discriminating more than one trillion smells (Bushdid et al. 2014).
- Cannabis essential oil is one of the world's most expensive essential oils, sometimes selling for about $30.00/mL, although market prices fluctuate. Agarwood (from an endangered Asian tropical tree, *Aquilaria malaccensis*), at about $40.00/mL, is often said to be the world's most expensive essential oil. "Absolutes" are similar to essential oils but are extracted from plants by solvents or fats, whereas essential oils are typically obtained by steam distillation. Expensive absolutes include champaca (from a southern Asian tree, *Michelia champaca*) at $80.00/mL, frangipani (from a tropical American tree or shrub,

Plumeria rubra) at $50.00/mL, and tuberose (from the Mexican perennial *Polianthes tuberosa*) at $50.00/mL.

- Humans are not only fascinated by the odors of plants but also by the smell of other humans. One of the more curious studies reported that underarms of men smell of cheese while those of women smell of onions (Troccaz et al. 2009). The onion odor of women's armpits was found to be due to high amounts of an odorless sulphur-containing compound, which underarm bacteria transformed into the onion-smelling chemical thiol. The cheesy odor of men's armpits was caused by an odorless fatty acid which became smelly when acted on by underarm bacteria.

10 Minor Uses

Cannabis sativa is an exceptionally versatile crop. Fiber (Chapter 7), oilseed (Chapter 8), and cannabinoid drugs (Chapter 13) are the main economic products. Essential oil, of minor significance, is discussed in Chapter 9. Other actual or potential uses are examined in this chapter. The following sequence of subchapters is arranged in decreasing order of probable potential usefulness.

BIOMASS

MERITS OF CANNABIS SATIVA AS A SOURCE OF BIOMASS

Biomass refers to material from living or recently living organisms, especially from plants, which is usually employed as an energy source, either burned to produce heat or converted to biofuel. Numerous plants are capable of generating considerable biomass, and *C. sativa* is one of them (Poiša et al. 2010; Figure 10.1). Concern over rising prices and ecological damage associated with the use of petrochemicals has led to attempts to reduce fossil fuel use by substituting biomass plants to produce energy. Most biomass is currently derived from wood, but as discussed in Chapter 16, dealing with sustainability, trees are a diminishing resource, and crops are being considered as new sources of biomass. Hempseed-based biodiesel is discussed in Chapter 8. Biodiesel is usually produced from edible oilseed crops (such as rapeseed) and bioethanol is usually manufactured from edible carbohydrate crops (such as maize and sugar cane). This is controversial, since using cropland to produce biomass instead of food can reduce the availability and increase the cost of food, especially in low-income nations. It has been argued that using crops that produce only inedible "lignocellulosic biomass" (such as fiber hemp) avoids the ethical problem, but it does not, since the land could be used for food production (including oilseed hemp). Rehman et al. (2013) explored the possibility of using harvested wild-growing hemp in Pakistan as a source of biomass, which certainly would be an ethical strategy. As discussed in Chapter 16, *C. sativa* is an especially sustainable, environmentally friendly plant, and so when grown as a crop for whatever purpose, possibly including biomass production, it is relatively benign to the planet and to people. Compared to other crops grown for energy, hemp is considered to be a reasonably efficient source (Finnan and Styles 2013).

It has been contended that hemp is notably superior to most crops in terms of biomass production, but Van der Werf (1994b) observed that the annual dry matter yield of hemp (rarely approaching 20 tonnes/ha) is not exceptional compared to corn, beet, or potato. Meijer et al. (1995) also noted that there are constraints to the biomass production of hemp. However, most hemp varieties have been selected for production of fiber, not for biomass. Hemp has been rated on a variety of criteria as one of the best crops available to produce energy in Europe (Biewinga and van der Bijl 1996). Hemp, especially the hurds, can be burned as is or processed into charcoal, methanol, methane, or gasoline through pyrolysis (destructive distillation). Hemp could be used to create cellulosic-based ethanol (Sipos et al. 2010; Kuglarz et al. 2016). González-García et al. (2012) showed that ethanol derived from hemp hurds under some scenarios could be practical. However, conversion of hemp biomass into fuel or alcohol is impractical in areas where there are abundant supplies of wood, and energy can be produced relatively cheaply from a variety of sources. Prade et al. (2012) concluded, "The main competitors for hemp are maize and sugar beets for biogas production and the perennial crops willow, reed canary grass and miscanthus for

FIGURE 10.1   Scenes illustrating considerable biomass production by hemp (from a Canadian medicinal marijuana plantation in Ottawa in 1971, described in Small et al. 1975).

solid biofuel production. Hemp is an above-average energy crop with a large potential for yield improvements."

Biogas

"Biogas" (especially methane) is produced in some countries from various feedstocks, particularly animal waste, crop residues, household organic waste, and sewage sludge. In Germany, maize has been used as a source of biogas (Rehman et al. 2013), and other crops have been considered for the purpose. Mallik et al. (1990) studied the possibility of using hemp for methane production and decided that it was unsuitable for this purpose. Pinfold Consulting (1998) concluded that while there may be some potential for hemp biomass fuel near areas where hemp is cultivated, "a fuel ethanol industry is not expected to develop based on hemp." Kreuger et al. (2011a,b) were more optimistic, considering hemp to be a potential source of biomass for biogas generators, based in part on their observation that steam pretreatment notably increased the conversion of hemp straw into methane.

Hemp Solid Fuel

Hemp can be burned directly for energy (Rice 2008), but there has been limited interest in this. However, because of its high biomass productivity, hemp is a potential feedstock for the production of solid biofuels such as briquettes and pellets (Prade et al. 2011; Aluru et al. 2013). Pelleted combustible material that can be used as fuel for pellet stoves and boilers represents a niche market. Today, fuel pellets are made almost exclusively from wood, although other biomass energy crops (such as cereal straw, miscanthus, switchgrass, and hemp) are being explored for the purpose (Kolarikova et al. 2013). Hemp produces relatively little ash when burned (often under 2%) and is comparable in corrosive effect to wood pellets (pellets from straw, miscanthus, and switchgrass can be relatively corrosive), and these are advantages for most pellet stoves currently marketed. However, Kolarikova et al. (2015) found that utilization of hemp for briquettes was not economically feasible.

NONSEED USE OF HEMP AS LIVESTOCK FEED

As noted in Chapter 8, hemp seed and its derivatives make excellent feed for animals. However, feeding entire plants is another matter because the leaves are covered with resin-producing glands. The herbaceous material (not the seeds) appears to have toxic potential if eaten in very large amounts. While deer, groundhogs, rabbits, and other mammals will nibble on hemp plants, mammals generally do not choose to eat hemp. Jain and Aroroa (1988) fed marijuana refuse to cattle and found that the animals "suffered variable degrees of depression and revealed incoordination in movement." Driemeier (1997) reported that four of five cattle died after consuming bales of dried marijuana leaves. Companion animals, especially dogs, occasionally are intoxicated as a result of consuming relatively small amounts of herbal marijuana but rarely are seriously harmed (foods such as brownies prepared with marijuana extracts can result in large amounts being consumed and consequent greater risk; see Chapter 12).

The EFSA Panel on Additives and Products or Substances Used in Animal Feed (2011), a comprehensive committee-based evaluation of hempseed as animal feed, came to the following evaluation (pertaining to low-THC hemp): The whole hemp plant (including stalk and leaves), due to its high fiber content, would make a suitable feed material for ruminants (and horses), and daily amounts of 0.5 to 1.5 kg whole hemp plant dry matter could likely be incorporated in the daily ration of dairy cows. However, due to observations that cows so fed secreted milk with THC and concern that other products (meat, eggs) could be similarly affected, the panel recommended that "whole hemp plant, hemp hurds, hemp flour (ground dried hemp leaves) should be placed on the list of materials whose placing on the market or use for animal nutritional purposes is restricted or prohibited."

Letniak et al. (2000) conducted an experimental trial of hemp as silage. No significant differences were found between yield of the hemp and of barley/oat silage fed to heifers, suggesting that fermenting hemp plants reduces possible harmful constituents.

ORNAMENTAL USE

Hemp has, at times in the past, been grown simply for its ornamental value. The short, strongly branched cultivar Panorama (Figure 10.2), bred by Iván Bósca, was commercialized in Hungary in the 1980s (*Journal of the IHA* 1994) and has been said to be the only ornamental hemp cultivar available. It has had limited success, of course, because there are very few circumstances that permit private gardeners to grow *Cannabis* as an ornamental today. By contrast, beautiful ornamental cultivars of opium poppy are widely grown in home gardens across North America, a very curious situation widely tolerated by the police and governments despite their illegality according to a strict interpretation of certain legislation. Tall fiber *C. sativa* has been employed in France as an ornamental maze (Figure 10.3). Doubtless, should it became legally permissible, many would grow hemp as an ornamental.

FIGURE 10.2 Panorama—the world's only ornamental hemp cultivar, with the breeder, Iván Bócsa. Photo courtesy of the late Professor Bócsa. According to De Meijer (1998, based on information provided by Bócsa), this arose as a back-cross hybrid between a globe-shaped dwarf mutant of a Lebanese drug strain and the monoecious cultivar Fibrimon.

FIGURE 10.3 Ornamental hemp maze in France. Photo by Barbetorte (CC BY 3.0).

HEMP AS A PROTECTIVE COMPANION PLANT

"Companion plants" are pairs of species, at least one of which benefits by being grown near the other. Sometimes, plant species are toxic to mobile insects, and their mere presence seems to safeguard nearby plants to some degree. Sometimes, plant species are allelopathic: toxic chemicals diffuse from them, particularly from the roots, and suppress nearby species, such as harmful soil organisms. As reviewed by McPartland (1997a), hemp near cotton and vegetable crops has been shown to protect them to some degree against certain of their pests, particularly nematodes, the reduction of these making the soil less threatening for subsequent different crops.

Crop rotation is a form of companion planting in which one species is deliberately planted in the same place as its benefactor grew the previous season. For most crops, rotation tends to reduce pests and diseases, in part because the unwanted organisms that build up on a given crop tend to have less success on the next season's different crop. As noted in Chapter 7, hemp is best alternated with several different crops in sequence over a period of years. Hemp benefits particularly from being planted where legume crops with nitrogen-fixing bacterial associates, such as alfalfa and clovers, have grown. In turn, hemp can benefit other crops in the rotation.

NATURAL PESTICIDES

McPartland (1997a) reviewed research on the pesticide and repellent applications of *Cannabis*. Powdered material and extracts of *C. sativa* have been used as antifeedants, repellents, and insecticides (Bouquet 1950; McPartland 1997b). Mukhtar et al. (2013) found that *C. sativa* is effective against nematodes. Gorski et al. (2009) found that hemp oil repelled aphids. There are numerous studies of the effects of crude preparations of cannabis on various classes of noxious organisms, but there is often insufficient evidence to attribute the effects to particular chemicals present. Nevertheless, dried plant parts and extracts of *Cannabis* have received rather extensive usage as homemade repellents in the past, raising the possibility that research could produce formulations of commercial value. Natural plant pesticides tend to be relatively benign to the environment and biodiversity compared to synthetics, so they are often welcomed in the marketplace. However, the commercial value of cannabis extracts is uncertain at present.

HEMP JEWELRY

Hemp jewelry combines colored hemp twine and (usually) colorful beads in the form of anklets, bracelets, necklaces, purses, and various other (usually female) accessories (see Figure 10.4). Wearing hemp jewelry has been popular among youth, frequently as an expression of "eco-chic." As expressed by Dvorak (2004), "Around college campuses and at concerts and village greens, hemp twine jewelry worn by the younger generation has become ubiquitous. Many people wearing it consider this to be a statement that they are for the environment and against cannabis hemp prohibition. Others simply wear it because it's cool." Complex knitting and knotting are often employed. Kits

FIGURE 10.4   Examples of hemp jewelry. Headbands by Totally Hemp (CC BY 2.0).

FIGURE 10.5 Gorgeous handbags made with hemp, exhibited in the Yunnan Nationalities Museum, Kunming, Yunnan, China. Photo by Daderot (released into the public domain).

for preparing hemp jewelry are widely available and so are preparation instructions on the Internet and in books (e.g., Baskett 1999; Lunger 1999). The recent interest in using hemp fiber for arts and crafts preparations mainly for women's attire is ironic, given that since ancient times, the Chinese have employed hemp to prepare astonishingly artistic items of the same nature (Figure 10.5).

HEMPSEED AS FISH BAIT

In some European countries, hempseed is considered to be an outstanding fish lure and is often sold in bait shops as dry, sterilized seeds. The seeds are usually prepared by boiling to the point that a hook will pass easily through them. Many anglers simply toss the intact hulled or ground-up seeds into the water to attract fish, using hooks baited with other materials to actually catch the fish. Instructional videos on the use of hempseeds for fishing are available online. In the United Kingdom, where hempseed is a popular fish bait, some fishermen with hempseeds intended for fishing have been arrested and charged with possession of a narcotic—an obvious misapplication of the law.

HEMPSEED AS A GROWTH MEDIUM FOR FUNGI AND OTHER MICROORGANISMS

Mycologists frequently employ boiled hempseed in sterile water to culture aquatic fungi, and indeed, this is the medium of choice for numerous water molds. Since hempseed is very attractive to many fungi and some other microorganisms that grow in water, hempseed is also often placed in natural aquatic (and even terrestrial) habitats for a period as "bait," and after the seeds have been colonized by these organisms, the material can be examined to determine exactly which species are present in the location.

HEMP AS AN AGRICULTURAL POLLEN BARRIER

One of the most curious uses of hemp, occasionally observed in Europe, is as a tall fence to physically prevent pollen transfer in commercial production of seeds. Isolation distances for ensuring that seeds produced are pure are considerable for many plants and are sometimes too large to be practical. At one point in the 1980s, the only permitted use of hemp in Germany was as a fence or hedge to prevent plots of beets being used for seed production from becoming contaminated by pollen from ruderal beets. The high and rather impenetrable hedge that hemp can produce was considered unsurpassed by any other species for the purpose. As well, the sticky leaves of hemp were thought to trap pollen. However, Saeglitz et al. (2000) demonstrated that the spread of beet pollen is only partly prevented by hemp hedges. Tall fiber varieties of hemp were also once used in Europe as wind-breaks, protecting vulnerable crops against wind damage. Although hemp plants can lodge (bend over permanently), on the whole, very tall hemp is remarkably resistant against wind.

EDIBLE SHOOTS (STEMS AND FOLIAGE)

The extensive edible uses of the seeds and their fixed seed oil and the possible use of seedlings as sprouts are discussed in Chapter 8. The flavoring use of the essential oil is discussed in Chapter 9. As noted in Chapter 4, honey bees collect pollen from *C. sativa*, and so some of this may end up in honey, although almost certainly in insignificant amounts. Recreational marijuana is widely incorporated into edible preparations (Chapter 12). However, eating the foliage, stems, or floral material is unpleasant and at least slightly toxic, and edible material for humans generally has been prepared by extracting THC (usually in fats such as butter when used in brownies and the like or in alcohol in liquid preparations), so that the marijuana itself is not eaten. Despite the lack of palatability and potential toxicity, there are statements from various authors (not well documented) that leaves, twigs, or flowers are occasionally eaten (Cheatham et al. 2009, pp. 38–39). As noted in Chapter 12, bhang is a traditional Asian beverage made with chopped cannabis foliage.

CURIOSITIES OF SCIENCE, TECHNOLOGY, AND HUMAN BEHAVIOR

- The use of plant biomass to generate fuels is widely regarded as an important environmentally beneficial way to reduce the consumption of fossil fuels and the consequent generation of greenhouse gases contributing to greenhouse gases and climate change. Until 2012, the European Union (EU) heavily subsidized crops such as hemp and flax that are regarded as good for the environment. In the 1990s, the EU provided a subsidy of more than $1000 per hectare (approximately $400/acre) for farmers who grew flax and hemp. In a classic example of how good-intentioned legislation can be abused, in 1998, thousands of hectares of flax were grown in Spain and Ireland, the subsidies were collected, and all of the crops were simply unharvested or burned in the fields, worsening environmental pollution.
- Supercapacitors are energy storage devices using activated carbon electrodes to provide quick bursts of power. They are used currently in braking systems for buses and fast-charging flashlights. Heating and chemically treating hemp phloem fibers have been found to produce a material with potential to store much more energy in supercapacitors than the activated carbon electrodes currently used (Wang et al. 2013).
- According to *Popular Mechanics* (1938), "Thousands of tons of hemp hurds are used every year by one large powder company for the manufacture of dynamite and TNT." Hemp hurds (the woody interior of hemp stems) are highly absorbent and so would have been useful to soak up the explosive chemicals in cylinders or sticks. However, the use of organic materials such as hemp hurds and sawdust to hold the explosives has generally been discontinued in favor of more stable absorbents.

11 Cannabis Chemistry: Cannabinoids in *Cannabis*, Humans, and Other Species

Plants produce thousands of chemical compounds, and *Cannabis sativa* is no exception. Chemical aspects of the stem fiber, oilseeds, and essential oil are dealt with in other chapters. This chapter deals with the cannabinoids, the chief chemicals of interest. As will be discussed, there is evidence of compounds that are structurally similar to the cannabinoids in a few other plants.

Much more importantly, there are analogous (structurally dissimilar but functionally similar) compounds, also considered to be cannabinoids, in humans and other animals, that affect metabolism in countless ways. Indeed, as discussed in Chapter 13, it is now clear that ingestion of plant and synthetic cannabinoids affects humans, for better or worse, by altering the body's metabolic engines, which normally employ internally produced cannabinoids. Also, in many plants, there are chemicals that are structurally dissimilar to the cannabinoids of *Cannabis* or humans that can also influence animal metabolism (including human physiology) by influencing the metabolic system that handles cannabinoids. However, this chapter is mostly concerned with the cannabinoids of *C. sativa*: where they are produced in the plant, variation of concentrations in different kinds of plant and in different growth circumstances, and their basic chemistry.

GLANDULAR TRICHOMES OF *CANNABIS*: THE PLANT'S DRUG FACTORIES

Most plant species have very small epidermal appendages termed "trichomes" on the aerial parts, widely considered to be protective against pathogens and arthropod herbivores (Levin 1973), although numerous other hypotheses have been proposed to explain their presence (Wagner 1990; Werker 2000; Theis and Lerdau 2003; Wagner et al. 2004). Trichomes are sometimes termed "hairs" because they are often hair-like, but most biologists reserve the term hair for animals. Nonglandular trichomes were discussed in Chapter 6; this section deals with glandular trichomes (which, like human glands, synthesize particular chemicals).

About 30% of flowering plants possess "glandular trichomes," producing secondary chemicals, usually at the tip of the structure, often in distinctive head-like containers (Dell and McComb 1978; Glas et al. 2012). (Secondary chemicals are organic compounds produced during metabolism, which are not directly involved in essential biological structures or in normal development or reproduction.) The substances manufactured are frequently known to serve the plant as protective agents but are also often immensely useful to humans as natural pesticides, food additives, fragrances, and pharmaceuticals (Duke et al. 2000). The psychoactive chemicals of *Cannabis* (cannabinoids, principally THC) are produced in specialized tiny secretory trichomes, which are almost always multicellular (the nonglandular trichomes of *C. sativa* are unicellular).

There is by no means agreement by botanists how multicellular secretory trichomes should be classified, either in plants generally or *Cannabis* specifically. The most important criterion for distinguishing classes of trichomes in *Cannabis* is stalk length, and size is next in importance.

As many as three classes of epidermal secretory glandular trichomes can be distinguished on the basis of basal stalk. The so-called stalkless or sessile type, which hardly resembles a hair-like structure, may have a very short stalk that is not visible as it is hidden under the gland head (Figure 11.1). These glandular trichomes tend to be comparatively small. The long-stalked

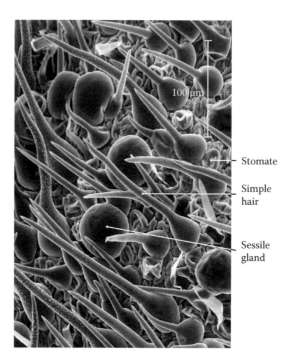

FIGURE 11.1 Scanning electron micrograph of the abaxial (lower) surface of a young leaf of *C. sativa*, showing simple unicellular hairs and stalkless (sessile) multicellular secretory glands (both representing kinds of "trichomes"). Prepared by E. Small and T. Antle.

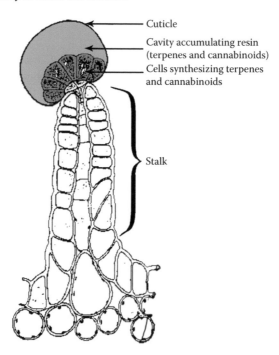

FIGURE 11.2 Diagram of longitudinal section of a long-stalked secretory gland of *C. sativa*. Resin containing cannabinoids is synthesized by the cells in the basal part (shown in red) of the more or less spherical head and accumulates in the cavity (shown in green) above these cells within the external membrane covering the head. Sometimes, the head breaks open and the resin seeps over the adjacent plant tissues. Adapted from Briosi and Tognini (1894).

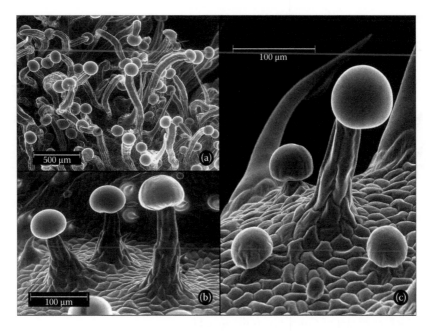

FIGURE 11.3 Scanning electron micrographs of secretory glands of the abaxial (lower, outer) epidermis of perigonal bracts (i.e., the single bract covering each female flower) of high-THC forms of *C. sativa*. (a) Dense concentration of long-stalked glands. (b) Three long-stalked glands. (c) A long-stalked secretory gland (center) around which are three short-stalked multicellular glands. Also shown is a nonglandular hair (a unicellular structure). Resin containing cannabinoids is synthesized in the spherical heads of the glandular trichomes. The perigonal bracts are the most intoxicating plant organ of high-THC forms of the plant. Prepared by E. Small and T. Antle.

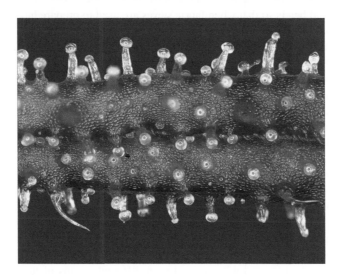

FIGURE 11.4 Light microscope photo, edge view, of a young leaf of *C. sativa*, cultivar FINOLA, covered by long-stalked secretory glandular hairs.

(Figures 11.2 through 11.4) and short-stalked (Figure 11.3c, in part) types are distinguished arbitrarily on the basis of stalk length and tend to be comparatively large. Additionally, a very distinctive kind of glandular trichome occurs on the anthers (Figure 11.5). In all of these cases, the essential part of the gland is a more or less hemispherical head, sometimes compared in size to the head of a pin. Inside the head at its base there are specialized secretory "disk cells," and above these

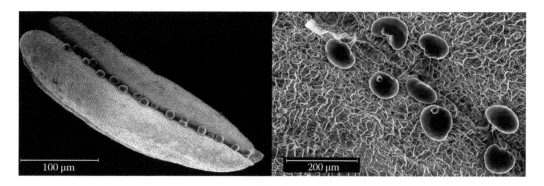

FIGURE 11.5   Scanning electron micrographs of short-stalked secretory glands on an anther of *C. sativa*. Left: A row of glands separating the pollen-containing segments. Right: Close-up of several of the glands. Prepared by E. Small and T. Antle.

there is a noncellular cavity where secreted resin is accumulated, enlarging the covering sheath (a waxy cuticle) of the head into a spherical blister (Figure 11.2). The resin is a sticky mixture of cannabinoids and a variety of terpenes. In marijuana varieties, the resin is rich in the cannabinoid THC, the chief intoxicant of *Cannabis*, as discussed later.

Happyana et al. (2013) found that the stalks of stalked glandular trichomes contained traces of cannabinoids, but whether this is the result of contamination is unclear, and it does seem that most, if not all, of the cannabinoids are synthesized in the secretory disk cells. Lanyon et al. (1981) showed that nearly all of the cannabinoids of the stalked glands occur in the noncellular cavity above the disk cells. Its sheath may eventually rupture, releasing resin onto the surface of the plant. Hot conditions seem to favor release of the resin, but apparently there has been selection for strains that retain resin within the gland heads so that when fabric sieves are used to prepare hashish (as described in Chapter 12), they will not become clogged with sticky resin. However, strains that produce extruded sticky resin have been favored when leather or hands are used to rub off the resin for hashish preparation (Clarke 1998a; McPartland and Guy 2004a).

Various authors (e.g., Clarke and Watson 2002; Mahlberg and Kim 2004) have characterized a narrowed portion of the top of the stalk, just below the base of the head, as an "abscission layer." In the living state, the gland heads always burst immediately when touched but do not readily fall off from the living plant, so just why stalked glandular trichomes develop a constricted area just beneath the gland heads is unclear adaptively. In no way is the "abscission layer" of cannabis stalked trichomes comparable to the abscission zone at the base of the foliage of deciduous trees or at the base of fruits that abscise at maturity (as described in detail in Chapter 3 for the fruit of *Cannabis*). There seems no obvious reason why dropping the heads is adaptive from the plant's perspective (which they simply do not do while fresh) but when the plants is dried, the gland heads do fall off very readily when agitated. This facilitates harvesting the heads for hashish preparation, and some strains may have been selected for ease of harvesting the heads for making intoxicating preparations.

The secretory glands differ notably in density on different organs of the plant (high concentrations occur on the lower surface of the young leaves, on young twigs, on the sepals, anthers [Figure 11.5], and especially on the perigonal bracts [Figure 11.3], where they are very dense and productive). Given this distribution, the glands seem to be protective of young and reproductive above-ground exposed tissues (the roots and achenes, which are not exposed, lack glands). Clarke (1998a) observed that marijuana varieties differ widely in the size of glands, but there is evidence that selection of high-THC forms has favored greater gland size, greater gland density, or both (Small and Naraine 2016b). Small and Naraine (2016b) found that a sample of elite pharmaceutical strains all had much larger gland heads than those of industrial hemp cultivars. Mahlberg and

Kim (2004) recorded that the cannabinoid content of the long-stalked glands they examined possessed about 20 times the cannabinoid content of the sessile glands. The glands of *Cannabis* have been described in detail by Potter (2009) and extensively examined by Mahlberg and associates (Hammond and Mahlberg 1977, 1978; Turner et al., 1980, 1981a,b; Mahlberg et al. 1984; Mahlberg and Kim 1991, 1992, 2004; Kim and Mahlberg 1995, 1997, 2003). It has been established that cannabinoids are synthesized within the secretory glands, not elsewhere, and transported to the glands (Sirikantaramas et al. 2005; Stout et al. 2012). By contrast, the alkaloid nicotine is synthesized in the roots of tobacco plants and transported to trichomes of the foliage, a phenomenon recorded in several other plants (Vivanco and Baluška 2012).

Comparatively small glands with very small heads (sometimes unicellular, typically less than 20 microns in diameter) and very short stalks (sometimes just two cells) often occur over much of the plant. Such glands are often termed "bulbous," in contrast to the larger glands described previously, which are termed "capitate" (meaning head-like). (Bulbous and capitate are not obviously distinguishable terms and can be misleading.) The small glands could simply have failed to develop into larger glands. They are so small that they cannot produce appreciable amounts of the cannabinoids. A long-stalked gland with a capitate secretory head is widely referred to as a "stalked-capitate trichome" and is somewhat reminiscent of a golf ball sitting atop a golf tee.

Most THC in cannabis drug preparations is located in the resin heads of the stalked glandular trichomes, which cover the protective perigonal bracts enveloping the pistils and seeds. Stalked glands are located mainly on the perigonal bracts in the inflorescence but also often occur on the veins of the underside of the leaves. Small and Naraine (2016b) found that after harvest, the resin heads shrink in diameter in exponential decay fashion under ambient room conditions, losing about 15% of their diameter in the first month, rising to 25% over the first year, 30% by 40 years, and 33% after a century (Figure 11.6). An equation accounting for the asymptotic curve descriptive of the progression of shrinkage was determined (original gland head diameter in microns = observed diameter) divided by ([0.5255 + 0.4745 multiplied by time in days to the power −0.1185]), so that if the age of a specimen is known, the original diameter of the gland heads in the fresh state can be extrapolated.

There is controversial evidence for trace amounts of cannabinoid production outside of the epidermal glands. Laticifers (latex-containing internal tissues or cells) occur in the foliage and stems (Zander 1928). These are of the unbranched, nonarticulated form, made up of an elongated secretory cell producing a kind of latex. Furr and Mahlberg (1981) reported that they detected cannabinoids

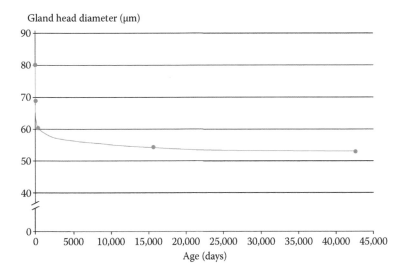

FIGURE 11.6 Time-shrinkage curve for gland head diameter, based on Small, E., Naraine, S.G.U., *Gene. Resour. Crop Evol.*, 63, 349–359, 2016.

in laticifers of *C. sativa*. However, because of the possibility of contamination, the presence of cannabinoids in the laticifers needs to be verified. Veliky and Genest (1972), Hemphill et al. (1978), Pacifico et al. (2008), Flores-Sanchez et al. (2009), and others found no production of THC in tissue cultures, suggesting that nonsecretory cells do not produce cannabinoids. However, some experiments have demonstrated production of cannabinoids in cell cultures of *Cannabis*, but in extremely limited amounts (Heitrich and Binder 1982; Hartsel et al. 1983; Loh et al. 1983; Mandolino and Ranalli 1998). Braemer and Paris (1987) showed that cell cultures could at least transform some cannabinoids to other cannabinoids.

CANNABINOIDS IN *CANNABIS SATIVA*

CLARIFICATION OF THE TERM "PSYCHOACTIVE"

In Chapter 13, the section "Pharmacological Terminology for Marijuana" discusses at length the various terms that have been applied to describe the psychological states produced by marijuana. This chapter provides information on the effect of several of the cannabinoids, the most significant chemicals in *C. sativa*. Before proceeding, it is important to clarify the meaning of "psychoactive," a term that is very widely employed for cannabis in a selective way. Etymologically, "psychoactive" should mean "psychologically active," i.e., significantly affecting mental status, which could include any of mood, emotion, perception, and cognition. In the context of discussing cannabis, a tradition has developed of employing "psychoactive" to refer exclusively to the euphoria (the "high" or intoxicant state) produced by marijuana, while ignoring other significant induced mental states, most particularly sedation, but also such other effects as anxiety (reduction or increase).

In reading most of the pharmacological and experimental cannabis literature, it is therefore important to consider how authors are employing the term "psychoactive." Most egregiously, the cannabinoid cannabidiol (CBD) is almost universally referred to as "nonpsychoactive," which indisputably is incorrect (i.e., although CBD will not induce marijuana intoxication, depending on dosage and context, it will modify psychological status). Burstein (2015) pointed out, "the structure of CBD was not completely elucidated until 1963. Subsequent studies resulted in the pronouncement that THC was the 'active' principle of *Cannabis* and research then focused primarily on it to the virtual exclusion of CBD. This was no doubt due to the belief that activity meant psychoactivity that was shown by THC and not by CBD. In retrospect this must be seen as unfortunate." Piomelli and Russo (2016) stated, "although cannabidiol is nonintoxicating, it certainly has antianxiety, antipsychotic, and even antidepressant effects, so properly they must be considered psychoactive."

CLARIFICATION OF THE TERM "DRUG"

Occasionally, kinds of *C. sativa* used for production of marijuana (always high in THC) are referred to as drug strains, in contrast to kinds grown as hemp (almost always low in THC but high in CBD). Indeed, Clarke and Merlin (2015) recommended this nomenclatural distinction. Restricting "drug strain" to high-THC *C. sativa* often implies that this kind of *Cannabis* is consumed for recreational or spiritual purposes. However, employing high levels of THC as a basis for defining "drug" kinds of *C. sativa* may reflect either a pejorative way of implying that marijuana is harmful or a mistaken notion that THC is the principal *Cannabis* cannabinoid of medicinal value. As detailed in Chapter 13, CBD has at least as much potential for use in drugs as THC and indeed is currently under exploration for treatment of many more medical conditions.

BASIC CHEMISTRY OF THE CANNABINOIDS

Cannabis contains an unusual class of terpenophenolic secondary metabolites, defined as "cannabinoids" (Mechoulam and Gaoni 1967). It was once believed that these natural plant cannabinoids constituted all cannabinoids, but over time, broader conceptions of "cannabinoids" have developed. Cannabinoids have become so significant that they are the subject of thousands of articles published annually (Bab 2011). More than 100 cannabinoids in *C. sativa* have been described (Grotenhermen and Russo 2002; ElSohly and Slade 2005; ElSohly 2006; Radwan et al. 2009; De Meijer 2014). According to ElSohly and Gul (2014), over 150 cannabinoids have been recorded for *C. sativa*.

Numerous publications discussing the kind of cannabinoids found in *C. sativa* state that they are unique to *Cannabis*, but this is incorrect. There are reports in the literature that cannabinoids occur in other plants—in the composite *Helichrysum* (Bohlmann and Hoffmann 1979; Lourens et al. 2008), in the liverwort *Radula* (Toyota et al. 1994, 2002), and in the legume *Glycyrrhiza foetida* Desf. (Raedestorff et al. 2012). Cannabinoid-like compounds have also been reported in the legumes *Amorpha* (Kemal et al. 1979), *Machaerium multiflorum* (Muhammad et al. 2001), and *Desmodium canum* (Botta et al. 2003). The fungus *Cylindrocarpon olidum* produces cannabiorci-chromenic acid (Quaghebeur et al. 1994).

However, virtually all specialists on the cannabinoids are of the view that they are more characteristic of *Cannabis* than any other plant, and the major cannabinoids of *C. sativa* occur only in this species. Additional chemical investigation is required to establish whether some of the cannabinoids of *C. sativa* that have been described occur as original metabolic products of the plant or are degenerative products or artefacts.

Delta-9-THC (Δ^9-THC, or simply THC) and CBD are the plant cannabinoids of most importance. THC is the principal intoxicant constituent of *C. sativa*, and CBD, which is not intoxicating, is the principal cannabinoid of hemp (nonintoxicating forms of *C. sativa*). Both compounds have numerous medicinal properties, as presented in Chapter 13. THC was isolated from *C. sativa*, molecularly characterized, and even synthesized in the laboratory of Raphael Mechoulam (Figure 11.7) in the mid-1960s (Gaoni and Mechoulam 1964). As detailed in other chapters, plants that have been selected for fiber and oilseed characteristics usually produce resin in the secretory glands, which

FIGURE 11.7   Professor Raphael Mechoulam, chemist at the Hebrew University of Jerusalem, who, along with colleagues, has conducted pioneering research on the cannabinoids of *C. sativa* and the endocannabinoids of humans. Photo provided by Dr. Mechoulam.

has no or limited amounts of THC, but high amounts of CBD. In Chapter 18, these are placed in *C. sativa* subsp. *sativa*. In contrast, plants that have been selected for intoxicating drug properties are generally high in THC and are placed in *C. sativa* subsp. *indica*. As discussed in Chapters 12 and 13, "sativa type" marijuana strains have no or very little CBD, while "indica type" marijuana strains frequently have substantial amounts of both THC and CBD (Figure 11.8).

In the living plant, the cannabinoids exist predominantly in the form of carboxylic acids (i.e., a *−COOH* radicle is attached to the molecule). These decarboxylate into their neutral counterparts (the molecules lose the acidic *−COOH* radicle, leaving an H atom), as shown for THC in Figure 11.9, under the influence of light, time (such as prolonged storage), alkaline conditions, or when heated, as occurs when marijuana is smoked or cooked (e.g., in brownies). Romano and Hazekamp (2013) noted that cooking marijuana in an oven at 145°C for 30 minutes substantially decarboxylated the material, with very little degeneration to cannabinol (CBN).

The more important cannabinoids are shown in Figure 11.10 (see the legend for abbreviations of the cannabinoid names). These have a basic 21-carbon skeleton (22 in the carboxylated forms). The designation Δ^9-THC employs formal chemical nomenclature for pyran-type compounds (the "dibenzopyran system," favored in North America; in an alternative nomenclature system often employed in Europe, the "monoterpenoid system," based on regarding the cannabinoids as substituted monoterpenoids, this is known as Δ^1-THC). The "Δ^9" in Δ^9-THC points out the presence of a carbon–carbon double bond (indicated by Δ) located between carbon atoms 9 and 10, as indicated by the exponent 9 in Δ^9 (Khan et al. 2012).

THC is the world's most popular illicit chemical and indeed the fourth most popular recreational chemical after caffeine, ethyl alcohol, and nicotine, all of which are addictive. Other THC isomers also occur, particularly Δ^8-THC, which is also euphoric. Δ^8-THC is much less abundant in *C. sativa*, occurring only in trace amounts if at all and is somewhat less potent than Δ^9-THC,

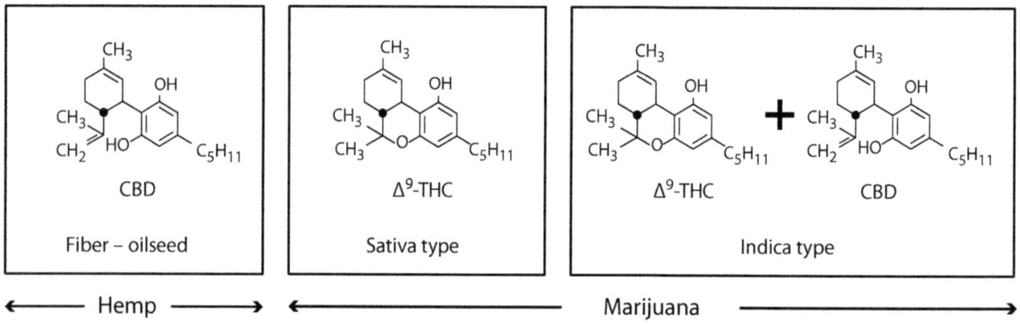

FIGURE 11.8 Three basic categories of cultivated plants of *C. sativa* based on predominant cannabinoids.

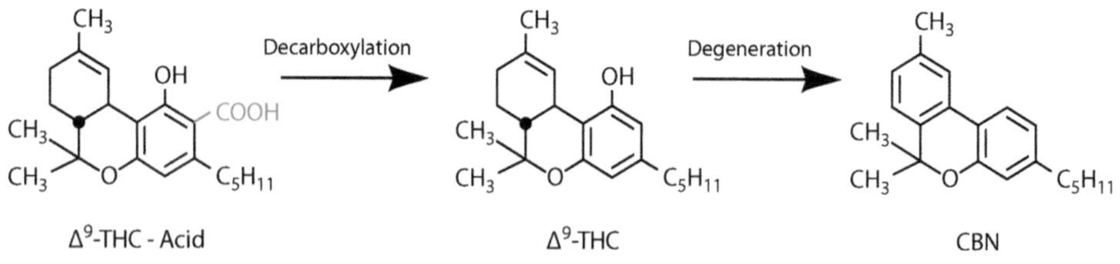

FIGURE 11.9 Spontaneous alteration of THC with time and/or storage conditions. In the living plant, Δ^9-THC is carboxylated (with a –COOH moiety attached to the benzene ring), and in this form (known as THCA), it is only marginally psychoactive. With mild heat (as applied when smoking or vaporizing marijuana), drying, or aging after harvest, Δ^9-THC-COOH decarboxylates to form CO_2 and Δ^9-THC, which is psychoactive. THC degenerates with time (especially at higher temperatures and when exposed to oxygen) to CBN, which has much reduced psychoactivity.

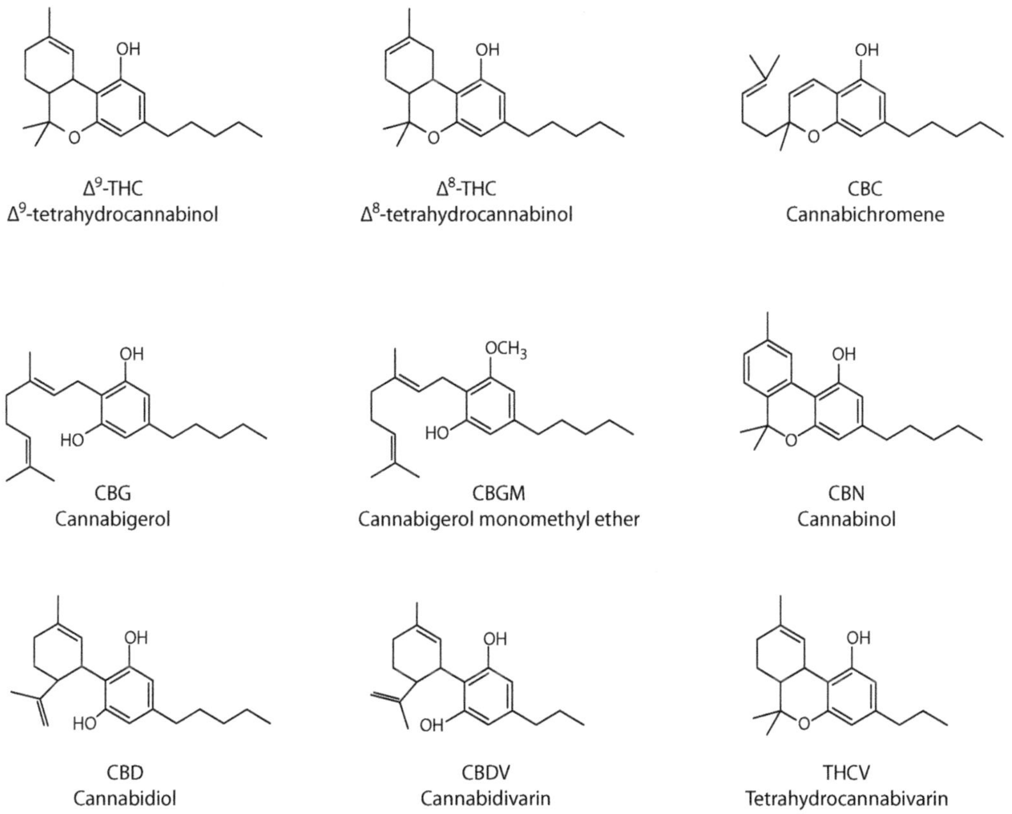

FIGURE 11.10 Chemical diagrams (decarboxylated forms) of several of the well-known cannabinoids. Δ⁹-THC is the chief intoxicant chemical and predominates in marijuana strains, while the isomer Δ⁸-THC, which is somewhat less intoxicant, is usually present in no more than trace amounts. CBN is a frequent degradation or oxidation product, usually not appreciably present in the fresh plant; it has some intoxicating ability. The remaining compounds shown are not intoxicant or at least not appreciably so and, except for CBD, are usually present in trace amounts or are absent. CBD is the chief nonintoxicant chemical and predominates in hemp strains. As described in Chapter 13, CBD has an astonishing variety of therapeutic properties. CBC is often detected in high-THC strains. CBG is the biosynthetic precursor (in the carboxylated form, as shown in Figure 11.11) of THC and CBD. It is more often observed in nonintoxicant strains than in high-THC strains. CBGM has been detected especially in populations from northeastern Asia. CBDV has been reported in populations from central Asia. THCV is usually present in trace amounts but occasionally in significant quantities, especially in populations from Asia and Africa.

although somewhat more heat-stable than Δ⁹-THC. CBN, the principal degeneration or oxidative breakdown product produced when Δ⁹-THC ages (Δ⁸-THC is notably more stable and persists in old material), is generally considered to have limited psychoactive potential—perhaps about 25% of the potency of THC (Piomelli and Russo 2016). Tetrahydrocannabivarin (THCV) "is certainly psychoactive, but rarely seen in high titer in commonly available *Cannabis* strains" (Piomelli and Russo 2016). The other molecules shown in Figure 11.10 are not euphoriant and, if present, almost always occur only in small concentrations.

DECREASE OF INTOXICANT ABILITY BY TRANSFORMATION OF THC TO CBN

THC converts gradually over time to other compounds, principally CBN, as illustrated in Figure 11.9. The ratio of THC to CBN is a crude indicator of the age since harvest of stored marijuana

(Ross and ElSohly 1997). CBN has been estimated to have 10% of the activity of THC (Pertwee 2008), and because it is much less psychoactive, marijuana preparations have a limited shelf life (Fairbairn et al. 1976; Lewis and Turner 1978; Narayanaswami et al. 1978; Harvey 1990; Trofin et al. 2011, 2012). High temperature, moisture (high humidity), light, and oxygen all contribute to gradual lowering of the quality (potency) of cannabis drug preparations, but storage in a dark, cool place with exclusion of air minimizes loss of activity (Clarke 1998a). It appears that the intact gland heads act as microcontainers, providing some protection against degradation of THC, presumably particularly against oxidation. Estimates of the shelf life of cannabis preparations vary considerably. Clarke (1998a) stated that at room temperature, 50% of THC is lost annually in hashish. Lindholst (2010) observed exponential decay of THC content in hashish over four years, with about a one-third reduction in the first year. However, Health Canada (2013) stated that only a third of the THC content of marijuana stored at room temperature is lost over a five-year period. Turner et al. (1973) measured the decomposition rate of THC in flowering tops. They stored the buds in darkness for two years, in glass jars with aluminum screw-caps. At room temperature (22°C), THC decomposed at a rate of 6.9% loss/year. In a refrigerator at 4°C, the loss dropped to 5.4% per year. In a freezer at −18°C, the loss decreased to 3.8% annually.

VARIATION OF CANNABINOIDS AMONG DIFFERENT KINDS OF CANNABIS PLANTS

There is a general inverse relationship in the resin of *Cannabis* between the amount of THC present and the amount of CBD. Whereas most drug strains contain primarily THC and little or no CBD, fiber and oilseed strains primarily contain CBD and very little THC.

There have been numerous studies of cannabinoid variation, mostly employing the predominance of either THC or CBD respectively as indicators of intoxicating marijuana kinds and nonintoxicating hemp kinds (for examples, Fetterman et al. 1971; Small and Beckstead 1973a,b; Small et al. 1975; Avico et al. 1985). Many publications have recognized "chemical phenotypes" based particularly on ratios of THC and CBD in the resin or on the presence of one of the less common cannabinoids. In Chapter 18, two subspecies are recognized using THC content for separation.

Cannabis sativa subsp. *sativa* has limited THC, and *C. sativa* subsp. *indica* has appreciable THC. A dividing line of 0.3% (dry weight content in the inflorescence or young infructescence) was established by Small et al. (1976) based on study of variation in several hundred populations and subsequently was adopted in the European Community, Canada, parts of Australia, and the U.S.S.R. as a criterion between cultivars that can be legally cultivated under license and forms that are considered to have too high a drug potential (in some countries, the allowable level is currently different). The 113th U.S. Congress enacted the Agricultural Act of 2014 ("farm bill," P.L. 113-79), which provided a statutory definition of "industrial hemp" as the plant *C. sativa* L. and any part of such plant with a Δ^9-THC concentration of not more than 0.3% on a dry weight basis. A level of about 1% THC is considered the threshold for marijuana to have intoxicating potential, so the 0.3% level is conservative, and some jurisdictions (e.g., Switzerland and parts of Australia) have permitted the cultivation of cultivars with higher levels. It is well known in the illicit trade how to screen off the more potent fractions of the plant in order to increase THC levels in resultant drug products. Nevertheless, a level of 0.3% THC in the flowering parts of the plant is reflective of material that is too low in intoxicant potential to actually be used practically for illicit production of marijuana or other types of cannabis drugs. While this criterion is in common use to separate adult plants, the ratio of CBD and THC often suffices to distinguish high-THC and low-THC plants as young as seedlings (Rowan and Fairbairn 1977; Broséus et al. 2010; De Backer et al. 2012). However, Vogelmann et al. (1988) found that the cannabinoids of extremely young seedlings were dominated by cannabichromene (CBC), and De Meijer et al. (2009a) also observed that CBC is often dominant in juvenile plants and young cuttings.

CBC is a frequent minor constituent of highly intoxicating strains of *C. sativa*, especially from Africa, and strains high in CBC have been selected for medicinal experimentation. De Meijer et al. (2009a) provided evidence that CBC is present in substantial amounts in juvenile plants and declines with maturation. These authors found plant variants in which CBC persisted into maturity and noticed that this is associated with a reduced presence of perigonal bracts and secretory glands. Potter (2009) recorded a greater presence of CBC in the small (nonstalked) secretory glands of the foliage than in the large (stalked) glands of the inflorescence.

Cannabigerol (CBG) rarely dominates the resin of *Cannabis* (Fournier et al. 1987). Some geographical races with minor or trace amounts of cannabinoids have been described, notably for cannabigerol monomethyl ether (CBGM) in some northeastern Asian populations, cannabidivarin (CBDV) in some populations from central Asia, and THCV in some collections from Asia and Africa.

CONVERSION OF NONINTOXICANT CBD TO INTOXICANT THC IN THE LABORATORY AND POSSIBLY IN THE HUMAN BODY

The nonintoxicant CBD can be converted to the intoxicant THC by acid catalysis (i.e., acidic conditions stimulate the reaction), a fact which has been known for many years (Adams et al. 1940, 1941). More recent research demonstrated that laboratory conversion could be carried out using a number of acidic reagents (Gaoni and Mechoulam 1966, 1968). Watanabe et al. (2007) found that gastric juice converted CBD to THC and other cannabinoids with limited intoxicant ability, suggesting that some CBD in edible forms of marijuana could be converted to THC in vivo (in animals). The possible conversion of CBD to THC raises the possibility that the illicit cannabis trade could take CBD from industrial hemp and use it to make THC or other intoxicating analogues. However, the chemical reactions required are elaborate and expensive, and there are no reports of illicit attempts to manufacture THC from CBD. A process to convert CBD to THC has been patented (Webster et al. 2008). The fact that stomach acid is capable of converting at least a small amount of CBD to THC in orally ingested edible forms of marijuana may be of more than academic interest, since this can influence both experimental and therapeutic results (Watanabe et al. 2007). However, "although CBD can turn into THC at very low pH, there is absolutely no evidence that this occurs in the human body. GW Pharmaceuticals did extensive pharmacokinetic studies with pure cannabidiol and none of those patients showed any serum levels of THC or its direct metabolites" (Ethan Russo, personal communication).

BIOSYNTHESIS AND GENETICS OF THE CANNABINOIDS OF *CANNABIS*

A pentyl side chain has the formula $-C_5H_{11}$. The biosynthetic pathways of the major cannabinoids with pentyl side chains (CBC, CBD, CBG, and THC) were established in the 1990s. The first event in the pentyl cannabinoid biosynthesis is the production of CBG, produced by condensation of a phenol-derived olivetolic acid and a terpene-based geranylpyrophosphate catalyzed by the enzyme geranylpyrophosphate:olivetolate geranyltransferase (Fellermeier and Zenk 1998). From CBG, Δ^9-THC, CBD, and CBC are synthesized, each by a specific synthase enzyme. The enzyme converting CBG to THC was clarified by Taura et al. (1995). The enzyme converting CBG to CBD was studied by Taura et al. (1996, 1997). An outline of the biosynthesis of the two most important cannabinoids, THC and CBD, is shown in Figure 11.11. For more complete analyses of cannabinoid biosynthesis, see Sirikantaramas et al. (2007), Flores-Sanchez and Verpoorte (2008), Van Bakel et al. (2011), and Gagne et al. (2012).

As emphasized by Hillig (2002), De Meijer et al. (2003), and De Meijer (2014), it is important to distinguish quantitative and qualitative aspects of cannabinoid inheritance. The absolute quantity of cannabinoids produced by an individual plant or by a population (on an acreage basis) depends on growth and developmental traits (such as size and proportion of tissues constituted by secretory glands), which are (a) probably determined polygenically, (b) unrelated to cannabinoid biosynthetic pathways, and (c) subject to strong environmental modification. Qualitative aspects, discussed in

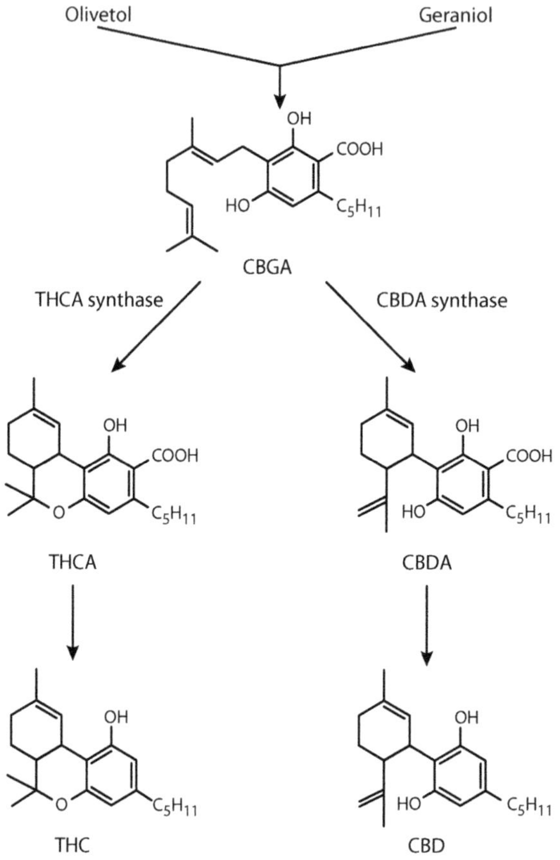

FIGURE 11.11 Biosynthetic pathways of THC and CBD, the predominant cannabinoids of *C. sativa.* CBGA = cannabigerolic acid, THCA = tetrahydrocannabinolic acid, CBDA = cannabidiolic acid (the carboxylated forms, respectively, of CBG, THC, and CBD). Decarboxylation (conversion of THCA to THC and conversion of CBDA to CBD) is not part of the biosynthetic pathway but occurs spontaneously with aging and/or heat.

the next paragraph, relate to the genetic control of genes influencing the relative amounts of the cannabinoids.

First-generation (F_1) hybrids between high-THC marijuana strains and high-CBD fiber cultivars are usually more or less intermediate between the parents. Small and Beckstead (1979) found that numerous first-generation hybrids were more or less intermediate in THC proportion. Beutler and Der Marderosian (1978) crossed a ruderal low-THC form and a race with higher THC and also found that the first-generation hybrids were more or less intermediate, although the progeny tended to fall into two groups with respect to THC proportion, one tending to the parent with the higher level, the other tending to the parent with the lower level. As expected for an outcrossing species, F_1 hybrids frequently show evidence of heterosis for various characteristics. Various authors have observed cannabinoid segregation ratios in F_2 generation hybrids (see literature citations in De Meijer et al. 2003), and as discussed in the next paragraph, this is due to allelic segregation.

Sytnik and Stelmah (1999) suggested that CBD and THC are controlled by closely linked but independent genes. Inheritance of the key cannabinoids THC acid (THCA) and CBD acid (CBDA) was found to be apparently determined by the allelic status at a single locus (referred to as B) (De Meijer et al. 2003; Mandolino et al. 2003; Pacifico et al. 2006). De Meijer et al. (2003) (cf. Mandolino and Ranalli 2002; Mandolino et al. 2003; Mandolino 2004) found evidence that THCA development in *C. sativa* is under the partial genetic control of codominant alleles. Allele B_D is postulated to encode CBDA synthase, while allele B_T encodes THCA synthase. This genetic model holds that plants in which CBDA is predominant have a B_D/B_D genotype at the B locus, plants in which THC is predominant have a B_T/B_T genotype, and plants with substantial amounts of both THCA and CBDA are heterozygous (B_D/B_T genotype). De Meijer and Hammond (2005) found that plants accumulating CBG have a mutation of B_D (which they term B_0) in the homozygous state that encodes for a poorly functional CBD synthase; and De Meijer et al. (2009b) selected a variant of this that almost completely prevents the conversion of CBG into CBD.

The hypothesis that the enzymes that produce THCA (THCA synthase) and CBDA (CBDA synthase) from the same precursor compound, cannabigerolic acid, are controlled exclusively by two alleles of the same gene was challenged recently by Weiblen et al. (2015). They found that THCA synthase and CBDA synthase are encoded by two separate but linked regions. THC-predominant plants simply have a nonfunctional copy of CBDA synthase, so they convert all cannabigerolic acid into THCA. Other evidence also indicates that other genes control the pathways to THCA and CBDA (Van Bakel et al. 2011; Onofri et al. 2015).

Shoyama et al. (2001) transferred the THC-synthase gene from *Cannabis* to a tissue culture of tobacco (*Nicotiana tabacum*), inducing it to convert supplied CBGA to THCA. The same research group bioengineered the enzyme THCA-synthase into yeast, *Pichia pastoris*, which also produced THCA upon feeding of CBGA (Taura et al. 2007). Subsequently, insect cell cultures and other microorganisms have been similarly induced to carry out this final biosynthetic step in THCA synthesis (Zirpel et al. 2015). This raises the prospect that transgenic tobacco (or indeed any other plant) could be smoked as a marijuana substitute! (Of course, an entire suite of enzymes would be required to achieve a total biosynthesis of THC starting from terpenes.)

QUANTITATIVE VARIATION OF CANNABINOIDS IN DIFFERENT PLANT PARTS

This section discusses relative quantitative levels of cannabinoids in different organs of *C. sativa*, i.e., the relative absolute amounts of resin cannabinoids. In most studies, THC has been measured. Populations of some cultivars have proven to be rather uniform in THC content, whereas in others, considerable variation among plants has been found (Mechtler et al. 2004), although populations generally strongly tend to be quite consistent in cannabinoid development.

Cannabinoid levels (concentrations) in the plant generally increase from the seedling stage to the flowering stage (Phillips et al. 1970; Latta and Eaton 1975; Turner et al. 1975; Small 1979b; Hemphill et al. 1980; Kushima et al. 1980).

Absolute cannabinoid content differs in different parts of the plant, increasing in the following order: large stems, smaller stems, older and larger leaves, younger and smaller leaves, and perigonal bracts covering the female flowers (and consequently covering the fruits). Epidermal secretory glands are present on all of the preceding structures, and their relative density, coupled with the relative size of the resin-containing trichome heads, clearly determines the relative presence of cannabinoids in different organs.

The males bear resin glands on their anthers (pollen-containing structures) and filaments (the stalks subtending the anthers) and so are associated with cannabinoids (Fairbairn 1972; Dayanandan and Kaufman 1976). The female flowers are devoid of resin glands (Small and Naraine 2016a). Fetterman et al. (1971) reported that the female flowers tested positive for cannabinoids, but it is erroneous to conclude that they actually contain cannabinoids, because the female flowers are devoid of secretory glandular trichomes. As detailed in Small and Naraine (2016a), resin heads frequently fall off the glandular trichomes of the perigonal bracts and stick to the stigmas, so that the female flowers appear to be rich in THC. There are reports of cannabinoids in minute amounts in seeds (excluding bracts) and roots, but this is also likely due to contamination, as the resin of the plant is easily transferred. THC and other cannabinoids have been reported in the pollen (Paris et al. 1975; Ross et al. 2005), but once again, this is likely the result of contamination from the secretory glands of the anthers (note Figure 11.5).

QUANTITATIVE VARIATION OF CANNABINOIDS IN RELATION TO LENGTH OF GROWTH PERIOD

The relative amount of resin or cannabinoids produced increases (whether expressed on a per plant basis or area basis) over the vegetative and early flowering periods. Photoperiodic adaptation usually determines how long a given race of *C. sativa* can grow in a particular location and hence how much resin (or THC, which is the usual desired cannabinoid) can be harvested. In an indoors, controlled environment, trial and error usually is employed to determine precisely how long a marijuana strain should be grown to most efficiently produce high-THC material (see discussion in Chapter 14). The density of secretory glandular hairs produced on the younger leaves and perigonal bracts and the mean maturity of the resin are what determine the quantity of resin or THC at a given time, and the objective usually is to identify the stage of maximum production before cannabinoid degradation becomes significant.

QUANTITATIVE VARIATION OF CANNABINOIDS IN RELATION TO ENVIRONMENTAL FACTORS

Several elementary considerations should be kept in mind in considering quantitative aspects of resin production in *C. sativa*. Bigger plants, generally speaking, produce a larger quantity of resin or THC, so factors that make for vigorous, healthy growth contribute to increased production. Crowding reduces photosynthetic efficiency, and as with many agricultural plants, trial and error has been employed to determine ideal planting densities. The growth and productivity of all plants are determined by supplies of essential factors, including light, warmth, water, nutritional elements, and CO_2—and of course, these necessarily limit the quantity of resin or THC produced. Plants near the equator have the luxury of extremely high sunlight, extremely long seasons, and relatively high temperatures, and so biotypes of cannabis capable of exploiting these conditions (or comparable conditions in a controlled interior environment) can become very large and productive (although not necessarily more productive on a time or area basis).

Of all factors determining the relative quantity of resin production in *C. sativa*, the supply of water has been most discussed. The geographical location of hashish production in Asia in areas of low atmospheric humidity or rainfall has been advanced as evidence that low atmospheric humidity contributes to increased resin production. Bouquet (1950) pointed out that primarily dry areas of Lebanon were devoted to hashish production because they seemed conducive to copious resin production, while the western mountainous side of Lebanon was less suitable because of humid sea winds. De Faubert Maunder (1976) similarly noted that the copious resin production needed for hashish preparation occurred only in a belt of countries from Morocco eastward, including the Mediterranean area, Arabia, the Indian subcontinent, and Indo-China, all areas with sparse rainfall and low humidity. However, the conclusion that rainfall or atmospheric humidity is *causing* the plants to produce more resin does not follow from these observations. As pointed out in Chapter 12, hashish production in Asia requires a dry outdoors environment because moist sticky material cannot be forced through the fine screens used for manufacturing hashish.

QUANTITATIVE VARIATION OF CANNABINOIDS IN RELATION TO STRESS

Stress makes for smaller plants with less biomass and, hence, a lower overall production of cannabinoids per unit area of land occupied. The question arises, however, whether the smaller plants differ in relative concentration of cannabinoids (especially THC) on a weight basis. Coffman and Gentner observed that smaller plants tended to have higher THC levels. Small et al. (1975) found that plants subjected to extreme deficiencies of nitrogen, phosphorus, and potassium were dwarfed (Figure 3.2) but developed comparable THC concentrations to control plants. Haney and Kutscheid (1973) demonstrated that wild hemp populations in Illinois were highest in cannabinoid concentrations when stressed, either by nutrient limitations or by drought, although shading did not have any measurable effect. Latta and Eaton (1975) found that Kansas wild hemp tended to produce increased percentages of THC in less favorable locations. Sharma (1975) observed greater density of glandular trichomes on leaves of *C. sativa* growing in xeric situations. Murari et al. (1983) noted that fiber cultivars grown in drier continental areas developed higher THC levels than the same cultivars grown in a moist maritime region.

The mechanism of how stress affects cannabinoid production is unclear—do stress factors influence the enzymes controlling the rate of production, do they alter the development of the secretory glands, or does stress merely alter the ratio of cannabinoids to other materials constituting the dry weight of the tissues? Stress tends to make the plants drop their lower leaves which are naturally low in THC, and so it is difficult to evaluate the effects of stress on a whole-plant basis.

QUANTITATIVE VARIATION OF CANNABINOIDS IN RELATION TO CHROMOSOME DOUBLING

Doubling chromosome number of plants (usually by treating seeds with the chemical colchicine) has had variable results with respect to cannabinoid content. De Pasquale et al. (1979) reported that colchicine treatment doubled THC concentration and halved CBD concentration compared to control plants. However, Bagheri and Mansouri (2015) found that in colchicine-induced tetraploids (i.e., with double the normal number of chromosomes), the density of trichomes was reduced and so was the concentration of cannabinoids. These authors also noted similar decreases of active chemicals accompanying doubling of chromosome number with respect to the essential oil of mint (*Mentha spicata* L.) and the glycosides of purple foxglove (*Digitalis purpurea* L.).

QUALITATIVE VARIATION OF CANNABINOIDS IN RELATION TO STAGE OF PLANT DEVELOPMENT

There is very limited evidence that cannabinoids differ *qualitatively* in different organs (e.g., that THC is higher in some organs while CBD is higher in other organs). However, as detailed previously in this chapter, CBC has been shown to be present in substantial amounts in some seedling and juvenile plants and to decline with maturation. Seasonal fluctuations in relative proportion of THC and CBD have also been observed (Phillips et al. 1970; Latta and Eaton 1975; Pate 1998a), with differences in male and female plants (Turner et al. 1975).

QUALITATIVE VARIATION OF CANNABINOIDS IN RELATION TO ENVIRONMENTAL FACTORS

NATURE VS. NURTURE IN THE DETERMINATION OF CANNABINOIDS OF *CANNABIS SATIVA*

In the early twentieth century, the idea became popular that the two principal products for which *C. sativa* is grown (fiber from the stems and intoxicating drugs from the foliage and flowering parts) were developed in the plant primarily in response to climate. As (erroneously) expressed by Abel (1980): "Depending on the conditions under which it grows, cannabis will either produce more resin or more fiber. When raised in hot, dry climates, resin is produced in great quantities and fiber quality is poor. In countries with mild, humid weather, less resin is produced and the fiber is stronger and more durable. It is because of these climate-related characteristics that most Europeans knew very little of the intoxicating properties of the cannabis plant until the nineteenth century." As noted in this book, resin quantity and quality are indeed somewhat modifiable by climate and other environmental variables, and the same is true for fiber quantity and quality. Additionally, the length of season available until days become shorter in the autumn plays a large role in plant development. However, the view that natural environmental variables are *more* determinative of the characteristics of the plants than their genetic constitution is quite false and indeed rather naïve. It is quite possible to produce considerable high-quality fiber or considerable high-quality resin in almost any climate that *C. sativa* can be raised, provided that the biotype grown is suitable.

A more subtle interpretation of the greater importance of environment compared to heredity also was also prevalent in the early twentieth century—that environment could induce changes in fiber or resin production over a few generations. As (erroneously) expressed, by Bouquet (1950), "Secretion of resin does not vary in different varieties of hemp, but depends largely on climatic conditions. Non-existent or negligible in Northern Europe, it becomes considerable in the same variety of hemp when grown in hot, dry climates" and "hemp cultivated in the plains gradually loses the property of supplying active resin." Similarly Hakim et al. (1986) stated that a low-THC English hemp variety when grown in the Sudan over two generations increased its concentration of THC. Such reports reflect the view that *C. sativa* strains suited for fiber production in temperate climates, when transplanted to hot, dry climates, would transform over generations into drug cultivars, and vice versa. In essence, this amounts to the rejected theory of "Lamarckian inheritance," which states that environmentally determined characteristics acquired by plants or animals can be passed down intergenerationally to their offspring. (Currently, the field of epigenetics has raised the possibility that some limited aspects of Lamarckian evolution may be valid.) Lamarckian inheritance is often exemplified by the idea that the large muscles acquired by a blacksmith pursuing his physically demanding trade would be passed on to his sons (whether or not they in turn exercised their muscles). As explained in Chapter 18, maintaining the purity of a strain of *Cannabis* requires stabilizing selection and protection from contaminating pollen, and the absence of these probably accounts for observations that *Cannabis* grown in a foreign location seemed to transform remarkably in a few generations. It is clear that although environment does influence the development of the characteristics of *Cannabis*, indeed of all organisms, strains selected for fiber or marijuana characteristics retain their capacities for such production so long as their gene frequencies are maintained. Of course, the ability of *Cannabis* to change genetically as a result of hybridization and selection should not be confused with the concepts of the environment directly altering or permanently changing the characteristics of plants.

ENVIRONMENTAL FACTORS INFLUENCING THE QUALITATIVE DEVELOPMENT OF CANNABINOIDS

Various environmental circumstances can modify, albeit relatively slightly, the qualitative cannabinoid content of the resin of *Cannabis* (i.e., relative proportions of given cannabinoids). Factors that have been examined include temperature (Bazzaz et al. 1975; Braut-Boucher 1980; Sikora et al. 2011), nutrient availability (Haney and Kutscheid 1973; Coffman and Gentner 1975, 1977; Bócsa et al. 1997), light intensity (Potter and Duncombe 2012), ultraviolet (UV) light intensity (Lydon et al. 1987; Pate 1994), light quality (Mahlberg and Hemphill 1983), and photoperiod (Valle et al. 1978). As noted in the previous discussion of cannabinoid development in response to stress, there has also been examination of the effects of limitations of water (atmospheric humidity or soil water) on content of THC, although interpretation of recorded observations is problematical.

Qualitative variation in cannabinoid production seems to be much more influenced by heredity than by environment. The range of relative THC concentrations in the resin, developed by populations of low-THC cultivars (those typically with no more than 0.3% THC) under different environmental circumstances, on the whole is limited, for the most part generally not varying more than 0.2 percentage points when grown in a range of circumstances, and usually less. The range of relative THC concentrations in the resin, developed by populations of high-THC strains under different circumstances, has not been examined nearly as extensively (because such strains are illicit in most countries) but seems to also follow the pattern of not varying greatly when a given strain is raised in different situations.

BREEDING FOR HIGH AND LOW LEVELS OF CANNABINOIDS

HIGH-THC STRAINS

THC (Figure 11.12) has been the subject of breeding to increase or decrease content in plants. Clandestine marijuana breeders, for several decades, have produced "improved" types of drug plants, and hundreds of selections have been named and offered in the illicit trade. Many named selections are described in Rosenthal (2001, 2004, 2007, 2010), Snoeijer (2002), Danko (2010), Grisswell and Young (2011), Oner (2011, 2012a,b, 2013a,b, 2014), and Backes (2014). Because of legal constraints, very few of these appear to possess protected status as accorded by national and international agreements governing registered cultivated varieties and intellectual property. In the Netherlands, some firms are (or were) authorized to distribute drug selections, and there have been some claims for property rights for these. In 1998, a pharmaceutical drug cultivar called Medisins was registered in the Netherlands by HortaPharm, one of the earliest officially recognized drug cultivars, followed by Grace, registered by GW Pharmaceuticals in 2004, both awarded plant breeders rights (Clarke and Merlin 2013). Pharmaceutical varieties developed in the Netherlands by HortaPharm BV were transferred to GW Pharmaceuticals, centered in the United Kingdom, which has plant breeder's rights to at least 30 to 40 selections (Anonymous 2006). GW Pharmaceuticals, the leading developer of cannabis drugs, is developing strains that predominantly produce one of the four major cannabinoid compounds (THC, CBD, CBC, and CBG), as well as varieties with mixed cannabinoid or terpene profiles (Clarke and Merlin 2013). Several GW Pharmaceuticals selections produce single cannabinoids at very high levels, ranging from over 50% to almost 100%, and one clone is cannabinoid-free (De Meijer 2014). Other private firms, especially in the Netherlands, have also selected "medicinal" lines with particular cannabinoid profiles as well as other attributes.

LOW-THC CULTIVARS

Breeding for low-THC cultivars in Europe has been reviewed by Bredemann et al. (1956), Sokora (1979), Bócsa (1998), Bócsa and Karus (1998), and Virovets (1996). Pacifico et al. (2006) were unable to detect cannabinoids in some plants of European fiber cultivars (USO-31 and Santhica 23). However, at present, no commercial cultivar seems to be 100% free of THC, although Holoborodko et al. (2014) point out that some are close to having none. THC content has proven to be more easily reduced in monoecious varieties, which are inbred, than in dioecious varieties, which are outbred. France has been particularly active in breeding low-THC hemp (Fournier 2000). As a strategic economic and political tactic, France has been attempting for several years to have the European Union (EU) adopt legislation forbidding the cultivation of industrial hemp cultivars with more than 0.1% THC, which would mean that primarily French varieties would have to be cultivated in Europe.

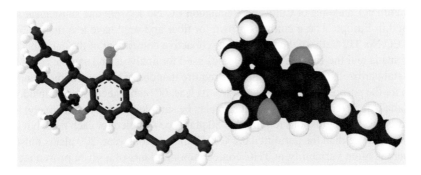

FIGURE 11.12   Molecular models of THC. Left: ball-and-stick model. Right: sphere model. Prepared by B. Brookes.

In theory, a simple way of making plants THC-free is to eliminate the capacity to produce any kind of cannabinoid. De Meijer et al. (2009b) noted that there are two ways of accomplishing this: (1) disrupt the morphogenesis of the glandular trichomes and (2) block one or more biochemical pathways crucial for the formulation of the cannabinoids. Gorshkova et al. (1988) reported on plants that lacked glandular trichomes and plants with odd glandular trichomes (with white heads), both types lacking cannabinoids, but a cultivar or selection in which all plants lack glandular trichomes has not been described. De Meijer et al. (2009b), based on selections from a fiber hemp cultivar (USO-31), discovered a genetic factor (termed a "knockout gene") that completely blocks cannabinoid biosynthesis in *C. sativa*, apparently functioning by preventing the conversion of the phenolic precursors of the cannabinoids into the cannabinoids.

THC TEST PROTOCOLS FOR INDUSTRIAL HEMP

This section provides information regarding maximum permitted THC limits in licensed cultivated hemp plants according to the Canadian and EU regulations, which are more precise than in other jurisdictions. Industrial hemp is widely required to have less than 0.3% THC, a criterion first proposed by Small and Cronquist (1976). As noted further in this section, this criterion was lowered to 0.2% in the EU. In Canada, cultivars demonstrating levels consistently much lower than 0.3% are afforded relief from some of the stringent sampling requirements. In the EU, cultivars that meet the standard have been eligible for subsidization payments. In both political regions, cultivars that are found to regularly exceed the limit are prevented from being cultivated in the future.

Canadian regulations are particularly demanding in terms of how confirmatory measurements need to be taken. The procedures require that samples be collected from the upper, reproductive parts of the plant ("the entire, fruit-bearing part of the plant shall be used as a sample…normally the top one-third of the plant") when the plants are beginning to produce mature seeds ("when the first seeds of 50% of the plants are resistant to compression"). The collection standard calls for screening the material through a sieve of mesh size 2 mm, which eliminates seeds and all but the smallest twigs and essentially represents leaves and floral (perigonal) bracts. In Canada, every grower and every cultivar grown (sometimes individual fields) of industrial hemp must be tested for THC content by an independent laboratory, and under the industrial hemp regulations, fields of hemp with more than 0.3% THC may be destroyed. Once a cultivar has established a proven record of developing less than 0.3% THC, the requirements for testing are reduced. A list of approved cultivars as of 2015 is available at http://www.hanfplantage.de/wp-content/uploads/2015/04/LOAC_2015_EN _-_Health_Canada_-_List_of_approved_Cultivars_Cannabis_Sativa.pdf. Importation of experimental hemp lines (i.e., other than the approved cultivars) requires special licenses (as well as phytosanitary clearance of imported shipments of seeds by the Canadian Food Inspection Agency), and the licenses have to provide an indication that the THC contents are low.

As pointed out by Da Porto et al. (2015), "Only varieties of industrial hemp named in the EU Approved Common Catalogue of Cultivars (Regulation EC No 1251/99 and subsequent amendments), can be planted in Europe. These cultivars grown for fiber and seed have less than 0.2% of Δ^9-THC (Regulation EC No. 1124/2008), the principal psychoactive constituent of the cannabis plant."

In both Canada and the EU, the methodologies used for analyses and sample collection for THC analysis are standardized. Canadian regulations require that one of the several independent laboratories licensed for the purpose conduct the analyses. At least 60 plants need to be sampled. Jurisdictions outside of Canada may require 30, 50, or more plants be sampled (usually collectively) to ensure that THC levels are low. EU regulations require that 50 plants be sampled for each field, but if the cultivar grown developed more than the permitted 0.2% THC the previous year, 200 plants must be sampled.

While there is often variability in THC content among plants of a given provenance, De Meijer et al. (1992) found that a sample of 20 provided a reliable approximation of average THC content. Mechtler et al. (2004) found that most hemp cultivars produce populations in which the THC level is very close in most plants, but occasional ones are significantly divergent.

In the EU, permitted THC levels in hemp were lowered to 0.2% from 0.3% in 2001, and there has been continuing lobbying by French hemp companies to lower the level to 0.1% (which would favor hemp cultivars of France, since they are typically very low in THC). In Ukraine, the maximum THC content allowed for industrial hemp varieties was 0.3% in 1973, 0.2% in 1980, and 0.15% in 1995. Callaway (2008) provides a critique of THC sampling protocols of the EU.

INFORMATIVE ONLINE WEBSITES REGARDING THC SAMPLING PROCEDURES FOR HEMP

Chapter 13 provides a guide to key websites dealing with THC test procedures mainly for high-THC material (for medical marijuana or for law enforcement regarding illicit material). There are no legislated limits for cannabinoid content for medical or recreational marijuana, although one can contemplate that in the future there may be.

For industrial hemp, the concern is that maximum legislated content of THC not be exceeded in plants growing in the fields and often additionally as well in resulting products (especially food). In this context, the following websites provide guides to how plant sampling and laboratory measurement of THC should be carried out.

- Health Canada industrial hemp technical manual—www.hc-sc.gc.ca/hc-ps/pubs/precurs/tech-man-eng.php

    This document applicable to Canada provides extensive information on how samples must be collected in the field, prepared for laboratory analysis, and chemically analyzed.
- EU Commission regulations regarding subsidization of farmers (code: 2009R1122 -EN-03.03.2010-001.001)—www.finola.fi/1122_2009.pdf

    Most of this 76-page document is irrelevant to hemp. The key information regarding plant sampling protocols and THC laboratory analysis procedures is found in Annex 1 (pages 68 and 69).

ADAPTIVE PURPOSE OF THE CANNABINOIDS

The natural adaptive functions in *C. sativa* of the abundant secretory glands, of the large volume of resin they produce, and of the cannabinoids in the resin have not been established, although there is no shortage of hypothesized explanations. The glands are rich in terpenes, which are very common in higher plants and are known to be protective against many harmful organisms, so this is certainly an important consideration. However, why the plant elaborates some of its terpenes into cannabinoids is not clear. Potter (2009) suggested that, consistent with the antioxidant property of the cannabinoids, their function might be to prevent oxidation of the monoterpenes present in the resin.

There is some evidence that drought, high light intensity, and high elevations (and therefore greater UV light) increase the release of exudate on the leaf surfaces, and this has led to the hypothesis that the resin is a protective sunscreen (Bouquet 1950 stated that the resin is an "insulating protective varnish" against high temperature and moisture loss). Pate (1983, 1994) hypothesized that THC is protective against UV-B radiation and that the much higher THC production in subtropical and tropical biotypes in contrast to the much lower production in temperate regions may be the result of selection, adapting the more southern races to the high UV-B radiation in the south. However, Lydon et al. (1987) concluded that "the contribution of cannabinoids as selective UV-B filters in *C. sativa* is equivocal." The glands and consequently the resin that is secreted are concentrated on the abaxial ("lower") side of the leaves (the same is true for the perigonal bracts in the inflorescence); it hardly makes sense for a sunscreen to be present on the shaded lower side of the foliage rather than the exposed upper side, and employing a resinous sunscreen seems quite speculative in view of the fact that plants commonly use several other strategies for reducing the intensity of solar radiation (see, for example, Small 2014a).

The cannabinoids appear to provide some protection against bacteria (Schultz and Haffner 1959; Radosevic et al. 1962; Van Klingeren and Ham 1976) and fungi (Misra and Dixit 1979; Pandey 1982; McPartland 1984). *Cannabis* extracts have controlled bacterial and fungal diseases in crops, livestock, and humans, as reviewed in McPartland (1997a) and McPartland et al. (2000).

Clarke and Merlin (2015) speculated that the "abscission layer" at the base of the large gland heads which contains most of the plant's cannabinoids, is adaptive in allowing gland heads to fall to the ground and release allelopathic constituents to repel competing vegetation. This hypothesis was disputed by Small (2015b). As pointed out in Small and Naraine (2016b), about 30% of flowering plants possess glandular trichomes producing secondary chemicals, usually at the tip of the structures, often in distinctive head-like containers, and there is an overwhelming majority view that these are primarily an adaptation protecting the plants against herbivorous animals, not against competing plants. In any case, living plants simply do not drop the resin heads. There is also a timing problem with Clarke and Merlin's hypothesis: *C. sativa* does not produce its inflorescences until late summer and early autumn, so it does not have a supply of stalked glands to drop until then. It hardly makes sense to schedule allelopathic activity so late in the season, when competing plants have completed most of their seasonal growth.

Cannabis sativa has minor allelopathic properties (Inam et al. 1989; McPartland 1997a; McPartland et al. 2000), and chemicals leached to the soil may inhibit competing plants, as suggested by Haney and Bazzaz (1970) (also see the discussion of allelopathy in Chapter 3). The foliage surface has many small glands, which unlike the much larger capitate glands of the inflorescence, do not burst when touched (their function also seems to be to discourage herbivores from eating the plant but not to additionally act as flypaper like the large stalked glands in the flowering part). Also unlike the larger inflorescence stalked glands, the glandular heads of which fall off with light agitation in dried material, the foliage glands of the leaves are extremely difficult to separate and remain rigidly attached to desiccated, dead leaves. This likely holds the key to the allelopathic nature of *C. sativa*. Allelopathy is very well known in the plant kingdom, especially as a result of toxic chemicals frequently leaching into the soil from the roots or other vegetative parts of the plant. Older, senescent foliage of *C. sativa* characteristically abscises and falls to the ground, and while the content of resin in the foliage is very much lower than in the inflorescence (Potter 2009), the fallen leaves may contribute some allelopathic protection. In any event, allelopathy of *C. sativa* is not particularly effective, since nearby weeds and grasses are usually superior competitors (unless *C. sativa* has grown large enough to shade out lower plants). Moreover, allelopathy is most effective in repelling seedlings, not established plants. *Cannabis sativa* is an annual and does not become large enough in the spring to produce a sufficient volume of resin to have any substantial effect on surrounding competing seedlings. (Whether overwintering litter from the plant can affect the next year's growth is another issue.)

Insects are by far the principal herbivores of plants, which employ many chemical defenses against them. Curiously, insects lack endocannabinoid receptors (see Chapter 13) and so do not respond to the cannabinoids in the same way as most animal groups. However, insects need not consume the resin for it to be repellent. There have been repeated suggestions that exuded resin could be a mechanical defense, ensnaring small insects like flypaper (e.g., Ledbetter and Krikorian 1975). "Touch-sensitive glandular trichomes" rupture when touched by an arthropod, rapidly releasing a sticky exudate, which can discourage, even kill, herbivorous insects (Krings et al. 2002; Figures 11.13 and 11.14). In living (but not dried) cannabis glands, the resin head readily ruptures when touched, suggesting that the released resin is indeed antiherbivorous (Small and Naraine 2016b). Potter (2009) noted that the high proportion of monoterpenes in the large inflorescence glandular hairs lowers the resin's viscosity compared to the foliage glandular hairs. Once the large gland heads are punctured by insect contact, the contents would more readily spread over the insect surface, and as the monoterpenes volatalise, the result is an increasingly viscous adhesive coating the insect. The acidic forms of THC (THCA) can kill insect cells, suggesting that cannabinoids are protective against insects, like the terpenes (Taura et al. 2009).

FIGURE 11.13   Possible modes of antiherbivorous action of the large stalked glandular trichomes. Left, resin secretions from the gland heads act like flypaper to trap a small insect. Right, a larger insect contacting the resinous secretions is discouraged by the sticky substance and perhaps also by its odor and taste. Prepared by B. Brookes.

FIGURE 11.14   Insect stuck on glandular stalked trichomes of a young leaf. Photo by Psychonaught (CC0 1.0).

All of the preceding does not provide a definitive answer to the question of why the cannabinoids have evolved, an issue that remains open to speculation (indeed, why other species in the Cannabaceae have secretory epidermal cells is equally unclear). Most secondary compounds are likely (a) metabolic waste products, (b) generalized antibiotics (acting against all harmful classes of organisms; see Pate 1994), or (c) evolutionary holdovers from ancestors in which the chemicals were adaptive. The cannabinoids probably fall within one or more of these categories.

EXPANDED DEFINITIONS OF CANNABINOIDS

The term "cannabinoids" has been expanded from its original meaning referring only to a unique class of compounds synthesized mostly by *Cannabis* and rarely by a few other plants. Some

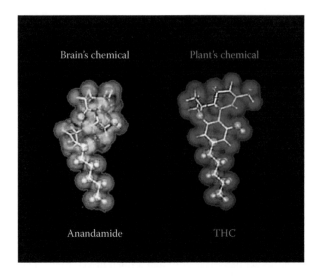

Brain's chemical   Plant's chemical

Anandamide   THC

FIGURE 11.15   U.S. National Institute of Drug Abuse display (public domain, modified), accompanied by the text: "THC's chemical structure is similar to the brain chemical anandamide. Similarity in structure allows drugs to be recognized by the body and to alter normal brain communication."

researchers also include in the term "cannabinoids" (a) chemically synthesized analogues ("synthetic cannabimimetics"; Ashton 2012) and (b) chemicals of quite different structure called "endocannabinoids" (endogenous cannabinoids), found in animals, including humans, which trigger the cannabinoid receptors, particularly those that function in neurochemistry, as described in detail in Chapter 13. For example, anandamide, the best known of the human body's natural mimics of THC, is functionally much like THC in its effects but is structurally very different (Figure 11.15).

The cannabinoids of the cannabis plant have occasionally been referred to as "classical cannabinoids" in contrast to synthetic cannabinoids of different structure, which are termed "nonclassical cannabinoids." Plant-derived cannabinoids (phytocannabinoids) collectively (whether from *C. sativa* or other plants) have been referred to as "exocannabinoids," in contrast to endocannabinoids. The term "phytocannabinoids," which was once restricted to the cannabinoids of *Cannabis* (Pate 1999), has been enlarged by Gertsch et al. (2010) as follows: "any plant-derived natural product capable of either directly interacting with cannabinoid receptors or sharing chemical similarity with cannabinoids or both." A range of chemical compounds produced by a range of organisms directly affect the cannabinoid receptors, for example, benzyl derivatives from the fungus *Eurotium repens* (Gao et al. 2011). Indirect effects on the cannabinoid receptors are illustrated by the pentacyclic triterpene, beta-amyrin, produced by *Protium*—Brazilian resinous trees (Simão da Silva et al. 2011), which inhibits the breakdown of 2-AG (Matos et al. 2013). Very curiously, beta-caryophyllene, a major compound of the essential oil of *C. sativa* (and many other plants), directly activates the CB_2 receptors (see Chapter 13), and thus, *C. sativa* produces two quite distinctive classes of phytocannabinoids. *N*-alkyamide in echinacea (*Echinacea* species) has also been shown to directly stimulate the CB_2 receptor system.

CURIOSITIES OF SCIENCE, TECHNOLOGY, AND HUMAN BEHAVIOR

- According to chemical nomenclature, Δ^9-THC is (6aR-trans)-6a,7,8,10a-tetrahydro6,6,9-trimethyl-3-pentyl-6H-dibenzo[b,d]pyran-1-ol. It is easy to see why such enormous names require shorter versions. The acronym THC has become universally understood for tetrahydrocannibal, although there are well over a hundred other meanings, such as Texas Historical Commission, The History Channel, and Terminal Handling Charge. CBD similarly also has other meanings, particularly "Convention on Biological Diversity."

- Rothschild and Fairbairn (1980) observed that the ovipositing butterfly *Pieris brassicae* was repelled more by an intoxicating Mexican strain of *C. sativa* than by a fiber Turkish strain, apparently because of the high THC content in the former.
- According to a widely circulated Internet hoax, in the 1990s, a mythical Florida biochemist, Professor Nanofsky, transferred the ability to produce THC into orange trees and circulated seeds of the transgenic "pot orange" in revenge against the U.S. Drug Enforcement Administration for some perceived harassment.

12 Nonmedical Drug Usage

This chapter provides a basic account of nonmedical cannabis drug consumption, mostly for recreational enjoyment, occasionally for religious or ritualistic purposes. An understanding of the nature of nonmedical usage to alter consciousness is needed to understand the context for employing *Cannabis sativa* for the authorized medical purposes documented in Chapter 13 and the recreational possibilities described in Chapter 15. There are hundreds of books providing instructions on the (usually illegal) preparation and use of cannabis products. Most of these masquerade as "educational" works and slyly provide disclaimers that laws should not be contravened. For a period in the late twentieth century, law enforcement authorities attempted to prohibit the sale and distribution of such volumes, but with an endless supply of articles now on the Web, this has proven futile. The fact is, despite the usually illegal status of cannabis, marijuana is being generated and used in huge quantities, and society in general has become quite knowledgeable about the subject. However, it is also true that a great deal of misinformation is in circulation.

RELIABILITY OF COUNTERCULTURE INFORMATION

Voluminous information on many aspects concerned with the nonmedical use of cannabis is available in drug counterculture/underground publications, both in print and on the Web. Such works typically advocate (explicitly or implicitly) the consumption of restricted (usually proscribed) drugs for recreational use and additionally often provide instructions for obtaining, preparing, or using such drugs. Most of these publications are authored anonymously or under pseudonyms. Particular publications may not be *illicit* (habitually, phrases like "Do not contravene any local laws!" are inserted, tongue-in-cheek); they may not entirely represent the *counterculture* (usually conceived of as representing a minority viewpoint); and they may not be *underground* (some jurisdictions are more permissive with respect to illicit drugs than others, and cannabis books that once could have led to prosecution are now being widely printed and circulated by the publishing industry). An appreciable proportion of the information in most of these publications is either invalid or seriously suspect. On the other hand, some of the content has been compiled as a result of careful observation, experimentation and/or inventiveness and is often mentioned in this book.

THE CANNABIS EXPERIENCE

Biphasic Effects

Cannabis and some of its constituent cannabinoids can generate "biphasic effects," a phrase that can mean "two phases" but in the case of cannabis refers to different responses: low doses produce symptoms or consequences opposite to those resulting from high doses. As noted by Ashton et al. (2005), many of the psychological effects of cannabis (particularly of THC) "are biphasic and bidirectional, depending on dose, mode of administration, environment, expectation, personality, degree of tolerance and other individual factors, as well as time-frame... Thus, acute effects in normal subjects can include euphoria or dysphoria, relaxation or anxiety, excitation followed by sedation, heightened perception followed by perceptual distortion, and increased motor activity followed by incoordination." The dual nature of cannabis was appreciated in ancient civilizations, particularly India and China. For example, the first century AD Chinese classic medical pharmacopeia *Ben Ts'ao* pointed out that cannabis was a benign treatment for numerous maladies, "but when taken in excess it could cause seeing devils" (Mechoulam 1986; Mechoulam and Parker 2013a).

Sequence of Psychological Impacts

Iversen (2000) recognized four stages or phases of cannabis intoxication: "buzz," "high," "stoned," and "come-down." During the short initial "buzz," the user may feel lightheaded or slightly dizzy, and tingling may occur in the extremities or elsewhere. During the "high," there may be euphoria, exhilaration, and unrestrained behavior such as giggling, laughter, talkativeness, and other forms of sociability. When "stoned," the user usually feels calm, relaxed, and happy and may feel as if in a dreamlike state. After several hours, the come-down period occurs, marked by diminishment of the sensations. Iverson's analysis is debatably applicable to most people, especially to the effects of certain marijuana strains, but (in common with many drugs, including alcohol) conforms to a sort of bell curve of a gradually increasing, peak, and gradually decreasing set of reactions.

Pleasant Psychological Effects

Recreational usage of marijuana, for at least the majority of regular consumers, is employed for fun (Figure 12.1). It usually produces mild euphoria (a "high"), relaxation, decreased anxiety, and reduction of inhibitions. There are often accompanying decreases in abilities to judge time and focus intellectually and induction of sleepiness and incoordination, but these symptoms are expected in experienced users and may even be welcome. Some sensory perceptions frequently become more intense, and these may include those associated with hunger, sex, music, and other pleasurable activities. Marijuana is usually consumed in a group setting (it is said to increase sociability), which contributes to the enjoyment of the experience. "Social setting and expectations have a great effect on the perceived state of 'intoxication' induced by marijuana" (Yazulla 2008).

FIGURE 12.1   Social intoxicant usage of marijuana. Prepared by B. Brookes.

UNPLEASANT PSYCHOLOGICAL SYMPTOMS

Because the drug is usually taken in an illegal setting, there may be anxiety or guilt associated with the possibility of being exposed. Most users today, however, have limited concerns for the legal status or possible harm related to marijuana and do not feel the shame and fear that are frequent with other illegal drugs. There is considerable variation in individual reaction, with personality playing a significant role, and some, particularly naïve users and those in stressful situations, can experience anxiety to the level of panic, paranoia, and psychosis (Gregg et al. 1976; Kalant 2001; Hall and Pacula 2003). Unpleasant psychological reactions occasionally occur and may include anxiety, depressed mood, dizziness, and panic attacks. Psychotic symptoms like delusions and hallucinations are rare but more likely at very high doses of THC. High doses can cause dysphoria (a profound state of unease or dissatisfaction), sensory distortion, and even hallucinations. As discussed in the next chapter, the particular constituents present in given samples of cannabis can result in different psychological effects.

COUCHLOCK

"Couchlock" (also couch lock and couch-lock) refers to a state of extreme lethargy (literally to the point of being unable to rise from one's couch) resulting from smoking marijuana. "Zapped," "zonked," and "zombied" are occasionally terms also employed to indicate profound physical tiredness induced by marijuana. Couchlock has often been attributed to the sedative effect of the considerable cannabidiol (CBD) in indica strains but seems to result from a sedative interaction of THC and strains high in the terpene myrcene (Russo 2011a). As noted in Chapter 2, myrcene contributes to the sedative effect of *Humulus*, the sister genus of *Cannabis*.

PHYSIOLOGICAL EFFECTS

Short-term noticeable physiological effects may include an increase in heart rate, a decrease in blood pressure when standing, dry mouth ("cotton mouth," "pasties"), lowering of body temperature, increased oxygen demand, reduced tear flow, reduced bowel movement, and delayed gastric emptying (Fišar 2009b).

"Red eye" (redeye, bloodshot eye, or conjunctival vasodilation) is another common effect of smoking marijuana (not to be confused with photographic redeye due to reflection of flash light from the retina). Red eye is redness of the sclera or white region of the eye (Figure 12.2) due to vasodilation of small vessels in the eye or occasionally simply from irritating marijuana smoke. Although red eye may result from smoking marijuana, the occurrence of pupil dilation, widely thought to occur, has been challenged (Weil et al. 1968; also see the next chapter).

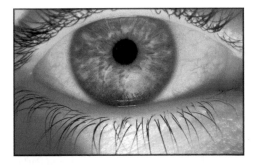

FIGURE 12.2   Red eye caused by smoking marijuana. Photo by Psychonaught (released into the public domain).

HEALTH RISKS

Although cannabis is an enjoyable diversion for most, there are potential hazards associated with nonmedical usage of marijuana. Of course, purchasing material from criminals and exposure to prosecution are dangers that can indirectly be harmful psychologically and physically. There is no shortage of studies showing harmful psychological and physiological effects of marijuana, but, as pointed out in the next chapter, essentially only studies intended to demonstrate negative effects of marijuana consumption have been permitted until very recently, and research funding has been predicated on documenting the harmfulness of marijuana.

As pointed out in Chapter 13, a range of opinion exists in the medical community regarding just how hazardous consumption of marijuana is with respect to *causing* particular illnesses, just as there is a range of opinion with regard to how beneficial medical marijuana is with respect to *alleviating* particular illnesses. As also noted in Chapter 15, the official position of many national governments (particularly in North America) is (or at least has been until recently) that herbal marijuana is a dangerous "narcotic" with absolutely no valid medical value.

Health risks mentioned in the context of using medical marijuana are discussed in the next chapter. Except for the special hazards experienced by patients with weakened immune or cardiac systems (examined in the next chapter), the risks of using medical marijuana are essentially the same as for healthy people using recreational marijuana. The specific risks associated with smoking, hazardous contaminants, pathogens, pregnancy, and lactation are also discussed in the next chapter. The majority of medical associations have issued statements noting the potential harmfulness of marijuana and usually also indicating the need for research before it is accepted for medical usage. For example, the American Psychiatric Association (2013) revised *Diagnostic and Statistical Manual of Mental Disorders* warns that use of marijuana can progress to "marijuana use disorder," a condition requiring treatment. Kepp and Raich (2014) list the position statements on the use of medical marijuana of numerous medical organizations. While most such evaluations are rather negative, as detailed in the next chapter, there are quite promising therapeutic applications under examination and development. The following comments regarding health risks are pertinent to the recreational usage of marijuana.

CONCERN FOR MENTAL STATUS

There are many publications documenting mental health concerns associated with marijuana, including the possibilities of addiction and mental illness (Fernández-Artamendi et al. 2011; Zvolensky et al. 2011; Bostwick 2012; Greydanus et al. 2013; Karila et al. 2014; Lev-Ran et al. 2014; Rubino and Parolaro 2014; Volkow et al. 2014). Cannabis is extraordinarily attractive to adolescents (Figure 12.3), and the possible effects on the mental health of the young are the leading health concern associated with recreational marijuana. Moore et al. (2007) wrote "there is now sufficient evidence to warn young people that using cannabis could increase their risk of developing a psychotic illness later in life."

Possibly the most human characteristic is our advanced degree of cognition—the brain's ability to acquire, store, and later retrieve new information. Numerous studies have concluded that marijuana is not completely without serious risk for mental status. "Clearly, the chief psychoactive component in cannabis, THC, produces acute cognitive disturbances in humans and animals, more profoundly affecting short-term than long-term memory" (Mechoulam and Parker 2013a). Gilman et al. (2014) studied the brains of young recreational users ranging in age from 18 to 25 years and found "that in young, recreational marijuana users, structural abnormalities in gray matter density, volume, and shape of the nucleus accumbens and amygdala can be observed." Some reports have found that chronic marijuana use over several years produces quite significant cognitive impairments; others dispute the validity of these studies (Mechoulam and Parker 2013a). Jager (2012) stated, "Taken together, studies on long-term effects of cannabis on cognition have failed to find proof of

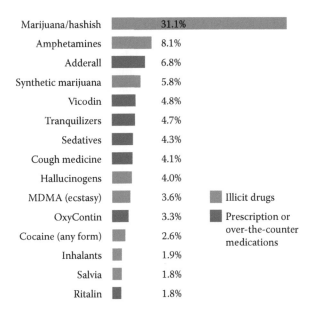

FIGURE 12.3 Annual use in 2014 (at least once) of recreational drugs by high school seniors (grade 12) in the United States. Sample size = 12,400. During the year, 38.7% of the students employed one of the illicit drugs shown at least once annually, 60.2% consumed alcohol at least once annually (37.4% at least once monthly), and 6.7% smoked tobacco daily. Based on information from the University of Michigan "Monitoring the Future" website (http://www.monitoringthefuture.org/), sponsored by the U.S. National Institute on Drug Abuse (NIDA).

gross abnormalities, but there is some evidence for mild cognitive impairments, particularly in the domain of memory and learning… The majority of recreational cannabis users does not experience serious adverse reactions and is able to regulate their use. However, a minority of frequent or long-term users will develop problems."

In a very extensive meta-analysis, Minozzi et al. (2010) stated, "We conclude that there is insufficient knowledge to determine the level of risk associated with cannabis use in relation to psychotic symptoms and that more information is needed on both the risks of cannabis use and the benefits of preventive interventions to support evidence-based approaches in this area." Radhakrishnan et al. (2014) concluded that there is danger of precipitating mental illness in those vulnerable to mental problems and risk of seriously aggravating mental conditions in those already suffering from mental problems. However, Zammit et al. (2008) were of the view that the degree of risk has not been clearly evaluated. A study by Phillips et al. (2002) suggested that marijuana use in a study group of patients at risk of developing psychosis did not increase that risk. Ksir and Hart (2016) reviewed studies of cannabis and psychosis and concluded that the research "suggests that cannabis does not in itself cause a psychosis disorder. Rather, the evidence leads us to conclude that both early use and heavy use of cannabis are more likely in individuals with a vulnerability to psychosis." Clearly, there is not a clear consensus on the potential serious harm that marijuana poses to human cognition.

CANNABIS DEPENDENCE

"Addiction" as a term has become less popular in medicine; "dependence" is the preferred term for extreme, excessive harmful usage. Nevertheless, severe, chronic dependence is usually called addiction. "Abuse" can be employed to refer to a lesser degree of overuse.

The risk of addiction from cannabis is considerably lower than that of numerous prescribed and illegal agents (Grotenhermen and Russo 2002; Guzmán 2003). While the addictive potential of cannabis is low compared to nicotine, alcohol, and some controlled drugs, there is general agreement that compulsive usage sometimes develops. "Physical dependence" on cannabis has not been demonstrated (Gordon et al. 2013), but psychological dependence on cannabis is often accepted as a genuine phenomenon (e.g., Norberg et al. 2012). Gardner (2014) stated, "It seems irrefutable that cannabis can be considered to be addictive at the human level."

The *Diagnostic and Statistical Manual of Mental Disorders*, published by the American Psychiatric Association headquartered in Washington, DC (fifth edition published in 2013), is the official analytical system for mental disorders in the United States. The *International Classification of Diseases* published by the World Health Organization (10th edition completed in 1992, 11th edition due in 2017) is the other major professional manual for mental disorders, although it covers all health conditions and is used mainly outside the United States. Both manuals accept "cannabis dependence" (cannabis use disorder) as a psychological condition requiring treatment. Budney et al. (2007) estimated that 9% of Americans who had ever used cannabis were psychologically dependent on it (cf. Crean et al. 2011). Degenhardt et al. (2013) estimated that on a worldwide basis, the figure is 20%.

The *International Classification of Diseases* (the standard diagnostic tool for epidemiology, health management, and clinical purposes; for details, see the World Health Organization's website at http://www.who.int/classifications/icd/en/) provides the following information: "Individuals who have cannabis dependence compulsively use the drug but do not usually develop physiological dependence, although frequently tolerance to the effects of cannabis has been reported by these individuals. Some users also reported withdrawal symptoms, although the symptoms have not usually been clinically significant. Frequently people with cannabis dependence use very potent cannabis over a period of months and sometimes years, and may spend significant time acquiring and using the substance. Cannabis dependence often interferes with family, work, school, or recreational activities. Individuals with cannabis dependence may also persist in using this drug although knowledge of physical or psychological problems may result."

The fifth edition of the *Diagnostic and Statistical Manual of Mental Disorders* lists 11 criteria for cannabis use disorder (cannabis dependence). These include cravings, giving up important life activities in order to use cannabis, continuing to use despite adverse physical or psychological problems caused or exacerbated by using, tolerance, withdrawal, and persistent unsuccessful efforts to quit. The validity of some of the criteria employed to define cannabis abuse has been disputed (e.g., Piontek et al. 2011). Clearly, there is subjective judgment associated with criteria based on personal choice to continue an activity that one judges to be pleasant, even if there are some unpleasant consequences. Some have argued that there is a "caffeine use disorder," as evidenced by the fact that some of the criteria for cannabis use disorder are analogous to those manifested by coffee addicts.

CANNABIS WITHDRAWAL SYNDROME

A "cannabis withdrawal syndrome" is said to occur in frequent users shortly after they quit, the symptoms lasting a week or more (Budney and Hughes 2006). The fifth edition of *Diagnostic and Statistical Manual of Mental Disorders* lists seven symptoms of withdrawal: anger, anxiety, depression, loss of appetite, restlessness, sleep difficulties, and physical symptoms that cause significant discomfort, including chills, fever, headache, stomach pains, sweats, and tremors. Gorelick et al. (2012) noted that the symptoms are often vague. The very existence of the phenomenon of a withdrawal syndrome has not been universally accepted (Jager 2012). It could be argued that there is a "caffeine withdrawal syndrome," as evidenced by the fact that some of the symptoms from being denied coffee or other caffeine-laced beverages are the same as those described for cannabis withdrawal syndrome. However, several studies have found that ceasing usage of marijuana has at least temporary physiological effects (Allsop et al. 2012; Fratta and Fattore 2013).

Marijuana as a "Gateway Drug"

One of the most frequently cited dangers claimed to be associated with marijuana is that it is a "gateway drug," leading to the use of other more potent and addictive substances of abuse. Marijuana is just one of several drugs examined in "stepping-stone" or "stairway" models accounting for progression of usage of a sequence of more serious drugs (Tarter et al. 2012). Kepp and Raisch (2014) assert, "there is considerable evidence that marijuana is a gateway drug to other illicit drugs." Joy et al. (1999) noted that it is not surprising that most users of other illicit drugs have used marijuana first, simply because it is the most widely used illegal drug and therefore the first one most people encounter. But since most drug users employed alcohol and nicotine before using marijuana (and indeed mother's milk before that), it is hardly deductive logic to assume that preceding usage is causally linked to subsequent usage. Joy et al. (1999) pointed out that there is no conclusive evidence that marijuana is in fact a gateway drug. Morral et al. (2002) stated: "Strong associations between marijuana use and initiation of hard drugs are cited in support of the claim that marijuana use per se increases youths' risk of initiating hard drugs (the 'marijuana gateway' effect)... Marijuana gateway effects may exist. Our results demonstrate, however, that the phenomena used to motivate belief in such an effect are consistent with an alternative simple, plausible common-factor model. No gateway effect is required to explain them. The common-factor model has implications for evaluating marijuana control policies that differ significantly from those supported by the gateway model." Evaluation that marijuana is a gateway drug is difficult as there are various alternative explanations possible of observed correlations between marijuana usage and subsequent usage of other drugs (Hall and Lynskey 2005; Fergusson et al. 2006; Cleveland and Wiebe 2008). Melberg et al. (2010) suggested that there are "two distinct groups; a smaller group of 'troubled youths' for whom there is a statistically significant gateway effect that more than doubles the hazard of starting to use hard drugs and a larger fraction of youths for whom previous cannabis use has less impact."

Concern for Lung Function

Pletcher et al. (2012) conducted a large-scale study of the harm of smoking marijuana upon lung function, and concluded that "Occasional and low cumulative marijuana use was not associated with adverse effects on pulmonary function." However, in the main, the medical profession has a very negative view of how smoking marijuana influences breathing. The American Lung Association (2015) issued the following statement: "Smoke from marijuana combustion has been shown to contain many of the same toxins, irritants and carcinogens as tobacco smoke... Marijuana smokers tend to inhale more deeply and hold their breath longer than cigarette smokers, which leads to a greater exposure per breath to tar. Secondhand marijuana smoke contains many of the same toxins and carcinogens found in directly inhaled marijuana smoke, in similar amounts if not more. While there is no data on the health consequences of breathing secondhand marijuana smoke, there is concern that it could cause harmful health effects, especially among vulnerable children in the home. Additional research on the health effects of secondhand marijuana smoke is needed. Smoking marijuana clearly damages the human lung. Research shows that smoking marijuana causes chronic bronchitis and marijuana smoke has been shown to injure the cell linings of the large airways, which could explain why smoking marijuana leads to symptoms such as chronic cough, phlegm production, wheeze and acute bronchitis... Smoking marijuana hurts the lungs' first line of defense against infection by killing cells that help remove dust and germs as well as causing more mucus to be formed. In addition, it also suppresses the immune system. These effects could lead to an increased risk of lower respiratory tract infections among marijuana smokers, although there is no clear evidence of such actual infections being more common among marijuana smokers. However, frequent marijuana-only smokers have more healthcare visits for respiratory conditions compared to nonsmokers. Studies have shown that smoking marijuana may increase the risk of opportunistic infections among those who are HIV positive, although it does not seem to effect the development of AIDS or lower white cell counts. Another potential threat to those with weakened immune systems is *Aspergillus*, a mold that can cause lung disorders. It can grow on marijuana, which if then smoked exposes the lungs to this fungus. However, it rarely causes problems in people with healthy immune systems." For additional information regarding the risks from *Aspergillus*, see the section "Microbiological Safety and Sterilization" in Chapter 14.

One of the least discussed risks associated with smoking marijuana is the unsanitary but widespread practice of sharing joints and bongs and the possibility of acquiring diseases such as hepatitis. It has been hypothesized that communal smoking could be one of the factors responsible for transmitting the human papilloma virus, linked to cancers of the throat and tongue (Zwenger 2009). Smoking marijuana has been alleged to have an inhibitory effect on the immune system, which could predispose users to infectious diseases. There are scattered case reports of *Aspergillus* infection in immunocompromised patients and even meningitis from passed joints. However, doses that are capable of producing immunosuppression in rodents are 50–100 times higher than usual human doses (Ethan Russo, personal communication).

SOCIETAL VS. INDIVIDUAL HEALTH

Euphoric drugs can be used for good or evil, but the harm that can result is not limited to physical and/or psychological damage to the individual user (Figure 12.4). Opium use was once so widespread in China that a substantial portion of the population became nonproductive, and a burden on the state. The same situation prevails today for khat (*Catha edulis* L.) usage in some parts of the

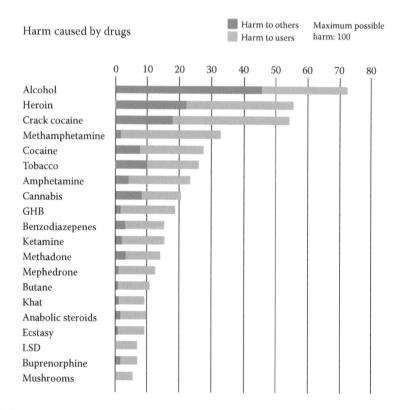

FIGURE 12.4 Ranking of harm to users and to society caused by drugs, in the United Kingdom, from Nutt, D.J., King, L.A., Phillips, L.D., *The Lancet*, 376, 1558–1565, 2010. Rankings are the opinion of drug-harm experts, using measures such as damages associated with health, drug dependency, economic costs, and crime. Figure downloaded by Tesseract2 (CC BY SA 3.0).

Middle East (Small 2004). Indeed, most proscribed illicit drugs harm not just the users but also are threats to the financial and physical welfare of many others. Alcohol and tobacco are obvious examples of legal (albeit controlled) harmful drugs that represent huge burdens on society. The same is true for sucrose (common table sugar) and other psychologically addictive foods that contribute to the obesity epidemic. The discussion of harmful effects of nonmedical uses of marijuana in this chapter is restricted to individual health, but it is well to remember that there are also potential effects on the collective welfare of society.

Driving Risks

Cannabis is a central nervous system depressant, and in some respects, the acute effects (i.e., at high dosages) resemble those of alcohol and other central nervous depressants. Cannabis can produce drowsiness, slower reactions, decreased memory (Figure 12.5), decreased attention, poorer psychomotor task performance, and poorer performance in driving. Asbridge et al. (2012) and Li et al. (2012) concluded that the risk of involvement in a motor vehicle accident increases approximately twofold after acute cannabis smoking. However, very experienced users seem able to compensate substantially, apparently tolerating the drug's actions well (Hart et al. 2001). Marijuana users operating vehicles seem to be less aggressive and more cautious than drunk drivers (Smiley 1999). (A joke that comes to mind: "A drunk driver will run a stop sign, a high driver will stop and wait for it to turn green.") Some of the impairment caused by cannabis is mitigated, since subjects appear to perceive that they are indeed impaired. Where they can compensate, they do, for example, by not overtaking, by slowing down, and by focusing their attention when they know a response will be required (Grotenhermen 2007). Experienced users appear able to develop physiological tolerance to the drug, substantially retaining driving ability (Grotenhermen and Müller-Vahl 2012). The combination of marijuana and alcohol or several illegal drugs is additive or synergistic, considerably increasing the risks associated with driving under the influence of drugs. Neavyn et al. (2014) recommended that users refrain from driving for 8 hours following a "high."

Hartman and Huestis (2013) reviewed the current literature on cannabis effects on driving and concluded, "drivers attempt to compensate by driving more slowly after smoking cannabis, but control deteriorates with increasing task complexity. Cannabis smoking increases lane weaving and impaired cognitive function. Critical-tracking tests, reaction times, divided-attention tasks, and lane-position variability all show cannabis-induced impairment. Despite purported tolerance in

FIGURE 12.5 The perils of driving while stoned. Prepared by B. Brookes.

frequent smokers, complex tasks still show impairment. Combining cannabis with alcohol enhances impairment, especially lane weaving."

Similarly, Sewell et al. (2009) concluded: "Marijuana smokers tend to compensate effectively while driving by utilizing a variety of behavioral strategies. Combining marijuana with alcohol eliminates the ability to use such strategies effectively, however, and results in impairment even at doses which would be insignificant were they of either drug alone. Epidemiological studies have been inconclusive regarding whether cannabis use causes an increased risk of accidents; in contrast, unanimity exists that alcohol use increases crash risk. Furthermore, the risk from driving under the influence of both alcohol and cannabis is greater than the risk of driving under the influence of either alone."

Whitehill et al. (2014) noted (for the United States): "The issue of marijuana-impaired driving is particularly salient for young drivers, for whom the combination of inexperience and substance use elevates crash risk. Youth younger than 21 are at the highest risk of involvement in a fatal motor vehicle crash. They are also the age group most likely to use marijuana. Nationally, cannabis was involved in 12% of fatal crashes among 16–20 year olds. College students are a population at increased risk of substance-related risk behaviors, such as impaired driving… Marijuana is second only to alcohol for substances most abused by this population… Findings of previous studies suggest that male students are twice as likely as female students to drive while high on marijuana and 20% more likely to ride with a marijuana-using driver."

RISK OF CONTAMINATION AND ADULTERATION OF STREET MARIJUANA

The quality of cannabis drugs purchased in the illicit market is often uncertain, and this is one of the chief reasons that many grow and prepare their own supplies. Although sometimes a harmless substance is sold as marijuana, often the consumer is offered a product that is significantly more harmful that medical-grade forms of the drug.

Herbal marijuana may be contaminated as a result of negligent cultivation, preparation, or storage techniques. As discussed in Chapter 13, this can introduce dangerous fungi, aflatoxins (toxic fungal metabolites), other microbes (particularly bacteria), pesticide residues, and heavy metals. Law enforcement in some countries has employed Paraquat herbicide to control illicit marijuana, notably in Mexico (Figure 12.13b), and there has been concern that imported marijuana could be contaminated (Landrigan et al. 1983). However, according to Barceloux (2012), "the high combustion temperatures in marijuana cigarettes destroys Paraquat; therefore there is no significant risk of Paraquat-induced pulmonary fibrosis from cannabis smoking." Illicit growers sometimes have little concern about the health risks of their customers and can produce chemically contaminated marijuana. They may use banned plant growth regulators to force early flowering and production of bigger more compact buds, such as paclobutrazol, or daminozide (Alar) which degrades into the dangerous chemical hydrazine (Upton et al. 2013). So-called "growth enhancers" whose chemical nature is uncertain may also have been employed. Sullivan et al. (2013) examined how the presence in marijuana of three commonly employed pesticides, bifenthrin, diazinon, and permethrin, as well as the plant growth regulator paclobutrazol, produced contaminants in the resulting inhaled smoke. Recovered residues were as high as 70%, "suggesting that the potential of pesticide and chemical residue exposures to cannabis users is substantial and may pose a significant toxicological threat." Hair (from humans or pets), although not particularly hazardous, is commonly found on street marijuana, reflecting the sloppiness of many illicit marijuana producers and sellers.

Solvent extracts ("hash oil," discussed later) can be contaminated with dangerous chemical residues. "Synthetic" or "fake" marijuana, discussed in Chapter 13, and chemically synthesized analogues of THC are usually quite hazardous products.

Adulteration refers to the deliberate inclusion of inferior materials in a product in order to mislead the purchaser into thinking that the quantity or quality is superior. A chief motivation for this is to make the appearance more attractive and/or to increase the weight. Adding sand, chalk particles, or tiny glass shards can make marijuana appear to have more of the desirable glistening trichome gland heads as well as increase the density. In the United Kingdom during the Victorian era, lead was a common adulterant, used for example to color cheese. Because street marijuana is sold by weight, some unscrupulous dealers have added lead particles to their offerings, resulting in poisoning consumers (Busse et al. 2008a,b). Especially dangerous is the practice of some dealers of adding dangerous drugs or plants to marijuana (McPartland 2008b; Upton et al. 2013).

RISK OF EXPOSURE TO INDOOR GROW-OP ENVIRONMENTS

Martyny et al. (2013) examined the environmental dangers to law enforcement personnel of entering illegal indoor marijuana growing operations. They did not find hazardous levels of volatile organic compounds, carbon dioxide, carbon monoxide, and common chemicals utilized by illicit growers, primarily pesticides and fertilizers, and none of these showed high toxicity. Airborne fungal spores, however, were of significant concern, and it was noted that "removal of the marijuana plants could potentially expose responders to levels of exposure consistent with those associated with mold remediation processes and that respiratory protection is advisable."

RISK TO DOGS

Man's best friend is in special danger from cannabis (Figure 12.6). The American Society for the Prevention of Cruelty to Animals Animal Poison Control Center reported that dogs account for 96% of their marijuana toxicity cases (Donaldson 2002). Ingestion of baked goods made with cannabis has resulted in sickness, even death of canines, which seem especially prone to being poisoned. Fitzgerald et al. (2013) reported that "The minimum lethal oral dose for dogs for THC is more than 3 g/kg. Although the drug has a high margin of safety, deaths have been seen after ingestion of food products containing the more concentrated medical-grade THC butter." (This report, based in part on a study of dogs dying from being tube-fed a large bolus of material causing aspiration and respiratory arrest, appears erroneous with respect to marijuana directly causing death in dogs. As CB_1 receptors are practically absent from the brain stem cardiorespiratory drive nuclei, true overdoses

FIGURE 12.6 Canine cannabis convention. Dogs are commonly poisoned by marijuana, especially edibles. Prepared by B. Brookes (a modification of the public domain "Dogs Playing Poker/A friend in need" by C. M. Coolidge).

have not been reported.) Meola et al. (2012) provided the following information: "Toxicosis in dogs can be caused by inhalation of the smoke, direct ingestion of the leaves, seeds, stems and flowers of the plant, ingestion of products laced with marijuana leaves, or ingestion of products made with concentrated THC or hashish oil. Clinical signs may be seen within 30–60 minutes after ingestion of marijuana. THC toxicosis in dogs can cause considerable morbidity. The most common reported clinical signs of marijuana toxicosis in dogs include central nervous system depression, ataxia [loss of control of bodily movements], mydriasis [pupil dilation], increased sensitivity to motion or sound, hyperesthesia [increased sensitivity to stimulation], especially ptyalism [excessive salivation], tremors, and the acute onset of urinary incontinence." A case of dermatitis in a dog resulting from moving into a residence that was previously used as a marijuana grow house was documented by Evans (1989).

THE GREAT DEBATE: IS CANNABIS RELATIVELY HARMFUL OR BENIGN?

Society has ferociously debated the merits of many scientific issues historically, such as biological evolution and vaccination, and such contentious debates continue to this day, exemplified by climate change and the comparative benefits of fad diets. Some debates are so esoteric and complex that they cannot be decided to the universal satisfaction of everyone (Figure 12.7), but in the fullness of time, disagreements concerning scientific facts can usually be settled. However, the debate concerning psychoactive cannabis (whether recreational or medicinal) is multifaceted, involving examination of not only its merits and risks in many independent respects but also human values concerning personal liberty and choice. It is easy to conclude when observing some debates (especially between politicians) that one or both parties are dishonest, foolish, deluded, prejudiced, and lacking in objectivity. As noted in this book, scientific methodology (at its best) controls these human weaknesses—but it is not possible to completely remove the human element from science, and particularly when scientific facts are not entirely clear, it is not surprising that there are very different perspectives. The following quotations representing such different perspectives of cannabis are intended simply to illustrate this phenomenon, *not* to summarize the best evidence or arguments, which are detailed

FIGURE 12.7 Left: "The School of Athens," representing philosophy, a fresco painted in 1509 in the Apostolic Palace in the Vatican by Raphael, considered to be his best painting. Right, detail, showing Plato (left) and Aristotle (right), considered to be among the greatest thinkers in history, and illustrating how the same reality can be viewed very differently by sincere, highly intelligent analysts. Photo credit: Web Gallery of Art; public domain.

elsewhere (especially in Chapter 13). As reflected by the predominance of negative views of the harmfulness of recreational marijuana, the *professional literature* is overwhelmingly dominated by those who regard marijuana as significantly harmful. A more balanced view (i.e., both for and against) is found in the media at large, but this book is most concerned with the views of informed individuals.

Representative Viewpoints That Cannabis Is Relatively Harmful

Hall and Degenhardt (2009) concluded, "The most probable adverse effects include a dependence syndrome, increased risk of motor vehicle crashes, impaired respiratory function, cardiovascular disease, and adverse effects of regular use on adolescent psychosocial development and mental health." Reece (2009) stated, "Chronic cannabis use is associated with psychiatric, respiratory, cardiovascular and bone effects. It also has oncogenic, teratogenic and mutagenic effects" (as discussed elsewhere, the contentions in the latter sentence are disputed). Hoch et al. (2015) concluded, "Various medical conditions can arise acutely after cannabis use, depending on the user's age, dose, frequency, mode and situation of use, and individual disposition; these include panic attacks, psychotic symptoms, deficient attention, impaired concentration, motor incoordination, and nausea. In particular, intense use of high doses of cannabis over many years, and the initiation of cannabis use in adolescence, can be associated with substance dependence, specific withdrawal symptoms, cognitive impairment, affective disorders, psychosis, anxiety disorders, and physical disease outside the brain (mainly respiratory and cardiovascular conditions)."

Hall (2014) concluded that chronic (long-term regular) recreational use is associated with the following hazards:

- Risk of developing a dependence syndrome (1 in 10 of all marijuana users, 1 in 6 of those starting in adolescence).
- Doubled risk of psychotic symptoms and disorders, especially with a personal or family history of psychotic disorders, and when use started in the mid-teens.
- Lower educational attainment by those beginning use as adolescents (causal link not established).
- Increased use of other illicit drugs by adolescent users (causal link not established).
- Intellectual impairment when use begins in adolescence and continues through young adulthood (reversibility of the impairment is unclear).
- Double the risk of schizophrenia or psychotic symptoms in adulthood when use begins in adolescence.
- Increased risk of chronic bronchitis.
- Probable increased risk of myocardial infarction from smoking in middle age.

Thompson and Koenen (2011) issued the following warning: "Predictable side effects of marijuana use include impaired judgment, cognitive impairment, impaired driving ability, hallucinations, early onset of psychosis in certain individuals, memory impairment, worsening of mood and anxiety disorders, and the risk of dependence. Individuals with major mental illnesses are especially vulnerable to the deleterious effects of cannabis. Smoking marijuana includes risks of rapid onset of intoxication as well as exposure to a variety of toxic and carcinogenic combustible products. Vaporization reduces exposure to some potential toxins such as carbon monoxide, but is unable to remove aluminum, ammonia, acetaldehyde, and other substances."

Hasin et al. (2015) concluded that in 2012–2013, 9.5% of U.S. adults (about 22 million) used marijuana in the past year, and nearly 3 of every 10 had a diagnosis of a marijuana use disorder. They also commented: "studies have shown that use or early use of marijuana is associated with increased risk for many outcomes, including cognitive decline, psychosocial impairments, vehicle crashes, emergency department visits, psychiatric symptoms, poor quality of life, use of other drugs, a cannabis-withdrawal syndrome, and addiction risk. Further, marijuana use disorders (abuse or dependence) are associated with substantial comorbidity and disability and are consequently of substantial public health concern." (See Meier et al. 2015 for a discussion of the difficulties of concluding that marijuana use among teenagers leads to permanent damage.)

Zeisser et al. (2012) stated: "Chronic cannabis use, generally referred to as a pattern that entails weekly or more frequent use, has been associated with an increased likelihood of cannabis dependence, chronic bronchitis and impaired respiratory function, psychotic disorders and impaired cognitive functioning as well as psychosocial effects such as impaired educational attainment in adolescents, and an increased likelihood of using other illicit drugs. Individuals who use cannabis may also experience acute adverse effects such as anxiety and panic, and an increased risk of motor vehicle crashes."

Representative Viewpoints That Cannabis Is Relatively Benign

Van Ours and Williams (2012) concluded: "Widespread use reflects the common belief that cannabis is not a particularly harmful drug. The weight of evidence supports this belief...the harms associated with cannabis use are much less serious than those associated with 'hard' drugs such as cocaine or heroin and may even be smaller than those associated with alcohol and cigarettes. And while it is generally acknowledged that there are risks associated with long term heavy use of cannabis such as respiratory diseases, cancer and perhaps psychotic disorders, only a small fraction of those who ever use cannabis actually become long term heavy users...for those who are not long term heavy users of cannabis, the physical and mental health effects of their cannabis use are likely to be small." (See comments in the next chapter regarding cancer risk from smoking.)

Nathan (2013) stated: "I am a father who worries about my kids getting sidetracked by cannabis before their brains have a chance to develop. But I am also a physician who understands that the negative legal consequences of marijuana use are far worse than the medical consequences... Alcohol, tobacco, marijuana, caffeine and refined sugar are among the most commonly used, potentially habit-forming recreational substances. All are best left out of our daily diets. Only marijuana is illegal, though alcohol and tobacco are clearly more harmful. In several respects, even sugar poses more of a threat to our nation's health than pot... If you still believe that cannabis should be illegal, then you must logically support the criminalization of alcohol and tobacco, with vigorous prosecution and even imprisonment of producers and consumers. Does that sound ridiculous? Then you must conclude that the only rational approach to cannabis is to legalize, regulate and tax it."

Graham (2014) wrote: "For proponents of the legalization of marijuana...legalizing pot means the market will be regulated, governments will reap the tax revenue, and drastically fewer people will be mired in the violence of the drug war and the injustices of the legal system. If public health suffers a bit as use of the substance increases, so be it... But what if the rise in marijuana smoking prompted by legalization brings more than just tolerable negative side effects? What if it is actually good for public health? A growing body of research suggests that marijuana may replace alcohol or hard drugs in many people's lives. Other recent studies suggest that looser restrictions on weed decrease traffic fatalities and even the suicide rate. That means the rising tide of legalization may mean more than just an acceptable but unfortunate societal burden—it may be a boon to public health."

The New York Times Editorial Board (2014) concluded: "It took 13 years for the United States to come to its senses and end Prohibition, 13 years in which people kept drinking, otherwise law-abiding citizens became criminals and crime syndicates arose and flourished. It has been more than 40 years since Congress passed the current ban on marijuana, inflicting great harm on society just to prohibit a substance far less dangerous than alcohol. The federal government should repeal the ban on marijuana... There are no perfect answers to people's legitimate concerns about marijuana use. But neither are there such answers about tobacco or alcohol, and we believe that on every level— health effects, the impact on society and law-and-order issues—the balance falls squarely on the side of national legalization. The social costs of the marijuana laws are vast. There were 658,000 arrests for marijuana possession in 2012, according to F.B.I. figures, compared with 256,000 for cocaine, heroin and their derivatives. Even worse, the result is racist, falling disproportionately on young black men, ruining their lives and creating new generations of career criminals. There is honest debate among scientists about the health effects of marijuana, but we believe that the evidence is overwhelming that addiction and dependence are relatively minor problems, especially compared with alcohol and tobacco. Moderate use of marijuana does not appear to pose a risk for otherwise healthy adults. Claims that marijuana is a gateway to more dangerous drugs are as fanciful as the 'Reefer Madness' images of murder, rape and suicide. There are legitimate concerns about marijuana on the development of adolescent brains. For that reason, we advocate the prohibition of sales to people under 21."

GLOBAL USAGE OF MARIJUANA

According to United Nations (2014): "In 2012, between 125 million and 227 million people were estimated to have used cannabis, corresponding to between 2.7 and 4.9 per cent of the population aged 15–64 years. West and Central Africa, North America, Oceania and, to a lesser extent, Western and Central Europe remain the regions with prevalence rates considerably higher than the global average. Over the past five years in North America, the largest cannabis herb market, prevalence rates have followed an upward trend." Cannabis users account for 80% of the illicit drug users in the world (Van Ours and Williams 2012). An analysis of consumption patterns and trends in Europe is provided by EMCDDA (2008, 2012). Estimates for the United States, the leader in usage, range up to 25% of the population. Marijuana has been claimed to be at least the fourth most valuable crop in America, outranked only by corn, soybeans, and hay (Small and Marcus 2002). As noted in Chapter 15, some authors claim that marijuana is the leading cash crop in the United States.

A BRIEF HISTORY OF NONMEDICAL PSYCHOTROPIC USAGE OF CANNABIS DRUGS

Natural drugs have been employed historically for three discernible purposes: spiritually or religiously, therapeutically, and recreationally as a euphoric. This chapter is mainly concerned with recreational and spiritual usage, and historical aspects related to medicinal usage are discussed in the next chapter. It is often difficult, perhaps impossible, to distinguish the three usages in ancient times because natural drugs were often utilized for more than one of these goals at the same time. The earliest recorded reference to euphoric use of *C. sativa* appears to date to about 5000 years ago, associated with Kurgan culture (the Kurgans were early people of the Caucasus region, known for elaborate burials mounds). A smoking cup with remnants of charred hemp seeds, associated with the Kurgans, who occupied Romania at that time, is suggestive of the flowering parts of *C. sativa* being combusted perhaps for euphoric ritualistic purposes. Kurgan incense burners have been hypothesized to have had the same function. It is speculative whether these early usages of cannabis actually were intended to produce intoxication, but it does seem that cannabis was used ritualistically in the Black Sea–Caucasus region (Sherratt 1991). The Scythians, noted next in this context, exemplify this tradition particularly well, and it seems that cannabis was also used for sacred purposes in Assyria, Babylon, and ancient Palestine (Rubin and Comitas 1975).

The ancient region of Scythia included a large area from the Ukraine to the borders of present-day India. The Scythians included nomadic Caucasoid tribes, wandering to the borders of modern Russia and China. Scythian culture thrived from the ninth to the third centuries BC. The fifth century BC Greek historian Herodotus described a Scythian funeral ceremony in which vapors from burning cannabis seeds (possibly entire fruiting heads, which could contain appreciable THC) were inhaled in small tents (Figure 12.8). Merlin (1972) provided Herodotus' account: "The Scythians then take this seed of hemp and, creeping under the mats, they throw it on the red-hot stones; and,

FIGURE 12.8 Artist's conception of a Scythian encampment, showing inhalation of smoke from cannabis being burned on braziers. The smaller tent shown is historically accurate; the larger tent is hypothesized. Prepared by B. Flahey.

being so thrown, it smolders and sends forth so much steam that no Greek vapor bath could surpass it. The Scythians howl in their joy at the vapor bath." Rudenko (1970) found archaeological evidence of a metal tripod censer with remnants of hemp seeds that Scythians of southern Central Asia had apparently employed during funeral rites. While the Scythian records have been interpreted as usage of marijuana to induce intoxication, this is uncertain. Plant materials are often burned ritualistically by various cultures without motivation to alter mental state. Nevertheless, in a rather reminiscent old practice in Poland, Russia, and Lithuania, hemp seeds were thrown on hot stones and the vapors inhaled in order to alleviate toothache (Benet 1975), suggesting relief of psychic stress as typically is induced by inebriants.

Practitioners of Buddhism and Shintoism historically often employed cordage or fabric made of hemp for ceremonial purposes (Olson 1997; Figure 12.11c), although this is not reflective of intoxicant usage. As late as the nineteenth century, there were cults and sects worshipping *C. sativa* in Africa (Benet 1975). Williams-Garcia (1975) described ritual usage of cannabis by an Indian tribal group in Mexico.

Touw (1981) reviewed evidence of shamanistic use in ancient China and suggested that the psychotropic properties of cannabis may have been known as early as five millennia ago there. Jiang et al. (2006) and Russo et al. (2008) documented a 2700-year-old grave, the Yanghai Tombs near Turpan, China, in which remains of apparently high-THC cannabis were detected, suggesting a possible ritualistic psychotropic purpose. (The DNA of this material was examined by Mukherjee et al. 2008, although the analysis is unclear with regard to relationships with modern varieties.)

Zoroastrianism, a monotheistic religion of Iran, was founded by the Prophet Zoroaster in ancient Persia approximately 3500 years ago and is still practiced by about three million devotees. Cannabis intoxication appears to have been a central activity in early Zoroastrian shamanic ecstasy (Mechoulam 1986).

Over the last millennium, cannabis consumption became more firmly entrenched in southern Asia from Afghanistan to India, than anywhere else in the world, both for medical and cultural purposes, some of which involved consumption of cannabis as an inebriant. Cannabis became intimately associated with religions of southern Asia (Aldrich 1977), and its sacred use in India predates written records (Hasan 1975). Not surprisingly, highly domesticated drug land races were selected there.

While *Cannabis* has been extensively used as an inebriating and medicinal drug for thousands of years in southern Asia and subsequently in the Near East (Figure 12.9), parts of Africa, and other Old World areas, widespread drug use simply did not develop in temperate region countries, where by contrast fiber hemp was raised. After the French war in Egypt and Syria (1798–1801), returning Napoleonic soldiers brought back knowledge of cannabis usage to France. Similarly, British physicians returning from India also introduced the intoxicant use of cannabis drugs to their homeland (see Chapter 13). In due course, the recreational use of cannabis became popular in Europe among intellectuals, who assembled in small "hashish clubs" in the nineteenth century. Most infamous of these was the "Club des Hashischins" of Paris (Figure 12.10), established around 1835, with monthly meetings in a hotel. The participants experimented with hashish and other drugs and included such famous French literary figures as Honoré de Balzac, Charles Baudelaire, Alexandre Dumas, Théophile Gautier, and Victor Hugo.

The use of cannabis for recreational, spiritual, and medicinal purposes was probably imported into the Americas by African slaves as early as the sixteenth century, becoming established in early times among low-income rural groups in South America. By the late nineteenth century, recreational marijuana usage had migrated to Mexico and the southern United States, where it remained a stigmatized drug associated with the poor and underprivileged, particularly with Hispanics and Blacks.

The use of *Cannabis* as a recreational inebriant in sophisticated, largely urban settings began substantially in the latter half of the twentieth century. In the 1960s, "hippies" made pilgrimages to Asia in search of enlightenment and established what came to be known as the "hippie trail" or "hashish trail" extending across Eurasia. Up until then, drug preparations of *Cannabis* were used predominantly as a recreational intoxicant in poor countries and the lower socioeconomic classes of developed nations. In the late 1960s, marijuana became associated with the rise of a hedonistic, psychedelic ethos, first among college students in the United States (Abel 1980; Booth 2004) and

FIGURE 12.9   "Les fumeurs de kiff" (hashish smokers) by Gabriel Ferrier (1847–1914). Public domain photo (Salon de Paris 1887, no. 908).

FIGURE 12.10   Artist's conception of the notorious nineteenth century Parisian "Club des Hashischins." Prepared by B. Flahey.

eventually over much of the world, with the consequent development of a huge international illicit market.

Tarter et al. (2012) noted that in the United States, "Policies aimed at curtailing substance use have been largely guided by ideology and political expediency effected primarily through the criminal justice system. The Eighteenth Amendment of the U.S. Constitution and the Volstead Act banning manufacture, transport and selling of alcohol beverages between 1919 and 1933, for example, culminated a long struggle spearheaded by the Anti-Saloon League, Prohibition Party, and Woman's Christian Temperance Union. Similarly, the first Director of the Federal Bureau of Narcotics, Harry J. Anslinger, demonized marijuana for primarily political reasons, namely to bolster the visibility, prestige and budget of the Federal Bureau of Narcotics." Cultivation, commerce, and consumption of drug preparations of *Cannabis* were also proscribed in most other countries during the twentieth century, but cannabis continues to contribute substantially to the current illicit drug problems of the world.

MODERN SPIRITUAL USAGE OF CANNABIS

Despite the extensive historical usage of cannabis for ritualistic purposes described previously, there is limited employment of marijuana today for religious usage. Information on Indian, particularly Hindu, religious usage of cannabis is presented in Bey and Zug (2004). Hindu devotees of Shiva believe that cannabis pleases this god (Acharya et al. 2014). Some Sikh festivals employ cannabis. Sufism has long employed cannabis. Of all current religious traditions, Rastafarianism in Jamaica is most associated with the use of marijuana, attributing divine power to the drug (Beaubrun 1983). Some American cults (an example is shown in Figure 12.11a and b) have taken the position that their use of cannabis is exempt from drug laws, but their claims have been rejected by courts. It should be kept in mind that a considerable amount of spiritual usage of cannabis was once concerned with the ritualistic use of hemp, not marijuana, as illustrated in Figure 12.11c.

FIGURE 12.11 Recent examples of spiritual use of cannabis. (a) Entrance sign of the First Church of Cannabis, founded in Indianapolis in 2015. Photo by Ayjazz (CC BY SA 4.0). (b) Mural inside the church, painted in the style of "The Creation of Adam," showing a joint being passed between hands. Photo by Janulus 144 (CC BY SA 4.0). (c) Japanese Shinto shrine with ceremonial rope made of hemp. Photo by Kamidana (CC BY 3.0).

CANNABIPHOBIA AND THE CULTURAL WAR ON MARIJUANA

During the early part of the twentieth century, marijuana was savagely villainized as a drug leading to extreme physical and mental degeneration (Figure 12.12). In particular, the 1936 American cult film *Reefer Madness* propagandized the evils of marijuana (described as "the plant with its roots in hell") in such an exaggerated and alarmist fashion—portraying users as homicidally depraved raving lunatics—that today it seems ridiculous. Nevertheless, since the latter half of the twentieth century and lasting until the present, marijuana has been a principal target of the "war on drugs" as declared in the United States by the Nixon administration but waged throughout the Western World. The use of cannabis was widely claimed to be associated with sexual permissiveness, dropping out of productive society, and a breakdown of culture and conventional morals. Young marijuana users were accused of developing an "amotivational syndrome," causing them to become alienated and unproductive, and cannabis was said to be a "gateway" inducement to harder drugs. Law enforcement has dedicated huge efforts to eradicating illicit material (Figure 12.13) and prosecuting and jailing millions of users.

GROWING PUBLIC ACCEPTANCE OF MARIJUANA

Despite substantial continuing condemnation of the use of marijuana, much of the public in Western nations has become tolerant or sympathetic to it. As pointed out by Leggett (2006):

"A sizeable share of the population in the world has experimented with cannabis and not experienced dramatic negative repercussions. It is widely understood that, unlike other drugs, one cannot die of a cannabis overdose and few people develop cannabis habits that force them into street crime or prostitution. Cannabis is not associated with violent behavior in many countries and its role in accidents is vague in the public mind. The stereotypical 'stoner' character has become celebrated in the popular media as harmless and somewhat endearing. Claims of purported medical benefits of cannabis have created the impression that cannabis is not only virtually harmless but that it can actually be beneficial to health." (See Graham 2014 cited earlier as asking: what if marijuana is actually good for public health?) As pointed out in Chapter 15, in the United States, a majority of the population has recently shifted to favoring the decriminalization of marijuana.

FIGURE 12.12 Lurid, mid-twentieth century, American governmental propaganda posters (in the public domain) demonizing marijuana as a catalyst for sexual deviance and psychosis. Ironically, such exaggerated warnings undermined the credibility of subsequent health cautions. Also ironically, there is substantial evidence that marijuana can indeed decrease inhibition and increase libido and sexual pleasure (Stuart et al. 2014).

THE GENETIC "IMPROVEMENT" OF MARIJUANA DUE TO LAW ENFORCEMENT

Ironically, law enforcement pressure has had the unintended effects of (1) driving marijuana production indoors, where it is harder to locate, and (2) increasing potency. Cannabis quality and yield efficiency have been greatly improved by breeders and cultivators, especially in the Netherlands and North America, since the early 1970s. Breeding has generated strains that are more potent, more productive, faster maturing, hardier, and more attractive to consumers. Yields have also been increased dramatically by improved cultivation techniques. The cultivation of elite female clones and the use of indoor production techniques that hide plants from the authorities (typically in bedrooms, basements, attics, closets, garages, or sheds) have been perfected. Growers are able to harvest up to six crops annually, with much greater or faster growth in smaller spaces than achieved previously.

Breeding of superior intoxicating strains of *C. sativa* has largely been done in a clandestine fashion because of the illegality of marijuana. However, by no means have marijuana breeders regarded themselves as engaging in a shameful activity, and indeed, many are proud of their achievements, often exhibiting photos of their best plants and buds on the Internet (usually under a pseudonym). There are also competitions for the most impressive strains, particularly in the Netherlands. Best known of these is the annual *High Times* (magazine) "Cannabis Cup" in Amsterdam (Smith 2012; Figure 12.14). The breeding achievements of motivated amateur horticulturists can be remarkable.

FIGURE 12.13 Law enforcement activities by the U.S. government to control the illicit use of *C. sativa*. (a) A seizure of about a ton of hashish in Afghanistan. Photo by isafmedia (CC BY 2.0). (b) Helicopter spraying of Paraquat herbicide on a field of marijuana. (c) Burning seized marijuana. (d) Clandestine indoor cultivation. (Photos b–d provided by the U.S. Drug Enforcement Administration.)

FIGURE 12.14 Awards table at the 27th *High Times* Cannabis Cup ceremony in 2014 in Amsterdam (CC BY 2.0).

In the eighteenth century, "gooseberry clubs" became popular in Britain, with the goal of giving prizes for the heaviest gooseberries (the fruit of *Ribes* species). Previously, wild gooseberries weighed only about 7 g and were about the size of a small pea, but the breeding efforts produced fruits resembling small apples and weighing as much as 57 g (Small 2013a).

The authorities attempting to suppress marijuana cultivation have been faced with the daunting problem of limited international control over distribution of seeds and knowledge. The Netherlands has been uniquely responsible for much of this situation, as it has been substantially free to develop marijuana strains and knowledge and to disseminate both throughout the world via the Internet. The information revolution has spread technical knowledge globally, while Web blogs and chat groups provide tips about every aspect concerned with acquiring, growing, preparing, and using marijuana.

FORMAL BOTANICAL NOMENCLATURE AND "STRAINS" OF *CANNABIS SATIVA*

Terms used in botanical classification are dealt with in Chapter 18, but one technical point bears mention here in order to correctly refer to genetic variations of marijuana plants. Article 2.2 of the current nomenclatural code for cultivated plants (Brickell et al. 2009), a legalistic document that governs names for cultivated plants, forbids the use of the term "strain" as equivalent to "cultivar" for the purpose of formal recognition. Very few marijuana strains satisfy the descriptive requirements for cultivar recognition, although many marijuana cultivars (mostly grown for fiber or oilseed rather than cannabinoids) do and by convention are denoted in single quotes. However, *Cannabis* strains are conceptually identical to *Cannabis* cultivars. Snoeijer (2002) treated *Cannabis* strain names as equivalent to cultivar names.

THE EVOLUTION OF HIGH-THC STRAINS

High-THC forms of *Cannabis* were initially selected many centuries (possibly millennia) ago, and during these early times, fairly primitive techniques were employed to make intoxicant preparations. Particularly in recent decades, a considerable understanding of the biochemistry and genetic control of cannabinoid metabolism has been achieved, and strains are now being generated that are rich in given cannabinoids for potential medicinal applications. Sophisticated techniques for breeding strains have been developed, including the generation of all-female lines (Chapter 4). Technologies (described in this chapter) have been created to collect and concentrate the THC-rich heads of the glandular trichomes, and this development may have resulted in the selection of strains in which the THC-rich trichome heads separate readily so that they can be collected easily.

In previous chapters, information has been provided on some of the ways that the characteristics of high-THC strains have evolved. Strains have been chosen that differ in architecture (Chapter 6) and cannabinoid profile (as noted in Chapter 11). Geographical biotypes have been found with one or more rare cannabinoids in unusually high presence (Chapter 11), which is probably the result of genetic drift (change in population genetics occurring in small populations simply by haphazard survival of unusual plants). A variety of different essential oil profiles seem to have been selected in high-THC strains (Chapter 9). There also seems to have been selection for concentration and distribution of the secretory glands, with very large densities of the glands and larger glands present on the floral bracts of some strains (Chapter 11). In response to demand for very high levels of THC, there has been selection for congested female inflorescences (production of numerous, well-formed "buds" being a recent quality criterion; Chapter 11).

Chapter 6 provided information on the evolution of shoot architecture in the two groups of high-THC plants ("indica type" and "sativa type"), and Chapter 3 provided information on how the seeds of domesticated plants (including the high-THC groups) have been modified by comparison with wild plants. This information is not repeated here.

The two basic kinds of high-THC plants (sativa type, characterized by very high THC levels, and indica type, characterized by moderate amounts of THC supplemented by noneuphoric CBD) are described next. They have become foundational breeding material for generating by hybridization a wide range of marijuana strains.

"SATIVA TYPE" AND "INDICA TYPE," THE TWO DOMESTICATED KINDS OF MARIJUANA PLANTS

Two discernibly different groups of high-THC cannabis plants were selected in Asia: "sativa type" and "indica type." The ancient distribution of these is shown in Figure 2.7, and in Figure 2.8, it is pointed out how the much more popular "sativa type" has been distributed in much of the world. In Figure 18.13, it is noted that the indica type probably arose from the sativa type and that extensive hybrids have been generated between the two kinds. The terms "indica" and "sativa" are widely employed, in the senses explained in this section, in innumerable books and websites providing instructions on how to (usually illegally) cultivate marijuana and more recently for medical marijuana.

Table 12.1 summarizes differences that have been alleged to distinguish the two kinds (no adequate statistically based study of differences has been published, and since hybrids between the two kinds dominate strains of marijuana currently grown, the two kinds are best considered as polar extremes connected by a continuous spectrum of intermediate forms). The two kinds are contrasted in Figures 12.15 through 12.17.

Strains of the sativa type tend to resemble European fiber cultivars, often being almost as tall although usually much more branched and tending to have relatively narrower leaflets. Sativa type strains characteristically have very high THC level in the cannabinoids and no or small amounts of

---

TABLE 12.1

Alleged Differences between the Two Basic Kinds of Domesticated Marijuana Plants

Group (Marijuana Trade Terminology)	Sativa Type	Indica Type
Early distribution area (see Figure 2.7)	Widespread (southern Asia)	Restricted (Afghanistan, Pakistan, northwestern India)
Seasonal adaptation	Relatively long (late-maturing), often in semi-tropical regions	Relatively short (early-maturing), adapted to relatively cool, arid regions
Height (under optimal growth conditions)	Relatively tall (2–4 m)	Relative short (1–2 m)
Habit	Diffusely branched (longer internodes); less dense, more elongated "buds"	Bushy (short internodes), often conical; very dense, more compact "buds"
Leaflet width	Leaflets narrow	Leaflets broad
Intensity of leaf color	Leaves lighter green	Leaves dark green
Length of season	Relatively late maturation	Relatively early maturation
Aroma (i.e., odor + "taste")	Relatively pleasant aroma (often described as "sweet")	Relatively poorer aroma (sometimes described as "sour" and "acrid")
Ease of detachment of heads from secretory glands (McPartland and Guy 2004a)	Variable	Easily detached
Presence of CBD	Little or no CBD	Substantial CBD
Alleged psychological effects	Relatively euphoric: a "cerebral high" promoting energy and creative thought (occasionally panic attacks in inexperienced users, or a drained feeling); recommended for daytime use	Relatively sedative: physically relaxing, producing lethargy ("couchlock"); recommended as a "nightcap" (cf. information regarding couchlock, in this chapter)

Note: Most of these differences are discussed in Clarke (1998a) and Clarke and Merlin (2013).

FIGURE 12.15   Contrast of the taller "sativa type" (above) and the shorter "indica type" (below) marijuana plants of *C. sativa*. Prepared by B. Flahey.

CBD. As pointed out in Chapter 18, usage of the term "sativa" to indicate extremely intoxicating (high-THC) plants is quite inconsistent with the tradition of employing the "epithet" (a word used in scientific names) taxonomically for nonintoxicant plants. Sativa type strains are extremely widespread in the illicit trade of Western nations.

Indica type strains tend to be short (about a meter in height) and compact under the often inhospitable conditions under which they are typically grown in Asia. They are often also highly branched, with large leaves and wide leaflets. The appearance is often reminiscent of a miniature, conical Christmas tree. Strains of this group characteristically have moderate levels of both THC and CBD in their cannabinoid profile. Like the sativa type, the indica type has historically been employed to produce hashish in southern Asia, particularly in Afghanistan and neighboring countries. Hashish is prepared by pooling collections from many plants, so individual plants may vary in proportions of cannabinoids (i.e., not all plants necessarily have moderate levels of both THC and CBD).

FIGURE 12.16   Contrast of the "sativa type" ([a]; note the tall stature and narrow leaflets) and the "indica type" ([b, c]; note limited stature and wide leaflets) marijuana plants of *C. sativa*. (a) Photographed at the U.S. Government marijuana production site at the University of Mississippi, Oxford (public domain photo). (b) Photo by Mr TM (CC BY 3.0). (c) Photo by otrs:2009060510011997 (CC BY 2.0).

FIGURE 12.17   A contrast of leaves of "sativa type" (left; narrow leaflets) and "indica type" (right; wide leaflets) marijuana plants. Photo by Transmitdistort (CC BY 3.0).

Clarke (1998a) and McPartland and Guy (2004a) interpreted indica type strains as having evolved in the cold, arid regions of Afghanistan and western Turkmenistan and explained their short height as an adaptation to the relatively short growing season. The relatively early-flowering nature of indica type strains is also an adaptation to a relatively short growing season.

Sativa type strains are very potent (higher in THC than most indica type strains) and hence more popular, although harder to grow indoors where room height is limited, because of their tallness. Hybrids between the two groups have proven to be well adapted to indoor cultivation and are progressively being marketed (Clarke and Watson 2006). Increasingly, strains with alleged percentages of each type are being sold.

There are varying descriptions in the literature about the contrasting psychological effects of indica type and sativa type strains (see, for example, Hazekamp and Fischedick 2012 and Smith 2012). These descriptions generally credit the high-THC sativa type with producing a more euphoric "high" and the lower-THC indica type with substantial CBD with producing a more subdued but attenuated (longer-lasting) experience, consistent not just with the lower THC content but more particularly with how CBD in marijuana substantially alters the effects of THC, as explained in Chapter 13. Erkelens and Hazekamp (2014) summarized the alleged effects as follows: "The sativa high is often characterized as uplifting and energetic. The effects are mostly cerebral (head-high), also described as spacey or hallucinogenic. This type gives a feeling of optimism and wellbeing, as well as providing a good measure of pain relief for certain symptoms... Sativa strains are generally considered a good choice for daytime smoking. In contrast, the indica high is most often described as a pleasant body buzz (body-high). Indica strains are primarily enjoyed for relaxation, stress relief, and for an overall sense of calm and serenity. They are supposedly effective for overall body pain relief, and often used in the treatment of insomnia; they are the late-evening choice of many smokers as an aid for uninterrupted sleep."

In Asia, strains of both kinds were often used to prepare hashish, but in most Western nations, they are predominantly employed to prepare marijuana. Traditional Asian hashish is typically rich in both the intoxicant THC and the noneuphoriant CBD, and indica type land races have been particularly selected for making hashish. By contrast, most high-THC sativa type cultivars have been selected just for THC, and indeed, most have limited or no CBD. An explanation for the presence of CBD in traditional hashish land races was offered by Clarke and Watson (2006): "Hashish cultivars are usually selected for resin quantity rather than potency, so the farmer chooses plants and saves seed by observing which one produces the most resin, unaware of whether it contains predominantly THC or CBD."

SELECTION FOR COLOR IN MARIJUANA STRAINS

The attraction that humans have for white or at least light shades of seeds was pointed out in Chapter 8. Another example of human preference for light hues is provided by the flowering parts of marijuana strains that have been selected by clandestine breeders in the last several decades. There appears to have been selection for strains developing whitish inflorescences (Figure 12.18). The immature stigmas of the female flowers are whitish, although becoming reddish or brown with age. High concentrations of female flowers in the inflorescence of marijuana strains is extremely desirable, since this increases potency, and because higher whiteness is reflective of more female flowers, selection for whiteness has been a simple way of selecting for higher potency and yield. The secretory glands responsible for producing THC are present in high density on the perigonal bracts, and these often glisten under strong light, also contributing to a whitish appearance of the female inflorescence. So-called "white strains" are very popular, as reflected by such names as White Diesel, White Fire, White Gold, White Haze, White Ice, White Label, White Queen, White Rhino, White Russian, White Skunk, White Widow, Early Pearl, Silver Haze, and X-Haze.

Humans are also fond of mutations that develop purplish foliage in domesticated plants, due to the prominence of anthocyanin pigments (e.g., Crimson King, a very popular variant of Norway

FIGURE 12.18   "Buds" of marijuana strains with notable development of white stigmas. Left: White Dwarf. Photo by Ankari80 (CC BY 3.0). Right: Photo by Psychonaught (released into the public domain).

maple; red (purple) cabbage). As is evident in Figure 12.19, when *C. sativa* is exposed to significant frost, it tends to become quite purple (or less green, since chlorophyll tends to degrade, revealing the anthocyanins), and sometimes, the same effect is noticed at high altitudes (perhaps related to high, damaging insolation), demonstrating a propensity for violet coloration. Often, purple coloration develops simply because of cultural conditions (Figure 12.19, right). Dewey (1913) found a purple-leaved mutation arising in Chinese hemp (a fiber biotype, not a marijuana strain), inbred for nine years in Kentucky, Minnesota, and Washington, DC. He named the inbred cultivar Kymington (based on Ky-Min-[Wash]-ington).

Purple coloration of the inflorescences of marijuana strains became quite attractive to consumers in the second half of the 1970s (Clarke and Merlin 2013; note Figure 12.20), many expressing the belief that such varieties are qualitatively superior. Examples of purplish strain names include Purple Bubba Kush, Purple Butter, Purple Cheese, Purple Diesel, Purple Dogg, Purple Erkle, Purple Haze, Purple Kush, Purple Maroc, Purple Monkey Balls, Purple Nepal, Purple Passion, Purple Pine, Purple Pineberry, Purple Power, Purple Pussy, Purple Snow, Purple Urkle, Purple Wreck, Grand Daddy Purple, Blackberry, Blueberry, Grape Ape, and Mendocino Purple. The development of purple coloration in foliage and/or stems occurs in some marijuana strains, likely reflecting past

FIGURE 12.19   Anthocyanin (purplish) coloration in *C. sativa*. Left: Purple color induced in foliage by exposure to frost in late autumn. Right: Purple color induced in the marijuana strain Bubba Kush by cultural conditions. Photo courtesy of Steve Naraine.

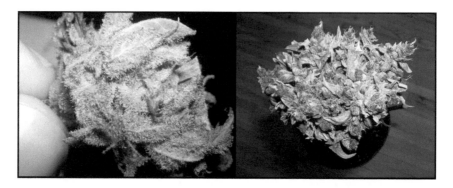

FIGURE 12.20 Marijuana strains of *C. sativa* illustrating selection of purple (anthocyanin) coloration under domestication. Left: A bud of Power to the Purple. Photo by Psychonaught, released into the public domain. Right: An inflorescence with numerous buds of Purple Haze. Photo by HansRoht (CC BY 3.0).

selection for expression of anthocyanin pigmentation, and this sometimes alarms illicit marijuana growers suspecting that their plants are diseased or haven't been cared for properly.

PLANT PRODUCTION

Information on outdoor cultivation of industrial hemp for fiber, oilseed, and essential oil is provided respectively in Chapters 7 through 9. Most illicit marijuana is also produced outdoors, although wind and rain can have detrimental effects on marijuana quality. For the most part, outdoor cultivation requirements for marijuana are similar to the requirements for industrial hempseed. Information on authorized indoor cultivation of medicinal marijuana is provided in the next chapter, where it is noted that over a hundred books, in addition to countless websites, provide detailed directions for the illicit cultivation of marijuana plants and consequent preparation of cannabis products. There is nothing fundamentally different in growing *C. sativa* for legitimate or illegitimate purposes, except for the needs for stealth and concealment when cultivating the plant illegally. Of course, this book is not intended to provide guidance on illegal cultivation.

DISTORTION OF BOTANICAL AND HORTICULTURAL TERMS BY THE MARIJUANA TRADE

Cannabis sativa is a plant and is most precisely described by the scientific terminology conventionally employed by biologists, agriculturists, and horticulturalists. For the past half-century, marijuana has been produced, traded, and employed mostly by people with limited knowledge, interest, and appreciation of "official" terminology and indeed who have often delighted in adopting terms that were unintelligible to conventional society. Unfortunately, some of these terms are ambiguous; i.e., they have one meaning scientifically and another meaning in the context of marijuana-specific street language. Frequently, technical botanical terms have been misinterpreted by the marijuana-using community (facetiously, one may ask why these intellectual lapses occurred). Sometimes, the differences in the meanings are subtle and require thought to understand exactly how a scientific term has been distorted. The terms in Table 12.2 especially often lead to misunderstanding.

TECHNOLOGIES FOR PREPARING CANNABIS DRUGS

Marijuana is consumed in a wide variety of forms, as discussed in this and the next chapter. These include several preparations that are smoked, edible formulations, skin patches, ointments, sprays, capsules, suppositories, and even sex lubricants for women. This chapter is concerned with formulations, apparatus,

TABLE 12.2

Examples of How "Marijuana Language" Has Distorted Correct Scientific Terminology

Term	Scientific Meaning	"Marijuana Language" Meaning
Bud	Meristem (growing point of a part of a plant, producing a stem, flower, or leaf; "eye" of tubers like potato)	Inflorescence (congested female flowering axis; see the discussion of why marijuana is not "flowers" in Chapter 1)
Indica	Part of the scientific name *Cannabis indica*, or the name *C. sativa* subsp. *indica*, conventionally employed to designate *all* cannabis plants that are rich in the intoxicating constituent THC	Employed to designate a distinctive class of intoxicating plants that have moderate levels of *both* THC and the nonintoxicating constituent CBD (see discussion of indica type in Chapter 18)
Pistil	Female portion of a flower (style + stigma + ovary)	Stigma (pollen-receptive part of a flower)
Pollen	Male fertilizing agent (functionally like animal sperm, although more complex)	Secretory heads of cannabis glands, collected by filtering techniques
Sativa	Part of the scientific name *C. sativa*, or the name *C. sativa* subsp. *sativa*, conventionally employed to designate all cannabis plants that are very low in the intoxicating constituent THC	Employed to designate intoxicating plants that have very high levels of THC and very low or no levels of the nonintoxicating constituent CBD (see Chapter 18)
Style	Transitional area of a female flower between the stigma (pollen receptive part) and ovary (seed-containing part); it is nonreceptive to pollen	Stigma (pollen receptive part of a female flower)
Trichomes	Small appendages on the surface of plants (includes "hairs"); in reference to cannabis, particularly the hairs tipped with resin-containing heads (i.e., both stalks and the resin-containing heads)	Resin-containing heads of cannabis stalked glandular trichomes

and methods used mainly for recreational purposes, while the next chapter is concerned with medical technologies, many of which are based on the same kinds of apparatus used for recreational marijuana.

Marijuana

"Manicured marijuana" is composed of flowering parts of the plant coupled with associated small leaves, prepared using intoxicant varieties. It is comparable in texture to smoking tobacco. Marijuana is conventionally prepared by (1) breaking up the dried flowering tops and eliminating all but the smallest twigs, (2) forcing the resulting material through a coarse screen, and (optionally) (3) crumbling. The result is a mixture of plant particles, including the tiny secretory trichome glands that contain most of the resin (some resin is smeared on plant particles during preparation). Up until the last two decades, in the Western world, marijuana often included a substantial content of seeds (which do not contain THC) and foliage (which contains limited THC, as illustrated in Figure 12.21, bottom). As a result, marijuana in the past usually contained no more than 5% THC, often less. Currently, marijuana rarely has seeds or larger leaves, and the THC content is at least 5%, sometimes as high as 25%. ElSohly et al. (2016) surveyed about 39,000 samples of cannabis confiscated by the American Drug Enforcement Administration between January 1, 1995 and December 31, 2014. The proportion of sinsemilla (seedless) samples increased, and (consistent with this) the potency of illicit cannabis plant material consistently rose from approximately 4% in 1995 to approximately 12% in 2014 (CBD content fell from approximately 0.28% in 2001 to <0.15% in 2014). Marijuana is sometimes referred to as "herbal-type" cannabis, in contrast to hashish, termed a "resin-type" form of cannabis.

The perigonal bracts subtending the female flowers are very rich in THC, and the market for marijuana has evolved toward the use of the unfertilized female inflorescences (which contain these bracts), i.e., the congested flowering branches, usually referred to as "buds," much less frequently termed "colas," illustrated in Figure 12.22a, b, and c. "Sinsemilla" is the most

FIGURE 12.21 Old-fashioned marijuana. Top: Sieving mature flowering tops with substantial amounts of foliage through a screen. Bottom: Grades of marijuana commonly encountered in the 1960s through the 1980s. Increasing quality is indicated by lesser content of twigs and seeds, which contain little or no THC. In the past, THC content of herbal marijuana rarely exceeded 5% dry weight.

frequent term, collectively referring to high-THC marijuana prepared mostly from the unfertilized female inflorescences. In the United Kingdom, sinsemilla is often called "skunk," a transfer of the word from the well-known strain Skunk #1. Similarly, "Kush" (part of the name of numerous marijuana strains) has become somewhat synonymous with high-grade marijuana in North America. Whole buds rather than those that are ground up are a favored commercial form of sinsemilla. Races with female marijuana plants have been selected to produce flowering heads with abundant flowers in tight heads. Buds have become much more popular as a sales item because they are usually a reliable indicator of high-grade marijuana (it is impossible to judge the quality of manicured marijuana without smoking it or measuring THC content). Contents of 10% to 20% THC are common in street grade bud. Rarely, 30% THC marijuana is found in illicitly sold material, although such high-potency material is claimed by some authorized medicinal marijuana suppliers (and can be achieved by careful trimming away of leaves). Buds are too large to smoke directly, so they are broken up into a tobacco-like consistency, often using a herb grinder (shown in Figure 12.22d).

TRADITIONAL HASHISH

Hashish (Figure 12.23c) is a relatively pure preparation of the resinous secretions of intoxicant varieties of the plant. As traditionally made in Asia, it is prepared by a variety of methods (see

FIGURE 12.22   (a, b) Marijuana "buds" and their processing for smoking. (a) Buds (unfertilized, congested, female inflorescences, with large numbers of perigonal bracts rich in secretory glands), increasingly popular since the 1980s. THC content typically ranges between 10% and 20%. Achenes ("seeds"), which do not contain cannabinoids, are not present. Often, the buds are manicured (the small unifoliolate leaves present are trimmed away with scissors) to additionally increase THC content. (a) Bud of the strain Blue Dream. Photo by Psychonaught (released into the public domain). (b) Buds of the strains Platinum Bubba on top and Skywalker OG on bottom. Photo by Coaster420 (released into the public domain). (c) Portions of buds in a container, the kind of commercial product that is currently the most popular form of cannabis. Photo by Cannabis Culture (CC BY 2.0). (d) Herb grinder, a device composed of two separable halves with sharp teeth or pegs that shred contained material when the halves are rotated. Originally used to shred herbs and spices for culinary purposes, they are now more frequently employed to shred bud into finely ground bits that burn evenly. Photo (public domain) by Liquid Splitter.

Clarke 1998a, Hamayun and Shinwari 2004, and Figure 12.23a and b) but is always a mixture of resinous herbal material collected from the female inflorescences of *C. sativa*. It is predominantly prepared by filtering cannabis material through very fine fabric screens (such as silk) or sieves. Sieving requires the plants be dried first, and because applying artificial heat is usually too expensive, an arid climate is essential (such as provided in regions of Afghanistan, Lebanon, Morocco, and Pakistan). Additional treatments following collection of the powdery sieved particles vary depending on region, but usually the material is compressed, sometimes gently heated, resulting in a solidified, sticky mass, typically pressed or rolled to form hardened resinous cakes. Hashish in illicit markets typically has a THC content of 5% to 25% (levels as high as 45% have been reported). Texture or consistency varies from putty-like to brittle or dusty. Colors encountered include sandy, reddish, and black (often due to oxidation, reflective of aging or manner of handling). Green color is often due to the presence of unwanted plant material or collection when the plants were immature. Adulterants, such as oils or bulking agents, are sometimes introduced.

FIGURE 12.23 Traditional methods of preparing hashish (mostly encountered in Asia). (a) Preparation of hashish by sieving through a fine silk screen. Drawn by B. Flahey. (b) Preparation of hand-rubbed hashish ("charas"), as once practiced in Asia. Secretory glands and resin rich in THC accumulate on the hands during prolonged manual contact with the plant and are scraped off. Drawn by B. Brookes. (c) Confiscated bricks and cubes of compressed hashish. Such preparations are primarily an Asian product and are currently often made from indica type races with more or less equal amounts of THC and CBD. THC contents generally range from 5% to 15%, dry weight.

In contrast to filtering or sieving (to produce "sieved hashish"), an alternative method of preparing hashish in Asia is to rub the female inflorescences by hand so that the sticky resin glands and secretions stick to the hands (Figure 12.23b) and are scraped off (to produce "rubbed hashish"). Similarly in the past, people dressed in leather brushed against the sticky inflorescences until resin accumulated on their garments, subsequently scraping off the resin (Bouquet 1950). The hand-rubbing technique has been mostly abandoned because it is so labor-intensive, although Clarke (1998a) wrote that hand-rubbed charas is sold in Himalchal Pradesh, Kashmir, Chitral, and Nepal. Abel (1980) stated that in Nepal, workers ran naked through the cannabis fields, and the sticky resin was scraped from their bodies. Reports of this rather gross practice are probably apocryphal, indeed likely mythical (Ethan Russo, personal communication), as anyone working in a field of *C. sativa* quickly learns how abrasive the foliage can be to bare flesh. Hamayun and Shinwari (2004) is an excellent anthropological study of traditional methods of producing hashish.

The abandoned hand-rubbing technique requires that sticky resin accumulate on the plant's surfaces. Stickiness is due to the secretory glands releasing terpene and cannabinoid secretions over the outer surface of the glands. Probably the agitation from wind tends to break some gland heads, which exude their resin. In very windy, dry, or cold environments, secretions tend to volatilize more readily, decreasing stickiness (terpenes volatilize readily, THC does not); by contrast, in hot, still environments (whether outdoors or under intense grow-lights), secretions appear to accumulate more readily, and the plant surfaces can become very sticky. It is unclear whether high-THC land races were selected that were particularly suitable for hashish preparation by the hand-rubbing technique, by virtue of tending to secrete resin readily rather than retaining it within the gland heads, but this seems plausible.

Traditional hashish is typically higher in THC content than traditional marijuana although buds can be higher in THC. Hashish is also more compact and retains THC levels longer (outer parts of a brick of hashish oxidize, but inner portions are relatively protected from oxygen and light). Hashish is more portable, transportable, and easier to conceal (both visually and with respect to odor). However, traditional hashish requires much more labor and land to produce than marijuana of equivalent psychoactive status. Its production is largely restricted to parts of Asia, and its export is mostly limited to Eurasia.

SOLVENT EXTRACTS

Hashish in the illicit trade may be made by the use of solvents—often a fire and explosion hazard for preparers. There are several counterculture guides on the preparation of such cannabinoid extracts (e.g., Gold 1973; Starks 1990). The products range from liquid form (with substantial solvent remaining) through thick oil (most solvent removed) and viscous or hard consistency, depending on the extent of distillation (Figure 12.24). A variety of terms, most of them slang, are applied, the most common of which include hashish oil (hash oil), butane hash (when prepared with butane as a solvent), liquid hashish, honey oil, wax, dabs, shatter (referring to a glass-like consistency that often snaps or "shatters"), budder, and nug runs. (Note that the word "oil" in these phrases does not necessarily indicate liquid form but seems to have been adopted because of viscous consistency and/or stickiness. "Hashish oil" may be liquid, semisolid, or tar-like. "Waxes" have the consistency of a sticky wax.) Such products may have a THC content of 20%–50% (levels exceeding 60% are rarely reported). Solvent-prepared hashish is usually too strong to consume directly and is normally cut (diluted) using tobacco or marijuana. Given the lack of quality oversight in illegal operations, these formulations often contain toxic residues and may be particularly dangerous. Nevertheless, there is considerable home preparation of hash oil, often using nonbud material, material trimmed away from the buds ("trim"), or remains ("shake") after filtering the more potent fractions. Romano and Hazekamp (2013) analyzed the comparative value of employing naphtha, petroleum ether, ethanol, and olive oil as solvents in preparing cannabis oil and noted that olive oil was the safest and cheapest.

FIGURE 12.24   Home-prepared chemical extracts of *C. sativa* rich in THC. Left: "Hash oil." Photo (public domain) by Erik Fenderson. Right: "Butane honey oil." Photo by Vjiced (CC BY 3.0). Preparations (usually illicit) such as these are often dangerous because of the possible presence of toxic chemical impurities, very high THC content, and deliberate contamination with dangerous drugs.

Solvents are also used to extract cannabinoids by the medical cannabis industry, and reputable products are often available in the form of liquids (especially alcoholic) containing considerable dissolved cannabinoids, as discussed in the next chapter.

ADVANCED NONSOLVENT TECHNOLOGIES FOR PREPARING CONCENTRATES OF GLAND HEADS

New technologies, not employing solvents, have been created in Western countries to produce preparations, best termed "resin powder," which are rich in the THC-containing resin glands (or their heads). Such concentrates are commonly termed "pollen," "crystal," "bubble hash," and "kief" and are also known by a variety of other names. The Asian tradition of using filters is employed, but the millipore screens commonly used have much smaller openings (50–150 μm in diameter), and the techniques utilized produce a material that is very much richer in presence of secretory glands, very much lower in presence of other herbal material, and (usually) higher in THC, by comparison with conventional Asian hashish. Clarke (1998a) is widely considered to be the "gold standard" on the topic of preparing potent marijuana by such nonsolvent methods (especially see the chapter "High-Tech Hashish-Making"). Although remarkably ingenious systems have been devised to separate and collect the cannabinoid-rich trichome gland heads of *C. sativa*, similar systems have been employed to collect the gland heads of its relative, hop (*Humulus lupulus*), employed as a flavorant and medicinal (Bishop 1966; Rigby 2000).

Cannabis "resin powder" is produced by sieving high-THC parts of the cannabis plant through very fine-pored screens. Crude grades of cannabis resin powder are in fact the basis of most traditionally produced hashish (hashish, in essence, is cannabis resin powder that has been very strongly compressed into massed material). However, in the last two decades, techniques and apparatus have become available that produce cannabis resin powder of exceptional potency (sometimes with about 50% THC). To date, highly potent resin powder has been produced as "connoisseur," counterculture, illicit, or quasi-legal drug products that are very expensive and available in limited supply. Such high-THC material is generated (1) from bud, to produce very high THC "gourmet" material and (2) from the "nonbud" (i.e., usually discarded), low-THC parts of marijuana plants, in order to salvage a high-grade of cannabis drug. The expense of such high-THC preparations, due to the high cost of preparation, is the chief factor limiting their popularity. In the illicit trade, the very large wastage factors mean that high-quality marijuana is generally not used for producing cannabis resin powder. Rather, resin powder is produced as a salvage operation based on waste material that otherwise is simply discarded. Substantial amounts of high-grade marijuana must be sacrificed to produce cannabis resin powder. For example, starting with material of 12% THC, 1 kg of material would have to be sacrificed to produce just 1 g of resin powder of 30% THC content (i.e., 99.9% of the starting material is "wasted"). The attraction is that relatively little material needs to be smoked. However, because of the difficulty in smoking the very small amount of material required to become "high," it is occasionally diluted with regular marijuana or tobacco. The product is powdery in nature, hence adaptable to dispersion in marijuana for purposes of increasing the THC content of the latter. Because it is so expensive, deterioration is a major concern, but resin powder can be stored long-term under appropriate (very cold, dry, dark) conditions for later use.

The terms "pollen" and "crystal" are currently widely applied to cannabis resin powder. Uncompressed resin powder is often referred to as "kief" (sometimes "kif"). Very-high-quality compressed resin powder is often known as "bubble hash" (Figure 12.25), an expression reflecting the frequent occurring of bubbling when the preparations are burned for consumption. The technologies described in the following concentrate the resin glands into a fine powder, reminiscent of plant pollen, and hence the term "pollen" was taken up as also designating the concentrated resin glands. The term "pollen" is used almost exclusively in the illicit drug counterculture cannabis community but is inappropriate since it is incomprehensible to most people. Nevertheless, when searching for information on cannabis resin gland preparations, the term "pollen" needs to be considered. Searches for "crystal" often produce information for crystal meth (amphetamine).

FIGURE 12.25 "Bubble hash," very potent forms of hashish (THC content has been claimed to sometimes exceed 50%) often prepared from resin powder. Photo at left by Andres Rodriguez, photo at right by J. Adams (both CC BY 2.0).

In the illicit drug counterculture, cannabis resin powder (which indeed is a powdery preparation) is compressed (e.g., by 5-ton hydraulic presses modified for the purpose), so that the preparation does have a superficial similarity to Asian hashish. Illicit drug counterculture publications use the term "hashish" to refer not only to traditional Asian styles of hashish but also to compressed cannabis resin powder, although the latter is different from traditional hashish. Clarke (1998a) refers to resin gland preparations produced by modern technologies as "high-tech hashish." When searching for information on cannabis resin powder, the term "hashish" needs to be employed.

Preparation of resin powder by modern techniques involves a combination of (1) very carefully regulated and limited application of force to separate secretory gland heads from the remainder of the plant materials and (2) the use of fine sieves (with very small pores). The sieves have holes 50–150 µm in diameter, the aperture size varied to separate the secretory glands from other plant materials. As noted in Clarke (1998a), marijuana varieties differ widely in gland size and so the filters used to produce resin powder should have pore sizes appropriate to the range of gland sizes. As noted in Small and Naraine (2016b), gland heads decrease in size with age, which can also affect the appropriate pore size required. Agitation and/or physical pressure is used to separate the glands and may be preceded by freezing to facilitate separation of intact glands. In "wet" techniques, dispersion in water is also employed, taking advantage of the principles that the cannabinoids are largely immiscible in water and differ in density from other parts of the plant. The result (potentially) is the production of a grade of material that is much richer in THC level than conventional marijuana. Portable handheld devices known as "kief boxes" are often used to transform small amounts of marijuana into a relatively crude grade of resin powder for personal use (Figure 12.26d).

The following are principal techniques employed to produce resin powder.

A. "Dry" technologies
  a. Vibration
    1. "Flat-screening": The simplest automated apparatus is a motor-driven shaker-sieve, preferably with both up-and-down as well as side-to-side motion. The material is placed in a container, the horizontal sieve forming the bottom of the container, and the container is kept in gentle motion for a limited period of time.
    2. "Drum-screening": Alternatively, a cylindrical container constructed of sieve material may hold the material, and the cylinder rotated, as in a conventional clothes dryer. Several designs are available. Drum-screening is considered to be preferable to flat-screening.

      With these kinds of devices, it is critical to limit the degree and period of agitation so that primarily the larger secretory gland-heads are separated. These are the first to separate. If agitation is too strong or continued for too long a period, the result is that additional plant materials pass through the pores of the sieve material, and the THC

FIGURE 12.26 Top: recent commercially available extraction systems for preparation of purified, high-THC concentrates of secretory glands starting with herbal material (leaves and flowers). (a) The "Pollinator" is a dry sifting machine. Herbal material is placed in the revolving drum, which is perforated with 150 μm holes. Resin glands are expelled through the holes and collect in the box containing the drum. (b) The "Bubbleator" is constructed like a small washing machine. Frozen herbal material is placed, along with ice water, in a series of bags that are perforated with holes of decreasing size that permit the resin glands to be expelled. These in turn are placed in the device, which agitates the bags for a period, and then the separated resin glands are purified by additional sieving, and dried. This device takes advantage of the insolubility of the resin in water and the brittleness of the glands when frozen. (c) The "Ice-O-Lator" is a similar but simpler apparatus, in which an agitating device is placed on top of a bucket. Detailed operating instructions are available at various websites. These devices may be considered to be illegal drug paraphernalia in some countries. Photographs courtesy of Mila and Chimed Jansen of the Pollinator Company. Bottom: (d) "Kief box" ("pollen box"). This is a small box fitted with a fine screen through which bud is gently sifted, allowing mostly secretory trichome gland heads to fall through the screen onto a collection plate. (e) Cannabis resin powder. Photo by Mjpression (CC BY 3.0).

level of the resulting resin powder is decreased. Moreover, by freezing the starting material just prior to sieving, the stalked glands become much more easily detached, facilitating separation of high-grade resin powder (illustrated in Figure 12.26d). The most widely advertised, sold, and used apparatus in the dry technology category is the Pollinator (Figure 12.26a), a device inspired by a clothes dryer.

b. Sonication (ultrasonic vibration)

Ultrasonic vibrators are an alternative to the use of motor-driven shakers. Commercial sonicator models, employing a liquid bath, are widely used to clean by shaking dirt off objects. Because the marijuana used as starting material must not contact liquid, the liquid either is simply not placed in the bath chamber or is first placed in a water-tight container. The resin powder collects at the base of the container or bath chamber. Commercially available devices specifically employing sonication for the production of cannabis resin powder do not seem to be available.

B. "Wet" technologies

Wet technologies exploit the fact that mature secretory glands are heavier than water (as well as the fact that the resin in the gland is basically not dissolvable in water), while most plant parts are lighter than water. When mixed with water, cannabis powder resin can thus be substantially separated. The principal marketed devices also utilize freezing to make the secretory glands more separable, combining this with filters and agitation.

The Bubbleator (Figure 12.26b) resembles a miniature washing machine. Bags made up of very fine-pored material ("bubble bags") are employed (a coarse-pored one with the material is placed inside a second bag with finer pores), and by varying the pore size of the bags and repeating the sifting process, it is possible to separate a series of resin powders of different THC levels.

The Ice-o-lator (Figure 12.26c) is one of the principal devices used. Material is placed in a bucket of cold (4°C) water to harden the resin glands and make them more easily separated. Agitation by a motor-driven mixer results in the resin glands separating. The denser resin glands sink while the less dense remaining parts of the plant float on the surface. A course screen (e.g., with hole size 187 μm) is used to skim off the floating materials, and a fine screen (e.g., with hole size 62 μm) is used to separate the resin glands from any finer particulate material that has sunk. Resin powder prepared with this apparatus is sometimes termed "iceolator hash" or "water hash."

Water Extracts ("Teas")

The word "tea" in common usage corresponds to two kinds of liquid extract in technical pharmacological literature. An "infusion" is a liquid solution extracting a compound of interest, prepared by soaking or steeping, usually in water. An infusion is usually made by pouring boiling water over herbaceous material and allowing this to steep (it can also be made by adding concentrated extracts to water). A "decoction" is an extract obtained by boiling in water (the strained liquor is called the decoction). In pharmacy, a decoction may be contrasted with an infusion, where there is merely steeping. Decoctions are often made using hard components such as roots and bark that are resistant to boiling. Both infusions and decoction are employed to prepare cannabis teas.

When cannabis is smoked, vaporized, or baked, the heat is sufficient to convert essentially 100% of the nonpsychoactive THC acid (THCA) to the psychoactive THC. When cannabis is placed in boiling water, only a small percentage of the THCA is converted to THC, so that cannabis tea is a comparatively less psychoactive way of consumption (unless one likes to drink a lot). Moreover, the amount of THC that can be dissolved in water is very low: Hazekamp et al. (2007) found that when pure THC is placed in boiling water, only 17% was solubilized. However, it is well known that THC is soluble in fats like milk, so adding some form of milk to the water greatly increases the THC dissolved. As noted later, in India, beverages prepared with cannabis often have milk added to extract the THC. Hazekamp et al. (2007) observed that if milk is added to cannabis tea, it stores well for several days, but if not added, most of the THC precipitates in only one day, so that the tea loses its potency.

In Jamaica, cannabis tea is used as a remedy for cold, fever, and stress (Hazekamp et al. 2007). In Europe (and occasionally in North America), packages of foliage of *C. sativa* are available (usually illegally) for preparing cannabis tea. The Office of Medicinal Cannabis of the Netherlands (http://www .cannabisbureau.nl/en/) provides the following instructions for preparing cannabis tea using marijuana:

- Boil 500 mL of water in a pan with the lid on.
- Add 0.5 g (about two teaspoons or one measuring scoop) of medicinal cannabis.
- Turn down the heat and let the tea simmer gently for 15 minutes with the lid still on the pan.
- Take the tea off the stove and pour it through a sieve.
- Keep the tea in a thermos flask if you plan to drink it the same day.

If you want to make tea for several days, use 1 g (about four teaspoons or two measuring scoops) of medicinal cannabis for 1 L of water. Then, after preparing the tea as described previously, add a package or teaspoon of coffee creamer powder to the warm tea. This will keep the active substances in the tea from sticking to the inside of the teapot or cup, reducing its effectiveness. Let the tea cool down and store it in the fridge. It will store for several days. You may reheat the refrigerated tea and can add sugar, syrup, or honey to improve its taste.

TECHNOLOGIES FOR SMOKING AND VAPING CANNABIS DRUGS

Representative traditional and novel methods of inhaling cannabis are discussed in the following presentation (in some jurisdictions, the materials illustrated are illicit). Regardless of smoking method, the very undesirable health effects of smoking are discussed in the next chapter, with emphasis on the relative desirability of inhalation modes that reduce the intake of toxins.

JOINTS AND BLUNTS

"Joints" (marijuana cigarettes; Figure 12.27a), also referred to as "reefers," "spliffs," "doobies," and by numerous other slang names, as indicated in Abel (1982), are the most widely employed method of smoking. Occasionally, "blunts" (marijuana cigars; Figure 12.27b) are prepared, although these are much too large for a single dosage. In Europe, cigarettes are frequently fashioned by combining tobacco with 0.1–0.3 g of marijuana, often using high-potency material since the tobacco occupies much of the cigarette; in North America, tobacco is infrequently employed and joints tend to be smaller. In the past in North America, marijuana was of lower potency, and up to 0.5 g was placed in a joint. The word "spliff" is sometimes used for a joint prepared with both cannabis and tobacco, but in the West Indies, where the term originated, and in North America, it normally designates a joint made only with marijuana.

FIGURE 12.27 Variations of marijuana cigarettes. (a) Hand-rolled marijuana cigarettes ("joints") and a regular tobacco cigarette for scale. Twisting the ends is common because marijuana lacks the packing qualities of tobacco. Note the smaller amounts that are typically smoked by comparison with tobacco. (b) A marijuana cigar ("blunt"), often prepared by replacing the tobacco in a conventional cigar with marijuana. Photo by iTopher (CC BY 2.0). (c) An unrolled joint with a rolled up piece of cardboard stock employed at the base so that the marijuana can be completely smoked. Photo by Erik Fenderson (released into the public domain). (d) A rolled up piece of cardboard as shown in (c). Photo by Erik Fenderson (released into the public domain).

Roach clips are devices employed to hold the lit butt of a joint, in order to avoid finger burns and stains. They may be as simple as a paper clip or tweezers. Roach clips are considered passé today in North America; an alternative technique is to use a piece of rolled-up business card (Figure 12.27c and d) inside the base of a joint so that it can be smoked completely.

Simple (Nonfiltering) Pipes

An impressive array of smoking devices are employed for marijuana. Very crude instruments suffice, for example, the "chillum" (Figure 12.28a), a simple, clay pipe employed in India and Jamaica. (While elementary in design, traditional usage requires an assistant to light the marijuana, while the other inhales the smoke through a wet cloth wrapped around the mouthpiece to cool the smoke and prevent inhalation of embers.) Often, makeshift crude instruments are fashioned out of all kinds of objects, such as hollowed-out apples (Figure 12.28b) and beer cans. Sebsi pipes (Figure 12.29),

(a)   (b)

FIGURE 12.28 Examples of simple marijuana pipes. (a) Earthen chillums displayed for sale in the city of Jorhat, India. Photo by Anupom sarmah (CC BY SA 4.0). (b) Apple pipe. Photo by Payman (CC BY SA 3.0).

FIGURE 12.29 Drawing of a sebsi pipe being smoked in North Africa. (Courtesy of Ebers, G., *Egypt: Descriptive, Historical, and Picturesque, Vol. 1*, Cassell & Company, New York, 1878. Public domain.)

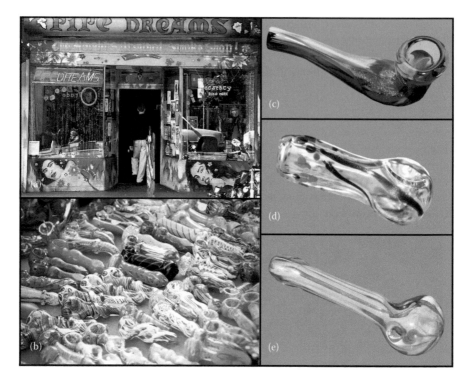

FIGURE 12.30 Artistic glass pipes employed for smoking marijuana. (a) A store in San Francisco-Haight & Ashbury-specializing in cannabis pipes. Photo by David Ohmer (CC BY 2.0). (b) Display of marijuana pipes. Photo by Charlie Gaddie (CC BY 2.0). (c) A general-purpose drug pipe. Photo (public domain) from the U.S. Drug Enforcement Administration. (d) A "spoon pipe" (so-named for its resemblance to a spoon). Photo by Todd Blaisdell (CC BY ND 2.0). (e) Note the ventilation hole ("carb," "choke," "shottie," "shot hole," "rush hole") in the bowl of this pipe, intended for thumb control; like the carburetor of a car, this serves to regulate burning and airflow by controlling access to air. Photo by TheChanel (CC BY 2.0).

which are popular in North Africa, especially in Morocco, are long-stemmed and have a small metal or ceramic bowl, both features that cool the smoke.

Multicolored handblown glass pipes have become quite popular (Figure 12.30) and are often works of art, sometimes designed to change color as the pipes heat up.

WATER PIPES

Based mostly on tobacco, different cultures have created complex devices to cool smoke by passing it through a water chamber (Figure 12.31a). These are known as water pipes, hookahs, hubble bubbles, nargils, and by other names. Such instruments modified specifically for cannabis consumption are known as "bongs." Most bongs have a carb (explained in Figure 12.30e) to clear smoke from the portion of the chamber above the water, but some bongs have a removable stem called a "slide," that has the same purpose. Today, a variety of instruments, often quite artistically designed and occasionally costing as much as thousands of dollars, are marketed. Often, instruments are constructed from household materials by individuals for personal use (Stone 2010). A "Rasta chalice" (tracing to Rastafarian religious use of marijuana) can be as simple as a hollowed out coconut with two holes, a smoking bowl inserted in one of the holes, its tube extending into the water placed in the coconut, and a drawtube inserted in the other hole. "Ice catchers" (ice bongs; Figure 12.32) incorporate indents in the instrument so that ice can be supported in the air flow column to cool the smoke (in addition to the water through which the smoke is bubbled). In past centuries, extraordinarily crafted hookahs were often

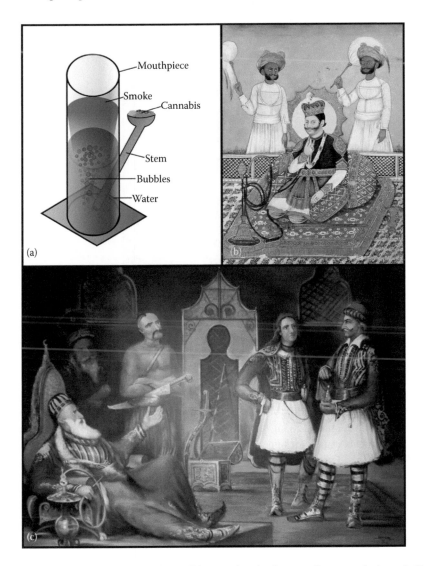

FIGURE 12.31   Hookahs for cooling smoke. (a) Diagram showing how marijuana smoke is cooled by a water pipe. Prepared by Christopher Thomas (CC BY 3.0). (b) A king of Nepal with a hookah (created in 1840 by an unknown Indian artist; public domain photo). Credit: The San Diego Museum of Art, Edwin Binney 3rd Collection Accession Number: 1990.177. (c) Ali-Pasha and his hookah. He was an eighteenth century Muslim Albanian ruler who served as a pasha for the Ottoman Empire. Photo by Dimitris Siskopoulos, of a mural in the Ali Pasha Museum, Epirus, Greece (CC BY 2.0).

prized possessions of Asian potentates (Figure 12.31b and c). As detailed in the next chapter, while water filtration reduces the amount of some harmful constituents, nevertheless, significant toxic compounds are inhaled when water pipes are employed to smoke cannabis. Some hookahs are designed to not only pass smoke through water but also through charcoal. However, waterpipes, even when fitted with solid filters, are ineffective at improving the THC/tar ratio in smoke (Gieringer 2001).

VAPORIZERS

Vaporizers are instruments designed to vaporize materials for inhalation. Unlike "smoking" (inhalation of "smoke"—a combination of vapors and combusted particles, invariably including hundreds of dangerous chemicals), the heat produced is sufficient to produce steam or vapor but ignition

FIGURE 12.32　An ice bong, for cooling smoke. Photo by Taschenkrebs, released into the public domain.

or burning does not occur. Commercial vaporizers specifically for inhaling the cannabinoids are available, particularly the popular "Volcano" series (Figure 12.33b). Personally constructed vaporizers have been made (Figure 12.33a). "Electronic cigarettes" (e-pen vaporizers, as shown in Figure 12.33c) are instruments usually designed to vaporize nicotine for inhalation but are also commonly employed for cannabis. Forms of cannabis used for vaporization are usually quite concentrated—resin or oil. E-pen vaporizers are increasing rapidly in popularity and could well become the most common instruments used for cannabis consumption. As detailed in the next chapter, vaporizers considerably reduce but do not entirely eliminate the intake of toxins experienced by smoking. Jensen et al. (2015) noted that E-cigarette liquids are typically solutions of propylene glycol, which can degrade to produce alarming levels of formaldehyde during vaping.

DABBING

"Dabbing" refers to a practice, largely conducted by a subculture of marijuana users, usually young, who employ cannabis concentrates (so concentrated that a "dab" suffices) to become very high very quickly. The technique was developed partly to efficiently use concentrates, which are easily ignited and wasted, and partly to get high rapidly. "Blasting dabs" is done by heating on a hot surface (Figure 12.34, right), often as simple as a real nail, or a similar structure termed a "nail" (often made of titanium, sometimes quartz or glass), and inhaling the vapor. Frequently, specialized instruments are available (Figure 12.34, left), as well as special gear such as "dab tools" (utensils for smoking concentrates) are employed. Blow torches (often specially designed) are typically used to heat nails or glass bongs (electronic nails have been developed to eliminate the need for fire). Some dabbing pipes look like traditional meth or crack cocaine pipes, and dabbing using torches has led to the practice being termed "the crack of pot." Like those who drink to become very drunk quickly, the intense highs desired by "dabbers" may be a sign of addiction or maladjustment (Loflin and Earleywine 2014). Because concentrates produced illicitly are often unsafe, and it is very easy to overdose, dabbing is dangerous (Gieringer 2015), especially for novices.

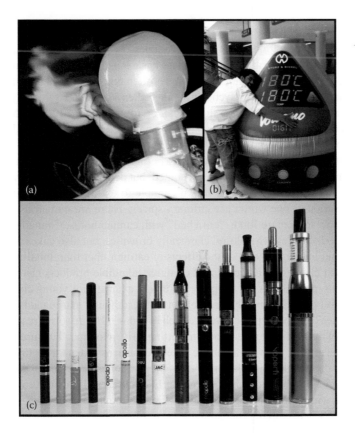

FIGURE 12.33 Vaporizers. (a) A vaporizer constructed for smoking cannabis. Photo (self-portrait) by Patrick Morris (CC BY 2.0; face obscured). (b) An exhibit of the popular Volcano vaporizer at a medical marijuana sales event in Toronto in 2013. Photo courtesy of Steve Naraine. (c) Different types of electronic cigarettes, adaptable for smoking cannabis resin and oil. Photo by Vaping 360.com (CC BY 2.0).

FIGURE 12.34 Dabbing of cannabis. Left: A dab rig. Photo by Steven Schwartz (CC BY 2.0). Right: Dabbing hash oil. Concentrated cannabis resin on the end of a metal poker is being applied to the cup of a glass "nail," which replaces the bowl of a bong. This nail has been heated using a butane torch, and applied resin will be almost instantaneously vaporized. The THC-rich vapor that results will be drawn through the hollow stem (seen in the center of the nail) into an attached smoking device (like a bong) and inhaled. Photo by DJ Colonel Corn (face obscured; CC BY SA 3.0).

EDIBLE CANNABIS

Marijuana can be ingested as food, often termed "edibles" (Wolkowicz 2012), such as illustrated in Figure 12.35. Eating cannabis to alter consciousness has a long history. India traditionally produced "bhang"—chopped, macerated cannabis leaves most often consumed as a beverage (typically with added milk), sometimes with other psychoactive ingredients added. To this day, preparing bhang is still practiced in India, where "bhang shops" can be found (Figure 12.36). "Majoon" is another traditional Indian marijuana foliage-based confection prepared with various foods and spices, some of which may also be psychotropic.

Humans are not well equipped to chew and digest herbal marijuana, which is not tasty. Consuming hashish is preferable in that less needs to be consumed, but depending on how it was prepared, there may be problems of toxicity and dosage. It is preferable to extract the cannabinoids as a solution and employ the solution as one would use culinary spices. There are numerous books and Internet sources detailing how to prepare foods "enriched" with cannabinoids. Cannabis is frequently consumed in baked goods in Western culture, classically brownies, but also cakes, cookies, and fudge. The slower onset and longer duration of the effects of eating rather than inhaling cannabinoids are discussed in the next chapter. THC content of commercial edible products has proven to be erratic in some jurisdictions, and edibles not properly stored are a hazard for young children.

As noted previously, THC is mostly insoluble in water, although marijuana is nevertheless sometimes made into a "tea" or other type of beverage such as sodas. THC dissolves readily in alcohol, so an alcoholic infusion of marijuana can be a route to getting THC into food. Tinctures are typically made by placing marijuana in glass jars with alcoholic beverages such as vodka for six to eight weeks to allow the cannabinoids to dissolve. A much faster mode of preparation is described in the next paragraph.

THC is fat-soluble, and by infusing it into a fat like butter, milk, or olive oil, cannabis can be incorporated into a wide range of edible products. Finely ground marijuana can be cooked with oil or butter (Figure 12.37), dissolving the cannabinoids in these fats. Cannabis butter ("cannabutter,"

FIGURE 12.35 Confiscated edible marijuana products, including "Allmy Joy," "Munchy Way," "Pot Tarts," "Stony Rancher," "Rasta Reece's," "Buddafingers," "Double Puff Oeo," "Keef Kat," "Budtella," "Puff-A-Mint Pattie," "Puffsi," "Bong's Root Beer," and "Toka Cola." Photos (public domain) from the U.S. Drug Enforcement Administration.

FIGURE 12.36 Bhang. (a–d) Preparing bhang in Punjab, India. (a, b) Macerating herbal cannabis with other ingredients. (c) Filtering macerated ingredients through cloth. (d) A glass of bhang. Photos by Marcusprasad (CC BY SA 4.0). (e) "Bhang eaters before two huts," dated ca. 1790. Credit: Edwin Binney 3rd Collection Accession Number: 1990.642, The San Diego Museum (public domain). (f) A bhang shop in Jaisalmer, Rajasthan, India. Photo by Tom Maisey (CC BY 2.0).

FIGURE 12.37 Examples of marijuana culinary arts. Left: Preparation of "cannabutter," a widely employed ingredient of edible cannabis. Photo by Realclark, released into the public domain. Right: Preparing marijuana brownies by cooking cannabis-infused butter with chocolate. Photo by Antoine (CC BY SA 3.0).

FIGURE 12.38 Scenes of THC-infused ice cream available in California. Photos courtesy of Cannabis Creamery (https://www.facebook.com/pages/Cannabis-Creamery/704824732895521?sk=info&tab=page_info).

"butteruana") can be quite green when prepared by combining macerated marijuana with butter because of the presence of extracted chlorophyll. "Recommendations range from 1/8 to 1/2 ounce [3.5–14 g] of marijuana per 1/4 cup [59 ml] of oil or butter. Cooking times range from 20 to 45 minutes. Shorter cooking times should release fewer flavonoids and minimize the grassy taste" (Earleywine 2010). The cannabinoid-infused butter or oil can be employed in many recipes, including ice cream (Figure 12.38).

As discussed in detail in Chapter 8, hempseed and hempseed oil are widely incorporated in edible products. Often outlets marketing cannabis products sell hempseed foods lacking THC as well as foods laced with THC, and sometimes, the THC-free products are there simply to augment sales because many people purchasing psychotropic foods are sympathetic to anything manufactured with cannabis.

PREFERRED MODES OF MARIJUANA CONSUMPTION

Based on a sample of over 4000 Americans at least 18 years of age who reported consumption of marijuana in 2014, Schauer et al. (2016) reported that: "Overall, 7.2% of respondents reported current marijuana use; 34.5% reported never use. Among current users, 10.5% reported medicinal-only use, 53.4% reported recreational-only use, and 36.1% reported both. Use of bowl or pipe (49.5%) and joint (49.2%) predominated among current marijuana users, with lesser use of bong, water pipe, or hookah (21.7%); blunts (20.3%); edibles/drinks (16.1%); and vaporizers (7.6%); 92.1% of the sample reported combusted-only marijuana use." Smoking, as pointed out in the next chapter, is deleterious to health, and likely, the noncombustion consumption of cannabis as edibles and by convenient e-pens will increase in the future.

FIGURE 12.39   Illicit, dangerous marijuana-like (noncannabinoid) preparations. Left: Anti-synthetic cannabinoid poster. Credit: Hawkes Bay District Health Board, New Zealand. Right: "Spice" and "K2," the principal marketed "fake marijuana" products. Photo (public domain) by U.S. Drug Enforcement Administration.

FAKE MARIJUANA

In recent years, "designer drug" mixtures of shredded plant material laced with chemical additives have been marketed as "legal marijuana" alternatives, under such names as Spice, K2, fake weed, Yucatan Fire, Skunk, and Moon Rocks (Dresen et al. 2010; Vardakou et al. 2010; Ashton 2012; Rosenbaum et al. 2012; Seelly et al. 2012; Thomas et al. 2014a; Figure 12.39). Well over 100 products are being marketed (Zawilska and Wojcieszak 2014), sometimes with, often not with chemicals related to the cannabinoids. Frequently with labels such as "not for human consumption" or "for aromatherapy only," the intent has been to provide a marijuana substitute that evaded current laws. These preparations have proven to be quite attractive to youth. Most jurisdictions have made these marijuana mimics illegal, as they have often resulted in sickness. Synthetic cannabinoids are sometimes used in such products (see the discussion of prescription "cannabimimetic" substances in the next chapter). Synthetic cannabinoids in samples of Spice products have sometimes shown an affinity for the CB_1 receptor (discussed in the next chapter) that is four to five times greater than natural THC (Vandrey et al. 2012). A wide variety of cheap toxic compounds has also been employed in fake marijuana.

It needs to be pointed out that by no means are all synthetic cannabinoids deleterious. Some are useful as research and therapeutic tools (Chiurchiù et al. 2015).

ETHICAL PERSPECTIVES OF DECRIMINALIZATION
AND LEGALIZATION OF RECREATIONAL MARIJUANA

There are endless publications arguing the merits for and against decriminalization or legalization of recreational marijuana from an ethical perspective. This book is not intended to take a position on the issue, but some of the arguments and considerations that are commonly raised are presented in the next paragraphs.

The following statement regarding the U.S. marijuana prohibition is representative of the viewpoint that recreational marijuana should be legalized: "Our marijuana laws are clearly doing more harm than good... Law enforcement agencies today spend many billions of taxpayer dollars annually trying to enforce this unenforceable prohibition. The roughly 750,000 arrests they make each year for possession of small amounts of marijuana represent more than 40% of all drug arrests. Regulating and taxing marijuana would simultaneously save taxpayers billions of dollars in enforcement and incarceration costs, while providing many billions of dollars in revenue annually. It also would reduce the crime, violence and corruption associated with drug markets, and the violations of civil liberties and human rights that occur when large numbers of otherwise law-abiding citizens are subject to arrest. Police could focus on serious crime instead. The racial inequities that are part and parcel of marijuana enforcement policies cannot be ignored. African-Americans are no more likely than other Americans to use marijuana but they are three, five or even 10 times more likely—depending on the city—to be arrested for possessing marijuana… Who most benefits from keeping marijuana illegal? The greatest beneficiaries are the major criminal organizations in Mexico and elsewhere that earn billions of dollars annually from this illicit trade—and who would rapidly lose their competitive advantage if marijuana were a legal commodity…. Like many parents and grandparents, I am worried about young people getting into trouble with marijuana and other drugs. The best solution, however, is honest and effective drug education" (Soros 2010).

The following statement is representative of the viewpoint that marijuana should not be legalized: "Marijuana is the most commonly abused illegal drug in the U.S. and around the world. Those who support its legalization, for medical or for general use, fail to recognize that the greatest costs of marijuana are not related to its prohibition; they are the costs resulting from marijuana use itself… Rapidly accumulating new research shows that marijuana use is associated with increases in a range of serious mental and physical problems. Lack of public understanding on this relationship is undermining prevention efforts and adversely affecting the nation's youth and their families. Drug-impaired driving will also increase if marijuana is legalized… Since legalization of marijuana for medical or general use would increase marijuana use rather than reduce it and would lead to increased rates of addiction to marijuana among youth and adults, legalizing marijuana is not a smart public health or public safety strategy for any state or for our nation" (DuPont 2010).

CURIOSITIES OF SCIENCE, TECHNOLOGY, AND HUMAN BEHAVIOR

- London's hosting of the 2012 Summer Olympic and Paralympic Games was associated with the construction of Britain's largest piece of public art, officially named "The ArcelorMittal Orbit." Located in Olympic Park in Stratford, London, it is 114.5 m tall (the Eiffel Tower is 324 m in height). The construction is eerily reminiscent of a hookah (Figure 12.40).
- Between the eleventh and thirteenth centuries, in the Middle East, a sect known as the Assassins—followers of Hasan-ibn-Sabah—dominated the Middle East through a reign of terror. Italian traveler Marco Polo (1254?–1324?) reported that the Assassins used a drug to rouse themselves to bloody deeds. In the nineteenth century, several European writers claimed that the word "Assassin" was derived from the word "hashish" and that this was the drug used by the Assassins. Thereafter, *Cannabis* was frequently associated with violence in many anti-marijuana stories, although marijuana is well known to induce sleepiness rather than hyperactivity and aggression.
- Intercouple violence in a marriage has been found to be less when one spouse is a marijuana smoker and even less when both partners are users (Smith et al. 2014).
- In the 1890s, several women's temperance societies recommended the recreational use of hashish instead of alcohol, in the belief that liquor led to wife-beating, while hashish just made people sleepy.
- The recreational use of cannabis in the United States is so widespread that about 10% of paper currency has been found to be contaminated with cannabinoid residues (Lavins et al. 2004).
- President Bill Clinton is famous for saying "When I was in England I experimented with marijuana a time or two, and I didn't like it. I didn't inhale." Clinton is believed to have had hemp beer ("Hempen Gold beer" manufactured by the Frederick Brewing Company

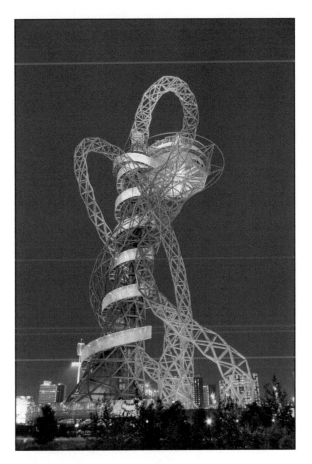

FIGURE 12.40   A giant hookah? Known as the ArcelorMittal Orbit, this huge sculpture and observation tower is located in London, England. Photo by BaldBoris (CC BY SA 2.0).

of Frederick, Maryland) served on February 15, 1999, on Air Force One, the presidential jet. Beer, of course, is usually flavored principally by hemp's cousin, hop. According to one reporter, the president "tasted but didn't swallow." President Barack Obama replied when asked if he had smoked marijuana, "I inhaled frequently…that was the point" (https://www.youtube.com/watch?v=cpBzQI_7ez8).

- "420" (4:20, 4/20, pronounced four-twenty) is a code reference to marijuana consumption, a term dating to the early 1970s in California. It has been suggested that the expression traces to 4:20 p.m., a time when school is over and students are free. Celebrations to advocate the legalization of marijuana (Figure 12.41) and other events based on marijuana are commonly scheduled for 4:20 during the day and on April 20 (i.e., fourth month, 20th day). The Colorado Department of Transportation replaced the frequently stolen Mile Marker 420 sign on I-70 east of Denver with one reading 419.99 to stop the thievery. The occasional expression "8:40" (twice 4:20) means that by 8:40 p.m., one has become twice as high as normally achieved at 4:20 p.m.
- The number "13" associated with the 13th letter of the alphabet, M, is occasionally used as code for "marijuana." For example, 13 on a motorcycle gang member's biker jacket sometimes stands for marijuana.
- In the eighteenth century, the term "sawbuck" was slang for a sawhorse, made by joining pieces of wood into an "X" shape to support boards at their ends while they were being cut. The first U.S. $10 bill bore the Roman numeral X, and consequently "sawbuck"

FIGURE 12.41 A four-twenty event in Boulder, Colorado, 4:20 p.m., April 20, 2010. Photo by Zach Dischner (CC BY 2.0).

became slang for the $10 bill, while "buck" became a reference to the dollar. In the mid-1900s, "sawbuck" became street slang, apparently originating in Chicago, for a 10-dollar bag of marijuana. Since the 1980s, the term has referred to a 10-dollar package of any street drug.

- Oxford Dictionaries declared "vape" to be its 2014 word of the year, including the verb ("inhale and exhale the vapor produced by an electronic cigarette or similar device") and noun ("an electronic cigarette or similar device; an act of inhaling and exhaling the vapor produced by an electronic cigarette or similar device"). Lake Superior State University (in northern Michigan), on the last day of 2014, issued its 41st annual "List of Words Banished from the Queen's English for Misuse, Overuse and General Uselessness," and "vape" was one of the 13 words included. Although "vape pens" are predominantly used for tobacco, they are increasingly being adapted for marijuana.

- "Budtender" (based on a combination of bud and bartender) refers to a worker who sells and is knowledgeable about the cannabis products in a medical marijuana dispensary or recreational marijuana shop. The qualifications of such personnel are suspect, and they have been referred to as "quasimedical vendors" (Kleiman 2015).

- In assessing the effects of smoking recreational marijuana, many researchers employ the unit "joint-year," where one joint-year equals smoking one joint/day for one year.

- "Bogarting" is a slang verb meaning to selfishly keep something completely for oneself, especially applied to joints (a Bogart is a person who hogs a joint). The term is based on actor Humphrey Bogart, who habitually kept a cigarette in his mouth.

- In the context of cannabis, a "hotbox" is a sealed room or vehicle in which the exhaled smoke from several pot smokers accumulates, so that the secondhand smoke reinforces the effects of the original smoke. The method was originally employed to avoid detection, but became a social activity. A "Jamaican hotbox" uses a bathroom in which the shower is turned on hot, so that steam also fills the room. It has been shown that the secondhand smoke in a hotbox can have appreciable intoxicating effect (Herrmann et al. 2015).

- Detecting illicit drugs (including cannabis) in urine has become a multibillion dollar industry, and in parallel, "fake urine" ("synthetic urine") businesses have developed to assist users of illegal drugs to pass these tests. (Of course, the simplest way most people avoid detection is to obtain clean urine from someone who is not using illicit substances and is a match for sex and age.) Typically costing $30.00 to $100.00, some of the products offered are claimed to be heatable to body temperature and sometimes are accompanied by portable warmers (since many labs check this to ensure genuineness). The "Whizzinator" is a prosthetic urine-delivery device that has been employed to simulate actually urinating into a collection bottle for "observed tests." The best formulations can be difficult to detect, closely simulating pH, creatinine, and the specific gravity of normal urine. Indeed, there is a sort of chemical warfare between the makers of fake urine, who keep improving their product, and detection laboratories, who are constantly seeking new ways to detect artificial urine. In some jurisdictions, such cheating is subject to specific criminal penalties.

13 Medical Marijuana: Theory and Practice

INTRODUCTION TO THE CONTROVERSY

"Medical marijuana" is also known as "medicinal marijuana," although some clinicians consider the word marijuana to be pejorative and prefer "medical cannabis." Medical marijuana is the most controversial therapeutic agent of modern times. The idea of employing it as a legitimate medicine is shocking to many (Figure 13.1). Marijuana is the world's most popular illegal drug, and no other plant is as extensively associated with crime and immorality. Its principal euphoriant ingredient, tetrahydrocannabinol (THC), is the most prevalent illicit chemical and indeed the fourth most common recreational drug after caffeine, alcohol, and nicotine. No other medicine is consumed by smoking, a practice that is increasingly viewed as antisocial and the antithesis of health promotion. It is curious indeed that, despite its sinful and deleterious reputation, a torrential demand for marijuana as a therapeutic agent has developed.

However, the present notoriety of marijuana obscures the fact that forms of cannabis have been employed as accepted, reputable drugs since ancient times. The illegality of cannabis during most of the twentieth century retarded research and development of modern products, both of a therapeutic nature and otherwise. In the last several decades, however, there have been great advances in the scientific understanding of how cannabis affects human physiology, and new products and technologies intended for therapeutic use have appeared. Even countries that unambiguously state that there is no legitimate use of cannabis whatsoever have nevertheless authorized the use of certain cannabis-based medicines. Long considered a "pariah drug," it is ironic that access to marijuana on the black market "allowed many thousands of patients to rediscover the apparent power of the drug to alleviate symptoms of some of the most cruel and refractory diseases known to humankind" (Robson 2005). Throughout the Western World, jurisdictions are authorizing access to various forms of cannabis, despite the majority scientific viewpoint that evidence supporting usage as a therapeutic agent is inadequate for most applications and the prevalent fear that that the increasing popularity of medical marijuana is a dangerous experiment with possibly harmful consequences for at least a susceptible proportion of individuals as well as for society in general. This chapter outlines what is and isn't known about medical marijuana, which is surely prerequisite to judge its utility and potential.

HOW MEDICAL DOES MARIJUANA HAVE TO BE TO QUALIFY AS "MEDICAL MARIJUANA?"

Perhaps the chief reason that medical marijuana is controversial rests in its dual capacities to be used or abused, both recreationally and medically. This is a frequent problem for psychotropic substances that produce temporarily desirable mental states—either euphoria or at least a dulling of pain. Consider the following curious distinction. Pharmacologically, an "elixir" is a sweetened, aromatic solution of alcohol and water containing medicinal substances. A "liqueur" is a sweetened aromatic solution of alcohol and water containing flavorants such as spices and herbs, which invariably have medicinal properties, so liqueurs effectively are elixirs. In fact, liqueurs such as Bénédictine have been employed both medicinally and recreationally. The point is that plant-based preparations almost invariably have the potential to be used either therapeutically or for other

FIGURE 13.1   The controversy over medical marijuana—a useful therapy or a dangerous drug of abuse? Prepared by B. Brookes.

purposes (legitimate or not), so their characterization as medical or nonmedical is problematical, at least to a degree.

In Chapter 18, concerning classification of *Cannabis*, the difficulties posed by "stereotypical thinking" are discussed. As pointed out in Chapter 18, stereotypical thinking is a rigid conceptualization of things as being of one nature and not another. Such inflexible thinking makes it difficult to view marijuana as having a dual nature, one of which is therapeutic. For many, conditioned by a century of vilification of marijuana as a dangerous narcotic, it is difficult to consider it only in the context of medicine.

In theory, "medical marijuana" can include all of the forms of recreational marijuana presented in the preceding chapter. Similarly, all of the "drug paraphernalia" (illicit or not) described in the last chapter could be considered to be "medical devices" (admittedly a stretch in some cases). In practice, legislative requirements, safety concerns, and common sense dictate what preparations qualify as medical marijuana and what devices are legitimate for medical cannabis consumption. The qualifications of health professionals to prescribe, recommend, administer, or oversee medical marijuana are based on jurisdictional and professional association regulations, and these considerations ultimately determine who merits treatment with medical marijuana. As discussed in the next section, the medical legitimacy of a considerable proportion of the medical marijuana trade is debatable.

Some authors use the phrase "medical marijuana" to refer just to herbal material; others include extracts as well as natural and synthetic cannabinoids. In this book, both "medical marijuana" and "medical cannabis" are employed to include all of the preceding. So-called "highless marijuana," a seemingly oxymoronic phrase ("highless cannabis" would be preferable), is discussed later.

MEDICAL MARIJUANA AS A PRETEXT FOR NONMEDICAL USE

In certain jurisdictions, it is clear that recreational usage has been the principal motivation for advancing the cause of medical marijuana. For some, the ultimate goal is political, commercial, or philosophical, not therapeutic. Indeed, the same contention, that a valid usage for *Cannabis sativa* was a stalking horse or wedge for a hidden illegitimate agenda, was advanced (with some truth) during the reintroduction of industrial hemp in many countries. Unfortunately, the legitimate advancement of medical marijuana has, to some extent, been hijacked by those seeking to legalize recreational usage. In some U.S. states, "the process of getting a 'recommendation' for medical

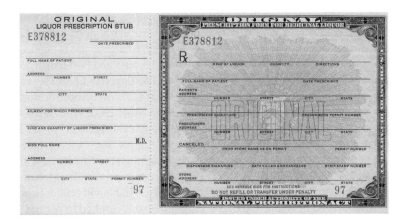

FIGURE 13.2   A U.S. government medicinal alcohol form (public domain) from the 1920s, used during the American Prohibition to acquire prescription alcohol, usually whiskey, ostensibly for strictly medicinal purposes.

marijuana is an open mockery" (Caulkins et al. 2012). At its worst, physicians with questionable ethics are providing prescriptions to "pretend patients" to be filled at "nonprofit" clinics motivated basically by financial gain (for additional discussion, see Chapter 15). This subterfuge is amusingly reminiscent of how "medicinal alcohol" was sometimes deceptively prescribed (Figure 13.2) during the period of American alcohol prohibition. Similarly, grape concentrates called "grape bricks," supposedly intended to prepare healthful fruit juice, were sold with the tongue-in-cheek warning: "Do not add to a jug of water and put in a dark place or it will ferment into wine." The unfortunate hypocrisy associated with medical marijuana should not detract from objectively evaluating its appreciable therapeutic values.

A BRIEF HISTORY OF MEDICAL USAGE OF CANNABIS

Many herbs were employed in past times for medical purposes, and in retrospect, it is obvious that, sometimes, the "potions" and "concoctions" employed were useless, even harmful. Often, however, crude herbs were genuinely effective because they contained medicinal ingredients. As documented in this chapter, *C. sativa* certainly contains medicinal compounds. Modern medicine tends to dismiss past "primitive" and "folk" medicinal practices, but this disrespectful view often obscures insightful knowledge that still has potential value for adopting or finding better therapies. Indeed, the careful interpretation of past medical usage of herbs is an important contribution toward improving medicine in the future. In the case of *C. sativa*, there are numerous reviews of the ancient history of cannabis for therapeutic purposes, notably, Merlin (1972), Abel (1980), Mechoulam (1986), Aldrich (1997), Gurley et al. (1998), Fankhauser (2002), Russo (2004a, 2007, 2014), and several articles in Russo and Grotenhermen (2006).

It is very difficult to identify the earliest written reference to therapeutic use of *C. sativa*, and often, in ancient times, medicinal usage merges with ritualistic, spiritual, or religious practices. As most modern historians appreciate, early history in general is often poorly documented, very difficult to interpret, and subject to the imaginations of both the ancients who recorded information and modern interpreters. Nevertheless, it is clear that cannabis was employed in major civilizations of the ancient world, including Assyria, Egypt, India, Greece, Rome, and the Islamic empire. Certainly, cannabis has a very long and extensive folklore associated with it, as well as many enthusiastic interpreters of that history. The phrase "medical marijuana" has exploded in popularity in recent decades, but in fact, cannabis has been used for medical purposes throughout most of the world, in the majority of human cultures, for most of recorded history.

Cannabis has been applied medicinally in folk medicine since antiquity in Asia. For several thousand years, it was employed to treat a wide variety of illnesses, particularly in traditional herbal medicine of China, Ayurvedic medicine of India, and Tibetan medicine. Analgesic use is implied from Chinese oral tradition allegedly dating to 2700 BC (Li 1973; Figure 13.3, left), and East Indian documents in the Atharva Veda dated at about 2000 BC (Gurley et al. 1998). The Egyptian Ebers Papyrus (Figure 13.3, right) described a plant called *shemshemet*, often interpreted as *C. sativa* because of allusions to its fiber and medicinal uses, although the accuracy of this is uncertain (Abel 1980; Wills 1998).

Assyria was a major Near East kingdom and empire from about 1250 BC to 612 BC. Assyrians employed cannabis as a psychoactive mind-altering drug as well as for medical purposes (Mechoulam and Parker 2013a). As *C. sativa* was spread through the Middle East and Africa over the last two millennia, medicinal usages were adopted, particularly in the Mohammedan world, especially in Persia and Arabia. Indeed, the word "hashish" is Arabian in origin. However, intoxication is strongly discouraged in Islam, and over the centuries, this resulted in periods of suppressing the use of cannabis (Nahas 1982). Zlas et al. (1993) recorded an interesting medical usage based on the skeleton of a girl about 14 years of age found in a fourth century gravesite near Jerusalem. She had apparently died in childbirth, unable to expel a full-term fetus, the skeleton of which was present in her mother's pelvis. Ashes in the tomb were found to have cannabinoids, suggesting that cannabis had been burned to produce vapors to promote uterine contraction and reduce labor pains, properties long attributed to the plant.

With the transfer of African slaves to South America in the seventeenth and eighteenth centuries, traditional medical (as well as recreational) usages were imported to the continent, particularly to Brazil. The same cultural diffusion seems to have occurred when African slaves were transferred to the Caribbean area. In the West Indies, where cannabis is used extensively, it may have been introduced by workers from India and elsewhere in Asia during the mid-1800s (Rubin and Comitas 1975; Wills 1998).

FIGURE 13.3 Alleged earliest exponents of the medical use of cannabis. Left: The legendary Shen Nung, Chinese deity of Medicine, Pharmacy and Agriculture, and China's mythical second emperor (presumably ca. 3500–2600 BC). A pharmacopoeia (medical text systematically providing information on curative prescriptions) attributed to him, which apparently mentions the therapeutic use of cannabis, is about 5000 years old. This illustration is copied from an old painting. The god is seated at the mouth of a cave, dressed in his traditional garb made from leaves, holding in his right hand a branch with leaves and berries, the virtues of which he has been demonstrating. Image from Wellcome Images/Wellcome Trust (CC BY 4.0). Right: The Ebers papyrus, a medical treatise from ancient Egypt, dated at 1550 BC but containing material from 5 to 20 centuries earlier. Passages in the document appear to indicate very early medical use of *C. sativa*. Photo by Einsamer Schütze (CC BY 3.0).

Medicinal usage of cannabis in prehistoric Europe was very limited in ancient times (Zuardi 2006). Scythian invaders are suspected of bringing some medicinal knowledge of cannabis from the Middle East to Europe more than two millennia ago (see the historical account in Chapter 12). Cannabis was employed medicinally in ancient Greece and Rome, as recorded in the *Herbal of Dioscorides* (ca. 40–90 AD; Greek physician, pharmacologist and botanist whose five-volume *De Materia Medica* was widely respected for 1500 years) and the records of Galen (129–ca. AD 200 to 216; Greek physician, surgeon and philosopher; prominent authority in the Roman empire who influenced Western medicine until the nineteenth century). For the first millennium in Europe, there was limited medicinal usage of cannabis, and while, subsequently, *C. sativa* was employed in various remedies, it appears that the species was grown almost exclusively for fiber hemp. Asian medicinal usage of high-THC cannabis was mostly ignored in Europe until the nineteenth century.

The French psychiatrist Jacques-Joseph Moreau (1804–1884; Figure 13.4, right), nicknamed "Moreau de Tours," observed the effect of hashish during his North African travels in the 1830s. He theorized that it produced psychosis and might be useful in treating mental illness (Zuardi 2006). Moreau's activities stimulated the "Club des Hashischins" of Paris (described in the previous chapter) to flourish in the 1850s, using hashish as a route to "esthetic self-realization" (Kalant 2001). Cannabis was seriously introduced into Western medicine in the first half of the nineteenth century, notably through William B. O'Shaughnessy (1809–1889; Figure 13.4, left). An Irish physician working for the British Crown in Calcutta in the 1830s, he recommended cannabis for pain, sedation, inflammation, vomiting, convulsions, and spasticity, remarkably reminiscent of several of the conditions currently being treated with marijuana. As a result of O'Shaughnessy's enthusiasm, extracts of cannabis were adopted into the American pharmacopoeia (official publications listing medicinal drugs, their effects, and their uses) and the British equivalent. The U.S. Pharmacopoeia listed cannabis from 1850 to 1942, available for such conditions as labor pain, nausea, and rheumatism (indeed, foreshadowing modern usage for alleviation of pain and spasticity). Cannabis preparations were listed in the British Pharmaceutical Codex of 1949. It was possible to order cannabis tinctures from Sears Roebuck catalogues of the 1900s. *Cannabis* drug preparations were extensively used in the West between the middle of the nineteenth century and World War II, particularly as a substitute for opiates and as antispasmodic, analgesic, hypnotic (sleep-inducing), and sedative agents (Mikuriya 1969). Cannabis was used to treat a very wide range of ailments, including

FIGURE 13.4 Physicians who pioneered in the use of cannabis in Western medicine. Left: William B. O'Shaughnessy. O'Shaughnessy popularized medical cannabis in Great Britain and was knighted for his contributions by Queen Victoria. Right: Jacques-Joseph Moreau. Moreau promoted examination of the psychiatric effects of cannabis, but by acquainting France with hashish, he also stimulated its recreational use. Public domain illustrations.

FIGURE 13.5   Nineteenth century medicinal cannabis bottles displayed in the Hash, Marijuana & Hemp Museum in Amsterdam. Photo by Didier le Ger (CC BY SA 3.0).

insomnia, headaches, anorexia, sexual dysfunction, whooping cough, and asthma. Orally adminis-tered tinctures, especially alcoholic, were particularly popular, with hundreds of brands in circula-tion (Fankhauser 2002; antiquecannabisbook.com; Figure 13.5).

Following the Second World War, medical use declined because of several developments: qual-ity limitations of available cannabis (such as variable potency, poor storage, and erratic absorption of fluid products); the introduction of new medications, including vaccines and alternative pain relievers; the development of hypodermic syringes allowing the injectable use of morphine; the use of synthetic analgesics and sedatives; and the progressive criminalization of cannabis. After the Second World War and until the end of the twentieth century, there was very limited authorized medical use, and the plant and its medicinal preparations fell into disgrace. Nevertheless, toward the end of the twentieth century, there was considerable unauthorized dispensing of marijuana to gravely ill people by so-called "compassion clubs" (Feldman and Mandel 1998), in addition to widespread self-medication using illegal street marijuana. In 1996, voters approved Proposition 215, making California the first American state to legalize the medicinal use of cannabis. In 2001, Canada became the first country in the world to adopt a federal system regulating the use of herbal marijuana for "medicinal purposes" (Fisher and Johnston 2002).

Currently, cannabis is frequently prescribed as a complementary or adjunct medicine, not a pri-mary or initial treatment for serious conditions. This, however, represents a great change over the virtual ban on medicinal usage several decades ago. Thousands of researchers have recently under-taken medical research on cannabis, and as explained in this chapter, the prospects for greatly increased medical usage are very promising.

NONCANNABINOID MEDICINAL COMPONENTS

While this chapter deals mainly with the medicinal significance of the cannabinoids of *Cannabis*, it may be noted that other constituents of the plant have medical importance. As noted in Chapter 8, the fixed oil of *Cannabis* (hempseed oil) is highly nutritional, sometimes serving as therapeutic nutritional supplements (particularly for cardioprotection by preventing platelet aggregation) and also as topical preparations for the treatment of skin conditions (especially for improving atopic dermatitis).

The noncannabinoid components in marijuana may also contribute significantly to potential therapeutic effects, and so any consideration of medicinal marijuana and of THC delivery systems needs to take this into consideration. As noted in Chapter 9, the essential oil of *Cannabis* may have some therapeutic potential. Potentiating interactions of the cannabinoids and various terpenes, as well as the 20 or so flavonoids that are present, have been hypothesized to modify synergistically the psychological and physiological effects of cannabis drugs (McPartland and Pruitt 1997, 1999; Clarke 1998c; McPartland 2001; McPartland and Mediavilla 2002; Russo 2011a). McPartland and Russo (2001) concluded, "Good evidence shows that secondary compounds in cannabis may enhance the beneficial effects of THC. Other cannabinoid and non-cannabinoid compounds in herbal cannabis or its extracts may reduce THC-induced anxiety, cholinergic deficits [affecting key components of the nervous system], and immunosuppression. Cannabis terpenoids and flavonoids may also increase cerebral blood flow, enhance cortical activity, kill respiratory pathogens, and provide anti-inflammatory activity." The phrase "entourage effect," first proposed by Ben-Shabat et al. (1998) and Mechoulam and Ben-Shabat (1999), refers to the collective physiological interaction of chemicals that naturally occur with endocannabinoids and modify their effects, normally in a beneficial manner, and it is clear that similarly, there are beneficial therapeutic effects from interactions among components of medical marijuana. However, Fischedick et al. (2010) indicated that aside from the established interaction of THC and cannabidiol (CBD), it is premature to draw hard conclusions about therapeutic drug effects among the cannabinoids and other compounds.

BIAS IN RESEARCH ON MEDICAL MARIJUANA

The medicinal efficacy of cannabis is extremely controversial and regrettably is often confounded with the issue of balancing harm and liberty concerning the proscriptions against recreational use of marijuana. There is a tidal wave of anecdotal reports and testimonials from people convinced, on the one hand, that marijuana has uniquely relieved their medical conditions, or on the other, that it led to their physical and mental degeneration. Scientists, physicians, academics, the legal and law enforcement professions, politicians, and the general public are all divided on the merits and misuses of medicinal marijuana. U.S. Senator Daniel Patrick Moynihan (1927–2003) famously said, "Everyone is entitled to his own opinion, but not his own facts." In fact, the facts about medical marijuana are not yet entirely clear, and bias against it has prevented the search for truth more dramatically than almost any other topic of interest to society.

Only studies intended to demonstrate negative effects of marijuana consumption have been permitted until recently, and research funding has been predicated on documenting the harmfulness of marijuana. Consistent with this research straitjacket, most scientists working on cannabis until recently have provided evidence in publications that it is harmful, often writing with a condemnatory enthusiasm that belies neutrality. Such departure from objectivity is often revealed in the mandatory "introduction" to scientific papers, where authors who have become hostile to cannabis will refer to its "abuse" rather than its "use."

It is important to acknowledge that bias with regard to cannabis may involve excessive support as well as excessive opposition. There is a tendency for scientists working on a subject to develop respect, even fondness for it, but when they become involved in financial investments, it is all too easy to cross an ethical line. As is well known, some academic physicians funded by the pharmaceutical industry have engaged in questionable exaggeration on behalf of their patrons, including misrepresentation, selective presentation of facts, and "ghost authorship" of publications actually prepared by corporate employees. McPartland (2009) analyzed publications concerned with the cannabinoid-based antiobesity drug rimonabant (discussed later) and found examples of (1) exaggeration of its efficacy and safety, (2) lack of criticism, (3) "disease mongering" (exaggerating the frequency and seriousness of an illness to expand the market for its treatment), and (4) failure to disclose sponsoring financial support.

The saying "money corrupts" is applied to many of the institutions of society, and unfortunately, it also applies to science, as noted in the previous paragraph. Very few scientists currently are self-financed or are supported by employers or granting agencies that are completely free of preferences regarding expected discoveries and conclusions. Virtually every scientist is constrained by finances, almost always severely, and when a particular subject is contentious, significant pressure by management or other controllers of resources is common. When an adversarial relationship develops between scientists and the sources of their funding, the latter almost always prevail.

Although scientists are rarely completely free to conduct science, fortunately, there is an almost universal understanding that scientific truth ultimately is advantageous for almost everyone. Science is indeed a search for truth, and scientists employ rigorous experimental designs and objective statistical tests to evaluate the merits of what they are investigating. So how could bias alter the truthfulness of their results? First, "It is a natural tendency of most researchers to overstate the strength of their own research in relation to their favorite causal hypotheses; this near-universal phenomenon among researchers is referred to generally as 'wish bias'" (Macleod and Hickman 2010; cf. Wynder et al. 1990). Basically, it amounts to making a mountain out of a molehill. Second, scientists are human, and indeed often blind to their frailties, and with the best of intentions, observations and analyses can be faulty. And third, it is all too easy to inflate the significance of one set of findings while downplaying contrary results—so those who were already convinced that marijuana is bad may tend to exaggerate its harm while those who were already convinced that marijuana is good may do the opposite.

The objectivity of scientific evaluation of the medicinal value of marijuana to date has indeed been questioned. Up until about the start of the twenty-first century, it was extremely difficult to undertake cannabis medical research. In the words of Hirst et al. (1998): "The...status of cannabis has made modern clinical research almost impossible. This is primarily because of the legal, ethical and bureaucratic difficulties in conducting trials with patients. Additionally, the general attitude towards cannabis, in which it is seen only as a drug of abuse and addiction, has not helped."

The respected journal *Nature* (2001) stated: "Governments, including the U.S. federal government, have until recently refused to sanction the medical use of marijuana, and have also done what they can to prevent its clinical testing. They have defended their inaction by claiming that either step would signal to the public a softening of the so-called 'war on drugs'... The pharmacology of cannabinoids is a valid field of scientific investigation. Pharmacologists have the tools and the methodologies to realize its considerable potential, provided the political climate permits them to do so." The ferocity of opposition to medical marijuana by the U.S. and other governments is indicated by numerous official documents (e.g., U.S. House of Representatives 2005; Drug Enforcement Administration 2011). This has led to a climate of fear in which it is difficult to conduct research that could contradict the politically sanctioned position. To this day, passionate advocacy for and against marijuana is hampering evaluation of medical benefit/risk evaluation.

Experimental design and statistical methods for controlling bias are available (Centre for Reviews and Dissemination 2009; Higgins and Green 2011; Higgins et al. 2011) but are infrequently employed in cannabis studies (Whiting et al. 2015).

BIAS AGAINST EUPHORIC MEDICINES

The psychopharmacological profession is enthusiastic about, one might even say addicted to, prescription psychotropics (mood, perception, and behavior enhancers and stabilizers). Antidepressants, tranquillizers, antipsychotics, stimulants, and other psychiatric drugs often seem to be overprescribed, particularly for the young (Sinclair 2012), the old (Hilmer and Gnjidic 2013), and women (Ettorre and Riska 1995). The issue is highly polarized, with camps for and against the position that there is overprescription, and indeed such disagreement serves to indicate how hard it is to decide on the pros and cons of mind-modifying drugs like marijuana. Regardless, the general goal of medicinal psychotropics is to make patients feel good—but not too good.

It is curious that while a chief purpose of medicine is the immediate relief of pain, the opposite of pain, pleasure, is rarely identified as a direct medical goal, and indeed, euphoria is frequently highlighted as an undesirable side effect of medication. This is particularly true for marijuana. One of the objectives of the search for synthetic versions of THC was to find compounds that would deliver the medical benefits without the "high." However, as pointed out by Cristino and Di Marzo (2014), "the euphoric effects of cannabinoids are not always an obstacle to their effective administration. Mood elevation... may be an important component of its effectiveness in patients with cancer or AIDS."

Notcutt (2004) commented with regard to medical cannabis: "there are many health professionals who perceive that a mild psychoactive effect from the drug is somehow wrong. This only seems to be of concern to those who do not treat patients in pain or distress... Elevating the mood of a patient whose life is miserable because of chronic, untreatable pain would seem to be a worthwhile goal. There is also a desire by some to find a form of cannabis that is stripped of psychoactive effects, but this goes against the experience with morphine. Any drug that has multiple sites of action within the central nervous system is unlikely to be obtained in a form that does not affect consciousness, mood or cognition in some way. No one has achieved this with opiates after a century of research."

THE ISSUE OF ADDICTION

Western medicine almost universally frowns on the use of medicinals to produce euphoria for fear that patients will become addicted to them. The basis for this is that "it is now well accepted that abusable drugs derive their addictive potential from activating the core pleasure/reward circuitry of the brain" (Gardner 2014). A common-sense exception is made for those who are terminally ill or at least suffering from extremely grave or debilitating conditions. The psychological and physical horrors of those addicted to narcotics are a popular theme in literature, and it is clear that for many individuals addiction is serious, even deadly. However, some patients complain that caregivers refuse to provide adequate amounts of painkillers because of fear that addiction (and possibly lawsuits or professional censure) will result. "The reason that patients with chronic pain are not given opiates when indicated is...the intolerance in our society to any ongoing use of opiates in non-terminal patients. This has been described as 'opiophobia' and is a major medical problem" (Trachtenberg 1994; the similar word "ophiophobia" refers to fear of snakes). It is important to distinguish the use of psychotropics in a medical setting from the abuse in a nonmedical context and to weigh the possibility of addiction in relation to the individual characteristics and context of patients. In the case of cannabis, as pointed out in Chapter 12, perhaps 9% of recreational users become addicted, but the health consequences pale in seriousness compared to the opiates.

THE ISSUE OF INCAPACITATION (REDUCTION IN ABILITIES) WHILE INTOXICATED

Western medicine is reluctant to employ pharmaceuticals that produce euphoria as a side effect because in this state patients cannot function as well as healthy individuals. An exception is made for those who are suffering chronic, incurable, extremely miserable, or debilitating conditions such as often associated with acquired immunodeficiency syndrome (AIDS) and cancer. As pointed out by Fitzcharles et al. (2014): "The overriding principle for any pain treatment is to maintain function, without sacrificing cognitive or psychomotor function, a concept clearly different from pain management for medical conditions predominantly requiring palliation." It is a remarkable fact that some powerful potentially incapacitating drugs of abuse can be used regularly and responsibly. Cocaine is a very serious addictive drug problem in North America, but coca leaf is responsibly employed as a stimulant in much of South America in the same way as coffee and tea are used (Small 2004). Alcohol is not employed as a medicine (although alcoholic beverages sometimes have health benefits), but the fact that many can drink responsibly also illustrates that some potentially harmful substances can be consumed appropriately by a substantial proportion of society. This is not to ignore the dangers of adding marijuana to the list of potentially dangerous substances that are permitted, but to point out that compromise rather than outright veto is possible.

TRANSFER OF BIAS AGAINST THC TO NONINTOXICATING CANNABINOIDS

THC is the predominant psychotropic cannabinoid of *C. sativa* and, were it not present, marijuana would not be intoxicating and all the legal problems associated with the plant and its preparations would not exist. Most cannabinoids are not intoxicating, so why not free the "innocent" cannabinoids from legal constraints? There are some reasons for retaining laws against cannabinoids (as is the case in many jurisdictions)—for example, they can be chemically converted to intoxicants (the conversion of CBD to THC was examined in Chapter 11), and some synthetic cannabinoids are quite dangerous. Nevertheless, for practical purposes, nonintoxicating cannabinoids from the plant could be sold as conventional pharmaceuticals or over-the-counter preparations (often classified as "herbals" or "supplements"). The stigma that THC bears because it is intoxicating has unfortunately been transferred to the essentially harmless cannabinoids. As described later, CBD is the most promising medicinal cannabinoid, with astonishing potential for therapeutic applications, and it is important for society to relieve this remarkable compound from the burdens of its more hazardous sibling, THC.

THE IMPORTANCE OF EXPERIMENTAL DESIGN IN EVALUATING MEDICAL MARIJUANA

Most public knowledge of cannabis, and indeed most literature about it, is based on anecdotal evidence—i.e., accumulated random observations and conclusions. Most of what people have learned and come to believe is in fact based on repeated observations, and useful "experience" is the result of recurrently observing the dependability of phenomena. But the fact that people disagree on many "facts" demonstrates that humans are often mistaken. Since there is so much disagreement about the merits of medical marijuana, it follows that scientific (statistical) evaluation, including appropriate scientific experimental design, is indispensable to evaluate the pros and cons of medical marijuana. Perhaps the chief biasing factor that needs to be controlled is expectation, as epitomized by the placebo effect (if you think something is effective, even if it isn't, there is probability it will seem useful). Particularly with respect to evaluation of drugs, prior belief about whether they will be effective or not has a powerful biasing effect (Peck and Coleman 1991; Kirk et al. 1998; Quitkin 1999). The placebo effect in cannabinoid research is as high as 70% (Consroe et al. 1991). To validly evaluate the merits of marijuana, rigorous placebo-controlled, double-blind studies (both the investigators and the subjects unaware of who is receiving the drug or a placebo) with adequately large samples for statistical evaluation are essential (Wright et al. 2012). Without blinding, in some experiments over 75% of participants in studies can be aware of whether they have received the drug or the placebo (Grant et al. 2012). Appropriately designed studies of cannabis are not yet sufficient to evaluate the usefulness of marijuana for many medical conditions, although unqualified claims like "marijuana cures cancer" are common.

THE PRECAUTIONARY PRINCIPLE IN RELATION TO MEDICAL MARIJUANA

The "precautionary principle" is a conservative approach, when a decisive majority view does not prevail concerning conflicting evidence on the opposite sides of a publically prominent issue in which there is plausible (perhaps uncorrectable) risk to human welfare or to some other critically important situation. The precautionary principle came to prominence in 1972 at the United Nations Conference on the Environment (in Stockholm), in regard to serious or irreversible harm to ecosystems. In essence, the so-called "principle" advocates inaction or maintenance of the status quo, in fear of extensive, catastrophic harm ("it's better to be safe than sorry"; "err on the side of caution"), but is often accompanied with a recommendation that additional evidence should be sought before an action is considered further.

The precautionary principle is a reaction to the common view that an activity is innocent until it is proven guilty, a regulatory viewpoint that is decidedly dangerous to public health. In the health sphere, the practice of medicine is conducted with exceptional caution, and most clinicians are very reluctant to adopt new treatments without quite convincing evidence. This is consistent with a fundamental precept of bioethics, captured in the phrase "First do no harm" (the Latin *Primum non nocere* or *Primum nil nocere*; popularly misconceived as part of the Hippocratic Oath).

As documented in this chapter, while there are many potential medical applications of cannabis, the medical community is divided on the merit of the majority of possible usages. This disagreement, in the perspective of the precautionary principle, has been employed as an argument against the use of medical marijuana. As pointed out in Chapter 15, in regard to the authorization of recreational marijuana, the precautionary principle is one of several relevant considerations. Moreover, as also examined in Chapter 15, since the issue of medical marijuana in some jurisdictions has been taken out of the hands of the medical profession and put to public vote, the precautionary principle is a critical factor for public health.

However, the precautionary principle must not be allowed to block medical progress and the availability of effective treatments with cannabis. Medical therapy is frequently and necessarily a question of balance of risk and benefit, and a literal interpretation of the precautionary principle could prevent treatment merely because there is risk. The practice of medicine is already very highly regulated, with appropriate professional evaluation of risk, and medical marijuana should be judged on the same basis as are all other developing therapies.

RECENT AUTHORIZED MEDICAL MARIJUANA PRODUCTION AND USAGE

In most countries, cannabis usage for medical purposes has remained prohibited. However, during the last several decades, there has been a momentous societal and scientific debate regarding the wisdom of employing cannabis drugs medically. This has led to experimental production and medical evaluation in many Western countries. Several European and Commonwealth countries and many states of the United States currently allow medical dispensation of marijuana, while Uruguay and several U.S. states recently have even permitted the sale of marijuana for recreational use by licensed vendors. The fact that some governments have authorized medicinal use does not necessarily indicate acceptance of its value but often reflects a desire for effort at medical evaluation, compassion for those who haven't found a better alternative, court decisions that have limited legislation, or simply political expediency.

In the United States, the National Institutes of Health (NIH) has a long-standing program of research into medicinal marijuana and for a period supplied a handful of patients with medicinal marijuana for treatment. In the early part of the twenty-first century, governmental authorities in the United States consistently rejected the authorization of marijuana for medical use (Anonymous 2006). The American Drug Enforcement Administration has been hostile to the medicinal use of *Cannabis* (Drug Enforcement Administration 2013), and for decades research on medicinal properties of *Cannabis* in the United States was in an extremely inhospitable climate, except for projects and researchers concerned with curbing drug abuse. Essentially, the only standard research marijuana for experimentation in the United States had to be obtained from the NIH marijuana farm at the University of Mississippi, and few were allowed to obtain research material. However, in 2014, a very large increase in the number of medical research projects in the United States was authorized, reflecting the belated recognition of the desirability of examining the safety and efficacy of medical marijuana.

PHARMACOLOGICAL TERMINOLOGY FOR MARIJUANA

People smoke cannabis because it pleasurably alters consciousness, producing euphoria and relaxation and intensifying ordinary sensory experiences. It also alters perception—for example of time—and can promote sociability, stimulating laughter and talkativeness. These manifestations are common to many social inebriants, but cannabis is fairly unique and is difficult to characterize. Moreover, individual response can differ considerably. For example, some people claim that marijuana increases creativity, others that it just makes them sleepy (note the discussion presented later that different kinds of marijuana may produce different effects).

The word "narcotic," often used to describe the psychological effects associated with marijuana, has been extensively and ambiguously employed in lay, legal, and scientific circles. The term is most widely used as an arbitrary juridical category—a narcotic is simply a substance or preparation that is associated with severe penalties because of real or alleged dangerous, addictive properties. "Legally, cannabis has traditionally been classified with the opiate narcotics, and while they may share some euphorogenic and analgesic properties, they are otherwise quite distinct pharmacologically" (Le Dain 1972). Etymologically, based on "narcosis," a narcotic would be expected to be a substance promoting sleep, and indeed, some use the term to characterize any drug that produces sleep, stupor, or insensibility. Both THC and CBD, at least one of which dominates the cannabinoids of most biotypes of *C. sativa*, have sleep-inducing properties at some dosage, albeit CBD is stimulative at low and moderate dosages (Piomelli and Russo 2016) and is sedative only at quite elevated doses (Carlini and Cunha 1981; Pickens 1981). Moreover, the terpene myrcene is common in *C. sativa* (especially in marijuana strains with appreciable CBD) and is sedative (Russo 2011a). Accordingly, the soporific property of cannabis provides some limited justification for referring to it as a narcotic, although it is by no means best known for its sedative properties. Nevertheless, the term "narcotic" is better known as characterizing an intoxicant than a sedative. Because "narcotic" is often used pejoratively, it is probably best avoided as descriptive of pharmacological effects. Although substances called narcotics are widely viewed as intrinsically evil, the world's leading controlled so-called narcotic crops have some legitimate, useful applications (Small 2004; Small and Catling 2009).

The pharmacological classification of cannabis is controversial. It has been characterized as a sedative-hypnotic-general-anesthetic like alcohol and nitrous oxide; a mixed stimulant-depressant; a mild hallucinogen, especially at higher doses; a "psychedelic," like LSD at very high doses; and as a separate category of psychic experience (Le Dain 1972). The following terms have been used to describe cannabis: psychedelic (mind-manifesting or consciousness-expanding), hallucinogenic (hallucination-producing), psychotomimetic (psychosis-imitating), illusinogenic (illusion-producing), and psychodysleptic (mind-disrupting); as noted in Le Dain (1972, p. 396), all of these terms are problematical. None of the terms is completely satisfactory to denote the euphoric psychological effects of marijuana in general and THC in particular.

There is little dispute that cannabis is a "psychoactive" drug (one altering sensation, mood, consciousness, or other psychological or behavioral functions). However, "psychoactive" is so broad it applies to a very wide variety of psychological states. Clearly, marijuana is popular (albeit largely illegal), employed primarily as a social inebriant and euphoriant. "Psychotropic," meaning mind-altering, is also widely used, but both "intoxicant" and "nonintoxicant" types of *Cannabis* can influence the mind by virtue of the properties of THC and CBD. "Hallucinogenic" is less appropriate since true hallucinogens are rarely produced. Psychotomimetic (mood-altering) is perhaps the most appropriate pharmacological term but is hardly definitive since it could be applied to numerous substances, including chocolate and caffeinated beverages. Although not a technical phrase, "mood enhancer" is sometimes applied to marijuana.

COMPARATIVE THERAPEUTIC VALUE OF PURE
CHEMICALS AND HERBAL MIXTURES

An important issue in the therapeutic use of medicinal substances is the relative wisdom of employing pure chemicals and herbal mixtures. The former have been described as "monomolecular bullets" (implying that they can be directed precisely to targeted symptoms or their underlying causes), while the latter have been characterized as "polymolecular shotgun shells" (implying that they cannot be directly targeted and may inadvertently cause damage).

"Pharmacognosy" used to be defined as the study of crude drugs. "Crude drugs" refers to materials obtained from natural sources, including plants, animals, and minerals. In practice, crude drugs originate mostly from plants and are "unrefined" beyond being dried and ground up. Crude drugs from plants represent a complex mixture of chemicals that they have produced. Today, the word "pharmacognosy" has been expanded to include all drugs from natural sources, including purified extracted molecules. The more inclusive word "pharmacology" refers to the study of all drugs. Human therapeutic use of unrefined plant materials is the foundation of all medicine, and remains the basis of most healing practiced in the world because crude plant drugs are far cheaper and easier to obtain than modern pharmaceuticals, and so are widely employed in the Third World. The most extensively utilized herbal medicine is Asian ginseng (the root of *Panax quinquefolius* L.), consumed by hundreds of millions, despite the fact that its medicinal value has been almost as contentious as that of marijuana (Small 2006). In the rich, technologically advanced Western World, most medicines are pure chemicals—single molecules extracted from species or synthesized. Marijuana is a crude drug (although purified extracts are not), and as noted in the following, this is the basis of much criticism of its value.

In the phrase "crude drug," the word "crude" simply means "in a natural or raw state, or not processed or refined." However, the word "crude" also means "marked by the primitive, gross, or elemental or by uncultivated simplicity or vulgarity." It is possible that the disfavor for crude drugs today is subconsciously being influenced by the latter, pejorative meaning. The fact is, crude plant materials in the form of vegetables and fruits are the mainstay of most human health.

In the prestigious report *Marijuana and Medicine: Assessing the Science Base* (Joy et al. 1999), the following statement is presented: "Defined substances, such as purified cannabinoid compounds, are preferable to plant products, which are of variable and uncertain composition. Use of defined cannabinoids permits a more precise evaluation of their effects, whether in combination or alone." Modern medicine has been said to prefer single-component "silver bullets" rather than multicomponent "herbal shotguns" (Spelman 2009). Kalant (2001) noted: "The history of drug therapy has been to a large extent one of progressive movement away from natural products of unknown or variable composition and potency, toward the use of pure active compounds of precisely known composition, stability, dosage and pharmacology."

Pharmacological drug research today is heavily slanted toward discovering or computer-designing compounds that affect cellular receptors, compensating for abnormal or inadequate genes at the root of diseases. Since medical conditions are often the result of combinations of causes, a single pharmaceutical "silver bullet" may not suffice, requiring a combination of silver bullets (or a designer pharmaceutical shotgun shell). In either alternative, the movement away from crude natural products remains.

A concern about all crude drugs is that even if they contain therapeutic chemicals, they may also have toxic compounds. Dupont (2000) characterized marijuana as "a complex chemical slush" with more than 2000 chemicals. While this seems alarming, it should be remembered that our vegetables, fruits, and cereals are also composed of thousands of chemical compounds, and the liver—the largest internal organ—is designed to cope with regular intake of small amounts of toxins.

Western-based medicine has indeed become reliant on single-molecule pharmaceuticals or at least carefully evaluated combinations of drugs (described as "polypharmacy"). Even with the resurgence of alternative (especially herbal-based) modalities, there is widespread disrespect (in the West) for traditional plant-based medicines because they are not precisely defined mixtures (although they are effectively polypharmaceutical in nature). However, the perspective that herbal (crude drug) preparations are inherently inferior is short-sighted. Many herbal products in Europe are standardized and have been clinically demonstrated to be efficacious in double-blind placebo-controlled trials (Tyler 1993a,b, 1996; Tyler and Foster 1996; Small and Catling 1999).

Defenders of herbal medicine often point out that there may be synergistic effects (which increase potency or other desirable effects) or mitigative effects (decreasing toxicity) because of interactions among the constituents of crude drugs and that, over time, humans have learned by trial and error the circumstances when these preparations are efficacious (Lewis and Elvin-Lewis 2003). Of course, research is required to examine the comparative merits of crude drugs, extracts, and synthetic analogues, and this is particularly true for *C. sativa*. Crude cannabis drugs (marijuana, hashish) are currently the main options exercised for medical use of *C. sativa*, and indeed, they are often chosen in preference to extracts and synthetic analogues by patients. It is very well known that extracted cannabinoids produce somewhat different effects from crude marijuana (Segelman et al. 1974; Fairbairn and Pickens 1981; Pickens 1981; Johnson et al. 1984; Wilkinson et al. 2003; Whalley et al. 2004; Ryan et al. 2006) and often do not satisfy patients as well as crude drugs, and this suggests that interactions of natural constituents are very important therapeutically (McPartland and Pruitt 1999; McPartland and Russo 2001; Russo and McPartland 2003; Russo 2014).

MEDICINAL IMPORTANCE OF COMBINING THC AND CBD

The previous paragraph points out the possibility that there are therapeutic interactions among the components of marijuana. In fact, a therapeutic interaction between THC and CBD dramatically validates such an occurrence. Although widely said to be "nonpsychoactive," it has long been appreciated that, at least at high dosages, CBD has sleep-inducing or sedative properties (Carlini and Cunha 1981), although at lower doses, it has alerting properties (Nicholson et al. 2004; Russo and Guy 2006). The reputation of "indica type" (high-CBD) marijuana strains for inducing sleep was demonstrated by a survey of patients at a dispensary, who chose such strains to treat insomnia (Belendiuk et al. 2015).

It is apparent that CBD antagonizes (potentiates, reduces, mitigates) undesirable side effects of THC. CBD ameliorates (in a therapeutic sense) the effects of THC, blocking anxiety provoked by THC, reducing psychotic experiences associated with high-THC marijuana, and attenuating memory-impairment effects of THC (Russo and Guy 2006; Zuardi et al. 2006, 2012; Mechoulam 2012; McPartland et al. 2015). The combination of THC and CBD is now appreciated to have medicinal advantages, although the nature of the interaction is not well understood (Bhattacharyya et al. 2010; Zuardi et al. 2012). Reducing the intensity of the THC experience is considered especially beneficial for inexperienced users, who may be subject to panic attacks, paranoia, delusions, or other disturbing symptoms on exposure to a high level of THC. Mechoulam and Parker (2013b) commented: "The reversal by CBD of some of the undesirable effects produced by pure THC or by cannabis with low levels of CBD and high levels of THC…strengthens the view that medicinal cannabis containing reasonably high levels of CBD is a better drug than cannabis with low levels of CBD or pure THC alone."

Sativex (Figure 13.6) is the most significant innovative cannabis-based medicine developed to date. The USAN name for the trade name Sativex is nabiximols. ("United States Adopted Names" [USAN] are unique nonproprietary names assigned to pharmaceuticals marketed in the United States.) Sativex is standardized in composition and is effectively a tincture. This cannabinoid-based analgesic, developed by the United Kingdom firm GW Pharmaceuticals (Guy and Stott 2005), exploits the advantages of combining approximately equivalent amounts of THC and CBD. Sativex is a buccal ("oromucosal") spray, applied under the tongue or inside the cheeks (never into the nose) using a nebulizer. The extracts in Sativex comprise about 70% THC + CBD, minor cannabinoids (5%–6%), terpenoids (6%–7%), sterols (6%), as well as other components (triglycerides, alkanes, squalene, tocopherol, carotenoids, and other compounds) derived from cannabis plants (Russo and Guy 2006).

In 2005, Sativex became the world's first licensed prescription medicine based on extracts of *C. sativa*, when it was licensed for usage in Canada. It is now licensed for various usages in numerous countries. It has been reported that some patients experience stinging or lesions (mouth ulcers)

FIGURE 13.6   Sativex, cannabinoid spray, an alcoholic solution with about equal amounts of THC and CBD, sprayed under the tongue and authorized for use in more than two dozen countries. Photos provided by GW Pharmaceuticals plc.

from the high alcohol content of Sativex (Scully 2007), and other relatively minor side effects have been reported. While it is clear that the interaction of THC and CBD is the principal medically beneficial interplay of cannabis constituents, it is quite possible that the additional components in Sativex and indeed in marijuana are also beneficial, either additively or interactively.

THE ACCIDENTAL MEDICINAL VALUE OF INDICA TYPE STRAINS

In southern Asia, the homeland of potent intoxicating cannabis preparations, sativa type land races used to produce both hashish and marijuana-type products have a cannabinoid profile typically dominated by THC, while indica type land races often used to produce hashish typically have a cannabinoid profile dominated by both THC and CBD. (This does not necessarily apply to single indica type plants, but rather to populations; Clarke 1998a.) For production of hashish, it is desirable that considerable resin be produced relative to plant tissues, and so hashish strains often seem to have been unconsciously selected for resin production at the expense of lowering THC content. The combination of THC and CBD has therapeutic advantages, as explained previously. It is unclear whether or not traditional (obsolete) medicines based on *C. sativa* took advantage of the naturally superior medicinal qualities of indica type strains with balanced THC and CBD. In principle, preparing marijuana from such strains should also be advantageous therapeutically, but halving the amount of THC (in order to have an equal amount of CBD) has the undesirable effect of increasing the amount of material that needs to be consumed (if absorbed by smoking).

OVERVIEW OF MEDICAL MARIJUANA PREPARATIONS

Medical marijuana is currently being dispensed in many jurisdictions. A variety of forms are available, as shown in Figure 13.7. Most preparations of cannabis described in Chapter 12 can be employed in a medicinal way, but each type has advantages and disadvantages, as discussed in this chapter. Illicitly produced cannabis is especially unsuitable for medical usage because of impurities and variability of cannabinoid content. Traditional hashish, in particular, is a sticky, contaminated mixture, which is very unlikely to be usable for medicinal purposes.

Cannabis is popularly consumed as smoked herbal material, and when experienced users become patients using cannabis therapeutically, there is a tendency to remain loyal to their traditional form

FIGURE 13.7   Medicinal marijuana scenes. (a) Medical marijuana van. Photo by Jamal Fanaian (CC BY 2.0). (b) Lollipops with infused cannabinoids. Such foods are made with fats (usually butter), in which THC readily dissolves. Photo by Berknot (CC BY SA 2.0). (c) Counter display of strains of buds. Photo by AudioVision— Public Radio (CC BY 2.0). (d) Confections prepared with infused cannabinoids. Photo by Compassion007 (released into the public domain). Although allowed or at least tolerated in some jurisdictions, the materials illustrated here are illicit in some circumstances or regions. Moreover, as discussed in the text, cannabis and associated advice on its use, acquired in many commercial dispensaries, may result in inferior treatment compared to care in a hospital or genuine medical clinic.

of consumption. Hazekamp et al. (2013) reported on a survey of preferences of almost 1000 patients employing cannabinoid-based medicines (CBMs), and found that "In general, herbal non-pharmaceutical CBMs received higher appreciation scores by participants than pharmaceutical products containing cannabinoids. However, the number of patients who reported experience with pharmaceutical products was low, limiting conclusions on preferences." Given the very small number of patients who had used pharmaceutical CBMs in the study of Hazekamp et al., it is possible that patients habituated to them would develop and maintain a preference for them rather than traditional marijuana.

In the illicit drug trade, cannabis preparations based on solvent extracts are unreliable and dangerous. However, alcoholic cannabinoid extracts were widely employed medically a century ago (Russo 2003a; Figure 13.8a and b), and solvent-extracts can be produced safely by authorized qualified personnel (Figure 13.8c). Sativex, obtained by supercritical liquid carbon dioxide extraction (Potter 2014), is the most successful commercial example. As illustrated in Figure 13.8d, a variety of balms, salves, and ointments based on extracted cannabinoids are currently being marketed.

Cannabis resin powder (Figure 13.9), described in Chapter 12, is a relatively well-defined product. The devices described in Chapter 12 for manufacturing cannabis resin powder could be used to produce a product suitable for medical consumption. Apparatus for the production of medicinal marijuana should be sterilizable, and it is not clear that available devices meet this criterion. There is no literature reporting the microbiological safety of the products produced by using the available resin powder devices and techniques, but appropriate precautions could be taken to ensure adherence to good manufacturing practice and safety in conformity with conventional medical products. Cannabis resin powder has the potential to be used like grated Parmesan cheese from a shaker bottle, although the natural stickiness of the material can make distributing the material difficult. Resin powder could be added to marijuana cigarettes ("reefers") to increase dosage, so less material

FIGURE 13.8 Medicinal preparations based on extracts of cannabinoids of *C. sativa*. (These may be illegal or require licenses depending on jurisdiction.) (a, b) Early twentieth century preparations (a, public domain photo; b, photo by E. Small). (c) Tincture accompanied by syringe for oral application. Photo by Stephen Charles Thompson (CC BY 3.0). (d) Cannabinoid-based preparations marketed in Colombia in 2016. (A) Extracts using lipids instead of hydrocarbons as solvents. (B) Lip balm. (C) Topical cream. (D) Alcoholic tinctures. Photo courtesy of Steve Naraine.

FIGURE 13.9 Resin powder. Photographs courtesy of Mila and Chimed Jansen of the Pollinator Company.

needs to be smoked. "Artificial reefers" have in fact been produced for medical research by adding solvent-extracted THC to cellulose-based "cigarettes." However, as noted in the following presentation, smoking cannabis is inconsistent with health promotion.

Purified cannabinoids and synthetic cannabinoids are often regarded as the ultimate path to responsible cannabis medications, but as noted in this chapter, they have not yet proven to be as consistently effective or acceptable to the majority of patients as cruder cannabis preparations.

MEDICAL MARIJUANA DRUG DELIVERY SYSTEMS

Aside from considerations of the inherent toxicity of cannabis is the issue of the relative safety (or harmfulness) of modes of drug delivery. Inhalation and oral ingestion are the most common methods of administration, but delivery can also be intrathecal (injected into the fluid surrounding the brain and spinal cord), intravenous, opthalmogic (ophthalmologic, i.e., by eye), rectal, vaginal, sublingual (under the tongue), or transdermal. The production of novel pharmacological products that deliver cannabinoids efficiently is often considered to be a key to the future acceptance of medical marijuana. New pharmacological products for delivery of cannabinoids may be based on suppositories, time release encapsulation, eye drops, nasal sprays, aerosols, topical ointments or creams, sublingual drops, transdermal patches, etc.

Safer Respiratory Systems

Conventional Smoking

"Smoking" is simply breathing in smoke (particulates and gases) from combustion. The heat from the fire in front of the herbal material that is set alight (located just back of the lit front of a joint or just underneath the burning material at the top of the bowl of a pipe) converts the THC-acid to free THC, as described in Chapter 11, and volatilizes the THC, which is inhaled with the smoke into the lungs. THC is very lipid-soluble, so it can easily cross the membranes of the alveoli (air sacs) and enter the blood in the pulmonary capillaries. The metabolized THC is rapidly carried to the heart and pumped to the brain.

One of the ways that cannabis is unique is that no other prescribed drug is administered by smoking (Mechoulam 2012). (Historically, in fact, tobacco was smoked occasionally for medical purposes.) The extremely serious health hazards of smoking tobacco are well known: bronchitis (inflammation of the mucous membranes lining the major airway leading to the lungs), emphysema (lung diseases damaging alveoli, leading to shortness of breath), lung cancer, heart disease, and numerous other disorders, including "bong lung" (Gill 2005). As a system for delivering the target chemical (nicotine in the case of tobacco), smoking of any herbal is likely to also deliver hundreds of toxins, and this unhealthy consequence is certain when marijuana is smoked (Tashkin 2002, 2005, 2013; Reid et al. 2010; Owen et al. 2014). Moir et al. (2008) observed that as well as finding numerous of the same toxic compounds as in tobacco, "ammonia was found in mainstream marijuana smoke at levels up to 20-fold greater than that found in tobacco. Hydrogen cyanide, NO, NO_x [NO_x is a generic term for the mono-nitrogen oxides NO and NO_2], and some aromatic amines were found in marijuana smoke at concentrations 3–5 times those found in tobacco smoke."

Macleod and Hickman (2009) concluded that "cannabis use is almost certainly harmful, mainly because of its intimate relation to tobacco use," and Kalant (2008) predicted that smoked marijuana had little future prospects for medicinal purposes. Many of the ingredients common to marijuana and tobacco smoke (including tar, carbon monoxide, aromatic hydrocarbons, hydrocyanic acid, oxides of nitrogen, acrolein, reactive aldehydes, several known carcinogens, and particulate matter such as naphthalene, dimethylphenol, and benzopyrene) are known to be toxic to respiratory tissue (Taylor 1988; Earleywine 2010). Cannabis smokers inhale more deeply than tobacco smokers, resulting in greater exposure to combusted material per inhalation, but most cannabis users consume fewer cigarettes. Unfortunately, marijuana is generally smoked with tobacco in Europe, parts of Asia, North Africa, Australia, and New Zealand, although not usually in North America (Leggett 2006). The American Lung Association (2015) issued the following statement. "The ALA encourages continued research into the health effects of marijuana use, as the benefits, risks and safety of marijuana use for medical purposes require further study. Patients considering using marijuana for medicinal purposes should make this decision in consultation with their doctor, and consider means of administration other than smoking."

Sullivan et al. (2013) compared the efficacy of filtration techniques on reducing pesticide content in absorbed smoke. A handheld glass pipe allowed between 60% and 70% of pesticide residues to be absorbed. An unfiltered water pipe allowed 42% to 60% absorption. A cotton-filtered water pipe allowed only 0.1% to 11% absorption. While filtration can significantly reduce toxic pesticide residues, this does not necessarily reflect how well other classes of toxins are removed.

The issue of harm to pulmonary function from recreational smoking of marijuana was discussed in the previous chapter. The harm from smoking medical marijuana should not be equated with the harm likely resulting from regular smoking of recreational marijuana. Joshi et al. (2014) concluded, "Medicinal use of marijuana is likely not harmful to lungs in low cumulative doses."

In the interests of harm reduction, it is preferable to utilize efficient systems that increase the proportion of cannabinoids taken up while decreasing exposure to numerous other volatilized substances. Smoking cannabis preparations with an increased proportion of THC is the most common way of achieving this. Thus, high-THC buds and hashish can be considered "safer," lowering the risk from the hazardous chemicals by reducing the amount of material that is smoked. Properly prepared hashish contains much higher levels of the cannabinoids than does marijuana (this is often not true in the illegal trade), and therefore, a smaller quantity needs to be consumed in comparison to marijuana. Accordingly, lesser amounts of toxins are absorbed in smoking, and at least *in this limited sense*, hashish is safer than marijuana. Whether smoking "safer" marijuana encourages unnecessary usage is another issue.

During the first decade of the twenty-first century, illicit marijuana has tended to increase in THC content in the United States, Netherlands, and United Kingdom (McLaren et al. 2008). The widespread criticism that, because cannabis products in the illicit trade have increased in potency (THC content) during the past 20 years (Licata et al. 2005; Cascini et al. 2012), they are more dangerous, tends not to be taken seriously by informed pharmacologists. This is not only because higher-potency material means less material needs to be smoked, but also because cannabis dosage is autotitrated (self-regulated) by experienced users (novices are less able to limit their consumption to a desired point of satiety). According to Leggett (2006), "Unlike other drugs, it is virtually impossible to 'overdose' on cannabis." (As noted later in the discussion of "dabbing," this is no longer true.) Similarly, when consuming alcoholic beverages, whether beer, wine, or liquors, many experienced users tend to self-dose up to a particular level of intoxication, and the different concentrations of alcohol present is of relatively limited importance. However, King et al. (2005) stated that "how far this parallel hold for cannabis is unknown," and clearly, alcoholism and alcohol-based death indicates that some drinkers cannot limit their intake. Di Forti et al. (2009) intimated that high-potency marijuana may increase the risk of psychosis, but their study design was inadequate to show a causal relationship.

It is possible to smoke extracts, but because these are so concentrated, in the illicit culture they are generally diluted with tobacco or less concentrated marijuana. Street-available extracts can be very dangerous, but properly made extracts can be free of contamination. In any event, it is far easier and safer to consume extracts by vaporization, described later in this chapter, than to attempt to smoke them.

Valsalva Maneuver and Prolonged Breath-Holding

Intuitively, breath-holding time and puff volume while smoking marijuana would seem to determine the amount of THC absorbed. Both assumptions have been questioned, particularly whether sustained maintenance of smoke in the lungs increases the resulting degree of intoxication. The deep inhalation and prolonged holding of breath—the usual technique of smoking marijuana—create a Valsalva maneuver. The Valsalva maneuver (named for the seventeenth century Italian physician A.M. Valsalva) results from attempting to forcibly exhale while keeping the mouth and nose closed, thus closing the windpipe. This impedes the return of venous blood to the heart and is sometimes used to diagnose cardiac conditions. It is also employed to assess a variety of other pathological conditions. In diving and air travel, the technique is often used to equilibrate pressure issues for the ears and sinuses. The Valsalva maneuver has been speculated to rarely cause pneumothorax (collection of air within the chest, causing a lung to collapse), pneumomediastinum (collection of air in the middle portion of the chest), and lung disease (Hii et al. 2008; Lee and Hancox 2011).

The common conviction that prolonged breath-holding of marijuana smoke enhances the effects of marijuana has been disputed (Zacny and Chait 1989, 1991; Azorlosa et al. 1995; Earleywine 2010). "Stoner" films regularly show the leading characters bravely holding in exaggerated puffs (they also feature poorly constructed plastic marijuana plants). The belief in breath-holding is so established that challenging it, as Earleywine (2010) pointed out, is likely to produce profound disbelief in the user community.

Water Pipes

Water pipes (devices to draw smoke through water; small contraptions are commonly called bongs, larger ones are hookahs; Figure 13.10; also see Figure 12.31) are widely employed by cannabis smokers in order to filter out toxins created by combustion and reduce pulmonary irritation. The heat of smoking can have negative effects, such as inflammation, and because water pipes cool smoke, they may be advantageous in this respect (Earleywine 2010). Water also removes gas-phase smoke toxins, such as ammonia, acetaldehyde, benzene, carbon monoxide, hydrogen cyanide, and nitrosamines, but is mostly ineffective against tars (polycyclic hydrocarbons) and particles in smoke (Bloor et al. 2008). Although THC is insoluble in water, the water nevertheless appears to trap some THC, leading to the need to smoke more when using a bong, with consequent increased intake of tars by comparison with smoking joints (Earleywine 2010). Gieringer (2001) stated, "studies have found that waterpipes and solid filters are ineffective at improving the THC/tar ratio in smoke." Regardless of smoking technique, because of incomplete decarboxylation of THCA, loss through exhalation, and destruction by pyrolysis, a maximum of about 30% of the THC in cannabis preparations is absorbed (Russo 2007).

Vaporization

A smokeless pulmonary technique now extensively used in the consumption of cannabis drugs employs vaporization or volatilization, i.e., heating to produce steam or vapor without ignition or burning. Because the temperature is kept below the point of combustion where pyrolytic toxic compounds are released, the production of irritating respiratory toxins is suppressed. Cannabinoids are quite volatile and will vaporize at temperatures much lower than the heat required to combust plant material. Vaporizers heat marijuana (usually buds or oil) to 180°C–190°C, vaporizing THC without burning the plant material, producing a mist rather than "smoke." These devices are likely to become especially widespread for medicinal usage. Vaporizers of the rechargeable battery-operated pen-type (electronic cigarettes or e-cigarettes; Figure 13.11, right), which can be used either with vials of hash oil or solid concentrates, have become common for discreet consumption of illicit marijuana because they are inconspicuous and produce limited odor. However, Etter (2015) found that the conventional larger electronic vaporizers (such as the Volcano shown in Figure 13.11, left) were more popular than electronic cigarettes. The purity of the material employed is key to the degree of safety in using vaporization. The delivery of THC to the bloodstream by use of vaporizers is considered comparable to that achieved by conventional smoking.

FIGURE 13.10   Elegant hookahs. Left: Turkish hookah. Photo by Solix (released into the public domain). Right: Hookahs on sale in Marakesh, Morocco. Such water pipes are particularly inappropriate for smoking medical marijuana because the dose is difficult to control. Photo by Just-a-cheeseburger (CC BY 2.0).

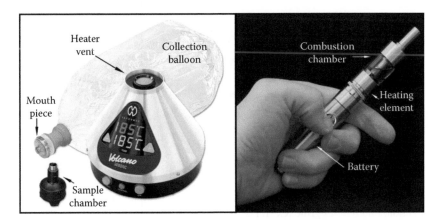

FIGURE 13.11   Vaporizers. Left: The Volcano vaporizer system for volatilizing the resin of *C. sativa*. This is the most extensively studied vaporizer available and has been used extensively in medical studies. A receptacle ("sample chamber") containing ground material is placed in the "heater vent" at the apex of the pyramid-like apparatus, where heat volatilizes the active components of the resin. The resulting gases are collected in the balloon, and the contents of this are inhaled. Such devices can greatly reduce the number of inhaled dangerous components that are present in marijuana (performance of the Volcano vaporizer was evaluated by Gieringer et al. 2004). Photo by Storz & Bickel (CC BY 3.0). Right: A portable electronic ("vape pen") vaporizer. Photo by Jonny Williams/www.ecigclick.co.uk (CC BY SA 2.0).

Ruchlemer et al. (2007) concluded that "vaporization of cannabis is a safe and effective mode of delivery of THC." Earleywine and Barnwell (2007) found comparatively reduced incidence of respiratory symptoms in users of vaporizers. However, inhaling the steam from vaporization does not eliminate all of the toxic materials produced. Low molecular toxic compounds such as ammonia tend to persist (Russo 2007; Bloor et al. 2008). Smith et al. (2015) studied the performance of the Volcano vaporizer and concluded: "Whilst the 'Volcano' device removes some toxic compounds from the smoke and reduces their inhalation by its user, it likely leads to enhanced ingestion of toxic ammonia known to result in neurobehavioral impairment." Grant et al. (2012) commented, "Vaporization is not a perfect solution since carbon monoxide is formed, but levels are significantly lower than with smoking."

ORAL CONSUMPTION

Cannabinoids can be consumed in foods (Figure 13.12). Because THC is lipophilic, orally consumed cannabis is absorbed better by the intestinal mucosa if some fat is ingested simultaneously (this is usually accomplished by adding a fatty liquid, such as cream to cannabis tea, or considerable butter when baked in brownies; animal lard and vegetable oils are also used). Brownies, candies, and cookies seem to be most popular. Teas (usually "decoctions," which are boiled preparations) and juices have proven useful, despite the notable limited solubility of THC in water. Chapter 12 provides information on how cannabinoids are added to common foods. Oral consumption in the form of foods or tinctures is a way of avoiding all lung problems, and during the nineteenth century, oral use was common both for medical and recreational use. However, becoming "high" from oral consumption is notoriously slow and comparatively unreliable. The speed of absorption of THC in the stomach depends on stomach contents and the presence of coingested drugs. Some degradation of THC by acids in the stomach and gut may occur.

Smoking produces effects within seconds to minutes, with a maximum from about 10 to 30 minutes and a duration of two or three hours, but with wide interindividual variation (Huestis 2005). THC can be detected immediately in plasma after the first puff. The rapid action of smoking is

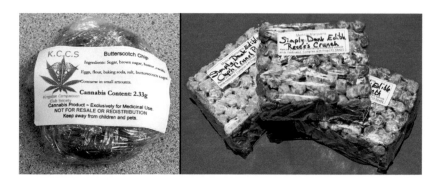

FIGURE 13.12 Edible medical marijuana preparations. Left: Cookie. Photo by Subvertc (CC BY 3.0). Right: Confection. Photo by Eggrole (CC BY 2.0).

due to THC being transported quickly to the brain. By contrast, eating does not produce effects for 30 minutes to three hours, and the effects are relatively prolonged, lasting five to eight hours or even longer. (Eating raw cannabis material that has not been heated to decarboxylate the acidic form of THC will produce only a minimal euphoric effect.) The slow action of orally ingested THC is due to its being transported from the stomach to the liver, where it is converted to 11-hydroxy-THC, a more potent and longer-lasting cannabinoid than THC. The more rapid onset and predictable decay of effects from smoking facilitates self-titration of dosing, in contrast to oral ingestion. Smoking and eating modes of metabolizing THC are contrasted in Figure 13.13.

SYNTHETIC ORAL CANNABIS DRUGS

A number of psychoactive analogues of THC have been synthesized and tested experimentally (Russo 2003b). Such analogues have been referred to as "cannabimimetic cannabinoids" and

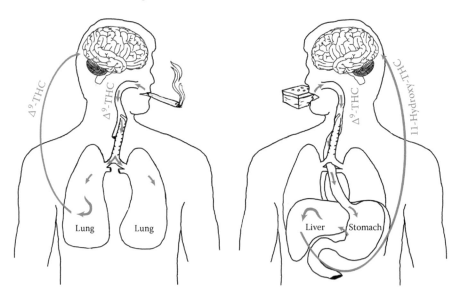

FIGURE 13.13 A contrast of the metabolism of inhaled and eaten marijuana. Left: Vaporized THC from smoking is carried to the lungs, where it is transported by blood vessels to the brain, exerting its psychoactive effects quickly (usually within 10 minutes). Right: Most orally ingested THC is transferred by blood vessels from the stomach to the liver, where it is converted to 11-hydroxy-THC, a more potent, longer-lasting metabolite, resulting in delayed (usually an hour or more) onset of psychoactive effects, which may last up to 24 hours and be stronger, less predictable, and less pleasant. Prepared by B. Brookes.

FIGURE 13.14 Synthetic ("cannabimimetic") cannabinoids. Left: Marinol (brand of dronabinol), a synthetic THC that has been prescribed for the control of nausea and vomiting caused by chemotherapeutic agents used to treat cancer and to stimulate appetite in AIDS patients. Right: Cesamet (brand of nabilone).

"syntho-cannabinoids." The following two (Figure 13.14) have been marketed commercially for several decades. Dronabinol is the synthetically manufactured (-)-trans-isomer of Δ^9-THC. Marinol is a dronabinol preparation, dissolved in sesame oil, provided as capsules. It is a registered trademark of Unimed Pharmaceuticals, Inc., a subsidiary of Solvay Pharmaceuticals, and is available in North America and some European countries. Nabilone is a synthetic derivative of Δ^9-THC with a slightly modified molecular structure. It is marketed under the name Cesamet, a registered trademark of ICN Canada Ltd., and is available in Canada, in the United States (through Valeant Pharmaceuticals International), and some European countries. These synthetic preparations of THC are expensive and are often considered to be less effective than simply smoking preparations of marijuana.

Rimonabant (the generic version of the trade pharmaceutical Accomplia; also trademarked as Zimulti) is a CB_1 cannabinoid receptor antagonist (discussed later in this chapter), marketed for

FIGURE 13.15 A cartoon parodying rimonabant, the once popular endocannabinoid-depressing weight-reducing pharmaceutical. Unlike THC, which stimulates CB_1 cannabinoid receptors and produces a desire to eat, rimonabant does the reverse, depressing appetite. During the brief period the chemical was authorized as a means of losing weight, it was widely and rather luridly advertised under various brand names in dozens of countries. Unscrupulous marketers still provide it by Internet sales. Prepared by B. Brookes.

several years as an appetite suppressant (Figure 13.15). By blocking endocannabinoid receipt in the brain, this synthetic chemical proved useful for weight loss and obesity control. Indeed, not only did the drug reduce weight and decrease waist size, it also increased high-density lipoprotein cholesterol (good cholesterol), decreased triglycerides, improved glucose tolerance and insulin levels, and seemed to help in cessation of smoking. Because of potential psychiatric side effects, including anxiety, depression, nausea, seizures, and (particularly) suicide, the drug was discontinued in Europe in 2006, the main region where it was sold (Bermudez-Silva et al. 2010).

Paradoxically, in view of the well-known ability of marijuana to stimulate appetite, Le Foll et al. (2013) showed that the prevalence of obesity is much lower in cannabis users compared to nonusers (see later discussion of obesity). It would of course be irresponsible to suggest that smoking marijuana is a reasonable way to lose excess weight.

DERMAL AND MUCOSAL DRUG ABSORPTION

"Skin" is a protective layer covering most of the outer body. Cannabinoids can be absorbed into the skin by topical (external) application; hence, concern has been expressed about the possible unintended absorption of THC through the skin because of the presence of contaminating THC in hempseed oil used in cosmetics, as described in Chapter 8. In fact, epidermal cannabinoid patches and gels are available commercially to deliberately absorb THC through the skin.

A different kind of protective layer, mucosal tissue, lines most internal body passages, notably occurring in the nose, lips, digestive tract, urinary system, genitals, and anus. These tissues are moist, secreting mucus to protect against infection. Mucosal absorption of cannabinoids is much more efficient than application to the skin (and is vastly more likely to produce intense psychoactivity from an overdose). Sativex (Figure 13.6), the cannabinoid-based analgesic supplied by the U.K. firm GW Pharmaceuticals, applies cannabinoids through mucosal tissues in the mouth. Ben Amar (2006) described Sativex as "a compromise between the inhaled and oral routes." The cannabinoids are diffused into tiny blood vessels, entering the bloodstream directly (like smoking), unlike oral consumption, which exposes the cannabinoids to the digestive tract and subsequently to the liver, with the resulting issues described previously. Suppositories are occasionally used as a form of THC absorption through the mucosal tissues in the rectum (see "Curiosities of Science, Technology, and Human Behavior" at the end of this chapter). Rectal absorption is more constant than oral absorption. A patent for the medical use of suppositories to administer THC was issued to ElSohly (1995). Vaginal application is also possible, and cannabinoid sprays for this purpose are being marketed, with claims that they produce extraordinary sexual sensation. Unless manufactured to precise standards and employed under the care of experienced health professionals, such products may be dangerous.

THC DOSAGES, ABSORPTION RATES, AND TIME COURSE OF EFFECTS

BIOAVAILABILITY

The relative amount of THC absorbed (bioavailability) during smoking has been claimed to be as high as 70%, depending on losses in combustion and sidestream smoke (Hall and Solowij 1998). Commonly, much of the THC is lost as smoke that is not inhaled and in vapor that is exhaled after inhalation. Bioavailability of inhaled THC is affected by volume inhaled, depth of inhalation into the lungs, and length of time that the breath is held (but note earlier contradictory viewpoints regarding this), retaining the smoke in the alveoli (Grotenhermen 2004a). The intensity of subjective effects is directly proportional to the puff volume and frequency (Azorlosa et al. 1992, 1995). In most circumstances, from 5% to 25% of the THC in marijuana is absorbed by smoking (Agurell et al. 1986; Holubek 2010; Strougo et al. 2008). Oral bioavailability ranges from 5%–20% of dose (Huestis 2005), usually less than 15% (Bowles et al. 2012).

DOSAGES

THC is very potent in humans, causing a "high" at a dose of 10 µg/kg through smoking, 30–50 µg/kg after intravenous injection, and 120 µg/kg from ingestion. A THC concentration in marijuana of approximately 0.9% has been suggested as a practical minimum level to achieve an intoxicant effect, but, as discussed later, CBD (the predominant cannabinoid of fiber and oilseed varieties) antagonizes (i.e., reduces) and potentiates (modifies) the effects of THC. Concentrations of 0.3% to 0.9% are considered to have "only a small drug potential" (Grotenhermen and Karus 1998).

Specifying the amount of marijuana that should be consumed in order to become high is subject to several difficulties. Experienced smokers are often capable of accessing about twice as much THC as casual users because they have developed a superior inhalation technique. Individuals differ in their response to marijuana, and heavy users can develop tolerance, so larger amounts may be necessary. Female patients with higher estrogen levels have been found to be relatively sensitive to the effects of medical cannabis (Lopez 2010). According to Leggett (2006), "a cannabis cigarette should not be considered a 'dose.' Consumption of an entire cannabis cigarette in a single sitting by a casual user would be rare, rather like a casual drinker consuming an entire bottle of wine. Cannabis of reasonable quality is actually more like spirits: just a few 'shots' is enough to produce intoxication." A "hit" (inhalation) or two of high-potency cannabis is effective in producing a "buzz," and several hits usually suffice during a session. Smoking 0.05–0.1 mg of high-quality cannabis (with 15% THC) is sufficient to become high (Leggett 2006), so a 0.5-mg joint would normally be consumed sequentially over several sessions by one person or simultaneously by several people. (The U.S. Drug Enforcement Administration employs 0.5 g of cannabis as representative of a cannabis cigarette.) Average consumption of 1 g/day is not unusual among regular users. For medical use or for those who choose to be more or less continuously intoxicated, consumption of as much as 5–10 g/day/person has been reported.

PHARMACOKINETICS

"Pharmacokinetics" is the study of the time course of drug absorption, distribution, metabolism, and excretion (more generally, "pharmacodynamics" is the study of how drugs affect the biochemistry and physiology of the body). Grotenhermen (2003) stated that "Pulmonary assimilation of inhaled THC causes a maximum plasma concentration within minutes, psychotropic effects start within seconds to a few minutes, reach a maximum after 15–30 minutes, and taper off within two to three hours. Following oral ingestion, psychotropic effects set in with a delay of 30–90 minutes, reach their maximum after two to three hours and last for about 4–12 hours, depending on dose and specific effect." Similarly, Abrams and Guzman (2014) observed that "When taken by mouth, there is a low (6%–20%) and variable oral bioavailability. Peak plasma concentrations occur after one to six hours and remain elevated with a terminal half-life of 20–30 hours. When consumed orally, delta-9-THC is initially metabolized in the liver to 11-OH-THC, also a potent psychoactive metabolite. On the other hand, when inhaled, the cannabinoids are rapidly absorbed into the bloodstream with a peak concentration in 2–10 minutes, which rapidly declines over the next 30 minutes. Thus, smoking achieves a higher peak concentration with a shorter duration of effect. Less of the psychoactive 11-OH-THC metabolite is formed." Grotenhermen et al. (2005) came to the following conclusions (cf. Armentano 2013). After smoking "typical medium to strong doses" of 15–20 mg THC, peak THC levels in blood occur 5–10 minutes after inhalation, and a waiting period of about three hours after smoking seems sufficient to reduce THC level to below a THC blood level of 5 ng/mL. Typical oral doses in social settings are in the 10–20 mg range, the effects occurring later than do those of smoking, usually peaking two to three hours after ingestion, and usually decreasing below the level of 5 ng/mL THC of blood in four hours.

BLOOD CONCENTRATIONS OF THC IN RELATION TO SAFE DRIVING

Grotenhermen et al. (2007) found that "Limited epidemiological studies indicate that serum concentrations of THC below 10 ng/ml are not associated with an elevated accident risk. A comparison of meta-analyses of experimental studies on the impairment of driving-relevant skills by alcohol or cannabis suggests that a THC concentration in the serum of 7–10 ng/ml is correlated with an impairment comparable to that caused by a blood alcohol concentration of 0.05%." However, it is difficult to reliably associate serum concentration and the degree of impairment as is done with alcohol (Zuurman et al. 2009). The state of Colorado, after authorizing the use of recreational marijuana, set a legal maximum limit for driving an automobile of 5 ng/mL THC of blood (many other states have zero tolerance).

THE ENDOCANNABINOID SYSTEM AS A BASIS FOR MUCH OF THE MEDICAL VALUE OF MARIJUANA

OVERVIEW OF THE ENDOCANNABINOID SYSTEM

The endocannabinoid system (often abbreviated as ECS, or shortened to eCB system) is a "command and control" set of biochemical communication regulatory mechanisms dealing with THC-like substances (endocannabinoids) produced naturally by the human body in general and the brain in particular, where receptors (discussed later) are concentrated in areas associated with thinking, memory, coordination, pleasure, and time perception. Like an automobile that has multiple systems requiring regulation (cooling, lubrication, electrical, etc.), the body has several systems to regulate physiological homeostasis, i.e., to maintain the orderly expression of different functional units to achieve a stable balance. For example, temperature, hormones, and energy levels of the body all require stabilization. The ECS is one of the important biological control systems. "The cannabinoid system helps regulate the function of major systems in the body, making it an integral part of the central homeostatic modulatory system—the check-and-balance molecular signalling networks that keeps the human body…healthy" (Aggarwal et al. 2009). "The modulation of the endocannabinoid system has therapeutic potential in a wide range of disparate diseases and pathologic conditions that affect humans, including neurodegenerative, kidney, and gastrointestinal diseases, pain, cancer, bone and cardiovascular disorders, obesity and metabolic syndrome, and inflammation, just to mention a few" (Horváth et al. 2012).

Virtually every organ system of the human body (Figure 13.16) is directly or indirectly influenced by the ECS. Moreover, as reviewed by McPartland et al. (2014) (also see McPartland 2008a), the system can be influenced by numerous factors: (1) pharmaceuticals such as analgesics (including acetaminophen, nonsteroidal anti-inflammatory drugs, opioids, and glucocorticoids), antidepressants, antipsychotics, anxiolytics, and anticonvulsants; (2) psychoactive substances, including alcohol, tobacco, coffee, and cannabis; (3) clinical interventions, including complementary and alternative medicine such as massage and manipulation, acupuncture, dietary supplements, and herbal medicines; (4) diet; and (5) lifestyle modification, including exercise and weight control. Given such extensive regulation of virtually all facets of the human body, and the numerous ways of modifying the controls, the significance of the ECS for human health can scarcely be exaggerated.

The human ECS is made up of cannabinoid receptors (notably CB_1 and CB_2), endocannabinoids (notably anandamide and 2-arachidonylglycerol [2-AG]), the biosynthetic precursors of the endocannabinoids, and the mechanisms (particularly the enzymes) involved in their biosynthesis and catabolism (inactivation) (Izzo et al. 2009). Information on these components of the ECS is presented in this section. As explained in the following discussion, the cannabinoids of *C. sativa* mimic the body's functionally similar molecules in binding to and activating (or neutralizing) tiny molecular receptor molecules embedded in the membranous surfaces of cells.

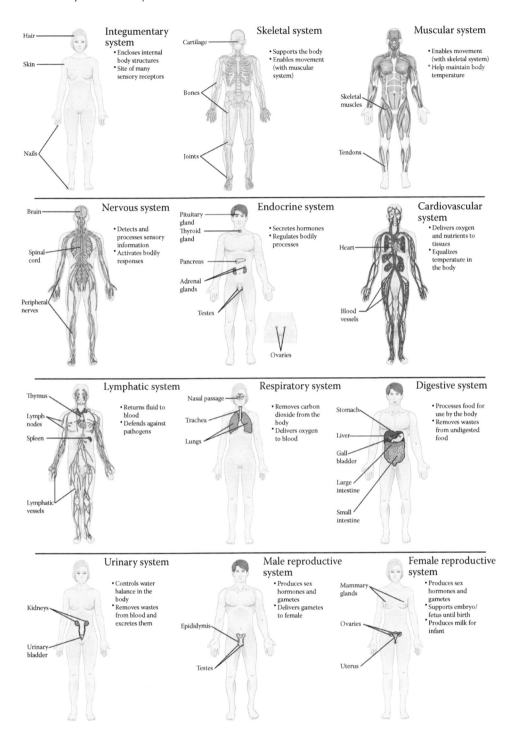

FIGURE 13.16 Human organ systems influenced by the endocannabinoid system. In fact, virtually all of the body's systems are influenced by the ECS. Deficiencies in the functioning of the ECS appear to be major causes of certain diseases, and the system seems to play at least a minor role in virtually every aspect of human health and sickness. Credit: Human Anatomy and Physiology, an OpenStax College resource. The OpenStax College name, OpenStax College logo, OpenStax College book covers, OpenStax CNX name, and OpenStax CNX logo are not subject to the creative commons license and may not be reproduced without the prior and express written consent of Rice University (CC BY 3.0).

There are well over 10,000 scientific publications dealing with the ECS, most of them very technical. The following discussion of the workings of the ECS and how it is affected by cannabinoids from the marijuana plants is a simplified overview. For a recent comprehensive technical review, see Mechoulam et al. (2014).

INTRODUCTION TO CELL RECEPTORS AND G PROTEIN-COUPLED RECEPTORS

The biochemistry of cannabinoid action in humans is best understood in terms of how they affect cell receptors. Receptors are proteins, located in the cell membrane or inside the cell, which respond to chemicals in their environment by altering metabolic functions within the cell. In the early 1970s, opiate receptors were discovered in the brain, which bind to molecules produced naturally by the body but also bind to morphine and other opiates that are consumed. Analogous to the discovery of opiate receptors, in the 1990s, a system was discovered in which cannabinoid cell receptors responded both to internally produced molecules and also the cannabinoids of cannabis that are absorbed. Receptors are usually named after the endogenous molecules that activate them, such as dopamine for dopamine receptors or serotonin for serotonin receptors.

"Cannabinoid receptors" are also called "endocannabinoid receptors." Chemically, molecules that bind to cellular receptors are called *ligands*; pharmacologically, chemicals contacting and activating receptors are *agonists*. Full agonists maximally activate receptors. Partial agonists do so partially. Neutral antagonists dock at receptors but do not activate them. Inverse agonists (antagonists) deactivate or prevent activation of receptors by displacing agonists and by suppressing "endogenous tone" and "constitutive activity" (both explained later in the section Endocannabinoid Receptors). Thus, a particular chemical docking at a receptor site can be stimulative or depressive. The different kinds of agonists have varying degrees of utility in manipulating the ECS. The use of an inverse agonist like rimonabant can have rather drastic effects, such as lowering the baseline activity of the ECS.

An analysis of terminology related to the ECS is presented by Di Marzo and De Petrocellis (2014), who attempt to provide more narrow definitions of some terms than usually encountered in the literature. For example, they dislike the term "cannabinoid receptor" mostly in use for the CB_1 and CB_2 receptors, since (as discussed later) cannabinoids affect other targets.

A particular chemical with drug properties, produced by a plant, may fortuitously mimic a natural chemical that the body employs to regulate cell functions through the receptor system. Plants produce a countless array of chemicals, and it is probably just coincidence that the cannabinoids of *C. sativa* just happen to have architectural (structural) features and ionic properties (distribution of electrical charges) that can activate the cannabinoid receptors of humans. In Figure 13.17, this kind of fortuitous relationship is conceptualized as analogous molecules with drug properties, whether originating from a plant or the human body, influencing the same cell receptor sites because of a sort of lock-and-key compatibility (receptors are analogous to locks; endocannabinoids and cannabinoids from *Cannabis* are analogous to keys). In the case of the ECS described here, the traditional lock-and-key metaphor is an oversimplification: different keys (chemicals) may have different degrees of fit into the lock, and cellular response depends on which kind of cell the lock is located.

The endocannabinoid receptors belong to a group called "G protein-coupled receptors" (illustrated in Figure 13.18). "G protein" is short for guanine nucleotide binding protein, the type of molecule that is normally attached but detaches when the receptor is stimulated, initiating a chain of biochemical events. G protein-coupled receptors are important in many diseases and are the target of about 40% of pharmaceuticals. This class of receptors includes numerous varieties of proteins that weave through cell membranes seven times (hence, they are also called seven-transmembrane receptors). G protein-coupled receptors function as intracellular molecular switches, possessing a unique binding pocket that "senses" or responds particularly to molecules docking in the pocket, and consequently, the external molecules transmit signals to the interior of the cell, triggering cellular responses.

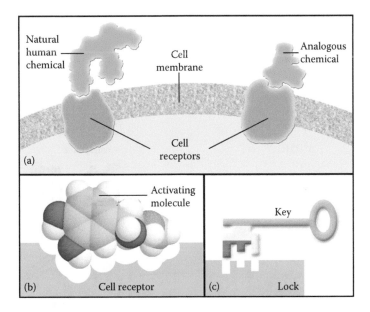

FIGURE 13.17   (a) Conceptual representation of how analogous drug chemicals from a plant and from the human body affect a given cell receptor because of three-dimensional "lock-and-key" compatibility. The plant's chemicals function as substitute "keys" and the cell receptors as the "locks." (b) Conceptual architectural fit of a chemical and its corresponding cell receptor. (c) Analogous fit of a key into a lock. Prepared by B. Brookes.

The mechanism of this "sensing" involves the docking causing a distortion of the protein that weaves seven times through the membrane, in turn producing a decoupling of the attached G protein inside the cell, which initiates biochemical events such as a flow of ions or the release of hormones, consequently influencing cell behavior. The type of detachable protein determines the kind of biochemical process that is triggered, and different types of cells can utilize a given receptor for different purposes by coupling to particular proteins. The body is known to possess thousands of G protein-coupled receptors, well-known types including dopamine, opioid, serotonin, and β-adrenergic receptors. Cannabinoid receptors are the most common kind of G protein-coupled receptor in the brain.

ENDOCANNABINOIDS

From the perspective of the human body, the cannabinoids of the cannabis plant are "exogenous" (i.e., derived externally), while its own cannabinoids ("endocannabinoids") are "endogenous" (originating internally). The endocannabinoids are derivatives of fatty acids and are relatively unrelated chemically to the cannabinoids of *C. sativa* (although the cannabinoids of *C. sativa* and the endocannabinoids of humans are all "lipids"). Although the chemical formulae of endocannabinoids are quite different from those of the cannabinoids of *C. sativa*, the three-dimensional structures of the two classes are thought to be quite similar, accounting for how marijuana cannabinoids can influence the human body's receptors.

There are two principal endocannabinoids. Arachidonylethanolamide or *N*-arachidonoyl-ethanolamine, mercifully nicknamed anandamide or AEA, exists particularly in the human nervous system and mimics the action of THC by influencing cannabinoid receptors. (The word anandamide was coined from "ananda," the Sanskrit word for "bliss" or "supreme joy," and "amide," meaning carboxylic acid derivatives.) Another chemical, later discovered to have similar effects on cannabinoid receptors, was coined 2-arachidonylglycerol (2-AG for short). Both of these endocannabinoids activate the CB_1 and CB_2 receptors discussed in the next section.

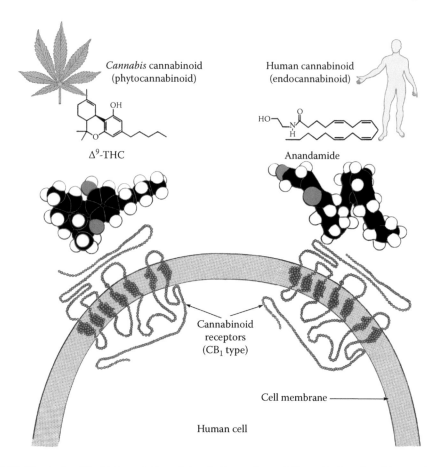

FIGURE 13.18   A simplified interpretation of the similar actions of *Cannabis*-based and human-based cannabinoids. Left: a molecule of Δ⁹-THC, the chief natural cannabinoid of *C. sativa*, is shown contacting and affecting a type CB₁ receptor embedded in the human cell membrane at bottom (note the characteristic structure of this polypeptide chain—a portion outside the membrane, winding through the membrane seven times, and a portion inside the membrane). Right: a molecule of anandamide, one of the chief natural endocannabinoids of the human body's internal cannabinoid system, similarly contacts and affects a type CB₁ receptor. The discovery that the cannabinoids of *C. sativa* affect (either positively or negatively) the human brain and other organs of the body through the internal endocannabinoid control system of the human body provides indisputable evidence that marijuana has medicinal properties (but not necessarily warranting medical usage). Drawn by B. Flahey.

Other known endocannabinoids (also capable of activating the CB₁ and CB₂ receptors) include 2-arachidonoyl glycerol ether, *O*-arachidonoyl ethanolamine (virodhamine), and N-arachidonoyl dopamine (Pacher et al. 2006). Lysophosphatidylinositol has been suggested to also be an endocannabinoid (it too activates CB₁ and CB₂ receptors and also seems to influence a putative cannabinoid receptor; Piñeiro and Falasca 2012).

Synthesis and release of endocannabinoids are induced "on demand" by physiological or pathological stimuli. Cannabinoid receptors are subsequently activated, resulting in adaptive (or in some disease conditions possibly inadaptive) responses.

ENDOCANNABINOID RECEPTORS

The brain and other organs have specific G protein-coupled receptors that recognize cannabinoids like THC as well as the body's endocannabinoids (Figure 13.18). While the receptors fortuitously respond to the cannabinoids from *C. sativa*, they appear to routinely function mainly in response to endocannabinoids, produced by the body's metabolism (Grotenhermen 2003, 2004a,b; Onaivi et al. 2005). Cannabinoid receptors have been identified in nerve terminals in the central nervous system, as well as in peripheral tissues, including sympathetic ganglia, dorsal root ganglia, adrenal glands, heart, lung, urinary bladder, reproductive tissues, gastrointestinal tissues, and immune cells. Cannabis drugs and extracts exert many of their biological functions through the receptors. Many of the potential therapeutic uses for cannabis drugs are related to the ways the drugs act on the cannabinoid receptors and how this influences human physiology (Joy et al. 1999; Onaivi et al. 2005b).

There are at least two types of cannabinoid receptors: CB_1 receptors, which are more numerous in the brain, particularly play neuromodulatory roles (i.e., they control nerve cell operations), and CB_2 receptors, which are best known for immunomodulatory functions (i.e., they control immune cell operations). The CB_1 receptor was discovered in 1991, and the CB_2 in 1993 (Pertwee et al. 2010). The CB_1 receptor contains 472 amino acids, while the CB_2 receptor has 360 (Sugiura et al. 2005). Only a few natural cannabinoids of *C. sativa* have been found to be capable of activating the receptors, of which the following (which produce euphoria in humans) are most significant. Δ^9-THC is a partial activator of both receptors but is more potent at CB_1 receptors than at CB_2 receptors (Pertwee et al. 2010). Δ^8-THC also activates the CB_1 receptors but is slightly less potent than Δ^9-THC, and it probably also activates CB_2 receptors (Pertwee and Cascio 2014). Cannabinol (CBN) binds less potently to both receptors than either Δ^9-THC or Δ^8-THC (Pertwee and Cascio 2014). In addition to these three cannabinoids, Δ^9-tetrahydrocannabivarin (THCV) affects both receptors. "No other phytocannabinoid investigated to date has been reported to activate CB_1 or CB_2 receptors with significant potency" (Cascio and Pertwee 2014).

Panlilio et al. (2015) summarized the activities of THC as follows: "The actions of THC at CB_1 receptors are considered to be critical for its psychoactive effects, but THC also acts at peroxisome proliferator-activated receptors and GPR55 receptors." CBD does not bind significantly either to CB_1 or CB_2 receptors, but there can be complex indirect effects (Pertwee 2008; Pertwee et al. 2010; Cascio and Pertwee 2014; Laprairie et al. 2015). CBD and indeed other nonpsychotropic (noneuphoric) cannabinoids have significant pharmacological actions on the human body that are independent of the ECS (Cascio and Pertwee 2014). (As pointed out elsewhere in this chapter, the psychotropic cannabinoids also have significant therapeutic potential independent of the ECS.)

The two kinds of receptors have somewhat different distributions, but collectively, they are in virtually all organs and body tissues. CB_1 receptors are the most abundant G-coupled receptors in the mammalian brain and they are the primary sites of action of THC taken into the human body. Human brains can contain as many as 100 billion neurons, which communicate electrochemically by means of neurotransmitters (chemical messengers), notably dopamine, serotonin, and glutamate. CB_1 receptors and the endocannabinoids (which are neurotransmitters) that affect them appear to play key roles in regulating the actions of the other neurotransmitters. Within the brain, the distribution of CB_1 receptors is consistent with the known effects of cannabinoids on cognition, memory, and motor function. The distribution of CB_1 receptors with respect to pain pathways in the brain, spinal cord, and on terminals of peripheral nervous system primary afferent neurons is also consistent with cannabinoid-induced analgesia. The male and female reproductive systems, bones, and indeed most body tissues are also home to CB_1 receptors, albeit in lower concentration than in nervous tissues. In the central nervous system, where CB_1 receptors are particularly densely distributed, they are responsible for such effects of marijuana as catalepsy (a nervous condition characterized by muscular rigidity and postural fixity), depression of motor activity, antinociception (decreased sensitivity to pain), analgesia, and feelings of well-being. In peripheral neurons, activation of the CB_1 receptors suppresses neurotransmitter (chemical messenger) release to the heart, bladder, intestines, and vas deferens. CB_1 receptors particularly affect brain functions and so are important in the medical consideration of psychiatric and memory diseases. CB_1 receptors also play key roles in regulating diseases related to pain, itching, muscle tone, and gastrointestinal functions such as digestion, secretion, and propulsion.

CB$_2$ receptors occur in nonneural tissue, especially in the spleen and immune system (hence their importance for anti-inflammatory functions), but to a lesser degree in the brain, pancreas, and liver. The distribution of CB$_2$ receptors on peripheral and central immune cells (such as in the spleen, thymus, tonsils, bone marrow lymph nodes, tonsils, and white blood cells) has been hypothesized to modulate immune effects of THC, through release of cytokines (proteins that are important in cell signaling, including some that act as chemical switches, turning certain immune cell types on and off).

CB$_2$ receptors particularly affect immunosuppression, and so are important in the medical consideration of numerous diseases and malfunctions of physiological systems, including inflammation (Croxford and Yamamura 2005; Basu and Dittel 2011). CB$_2$ receptors are also thought to affect disorders characterized by development of scar tissue (fibrosis), such as in liver cirrhosis, and to affect certain heart and kidney disorders. Indeed, an astonishing range of human maladies ranging from cardiovascular, gastrointestinal, liver, kidney, neurodegenerative, psychiatric, bone, skin, autoimmune and lung conditions, as well as pain and cancer, have been correlated with changes in endocannabinoid levels and/or CB$_2$ receptor expression (Pacher and Mechoulam 2011).

The expression of more or fewer receptors on cells is influenced by stresses. At least in some circumstances, increased numbers of receptors seems adaptive. Disease conditions may be accompanied either by more or by fewer cell receptors. A particularly interesting example is the study of Hirvonen et al. (2012), who demonstrated that chronic smoking in people was correlated with downregulation (decrease) of cannabinoid CB$_1$ receptors in certain brain regions but that this could be reversed by cessation of smoking.

Cannabinoid receptors (indeed, other classes of receptors) may be active to a degree despite the absence of cannabinoids (either endogenous or exogenous)—termed "constitutive activity"—and this may be difficult to distinguish from "endogenous tone," the relative efficacy or degree of functional balance of the ECS (for discussion, see Howlett et al. 2011).

For many years, CB$_1$ and CB$_2$ receptors were thought to have quite separate distributions, but more recently, CB$_2$ receptors were found to occur in brain tissues (Pamplona and Takahashi 2012).

There is evidence that there are additional cannabinoid receptors (which could be labeled CB$_3$, etc.), but their status is not yet certain (Pamplona and Takahashi 2012). Prominent among these candidate cannabinoid receptors is GPR55, a seven-transmembrane G protein-coupled receptor, which however does not meet current criteria employed to define cannabinoid receptors (Cabral et al. 2015). Transient receptor potential vanilloid type 1 (TRPV1; aka the capsaicin receptor) is also often viewed as an ion channel (rather than a G-protein-coupled) cannabinoid receptor (Tóth et al. 2009).

Intercellular Communication Involving Endocannabinoids

Nerve cells (neurons) of the brain and other parts of the nervous system communicate with each other by transmission of chemical or electrical messages across very shallow interneuron gaps. Such messages regulate and coordinate key functions such as appetite, memory, movement, and pain. A variety of messaging systems exist, and usually the stimulus arises entirely in one nerve cell and is transmitted directly to an adjoining nerve cell at a synapse. (Some employ the term "synapse" to refer to the gap [synaptic cleft] across which messages flow between nerve cells; more properly, the term also denotes the working parts of the adjacent neurons, i.e., the presynaptic and postsynaptic portions, as well as the synaptic cleft.)

The function of the ECS in the central nervous system is to dampen release of neurotransmitters. The receiving neuron has the capability of suppressing the message from its partner neuron (a so-called "retrograde feedback" or "upstream" or "backward arrangement"). As shown in Figure 13.19, the receiving (postsynaptic) membrane of a nerve cell synthesizes endocannabinoids (from phospholipids), which migrate across the synaptic gap to the transmitting (presynaptic) end of an adjoining nerve cell. Activation of the presynaptic CB$_1$ receptors decreases intracellular signaling ions (Ca^{2+}) which are required for release of stored intercellular cationic neurotransmitters from

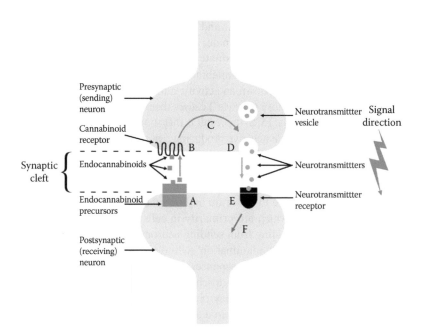

FIGURE 13.19 A simplified representation of endocannabinoid-based retrograde chemical message transmission at a synapse between nerve cells. Effectively, the endocannabinoid receptor (type CB$_1$) is functioning as a dimmer switch or surge protector, responding to chemical signals from the receiving cell, instructing the sending cell to dampen down its signal. (A) The receiving neuron synthesizes endocannabinoids from lipid precursors in the membrane of the postsynaptic neuron and these migrate across the synaptic cleft (gap) between the nerve cells to cannabinoid receptors of the transmitting neuron. (B) The endocannabinoids communicate with the presynaptic cannabinoid receptors, which are activated. (C) Activation of the CB receptors releases chemicals, decreasing signaling ions (Ca^{2+}) which are required for release of stored neurotransmitters from vesicles in the presynaptic neuron. (D) This suppresses the supply of neurotransmitters migrating to the postsynaptic neuron across the synaptic cleft and decreases the activation of receptors on the postsynaptic neuron. (E) Decreased activation of neurotransmitter receptors decreases triggering of cellular activity (F) in the postsynaptic neuron. Drawn by B. Brookes (after Velasco, G., Sánchez, C., Guzmán, M., *Natl. Rev. Cancer*, 12, 436–444, 2012).

vesicles in the presynaptic neuron. In short, this sequence of events suppresses the chemical signal between nerve cells. (For a technical account, see Reggio 2005.)

Most "classical" neurotransmitters (examples include acetylcholine, dopamine, and serotonin) are stored in vesicles in the presynaptic portion of a transmitting neuron, awaiting need (as shown in Figure 13.19). The endocannabinoid anandamide is not stored in the receiving neuron but is synthesized in the postsynaptic end of the receiving neuron and released on demand following stimulation. (The other major endocannabinoid, 2-AG, is almost universally thought to also follow this retrograde arrangement, but this awaits confirmation; Mechoulam and Parker 2013a). Mechoulam (2004) noted, "In many respects, the endocannabinoids differ from the classical modulatory molecules. They are not stored waiting for a customer. They are both formed and released on demand. They seem to act mostly (but not exclusively) within their anatomical vicinity and their actions are mediated (again mostly) by presynaptic, rather than by postsynaptic receptors, as with the classical neurotransmitters."

It appears that the postsynaptic neuron recognizes that the presynaptic neuron is sending too much signal and sends endocannabinoids to the former to decrease or halt transmission. Hyperstimulation of a postsynaptic neuron can degrade and even kill it (a phenomenon termed "excitotoxicity"). The endocannabinoid receptors have been compared to a dimmer switch, limiting or stimulating the amount of neurotransmitter released, thereby affecting how messages are sent, received, and processed. An illustration of how this might be useful is provided by the following example. Neuropathic or nerve-based pain is associated with many diseases (including diabetes, human immunodeficiency virus [HIV]/AIDS, and multiple sclerosis [MS]). The pain is produced by several neurotransmitters, including glutamate. To relieve excessive pain, the endocannabinoids act to limit glutamate release (too much glutamate can cause neurons to die). As noted later, such needless pain may be one of the targets of cannabinoid therapy.

The ECS is not limited to damping down an activity, although this seems to be its major function. Stimulation of some neurons ("inhibitory neurons") causes them to dampen down an activity, and when such neurons are damped down by the ECS, the result is that some activity is stimulated (such "synaptic disinhibition" is due to a sort of double negative producing a positive effect). Thus, the ECS is endowed with flexibility ("synaptic plasticity") facilitating complex mental functions such as learning and memory.

The previous discussion has concentrated on how CB_1 receptors suppress excessive stimulative activity from one cell to another to prevent damage to that receiving neuron. This is the main function of CB_1 receptors—to prevent damage to the brain and peripheral nerves from uncontrolled stimulation. Analogously, CB_2 receptors, which are primarily in cells associated with the immune system, prevent damage by suppressing signaling from sending neurons to immune cells. Immune cells have varied functions but often stimulate "inflammation" while contributing to a fight against infection. Normal inflammation is largely a localized response, where the body (particularly its immune cells) is trying to fight off infective agents, respond to chemical irritants, or remove dead or dying cells that have been injured. The immune system may produce protective compounds (a sort of chemical warfare) to ward off germs, which is desirable to a degree, but if done excessively, the result may be damage to innocent body tissues. (Autoimmune diseases, as discussed later, illustrate the danger of excessive inflammation.) Excessive inflammation can cause tissue damage, pain, and retardation of healing. Physicians conventionally prescribe pharmaceuticals to dampen down excessive inflammation (such as corticosteroids and nonsteroidal anti-inflammatory drugs [NSAIDS]), and one day, they may prescribe cannabinoids to mimic the body's natural inflammation-reducing endocannabinoids. A highly simplified interpretation of how the CB_1 and CB_2 receptors function as signal suppressors is shown in Figure 13.20.

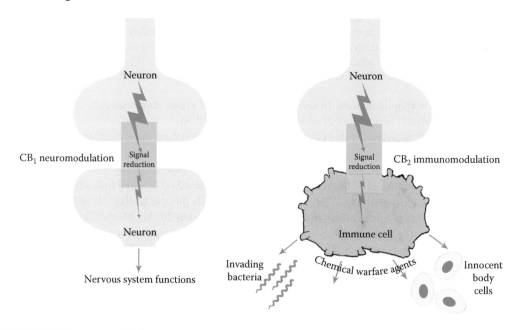

FIGURE 13.20 A simplified comparison of cannabinoid receptor signal reduction to prevent damage. Left: CB_1 neuromodulation of signal between a sending neuron and a receiving neuron, to prevent damage to the latter. Right: CB_2 immunomodulation of signal between a sending neuron and a receiving immune cell, to regulate inflammation to a level sufficient to kill invading microorganisms but not healthy body cells. Note the parallelism of the endocannabinoid modulation in the two circumstances. Prepared by B. Brookes.

The Broad Range of Compounds Affecting the Endocannabinoid System

In the section "Expanded Definitions of Cannabinoids" in Chapter 11, it was pointed out that some researchers have expanded the meaning of cannabinoids to include not only the cannabinoids of the cannabis plant, the endocannabinoids of humans (and other animals), but also chemicals that do not directly interact with the endocannabinoid receptors but do influence endocannabinoid function. Higher (flowering) plants do not have ECSs. However, a considerable number of chemicals produced by higher plants have been discovered to influence the CB receptors of humans (Gertsch et al. 2010). Anandamide, the first-discovered endocannabinoid in humans (Devane et al. 1992), critically affects brain functioning, and THC exerts its effects by substituting for it. Anandamide's tone (functionality) is affected by N-linoleoylethanolamide and N-oleoylethanolamide, which are found in a number of plants, most interestingly in cacao (*Theobroma cacao* L.), the source of chocolate, partially supporting the intuitive belief of many that the euphoric experiences from consuming chocolate and marijuana have some similarities (these chemicals do not directly affect the CB receptors but exemplify indirect effects). (The major endocannabinoid anandamide occurs in milk and therefore could be present in milk chocolate, but not in amounts that could produce observable psychotropic effects; Di Marzo et al. 1998). Gertsch et al. (2010) provide other examples of plant constituents that directly or indirectly affect CB receptors. These authors point out that THC is the most potent phytocannabinoid activator of the CB_1 receptor yet discovered.

Animals have chemicals that can influence ECSs of unrelated species. The stimulating molecules produced within a given species that regulate (activate or deactivate) its own ECS are in many cases capable of influencing the ECS of quite unrelated species.

In addition to anandamide and 2-AG, the two well-known primary endocannabinoids, the human body produces a range of chemicals that on their own do not affect the CB receptors but boost the effects of anandamide and 2-AG (a demonstration of the "entourage effect" discussed previously in this chapter). Some endocannabinoid specialists include these chemicals under the term "endocannabinoid."

Effects of Cannabinoids on Nonendocannabinoid Physiological Systems

"The current knowledge about cannabinoid receptors has been associated with mediating most of the psychoactive effects of marijuana, other neurobehavioral alterations, and the bulk of the cellular, biochemical, and physiological effects of cannabinoids" (Onaivi et al. 2005b). Nevertheless, while the ECS appears to be the primary way that cannabinoids directly exert physiological effects in humans, cannabinoids also affect other ("off target") receptor systems (particularly vanilloid receptors). The receptor TRPV1, known for reacting to capsaicin, the active ingredient of chile pepper, also responds to anandamide and CBD, although not to THC (Cabral et al. 2015). As noted earlier, it has been proposed to be part of the ECS.

Cannabinoids can also influence metabolism through nonreceptor mechanisms. *Cannabis sativa* and several of its cannabinoids have been shown to be antibacterial (Ferenczy et al. 1958; Kabelik et al. 1960; Radosevic et al. 1962; Gal and Vajda 1970; Farkas and Andrassy 1976; Van Klingeren and Ten Ham 1976; ElSohly et al. 1982; Appendino et al. 2008). Cannabinoid acids (precursors of the neutral cannabinoids) have been demonstrated to be antibiotic and were employed in veterinary medicine in Czechoslovakia in the 1960s (Izzo et al. 2009).

Medical Importance of the Endocannabinoid System

The ECS is of critical importance to comprehension of the value of medical marijuana. It is of paramount importance to human physiology and welfare, regulating numerous life-sustaining functions such as memory, perception, feeding behavior, digestion, blood pressure, body temperature, fertility, bone density, and disease resistance. The enthusiasm that many medical researchers are currently displaying for the significance of the ECS can scarcely be exaggerated. Di Marzo (2008a) stated,

"Perhaps no other signalling system discovered during the past 15 years is raising as many expectations for the development of new therapeutic drugs encompassing such a wide range of potential strategies for treatments." Russo (2004b) suggested that endocannabinoid deficiency accounts for therapeutic benefits of cannabis in migraine, fibromyalgia, irritable bowel syndrome, and other treatment-resistant conditions, a hypothesis supported by Smith and Wagner (2014).

Robson (2014) concluded that "The endocannabinoid system has emerged as an important physiological system and plausible target for new medicines. Its receptors and endogenous ligands play a vital modulatory role in diverse functions including immune response, food intake, cognition, emotion, perception, behavioural reinforcement, motor co-ordination, body temperature, wake/sleep cycle, bone formation and resorption, and various aspects of hormonal control. In disease it may act as part of the physiological response or as a component of the underlying pathology." McPartland et al. (2014) reviewed the literature, confirming that human ECS deficiencies have been implicated in schizophrenia, migraine, MS, Huntington's disease, Parkinson's disease (PD), irritable bowel syndrome, anorexia, and chronic motion sickness. Pacher et al. (2006) observed that because the ECS has indeed been found to be altered in diseases such as MS, spinal cord injury, neuropathic pain, cancer, atherosclerosis, stroke, myocardial infarction, hypertension, glaucoma, obesity, and osteoporosis, it is appropriate to search for new therapeutic strategies aimed at restoring normal system functionality. Among the new medications anticipated are analgesics, antiallergens, antidepressants, antiemetics (preventatives of nausea and vomiting), anti-inflammatories, antineoplastics (to inhibit or prevent tumors or malignant cells), appetite modulators, bronchodilators, immunosuppressants, muscle relaxants, and neuroleptics (antipsychotic or anesthetic drugs).

Yazulla (2008) wrote, "There is great interest in endocannabinoids for their role in neuroplasticity [the capacity of the nervous system to develop new nerve cell connections, allowing the brain to compensate for injury and disease and to adjust activities in response to new situations or environmental changes] as well as for therapeutic use in numerous conditions, including pain, stroke, cancer, obesity, osteoporosis, fertility, neurodegenerative diseases, MS, glaucoma and inflammatory diseases, among others." Pacher and Kunos (2013) stated, "the endocannabinoid system holds therapeutic promise for a broad range of diseases, including neurodegenerative, cardiovascular and inflammatory disorders; obesity/metabolic syndrome; cachexia (wasting syndrome); chemotherapy-induced nausea and vomiting; and tissue injury and pain, amongst others. However…a better understanding of the pathophysiological role of the endocannabinoid system is required to devise clinically successful treatment strategies."

EXTENSIVE ACTIVITY OF THE ENDOCANNABINOID SYSTEM LIMITS TARGETED THERAPY

"The broad spectrum activity of the endocannabinoid signalling in the central nervous system is… the main risk in the development of novel therapies based on this system, as it may be difficult to find pharmacological approaches for a specific disorder that do not affect other brain processes and produce important side effects" (Fernández-Ruiz et al. 2014a; cf. Di Marzo 2009). This problem is illustrated, as discussed later, by the commercial development of the CB_1 receptor antagonist rimonabant for weight control, which had to be discontinued because of unacceptable side effects. "Compared to other neuroactive compounds, cannabinoids are exceedingly pleiotropic in their activity, while their receptors are part of complex neural webs whose manipulation is difficult to predict in terms of in vivo effects. Although the debacle of the CB_1 inverse agonist rimonabant well exemplifies these difficulties, further investigations in this exciting field are strongly needed" (Appendino et al. 2009).

EVOLUTION OF THE ENDOCANNABINOID SYSTEM AND POTENTIAL FOR NONHUMAN PATIENTS

Cannabinoids from cannabis are not only of actual and potential medical value to humans but may be useful for veterinary treatment of other animals. Assessment of this possibility requires clarification of the evolution of the cannabinoid system among living things. As noted previously, *Cannabis* cannabinoids can influence metabolism without affecting cannabinoid receptors, but in the main, their value is presumed to be determined by their influence on these receptors. Accordingly, the presence of cannabinoid receptors in species determines at least in substantial part the potential value of cannabinoids to them.

Endocannabinoid-like molecules are synthesized in many plants and animals, indicating that the ability to manufacture them likely arose very early in evolution, possibly in the common ancestor of plants and animals (Elphick and Egertová 2001, 2015). Pacioni et al. (2015) found anandamide in truffles, members of the ascomycete fungal group. However, the capacity to employ such molecules in conjunction with receptors is restricted to certain groups of species. Insects, the largest group of animals, lack endocannabinoid receptors, although some, such as the fruit fly (*Drosophila melanogaster*) and the honey bee (*Apis mellifera*), can synthesize the endocannabinoids anandamide and 2-AG (McPartland et al. 2001; Jeffries et al. 2014). Facetiously, it seems, therefore, that insects are unable to become intoxicated by THC (Figure 13.21). Endocannabinoid receptors appear to be confined to a large, advanced lineage of animals termed the "deuterostomes" (formally termed Deuterostomia), which includes some invertebrates, such as sea squirts and nematodes, and the vertebrates—amphibians, birds, fish, mammals, and snakes (Elphick and Egertová 2001; McPartland 2006; McPartland et al. 2006; Elphick 2012). It would seem therefore that all deuterosomes could get high on marijuana (Figure 13.22).

Molecular (nucleic acid) coding for the CB$_1$ and CB$_2$ receptors is thought to have originated by an evolutionarily ancient duplication of a common ancestral gene (Cottone et al. 2013). Some primitive organisms, including sea squirts (such as *Ciona intestinalis*), have been found to possess this presumed ancestral gene, and their ancestry suggests that it evolved about 600 million years ago (MYA) (Viveros et al. 2008). Some primitive deuterosomes have just one receptor, which is genetically only partly homologous to human receptors. McPartland and Guy (2004b)

FIGURE 13.21   The hookah-smoking caterpillar from Lewis Carroll's *Alice's Adventures in Wonderland*. Some have contended that Carroll intended to suggest that the hookah contained cannabis, but this seems very unlikely. In any event, as noted in the text, insects lack cannabinoid receptors and are presumed to be unable to become intoxicated by smoking cannabis drugs. Drawing by John Tenniel, published in Carroll, L., *Alice's Adventures in Wonderland*, Macmillan, London, 1865. Colored by Mrwalletpants (CC BY 2.0).

FIGURE 13.22 Vertebrate animals as pictured here have been shown to possess cannabinoid receptors and therefore presumably can become high from marijuana. More importantly, their illnesses are potentially treatable with cannabinoids. Prepared by B. Brookes.

(cf. McPartland 2004) concluded that (1) endocannabinoids evolved before CB receptors, perhaps 2000 MYA; (2) they evolved independently multiple times; (3) the primordial CB receptor probably evolved about 790 MYA; (4) the CB_1–CB_2 duplication event occurred either prior to the origin of deuterostomes 590 MYA or prior to the divergence of fish and higher vertebrates 400 MYA; and (5) the vanilloid receptor may predate CB receptors but its affinity for anandamide is a recent acquisition, evolving after the appearance of mammals 300 MYA. McPartland et al. (2007a) proposed that cannabinoid receptors initially coevolved with a fatty acid ester ligand similar to anandamide in ancestral metazoans, and affinity for AEA evolved later. McPartland et al. (2007b) found that the cannabinoid system evolves conservatively, indicative of its importance for survival.

Gertsch et al. (2010) speculated that dietary contact with phytocannabinoids during mammalian evolution may have played a beneficial role in adapting species for survival. McPartland and Guy (2004a) extensively examined adaptive and coevolutionary hypotheses between humans and plant constituents that affect the human ECS. DiPatrizio and Piomelli (2012) hypothesized that the endocannabinoids play a critical role, in mammals at least, in providing a pleasure response to eating sweet and fatty food and thus were evolutionarily critical in accumulation of food energy for survival.

THE ENDOCANNABINOID SYSTEM IN RELATION TO THE LACK OF OVERDOSE MARIJUANA MORTALITY

LD_{50} (MEDIAN LETHAL DOSE)

In toxicology, the "median lethal dose," LD_{50} ("lethal dose, 50%"), is the relative amount required to kill half the members of a tested population after a specified period. The LD_{50} is a standard measurement of acute toxicity, widely employed to indicate how dangerous (or innocuous) a drug or substance is. Clarke and Pate (1994) stated that the LD_{50} "for orally ingested THC is approximately 1 g/kg of body weight. Simply interpreted, this means an average sized human would have to consume 50–100 g of pure THC to reach the LD_{50} level. Since high-potency *Cannabis* contains approximately 10% THC, a person would have to eat at least 500–1000 g of this marijuana before having a 50% chance of death." Annas (1997) estimated that the LD_{50} for marijuana is around 1:20,000 to 1:40,000, which would require smoking about 680 kg (1500 pounds) in 15 minutes.

THERAPEUTIC INDEX

The therapeutic index (TI; also called the therapeutic window, safety window, and therapeutic ratio) is a comparison of therapeutic to toxic effects (the latter measured by the LD_{50}, described previously). Clarke and Pate (1994) noted that "even accounting for pyrolytic decomposition and smoke loss, there is a several-thousandfold difference between an effective dose of THC and a potentially lethal one!" Loewe (1946) compared the ratio at which THC caused psychoactivity in mice to the LD_{50} (i.e., the TI of THC) and determined the TI to be 40,000:1. In comparison, the TI of morphine is 70:1, the TI of ethanol is 10:1, and the TI of digoxin is 2:1. Other common nonprescription drugs, such as aspirin, have similar relatively narrow margins of safe use. (Note in the discussion of "Dabbing" in Chapter 12 that this innovation facilitates overdosing.)

COMPARATIVE OVERDOSE LETHALITY OF CANNABIS AND OTHER DRUGS

The safety margin of crude marijuana, at least in the short-term, is impressive compared to other causes of mortality (Table 13.1). There is no authenticated example of death directly from an

TABLE 13.1
Causes of Death in the United States (2013 Data)

Cause of Death	Number
All causes	2,596,993
Major cardiovascular diseases	796,494
Malignant neoplasms (cancer)	584,881
Lung diseases (especially COPD[a], which includes emphysema and bronchitis)	149,205
Accidents (unintentional injuries)	130,557
Motor vehicle accidents (subset of total accidents, above)	35,369
Alzheimer's disease	84,767
Diabetes mellitus	75,578
Influenza and pneumonia	56,979
Kidney problems	47,112
Drug-induced deaths (legal and illegal)	46,471
Intentional self-harm (suicide)	41,149
Septicemia	38,156
Chronic liver disease and cirrhosis	36,427
Alcoholic liver disease (subset of chronic liver disease, above)	18,146
Injury by firearms	33,636
Alcohol-induced deaths	29,001
Parkinson's disease	25,196
Pneumonitis (lung inflammation, especially due to choking on ingested material, vomiting into the lungs, smoking, drugs, etc.)	18,579
Homicide	16,121
Viral hepatitis	8157
HIV disease	6955
Cannabis	0[b]

Source: Based on Drug Wars Facts, http://www.drugwarfacts.org/cms/Causes_of_Death.

Note: Drug-induced death categories are in bold (diseases due to smoking also represent a form of voluntary drug-induced mortality).

[a] COPD, chronic obstructive pulmonary disease.

[b] However, coroners occasionally rule that deaths are due to cannabis consumption. In Britain, Official British government statistics listed five deaths from cannabis in the period 1993–1995, although the House of Lords Select Committee on Science and Technology (1998) reported that they had died of inhalation of vomit, not directly from cannabis. Possibly thousands have died indirectly by associating with the illegal cannabis trade.

overdose of marijuana (Gable 2006). "Fatal overdose with cannabis alone has not been reported" (Grant et al. 2012). By contrast in the United States, the prescription drug mortality rate is higher than the death rate from illicit drugs combined, and drug overdose mortality exceeds mortality from motor vehicle accidents (ASTHO 2008). (It should be kept in mind that potentially deadly pharmaceuticals frequently are also lifesavers; aspirin kills hundreds in the United States annually, but probably saves hundreds of thousands.)

Until recently, the term "opioids" designated synthetic (chemically synthesized) molecules that emulate the natural opium alkaloids from the opium poppy plants. Today, the word "opioids" usually also refers to any of (1) natural narcotic alkaloids ("opiates") from the opium poppy plant, (2) "semisynthetics"—the natural alkaloids that have been chemically transformed (heroin is transformed opium), and (3) chemicals that are entirely synthesized but have similar structures and effects to natural opiates.

Opioid drugs (especially prescription analgesics, such as oxycodone, hydrocodone, and methadone) represent a principal cause of drug overdose deaths. In the United States, in 2013, there were 16,235 deaths involving prescription opioids, as well as 8257 heroin-related deaths, making opioids the leading cause of combined legal and illegal drug-induced deaths (Table 13.1). There are numerous opioid receptors in the brainstem, which controls breathing, so opioid drugs have high potential to stop respiration. By contrast, although there are cannabinoid receptors in the brainstem, there are none in the medullary respiratory centers, contributing to the relative safety of cannabis, at least with respect to the possibility of suppression of breathing.

This section should not be misinterpreted to imply that cannabis is necessarily safe, since long-term recreational consumption has been associated with several hazards (Chapter 12) and medical usage also requires cautions (discussed in this chapter). Although it is often said that marijuana has never killed anyone, this is not quite true, since simple association with the drug may indirectly be fatal. One of the more gruesome illustrations of this is the practice of "body packing" marijuana in the colon, recorded several times in Jamaica for the purpose of smuggling. This has led to colonic perforation and death (Cawich et al. 2010). However, "There is insufficient evidence...to assess whether the all-cause mortality rate is elevated among cannabis users in the general population" (Calabria et al. 2010).

CANNABINOIDS AS A POSSIBLE BASIS FOR ELIMINATING THE NEED FOR HERBAL MEDICAL MARIJUANA

In criticizing crude marijuana, Bostwick (2012) stated, "Exogenous plant-derived THC is a sledge-hammer compared with anandamide's delicate chisel, the former causing marked disruption of neuronal signaling and circuit dynamics in the finely tuned endogenous system" and "Recently, researchers have stated that the power of new pharmacologic products will obviate the need for botanical cannabis." Certainly, the growing knowledge of how particular cannabinoids (in the broadest sense) affect the ECS specifically and the body's metabolism in general is likely to lead to the development of pharmaceuticals precisely keyed to given illnesses. Nevertheless, as noted previously, the value of medicinal herbal preparations is too readily dismissed by much of the medical profession.

MEDICAL CONDITIONS FOR WHICH CANNABIS HAS OR MAY HAVE VALUE

As noted earlier, cannabis (broadly interpreted here as including herbal marijuana, extracted cannabinoids, and synthetics) has a very long history of medical applications. Notwithstanding this record, at the present time, medical usage of marijuana requires reexamination in the light of present medical knowledge. Indeed, although by no means have the possible medical applications of cannabis been adequately examined, although a mountain of recent information has been accumulated. PubMed (short for Public Medicine) is a massive online guide to publications in biomedical and life science publications, the most important comprehensive indexing of the world's medically significant professional literature. In early 2015, a PubMed search for scientific journal articles published in the last 20 years containing the word "cannabis" produced almost 9000 articles. Adding the word "cannabinoid" increased the number of publications to about 21,000. Over 95% of these deal directly or indirectly with medical aspects of cannabis. In recent years, thousands of such publications have appeared annually. Marijuana affects virtually every organ system of the body (Yazulla 2008), so it is not surprising that the literature on medicinal aspects of marijuana is extremely extensive, indeed rather overwhelming. The medical literature on cannabis expresses a wide range of judgments on the relative value and harm of cannabis for various conditions, indicative of the fact that there is not yet a medical consensus regarding its values.

As noted by Gordon et al. (2013), "The peer reviewed literature is the most objective means to examine purported and realized health effects of marijuana." The following presentation cites and often quotes the conclusions and evaluations of key publications and comprehensive recent reviews (frequently meta-analyses, i.e., overviews of previous studies) of the value of cannabis, particularly its components, for treating illnesses often claimed to be improved by herbal marijuana or cannabinoids.

Lambert (2007) stated, "Based on the diverse components of the endocannabinoid system and taking into account the wide distribution of the molecular targets, a therapeutic endocannabinoid-based strategy could be useful in numerous and diverse pathological conditions: mood and anxiety disorders, movement disorders such as Parkinson's and Huntington's diseases, inflammatory and neuropathic pains, multiple sclerosis and spinal cord injury, cancer, atherosclerosis and cardiovascular diseases, stroke, glaucoma, obesity and metabolic syndrome, addictions (tobacco, alcohol, drug of abuse), and osteoporosis are among the most cited in the literature." Grotenhermen (2003) noted that "Properties of cannabis that might be of therapeutic use include analgesia, muscle relaxation, immunosuppression, sedation, improvement of mood, stimulation of appetite, antiemesis, lowering of intraocular pressure, bronchodilation, neuroprotection and induction of apoptosis in cancer cells."

Ware et al. (2005), based on a survey of about 3000 patients in the United Kingdom, found the following frequencies of usage: chronic pain, 25%; MS, 22%; depression, 22%; arthritis, 21%; and neuropathy (peripheral nerve problems, often causing weakness, numbness, and pain, usually in the hands and feet), 19%. Walsh et al. (2013), on the basis of a survey of 628 Canadians, found that patients most commonly reported using cannabis to treat symptoms associated with sleep, pain, and anxiety. That is, most patients employ medical marijuana for symptomatic relief of pain and physical discomfort. (Note that "patients" in such surveys self-identify as such but often self-medicate without guidance by medical professionals.)

The Canadian regulations on "access to cannabis for medical purposes," established by Health Canada in July 2001, recognized two categories of patients eligible for medical cannabis. Category 1 required at least one of the following symptoms: severe pain and/or persistent muscle spasms from MS, from a spinal cord injury, from spinal cord disease; severe pain, cachexia, anorexia, weight loss, and/or severe nausea from cancer or HIV/AIDS infection; severe pain from severe forms of arthritis; or seizures from epilepsy. Category 2 covered debilitating symptom(s) of medical condition(s), other than those described in Category 1. In theory, the range of symptoms listed allowed considerable latitude for physicians to prescribe cannabis.

Alzheimer's Disease and Other Forms of Dementia

Dementia refers to pathological decline in mental abilities, including memory, thinking, and social skills severe enough to interfere with daily functioning. Walther and Halpern (2010) provided the following summary of dementia: "The prevalence rates vary among the different types of dementia. Alzheimer's disease is the most common dementia, accounting for 50%–60% of all cases. Prevalence rates increase with age. In Parkinson's disease the risk for developing dementia is increased six-fold. Approximately 30% of stroke survivors develop post-stroke dementia. Far lower prevalence rates are documented for Huntington's disease, which is frequently associated with dementia."

Alzheimer's disease is characterized by a progressive deterioration of cognition and memory because of deposition of protein plaques in the brain that apparently interfere with communication among brain cells. The disease is correlated with inflammation and death of neurons. No effective treatments are available. Fagan and Campbell (2015) noted that "Alzheimer's disease (AD) is a complex age-related neurodegenerative disease characterized by the progressive loss of memory and cognitive function. Approximately 36 million people worldwide suffer from AD and, with an increasingly aged global population, that number is estimated to triple by 2050. Symptoms of the disease include memory loss, difficulty with abstract thinking and completing familiar tasks, confusion, spatial and temporal disorientation, problems with speech, and altered mood. Deterioration in the health of patients with AD leads to death within three to nine years."

Because endocannabinoids are established to be important in brain function, there has been interest in exploring the possible therapeutic role of cannabis (Bonnet and Marchalant 2015). Eubanks et al. (2006) noted that "THC and its analogues may provide an improved therapeutic for Alzheimer's disease." Ramírez et al. (2005) found indications that "cannabinoid receptors are important in the pathology of AD and that cannabinoids succeed in preventing the neurodegenerative process occurring in the disease." González-Naranjo et al. (2013) observed that because Alzheimer's is multifaceted, cannabis may be advantageous in therapeutically addressing different aspects. In this vein, Bedse et al. (2015) stated, "The endocannabinoid (eCB) system appears to be a promising therapeutic target as it has the ability to modulate a range of aspects of AD pathology. At a first glance, it is striking that cannabinoids like delta-9-tetrahydrocannabinol (Δ^9-THC), known to impair memory, could be beneficial in AD. However, augmentation of eCB signaling could reduce excitotoxicity, oxidative stress, and neuroinflammation and thus could alleviate symptoms of AD."

Although cannabinoids have been hypothesized to have value for treating brain aging, authoritative confirmation of this has not been obtained (Campbell and Gowran 2007; Krishnan et al. 2009; Walther and Halpern 2010). Fernández-Ruiz et al. (2014b) stated, "Based on the potential shown by Δ^9-THC and CBD in experimental AD, the combination of both phytocannabinoids (i.e., Sativex) may be useful for the treatment of AD patients…the only clinical studies performed so far have been a small number that used dronabinol, an oil-based solution of synthetic Δ^9-THC, which was found to ameliorate only some AD-related symptoms." Similarly, Fagan and Campbell (2015) commented, "targeting this system has become a viable therapeutic approach for AD but further clinical studies elucidating the efficacy of cannabinoid treatment are required."

Amyotrophic Lateral Sclerosis (Lou Gehrig's Disease)

Amyotrophic lateral sclerosis (ALS; also known as motor neuron disease) is an uncommon, progressive, incurable condition that kills nerve cells in the brain and spinal column, paralyzing and ultimately killing the patient. Symptoms include stiff muscles, muscle twitching, and weakness associated with muscle wasting, and eventually difficulty speaking, swallowing, and breathing, usually resulting in death in three or four years. ALS is the third most common neurodegenerative disorder after Alzheimer's disease and Parkinson's disease. It occurs in about one person in 400,000 and is somewhat more prevalent in males. One form of the disease (familial ALS), responsible for 5%–10% of cases, is due to mutation in specific genes, at least 15 of which have been identified. It is suspected that numerous factors other than (just) mutative causative genes are involved in the remaining cases.

Carter and Rosen (2001) stated that marijuana should be considered in the pharmacological management of ALS because it has properties that may be therapeutic for this disease, including analgesia, muscle relaxation, bronchodilation, saliva reduction, appetite stimulation, and sleep induction, as well as strong antioxidative and neuroprotective effects, which might prolong neuronal cell survival. Weber et al. (2010) noted that "many patients with ALS experience cramps during the course of the disease but so far, none of the medications used has been of proven benefit" and in an experimental study found that THC did not provide relief for patients. Carter et al. (2010) noted that "Based on the currently available scientific data, it is reasonable to think that cannabis might significantly slow the progression of ALS, potentially extending life expectancy and substantially reducing the overall burden of the disease." De Lago et al. (2015) stated that "cannabinoid medicines may serve as novel therapy able to delay/arrest neurodegeneration in ALS." Chiurchiù et al. (2015) stated, "there is increasing evidence that cannabinoids and manipulation of the cannabinoid system may have therapeutic value in ALS." Nevertheless, Fernández-Ruiz et al. (2014b) concluded that "too few clinical data have yet been generated to allow any firm conclusions to be drawn."

Anorexia and Appetite Loss

Anorexia refers to loss of appetite (or more technically as loss of desire to eat despite caloric deprivation), which may be due to physical and/or psychological causes. The condition is a side effect of many chemotherapy agents and a component of cancer and AIDS pathophysiology. Anorexia nervosa is a psychological aversion to eating, often coupled with a distorted self-image and extreme desire to be thin, much more common in women as a result of pressures on females to be slim in Western culture. The disease is dangerous as it often leads to starvation. Cachexia (also known as wasting syndrome) is progressive, involuntary, extreme weight loss associated with chronic wasting disorders. Patients suffering from advanced stages of cancer or HIV infection often exhibit progressive weight loss, often exacerbated by chronic diarrhea. "Anorexia, early satiety, weight loss and cachexia are prevalent in late stage cancer and advanced HIV disease, but there are few effective treatments" (Grant et al. 2012).

"Endocannabinoids and endocannabinoid-related compounds are involved in food-related reward and suggest a dysregulation of their physiology in anorexia nervosa" (Monteleone et al. 2015). Cannabis has been shown to stimulate appetite, as well as reduce associated nausea and vomiting. Dronabinol (synthetic THC) therapy was found to be of some value in treating the condition (Andries et al. 2014). Cristino and Di Marzo (2014) concluded, "much progress has been made in the understanding of the mechanisms through which Δ^9-THC affects food intake and energy metabolism [but] the application of this knowledge to the development of novel, efficacious and safe treatments for cachexia and, particularly, anorexia, has unfortunately lagged behind."

Arthritis and Rheumatism

Arthritis refers to any of over 100 forms of joint inflammation. The most common type is osteoarthritis, a degenerative disease resulting from wear-and-tear damage to cartilage covering bones in joints, especially related to the hands, hips, knees, and spine. As with many diseases, should it prove that deficiencies in the ECS are partly responsible, supplying cannabinoids might alleviate the condition. La Porta et al. (2014) stated, "Currently, the therapeutic approaches for osteoarthritis are limited as no drugs are available to control the disease progression and the analgesic treatment has restricted efficacy. Increasing evidence from preclinical studies supports the interest of the endocannabinoid system as an emerging therapeutic target for osteoarthritis pain... The ubiquitous distribution of cannabinoid receptors, together with the physiological role of the endocannabinoid system in the regulation of pain, inflammation and even joint function further support the therapeutic interest of cannabinoids for osteoarthritis. However, limited clinical evidence has been provided to support this therapeutic use of cannabinoids, despite the promising preclinical data."

Rheumatism refers to diseases characterized by inflammation and pain in the joints, muscles, or fibrous tissue, especially rheumatoid arthritis, a chronic progressive disease producing inflammation in the joints and resulting in painful deformity and immobility, especially in the fingers, wrists, feet, and ankles. Rheumatoid arthritis is an autoimmune disease—caused by the immune system inappropriately attacking the lining of the joints. It affects joints symmetrically—on both sides of the body—while osteoarthritis typically affects one side, for example, one knee or hand. The hope has been that cannabis, through the ECS, could alleviate the inflammation and pain (via the CB_1 receptors) as well as reduce the counterproductive immune activity (via the CB_2 receptors) associated with rheumatoid arthritis.

Fitzcharles et al. (2014) offered the following comments with respect to the use of herbal cannabis for the treatment of rheumatic diseases: "Available drugs generally offer a modest effect only... It is therefore understandable that patients will continue to seek other remedies to reduce symptoms. Rheumatic disease patients commonly use complementary and alternative medicine, and with increasing advocacy for legalization of herbal cannabis as a recreational drug, cannabis may be perceived as a safe treatment option... Medical marijuana has...never been recommended by any rheumatology group worldwide for symptom relief in rheumatic conditions... To date, there is no formal study examining the efficacy or adverse effects of herbal cannabis in rheumatic diseases... Taking all factors into consideration, health care professionals should currently dissuade rheumatology patients from using herbal cannabis as a therapy." Echoing this viewpoint, Kalant (2015) commented, "the efficacy and safety of cannabis have not yet been demonstrated in well-designed clinical trials in patients with chronic rheumatic and arthritic conditions." Noting that "severe arthritis" is the most common reason that Canadians are authorized to possess medicinal herbal cannabis, Fitzcharles (2015) commented, "Preclinical animal studies as well as studies of osteoarthritis and rheumatoid arthritis joints provide evidence that the endocannabinoid system is locally activated in response to pain stimuli in arthritis, and functions as a pain modulator...it is understandable that cannabinoids have a potential for therapeutic relief, thereby calling for more scientific study of the effects of herbal cannabis in rheumatic conditions. The scientific research has, however, not yet translated into clinical application."

However, Blake et al. (2006) found that Sativex, described in the preceding discussions, significantly alleviated pain from rheumatoid arthritis. "In rheumatoid arthritis a significant reduction in disease activity was...noted, which is consistent with pre-clinical work demonstrating that cannabinoids are anti-inflammatory" (Lynch and Campbell 2011; cf. Lynch and Ware 2015). Dunn et al. (2012) commented, "Cannabinoids have been shown to reduce joint damage in animal models of arthritis." Gui et al. (2015) noted, "Increasing evidence suggests that the endocannabinoid system, especially cannabinoid receptor 2 (CB_2), has an important role in the pathophysiology of rheumatoid arthritis... In particular, specific activation of CB_2 may relieve rheumatoid arthritis."

Brain Injury

Traumatic brain injury refers to brain dysfunction caused by an outside force, usually a violent blow to the head, but not infrequently caused by an object penetrating the skull, such as a bullet. Mild injury may temporarily disrupt brain cells, but serious traumatic brain injury can result in long-term complications or death. "Traumatic brain injury (TBI)...affects an estimated 1.7 million people per year in North America. TBI patients often contend with persistent challenges including problems with learning and memory...studies suggest that the activation of the cannabinoid 1 receptor after a traumatic brain injury could be beneficial" (Arain et al. 2015).

"Traumatic brain injury (TBI) represents the leading cause of death in young individuals. It triggers the accumulation of harmful mediators, leading to secondary damage, yet protective mechanisms are also set in motion... There is a large body of evidence showing that endocannabinoids are markedly increased in response to pathogenic events. This fact, as well as numerous studies on experimental models of brain toxicity, neuroinflammation and trauma supports the notion that the endocannabinoids are part of the brain's compensatory or repair mechanisms" (Shohami et al. 2011).

Nguyen et al. (2014) compared the mortality of patients who had suffered severe traumatic head (brain) injury in relation to analyses of presence or absence of THC in their urine when admitted to hospital. The results bordered on the incredible: patients who had been using marijuana survived 80% more often than those who had not! This seems to represent a neuroprotective effect related to components in marijuana, presumably at least in part related to the role of CB receptors in reducing inflammation. With respect to brain injury associated with birth, Fernández-López et al. (2013) noted, "Numerous experimental studies have proven that the modulation of the endocannabinoid system...has beneficial effects during the acute and recovery phases after perinatal brain injury."

Similarly, Arain et al. (2015) commented, "studies provide strong support for the...therapeutic potential of modulating cannabinoid signaling to improve outcomes after a TBI."

Cancer

Cancer includes a group of diseases characterized by abnormal cell multiplication and consequent tissue damage, often accompanied by metastasis—spreading of such activity to other parts of the body from the area of origin. Most cancers produce tumors, described as "malignant" in contrast to "benign" tumors, which are not metastatic. Noncancerous cells undergo genetically programmed cell death (apoptosis), the body's way of regrowth and renewal. Malignant cells, however, have become mutated, escaping the gene control that prevented their continual division and unchecked development. Malignant growths are immortal (at least until their host dies or is cured), invading surrounding tissues, stimulating the production of blood vessels (angiogenesis) for their sustenance, and colonizing distant parts of the body. More than 14 million new cases of cancer and more than eight million deaths from cancer are reported annually on a world basis. Breast cancer represents about 30% of newly diagnosed cancers annually (Caffarel et al. 2010). Cancers may be caused by environmental factors such as chemicals, radiation, and infectious agents or can be the result of hereditary mutations. Cancer is the world's leading cause of death (followed by heart disease and stroke) and has the largest economic impact from premature death and disability of all causes of death worldwide (John and Ross 2010). The U.S. National Cancer Institute (2015) maintains a website dedicated to information on the use of cannabis in the treatment of cancer.

Almost a fifth of cancer mortality is due to smoking tobacco. However, as discussed earlier, claims that smoking marijuana produces cancer have limited validity. Zhang et al. (2015), based on a statistical analysis of the pooled data for six case-controlled studies, concluded that there is "little evidence for an increased risk of lung cancer among habitual or long-term cannabis smokers, although the possibility of potential adverse effect for heavy consumption cannot be excluded." Bowles et al. (2012) commented, "Medical cannabis remains a paradox in many ways. Cannabis smoke may be carcinogenic but it has been difficult to conclusively link cannabis use and cancer development epidemiologically, and cannabinoids have shown some promise as anti-cancer therapies." Huang et al. (2015) indicated that while smoking has not been statistically linked to lung cancer, there may be linkages to other cancers.

Marks et al. (2014) concluded that "The associations of marijuana use with oropharyngeal and oral tongue cancer are consistent with both possible pro- and anti-carcinogenic effects of cannabinoids. Additional work is needed."

Massi et al. (2013) wrote: "Over the past years, several lines of evidence support an antitumourigenic effect of cannabinoids including Δ^9-tetrahydrocannabinol, synthetic agonists, endocannabinoids and endocannabinoid transport or degradation inhibitors... Several cannabinoids have been shown to exert anti-proliferative and pro-apoptotic effects in various cancer types (lung, glioma, thyroid, lymphoma, skin, pancreas, uterus, breast, prostate and colorectal carcinoma) both in vitro and in vivo. Moreover, other antitumourigenic mechanisms of cannabinoids are currently emerging, showing their ability to interfere with tumour neovascularization, cancer cell migration, adhesion, invasion and metastasization."

Cannabinoids have been demonstrated in laboratory animals and cultures, and in some case reports in humans, to inhibit tumor growth or proliferation (Guzmán 2003; Caffarel et al. 2010; Guindon and Hohmann 2011; Abrams and Guzman 2014; Velasco et al. 2014; Solinas et al. 2015; Zogopoulos 2015), and these anticarcinogenetic effects have been widely and misleadingly employed in assertions that "marijuana cures cancer." Romano and Hazekamp (2013) discussed claims that cannabis oil cures cancer. Fowler (2015) examined such claims and the resulting persuasion of many that cannabis is currently a validated therapy to reduce or eliminate cancers, permanently or for some test period such as five years. He concluded that "the current preclinical data do not yet provide robust evidence that systemically administered Δ^9-THC will be useful for the curative treatment of cancer. There is more support for an intratumoral route of administration of higher doses of Δ^9-THC. CBD produces effects in relevant concentrations and models, although more data are needed concerning its use in conjunction with other treatment strategies." Kramer (2015) was of a similar viewpoint but nevertheless concluded, "cannabis and cannabinoid pharmaceuticals can be helpful for a number of problems, including many affecting patients with cancer." THC has been demonstrated to promote apoptosis in malignant conditions, and "this is obviously a fertile area for further research" (Russo 2003a). There is near-unanimous agreement that cannabis has antitumor potential and deserves to be examined for the development of anti-cancer therapies (Sarfaraz et al. 2008). With respect to directly treating tumors of human patients, Cridge and Rosengren (2013) commented, "Overall, the cannabinoids may show future promise in the treatment of cancer, but there are many significant hurdles to overcome."

As noted later in the discussion of CBD, this cannabinoid appears to have particular promise for treating cancer.

Brain cancer is a leading cause of cancer mortality, and "gliomas," which include all tumors arising from the gluey or supportive tissue of the brain, represent 30% of all brain tumors and 80% of all malignant tumors. Torres et al. (2011) found that THC had some effect on controlling cultured malignant gliomas. "Positive effects of cannabinoids in gliomas have attracted tremendous interest for the possibility that cannabinoids may serve as a novel form of chemotherapy for patients. However, the issue has progressed very slowly in the clinical area" (Fernández-Ruiz et al. 2014a). Foroughi et al. (2011) discussed two cases of regression of gliomas, possibly related to consumption of marijuana.

The main cancer-related therapeutic potential of cannabis at present seems to lie in alleviating the side effects and symptoms of cancer and its treatments. "The palliative effects of *Cannabis sativa* (marijuana), and its putative main active ingredient, Δ^9-tetrahydrocannabinol (THC), which include appetite stimulation, attenuation of nausea and emesis associated with chemo- or radiotherapy, pain relief, mood elevation, and relief from insomnia in cancer patients, are well-known" (Pacher 2013). "Cannabinoids are well-known to exert palliative effects in cancer patients, and their best-established use is in the inhibition of chemotherapy induced nausea and vomiting" (Velasco et al. 2012; cf. Machado Rocha et al. 2008).

On the whole, there is variable but enthusiastic support for the use of cannabis for aspects of cancer treatment. "The cannabinoids provide unquestionable advantages compared to current antitumoural therapies" (Bifulco et al. 2006). "Cannabinoids can palliate some cancer symptoms but it is unclear how effective they are compared to or combined with conventional therapies, or even whether cannabis, purified cannabinoids, or synthetic cannabinoids are more effective" (Bowles et al. 2012). "Medicinal cannabis is an invaluable adjunct therapy for pain relief, nausea, anorexia and mood modification in cancer patients" (Ruchlemer et al. 2015). "In addition to treatment of nausea and anorexia, cannabinoids may be of benefit in the treatment of cancer-related pain, possibly in a synergistic fashion with opioid analgesics" (Abrams and Guzman 2014).

CARDIOVASCULAR DISEASES

"Heart disease" (understood to include blood vessel disease) is a leading killer of men and women, responsible for over 17 million deaths annually throughout the world. Major problems include coronary heart disease (heart attacks), cerebrovascular disease (stroke), raised blood pressure (hypertension), peripheral artery disease, rheumatic heart disease, congenital heart disease, and heart failure.

The ECS and especially CB_1 receptors have important influences on heart function and blood circulation. "Activation of CB_1 receptors in the brain induces cardiovascular stress response, which increases heart oxygen consumption and lowers blood flow through coronary arteries" (Fišar 2009a). Modulating the functions of the ECS is considered to have therapeutic potential for cardiovascular disorders (Cunha et al. 2011). As discussed later in the section on cautions regarding the medical uses of cannabis, marijuana consumption is accompanied by a small risk of precipitating adverse cardiovascular incidents in susceptible individuals.

Russo (2015) reviewed cardiovascular risks of synthetic and natural cannabinoids and concluded: "To date, it appears that ultra-low THC and therapeutic phytocannabinoid dosing are cardioprotective, while supra-therapeutic recreational doses pose cardiovascular risks, and hyper-CB_1 stimulation by potent full agonists is distinctly dangerous to the heart. It remains for society to ascertain how science-based education may lower such risks and help potential consumers avoid a perilous misadventure in pharmacological roulette."

Atherosclerosis

Atherosclerosis is a disease of the arteries characterized by the deposition of plaques of fatty material on their inner walls, leading to decreased blood flow. It causes heart attacks, stroke, and peripheral vascular disease. "Atherosclerosis and its major acute complications, myocardial infarction and stroke, are the leading causes of death and morbidity worldwide... Growing evidence suggests that an overactive endocannabinoid-CB_1 receptor signaling promotes the development of cardiovascular risk factors" (Steffens and Pacher 2015). "The precise role of endocannabinoid system in atherosclerosis has not yet been fully recognized...cannabinoids can be linked...to useful effects on cardiovascular system, e.g., they have protective role in progression of atherosclerosis" (Fišar 2009a).

Hypertension

Hypertension ("high blood pressure") refers to abnormally high arterial blood pressure. It can damage the arteries, brain (leading to strokes), heart (leading to heart attacks), eyes, kidneys, and other organs. "Endogenous cannabinoids have no distinguished function in cardiovascular regulations under normal conditions, but they are involved in cardiovascular regulation in hypertension when they may hold down elevating of blood pressure through activation of CB_1 receptors" (Fišar 2009a).

Ischemia

Ischemia refers to an inadequate blood supply to an organ or part of the body, especially the heart muscles. "Increasing of static heartbeat, as far as about 60% during first 30 minutes after smoking cannabis, can be dangerous for men with cardiovascular disorder. Effects of THC on cardiovascular system are marked and are mediated largely by CB_1 receptors in blood cells and heart... Cannabinoids may contribute to cardiovascular collapse connected with myocardial infarction" (Fišar 2009a). However, Rajesh et al. (2010) found that CBD has protective cardiac effects. Similarly, Waldman et al. (2013), based on a study of mice, found that "a single ultra-low dose of THC before ischemia is a safe and effective treatment that reduces myocardial ischemic damage." With respect to stroke, Hillard (2008) commented: "Overall, the available data suggest that inhibition of CB_1 receptor activation together with increased CB_2 receptor activation produces beneficial effects" and "there is every reason to believe that the ECS contributes in some way to the injury produced by ischemia. In light of these facts, the lack of a clear picture of the precise role of this system, is on one hand, puzzling. The answer at the moment seems to be that we do not completely appreciate the complexity of ECS signaling and that its manipulation using global pharmacological approaches has led to incorrect conclusions. Future studies using more selective tools will certainly help to clarify this important and fascinating role of the ECS."

DIABETES

Diabetes is a set of diseases in which high blood glucose (blood sugar) develops because insulin production is inadequate or because the body does not respond properly to insulin. Insulin is a hormone necessary to convert sugar, starch, and other foods into energy. In type 1 (once referred to as "juvenile diabetes," affecting about 10% of diabetics), the pancreas does not produce enough insulin; in type 2 (once called "adult-onset diabetes"), the body's ability to use insulin is defective. "Most diabetic complications are associated with pathologic alterations in the vascular wall; the most common macrovascular complication of diabetes is atherosclerosis, which increases the risk of myocardial infarction, stroke, and peripheral artery disease... Diabetic complications have tremendous physical, emotional, and economic impact because diabetes is the leading cause of kidney failure, nontraumatic lower-limb amputation, and new cases of blindness among adults in the United States" (Horváth et al. 2012). Several hundred million people in the world suffer from diabetes. Obesity (discussed later) is the leading risk factor for insulin resistance. Poor eating habits are resulting in an increase in diabetes worldwide.

Several studies have shown a correlation between cannabis use and insulin levels. Cannabinoid receptors are known to occur in the pancreas and to function in regulating glucose and insulin. An epidemiological study showed that smoking marijuana is associated with a two-thirds reduction of incidence of diabetes (Jancin 2010). Alshaarawy and Anthony (2015) commented. "Recently active cannabis smoking and diabetes mellitus are inversely associated... Current evidence is too weak for causal inference...on a possibly protective (or spurious) cannabis smoking-diabetes mellitus association suggested in prior research." Frisher et al. (2010) noted, "Herbal cannabis use has been linked to harms and benefits for diabetic patients" and "The potential risks and benefits for diabetic patients remain unquantified at the present time. Cannabinoids appear to affect biochemical pathways associated with diabetes but it is too early to say whether this will lead to new treatments." Di Marzo (2008b) reported that CB_1 receptor activation impairs plasma glucose clearance, while CB_2 receptor activation promotes glucose clearance. Penner et al. (2013) observed lower fasting insulin levels in cannabis users. Also see the information presented later for CBD and THCV.

EPILEPSY

Epilepsy, which affects about 1% of adults and 2% of children (Blair et al. 2015) and 65 million people worldwide (Rosenberg et al. 2015), is a serious disorder of the nervous system, caused by abnormal electrical activity in the brain, leading to periodic seizures or convulsions. Dravet syndrome is a severe form of childhood epilepsy. Epilepsy is characterized either by mild, episodic loss of attention or sleepiness (petit mal) or by severe convulsions with loss of consciousness (grand mal). Williams et al. (2014) provide the following information: "Epilepsy is not a single disease, but encompasses a diverse family of disorders, all involving an abnormally increased predisposition to seizures." While there are dozens of drugs for epilepsy, "approximately 30% of the epileptic population experience intractable seizures regardless of antiepileptic drug treatments used." "Of all patients with epilepsy, approximately 50% will ultimately become refractory to drug treatment." "All existing antiepileptic drugs are associated with numerous side effects."

Friedman and Devinsky (2015) observed, "Patients need new treatments that control seizures and have fewer side effects. This treatment gap has led patients and families to seek alternative treatments. Cannabis-based treatment for epilepsy has recently received prominent attention in the lay press and in social media, with reports of dramatic improvements in seizure control in children with severe epilepsy. In response, many states have legalized cannabis for the treatment of epilepsy (and other medical conditions) in children and adults."

"Anecdotal evidence suggests that cannabis might have therapeutic value for epilepsy. In the fifteenth century, the epileptic son of a caliphate counsellor in Baghdad was treated with cannabis" (Mechoulam 1986). More recently, the use of cannabis in treating epilepsy came to widespread public attention when the CNN television network featured the successful treatment of Charlotte Figi, a five-year-old girl with epilepsy, with a CBD-rich strain of cannabis called Charlotte's Web (Maa and Figi 2014; Lorentzos and Webster 2015). Her seizures, before treatment occurring every half hour, essentially ceased. Note the comments in the following regarding the value of CBD for epilepsy.

Because the ECS inhibits electrical transmission, its potential role in serving to inhibit the undesirable electrical activity accompanying epilepsy seems promising. Hofmann and Frazier (2013) argued that the ECS has significant potential as a target for future development of antiepileptic therapies, based on three main contentions: "Endocannabinoid mediated systems are intimately involved in moment to moment regulation of neuronal excitability throughout many areas of the

CNS. Significant aspects of endogenous cannabinoid signaling systems are altered in a wide range of epileptic conditions. External modulation of cannabinoid mediated systems can prevent or modulate important aspects of epileptiform activity in a wide range of in vitro and in vivo models."

Blair et al. (2015) stated, "Basic scientific research studies that target different components of the brain ECS with specific pharmacological agents have shown great promise toward the development of novel therapeutic strategies for the control and possible prevention of epileptic seizure disorders." Di Maio (2013) found evidence based on animal research that the CB_1 receptor is a promising target for research on epilepsy. However, Gloss and Vickrey (2012) and Koppel et al. (2014) concluded that cannabis is of unknown efficacy in treating the symptoms of epilepsy. Rosenberg et al. (2015) reviewed the literature relating cannabis and epilepsy and stated, "For over a millennium, pre-clinical and clinical evidence have shown that cannabinoids such as CBD can be used to reduce seizures effectively, particularly in patients with treatment-resistant epilepsy. However, many questions still remain…regarding the mechanism, safety, and efficacy of cannabinoids in short- and long-term use." Reddy and Golub (2016) commented, "CBD is anticonvulsant, but it has a low affinity for the cannabinoid CB_1 and CB_2 receptors; therefore the exact mechanism by which it affects seizures remains poorly understood. A rigorous clinical evaluation of pharmaceutical CBD products is needed to establish the safety and efficacy for the treatment of epilepsy. Identification of mechanisms underlying the anticonvulsant efficacy of CBD is additionally critical to identify other potential treatment options." Tzadok et al. (2016) surveyed 74 children and adolescents with intractable epilepsy, who received CBD treatment, and described the effects as "highly promising" but noted that well-designed clinical trials using enriched CBD are needed. GW Pharmaceuticals is currently testing the value of Epidiolex, an experimental cannabis-based product, to treat epilepsy. Also see the comments further in this chapter below on the possible anticonvulsant value of CBD for treating epilepsy.

FIBROMYALGIA

Fibromyalgia is a central nervous system disorder, the second most common rheumatic disorder behind osteoarthritis. Fibromyalgia affects 2% of the population of North America, females much more commonly than males. It is characterized by chronic widespread musculoskeletal pain and heightened sensitivity to pressure in multiple tender points in the neck, spine, shoulders, and hips. An associated syndrome of symptoms may include fatigue, insomnia, joint stiffness, and other physical problems. There may also be accompanying psychiatric conditions such as depression and anxiety. Fibromyalgia does not respond well to conventional pain therapies, and in the past, it was often believed that sufferers were merely hypochondriacs whose pain was "in their heads."

Ste-Marie et al. (2012) found that fibromyalgia patients tended to self-medicate with marijuana. Schley et al. (2006) reported that a sample of fibromyalgia patients reported significant benefit from cannabis. Skrabek et al. (2008) also reported cannabis provided relief from fibromyalgia. Fiz et al. (2011) pointed out, "There is little clinical information on the effectiveness of cannabinoids in the amelioration of FM symptoms." In their own experimental study, Fiz et al. (2011) found that "the use of cannabis was associated with beneficial effects on some FM symptoms." Hendrickson and Ferraro (2013) reviewed the relationship of cannabis and fibromyalgia and concluded, "Evidence suggests incorporating cannabinoids into an approach including behavioral/lifestyle adaptations and cognitive behavioral therapies may optimize overall benefits for many sufferers."

GASTROINTESTINAL DISEASES

As noted in the earlier discussion of the ECS, the body's endocannabinoids control energy balance and metabolism, which includes feeding behavior and digestion. The gut is of course the site of energy extraction from food, and its orderly or disordered operation can be expected to be mediated by the ECS. Cannabinoid receptors are involved in such functions as secretion of stomach acid and digestive enzymes, contraction of the pyloric valve regulating food passage between the stomach and small intestine, moving food through the intestines by gut contractions, detecting visceral pain, and regulating appetite and vomiting.

There are numerous debilitating conditions of the gastrointestinal tract, including nausea, vomiting, ulcers, irritable bowl, Crohn's disease, and reflux disease. Inflammatory bowel disease is an incurable condition affecting millions in industrialized countries. Irritable bowel syndrome is a common disorder characterized by poor bowel function and abdominal discomfort. Crohn's disease is an inflammatory bowel disease usually confined to the lower end of the small intestine. It can lead to ulcers, bleeding, and scar formation, which may produce intestinal blockage, cramps, spasms, nausea, vomiting, loss of appetite and weight, severe diarrhea, and rectal bleeding. Paralytic ileus is a long-lasting inhibition of gastrointestinal motility, which can produce infection, gas, and a range of serious conditions. Gastroesophageal reflux disease is a digestive disorder affecting the lower esophageal sphincter (the muscle connecting the esophagus to the stomach). When the sphincter is weak or relaxes inappropriately, acidic gastric contents flow into the esophagus, damaging it.

Cannabis has a centuries-long history of use for gastrointestinal disorders, including emesis (vomiting), gastric ulcers, abdominal pain, gastroenteritis, diarrhea, and intestinal inflammation (Duncan and Izzo 2014). Di Marzo and Piscitelli (2011) commented, "the endocannabinoid system and related emerging signalling system may play a fundamental role in the control of all aspects of gastrointestinal physiology and pathology...strategies that either enhance or curb the activity of the ECS might be both employed for future therapies targeting various GI disorders." Todaro (2012) stated, "Today, the standard of care for prevention of CINV [chemotherapy-induced nausea and vomiting] for highly and moderately emetogenic chemotherapy is a 5-HT3 receptor antagonist, dexamethasone... With the approval of safer and more effective agents, cannabinoids are not recommended as first-line treatment for the prevention of CINV and are reserved for patients with breakthrough nausea and vomiting."

Slatkin (2007) wrote, "Improvement in antiemetic therapy across the entire spectrum of chemotherapy-induced nausea and vomiting will involve the use of agents with different mechanisms of action in concurrent or sequential combinations, and the best such combinations should be identified. In this effort, the utility of the cannabinoids should not be overlooked." Anecdotal and scientific evidence suggest that cannabis use may have a positive impact in patients suffering from inflammatory bowel disease (Borrelli et al. 2013). CB_1 receptors in the brain and in the gut regulate gut motility, and both receptor types attenuate gut inflammation, and these considerations may explain the therapeutic role of cannabis in gastrointestinal disorders (Duncan and Izzo 2014). Di Carlo and Izzo (2003) observed that cannabis has potential "for the treatment of a number of gastrointestinal diseases, including nausea and vomiting, gastric ulcers, irritable bowel syndrome, Crohn's disease, secretory diarrhoea, paralytic ileus and gastroesophageal reflux disease." There is some clinical evidence that cannabis is therapeutic for Crohn's disease (Naftali et al. 2011, 2013, 2014; Schicho and Storr 2014). At least in experimental animals, THC has been shows to reduce esophageal sphincter relaxation and decrease acid production, both of which may benefit gastroesophageal reflux disease (Caraceni et al. 2014).

Cannabis has proven useful in controlling chemotherapy-induced nausea and vomiting (Duran et al. 2010; Parker et al. 2011; Bowles et al. 2012; Borgelt et al. 2013; Rock et al. 2014), although long-term use can produce "cannabinoid hyperemesis syndrome" (Nicolson et al. 2012). Cannabinoid hyperemesis (a term coined in 2004) has been described as a rare effect of chronic marijuana use characterized by severe, cyclic nausea, vomiting, and abdominal pain, and (very oddly) marked by compulsive hot-water bathing for temporary symptom relief (Allen et al. 2004; Venkatesan et al. 2004; Wallace et al. 2011; Simonetto et al. 2012).

GLAUCOMA

Glaucoma is a group of eye conditions that damage the optic nerve, which is essential for good vision. Glaucoma is generally an age-related chronic eye condition causing progressive loss of peripheral vision. The damage is often caused by increased intraocular pressure (IOP), which can lead to blindness if not treated. Indeed, glaucoma is the world's leading cause of irreversible blindness.

It is well known that smoking marijuana can affect vision and may be useful for treating several degenerative and inflammatory retinal diseases (Xu and Azuaro-Blanco 2014). Marijuana can produce "red eye" ("conjunctival vasodilation" or "corneal vasodilation"; Figure 12.2) (Yazulla 2008). Anecdotal reports in the early 1990s indicated that Jamaican fishermen smoked marijuana to improve their vision for fishing at night. In fact, there is some evidence that marijuana enhances dim light vision (Russo et al. 2004). In the Caribbean area, Canasol, an alcoholic extract of *C. sativa*, is used to treat glaucoma. Canasol originated from research at the University of the West Indies and in private ophthalmic clinics (West 1997).

Several of the cannabinoids can reduce IOP and therefore may have utility in treating glaucoma. Some noncannabinoid components of *Cannabis* also may have this capacity. Nucci et al. (2008) stated, "The identification of plant-derived, endogenous, or synthetic CBs capable of interacting with the intraocular endocannabinoid system could open new perspectives for the treatment of glaucoma." Tomida et al. (2004) wrote, "Cannabinoids have the potential of becoming a useful treatment for glaucoma, as they seem to have neuroprotective properties and effectively reduce intraocular pressure. However, several challenges need to be overcome, including the problems associated with unwanted systemic side effects (psychotropic, reduction in systemic blood pressure), possible tolerance, and the difficulty in formulating a stable and effective topical preparation." The same information was given by Xu and Azuaro-Blanco (2014). Samudre et al. (2014), based on studies of rats, stated, "Topically applied cannabinoids are effective agents that reduce IOP and confer neuroprotection and are prime candidates for potential glaucoma treatment."

The American Glaucoma Society issued the following position statement (Jampel 2010): "Although marijuana can lower the IOP, its side effects and short duration of action, coupled with a lack of evidence that its use alters the course of glaucoma, preclude recommending this drug in any form for the treatment of glaucoma at the present time." Buys and Rafuse (2010), on behalf of the Canadian Ophthalmological Society, were equally critical. Sun et al. (2015) stated, "Currently, the deleterious effects of marijuana outweigh the benefits of its IOP-lowering capacity in most glaucoma patients. Under extremely rare circumstances, a few categories of glaucoma patients may be potential candidates for treatment with medical marijuana."

HEADACHE

Cannabis is "used by patients for relief of headache generally without physician recommendation. There is much anecdotal support and experimental evidence for this use but no good clinical trials for headache" (McGeeney 2013). "The literature suggests that the medicinal use of cannabis may have a therapeutic role for a multitude of diseases, particularly chronic pain disorders including headache. Supporting literature suggests a role for medicinal cannabis and cannabinoids in several types of headache disorders including migraine and cluster headache" (Baron 2015).

Migraine

Migraine refers to recurrent moderate to severe headaches associated with other symptoms. The headache typically affects half of the head, pulsates, and can last three days. Additional symptoms may include nausea; vomiting; sensitivity to light, sound, or smell; and an "aura" (a visual or other kind of sensory premonition of headache, occurring in about 10% of patients). The pain is generally increased by physical activity. Before puberty, boys are more affected than girls, but afterward, two to three times more women suffer migraines than men.

More than a century ago, Sir William Osler, perhaps the finest physician of his era, recommended cannabis as the "most satisfactory remedy for migraine" (Osler and McCrae 1915). Today, there is appreciable support for the thesis that cannabis can be therapeutic for migraine (Russo 1998, 2001, 2004b; McGeeney 2013). Greco and Tassorelli (2015) concluded, "Migraine has numerous relationships with endocannabinoids, and a deficiency in the endocannabinoid system has been hypothesized to underlie the pathophysiology of migraine. However biochemical studies providing a scientific basis for the potential efficacy of (endo)cannabinoids in migraine are, so far, limited."

HIV AND AIDS

HIV originated in west-central Africa during the late nineteenth or early twentieth century. HIV infection degrades the immune system, increasing susceptibility to diseases. The severe stages are referred to as AIDS and are often accompanied by extreme weight loss, pneumonia, and Kaposi's sarcoma, a type of cancer. If untreated, death is common within 10 years. A cure has not yet been formulated, but antiretroviral treatment can substantially suspend the condition.

HIV/AIDS has had a great impact on society, both as an illness and as a source of discrimination. (Werner 2001 discussed the difficulties of advocating marijuana to alleviate the effects of AIDS, since both were the subjects of prejudice.) The disease also has significant economic impact.

Patients with these conditions are one of the largest groups using cannabis for medicinal purposes. "Reasons for smoking cannabis cited by patients include countering the nausea, anorexia, stomach upset, and anxiety associated with the disease and with antiretroviral therapy" (Hazekamp and Grotenhermen 2010). Various studies reviewed by Hazekamp and Grotenhermen (2010) indicate that cannabis is useful for stimulating appetite and relieving pain in HIV/AIDS patients. Lutge et al. (2013) stated, "The use of cannabis (marijuana), its active ingredient or synthetic forms such as dronabinol has been advocated in patients with HIV/AIDS, in order to improve the appetite, promote weight gain and lift mood. Dronabinol has been registered for the treatment of AIDS-associated anorexia in some countries. However, the evidence for positive effects in patients with HIV/AIDS is limited, and some of that which exists may be subject to the effects of bias." Ellis et al. (2009) stated, "Neuropathic pain in HIV is an important and persisting clinical problem, affecting 30% or more of HIV-infected individuals. Although combination antiretroviral therapy has improved immunity and survival in HIV, it does not significantly benefit neuropathic pain… Our findings suggest that cannabinoid therapy may be an effective option for pain relief in patients with medically intractable pain due to HIV."

HUNTINGTON'S DISEASE

Huntington's disease (Huntington disease) is an inherited chronic condition in which nerve cells in the brain break down over time, deteriorating muscle coordination and mental capacities and leading to behavioral symptoms. Huntington's disease was once called Huntington's chorea, as chorea (involuntary movement) usually develops. The condition affects about 1 in 10,000 individuals, symptoms typically beginning in middle age (Casteels et al. 2015).

Regarding therapy with cannabis, Armstrong and Miyasaki (2012) wrote that "information is insufficient to recommend long-term use, particularly given abuse potential concerns." Similarly, Kluger et al. (2015) also found that "data are insufficient to draw conclusions regarding Huntington's disease." A clinical trial with Sativex failed to slow disease progression (Fernández-Ruiz et al. 2014b). Koppel et al. (2014) concluded that cannabis is of unknown efficacy in treating the symptoms of Huntington disease. Pietropaolo et al. (2015), based on mice studies, suggested that "prolonged administration of cannabinoid receptor agonists could be an appropriate strategy for selectively improving motor symptoms and stimulating neuroprotective processes in HD patients."

INSOMNIA

Insomnia is "characterized by a night complaint of an insufficient amount of sleep or not feeling rested after the habitual sleep episode. Insomnia is often associated with feelings of restlessness, irritability, anxiety, daytime fatigue and tiredness" (Murillo-Rodríguez et al. 2014). "Fifty-eight percent of adult Americans have reported symptoms of insomnia a few nights a week or more. The staggering prevalence of insomnia and the well-known complications of poor sleep quality, such as its effect on productivity, mental health, and cardiac and endocrinologic function, suggest the need for effective treatment of this spectrum of disorders" (Tringale and Jensen 2011).

Cannabis has historically been employed to treat sleep difficulties, and because it has sedative and anxiety-reducing properties (Nicholson et al. 2004; Bergamaschi et al. 2011), there is reason to believe that it has potential to treat insomnia. THC has sleep-inducing properties, while low-dose CBD can be alerting or neutral and high-dose CBD is sleep-inducing (Russo et al. 2009). CBD may also reduce symptoms that interfere with sleep. As noted in Chapter 9, dealing with essential oil, some of the volatile terpenes present may also play a role in inducing sleep. The pain-relieving properties of cannabis may also assist in promotion of sleep for some patients. Ware et al. (2010a) found that nabilone was effective in improving sleep in patients with fibromyalgia. Tringale and Jensen (2011) noted, "a significant decrease in reported time to of sleep after the use of marijuana in both those with and those without reported sleep difficulties" and that their "result is supported by recent findings concerning the endocannabinoid system, as well as voluminous anecdotal evidence." Patel et al. (2014) pointed out that THC increases melatonin production (melatonin is synthesized in the pineal gland at night, and contributes to sleep), and "could contribute to the well-known crash, or sleepiness, experienced after a bout of cannabis use."

LIVER DISEASE

Numerous maladies affect the liver, particularly cirrhosis or scarring, which compromise the liver's ability to remove toxins from the blood. Excessive alcohol consumption, nonalcoholic fatty liver disease, and hepatitis B and C are chief causes. More than 1.5 million die of liver diseases annually worldwide.

CB_1 receptors are common in the normal liver, and CB_2 receptors develop in the cirrhotic liver. Parfieniuk and Flisiak (2008) commented, "controlling CB_1 or CB_2 signalling appears to be an attractive target in managing liver diseases… Unquestionably, influencing endocannabinoid signaling may have a beneficial effect on delaying or even reversing hepatic fibrosis." The same viewpoint was expressed by Izzo and Camilleri (2008). Similarly, Mallat et al. (2011) observed that "Recent findings have revealed a role of endocannabinoids and their receptors in the pathogenesis of several key steps of acute and chronic liver injury, therefore identifying pharmacological modulation of cannabinoid receptors as an attractive strategy for the management of morbidity related to liver injury." As noted later in this chapter, there is evidence that CBD may be therapeutic for alcohol-induced liver damage.

MORNING SICKNESS

"Morning sickness," despite its name, affects pregnant women at any time of day or night. This condition is experienced by 50%–90% of pregnant women, usually early in pregnancy. It is characterized by nausea and/or vomiting and heightened sensitivity to smells and flavors. Rarely (1%–2% occurrence), a life-threatening form of vomiting referred to as hyperemesis gravidarum (hyperemesis syndrome, described previously) requires hospitalization. As noted in the section on gastrointestinal diseases, cannabis is effective against nausea and vomiting.

Cannabis has been employed historically to treat morning sickness, a use that persists among Rastafarian women in Jamaica (Westfall et al. 2006). Westfall et al. (2006) surveyed women who had employed cannabis for morning sickness and found that 92% considered it effective. However, as discussed in the section on cautions regarding medical marijuana, there is concern that cannabis might have effects on the fetus (although, as discussed by Russo 2002, cannabis has been employed extensively in historical times to treat female reproductive issues). The health implications of prenatal exposure to cannabis remain unclear. "Although it is the most widely used recreational drug in pregnancy...conclusive evidence is lacking with regards to immediate health effects of prenatal cannabis exposure, such as infant birthweight and rates of perinatal mortality and morbidity. In some studies, maternal cannabis use appears, at face value, to be associated with lower birth weight and higher rates of premature delivery; in others, birth outcomes are comparable to those of non-exposed infants" (Westfall et al. 2006). Hill and Reed (2013) examined evidence regarding the effects of marijuana exposure during pregnancy and breast-feeding and concluded that "The current evidence suggests subtle effects of heavy marijuana use on developmental outcomes of children. However, these effects are not sufficient to warrant concerns above those associated with tobacco use." Metz and Stickrath (2015) stated, "Further research is needed to provide evidence-based counselling of women regarding the anticipated outcomes of marijuana use in pregnancy. In the meantime, women should be advised not to use marijuana in pregnancy or while lactating."

MULTIPLE SCLEROSIS AND SPASTICITY

Multiple sclerosis (MS) is a neurodegenerative disease characterized by spasticity (uncomfortable and distressing muscle spasms and limb stiffness), painful muscle cramps and stiffness, reduced mobility, chronic pain in the arms and legs, tingling or pricking of the fingers and toes, ataxia (inability to control muscle movements), tremors, and intestinal and urination problems. "Multiple sclerosis (MS) is the commonest physically disabling neurological condition in young adults, with a prevalence between 50 and 200 per 100,000, depending on ethnic and geographical factors" (Novotna et al. 2011). MS is commonest in northern Europe and in regions colonized by white northern Europeans such as North America and is more frequent in females (Pryce and Baker 2014).

The axon (extension of a nerve cell that carries messages) is wrapped in a layer of myelin, which, like the coating around an electrical wire, insulates and promotes nerve transmission. MS produces inflammation that degrades myelin in the central nervous system neurons, as well as other features of the nerve cells. MS is generally believed to be an autoimmune condition. It is thought that cannabinoids may be therapeutically effective in treating MS because they are immunomodulatory.

In a survey of patient perceptions of the value of marijuana for treating MS, Page and Verhoef (2006) noted, "The perceived benefits of use were consistent with previous reports in the literature: reduction in pain, spasms, tremors, nausea, numbness, sleep problems, bladder and bowel problems, and fatigue and improved mood, ability to eat and drink, ability to write, and sexual functioning. Adverse effects included problems with cognition, balance, and fatigue and the feeling of being high. Although participants described risks associated with using marijuana, the benefits they derived made the risks acceptable."

There is considerable support for the use of cannabis to treat MS, and forms of it have been authorized for treatment of MS in numerous countries. "Cannabinoids...are effective in treating neuropathic pain in MS" (Iskedjian et al. 2007). "In clinical trials, more patients have been treated with cannabinoids for MS then for any other indication" and "MS is one of the few conditions where long-term extension studies have been performed with cannabis-based medicines" (Hazekamp and Grotenhermen 2010). "Undoubtedly, the most promising clinical use of cannabinoids concerns MS" (Chiurchiù et al. 2015).

Spasticity (and sometimes tremors) accompanies spinal cord injury as well as MS. There are indications that cannabis can indeed improve the symptoms of both MS and spinal cord sufferers (Hazekamp and Grotenhermen 2010). "MS results from disease that impairs neurotransmission and this is controlled by cannabinoid receptors and endogenous cannabinoid ligands. This can limit spasticity and may also influence the processes that drive the accumulation of progressive disability" (Baker et al. 2012). "An increasing amount of evidence is now emerging to confirm anecdotal reports of symptomatic improvement, particularly for muscle stiffness and spasms, neuropathic pain and sleep and bladder disturbance, in patients with MS treated with cannabinoids" (Zajicek and Apostu 2011). "We found evidence that combined THC and CBD extracts may provide therapeutic benefit for MS spasticity symptoms" (Lakhan and Rowland 2009). "Data from animal models of MS and clinical studies have supported the anecdotal data that cannabis can improve symptoms such as limb spasticity, which are commonly associated with progressive MS" (Pryce and Baker 2012). Corey-Bloom et al. (2012) "saw a beneficial effect of inhaled cannabis on spasticity among patients receiving insufficient relief from traditional treatments. Although generally well-tolerated, smoking cannabis had acute cognitive effects." Oreja-Guevara (2012) found that cannabis has "shown a clear-cut efficacy to reduce spasticity and their associated symptoms in those patients refractory to other therapies, with a good tolerability/safety profile." Koppel et al. (2014) reported that cannabis is "probably effective" in treating spasticity. Kluger et al. (2015) concluded that cannabis likely has no benefit for tremor in MS. The THC/CBD oromucosal spray Sativex (nabiximols) has been extensively employed to treat MS. "Real-life data confirm nabiximols as an effective and well-tolerated treatment option for resistant multiple sclerosis spasticity in clinical practice" (Flachenecker et al. 2014). Centonze et al. (2009) found that some parameters measuring the severity of MS were not altered by Sativex.

Neurodegenerative Diseases

Neurodegenerative diseases are incurable, debilitating conditions caused by progressive death or degeneration of neurons of the nervous system. This produces ataxias (problems with movement or motor functions) and dementias (decline in cognitive or mental functioning). "The treatment of neurodegenerative disorders is possibly one of the major challenges for biomedical research in the present century in developed countries, given the present and future increases in expected longevity in these countries, which will give greater opportunity for these disorders to appear" (Fernández-Ruiz et al. 2014a). Inflammation appears to be a common denominator among diverse neurodegenerative diseases.

Scotter et al. (2010) noted, "There has been anecdotal and preliminary scientific evidence of cannabis affording symptomatic relief in diverse neurodegenerative disorders. These include multiple sclerosis, Huntington's, Parkinson's and Alzheimer's diseases, and amyotrophic lateral sclerosis." The neuroprotective and anti-inflammatory properties of the cannabinoids (particularly of CBD) "may represent a very promising agent with the highest prospect for therapeutic use" (Iuvone et al. 2009). Rosales-Corral et al. (2015) noted, "Research on cannabinoids has been growing significantly in the last five years. More than fifty percent of this research corresponds to 'cannabinoids and brain,' particularly about neurodegeneration."

The American Academy of Neurology sponsored a review of all studies of medical marijuana published from 1948 through 2013 for evidence of efficacy in treating symptoms of MS, Huntington's disease, epilepsy, Parkinson's disease, cervical dystonia (a rare painful condition characterized by involuntary contraction of the muscles of the neck and shoulder, causing twisting and tilting of the head), and Tourette's syndrome. The review concluded that cannabis appears to help alleviate spasticity and central or spasm-related pain and some other MS symptoms, but there is little evidence of efficacy in treating epilepsy or movement disorders, and there are serious concerns about side effects (Koppel et al. 2014; Kurt 2014).

Inflammatory responses in the brain have become a central area of research to address numerous degenerative diseases of the brain (De Ceballos 2015). Chiurchiù et al. (2015) reviewed the mechanisms of brain functions in relation to how cannabis might be used therapeutically to address neuroinflammatory diseases, including Alzheimer's disease, multiple sclerosis and amyotrophic lateral sclerosis. Different classes of cells are involved in the protective inflammatory response by the nervous system and in the remainder of the body. Outside of the nervous system, most of the body's protective cells that combat invading pathogens and eliminate damaged cells are macrophages and lymphocytes, which arise in the blood system. However, particularly important to the protection of the central nervous system are "microglial cells," colorfully described by Chiurchiù et al. (2015) as protective soldiers, defending the brain in particular from infection, tumors, ischemia, trauma, and neurodegeneration. These are thought to remain dormant in the healthy brain but (like blood immune cells) become activated in response to damage or infection. Microglial cells normally appear to lack CB receptors, but when the nervous system experiences local inflammation, they develop high levels of CB_2 receptors, suggestive of a critical role in neuroinflammatory conditions. It is not surprising that CB_2 receptors are localized in protective brain cells, since outside the nervous system, they are widely present in immune system cells.

Zogopoulos et al. (2015) summarized the current status of therapeutic possibilities as follows: "Over the last years, considerable progress has been made in understanding the role of endocannabinoids in preventing or reducing the effects of progressive neurodegenerative diseases. The ECS has been shown to mediate neuroprotection in many neurological and psychiatric disorders including pain, schizophrenia, anxiety, depression, Parkinson's disease, Alzheimer's disease, Huntington's chorea, MS, amyotrophic lateral sclerosis and epilepsy. It also has neurotrophic and neuroprotective effects in cerebral ischemia (stroke) and traumatic brain injury. The endocannabinoid system represents a local messenger between the nervous and immune system and is obviously involved in the control of immune activation and neuroprotection. Manipulation of endocannabinoids and/or use of exogenous cannabinoids in vivo can constitute a potent treatment modality against inflammatory disorders."

See the sections in this chapter regarding the value of cannabis for each of Alzheimer's disease, Huntington's disease, multiple sclerosis, Parkinson's disease, Tourette's syndrome, and brain injury.

NEUROPATHY

"Peripheral neuropathy" (usually implied when the word neuropathy is used) refers to impaired sensation, movement, or body functions caused by damage or diseases affecting nerves. Carpal tunnel syndrome is a familiar example. "Chronic neuropathic pain affects 1%–2% of the adult population and is often refractory to standard pharmacologic treatment" (Ware et al. 2010b). "Painful peripheral neuropathy comprises multiple symptoms that can severely erode quality of life. These include allodynia (pain evoked by light stimuli that are not normally pain-evoking) and various abnormal sensations termed dysesthesias (e.g., electric shock sensations, 'pins and needles,' sensations of coldness or heat, numbness, and other types of uncomfortable and painful sensations). Common causes of peripheral neuropathy include diabetes, HIV/AIDS, spinal cord injuries, multiple sclerosis, and certain drugs and toxins. Commonly prescribed treatments come from drugs of the tricyclic and selective serotonin reuptake inhibitor (SSRI) antidepressant classes, anticonvulsants, opioids, and certain topical agents. Many patients receive only partial benefit from such treatments, and some either do not benefit or cannot tolerate these medications" (Grant 2013).

As noted under the section titled "Pain," cannabis appears helpful for pain relief accompanying various diseases. "The analgesic effects of herbal cannabis, derived from the dried leaves and flowers, have been most studied in neuropathic pain conditions" (Fitzcharles et al. 2014). "Cannabinoids show promise for treatment of neuropathic pain in humans either alone or as an add-on to other therapeutic agents" (Rahn and Hohmann 2009). Wilsey et al. (2013) concluded that "vaporized cannabis, even at low doses, may present an effective option for patients with treatment-resistant neuropathic pain." Grant (2013) stated, "data suggest, on balance, that cannabis may represent a reasonable alternative or adjunct to treatment of patients with serious painful peripheral neuropathy for whom other remedies have not provided fully satisfactory results." Gutierrez and Hohmann (2011) concluded: "Cannabinoids effectively suppress neuropathic pain in preclinical and clinical studies. The challenge is to balance analgesic efficacy with the presence of adverse side effects (i.e., dizziness or sedation).... In clinical settings, cannabinoid-based medications show therapeutic efficacy, and side effects are tolerable or no worse than those of conventional neuropathic pain medications (e.g., gabapentanoids, tricyclic antidepressants and anticonvulsants). Thus, cannabinoids show promise for managing neuropathic pain that is intractable to conventional treatments or as an adjunct to existing medications. Clinical trials are needed to evaluate the long-term effectiveness and safety of these compounds in humans."

Obesity

Obesity is the result of addictive behavior but has become so normalized that it doubtfully qualifies as a psychiatric condition. However, in 2015, the American Medical Association voted to classify obesity as a disease, albeit not specifically psychiatric in nature. The world is experiencing a fat epidemic: 1.2 billion are overweight and an additional 475 million are obese (World Obesity Federation 2015). Although imperfect as an indicator of degree of overweight status, body mass index (BMI), a measure of relative weight to height, is widely used. BMI = weight in kg divided by height in m^2. "Overweight" or preobesity is defined as a BMI of 25–29.9 kg/m^2, while "obesity" is defined as a BMI >30 kg/m^2. In the Western world, half of people are overweight, one in six is obese. Over 200 million school-age children are overweight, causing the present generation to have a shorter predicted lifespan than their parents. Obesity is more common globally that undernutrition and is one of the most important public health issues. Excess body weight in the form of accumulated fat can severely impair health. Many major chronic illnesses characterize modern industrialized societies, notably heart disease, hypertension, adult-onset diabetes, some kinds of cancer, and dental caries, and overeating is substantially responsible.

The ECS is an important regulator of eating. It is well known that marijuana stimulates the appetite, an effect caused by stimulation of the CB_1 receptors in hypothalamic feeding circuits in the brain (Williams et al. 2015). "Obesity seems to be a condition associated with a pathological overactivation of the endocannabinoid system" (Viveros et al. 2008). Accordingly, suppressing the receptor responsible for overeating would be expected to reduce appetite. Indeed, Trillou et al. (2004) found that knockout (genetically engineered) mice in which the CB_1 receptors were inactivated were quite resistant to obesity (Figure 13.23) and had notably improved energy metabolism, less tendency for fat deposition, and improved insulin levels. As noted earlier, the synthetic cannabinoid rimonabant, a CB_1 cannabinoid receptor suppressor, was licensed for a period as a weight-loss pharmaceutical for humans. It was sold in over 50 countries but was discontinued because of side effects such as suicidal thoughts and acute depression and was withdrawn from the legal market in 2009. The problem is that CB_1 receptors control many metabolic functions, and suppressing them to therapeutically control one function can negatively affect other vital activities. Illustrative of how deleterious suppression of the CB_1 receptors can be is the observation that feeding rimonabant to newborn mice results in their cessation of suckling and starving to death (Fride 2004).

FIGURE 13.23 The influence of CB_1 receptors on obesity. Mice genetically engineered to be deficient in CB_1 receptors (left) resist obesity even on high-fat diets. Prepared by B. Brookes.

In the experiment of Trillou et al. (2004) referred to previously, feeding efficiency of the mice lacking normal CB_1 receptors was lower (i.e., the mice extracted less energy from the same amount of food by comparison with normal mice), which would not be adaptive in the wilderness but is desirable for weight control. In nature, food is almost always in short supply most of the time, and evolution has favored metabolic systems that efficiently extract energy from food, maximizing storage of fat in the body's fat cells in order to survive periods of famine. The ECS is one of nature's ways of promoting food energy storage in a world where the supply of food is erratic and limited. Food is so abundant today that the ECS is not designed to limit its consumption (facetiously, it can be blamed for obesity).

Curiously, although cannabis stimulates the appetite, cannabis use is negatively correlated with obesity. Le Strat and Le Foll (2011) found that obesity is a third lower in regular marijuana smokers compared to abstainers. Penner et al. (2013) observed that cannabis users have smaller waist circumference (as well as higher levels of high-density lipoprotein, the desirable kind of cholesterol). The explanation for the net appetite-suppressing effect of cannabis may lie in the activation of not just CB_1 receptors but also of other appetite-regulating systems by components of marijuana.

OSTEOPOROSIS

Bone is a specialized tissue that continuously undergoes renewal and repair known as "bone remodeling." Specialized cells called osteoblasts generate new bone, while other specialized cells called osteoclasts dispose of old bone. Imbalance in bone formation and bone resorption is the main cause of bone disorders. Osteoporosis is characterized by gradual deterioration of bone mass, increased bone fragility, and increased risk of fracture. In developed nations, 2% to 8% of males and 9% to 38% of females are affected. Several other skeletal disorders such as rheumatoid arthritis are also characterized by bone deterioration. Bone fractures represent one of the main causes of death among elderly patients, and given the aging populations in industrialized countries, the disease is of increasing concern. Osteoporosis is often caused by estrogen deficiency (in postmenopausal women) and glucocorticoid treatment (glucocorticoids are widely used for chronic treatment of such conditions as bronchial asthma and skin diseases). Women in Western countries are more likely to die of a hip fracture than of breast cancer. Men suffering from prostate disease may experience associated bone problems.

Endocannabinoids, receptors (both CB_1 and CB_2), and their metabolizing enzymes are present in the skeleton and are the subject of ongoing research (Idris et al. 2009; Bab and Ofek 2011; Rossi et al. 2011; Bab 2012). CB_2 receptors are expressed in osteoblasts and osteoclasts; they stimulate bone formation and inhibit bone resorption (Bab et al. 2009). "There is a steadily growing body of evidence suggesting that the skeletal endocannabinoid system plays an important role in the regulation of bone mass in health and in disease... Future studies should...greatly enhance our understanding of the role of the skeletal endocannabinoid system in bone pathologies and encourage the development of cannabinoid-based therapy aimed at providing both anti-resorptive and anabolic effects in bone" (Idris 2010). An unanswered question is whether recreational use of marijuana is correlated with effects on the skeleton (Idris and Ralson 2010).

Aside from bone fractures due to osteoporosis, fractures in general are extremely common injuries. Kogan et al. (2015), in an experiment on fracture healing in rats, reported that "Many fractures heal by a process known as endochondral ossification. In this process initial bridging across the fracture gap is made by a cartilaginous callus that mineralizes, and is subsequently resorbed and replaced by a bony callus. The bony callus is further remodelled to form mature bone that is similar to the prefracture tissue...our results suggest that the CBD-induced stimulation of fracture healing occurs during the later phases of healing (after six weeks)."

PAIN

Pain is the most common symptom of illness and indeed constitutes an illness. Pain may be chronic (long-lasting, usually at least months) or acute (beginning suddenly, often from an injury and typically relieved when the injury heals or is treated). Analgesia—relief from pain—is the most frequent reason people seek medical treatment. Notcutt (2004) stated, "Chronic pain affects approximately 1 person in 12. For those who are over the age of 65 the incidence rises beyond 1 in 4. All the current analgesics have their problems... Sadly, having tried all the therapeutic options, many patients are left to continue their lives with inadequate or minimal control of pain."

There is a close relationship between pain signaling and the CB receptors (Woodhams et al. 2015). CB_1 receptors occur in the central and peripheral nervous systems concerned with pain. Inflammation produces considerable pain, and CB_2 receptors, which occur especially in immune system cells, seem to be particularly involved with inflammatory-related pain. Acetaminophen is the most popular drug for relief of pain and fever. It is known to produce analgesia through the CB_1 receptors (Mallet et al. 2008).

"'Severe [chronic] pain' is the most common reason for medicinal herbal cannabis use, with arthritis and musculoskeletal pain cited as the most prevalent specific medical condition" (Fitzcharles et al. 2014). Hazekamp and Grotenhermen (2010), Aggarwal (2013), and Russo and Hohmann (2013) reviewed studies reporting that cannabis relieves pain from a variety of sources.

Martín-Sánchez et al. (2009) were of the view that "Currently available evidence suggests that cannabis treatment is moderately efficacious for treatment of chronic pain, but beneficial effects may be partially (or completely) offset by potentially serious harms. More evidence from larger, well-designed trials is needed to clarify the true balance of benefits to harms." A similar cautious view was expressed by Bowles et al. (2012): "The data supporting cannabinoids for pain relief have been mixed." However, as illustrated by the following quotations, many researchers are enthusiastic about the potential value of cannabis as an analgesic. "There are well-documented reports of cannabis use leading to reduced headache, migraine and post-surgery pain" (Frisher et al. 2010). "There is evidence that cannabinoids are safe and modestly effective in neuropathic pain with preliminary evidence of efficacy in fibromyalgia and rheumatoid arthritis" (Lynch and Campbell 2011). "One of the most promising therapeutic uses of phytocannabinoids in humans is their employment as pain killers" (Costa and Comelli 2014).

The comparative analgesic value of opioids and cannabinoids is of particular interest. Carter et al. (2015) stated, "The field of pain medicine is at a crossroads given the epidemic of addiction and overdose deaths from prescription opioids. Cannabis and its active ingredients, cannabinoids, are a much safer therapeutic option...when used appropriately, cannabis is safe and effective for many forms of chronic pain and other conditions, and has no overdose levels. Current literature indicates many chronic pain patients could be treated with cannabis alone or with lower doses of opioids." The use of cannabis to complement opiates is a promising possibility. The endogenous opioid system produces pain-relieving endorphins, and the ECS may interact with the opioid system in effecting pain relief. "There are emerging data suggesting that cannabinoids augment opiates" (Bowles et al. 2012; also see Johnson et al. 2010, 2013; Abrams et al. 2011; Portenoy et al. 2012). Lucas (2012) also observed that "There is a growing body of evidence to support the use of medical cannabis as an adjunct to or substitute for prescription opiates in the treatment of chronic pain." Carter et al. (2011) noted, "Long-term drug safety is an important issue in palliative medicine. Opioids may produce significant morbidity. Cannabis is a safer alternative with broad applicability for palliative care." Burns and Ineck (2006) observed that "Chronic pain often requires a polypharmaceutical approach to management, and cannabinoids are a potential addition to the arsenal of treatment options."

Parkinson's Disease

Parkinson's disease (PD) is a chronic, progressive disorder of the central nervous system that affects movement and coordination. It is characterized by a fixed inexpressive facial expression, tremors, slowing of voluntary movements, gait with short accelerating steps, peculiar posture, and muscle weakness. Most patients are over 50. The condition is caused by death (for unknown reasons) of dopamine-producing neurons in the midbrain (dopamine is an important neurotransmitter, reduced levels interfering with coordination and motor function).

"Potential use of cannabinoids in PD is controversial" (Venderová et al. 2004). Carroll et al. (2004) found that "orally administered cannabis extract resulted in no objective or subjective improvement in dyskinesias [involuntary movement disorders] or parkinsonism." Koppel et al. (2014) concluded that cannabis is "probably ineffective" in treating PD. By contrast, Lotan et al. (2014) observed significant improvement after treatment in tremor, rigidity, bradykinesia (slow movement), sleep, and pain and stated that their study "suggests that cannabis might have a place in the therapeutic armamentarium of PD." Fernández-Ruiz et al. (2014b) stated, "it seems obvious that phytocannabinoids constitute potential novel neuroprotective therapies for PD… The phytocannabinoid Δ^9-THCV emerges as an interesting compound to be used alone or in combination with CBD." Gómez-Gálvez et al. (2015) found evidence of elevated levels of CB_2 receptors in mice models of Parkinson's as well as in postmortem patients, suggesting that this is an adaptive response that may be harnessed to alleviate the disease. Russo (2007) emphasized that prolonged administration of cannabis may be necessary before improvements in PD symptoms become evident. Concannon et al. (2015) concluded, "There is mounting evidence to suggest that the endocannabinoid system may have multiple therapeutic benefits in the treatment of PD. Evidence suggests that cannabinoid drugs have the potential to alleviate the symptoms of the condition as well as to ameliorate the debilitating motor side effects associated with current pharmacotherapies. Moreover the endocannabinoid system may also have the potential to modify the progression of the disease…our expanding knowledge of the endocannabinoid system gives cause for optimism for improved understanding and treatment of PD."

Psychiatric Disorders

Although estimates are uncertain, between one-quarter and one-third of the world's population, at some point in their lives, are believed to suffer from some form of mental disorder. Anxiety, depression, and schizophrenia are leading mental illnesses in Western countries. Electrical and neurotransmitter activities in the brain are important in determining healthy mental status, and abnormal symptoms are often correlated with abnormal nerve cell functioning. Because the ECS is vital in the health and pathology of nerve cells, cannabis has potential therapeutic value in maintaining or restoring health. However, there is hardly unanimity about the role of cannabinoids. As pointed out by Fride and Russo (2005), "The history and science of cannabis and cannabinoids and their relation to mental health and disease are fraught with controversy, ambivalence, and contradictory claims." Nevertheless, as noted by Patel and Hillard (2009), "preclinical and clinical data suggest a role for endogenous cannabinoid signaling in the modulation of anxiety and depression… These data provide evidence that ECS serves in an anxiolytic, and possibly anti-depressant, role. These data suggest novel approaches to treatment of affective disorders which could include enhancement of endogenous cannabinoid signaling."

Leweke and Koethe (2008) concluded that investigation of the ECS in relation to psychiatric conditions "provides a promising target for novel pharmacotherapeutic interventions." Similarly, Crippa et al. (2010) stated, "The cannabinoid system is a promising target for novel therapeutic interventions in psychiatry. Cannabinoids may be greatly useful in this field; however, additional controlled trials are still required to confirm these findings and determine the safety of these compounds." Less optimistically, Wyrofsky et al. (2015) commented "The endocannabinoid (eCB) system plays an important role in the control of mood, and its dysregulation has been implicated in several psychiatric disorders. Targeting the eCB system appears to represent an attractive and novel approach to the treatment of depression and other mood disorders. However, several failed clinical trials have diminished enthusiasm for the continued development of eCB-targeted therapeutics for psychiatric disorders, despite the encouraging preclinical data and promising preliminary results obtained with the synthetic cannabinoid nabilone for treating post-traumatic stress disorder."

Addiction

Addiction refers to obsessive thinking about and compulsive consumption of something despite resulting negative consequences (De Luca and Fattore 2015). Pathological compulsions for danger, drugs, food, gambling, and sex are common.

"The endocannabinoid system appears to be strongly associated with the neurobiological processes that underlie addictive disorders" (Fernández-Ruiz et al. 2014a). Cannabinoid CB_1 receptors are present in areas of the brain involved in reward processes, and when activated, they generate rewarding effects (De Luca and Fattore 2015). "Cannabinoids and endocannabinoids appear to boost the rewarding effects of addictive drugs, including alcohol, nicotine, cocaine, amphetamines, and opiates, suggesting that the endocannabinoid system may represent an important target for the treatment of addictive disorders" (De Luca and Fattore 2015). Although habit-forming drugs such as opioids, cocaine, tobacco, and alcohol are known to affect the CB_1 receptors in laboratory animals (CB_2 receptors have been found to play a role as well in cocaine addiction), therapeutic applications to humans have not yet been formulated (Fernández-Ruiz et al. 2014a).

Cannabis induces dependence less readily than the majority of other illicit drugs. Nevertheless, about 9% of marijuana users become dependent, and dependence increases up to 17% among those who initiate use at a young age (De Luca and Fattore 2015). It has been hypothesized that cannabis dependence could be addressed therapeutically by targeting the ECS (Clapper et al. 2009), a rather ironic scenario of using cannabinoids to cure cannabinoid addiction.

Anxiety

Anxiety drugs are among the top-selling prescription pharmaceuticals. Crippa et al. (2009) noted that although cannabis users have a high prevalence of anxiety disorders and, conversely, patients with anxiety disorders have relatively high rates of cannabis use, it is unclear if cannabis use increases the risk of developing anxiety disorders. "Anxiety disorders are the most common psychiatric disorders in the general population, and there is a particularly high incidence of cannabis use in patients with symptoms of anxiety and anxiety disorders…preclinical studies support the hypothesis that low doses and infrequent exposure to cannabis constituents can reduce feelings of anxiety and stress but that chronic use of large amounts has the opposite effect and could contribute to the development of anxiety and other psychiatric disorders" (Patel et al. 2014). The biphasic effects of cannabis (whether contributing to or alleviating anxiety) were examined by Rey et al. (2012). "Several studies have demonstrated a positive impact of CBD alone on anxiety in patients with social anxiety disorder" (Curran and Morgan 2014). Also see the information for posttraumatic stress disorder (PTSD) presented later in this section.

Autism Spectrum Disorder

Autism spectrum disorder, often simply called autism, includes a set of neurodevelopmental disorders characterized by complications in social interaction, impaired communication, and restricted, repetitive, stereotyped behaviors or interests (see Miles 2011 for a review). Asperger's syndrome (Asperger syndrome) is a form of autism on the higher-functioning end of the autism spectrum (IQs typically are in the normal to very superior range). No single psychopharmacologic treatment is applicable to autism, and concern has been expressed about the many toxic pharmaceutical medications that have been employed to treat the symptoms. Kurz and Blaas (2010) noted: "autistic children often show aggression against others and self-injurious behaviour, [and] also have sleep problems and eating disorders. Early infant autism affects 1 of 2000 children, with boys affected three times more often than girls. Autism does not equate with mental retardation, but intelligence is frequently limited (intelligence quotient (IQ) below 70). One quarter of autistic children achieve good results on IQ tests, termed 'high functional autism.' The cause of autism is still not fully explored, but seems to be multifactorial (including genetic, environmental and neurobiochemical disorders)… To date there have been no reports of the use of cannabinoids in autism. However, in internet blogs and discussion forums there are many reports of parents who have tried THC for their autistic children, but without medical monitoring and inappropriate administration." Grinspoon (2010) argued for the right of parents to conduct such experimentation in the light of desperation to alleviate their child's condition.

Onaivi et al. (2011), Busquets-Garcia et al. (2013), and Földy et al. (2013) suggested (based on rodent data) that the cannabinoid system is implicated in the development of autism. Khalil (2012) stated: "There has been a massive growth of public awareness and research funding around autism spectrum disorders (ASD) over the past 10 years... CBD and delta-9-THC may help in improving symptoms of ASD by their sedative, antipsychotic, anti-convulsant and tranquilizing effects." Siniscalco et al. (2013) observed a much higher development of CB_2 receptors in autistic children compared to normal children, suggesting the involvement of the ECS in autism development or maintenance.

Attention Deficit/Hyperactivity Disorder

Attention deficit/hyperactivity disorder (ADHD) is a neurodevelopmental psychiatric disorder, affecting about 40 million worldwide. ADHD is noted for lack of attention (distractability) and control of inhibitions, hyperactivity, and impulsiveness. It is the most common psychiatric condition in children and adolescents, affecting about three times as many boys as girls, beginning between the ages of 6 to 12 and often impairing school performance. The condition tends to become more manageable in adults, with a prevalence rate of 4.4% (McRae-Clark et al. 2010). The diagnosis and treatment have been controversial, especially insofar as it has involved prescribing drugs for children. The use of marijuana to treat children is particularly controversial because of the potential adverse effects on the developing brain.

Marijuana has been reported to be the most commonly used drug by adults with ADHD (McRae-Clark et al. 2010). Pedersen (2014), in a study of people who had self-diagnosed an illness and purchased marijuana for medical purposes in Norway, found that treatment of ADHD was the most commonly reported justification. Indeed, ADHD is often alleged to be the rationale for employing medical marijuana, although controlled studies of its efficacy are absent. However, there is at least an intuitive reason for using cannabis to treat ADHD. Wiskerske and Pattij (2015) pointed out, "Given its widespread and abundant expression in the brain and its known important role in other executive functions such as working memory, attention, time estimation, and behavioral flexibility, it should come as no surprise that the endocannabinoid system mediates impulsive behavior...the endocannabinoid system emerges as an interesting pharmacotherapeutic target to ameliorate impulsivity in psychopathology and neurological disorders." Strohbeck-Kuehner et al. (2008) provide an anecdotal report of improvement of ADHD in an individual treated with cannabis.

Bipolar Disorder

Bipolar disorder (bipolar affective disorder, manic-depressive illness) is a mental condition characterized by recurrent alternations of elevated mood (mania, characterized by expansive mood, decreased sleep, increased energy, and impulsive behavior) and depressed mood (depression, discussed next). About 3% of the world population develops bipolar disorder at some period of their lives. People with bipolar disorder frequently consume marijuana, surveys often showing a greater prevalence than for other misused drugs (Braga et al. 2015).

Ashton et al. (2005) commented, "Both THC and CBD have pharmacological properties that could be therapeutic in patients with bipolar affective disorder." Braga et al. (2015) stated, "advances in neurobiological research have facilitated greater understanding of the role of the endocannabinoid system in bipolar disorder and how cannabinoid modulators may serve as a treatment for the illness. To date, a few animal studies and anecdotal reports have suggested potential therapeutic effects. It should be noted that at this time, no definitive evidence suggests cannabis may be beneficial in any way."

Depression

Unlike the ups and downs that everyone experiences, the symptoms of clinical depression are severe. "Depression is one of the most common mental illnesses with a lifetime prevalence of about 15%–20%, resulting in enormous personal suffering, as well as social and economic burden. The major depressive disorder is characterized by episodes of depressed mood lasting for more than two weeks often associated with feelings of guilt, decreased interest in pleasurable activities and inability to experience pleasure (named anhedonia), low self-esteem and worthlessness, high anxiety, disturbed sleep patterns and appetite, impairment in memory and suicidal ideation" (Micale et al. 2015).

Denson and Earleywine (2006) surveyed marijuana users and suggested that "adults apparently do not increase their risk for depression by using marijuana." Patel et al. (2014) found that "studies examining the ability of cannabis and cannabinoids to reduce depression have yielded contradictory findings."

Posttraumatic Stress Disorder

PTSD is a pathological response to experiencing or witnessing severe traumatic events, such as combat, disaster, sexual abuse, violent crime, or accidents. It is characterized by mentally reexperiencing the event (frequently as flashbacks or nightmares) and often behavioral avoidance, hypervigilance, and anxiety. The condition appears to represent a situation where memories are best erased. The amygdala of the brain is important for emotional learning and memory, and endocannabinoids potentially could represent a key route to extinguishing unpleasant memories from this region of the brain. Elevated usage of marijuana has been associated with PTSD (Cougle et al. 2011).

Passie et al. (2012) stated, "Evidence is increasingly accumulating that cannabinoids might play a role in fear extinction and antidepressive effects…studies are warranted in order to evaluate the therapeutic potential of cannabinoids in PTSD." Patel et al. (2014) pointed out that "Data suggest a strong association between PTSD and cannabis use, and that subjects with PTSD use cannabis to reduce PTSD symptom severity. However, whether long-term outcomes are improved or worsened by cannabis use in PTSD patients remains to be determined." Gabbay et al. (2015) commented, "Substantial evidence suggests that endocannabinoid signaling is deficient in PTSD, thereby recommending the endocannabinoid system as a therapeutic target for that disorder. The endocannabinoid system is also a target for treatment of anxiety disorders, emotion dysregulation, depression, and cannabis use disorder, all of which are associated with PTSD. Moreover, as preclinical evidence suggests that the endocannabinoid system regulates fear memory processing and extinction, enhancing endocannabinoid function may ameliorate core deficits in PTSD. However, multiple challenges confront the effort to develop endocannabinoid-targeting therapies." Greer et al. (2014), based on a survey of patients, concluded, "Though currently there is no substantial proof of the efficacy of cannabis in PTSD treatment, the data reviewed here supports a conclusion that cannabis is associated with PTSD symptom reduction in some patients, and that a prospective, placebo-controlled study of cannabis or its constituents for treatment of PTSD is warranted." Neumeister et al. (2013) used a radioactive tracer detected by positron emission tomography to examine the brains of patients with PTSD (designated TC) as well as controls (HC). They reported: "Anandamide concentrations were reduced in the PTSD relative to the TC (53.1% lower) and HC (58.2% lower) groups… These results suggest that abnormal CB_1 receptor-mediated anandamide signaling is implicated in the etiology of PTSD, and provide a promising neurobiological model to develop novel, evidence-based pharmacotherapies for this disorder."

Schizophrenia

"Schizophrenia is a common psychiatric disorder characterized by impairments in the perception or expression of reality. Schizophrenic symptoms are subclassified into positive (or productive) symptoms such as delusions, auditory hallucinations, and thought disorder, and negative (or deficit) symptoms such as blunted affect and emotion, poverty of speech, anhedonia [inability to experience pleasure in normally pleasurable acts], and lack of motivation. For many patients the prognosis is poor with incomplete recovery and significant illness" (Müller-Vahl and Emrich 2008).

Although "there is substantial evidence that heavy cannabis abuse in healthy persons is a risk factor for the clinical manifestation of schizophrenia, CBD may be effective in the treatment of patients suffering from acute schizophrenia" (Müller-Vahl and Emrich 2008). The same view has been suggested by Leweke et al. (2012), Bossong et al. (2014), Parolaro et al. (2014), and Robson et al. (2014). Rohleder and Leweke (2015) concluded, "There is a strong relationship between the ECS, cannabis use, and schizophrenia. On the one hand, the ECS has to be regarded as part of the pathophysiology of schizophrenia, while on the other hand, cannabis use may contribute to weaken this system, a system that most likely plays a protective role in the neurobiology of this disease. Targeting this mechanism may become a viable new approach to the treatment of schizophrenia."

Skin Conditions

The skin is the largest organ of the body, its apparent simplicity belying astonishing complexity and importance. Skin has numerous functions, including protecting the body from microbial invasion, allergens, ultraviolet exposure, water, and chemicals; sensing heat, cold, touch, pressure, vibration, pain, itch, and injury; producing hair; controlling temperature (using sweat glands to control temperature; employing sense organs that regulate blood constriction to control skin temperature; containing fat that acts as insulation); and synthesizing hormones (best known of which is vitamin D).

CB_1 and CB_2 receptors occur extensively in the skin (Ständer et al. 2005), and its ECS influences various biological processes, including cell proliferation, growth, differentiation, programmed cell death, and hormonal functions (Bíró et al. 2009). Cannabinoids have been shown to have anti-inflammatory effect on skin (Tubaro et al. 2010). Pathological skin conditions and diseases that are potentially influenced by the ECS include general symptoms such as itch and pain, tumors, as well as the following conditions (Bíró et al. 2009).

Acne and Seborrhea

Acne (acne vulgaris), one of the most common skin conditions, is characterized by increased production of sebum (an oily product of the sebaceous glands, which waterproofs the skin) and inflammation of the sebaceous glands, induced or aggravated by factors such as stress and diet and common in adolescence. Seborrhea is an inflammatory skin condition particularly affecting areas of skin enriched by sebaceous glands. It is similar to acne, and indeed, acne and seborrhea are the most common dermatological diseases.

The sebaceous glands have endocannabinoid receptors, as do various skin cells, and there has been interest in the possibility that research, particularly with CBD, might produce a cannabinoid-based medicine to treat acne (Gardner 2010). Oláh et al. (2014) concluded that "CBD has potential as a promising therapeutic agent for the treatment of acne vulgaris."

Alopecia

Alopecia is a pathological hair loss particularly affecting the scalp (effluvium is a form of alopecia characterized by diffuse hair shedding).

Smoking marijuana has been claimed to reduce levels of plasma testosterone (e.g., Okosun et al. 2014), the principal androgen of men (androgens include any hormones that stimulate male reproductive traits and behavior). Consistent with this, Purohit et al. (1980) found that forms of cannabis acted as androgen antagonists, suggesting to some that baldness in men (which is exacerbated by male hormones) could be alleviated by consuming marijuana. Curiously, however, a debate is common on the Internet about whether or not consuming marijuana causes, rather than prevents, hair loss, a question that has not been decided.

The subject of hair loss is a great psychological stress to many, and cannabis has become intimately associated with this subject. Forensic departments frequently test for the illicit consumption of marijuana by sampling the cannabinoid content of hair samples, reflecting the fact that cannabinoids accumulate in hair, but it is unclear if this influences hair growth. Facetiously, smoking marijuana joints (especially roaches, i.e., the butts after most of the joint has been consumed) has caused hair loss by setting moustaches ablaze.

Dermatitis

Dermatitis is an inflammation of the skin, caused by various factors such as allergens ("allergic dermatitis"), infections, eczema ("atopic dermatitis"), external compounds ("contact dermatitis"), and so on.

Karsak et al. (2007) noted, "Allergic contact dermatitis affects about 5% of men and 11% of women in industrialized countries and is one of the leading causes for occupational diseases. In an animal model for cutaneous contact hypersensitivity, we show that… cannabinoid receptor antagonists exacerbated allergic inflammation, whereas receptor agonists attenuated inflammation. These results demonstrate a protective role of the endocannabinoid system in contact allergy in the skin and suggest a target for therapeutic intervention." Campora et al. (2012) found that "The endocannabinoid system and cannabimimetic compounds protect against effects of allergic inflammatory disorders in various species of mammals…this system may be a target for treatment of immune-mediated and inflammatory disorders such as allergic skin diseases."

Hirsutism

Hirsutism refers to excessive and increased hair growth (especially in women) on body regions where the occurrence of hair normally is minimal or absent.

Telek et al. (2007) found experimentally that they could inhibit hair growth by application of anandamide or THC.

Pruritis

Pruritis is a severe itching, which can be due to several causes.

Endocannabinoids can activate itch receptors (Bin Saif et al. 2011), suggesting the potential for therapeutic manipulation of the ECS to reduce itching. In a preliminary study, Neff et al. (2002) found THC alleviated itching in several patients with the liver disease cholestatic jaundice.

Psoriasis

Psoriasis is a chronic, autoimmune skin disease characterized by excessive production of skin cells and skin inflammation.

Wilkinson and Williamson (2006) concluded that there is "a potential role for cannabinoids in the treatment of psoriasis."

Scleroderma

Scleroderma (systemic sclerosis) is a chronic autoimmune disease characterized by diffuse fibrosis (accumulation of connective tissue), degenerative changes, and vascular abnormalities in the skin, joints, and internal organs.

Balistreri et al. (2011) found that a synthetic cannabinoid is capable of preventing skin fibrosis in a mouse model of scleroderma.

TOURETTE'S SYNDROME

Gilles de la Tourette's syndrome is a childhood-onset neurobehavioral complex of problems including motor and verbal tics and behavioral and cognitive disorders such as obsessive-compulsive behavior. Motor tics are recurrent, semivoluntary brief movements. Verbal tics may be meaningless utterances, rapid and involuntary repetition, and uncontrollable obscene language. About 0.4% of people develop Tourette's—four times as many males as females. The severity of the condition usually abates in adulthood.

Koppel et al. (2014) concluded that cannabis is of unknown efficacy in treating the symptoms of Tourette's syndrome. Other researchers have provided more optimistic evaluations. Curtis et al. (2009) stated that "Cannabinoid medication might be useful in the treatment of the symptoms in patients with Tourette's syndrome." Hazekamp and Grotenhermen (2010) concluded that studies have shown that THC has promising effects on tics and behavioral problems associated with Tourette's syndrome. Kluger et al. (2015) stated, "clinical observations and clinical trials of cannabinoid-based therapies suggest a possible benefit of cannabinoids for tics." Müller-Vahl (2013) noted that "Available data…consistently provide evidence for beneficial effects of cannabinoid-based medicines in the treatment of tics in patients with Tourette's syndrome. In addition, there is some evidence that cannabinoid-based medicines may also improve associated behavioral problems such as obsessive compulsive behavior, attention deficits, impulsivity, and self-injurious behavior in this group of patients… It can be assumed that beneficial effects of cannabinoid-based medicines in Tourette's syndrome are caused by modulations of the CB_1 receptor system, rather than unspecified effects such as sedation or decreased general activity." Müller-Vahl (2013) stated that "THC is recommended for the treatment of Tourette's syndrome in adult patients, when first line treatments failed to improve the tics. In treatment resistant adult patients, therefore, treatment with THC should be taken into consideration."

MEDICAL APPLICATION OF SPECIFIC CANNABINOIDS

Apart from the medical use of herbal marijuana, increasing efforts are underway to establish the efficacy of specific cannabinoids. THC, CBD, and CBN are the most studied cannabinoids. Howard et al. (2013) commented, "Currently available cannabinoids all contain the psychoactive constituent of *Cannabis sativa*, Δ^9-tetrahydrocannabinol (Δ^9-THC) or a synthetic analogue. They are generally less effective or less well tolerated than alternative drugs and are relatively expensive… Their analgesic effect is modest." Hill et al. (2012) stated: "Whilst Δ^9-THC is the most prevalent and widely studied pCB [phytocannabinoid], it is also the predominant psychotropic component of cannabis, a property that likely limits its widespread therapeutic use as an isolated agent. In this regard, research focus has recently widened to include other pCBs including CBD, cannabigerol (CBG), Δ^9-tetrahydrocannabivarin (Δ^9-THCV) and cannabidivarin (CBDV), some of which show potential as therapeutic agents in preclinical models of CNS disease. Moreover, it is becoming evident that these non-Δ^9-THC pCBs act at a wide range of pharmacological targets, not solely limited to CB receptors. Disorders that could be targeted include epilepsy, neurodegenerative diseases, affective disorders and the central modulation of feeding behaviour."

TETRAHYDROCANNABINOL

There is wide acceptance that THC is the main component of marijuana responsible for euphoria and mood elevation, results that may be medically advantageous for treating mood disorders. More than 100 metabolites of THC have been identified, notably 11-hydroxy-delta-9 THC (11-OH-THC). The latter produces analgesia and other psychoactive results, as well as muscle relaxant, appetite stimulant, and antiemetic (nausea reduction) effects. THC has been used to treat spasticity from spinal injury or MS, pain, inflammation, insomnia, asthma, loss of appetite, and other conditions (Izzo et al. 2009; Russo and Hohmann 2013). THC is a strong analgesic (Russo and Hohmann 2013) and can treat pain untouched by morphine-related analgesics. THC reduces intraocular pressure and has been used to treat glaucoma. Other properties of THC include anti-inflammation, bronchodilation, and antioxidation effects. THC also affects bone remodeling, fertility, short-term memory, tumor growth, and motor coordination (Mechoulam 2002; Iversen 2000), although these abilities have not been translated into therapies. THC is the main promoter of hyperphagia (hunger, producing the "munchies," traditionally satisfied with sweets and snacks), but other cannabinoids of *C. sativa* may also play a role (Farrimond et al. 2011).

CANNABIDIOL

CBD has been found to affect physiology in numerous ways and is considered to have pharmacological value for treating more pathological conditions than any other cannabinoid. This extraordinarily versatile cannabinoid has been *claimed* to have potential for therapeutic use for treatment of numerous diseases and symptoms, including Alzheimer's disease, arthritis, cancer, cerebral ischemia, congestion, convulsion, cough, diabetes, dystonia, epilepsy, hepatitis, inflammatory diseases, nausea, obesity, PD, skin diseases, and several psychological disorders including ADHD, PTSD, and psychosis (Zuardi et al. 2006; Zuardi 2008). Fernández-Ruiz et al. (2013) commented, "CBD acts in some experimental models as an anti-inflammatory, anticonvulsant, anti-oxidant, anti-emetic, anxiolytic and antipsychotic agent, and is therefore a potential medicine for the treatment of neuroinflammation, epilepsy, oxidative injury, vomiting and nausea, anxiety and schizophrenia, respectively. The neuroprotective potential of CBD, based on the combination of its anti-inflammatory and anti-oxidant properties, is of particular interest." As noted in the following, for several conditions, there has been experimental investigation of the efficacy of CBD.

CBD is credited with analgesic, anticonvulsant, antiemetic, antiepileptic, anti-inflammatory, muscle relaxant, anxiolytic (anxiety-reducing), neuroprotective, antioxidant, and antipsychotic activity (Russo and Guy 2006; Mechoulam et al. 2007). CBD also results in reduction of intraocular pressure. "CBD produces its biological effects without exerting significant intrinsic activity upon cannabinoid receptors. For this reason, CBD lacks the unwanted psychotropic effects characteristic of marijuana derivatives, so representing one of the bioactive constituents of *Cannabis sativa* with the highest potential for therapeutic use" (Scuderi et al. 2009). As stressed elsewhere in this review, CBD antagonizes the psychotropic effects of THC and so can be employed for moderating effects such as anxiety, intoxication, hunger, sedation, and increase in heart rate produced by THC (Russo et al. 2015). CBD also reduces the appetite-stimulating effect of THC (Morgan et al. 2010). Izzo et al. (2009) stated, "CBD exerts several positive pharmacological effects that make it a highly attractive therapeutic entity in inflammation, diabetes, cancer and affective or neurodegenerative diseases" and "CBD has an extremely safe profile in humans, and it has been clinically evaluated (albeit in a preliminary fashion) for the treatment of anxiety, psychosis, and movement disorders. There is good pre-clinical evidence to warrant clinical studies into its use for the treatment of diabetes, ischemia and cancer." Martin-Santos et al. (2012) concluded that "in healthy volunteers, THC has marked acute behavioural and physiological effects, whereas CBD has proven to be safe and well tolerated."

One of the more interesting studies of the potential practical value of CBD is that of Yang et al. (2014), who found that CBD protects the liver from damage from acute alcohol drinking, at least in mice.

Hemp strains are relatively high in CBD and low in THC, and so their resin is a natural source for extracted CBD. Oilseed hemp strains have often been selected to produce a congested head (to facilitate collection of seeds), but if grown in a seedless (sinsemilla) form, the flowering head could be used as "bud" material for smoking. Some medicinal strains of *C. sativa* have been selected for very high production of CBD coupled with very low THC. The Tikun Olam company in Israel developed the strain Avidekel, reportedly producing a product (debatably called "highless marijuana") containing 15.8% CBD and only traces of THC (see Chapter 8 for additional discussion of highless marijuana).

The following is a selection of conditions found to be beneficially affected by CBD. In fact, there are indications that dozens of illnesses might benefit from this remarkable cannabinoid.

Arthritis

Based on rodent studies, Malfait et al. (2000) found "that CBD, through its combined immunosuppressive and anti-inflammatory actions, has a potent anti-arthritic effect." "CBD exerts anti-arthritic actions through a combination of immunosuppressive and anti-inflammatory effects" (Izzo et al. 2009). Additional information was provided previously, in the general discussion of arthritis.

Cancer

Although as discussed previously, marijuana is not a cure for cancer, there is evidence that cannabinoids "are promising regulators of malignant cell growth" and that data "support the clinical testing of CBD against… carcinoma" (De Petrocellis et al. 2013; cf. Pacher 2013). "The non-psychoactive plant-derived cannabinoid CBD exhibits pro-apoptotic and anti-proliferative actions in different types of tumours and may also exert anti-migratory, anti-invasive, anti-metastatic and perhaps anti-angiogenic properties. On the basis of these results, evidence is emerging to suggest that CBD is a potent inhibitor of both cancer growth and spread" (Massi et al. 2013).

Diabetes

Weiss et al. (2008) found that CBD was useful in preventing diabetes in experimental mice. Izzo et al. (2009) commented, "CBD exerts beneficial actions against diabetes and some of its complications (e.g., retinal damage). The anti-inflammatory, antioxidant and neuroprotective actions of CBD could contribute to these protective effects." Di Marzo et al. (2011) conjectured that CBD may be useful in reducing the damage from diabetes in humans. Also see the information for diabetes presented earlier.

Epilepsy

The therapeutic value of CBD for epilepsy is uncertain (Zuardi 2008). Jones et al. (2010), Porter and Jacobson (2013), Cilio et al. (2014), Devinsky et al. (2014), and others have noted that CBD is anti-convulsant and deserves to be evaluated for its ability to treat the disease. Welty et al. (2014) stated, "At this time, there does seem to be a growing body of basic pharmacologic data suggesting there may be a role for CBD, especially in the treatment of refractory epilepsy…further clinical research should be wholeheartedly pursued."

Inflammatory Bowel Diseases

CBD is anti-inflammatory (Bowles et al. 2012) and so has potential value for treating inflammatory diseases. Commenting on the value of CBD for treating inflammatory bowel diseases, Esposito et al. (2013) stated that it "possesses an extraordinary range of beneficial effects that may slow the course of the disease, ameliorate symptoms and potentially increase the efficacy of the drugs actually available for the therapy of invalidating gut disorders such as ulcerative colitis or Crohn's disease."

Nausea and Vomiting

Rock et al. (2012) commented, "CBD acts in a biphasic manner, such that low doses suppress toxin-induced vomiting but high doses potentiate or have no effect on vomiting."

Neurodegenerative Diseases

"CBD is a well-known antioxidant, exerting neuroprotective actions that might be relevant to the treatment of neurodegenerative diseases, including Alzheimer's disease, Parkinson's disease and Huntington's disease" (Izzo et al. 2009).

Pain

Costa et al. (2007), on the basis of rodent studies, found that "Cannabidiol, the major psycho-inactive component of cannabis, has substantial anti-inflammatory and immunomodulatory effects" against pain.

Psychiatric Disorders

CBD has been found to be anxiety-reducing and has significance for this and other psychological problems, including depression and psychosis (Campos et al. 2012). "Experimental results suggest that it exerts antipsychotic actions and is associated with fewer adverse effects compared with 'typical antipsychotics'" (Izzo et al. 2009). "Most clinical studies with normal subjects or schizophrenic patients suggest that CBD has antipsychotic properties" (Campos et al. 2012). "Evidence suggests that CBD can ameliorate...symptoms of schizophrenia" (Deiana 2013). Leweke et al. (2012) found that CBD "exerts clinically relevant antipsychotic effects that are associated with marked tolerability and safety, when compared with current medications." Campos et al. (2012) concluded, "CBD is a safe compound with a wide range of therapeutic applications, including the treatment of psychiatric disorders. These findings make this drug an attractive candidate for future clinical use" and "this natural cannabinoid should be considered as a potential approach for the treatment of mood disorders." Rock et al. (2012) commented that CBD "has been shown to protect against cerebral ischaemia, inflammation, anxiety and, most recently, depression and even addiction." De Mello Schier et al. (2014) noted that "Anxiety and depression are pathologies that affect human beings in many aspects of life, including social life, productivity and health. Cannabidiol is a constituent non-psychotomimetic of *Cannabis sativa* with great psychiatric potential, including uses as an antidepressant-like and anxiolytic-like compound." Schubart et al. (2014) stated, "Although cannabis use is associated with an increased risk of developing psychosis...evidence from several research domains suggests that CBD shows potential for antipsychotic treatment." Prud'homme et al. (2015) reviewed the literature on addictive behavior in relation to CBD and concluded that "preclinical studies suggest that CBD may have therapeutic properties on opioid, cocaine, and psychostimulant addiction, and some preliminary data suggest that it may be beneficial in cannabis and tobacco addiction in humans. Further studies are clearly necessary to fully evaluate the potential of CBD as an intervention for addictive disorders." Curran and Morgan (2014) consider CBD to be "a candidate treatment for disorders of pathological fear memory, such as posttraumatic stress disorder and phobias." "CBD bears investigation in...neuropsychiatric disorders, including anxiety, schizophrenia, addiction, and neonatal hypoxic-ischemic encephalopathy. However, we lack data from well-powered double-blind randomized, controlled studies on the efficacy of pure CBD for any disorder" (Devinsky et al. 2014).

CANNABINOL

CBN (recall that this is considered to be a degenerative artefact of THC) is weakly psychotropic (Russo 2007). It has been shown experimentally to have anticonvulsant and anti-inflammatory effects, ability to promote skin and bone growth, decrease heart rate and intestinal motility, and inhibit platelet aggregation and is considered to have some medical potential (Izzo et al. 2009).

OTHER NATURAL CANNABINOIDS

All cannabinoids may have pharmacological value, but in addition to those mentioned previously, the following, which are relatively common in some strains, are most often mentioned as having medical potential (for discussion, see Russo 2011a).

Tetrahydrocannabivarin

Δ^9-THCV, a nonpsychoactive analogue of THC, can block cannabinoid receptors. "THCV seems to be a promising therapeutic compound because it has been shown to behave as a CB_1 receptor antagonist; at the same time, it activates CB_2 receptors, thereby decreasing inflammation and oxidative stress" (Horváth et al. 2012). Similarly García et al. (2011) observed, "Given its antioxidant properties and its ability to activate CB_2 but to block CB_1 receptors, $\Delta9$-THCV has a promising pharmacological profile for delaying disease progression in Parkinson's disease and also for ameliorating parkinsonian symptoms." McPartland et al. (2015) reviewed medical research on THCV and emphasized that its effects on CB receptors are complex, and it may demonstrate contradictory actions in in vitro (test tube) and in vivo (living animal) experiments. The following rodent studies suggest that it has therapeutic value. THCV has the capacity to reduce the intoxication of THC and has been shown to have anticonvulsant and anti-inflammatory properties (Bolognini et al. 2010), and to be capable of inducing weight loss in diabetic mice (Wargent et al. 2013). Hill et al. (2010), based on rat studies, demonstrated that THCV suppresses epileptic and seizure activity. Horváth et al. (2012) suggested that THCV may have value in treating diabetes. Cascio et al. (2015) found that THCV has apparent antipsychotic effects and that it "has therapeutic potential for ameliorating some of the negative, cognitive and positive symptoms of schizophrenia."

Cannabigerol

The nonpsychotropic cannabinoid CBG has some sedative, antibiotic, antifungal, antidepressant, and antihypertensive effects and decreases intraocular pressure. Borrelli et al. (2013), based on a study demonstrating therapeutic effects of CBG on inflammatory bowel disease in mice, suggested that it should be examined experimentally in patients. De Petrocellis et al. (2008) discussed the possibility of CBG having therapeutic effects on prostate cancer. Based on an in vivo (cell culture) study, Borrelli et al. (2014) found that CBG hampers colon cancer progression and inhibits the growth of colorectal cancer cells, and they recommended that CBG should be considered in the prevention and cure of colorectal cancer.

Cannabichromene

Cannabichromene has been found to exert anti-inflammatory, antimicrobial, antifungal, antidepressant, sedative, and analgesic activity (Izzo et al. 2009, 2012), based mostly on rodent studies.

CAUTIONS REGARDING MEDICAL MARIJUANA

Extensive negative effects of cannabis usage have been documented (as noted in Chapter 12). The variety of behavioral, psychological, and physical symptoms that are sometimes associated with the consumption of cannabis drugs are reviewed extensively in the references cited previously, as well as in the previous chapter. The following highlights some of the more significant concerns in a medical context.

The effects of cannabis on reproduction have not been completely evaluated. Toxicological studies have indicated that (in acute application studies in laboratory animals) cannabinoids can reduce the weight of sex organs and increase the weight of liver and adrenal glands, possibly through effects on sex hormones. Cannabis use during pregnancy has been alleged to slightly reduce birth weight of the baby (Hall 2014), although such studies rarely control for use of alcohol or cigarettes or check nutritional status. THC is transmitted in breast milk and can cross the placenta, reaching concentrations in the fetus of 10% to 30% of maternal levels (Holubek 2010). Accordingly, use during pregnancy and nursing seems (especially) contraindicated (but see information in the previous section dealing with morning sickness).

Cannabinoids are very lipid soluble and accumulate in fatty tissue throughout the body. They are released very slowly and so can persist in people for more than a month after consumption. This does not necessarily suggest toxicity but needs to be considered in relation to long-term use and dosage.

Wang et al. (2008a) reviewed adverse effects of the medical use of cannabis. Of about 5000 adverse effects, most were not serious. As noted previously, death in humans due directly (i.e., short-term) to cannabis has not been reliably determined. Dizziness was the most commonly reported nonserious occurrence. Of the 164 serious events, the most frequent was relapse of MS, vomiting, and urinary tract infection. These authors concluded that "the risks associated with long-term use were poorly characterized in published clinical trials and observational studies." Grant et al. (2012) noted the following risks in the medicinal use of cannabis: "Over the longer term cannabis may have unwanted systemic and psychoactive adverse effects that must be taken into consideration in chronic pain populations, who have high rates of co-occurring medical illness (e.g., cardiovascular disease) and co-morbid psychiatric and substance use disorders. In general these effects are dose-related, are of mild to moderate severity, appear to decline over time, and are reported less frequently in experienced than in naïve users. Reviews suggest the most frequent side effects are dizziness or lightheadedness (30%–60%), dry mouth (10%–25%), fatigue (5%–40%), muscle weakness (10%–25%), myalgia (25%), and palpitations (20%). Cough and throat irritation are reported in trials of smoked cannabis. Tachycardia and postural hypotension are infrequent but caution is warranted in patients with cardiovascular disease, and possibly younger adults who intend to embark on very vigorous physical activity. At higher doses, sedation and ataxia with loss of balance are frequent. Participants in some but not all studies report euphoria…There can be adverse psychiatric side effects. THC intoxication and euphoria can be disturbing, particularly to elderly patients. Anxiety and panic attacks occur, as do frank psychotic reactions (principally paranoia)." Borgelt et al. (2013) concluded, "Safety concerns regarding cannabis include the increased risk of developing schizophrenia with adolescent use, impairments in memory and cognition, accidental pediatric ingestions, and lack of safety packaging for medical cannabis formulations" and "Extreme caution should be used in patients with a history of cardiovascular disease or mental disorders and in adolescents." The small but real possibilities of cannabis usage triggering stroke or myocardial infarction are discussed by Mittleman et al. (2001), Mukamal et al. (2008), Renard et al. (2012), Wolff et al. (2013), and Thomas et al. (2014b) (also see the discussion of cardiovascular diseases presented earlier). Callaghan et al. (2013) reported that lifetime smoking of marijuana greater than 50 times was associated with more than a twofold risk of developing lung cancer (a conclusion disputed, for example, in Tashkin 2013). Risk of genotoxicity (genetic damage, including mutation) and carcinogenicity is considered to be low. "Although there have been suggestions that cannabis has adverse long-term effects on pregnancy, the immune system, fertility, and cognition, the preponderance of available evidence suggests that these are far less severe than originally thought" (Iverson 2002). "The evidence suggests that it is more likely than not that cannabis use precipitates schizophrenia in vulnerable persons… There is also some evidence that cannabis use is associated with increased likelihood of relapse to psychosis among those who have developed a psychotic disorder" (Degenhardt et al. 2013). In addition to the hazards reviewed here, see the information on "Health Risks" concerning recreational use, presented in the previous chapter.

LIMITATIONS OF EXPERT MEDICAL GUIDANCE IN THE USE OF MEDICAL MARIJUANA

Russo et al. (2015) summed up limitations on prescribed cannabis as follows: "Physicians often lack training in using botanical medicines, and endocannabinoid physiology is still absent from most medical school curricula. Many legal cannabis patients receive permission to use cannabis from their physician, but must rely on formula selection and dosing instructions provided by cannabis growers or dispensary staff with little training or experience."

ALLERGIES DUE TO *CANNABIS SATIVA*

"Allergy to marijuana is generally considered to be rare" (Tessmer et al. 2012). Nevertheless, as discussed in the following, a variety of allergic problems have been recorded (note Tennstedt and Saint-Remy 2011).

Atmospheric "aeroallergens" are a major source of allergies for people. Hemp pollen is widely distributed (as discussed in Chapter 4), and although not among the important sources of hay fever, it is a significant allergen for some people (Maloney and Brodkey 1940; Lindemayr and Jager 1980; Freeman 1983; Tanaka et al. 1998; Stokes et al. 2000; Singh and Kumar 2003; Mayoral et al. 2008;

Swerts et al. 2014). As with other air-borne pollen, inhalation of pollen of *C. sativa* has been observed to cause allergic rhinitis, asthma, and conjunctivitis. Wild plants and hemp plantations, not cultivated marijuana, are likely to be the major sources of outdoor *C. sativa* pollen. Sinsemilla marijuana is generated in the absence of male plants, so it should not be contaminated with pollen. Dust resulting from processing plants of *C. sativa* can also cause occupational allergic reactions (both pulmonary and cutaneous) to develop, both for hemp and marijuana, although not commonly. Skin (cutaneous) allergies (urticaria, pruritus) from handling plants have been recorded (Majmudar et al. 2006; Williams et al. 2008; Tessmer et al. 2012; Rojas Pérez-Ezquerra 2015). Several authors have suggested that cannabinoids, particularly THC, could directly cause allergies (reviewed in Ocampo and Rans 2015), but this has not been verified.

Pin prick allergy tests using macerated preparations of *C. sativa* have been employed to determine the proportion of people in various areas of the world who have become sensitized to the plant, since increasing cultivation and usage would suggest that there is increasing exposure to it. Larramendi et al. (2013) found that 8% of a sample of people in Spain were sensitized to cannabis. Sensitization to cannabis can take place indirectly through the phenomenon of cross-reactive sensitization—i.e., by exposure to a material which not only induces sensitivity to itself but also to cannabis. Armentia et al. (2011) found that an astonishing 92% of patients sensitive to tomato were also sensitive to cannabis (cf. De Larramendi et al. 2008). It has been suggested that the increasing use of cannabis has induced sensitivity to common foods (Ebo et al. 2013).

IS MARIJUANA A BONA FIDE MEDICINE? TENTATIVE CONCLUSIONS

For better or worse, marijuana is being dispensed as a therapeutic agent (Figure 13.24), albeit only in some jurisdictions. This section reviews its relative legitimacy for various conditions.

Swiss physician and alchemist Paracelsus (1493–1541; Figure 13.25) stated: "Everything is a poison. The difference between a poison and a medicine depends on the dose." This profoundly insightful observation is taught to every medical student, but knowing that marijuana has both toxic and therapeutic properties doesn't settle the simple question of whether it is a good medicine or a treatment of choice. As noted previously, marijuana and its constituent cannabinoids can alter human physiology in ways that may be medically advantageous for curing certain pathological states. However, not all medicines are equally efficacious for all diseases or for all people or for all circumstances, and the wisdom of utilizing cannabis for any particular person or malady, instead of alternative treatments, is unsettled. Unfortunately, medicinal aspects of cannabis are intertwined

FIGURE 13.24 Medical marijuana—a new alternative to established therapies, requiring evaluation using modern techniques. Prepared by B. Brookes.

FIGURE 13.25 Paracelsus, the "Father of Toxicology," whose observations bear on the medicinal nature of marijuana. This public domain painting is one of many copies of the lost original by artist Quentin Matsys (1466–1530).

with its popularity as a recreational euphoric and symbol of personal freedom on the one hand and, on the other, concern over the harmful effects of dangerous drugs on individuals and society. It is embarrassingly clear that in the name of legitimate medicine, there is widespread prescribing of medical marijuana simply for personal enjoyment. It is equally clear that the harmful effects have been so exaggerated by establishment authorities that trust in research findings has been eroded.

Hazekamp and Pappus (2014) stated: "According to some, herbal cannabis, also known as marijuana, is a substance whose abuse potential is well documented, but whose benefits are poorly characterized. However, this view overlooks the fact that the harmfulness of cannabis abuse is not as widely accepted as often assumed, and that some therapeutic effects claimed by patients are, in fact, clinically supported and sometimes even produced by registered [cannabis] medicines... As a result, it seems hard to reach any comfortable consensus on where the line may be drawn between the appropriate medical use and the abuse of this plant. Instead, what we observe is an interesting polarization of opinions on cannabis." Mendizábal and Adler-Graschinsky (2007) observed, "As in the case of the history of the therapeutic use of opioids, cannabinoid research is still the focus of legal and moral controversy, an issue that has powerfully contributed to the delay in the clinical application of these drugs."

As noted in this chapter, there is experimental evidence that cannabis drugs are clinically useful for alleviating nausea, vomiting, and appetite problems, especially following radiation therapy and chemotherapy, notably for cancer and AIDS patients; for relieving the tremors and/or spasticity associated with MS and other neurodegenerative conditions; and for pain relief. As well, there is substantial anecdotal evidence that cannabis drugs are potentially useful for a host of other conditions and symptoms. However, Turcotte et al. (2010) concluded, "It is important that cannabinoids not be considered 'first-line' therapies for conditions for which there are more supported and better-tolerated agents. Instead, these agents could be considered in a situation of treatment failure with standard therapies or as adjunctive agents where appropriate." At present, cannabis is not considered a first-line therapy for any medical condition. Wilkinson and D'Souza (2014) stated: "Medical marijuana differs significantly from other prescription medications. Evidence supporting its efficacy varies substantially and in general falls short of the standards required for approval of other drugs by the US Food and Drug

Administration (FDA). Some evidence suggests that marijuana may have efficacy in chemotherapy-induced vomiting, cachexia in HIV/AIDS patients, spasticity associated with multiple sclerosis, and neuropathic pain. However, the evidence for use in other conditions—including posttraumatic stress disorder, glaucoma, Crohn disease, and Alzheimer disease—relies largely on testimonials instead of adequately powered, double-blind, placebo-controlled randomized clinical trials. For most of these conditions, medications that have been subjected to the rigorous approval process of the FDA already exist." Hill (2015) surveyed the medical literature on cannabis and concluded, "Use of marijuana for chronic pain, neuropathic pain, and spasticity due to multiple sclerosis is supported by high-quality evidence… Medical marijuana is used to treat a host of indications, a few of which have evidence to support treatment with marijuana and many that do not. Physicians should educate patients about medical marijuana to ensure that it is used appropriately and that patients will benefit from its use."

Whiting et al. (2015) examined the literature on cannabis with respect to benefits and adverse effects. Specifically examined were randomized clinical trials dealing with nausea and vomiting due to chemotherapy, appetite stimulation in HIV/AIDS, chronic pain, spasticity due to MS or paraplegia, depression, anxiety disorder, sleep disorder, psychosis, glaucoma, and Tourette's syndrome. They concluded that "There was moderate-quality evidence to support the use of cannabinoids for the treatment of chronic pain and spasticity. There was low-quality evidence suggesting that cannabinoids were associated with improvements in nausea and vomiting due to chemotherapy, weight gain in HIV infection, sleep disorders, and Tourette syndrome. Cannabinoids were associated with an increased risk of short-term adverse effects."

Friedman and Devinsky (2015) reviewed the medical literature relating cannabis and epilepsy and concluded "The use of medical cannabis for the treatment of epilepsy could go the way of vitamin and nutritional supplements, for which the science never caught up to the hype and was drowned out by unverified claims, sensational testimonials, and clever marketing." Cannabis is currently being similarly advocated for dozens of illnesses, without adequate demonstration of effectiveness, and it is inevitable that some conditions will not prove to be usefully treated with it.

Despite thousands of research studies and the recent publication of dozens of comprehensive science-based reviews regarding the pros and cons of medical marijuana, there is not yet a consensus that any particular cannabis-based treatment is preferable to other available therapies, at least for the majority of patients. Testimonials to the efficacy of medical marijuana are widespread, but there are also cautions regarding possible significantly deleterious effects, particularly on youth and psychologically susceptible individuals.

However, there is a consensus that further research is required. Almost every recent published analysis of the medical nature of cannabis (including this book) concludes that more research is needed to assess the relative costs/benefits of marijuana employed for therapeutic purposes. Indeed, study is required before it can be concluded indisputably that cannabis is the most appropriate therapy for any particular condition or person.

Aggarwal et al. (2009) nicely evaluated the controversy over medical marijuana: "Arguably cannabis is neither a miracle compound nor the answer to everyone's ills. Yet it is not a plant that deserves the tremendous legal and societal commotion that has occurred over it." At least a minority of medical marijuana patients and their caregivers become so convinced of the value of cannabis that further discussion becomes fruitless; conversely, the view that marijuana is an unredeemable drug of abuse remains prevalent. Given the controversies and emotion associated with marijuana, it is probably impossible to state an evaluation of its current and potential medicinal value that will satisfy a majority of society or even the most informed minority. Nevertheless, the following tentative conclusion is offered. Medicinal cannabis and technologies for its usage currently appear to possess modest value for treating some illnesses but may have much greater potential for alleviating a wide variety of ailments. The endocannabinoid system of the body has such profound significance for human health that its medical manipulation may one day, with considerable probability, have outstanding therapeutic value. The persisting conviction that accepting marijuana as a legitimate medicine compromises society's "War on Drugs" remains a roadblock to research. Smoking and the consequent exposure to hundreds of toxic chemicals is universally viewed as extremely undesirable medically, and the production of new pharmacological products that deliver cannabinoids efficiently is a key to the future acceptance of medicinal marijuana.

MEDICAL-ETHICAL PERSPECTIVES OF LEGALIZATION OF MEDICAL MARIJUANA

There are endless numbers of publications arguing the merits for and against allowing the usage of medical marijuana from an ethical perspective. Aside from perspectives based exclusively on "human rights," the viewpoints of physicians and medical ethicists are particularly important because they are actually confronted with a wide variety of challenging situations. Thompson and Koenen (2011) noted: "There are compelling arguments both for and against the use of medical cannabis. Those who support its medical use argue that marijuana can be an effective medication to reduce suffering for patients who have exhausted all other means of treating a condition. Those who argue against the medical use of marijuana cite the lack of data on its safety and efficacy, an ever-expanding list of conditions that the drug is purported to treat, and fear that recommendations for medical marijuana are a physician's blessing for drug abuse."

The following statements are representative of the viewpoint that medical marijuana should not be available:

> There is some evidence to support the use of marijuana for nausea and vomiting related to chemotherapy, specific pain syndromes, and spasticity from multiple sclerosis. However, for most other indications…such as hepatitis C, Crohn disease, Parkinson disease, or Tourette syndrome, the evidence supporting its use is of poor quality… Evidence justifying marijuana use for various medical conditions will require the conduct of adequately powered, double-blind, randomized, placebo/active controlled clinical trials to test its short- and long-term efficacy and safety… Since medical marijuana is not a life-saving intervention, it may be prudent to wait before widely adopting its use until high-quality evidence is available to guide the development of a rational approval process. (D'Souza and Ranganathan 2015)
>
> Empirical and clinical studies clearly demonstrate significant adverse effects of cannabis smoking on physical and mental health as well as its interference with social and occupational functioning. These negative data far outweigh a few documented benefits for a limited set of medical indications, for which safe and effective alternative treatments are readily available. If there is any medical role for cannabinoid drugs, it lies with chemically defined compounds, not with unprocessed cannabis plant. Legalization or medical use of smoked cannabis is likely to impose significant public health risks, including an increased risk of schizophrenia, psychosis, and other forms of substance use disorders. (Svrakic et al. 2012)

The following statement is representative of the viewpoint that medical marijuana should be available:

> Recent studies have shown that medical marijuana is effective in controlling chronic non-cancer pain, alleviating nausea and vomiting associated with chemotherapy, treating wasting syndrome associated with AIDS, and controlling muscle spasms due to multiple sclerosis. These studies state that the alleviating benefits of marijuana outweigh the negative effects of the drug, and recommend that marijuana be administered to patients who have failed to respond to other therapies… After reviewing relevant scientific data and grounding the issue in ethical principles… there is a strong argument for allowing physicians to prescribe marijuana. Patients have a right to all beneficial treatments and to deny them this right violates their basic human rights. (Clark et al. 2011)

CURIOSITIES OF SCIENCE, TECHNOLOGY, AND HUMAN BEHAVIOR

- The medical use of cannabis as a suppository dates back to ancient Egypt (Russo 2003a). Today, chemical preparations of THC (hemisuccinates; THC itself is poorly absorbable rectally) are widely used in medical cannabis suppositories, providing good bioavailability and sparing patients the problems associated with oral and pulmonary application (Walker et al. 1999). The subject is "the butt of puns," including "the posterior is superior," "back-door medicine," and "bend over, it's time to get high."

- In ancient China, priest-doctors used hemp stalks into which snake-like figures were carved as an aide to curing the sick. The sickbed was pounded with the stalks in an attempt to drive away the demons responsible for the sickness.
- A belief that has persisted to modern times in the Himalayan region is that if a cobra is killed and buried, and *Cannabis* seeds are grown on the site, they will yield extremely potent forms of marijuana, which can be used for medicinal purposes, especially to treat tuberculosis.
- Hempseed was commonly used in folk medicine in Europe. In some parts of Eastern Europe, doctors instructed patients whose gums and teeth were thought to be infested with worms to inhale hempseed fumes so that the parasites would become intoxicated and fall out (Benet 1975).
- On Saint John's Eve in Europe, farmers would feed hemp flowers to their livestock in the belief that this protected the animals from evil and sickness.
- In the Balkans, an ancient folk ritual intended to cleanse people of diseases, and practiced until the early part of the twentieth century, was to run through a circle of burning hemp.
- In eighteenth century Britain, hempseed oil in milk was used to remedy venereal diseases.
- Free samples of medical *Cannabis* preparations were offered to visitors to the 1876 American Centennial Exhibition in Washington, DC.
- Sir Richard Reynolds, personal physician of Queen Victoria (1819–1901), is widely claimed to have prescribed an alcoholic tincture of *Cannabis* to treat her severe menstrual cramps (a report that has been disputed).
- As noted by Benet (1975), in central Asia in the early twentieth century, marijuana oint-ments were employed to alleviate the pain of wedding night defloration and "to shrink the vagina and prevent fluor alvus" (a disease caused by infection or inflammation of the genitals, commonly suffered by women). Male genitals also benefitted from comparable preparations of marijuana—in youth, to relieve the pain of circumcision and, in old age, as an aphrodisiac.
- Endocannabinoid is often abbreviated ECB. ECB is also a common abbreviation or acro-nym for many names, most often "European Central Bank" (the central banker for coun-tries in which the Euro is the standard currency). Among the more amusing meanings of ECB are "East Coast Bozoism" ("a syndrome that is seen in politics, economics, and business seen within DC and other eastern elite and power circles") and the "England and Wales Cricket Board" (which explained that EWCB is too long, "because the public don't identify with four initials as readily as with three").
- "Cloninger type 1 alcoholics" are characterized by anxiety-prone temperaments and late onset of alcohol problems, while type 2 alcoholics (about 20% of alcoholics) are strongly influenced by heredity and exhibit teenage-onset heavy drinking and antisocial behav-ior such as impulsive risk-taking and violent behavior. Based on brain postmortems, Lehtonen et al. (2010) found that anandamide levels in type 1 alcoholics were significantly low, but significantly high in the rarer type 2. These results suggest that most alcoholics might be better advised to use cannabis to compensate for anandamide deficiency rather than employing alcohol as their drug of choice. Curiously, in the early twentieth century, *C. sativa* was "used for the cure of chronic alcoholics in central Asia quite successfully" (Benet 1975). Russo (2011a) discussed the possibility that certain terpenes in marijuana can serve as short-term antidotes to alcohol intoxication.

14 Medical Marijuana: Production

THE NEED FOR HIGH STANDARDS OF PROFESSIONALISM

Gardening is the most popular leisure activity in the world. It is relatively easy to cultivate plants outdoors, somewhat more difficult indoors. Most people can grow plants of reasonable but rarely outstanding quality in both circumstances. Professional horticulturalists, however, employ sophisticated equipment and techniques to maximize yield, quality, and efficiency of production. Many illicit growers have in fact acquired a remarkable degree of skill and knowledge, and are in demand currently to oversee large commercial marijuana production enterprises that are being established, as shown in Figure 14.1. However, a very high degree of horticultural professionalism is now essential to compete in the legal marijuana production industry, and self-taught amateurs experienced in illicit clandestine techniques do not necessarily have the background that best suits a legal industry.

In particular, production of plants for consumable purposes, especially for medicinals, requires that the material be generated using "good agricultural practice" (GAP). GAP is variously defined but is generally understood to include responsible principles of cultivation, harvesting, and processing that result in safe products while giving due consideration to social and economic conditions, as well as environmental sustainability. As reviewed in Chapter 12, illicitly produced street marijuana is frequently contaminated or adulterated, sometimes dangerously so. As reviewed in Chapter 16, illicitly produced street marijuana is typically produced with flagrant disregard for the safety of the environment. While the increasing legalization of cannabis offers those who have engaged in criminal generation of marijuana the opportunity to engage in legitimate production, the industry needs to engage personnel with high ethical standards.

FUNDAMENTALS OF MARIJUANA PLANT PRODUCTION

OVERVIEW OF PRODUCTION

Techniques for the generation of high-grade marijuana have now become more or less standardized—in governmental and authorized private sector production, as well as in the illicit trade. There are well over a hundred books that provide directions for the cultivation of marijuana plants and consequent preparation of cannabis products. There is also an enormous "grey" literature (i.e., not like the kind of reputable documents produced by accepted authors and found in most libraries), much of it on the Internet. Despite the counter-culture nature of most such publications (really intended for the illicit trade) and the presence of some inaccurate and contradictory information, these sources are generally useful. Particularly competent or informative guides to aspects of marijuana cultivation include Clarke (1977, 1981), Drake (1979), Carver (1997), Frank (1997), Rosenthal et al. (1997), Rosenthal (1998), Green (2003), and Cervantes (2006, 2015). Note that a number of techniques (described in Chapter 6) are employed by illicit growers to maximize productivity in very small grow rooms, but these methods are of limited usefulness in professional large-scale facilities. A brief review of pertinent information on current production of marijuana follows.

Seedless marijuana is known in the illicit trade as "sinsemilla," Spanish for seedless. Since seeds are virtually without tetrahydrocannabinol (THC) and serve to adulterate or reduce the quality of the product, this generally means that seedless material is comparatively potent. Sinsemilla marijuana has become the standard form of intoxicating herbal cannabis. Female plants are cultivated in the absence of males, so they are protected against receiving pollen and do not develop seeds. As discussed in Chapter 11, males have lower levels of cannabinoids and are less productive than females. More importantly, because seeds are virtually devoid of cannabinoids, their presence

FIGURE 14.1 One of the growth facilities of GW Pharmaceuticals (U.K.), using a combination of natural and artificial lighting. Reproduced with permission of GW Pharmaceuticals plc.

substantially dilutes the female plant's production of cannabinoids, the chief desired chemicals. As noted in Chapter 4, sex expression is variable in *Cannabis sativa*, and it is necessary to check plants to ensure that some male flowers do not develop.

For the most part, once desirable female plants have been identified, they are reproduced asexually, yielding identical (i.e., cloned) individuals. The use of cloned plants contributes to the production of a uniform and predictable grade of marijuana. Branch cuttings (with a meristem, i.e., a growing point) are harvested, rooted, and grown to a desirable size and then are induced to flower by increasing the dark period of the 24-hour daily cycle. (As noted in Chapter 5, most strains of *C. sativa* are composed of short-day plants, photoperiodically adjusted to initiate flowers in the late summer and early autumn when day length shortens. As also noted, daily exposure to 13–14 hours of darkness causes most marijuana strains to come into flower.) In the absence of male plants producing pollen that normally fertilizes the female flowers, resulting in the production of seeds, the flowering branches continue to abundantly develop female flowers, each with a perigonal bract covered by secretory glands producing cannabinoids. The flowering branches become strongly covered with flowers (at which point the congested terminal branch systems are known in the illicit trade as "bud"). While much of the plant could be collected to produce marijuana, increasingly only the buds are collected for marijuana (although much of the above-ground plant is also used to extract cannabinoids).

Harvest time is based on judgment of when a maximum concentration of THC has developed. The secretory glands mature as the flowers mature, overmature glands tending to have decreased levels of THC. In the illicit cannabis drug trade, it is well known that marijuana plants need to be harvested when many of the glands are mature (which occurs when many of the flowers are mature) but while there are not too many overmature glands present. Some growers will harvest the terminal inflorescence first, since it matures relatively quickly, and allow the slower maturing lower

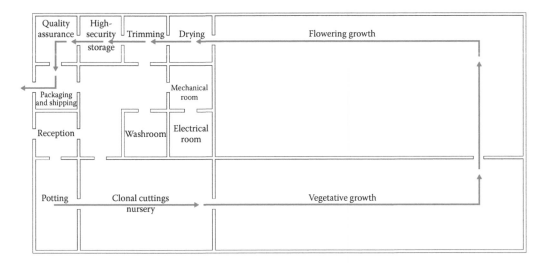

FIGURE 14.2 Simplified layout of a professional marijuana production facility. The red arrows indicate the sequence of production and processing. Designed by Steve Naraine.

flowering portions more time to develop (usually about an extra week). Should the ideal harvest period be exceeded, more material will be generated (because of plant growth), but that material will be lower in THC because of gland deterioration.

After harvest, plants are dried for several days, typically in a separate room. The buds may be individually picked and the larger leaves trimmed away with scissors (a labor-intensive procedure providing considerable employment; trimmings are frequently kept as raw material for extracts). Subsequently, the buds can be pulverized or screened, although intact trimmed buds usually fetch a premium price. The flowering material from the dried plants can simply be stripped off, followed by pulverizing. The final product is placed in packages of standard weight, or ground material can be made into cigarettes.

Large-scale greenhouse operations for ornamental, edible, and medicinal crops are normally automated to reduce labor and ensure strict control of schedules and environmental variables. This includes automatic implementation of regimes for lighting, ventilation, and fertigation (combining irrigation and liquid-soluble fertilization). A simplified plan showing essential rooms required for a professional marijuana production facility is shown in Figure 14.2.

As with all commercial endeavors in which several factors contribute to the final product, maximum profits are achieved by a compromise of inputs. Marijuana production efficiency primarily depends on an economic combination of light intensity, planting density, and strain of marijuana.

PLANTING DENSITY AND YIELDS

Whether indoors or outdoors, density of planting affects yield of marijuana harvestable from a given plant or from a given area. The best planting density for a given strain needs to be determined by trial and error, but the information presented in the following provides general guidelines. In most circumstances, the area available to growers is limited, and yield per unit area is of far greater importance than the amount of material harvested from a given plant. Yield is best expressed in terms of usable dried marijuana produced per unit area; in practice, the "dry weight" (air-dry weight) of marijuana can range from 10% to 15% of "fresh" or "green" weight.

Indoor Yields

For competitive reasons, commercial producers of marijuana usually do not reveal yield data. Vanhove et al. (2011) found that yield/m^2 did not differ between densities of 16 plants/m^2 and

20 plants m² (of course, the lower density produced larger plants). Toonen et al. (2006) surveyed yield of illicitly grown indoor sinsemilla (seedless) *C. sativa* in the Netherlands and found that based on a mean plant density of 15/m² and an illumination of 510 watts/m², the mean yield for a single harvest of buds was 33.7 g/plant or 505 g/m². Toonen et al. (2006) noted that four to six indoor harvests per year are possible, and the THC level of buds was about 15%. They recommended a density of 32 plants/m² for optimal yield. Also based on the Netherlands indoor plants, Leggett (2006) suggested average relative ratios of wet, dry, and bud material of 10:3:1 (i.e., wet weight of plant:dry weight of plant:dry weight of harvested buds). Bedrocan, the sole authorized supplier of medical marijuana to the Netherlands government since 2001, reported an average yield of 315 g/m² for their strain Bedrocan and 251 g/m² for their strain Bedica, employing a plant density of 2.33 per m² and a light intensity of 423 W/m² (Vanhove et al. 2011). Bedrocan's Bedrocan strain is a sativa type variety, while their strain Bedica is an indica type, the former longer-flowering, contributing to the higher production. The yield figures in Caulkins et al. (2012) and Hawken and Prieger (2013) vary from 325 to 430 g/m²/harvest.

In summary, the literature indicates that yields usually range from about 250 to about 500 g/m²/harvest, with about four harvests expected annually.

Outdoor Yields

Leggett (2006) suggests a representative figure of 100 g/m² (or 1 tonne/ha) for yield of outdoor crops, although five times as much is possible. In most temperate region locations, only one crop can be produced during the year, but two or three are possible in subtropical and tropical areas. (The length of the daily dark period is a major obstacle to growing most strains of marijuana outdoors near the equator. At the equator, the dark and light periods are about 12 hours long, but this very short dark period normally induces marijuana strains to flower, so near the equator, most strains would flower prematurely. Only day-neutral strains would be suitable [see Chapter 5].) Because several indoor crops can be grown in a year, and productivity can be manipulated by various techniques, indoor production can be 15–30 times as much as outdoor production for a given area (Leggett 2006). Nevertheless, as for most annual crops, outdoor cultivation is much more economical when expressed on a cost/gram basis, although for security and quality control of production, indoor cultivation is usually necessary for licensed marijuana production in most countries. Compared to most major field crops, a relatively small outdoor area is needed to satisfy the needs of a large consumer population (Small 1971).

Governmental Production Quotas

Most growers of marijuana benefit from high efficiency of production. (In some jurisdictions, private citizens are allowed to grow a small number of plants, in which case efficiency of production is not much of a concern.) Illicit growers of cannabis particularly need to maximize yield from a given area because the larger the area utilized, the greater the chance of detection. Authorized commercial growers of cannabis also need to maximize yield from a given area because governments often limit the area that may be used to cultivate plants. For commercial growers, higher production per unit area may have some detrimental aspects that lower profits. High plant densities can be conducive to pests and diseases and make handling (tying branches to supports, trimming away damaged branches, harvesting) difficult. Constructing very small walkways between benches allows more space for plants but less for handlers. Sometimes, misguided policies have authorized private sector growers to produce only a limited number of plants (e.g., 100), in which case they have often been carefully nurtured to develop into giants so that each individual would produce as much as 100 typical indoor marijuana plants. This can be achieved by manipulating the photoperiod to keep the plants vegetative and allowing them to grow for long periods under luxuriant conditions. When jurisdictions limit grower production by quotas, factors contributing to yield need to be specified. A professional marijuana production facility necessarily requires space for noncultivation facilities (washrooms, record rooms, storage, etc.) and space that directly supports cultivation (power, heating, etc.) and processing (drying, manicuring, packaging, etc.). Moreover, "growing space" can include areas dedicated to establishing seedlings or cuttings, areas for vegetative growth, and areas for flowering plants (Figure 14.2). Since the last-mentioned is the key stage of production, a sensible quota system could be based exclusively or mainly on the area in which plants are in flower.

GROWTH MEDIA

All of the techniques presently used to grow horticultural plants in commercial greenhouses can be employed to grow marijuana plants. These include hydroponics (roots growing in water but supported by an inert medium) and aeroponics (exposed roots growing in air are misted with nutrient solution), both typically conducted in large tanks housing numerous plants. The use of so-called "soilless" media, both inorganic (such as gravel or rockwool) and organic (such as coir, i.e., coconut husk), effectively represents hydroponic culture, even if the plants are placed in individual pots, since all nutrients are added in solution. The appropriate pH for nutrient solutions employed for hydroponic cultivation is lower (typically 5.2–5.8) than is optimum for culture in soil (6.5–7.2). A mystique about the alleged superiority of hydroponic growth of marijuana has developed. However, hydroponic production of marijuana does not appear to be notably advantageous over conventional soil-based methodology, and most growers do not employ purely water-based production systems. "Such systems do not appear to increase cannabis productivity or potency" (Potter 2014) and require considerable cultivator knowledge and experience. A possible advantage is that hydroponic growth can lessen the possibilities of microbiological and chemical contamination that sometimes comes from soil cultivation, but these potential problems always require diligence, whatever cultivation techniques are employed.

Traditional culture in pots of soil can produce entirely satisfactory results and is the principal method employed. Indoor plants are grown in pots of compost and soil (sometimes referred to as "organic," although this hardly defines organic agriculture as it is usually understood). Most professional marijuana growers prepare their own soil mixes or purchase commercial preparations. Peat-based mixes are commonly employed because of their water-retentive capacity, but lime is often added to reduce the natural acidity of peat. Sterilization of growth media prior to planting is highly advisable to reduce disease. Professional growers add mineral nutrients on a need basis, based on soil analysis. Whatever type of pot culture is adopted, horticulturalists often "flush" accumulated salts and minerals away with running water, typically after a substantial number of chemical nutrient fertilizations, often in the last two weeks prior to harvest.

Fertilizer mixtures alleged to be formulated specifically for marijuana plants are sometimes available commercially. "Growth boosters" (often plant growth hormones, sometimes preparations alleged to enhance root health) are also sometimes available, but the phrase "caveat emptor" should be remembered. Fertilizers based wholly or in part on animals (manure, bone meal, and blood) may harbor pathogenic bacteria, and although the risk is low, production of medicinal products requires extreme care. "Aquaponics" combines conventional "aquaculture" (raising aquatic animals such as snails, fish, crayfish, or prawns in tanks), taking advantage of the animal excretions to fertilize the plants. Because the excreta could be toxic, production of marijuana by such system would require study of safety aspects, although vegetables are often aquaponically grown. As discussed later, medical marijuana production is usually required to meet health standards of good agricultural practices and good manufacturing practices.

VEGETATIVE PROPAGATION

Once elite female plants are obtained or selected, they are maintained indefinitely by keeping them in a vegetative state (typically at 18 hours of light daily), although as discussed in Chapter 5, some strains are completely or partly insensitive to photoperiodic regime and will flower regardless of the grower's wishes. Occasionally, some female plants will prove to have a tendency to produce some male flowers, and these should be removed (although such unique plants are useful for breeding).

The art of reproducing horticultural plants by cuttings is well established, but different species tend to have somewhat different requirements. Numerous guides provide protocols for propagating *C. sativa* by cuttings (e.g., Chandra et al. 2010b; Upton et al. 2013). Vigorous "mother plants" (sometimes "motherplants") are employed as the starting material, and cuttings are harvested serially as needed. Mother plants can be maintained in the vegetative state for 20 or more years (Clarke and Watson 2006). However, new mother plants are usually established every few months from cuttings, as the old ones gradually deteriorate from all the cuttings taken (or sometimes from diseases that infect the cut stems). Typically, a short branch with two or three leaves is cut off, at a 45° angle below a node (a joint on the stem where a leaf or leaves arise) (a sterile blade is recommended), and about 2 cm of the base is dipped in a rooting hormone (such as indole-3-butryic acid). The cuttings may be grown in coco fiber, perlite, rock wool, sand, vermiculite, or some other rooting medium or mix (preferably sterile or sterilized), in small individual peat pots or large collective trays. They are placed so that at least one node at the base of the cutting is in the rooting medium. Plants are irrigated so that the rooting medium remains moist (hydroponic culture is alternatively practiced). The environment is kept moist (ca. 50% humidity), warm (ca. 25°C), and under diffused light. A small plastic tent is often fashioned over cuttings to maintain a desired humidity. Adequate rooting typically occurs in two to three weeks, and several weeks later, the young plants are placed in larger pots and transferred to a growth room or greenhouse.

Advanced biotechnological techniques have been employed to mass propagate excised apical meristems rather than whole cuttings (Wang et al. 2009; Upton et al. 2013) and even to encapsulate these as "synthetic seeds" for field planting (Chandra et al. 2010b; also see Chapter 4).

LIGHTING

Light is the most critical requirement for plant growth (for a review of the physiological roles that light plays, see Chen et al. 2004). Different kinds of lights have been employed for indoor production of marijuana. The advantages and disadvantages of the most popular kinds of lighting are examined in this section. The efficiency that different lamps convert electricity to light is a critical cost consideration. Large illicit grow-ops often steal electricity (Chapter 16) because it is so expensive and to avoid detection by the police, who often look for abnormally large consumption in residential areas.

Light Intensity

Cannabis sativa is naturally adapted to grow in full sunlight, and increasing light intensity strongly promotes photosynthesis and growth (Lydon et al. 1987; Chandra et al. 2008). Most marijuana strains originate from latitudes south of north latitude 40°N (indeed, generally south of 30°N) and so are adapted to very high light intensities. Chandra et al. (2010a) found that clonally propagated plants produced higher THC levels outdoors in Mississippi than the corresponding genetically identical (cloned) indoor plants and attributed this to the light intensity outdoors—slightly more than twice as intensive as the indoor situation. Light intensity in indoor facilities is normally the key factor limiting productivity. To increase lighting intensity, artificial light sources can be positioned close to the canopy of growing plants. However, because of the heat produced by most lighting systems, they cannot be placed too closely, although air-cooled high-intensity lamps can be placed relatively close to the plants.

For security reasons, indoor production of marijuana is often carried out using only artificial light (Figure 14.3). Potter (2014) described the production system (in England) of GW Pharmaceuticals, using a greenhouse system with supplemental illumination providing approximately half the light energy required during the year (Figure 14.1). This employed lighting supplying 55 W/m², which is well above levels in U.K. glasshouses producing food or ornamental crops.

The "strength" of light can be measured in various ways. "Light intensity" (measured in Lux or Lumens) is a total measure of solar radiation, but less than half of this includes photosynthetically active radiation (between wavelengths 400 and 700 nm). From the perspective of plant photosynthesis,

FIGURE 14.3   Medicinal marijuana production by Bedrocan, the sole authorized national supplier of medical marijuana to the Netherlands government since 2001. This growth room employs only artificial light. Photo courtesy of Bedrocan.

light is best measured as photon flux density, a measure of light energy (in moles) received per unit of time per unit of area. Chandra et al. (2008) reported that *C. sativa* benefits from light up to a photon flux density of 1500 μmol/m²/s. For practical reasons, however, yield data (discussed later) are often presented in terms of weight produced/watts utilized, since lamps are measured in wattage, and expressing the weight of marijuana produced in relation to wattage is a direct measure of cost/unit.

Cannabis plants can tolerate much more light intensity than humans can, and so while the plants will grow better under extremely high interior lighting, workers can find the lights too bright for comfort. In any event, after moderately high lighting is provided, progressive increases of light intensity produce diminishing increases of growth.

A productivity of 1 g of marijuana output per watt of light input is often considered to be a desirable level of efficiency. Interior lighting regimes for cannabis grow rooms in the Netherlands commonly utilize 270 to 600 W/m², and over this range, Potter and Duncombe (2012) observed the following: (1) THC level does not change significantly, (2) weight and proportion of flowering material (hence production of sinsemilla buds) increase by about one-quarter, and (3) efficiency (yield/area) decreases by about four-fifths.

Vanhove et al. (2011) stated that "Overhead lights should provide at least 54,000 lumens per m². Nowadays, these light intensities are amply achieved by high-pressure sodium or metal halide lamps of either 400 W or 600 W." One lamp can be positioned over the center of each square meter. Vanhove et al. (2012) noted that 600-W bulbs were the most popular choice of illicit growers in Belgium (although commercial growers may choose to use larger lights).

Light Quality

As noted in the following discussion, high-pressure sodium and metal halide are popular choices for growing marijuana, although the former is somewhat deficient in the blue end of the spectrum.

The blue portion of the photosynthetic spectrum has been observed to tend to suppress growth elongation of some marijuana strains, so that a choice of a lamp rich in blue could promote shorter plants with more compact buds. Potter and Duncombe (2012) found that ultraviolet light increased THC concentration, but insufficiently to warrant attention.

Fluorescent Lighting

Fluorescent light bulbs furnish light efficiently in terms of converting electricity to light energy, but conventional bulbs emit limited light so it is difficult to provide the high light intensity that benefits the growth of *C. sativa*. Full-spectrum fluorescent bulbs provide balanced red and blue light to promote photosynthesis and are preferable to cool-white bulbs. Fluorescents are often used for establishing cuttings and germinating seeds because they generate sufficient light for growth while producing little heat (ballasts, which do produce heat, can be located some distance away from the plants). Compact fluorescent bulbs (with self-contained ballasts) are often used for growing marijuana illicitly in small rooms because they produce relatively little heat and use relatively little electricity.

High-Pressure Sodium Lighting

High-pressure sodium lighting (HPS) produces light efficiently, and the bulbs tend to last twice as long as metal halides. The ballasts produce considerable heat, and cooling may be necessary. Recent marijuana cultivation guides commonly recommend a light regime of 600 W/m^2 using HPS. The height of the lights above the tops of the plant canopy needs to be kept constant (30–60 cm has been recommended) by raising the lights as the plants grow (otherwise, the highest part of the plants will be scorched). Higher wattage (such as 1000 W) bulbs require a meter or more distance from the top of the plants. The color spectrum of HPS is somewhat stronger in the orange/red portion of the spectrum, somewhat deficient at the blue end, and therefore better for vegetative growth than for the flowering phase, but this is rarely considered important.

Metal Halide Lighting

Metal halide lamps produce less quality light (balance of wavelengths required for photosynthesis) and tend to fade faster over long periods than high HPS does (metal halide lamps are typically useful for up to 10,000 hours, HPS, for up to 18,000 hours).

Light-Emitting Diode Lighting

Light-emitting diode (LED) lamps, compared to high-intensity discharge lamps, are longer lasting (lifetimes can exceed 100,000 hours), smaller, adjustable for light intensity (dimmable) and quality, and extremely efficient in converting electricity to light (Yeh and Chung 2009). LEDs can produce three times more light output per watt of input power on an area-equivalent basis compared to high-intensity discharge lamps (Morrow 2008). LEDs have a low surface temperature, so they can even be placed inside a crop canopy without injuring the plants. While LEDs produce light with drastically reduced power consumption and heat generation, they are expensive to purchase. Until recently, it has been difficult to achieve the high intensity needed for good growth of marijuana, and the technology has not been practical for commercial production. Most LED lighting currently available is unsuitable for growing marijuana indoors, but LED lighting has prospects of capturing an appreciable share of the market in the near future.

Recommended Lighting

By a considerable margin, high-pressure sodium and metal halide are the most popular choices for growing *C. sativa* under artificial light, and they are sometimes combined to promote color balance (Figure 14.4). Both are classified as "high-intensity discharge" lamps.

FIGURE 14.4 Medical marijuana growth facility of Tweed Inc. (Smith Falls, Ontario, Canada). The light installation shown here features alternating high-pressure sodium and metal halide lamps in rows between ventilation pipes to regulate temperature. Photo courtesy of Steve Naraine.

TEMPERATURE

Marijuana strains are generally adapted to grow well at quite warm temperatures, generally at least 25°C, often performing best under daylight temperatures of about 30°C. Required lighting elevates temperatures in indoor growth facilities. Higher temperatures produce better growth for many strains but are uncomfortable for workers and may encourage the growth of insects and spider mites. Because cooling is expensive, in practice growth facilities are kept fairly warm.

ATMOSPHERIC GROWTH CONDITIONS

Humidity

Outdoors, *C. sativa* tolerates a range of humidity, depending on strain. Cuttings are generally raised under elevated humidity (about 75%) since their developing root systems are too weak to absorb much water. Until cuttings are well rooted, growers often mist their foliage with a water spray. Grow rooms and greenhouses easily develop very high levels of humidity, and if prolonged, this encourages fungal infection. Air exchange is essential to maintain moderate humidity—about 50% to 60%. Fans are employed to promote uniform conditions, with the additional benefit that wind buffeting promotes stronger (and sometimes shorter) stems.

Air-Borne Contaminants in Relation to Stickiness

Marijuana plants often develop stickiness on the surface of the flowering parts. Probably some strains differ in their tendency to produce stickiness. Stickiness is due to terpene secretions from the gland heads of the stalked trichomes that are present (as described in Chapter 11). There is limited secretion from intact gland heads, but buffeting by wind, ventilation, or handling releases some resin. In very windy, dry, or cold environments, secretions tend to volatilize more readily, decreasing stickiness; by contrast, in hot, still environments (whether outdoors or under intense grow-lights), secretions appear to accumulate more readily, and the plant surfaces can become very sticky. The degree of stickiness depends in part on the atmospheric conditions provided—whether grown outdoors, in glass houses, or in growth chambers. There are potential problems associated with the degree of stickiness developed. Plants grown outdoors are particularly subjected to wind-borne particles (from soil, animals, etc.) and insects, which represent contaminants. Quite aside from the natural stickiness of the plants, some contaminants such as insect and bird excreta are themselves sticky and adhere tightly to the plants. Depending on ventilation system, glass house plants are generally subjected to a lower shower of contaminants, and plants in small chambers are usually susceptible to the lowest amount. It can be virtually impossible to separate tiny particles adhering to harvested buds, and stickiness also makes handling of ground-up material difficult. Care needs to be taken to minimize the presence of wind-borne materials. Filtering of circulated air is clearly desirable, as well as monitoring of the effects of temperature and humidity on the development of stickiness.

Carbon Dioxide Concentration

Like most plants, growth rate in *C. sativa* is increased by elevating CO_2 levels (Chandra et al. 2011a). Some growers employ CO_2 generators, fueled by natural gas or propane, to raise indoor CO_2 levels and accelerate plant productivity. Concentrations of CO_2 are often raised to four times natural levels (ca. 1600 ppm). As noted in Chapter 16, although releasing CO_2 might seem to contribute to atmospheric pollution, the speeding up of plant growth (and therefore the trapping of carbon within the plants) might, on balance, be beneficial for the environment.

INSECT CONTROL

It should be possible to produce marijuana plants organically, without employing any biocides (including fungicides), although some greenhouse pests, such as aphids, spider mites (Figure 14.5), thrips, and whiteflies are difficult to control. Biocontrol measures using beneficial insects and other

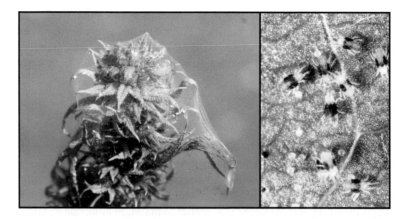

FIGURE 14.5 Left: Damage to indoor-grown plant of *C. sativa* caused by the twospotted spider mite (*Tetranychus urticae* Koch). Spider mites (especially the genus *Tetranychus*) are perhaps the most serious invertebrate pest of indoor *C. sativa*. Photo by Whitney Cranshaw, Colorado State University, Bugwood.org (CC BY 3.0). Right: Twospotted spider mite (adults and eggs). Photo by CSIRO (CC BY 3.0).

invertebrate predators (such as nematodes and predatory mites) that consume or parasitize invaders, and some organic insecticides, are preferable to employing synthetic pesticides. Applications need to be completed well in advance of harvest to prevent contamination.

HARVEST AND DRYING

Plants are harvested when it is judged that maximum cannabinoids have developed. This stage can be assessed by chemical analysis or more frequently is estimated as the point that numerous female flowers have developed and the stigmas of an appreciable proportion (at least 75%) of them have recently turned from white to brownish or orange. Outdoor plants are best harvested during a rainless period, so that the plants aren't soaked, but timing depends on Mother Nature. Plants are traditionally cut off at the base of the main stalk, just below the lowest branch. Different options are available for drying plants, but commonly, they are simply hung on cord (Figure 14.6). Curiously, the modern Chinese character ma, 麻, is based on a Zhou Dynasty bronze script ideograph for cannabis or hemp, showing 林 (plants) drying in a 广 (shed).

There is widespread opinion that slow drying or "curing" (up to a week) of harvested plants in a moderately dry, well-ventilated (and/or dehumidified), dark environment maintained at 25°C–30°C produces the best marijuana. However, some production systems conduct drying at up to 40°C for only 15 hours (Upton et al. 2013). An industrial grade ventilation system is normally required for large-scale production. At the conclusion of drying, the foliage should be crisp and brittle enough to be crumbled and the floral material should pull away readily from the stems. Fresh plants may have a moisture content of about 80%, which may dry down to about 15%. If material is to be stored before processing, it should be in an area maintained at low humidity, cool temperature, and in darkness. Marijuana should be dried to 5%–10% before packaging (8% has been recommended, although some consumers prefer a "wetter" product).

Alternative to harvesting entire plants, they may be kept growing while harvesting individual buds or portions of branches as they mature (rather like harvesting tomatoes from a given plant as it sequentially matures the fruit). The uppermost part of the main stalk usually matures first. (Home growers and sometimes commercial growers sometimes prune away lower branches in the expectation that this will produce larger terminal inflorescences.)

FIGURE 14.6 Marijuana branches covered with buds hung up to dry on a cord. Photo by Cannabis Training University (CC BY SA 3.0).

FIGURE 14.7   Hand-manicuring marijuana. Photo courtesy of Bedrocan (Netherlands).

PROCESSING

Once dried, the foliage and floral material is stripped from the stem tissues (stalk and twigs). The stems are routinely discarded since they are almost devoid of cannabinoids. For chemical extraction of cannabinoids, all of the remaining material can be employed. For production of ground-up (mani-cured) marijuana, the floral and other tissues in the flowering stem (mostly the perigonal bracts and smallest leaves) are screened. Loose gland heads tend to fall off and accumulate in the collection container, and since these are very concentrated in cannabinoids, a protocol needs to be established to maintain both herbal parts and the gland heads in a standard fashion in the final product.

Buds are increasingly desired in the marketplace and are often processed by hand (either trimmed or crumbled), a rather labor-intensive process (Figure 14.7). For sales presentation, the smallest leaves (with lesser levels of cannabinoids) are often trimmed away from the buds with scissors or machines. Machine trimming is much faster but less complete than hand trimming and is often employed as a first step followed by a final hand processing. The principal purpose of this practice is to present a product along with a claim that it has a very high level of cannabinoids, especially THC. Such trim-ming is best done before the buds are well dried, as the cannabinoid-rich trichomes tend to drop away with handling because well-dried buds are brittle. The "trim" or "skuff" (material removed from the buds) is sometimes discarded, but efficient systems can use this for solvent extraction of cannabinoids.

STORAGE

Since oxygen and light degrade THC, to retain quality, the marijuana should be protected from air (in tightly sealed containers) and maintained in the dark. Recommended storage temperatures are as fol-lows: short-term, 18°C–20°C, and long-term, −20°C (Upton et al. 2013). Handling should be kept to a minimum, since the trichome gland heads containing the cannabinoids are easily separated when mari-juana is dried and brittle. While material is frozen, it is especially prone to dropping the gland heads.

QUALITY CONTROL

Aside from the argument that marijuana is inherently harmful, medical marijuana needs to be pro-duced with considerable quality control, adhering to general Good Manufacturing Procedures and Best Practices.

MICROBIOLOGICAL SAFETY AND STERILIZATION

Plant material typically carries microorganisms on the surface. Illicit cannabis drugs may be significantly contaminated microbiologically, for example, as a result of using human excrement as fertilizer or simply from handling material with unwashed hands. Salmonella has occasionally been detected. Verweij et al. (2000) examined many samples of marijuana and found that they were heavily contaminated by fungal spores. The fungus genus *Aspergillus* occurs worldwide in water, soil, and air and can cause infection when its conidia (asexual spores) are inhaled. Sufficient heat can destroy the spores, but smoking often does not provide enough heat. Smoking marijuana contaminated with *Aspergillus* fungi and other microorganisms can cause significant, even fatal, lung diseases in patients with weakened health (Chusid et al. 1975; Llamas et al. 1978; Kagen 1981; Kagen et al. 1983; Karup et al. 1983; Schwartz 1985; Sutton et al. 1986; Hamadeh et al. 1988; Denning et al. 1991; Levitz and Diamond 1991; Marks et al. 1996; Szyper-Kravitz et al. 2001; Cescon et al. 2008; Gargani et al. 2011). Given that patients are often immunocompromised, certified medical cannabis needs to adhere to high production standards.

In Canada, the Netherlands, and other jurisdictions, officially produced medical marijuana conforms to very strict quality standards and is gamma-irradiated to sterilize coliform microbial infection (Cannabis Health 2005). According to Russo (2011a), "the safety of this technique for a smoked and inhaled product has never been specifically tested" (and as noted in Chapter 9, irradiation eliminates terpenes with possible medicinal value). Ruchlemer et al. (2015) explored sterilization of medical marijuana by autoclave, plasma hydrogen peroxide (H_2O_2), and ethylene oxide gas as alternatives, since nuclear facilities to irradiate marijuana are rarely available. They recommended H_2O_2 plasma, which has become a commonly used germicidal method.

MONITORING AND CONTROLLING THE PRODUCTION OF STANDARDIZED HERBAL MARIJUANA

Aside from microbiological safety, a professional medicinal marijuana analytical laboratory will be equipped to check representative samples for cannabinoid content, moisture level, and the presence of toxins of nonbiological origin such as heavy metals and pesticides. Marijuana can be contaminated with pesticide residues, heavy metals (absorbed from the soil or from wastewater used for irrigation), or by the occasional deliberate adulteration with (other) illicit drugs (McPartland 1994, 2002; McPartland and Pruitt 1997; McPartland et al. 2000). The underground literature sometimes describes the use of toxic growth-enhancing chemicals, and these may contaminate street marijuana. As pointed out by Sullivan et al. (2013), "pesticide toxicity…can pose substantial threats to immunocompromised patients or patients with other conditions, such as diseases of the liver, that may intensify the toxicological effects of pesticide exposure. Additionally, during heating pyrolysis products from the plant material form a highly complex mixture of products, many of which may interact with the pesticides or pyrolysis products of the pesticides forming more toxic materials, or highly toxic pyrolysis products may form from the pesticide residues alone." See Chapter 12 for additional discussion of toxic compounds potentially associated with marijuana.

Both for authorized medicinal research and for licensed dispensation of medicinal marijuana, it is critical to supply material of standardized and uniform THC and CBD content. Many patients using medical marijuana are in fact experienced users, who will judge the acceptability of the product in part on the basis on what they have consumed in the past. For such sophisticated users, familiar appearance and organoleptic qualities can be important indicators of product acceptability. Qualities valued by marijuana users include overall appearance, color, smell, humidity, grind size, and smoking characteristics such as burn rate, hotness, harshness, and taste (Ware et al. 2006). Accordingly, laboratory technicians may be expected to judge appearance and odor of the product, at least for consistency, since these qualities differ among strains.

Medical marijuana is mostly provided today in the form of whole (buds) or granulated (finely ground) material. The THC level of the final product is determined by (1) the natural genetic capacity to produce THC of the plants employed; (2) the environment in which the plants were raised; (3) the environment in which the plants were harvested, dried, and stored; and (4) the exact parts of the plant sampled. All of these variables need to be standardized and controlled. Medical marijuana currently marketed under license by authorized sources typically contains 10%–30% THC, the higher levels often considered to be a chief selling point.

Medical marijuana may be provided in the form of "manicured" material, prepared from dried "bud" by crumbling, screening, and/or cutting. The manicuring process needs to be standardized and controlled. Critical variables that need to be considered include the following:

1. Drying technique and humidity during manicuring. Overly dried material and/or material that is manicured in a very dry environment, upon crumbling and/or screening, tends to result in the microscopic glands (that contain the THC) breaking away from the remaining herbal material, and depending on the collection procedure, these may not be present in the final product. The results may be lowered THC content or marijuana with a variable THC content.
2. Screen (pore) size. Smaller screen size filters out larger parts, particularly twig material, which is very low in THC and results in a higher THC content of the final product.
3. Force applied during crumbling and or screening. A larger force hastens production but (depending particularly on how dry the plants are, and the ambient humidity) may tend to break away the microscopic glands and smear the contained resin over other parts of the herbal mixture, which, after screening, may alter THC content.
4. Handling of the product, which needs to be minimized. In (dried, prepared) marijuana, a proportion of the secretory glands (which carry most of the THC) separate from the epidermis of the plant material, and with increased handling, a larger percentage of the glands separate. This can result in considerable heterogeneity of marijuana preparations, the glands tending to sift to the bottom of collected material. Also, such variability can make it difficult to produce authorized material of homogeneous THC composition.

Unfortunately, even when marijuana is generated using genetically uniform plant material (i.e., clones), because of variations in plant growth conditions, preparation of material, and storage, differences in THC content develop. By standardizing (insofar as possible) plant growth conditions, preparation of material, and storage, material with a relatively predictable THC content can be generated, but nevertheless, from time to time, material of lower than desired THC content might be produced, requiring adjustment of THC content by mixing of batches of known potency.

INFORMATIVE ONLINE DOCUMENTS AND WEBSITES REGARDING PRODUCTION OF MEDICAL MARIJUANA

- Canadian government portal for medical marijuana—http://www.hc-sc.gc.ca/dhp-mps /marihuana/index-eng.php. This is one of the most extensive websites dealing with practical aspects of producing medical marijuana. It includes information on physical security measures, good production practices, testing specifications, and advice on packaging, labeling, and shipping.
- The American Herbal Products Association medical marijuana draft guidelines. The following four documents, cited in the Literature Cited, are available at http://www.ahpa.org /Default.aspx?tabid=267:
  - *The American Herbal Products Association draft guidelines for laboratory operations* (American Herbal Products Association 2014a).

- *The American Herbal Products Association draft guidelines for dispensaries* (American Herbal Products Association 2014b).
- *The American Herbal Products Association draft guidelines for manufacturing, packaging, labeling, and holding operations* (American Herbal Products Association 2014c).
- *The American Herbal Products Association draft guidelines for cultivation and processing operations* (American Herbal Products Association 2014d).
- Americans for Safe Access—http://www.safeaccessnow.org/. Sympathetic to the expansion of medical marijuana; provides information on a variety of related topics; has sponsored an industry certification program for cultivators and dispensaries.
- The Office of Medicinal Cannabis, Netherlands—http://www.cannabisbureau.nl/en/. Provides information on regulations in the Netherlands, as well as medical usage guidelines for patients and doctors. Note that only one supplier with a very limited number of strains is authorized officially to provide medical marijuana in the country.
- Scholten, W.K. 2003. Guidelines for cultivating cannabis for medicinal purposes. *Journal of Cannabis Therapeutics* 3: 51–61. http://www.cannabis-med.org/data/pdf/2003-02-4_0.pdf.
- *Recommended methods for testing cannabis: manual for use by national narcotics laboratories.* (United Nations 1987). http://www.unodc.org/documents/scientific/ST-NAR-40-Ebook.pdf.
- Daley, P., Lampach, D., and Sguerra, S. 2013. *Testing Cannabis for contaminants.* Botec Analysis Corporation I-502, Project 430-1a. Los Angeles, CA: Botech Analysis Corporation. 65 pp. http://liq.wa.gov/publications/Marijuana/BOTEC%20reports/1a-Testing-for-Contaminants-Final-Revised.pdf.
- *WHO guidelines on good agricultural and collection practices (GACP) for medicinal plants.* Geneva: World Health Organization. 2003. 72 pp. http://whqlibdoc.who.int/publications/2003/9241546271.pdf.
- *WHO guidelines on good manufacturing practices (GMP) for herbal medicines.* Geneva: World Health Organization. 2007. 92 pp. http://apps.who.int/medicinedocs/documents/s14215e/s14215e.pdf.
- *Quality control methods for herbal materials.* Geneva: World Health Organization. 2011. 187 pp. http://apps.who.int/medicinedocs/documents/h1791e/h1791e.pdf.
- *Recommended methods for the identification and analysis of cannabis and cannabis products (revised and updated).* New York: United Nations. 2009. Laboratory and Scientific Section, United Nations Office on Drugs and Crime, Vienna. 52 pp. https://www.unodc.org/unodc/en/scientists/recommended-methods-for-the-identification-and-analysis-of-cannabis-and-cannabis-products.html.
- Upton, R., Craker, L., ElSohly, M., Romm, A., Russo, E., and Sexton, M. eds. 2013. *American herbal pharmacopoeia: Cannabis inflorescence: Cannabis spp.: standards of identity, analysis, and quality control.* Scott's Valley, CA: American Herbal Pharmacopoeia. 63 pp. http://www.pdfsdocuments.com/american-herbal-pharmacopoeia-cannabis-inflorescence.pdf.
- GW Pharmaceuticals—http://www.gwpharm.com/. Website of the world's largest manufacturer of cannabinoid medicines, provides information on a variety of medicinal marijuana topics. The PhD thesis of Potter (2009) is presented, and this provides important information on the commercial cultivation of marijuana.

SECURITY CONSIDERATIONS

Compared to almost all other crops, relatively little marijuana needs to be generated for authorized purposes (Small 1971), but because it is a high-value material in considerable demand, strong security is essential (Figure 14.8). Cannabis has very high value, and stolen material has considerable

FIGURE 14.8   High security required for marijuana. Prepared by B. Brookes.

abuse and harm potential. Marijuana continues to be a favorite trade commodity of criminals, who can be quite ruthless in robbing, intimidating, and even eliminating legitimate competition. The authorized marijuana business can be lucrative, and the wealth generated also invites unwelcome attention from the criminal element. At present, production of medicinal marijuana by the private sector occurs mostly in indoor facilities that are very well protected by guards, locked doors, and monitoring cameras. Storage is in very heavy vaults, and elaborate protocols are in place to prevent unauthorized access. Most of the cost of marijuana is due directly to the needs for security. Governments that authorize the production of medicinal marijuana are very insistent on adequate security, and breaches of requirements are not tolerated. The website of the Canadian government, cited previously, is particularly informative on security measures.

BIOTECHNOLOGICAL APPROACHES OF POTENTIAL IMPORTANCE TO MEDICINAL MARIJUANA

IN VITRO PRODUCTION

Techniques for the "test tube" culture of plant cells, tissues, organs, and young plants in aseptic artificial media have been developing since the 1940s. Termed "micropropagation," this is essentially a form of clonal reproduction and can be a means of producing a large number of genetically identical plants in a short time. As a breeding technique, it is invaluable for preserving mutations, especially those that are (genetically) recessive. The responses of plant cells and tissues to various artificial culture media and hormone treatments (required to induce cell differentiation, organ production, and plantlet production) differ considerably among plant species. Some species have proven to be very difficult to reproduce by tissue culture, and until recently, this has included *C. sativa* (Hemphill et al. 1978; Loh et al. 1983; Mandolino et al. 1996; Mandolino and Ranalli 1998). As described in Chapter 4, it is now possible to produce "artificial seeds" of *C. sativa*: very young plantlets packaged with fertilizer and water in a gel, which can be planted just like real seeds to produce new plants.

The most elementary kind of test tube culture is simple cell culture. In some cases, plant cell cultures are sufficient for the commercial production of certain chemicals. However, a frequent problem encountered is that the production of compounds that occur in the mature plants of a species simply do not occur or occur in very low amounts when the cells are cultured. This has proven to be the case in *Cannabis*. Veliky and Genest (1972) found no production of THC in cell cultures. There have been some experiments that demonstrated production of cannabinoids in cell cultures of *Cannabis*, but in extremely limited amounts (Heitrich and Binder 1982; Hartsel et al. 1983; see review of Mandolino and Ranalli 1998). There remains a reasonable potential of utilizing in vitro production techniques for the test tube production of cannabinoids, although the technology remains to be developed.

GENETIC ENGINEERING

Techniques for introducing genes from quite unrelated organisms into given species have been developing since the 1980s. There have been remarkable successes in creating several new kinds of recombinants, and some are in commercial production today, although genetically engineered organisms are controversial. There appears to be a widespread sentiment against the production of genetically engineered hemp strains among growers of industrial hemp (particularly for the hemp edible oilseed industry, there is a strong view that the plants should remain "natural"). Nevertheless, it is interesting to speculate on some possibilities (cf. Watson and Clarke 1997). The production of natural cannabinoids in large amounts is at present known only in *C. sativa*, where it occurs almost completely in the epidermal secretory glands of the plant. Perhaps genetic engineering could one day succeed in transferring the ability to produce cannabinoids to other plants. This might seem like an academic exercise, but for various reasons, such interspecies transfers are often preferable (Small 2004). An important practical aspect of genetic engineering is the clarification of the genes of a plant under study, and this can lead to manipulation of the genome so that the plant becomes more useful. Modern techniques have made it possible to prepare detailed genetic maps, and in many cases, some of the specific functions of mapped genes have been clarified. In the case of *Cannabis*, this work is in its infancy. In the future, genetic engineering of *Cannabis* may greatly facilitate the production of plants with extremely well-defined characteristics that are very desirable for purposes of industry, medicine, and law enforcement (De Meijer 2014).

CURIOSITIES OF SCIENCE, TECHNOLOGY, AND HUMAN BEHAVIOR

- According to Leggett (2006), a hectare of outdoor marijuana plants could supply 10,000 "light users" with a daily dose for a year, and about 160 km^2 (100 square miles) could similarly provide a year's supply for all of the world's marijuana smokers.
- The Eden Project is a large-scale environmental complex near St Austell, Cornwall, England (Figure 14.9). The complex includes several transparent domes, the largest of which, the Tropical Biome, is claimed to be "the world's largest greenhouse." It covers 1.56 ha and measures 55 m in height, 100 m in width, and 200 m in length. As noted in the text, a concentration of 32 cannabis plants/m^2 has been recommended for optimal yield, and at this density, the Tropical Biome greenhouse could hold 640,000 plants. Commercial greenhouse complexes often link greenhouses together to occupy even larger acreages.
- The Crystal Palace was a giant glass and cast iron structure erected in Hyde Park, London, England, to house the Great Exhibition of 1851 (Figure 14.10). The building was 41 m high, with 71,794 m^2 on the ground floor and a total of 92,000 m^2 of exhibition space. The Crystal Palace was redesigned and moved to Sydenham Hill in 1854, where it burned down in 1936. At a concentration of 32 plants/m^2, it could have held three million cannabis plants.

FIGURE 14.9   The geodesic domes at the Eden Project in England. Photo courtesy of Jürgen Matern (CC BY SA 2.5).

FIGURE 14.10 The Crystal Palace in London, England. From Nash, J., Haghe, L., Roberts, D. *Dickinson's Comprehensive Pictures of the Great Exhibition of 1851*, Dickinson Brothers, London, U.K., 1852 (1854).

15 The Commercial Marijuana Revolution

A SEA CHANGE FOR BETTER OR WORSE

The early twenty-first century is a watershed period, witnessing an astonishing transformation of the status of high-THC cannabis from an entirely black market industry to considerable "legitimacy." The 11th edition (2012) of *Merriam-Webster's Collegiate Dictionary* defines "legitimate" as "(1): to give legal status or authorization to; (2): to show or affirm to be justified; (3): to lend authority or respectability to." This chapter discusses marijuana from the point of view of (1), i.e., as a legalized commodity. Whether cannabis commercialization is "justified" (sense 2) or is "respectable" (sense 3) remains a matter of controversy. Although many are inflexibly opposed to any softening of current cannabis legislation, there is a growing viewpoint that marijuana prohibition has been a multitrillion dollar failed social experiment and needs to be replaced with a rational management approach. This chapter is mainly concerned with managing the factors that bear on maximizing benefits while minimizing risks associated with commercialized marijuana—but in the real world rather than an idealistic context. Life is frequently a choice between flawed alternatives, necessitating management of the winds of change by adapting to their direction.

The term "cannabusiness" ("cannabiz" for short) is often encountered currently, indicating commerce in medical and/or recreational cannabis. However, in the recent past, the term has also been employed to include the industrial hemp industries, which were discussed in detail in earlier chapters.

The legalization of marijuana commerce is occurring only in certain locations, but the fact that this includes some states of the United States, the leader in establishing commercial trends, suggests that this trend will continue. Indeed, there is a veritable tidal wave of new legal commercial activities related to cannabis. Growth in medical and recreational aspects concerning cannabis is so explosive that the developing commerce has been described as a lucrative "green gold rush." Despite persisting stigma and legal concerns attached to marijuana, investment capital is rapidly being directed to the authorized medical cannabis industries and even to the developing recreational white market. Marijuana has become increasingly available without significant restrictions in some places, and resulting quasi-legal activities have often resulted in extensive confusion among law enforcement, businesses, and consumers. Clear regulations need to be based on consideration of the best public health interests of society at large.

This chapter addresses the commercial and associated regulatory aspects of *both* medical and recreational cannabis. Although significantly different considerations apply, from a business perspective, the two areas are becoming intertwined. Production of marijuana is much the same, whether for medical or recreational usage, and there are common concerns about security, regulations, business models, safety, and consumer acceptance. Moreover, as discussed later in this chapter, the potentially much more profitable recreational market is the long-term target of many who have invested in the development of medical marijuana.

COMMERCIAL OBSCURATION OF THE DISTINCTION BETWEEN MEDICAL AND RECREATIONAL MARIJUANA

Chapter 12 discussed the nonmedical (largely recreational) use of marijuana, and Chapter 13 reviewed the medical aspects. The distinction between medical and recreational use is straightforward:

recreational cannabis is employed to get high, while medical cannabis is used to get healthier. Other mind-affecting substances are similarly used, but like marijuana, until recently, the recreational usage remains entirely illegal (later in this chapter, opioids and laughing gas are discussed as guiding parallel business/regulatory examples). Prescription pharmaceuticals are potentially quite toxic and so require highly qualified professionals to recommend, formulate, dispense, and sometimes also to administer, and cannabis is increasingly being recognized as deserving of such status. However, in many jurisdictions, allegedly "medical" marijuana is being peddled in establishments (debatably labeled as "dispensaries") by advisers with about the same limited credentials as clerks in "health food" establishments. Much worse, as noted next, some doctors are clearly betraying their profession by aiding those who simply wish to use marijuana recreationally.

PHYSICIAN GATEKEEPERS FOR HIRE TO PROVIDE LEGAL ACCESS TO PSEUDOPATIENTS

A questionable aspect of the establishment of medical marijuana dispensation in some jurisdictions is the virtual authorization of bogus "patients" with fictitious or marginal conditions to determine their own consumption protocol—i.e., to self-medicate (Hazekamp and Pappus 2014; Figure 15.1). In this laissez-faire environment, one of the less savory employment niches that has appeared with the increasing popularity of medical marijuana is the need for prescriptions (or "recommendations" or "approvals," depending on local regulations) from medical professionals, simply to obtain material. While most physicians are ethical and will not prescribe marijuana without justification, there are some who are incentivized simply by the opportunity for lucrative fees. In several of the U.S. states that have authorized medical marijuana, it has been embarrassingly clear that the bar to obtain scripts for marijuana is very low and that many in perfect health have been authorized to smoke pot for trumped-up illnesses. This is evident by studies in California showing that "patients" at medical marijuana clinics are overwhelmingly young white men experienced in the use of recreational marijuana (Reinarman et al. 2011). Regan (2011) wrote that "California doctors have the latitude to prescribe marijuana for pretty much anything. Although California law suggests doctors recommend marijuana for patients over the age of 18 who are suffering from a specific set of diseases, the law also includes a provision allowing pot for 'any other illness for which marijuana provides relief.' So the reality is, anyone who wants pot can get a 'recommendation' for marijuana. Headaches, anxiety, trouble sleeping, you name it…recommending marijuana has become a thriving business in and of itself, as the multitude of doctors' advertisements in the back

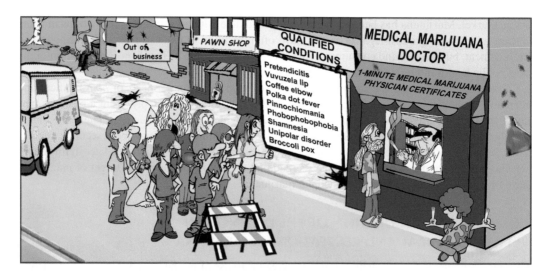

FIGURE 15.1   Medical marijuana malpractice. Prepared by B. Brookes.

of the pot magazines' 'yellow pages' suggest. Doctors typically charge a $200 consultation fee, and I've never heard of anyone being denied a recommendation. There are currently an estimated four hundred thousand medicinal marijuana patients in the state." Caulkins et al. (2012) described marijuana doctors as follows: "The physicians who write most of the recommendations openly advertise, promising not to diagnose and treat illness but simply to provide a recommendation. Some even advertise that the visit is free unless it results in a recommendation. Others have advertisements, or even storefronts, that list a range of conditions for which they will write marijuana recommendations. It's hard to take this seriously as the practice of medicine."

Prescription Drugs vs. Over-the-Counter Herbals

In Chapter 13, the debate between medical use of purified pharmaceutical chemicals and crude (herbal) drugs was presented, and it was argued that herbal drugs in the hands of qualified professionals can be as effective as pure drugs that can only be obtained by prescription. However, unlike prescribed pure drugs, which are necessarily best managed by highly educated and experienced personnel, numerous herbal preparations, such as echinacea, garlic, ginkgo, ginseng, and valerian, are typically dispensed by clerks who are personable and enthusiastic but often dangerously uninformed. There are emerging markets for both prescription and over-the-counter marijuana, and this is another important factor muddying the distinction between medicinal and recreational cannabis.

From an investment perspective, there are very few enterprises as lucrative as the pharmaceutical industry. This is because patented products (or manufacturing processes contributing to them) that are either uniquely effective or popular can result in enormous long-term profits. Traditional herbal preparations (raw plant materials) cannot be patented but can often earn appreciable revenues from sales in health food stores and outlets for over-the-counter (nonprescription) drugs and use by alternative health practitioners as well as those who subscribe to self-medication. Herbal marijuana, however, is in a class by itself, currently and for the foreseeable future much more popular and profitable in the burgeoning medicinal marijuana herbal sector than the patented formulations presently offered by pharmaceutical companies. "Cannabinoid-based medication—mainly based on delta-9-tetrahydrocannabinol (Δ^9-THC) and cannabidiol (CBD)—has not been able to significantly reduce the worldwide use of [herbal] cannabis as a medicine. Cannabis seems…on the one hand too potent to be regulated as an herbal (or alternative) medicine, on the other hand too herbal to be regarded as conventional medicine" (Hazekamp and Pappus 2014).

THE EVOLUTION OF PHYSICIAN ACCEPTANCE OF MEDICAL CANNABIS

Chapter 13 details research on medical aspects of cannabis, summarizing majority opinions on the merits of marijuana in treating numerous illnesses. Regardless of the assessments of medical researchers (who, like all scientists, sometimes have exaggerated evaluations of the adequacy of their findings), practicing clinical doctors tend to be very conservative and reluctant to adopt new treatments without very convincing evidence. The community of physicians who actually treat patients represents the most critical filter limiting the success and failure of medicinal marijuana, not just as therapy but as the subject of commerce.

Softening of Medical Opposition

Despite the current vociferous opposition to medical use of marijuana by many physicians, Adler and Colbert (2013) reported on the surprising result of a survey of medical professionals (*New England Journal of Medicine* 2013). Readers were asked to vote on two opposing positions: (1) Marijuana should be used medically "only when conservative options have failed for fully informed patients treated in ongoing therapeutic relationships" or (2) "there is little scientific basis for physicians to endorse smoked marijuana as a medical therapy." The majority (76% of 1446 votes, mostly from North America) were cast in favor of marijuana.

MEDICINE BY POPULAR VOTE

Pharmaceutical companies incessantly advertise their wares on the popular media, clearly to persuade patients to influence their doctors to prescribe them, although physicians are obviously much more qualified than their patients to choose the most appropriate therapies. In the case of therapeutic forms of cannabis, physicians are often being overwhelmed by the enthusiasm of patients to be treated with cannabis. Moreover, the demand for legislation allowing medical cannabis to be available in some jurisdictions has been determined not primarily by medical opinion but by popular demand. Although in theory, the wisdom of legalizing medical marijuana should be evaluated principally by the medical profession, in practice, prevailing societal demand is proving to be determinative, a phenomenon that has been termed "medicine by popular vote" (Voth 2001; Bostwick 2012).

PROFIT-DRIVEN VS. HEALTH-DRIVEN MEDICAL MARIJUANA

While business people and sometimes even corporations can be extremely ethical and conscientious, there is perhaps an inevitable amorality to most profit-centered ventures. Already, some newly created medical marijuana establishments are peddling their wares with the same exaggerated rhetoric by which cars and detergent are advertised. As noted in Chapter 13, normal human stress has been excessively medicalized, contributing to an epidemic of mood-altering drugs, and some physicians are adding marijuana to the list of needless prescriptions. At present, governmental regulations are highly restrictive in most jurisdictions, limiting the possibility of harm. All things considered, marijuana is so prevalent that if it were seriously toxic in the limited amounts it is usually consumed, its harm for the majority of adults would have been revealed by now. Nevertheless, wisdom dictates that a great deal of caution needs to be exercised with respect to both personal and societal health issues as marijuana becomes increasingly legitimatized.

PATIENT FINANCIAL REIMBURSEMENT

The cost of medicines, in the main, is eligible for subsidization from public and private insurance systems. Medical marijuana and cannabinoid-based pharmaceuticals (to say nothing of paraphernalia) could be eligible under insurance schemes, although many will find the very idea curious. A basic problem is distinguishing actual patients from those posing as such to obtain recreational drugs, and this issue is not easily solved since marijuana intergrades between a genuine medicine and a recreational intoxicant. Unfortunately, many real patients live on small budgets, and because of the lack of funds to purchase sometimes expensive pharmaceutical-grade marijuana and cannabinoid drugs, those without insurance may feel forced to buy unreliable street material. Philanthropy in the private sector is limited (except for public relations purposes), and a common response when a costly medicine is in large demand is for nongovernmental organizations and/or the state to achieve savings by large-scale purchases. At present, medical marijuana "clubs" are popular in some places, contributing to limiting costs. Perhaps, unlike any other substance, the threat of street-available marijuana remains a brake on cost inflation of herbal marijuana.

MEDICAL VS. RECREATIONAL MARIJUANA MARKETS

Medical marijuana inherently suggests the nobility of the medical profession, while recreational marijuana has its roots in the illicit, evil, and dangerous underworld. Medical marijuana is now authorized in many countries and is produced according to stringent quality standards. In principle, exactly the same material can be employed as recreational marijuana, albeit far fewer locations authorize this at present. Although most investment in cannabis is ostensibly in the field of medical marijuana, there is widespread realization that for most companies, the profit potential is much larger in the area of recreational marijuana. Accordingly, major players are strategizing to invest in medical marijuana with the long-term goal of expanding into a recreational market. Recreational marijuana has a negative image that can dissuade investment, so one tactic ("serious and studious") that is advantageous is to associate the company with legitimate medical research, pharmacological representatives, and bona fide medical dispensers and to create or sell products with at least some proven efficacy. Companies of this nature tend to market marijuana strains with very conservative names. The opposing strategy is to target an audience that views recreational marijuana sympathetically (even if it is only available in the form of medical marijuana). Such companies tend to offer marijuana strains with ostentatious street names, a variety of edibles, smoking apparatus (especially bongs), and T-shirts with psychedelic logos. Companies specializing in medical marijuana are resentful of those promoting recreational marijuana aspects, since it degrades the seriousness of the industry.

THE ROLE OF "BIG TOBACCO"

Smoking has been widely practiced for centuries (Figure 15.2), but at least in Western nations, tobacco is a dying industry, the result of its horrific effects on human health. Smoking tobacco is estimated to kill six million people worldwide annually (Barry et al. 2014). Tobacco companies are anxious to diversify into more viable products (notably, for a period there were attempts to extract proteins from tobacco leaves to prepare edible "tobacco burgers"). Facetiously, the concerns regarding dangers to human health and the repugnant aspects of peddling a socially condemned product that would repel many potential investors might actually be attractive to the tobacco sector. For years, there has been suspicion that tobacco-based interests have been examining the possibility of investing in legitimized herbal marijuana production and distribution. Given its expertise in producing nicotine-based smoking products and paraphernalia, the tobacco industry is well preadapted to establishing "Big Marijuana." "Legalizing marijuana opens the market to major corporations, including tobacco companies which have the financial resources, product design technology to optimize puff-by-puff delivery of a psychoactive drug (nicotine), marketing muscle, and political clout to transform the marijuana market" (Barry et al. 2014). The recent consumer emphasis on vaping tobacco rather than smoking

FIGURE 15.2   Classical illustrations of tobacco smoking. Left: "Smoking club" (public domain illustration) from Fairholt, F.W., *Tobacco, Its History and Associations*, Chapman and Hall, London, U.K., 1859. Right: Boy dressed in a Napoleonic military costume, smoking a pipe (nineteenth century American trade card). Photo by oaktree_brian_1976 (CC BY 2.0).

it is perfectly aligned with the same trend for marijuana. Very large scale, as exemplified by corporatization, may well be a prerequisite to future success of the recreational marijuana industry, and certainly tobacco corporations have the size, capitalization, and perspective to dominate a free-market marijuana-based economy. The tobacco industry is old and conservative, which may explain its low profile to date on the subject, but there is evidence of its attention to marijuana (Barry et al. 2014), and it would not be surprising for tobacco interests to be surreptitiously venturing into cannabis.

PROFESSIONAL NICHES IN THE DEVELOPING MARIJUANA BUSINESS SECTOR

The illicit marijuana market has spawned legitimate or at least quasi-legal employment for many. Of course, the millions of individuals who have been caught running afoul of the law have needed legal representation, and monumental investment has gone into policing and imprisonment. On the borders of legality, stores specializing in hydroponics, lights, and similar cultivation equipment required to grow the plants have benefitted, as have "headshops," organizers of marijuana festivals, and writers specializing in counterculture publications. These profitable fringe activities have convinced many that legalized marijuana can be developed by rebels and misfits.

However, legalized marijuana production and marketing are definitely not industries to be pursued by amateurs. Much more so than most other business activities, there are very demanding needs to prepare extensive applications for approval, to keep accurate records, to tolerate inspections, and to regularly file detailed reports. Working with several aspects of marijuana can be dangerous, both from the criminal element and by inadvertently failing to comply with often ambiguous regulations. There are stringent requirements best satisfied by dedicated professionals. Many are finding employment in ancillary services allied to the emerging marijuana industry (Figure 15.3) or are simply benefitting indirectly from it (Figure 15.4). Particularly in demand are commercial property and leasing managers, insurance specialists, security

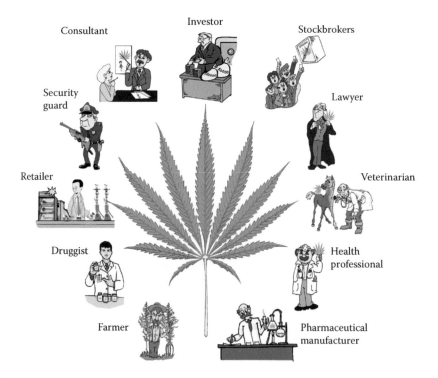

FIGURE 15.3 Occupations that have become associated with the rise of authorized medical and/or recreational marijuana. Prepared by B. Brookes.

FIGURE 15.4   An example of indirect benefits (in this case, to a junk food store) from the authorized marijuana industry. Prepared by B. Brookes.

personnel, financial and banking institutions, legal advisers, tax specialists, accountants, public relations and advertising representatives, horticulturalists, retailers, and media-savvy Web designers. With the expansion of professionals dedicated to facets of the marijuana industry, there is a parallel need for the establishment of trade associations and advisory councils to ensure that production and sales follow best practices and that ethical standards are maintained.

OPPOSITION TO COMMERCIALIZATION OF RECREATIONAL MARIJUANA

Any venture into the commercialization of marijuana needs to take account of the persisting social, moral, scientific, and political opposition. A century of prohibitionist policies and a negative image mean that many, indeed most, investors are likely to shy away from involvement.

MEDICAL OPPOSITION TO RECREATIONAL MARIJUANA

The most important source of opposition to recreational marijuana is a substantial proportion of the medical profession. The following statement succinctly summarize concerns (also see Chapters 12 and 13):

"Owing to limited funding and study opportunities, marijuana and health professionals and policymakers do not yet know the full scope of the effects of marijuana use; however, the evidence demonstrates that the regular consumption of marijuana does increase the risk of physical and mental health problems… The general consensus among substance abuse professionals is that underage marijuana use can be both dangerous and addictive. Studies show that teenagers' use can lead to negative physical, psychological, and behavioral consequences, such as chronic cough and bronchitis, memory deficits, and a loss of up to 8 points in I.Q. Additional public health issues… as a result of legalization include… acute health effects from contaminated marijuana products, the safety of edible marijuana products, accidental poisoning of young children from edible products, use among pregnant and breast-feeding women, secondhand smoke… substance abuse, potential impaired driving."

(Hickenlooper 2014; the governor of Colorado, a state which contradictorily has authorized not just medical but also recreational marijuana)

POLITICAL OPPOSITION TO RECREATIONAL MARIJUANA

In most of the Western world, marijuana remains a prohibited drug, often with exceptions for medical usage. Most governments are hostile to changing the status of cannabis, continuing to devote large resources to antimarijuana education and law enforcement. However, in several countries, court decisions based on personal rights, and referenda, have led to softening of laws restricting marijuana. In Canada, the government was elected in 2015 on a policy of authorizing the use of recreational marijuana. In the United States, the national government remains opposed to recreational use, although initiatives in some individual states have led to recreational marijuana becoming available.

As reviewed in earlier chapters, a majority of scientists and the public now accept that marijuana is not as deleterious as once believed, although not without significant harm, and that reformation of legislation is desirable to permit regulated usage. The governments of many nations have sponsored studies of the wisdom of laws limiting the use of marijuana (and indeed of other "drugs of abuse"), based particularly on harm potential, but these analyses have usually been hesitant to make decisive recommendations, although often reinforcing the view that marijuana legislation requires updating and emphasizing that further research is necessary. The fundamental issue to be addressed is how permissive or restrictive revised legislation should be.

BADMOUTHING THE COMPETITION

Curiously, many developing sectors of the nascent marijuana industry are very critical of each other, an obvious badmouthing of the competition in order to protect market share. For example, small dispensaries are horrified at the prospect that chain drug outlets could dominate retail aspects; medical dispensaries are opposed to the legalization of recreational marijuana because it reduces their profit potential and cheapens their image; and pharmaceutical-based companies, interested in cannabinoid drugs, see cheap mass-produced herbal cannabis as a severe limitation on future profits. Probably the most ironic objection to legitimizing marijuana is by the criminal trade, which faces the prospect of losing a substantial part of its drug-based income.

GROWING SOCIETAL ACCEPTANCE OF RECREATIONAL MARIJUANA

In a democracy, public opinion is the final arbiter of all issues, not scientists, politicians, or any other influential representatives of the people. It has been estimated that about 4% of the world's population consumes cannabis annually (Leggett 2006). However, in some countries, the majority of young people have smoked it, and there are strong indications of growing social acceptance and cultural normalization. As noted later, recreational marijuana is quite freely available in the Netherlands without fear of legal consequences, but this is very unusual. The critical development that is being closely observed is recent establishment of recreational marijuana industries in states like Colorado and Washington. This is regarded as a large-scale social experiment, and given the importance of the United States in influencing modern trends, it is possible that the results could either stimulate or discourage wider acceptance in the Western world. A key analysis of the possible future of recreational marijuana is provided by Galston and Dionne (2013). They note that in less than a decade, public opinion has shifted dramatically toward support of marijuana legalization and that 48% of Americans have personally used marijuana. Furthermore, a slim majority of Americans believe that it is less harmful than alcohol to individuals and society (by contrast, over 75% accept medical marijuana as beneficial). However, Galston and Dionne note that although support for legalization is growing, a substantial number of Americans are strongly opposed, and those who are in favor of legalization do not consider consumption harmless but rather view criminalization as harmful. Indeed, in Western countries, there has been a general trend to decriminalize drug abuse. Galston and Dionne concluded that "Over the long run, the attitudes of Americans…on the

"In a free and democratic society, which recognizes fundamentally but not exclusively the rule of law as the source of normative rules and in which government must promote autonomy as far as possible and therefore make only sparing use of the instruments of constraint, public policy on psychoactive substances must be structured around guiding principles respecting the life, health, security and rights and freedoms of individuals, who, naturally and legitimately, seek their own well-being and development and can recognize the presence, difference and equality of others."

Report of the Canadian Senate special committee on illegal drugs (2002)
http://www.parl.gc.ca/content/sen/committee/371/ille/rep/summary-e.htm

"For the first time in the more than four-decade history of polling on marijuana issues, a Pew poll in 2013 showed that a majority of people in the United States supported legalization. The sharp contrast between the growing support for legalization, which outstrips the percentage of the population currently using marijuana, must reflect some degree of a rejection of paternalism... some portion of the population must be rejecting the paternalistic decision of regulators to remove a person's autonomy when it comes to choosing to use marijuana."

Friedman (2014)

question will be shaped by whether the various experiments with legalization, decriminalization, and the use of marijuana for medical purposes are deemed successes or failures."

THE NEED FOR CAUTION IN LEGITIMIZING RECREATIONAL MARIJUANA

Any material that produces a euphoric state will be used to excess by some individuals. Alcohol and tobacco, which are physically addictive for many, have had devastating effects on human health. "Indeed, if alcohol were a newly formulated beverage, its high toxicity and addiction potential would surely prevent it from being marketed as a food or drug" (Gable 2006). Sugar, which is also psychologically addictive, is responsible for much of the current obesity epidemic. Clearly, some addictive substances are so dangerous that extreme efforts are justified to prevent or reduce their usage, at least in most circumstances, but just as clearly, in a free society, a majority of adults cannot be prevented from choosing to consume what they wish. The issue that needs to be resolved is what level of loosening of control over cannabis is advisable.

While this book is not intended to judge the wisdom of increasing authorized consumption of marijuana, either for medicinal or recreational purposes, it would be remiss to ignore the likelihood that some people are likely to be harmed. As discussed in Chapter 12, cannabis can produce anxiety, panic, and even paranoia in naïve users and anxious subjects and has significant potential for inducing psychological harm in adolescents and people with serious psychological vulnerabilities or suffering from mental illness. As also discussed in Chapter 12, cannabis additionally has potential to lower performance in handling equipment and automobiles and in conducting activities demanding alertness and manual dexterity, so inappropriate use could contribute to accidents and lowered productivity. Some additional issues are discussed in the following.

CHRONIC CONSUMPTION

Based on surveys in several countries, it appears that about 14% of illicit cannabis users consume marijuana daily (Leggett 2006). In the United States, about 12% of drinkers report imbibing eight or more alcoholic drinks in the past week—averaging more than one per day. In both cases, such high consumption is related to concerns about negative health effects. Often, regular usage of an inebriating drug is correlated with development of tolerance or insensitivity, so that higher amounts need to be consumed in order to achieve the same psychological state. Accordingly, the drive to consume in regular users can be exacerbated. Whether one's social drug is marijuana or alcohol, it appears that a very high proportion of usage (and potential profits) is concentrated in fairly small subpopulations. About half of alcohol profits come from one-quarter of alcohol drinkers, and one wonders about the sincerity of "Big Alcohol" advertising that their customers should "drink responsibly." A very large proportion of booze and virtually all cigarettes are consumed by addicts. In the same vein, Werb et al. (2012) estimated that over 90% of illicit cannabis in British Columbia was purchased by daily marijuana users. The relative harm of marijuana, tobacco, and alcohol is frequently debated, but there is agreement that excessive usage of drugs is undesirable, and adding marijuana to the set of legal social drugs may be particularly hazardous for those with a propensity to develop very high usage.

EFFECTS OF INCREASED AVAILABILITY OF AUTHORIZED CANNABIS

Increased allowable usage of marijuana is occurring especially in North America. Although authorization of medical usage, as well as decriminalization and permitted recreational usage, are very recent and by no means universal, there have been several studies of the associated social and medical effects. This information is relevant to business aspects because of the associated challenges faced by the cannabis industry in the future—in persuading the public, politicians, the medical profession, and indeed other business interests of the legitimacy of cannabis commercial activities.

Effects on Prevalence of Usage

Gorman and Huber (2007), on the basis of arrest and emergency hospital data, concluded that "consistent with other studies of the liberalization of cannabis laws, medical cannabis laws do not appear to increase use of the drug." In a study of adolescent usage, Lynne-Landsman et al. (2013) found that medical marijuana laws "have not measurably affected adolescent marijuana use in the first few years after their enactment. Longer-term results, after MMLs are more fully implemented, might be different." Nevertheless, in U.S. states with medical marijuana laws, rates of marijuana usage increased (Cerdá et al. 2012; Harper et al. 2012). However, whether liberalization of laws *causes* increased usage is unclear. Cerdá et al. (2012) noted that "Future research needs to examine whether the association is causal, or is due to an underlying common cause, such as community norms supportive of the legalization of medical marijuana and of marijuana use." From a commercial perspective, increased usage is desirable simply to increase the customer populations, but from a public health perspective, the opposite is often true.

Effects on Health

Cerdá et al. (2012) found that marijuana abuse/dependence was more prevalent in states with relaxed availability of marijuana, but suggested that this was accounted for simply by higher rates of use. As noted in the next section, it is conceivable that if marijuana becomes more available legally, the usage of more deleterious substances (debatably including alcohol) might decrease because of substitution.

Effects on Crime Rate

By definition, if marijuana becomes legal, the crime rate will drop! Indeed one might predict that overdosing on other illicit substances might decrease if marijuana became more accessible legally. Contrary to expectation that the legalization of marijuana for medical purposes poses a danger to public health in terms of exposure to violent crime and property crimes, Morris et al. (2014) found no evidence of increased crime in 11 states that had authorized the establishment of medical marijuana dispensaries. Similarly, Shepard and Blackley (2016) found that "There is no evidence of negative spillover effects from medical marijuana laws (MMLs) on violent or property crime. Instead, we find significant drops in rates of violent crime associated with state MMLs." However, Chu (2014) observed that medical marijuana authorization was associated with greater arrests for unauthorized (recreational) usage. Kepple and Freisthler (2012) found that the density of medical marijuana dispensaries seemed unassociated with crime rates and suggested that measures dispensaries take to reduce crime (such as doormen and video cameras) may deter offenders.

FINANCIAL GUESTIMATES OF THE POTENTIAL MONETARY VALUE OF RECREATIONAL MARIJUANA

Black market commodities are highly inflated in cost, but nevertheless, the value of the illicit marijuana market is staggering. Gettman (2006) stated: "Domestic marijuana production has a value of $35.8 billion, more than corn and wheat combined, easily making it America's largest and most lucrative cash crop.… Marijuana is the top cash crop in 12 states, one of the top 3 cash crops in 30 states, and one of the top 5 cash crops in 39 states. The domestic marijuana crop is larger than cotton in Alabama, larger than grapes, vegetables and hay combined in California, larger than peanuts in Georgia, and larger than tobacco in both South Carolina and North Carolina."

Miron and Waldock (2010) estimated that, in the United States, about $9 billion would be gained annually from legalizing marijuana, due to savings in government expenditures of enforcement of marijuana prohibition.

In 2012, more than 300 economists, including three Nobel laureates, signed the following petition, suggesting that legalization of marijuana, combined with taxation, could produce the equivalent of about $14 billion in revenue for the U.S. government:

AN OPEN LETTER TO THE PRESIDENT, CONGRESS, GOVERNORS, AND STATE LEGISLATURES

We, the undersigned, call your attention to the attached [2005] report by Professor Jeffrey A. Miron, *The Budgetary Implications of Marijuana Prohibition* [http://www.prohibitioncosts.org/mironreport/]. The report shows that marijuana legalization – replacing prohibition with a system of taxation and regulation – would save $7.7 billion per year in state and federal expenditures on prohibition enforcement and produce tax revenues of at least $2.4 billion annually if marijuana were taxed like most consumer goods. If, however, marijuana were taxed similarly to alcohol or tobacco, it might generate as much as $6.2 billion annually.

Economist Stephen Easton (2009) wrote (in "Bloomberg Business/Debate Room" blog, not available):

The current prohibition on marijuana consumption exactly parallels the 1920s alcohol prohibition… like booze during Prohibition, this substance, marijuana, is the easy revenue of organized crime, contributing tens of billions of dollars to growers, who commit a variety of bad acts both at home and abroad. How much money is made from this single illegal substance? In fairness, nobody knows for sure… total spending on marijuana may add up to $45 billion to $110 billion a year. What about possible tax revenue? From Canada we've learned that the production cost of [government-sponsored] marijuana is roughly 33¢ a gram. [Costs are higher, as noted later.] Currently, U.S. marijuana consumers pay at least $10 per gram retail for illegal marijuana. If the cost of retailing and distribution is the same as for legal tobacco cigarettes, about 10¢ a gram, then selling the (legal) product at exactly the same price as on the street today ($10 per gram) could raise $40 billion to $100 billion in new revenue. Not chump change. Government would simply be transferring revenue from organized crime to the public purse.

These descriptions of the size of the illegal and potentially legal marijuana industry for the United States are subjective and have been strongly disputed for several years. A principal point of contention is that the negative costs (for example, from intoxicated driving) are not considered. Moreover, because of the criminal, clandestine nature of most trade in marijuana at present, it is extremely difficult to evaluate the quantities of cannabis produced, trafficked, and consumed— traditional parameters of a market analysis. Despite these limitations, almost certainly potentially gigantic revenues are possible. Table 15.1 suggests possible comparative potential global earnings

TABLE 15.1

Estimates of Potential Annual World Economic Value of Categories of *Cannabis sativa*

Category	Value
Ornament	Thousands
Phytoremediation	10s of thousands
Biomass	100s of thousands
Fiber	10s of millions
Oilseed	100s of millions
Pharmaceuticals	Billions
Medical herbal marijuana	10s of billions
Recreational marijuana	100s of billions

from both industrial and drug kinds of cannabis—estimates that necessarily are subject to very large errors.

While the current explosive interest in marijuana legitimization (whether exclusively for medicine, a general decriminalization, or even comprehensive legalization) is being driven by sincere libertarians and believers in its potential medical benefits, financial considerations are often determinative in human affairs (put simply, "money talks"). Benefits from marijuana taxes to local governments may be as high as 100%, reflective of the tendency to impose heavy "sin taxes" on morally questionable commodities (often tolerated on the grounds that the funds will be used for noble purposes such as school and hospital construction). The ultimate economic impact of partial or complete legalization cannot be reliably predicted. Most marijuana is traded on the illicit market, and how this relates to a future legal market is uncertain. Nevertheless, there is a general consensus that medical marijuana is a multibillion dollar enterprise, and the value of recreational marijuana could be astronomical (Table 15.1).

The "Drug War" insofar as it has involved marijuana has clearly been very costly to the public purse, with huge expenditures dedicated to law enforcement (Duke 1995; see Collins 2014 for an extensive analysis of the drug war). Incarcerating many individuals for infractions involving marijuana not only involves considerable overhead costs but also removes them as potentially productive citizens during their imprisonment and often subsequently limits the employability of ex-convicts. Moreover, prohibition of any commodity demanded by a segment of society has huge costs resulting from criminal activities, such as street violence, bribery of public officials, tax avoidance, and health problems. Nevertheless, it is simplistic to think that removing the marijuana black market will substantially lower societal costs: it is possible that should criminals no longer be able to benefit from trade in marijuana, they will simply turn to other illicit commodities that are equally or even more deleterious to society.

ECONOMIC EFFECTS OF LEGITIMIZING RECREATIONAL MARIJUANA ON THE ILLEGITIMATE MARIJUANA INDUSTRY

For good or evil, regions that have been centers of production and trade in illicit marijuana will suffer economically should legalization occur. Producers of illegal marijuana range from small-scale cottage industry (often conducted by students or "mom and pop" entrepreneurs) to large-scale operations sponsored by organized crime. The distortion on local economies can be spectacular, affecting a wide range of businesses, especially those selling expensive goods and services. In northern California's Emerald Triangle (overlapping Humboldt, Mendocino, and Trinity counties), it has been estimated that marijuana accounts for up to two-thirds of economic activity in the region, and property values have become highly dependent on their marijuana-growing potential

(Regan 2011). As a result, "there are an awful lot of young people driving BMW's, Mercedes, and tricked-out trucks. Local fast-food restaurant owners complain they can't find high school students willing to work in their restaurants because there's so much more money to be made in the marijuana trade" (Regan 2011).

COSTS OF PRODUCTION OF MARIJUANA

Chapter 14 presented information on the production of medical marijuana indoors, and in particular documented yields on an area and harvest basis (keep in mind that indoor cultivation typically produces four harvests annually). To summarize the key production efficiency data presented in Chapter 14, the literature indicates that yields (indoor production, weight/area) usually range from about 250 to about 500 g/m^2/harvest, with four harvests expected annually.

Leaving aside the special costs for security, if marijuana could be produced outdoors like hemp, the cost of production would be dirt cheap. Gieringer (1994) stated that "In an untaxed free market, cannabis ought to be as cheap as other leaf crops. Bulk marijuana might reasonably retail at the price of other medicinal herbs, around $0.75–$1.50 an ounce. Premium cured and manicured sinsemilla buds might be compared to fine teas, which range up to $2 per ounce, or to pipe tobacco, which retails for $1.25–$2.00." (These figures are unrealistically low, as noted in the following paragraphs.)

The cost of marijuana is substantial when grown in greenhouses (with or without supplemental lighting), but given that ornamentals and vegetables are often grown out-of-season in greenhouses, marijuana could equally be produced relatively cheaply. The cost is magnified at least 10 times when marijuana is produced indoors with just artificial lighting, as is usually done currently. The major cost factor is electricity, particularly for lighting, but also for climate control. Stringent security requirements at present usually dictate indoor growing with no or very limited natural light, but should greenhouse cultivation be permitted, the cost of production would diminish substantially, and should outdoor growth be allowed, the costs would decrease dramatically.

Economies of scale in production of cannabis are discussed by Hawken and Prieger (2013), who point out that large-scale operations (at least 200 m^2 or several thousand square feet) are much more efficient than small-scale facilities, and volume discounts for large consumers of electricity may also be significant. They also note that when the goal is production of trimmed buds, trimmers may constitute about 50% of total labor (although machine trimming can lower the burden). Capital investment costs in physical facilities are discussed by Caulkins (2010) and Hawken and Prieger (2013). Except for the special requirements for security, the facilities necessary are comparable to what is needed for indoor crops in general, although medicinal plants are a more reasonable basis for comparison than are flowers or vegetables.

Hawken and Prieger (2013) indicate that marijuana can be produced under artificial illumination for approximately $2.00 to $2.25 per gram, the costs decreasing when investments are prorated for longer periods and for about half these figures when produced in greenhouses (using mostly natural light). Retail costs for medical marijuana usually range from $10.00 to $20.00 per gram. In the United States, black market (illicit) marijuana averages about $13.00/g. In some jurisdictions, licensed medical marijuana is available at cheaper prices than the illegal street counterpart!

In Canada, production costs have been claimed to be as low as U.S.$2.20 per gram, employing natural light greenhouses, which, as noted previously, are substantially more economical than growth rooms using artificial light (Koven 2015).

ELEMENTARY BUSINESS HAZARDS: WILL THE "POT BOOM" BECOME A "POT BUBBLE?"

There is phenomenal enthusiasm for investment in marijuana, the market amusingly described by phrases such as smoking hot, high trade, mind-blowing profit, growth industry, budding business, and green gold rush. Many expect that just as dot.com billionaires were created during the computer revolution, so marijuana moguls are in line for great wealth (Figure 15.5). It is well to be reminded that most new market commodities go through a life cycle, with several phases. Figure 15.6 illustrates in a hypothetical fashion the relationship of profitability and time for many new (especially patent-free) products in a free market economy, the situation that prevails for most traditional marijuana products, especially herbal forms. Phase 1 is the period of investment in research and development, necessary to bring a new product to the point of profitability (most new products in fact do not survive this foundational period). Most classes of recreational marijuana and associated paraphernalia are well beyond phase 1, but many forms of pharmaceutical preparations are very new or being researched. During phase 2, the market expands, along with profitability. Phase 3 is a stable period of profitability, at the end of which decline in profits occurs (phase 4). The most common cause of the decline in profitability is copycat competition from those who have observed the profitability of the item.

FIGURE 15.5 Private sector investment in the exploding legalized marijuana industry is rapidly producing fabulous wealth, as in the early growth period of the dot.com Internet barons. Prepared by B. Brookes.

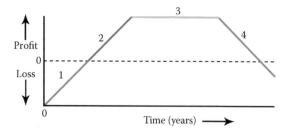

FIGURE 15.6 Simplified product life cycle, showing the phases of profitability frequently associated with new market offerings. See text for descriptions of the four phases and how they relate to the developing marijuana industry.

The more lucrative a given product becomes, the faster competition develops. A frequent consequence of such competition is oversaturation of the market, the generation of surpluses, and business failures of those who remained dependent on the sale of the once-profitable item. Developing new products (phase 1) is extremely risky but offers the greatest potential rewards because the inertia from having a large head start may allow one to rapidly capture a large proportion of the potential market and to hold that market for a long period. New pharmaceutical products, new manufacturing processes, and new applications, most of which are associated with intellectual property rights, are the riskiest but potentially the most profitable aspects of the marijuana industry. Provided that competition is still limited, the safest strategy is copying the example of those who recently established marijuana businesses centered on production, manufacturing, and distribution of herbal and extracted forms of marijuana based on public domain techniques. The most common mistake is belated investment in a popular commodity that has demonstrated sustained high profitability but for which the market is rapidly becoming oversaturated. In some areas of Los Angeles, medical marijuana dispensaries outnumber McDonald restaurants (more than 1000 dispensaries are claimed to be present in the city), and such proliferation of "pot shops" has developed in several North American cities (Figure 15.7), including Vancouver, British Columbia.

The laws of supply and demand and market timing ultimately determine profitability, and legal marijuana as a commodity is in the enviable position of having limited supply and enormous potential demand. However, this situation invites competition, rapid overinvestment, and the generation of an oversupply resulting in a depression of the market. In this regard, the history of *Cannabis sativa* introduction in Canada for industrial hemp is instructive. Authorization to grow licensed industrial hemp began in 1998. For several years thereafter, hundreds of licenses were acquired, mostly by idealistic individuals with limited experience in farming

FIGURE 15.7   The marijuana dispensary bubble. Top: Before; bottom: after. Prepared by B. Brookes.

and business. The result was widespread crop and business failures, bankruptcies, and an enormous glut of hempseed and hemp fiber that depressed the market, making industrial hemp unprofitable in the country for years. A parallel situation has developed in Canada for medical marijuana. The production of medicinal marijuana was authorized by just one major company in Canada from 2002 to 2013. As of 2015, any company satisfying rather demanding regulations can produce medical marijuana, and there have been more than 1000 license applications (although only a few dozen have been approved). The frenzy of new entrants was accompanied by marijuana companies issuing stock offerings, and the resulting giant marijuana stock bubble that was formed in 2014 collapsed in the same year (Koven and Pett 2015). However, with the election of a new government expected to authorize recreational marijuana, cannabis has once again become a hot commodity in the Canadian marketplace.

In the United States, there has been a frenzy of stock offerings in marijuana companies, and as of 2016, there have been associated reports of plummeting stock values. The British firm GW Pharmaceuticals, specializing in compounded cannabinoid extract preparations, is quite exceptional in its success (reportedly valued at $2 billion) and its stock offerings have been very profitable.

POSSIBLE SCENARIOS OF SUCCESS OR COLLAPSE
OF SECTORS OF THE MARIJUANA MARKET

HERBAL

Law enforcement agencies have often assigned a value of $1000.00 to a single marijuana plant. While this is an exaggeration, it is clear that on the black market, marijuana has a very large value. Even when produced legally, marijuana is an exceptionally valuable commodity. The National Institute on Drug Abuse, the only federally legal supplier in the United States, charges researchers $1525.00 per kg, or $7.00 per joint (Reardon 2015). The elevated price is necessary because production costs are very high (primarily because of the high costs of security and indoor production and the need to conform to quality standards). The very high price for the U.S. national supply of marijuana is artificial, a result of a tightly controlled market. In reality, marijuana plants can be grown and turned into product very cheaply, a circumstance in which the possibility of a market crash exists should regulations permit free competition.

At present, growing marijuana (even for personal consumption) is almost universally illegal, and this very strongly inhibits home cultivation. However, should recreational marijuana become widely available, it is likely that attitudes toward personal cultivation of plants will change. Were home cultivation to be widely permitted, it could drastically curtail commercial market demand for herbal forms of the plant, or at least suppress profitability. One may well ask why people do not grow their own tobacco plants, but this is probably because preparing high-quality tobacco is a very demanding art and science, while almost any dummy can raise a pot of pot, especially since equipment for cultivation is accessible via the Internet and from "grow-shops." Moreover, a very high percentage of cannabis growers cultivate the plant not primarily for profit but for enjoyment and ideology (Hammersvik et al. 2012). According to Leggett (2006): "With the possible exception of some amphetamine-type stimulants, cannabis is the only drug where the entire market chain, from production to consumption, can be contained in a single individual."

EXTRACTS

There are numerous financial risks associated with the development of new pharmaceuticals, particularly those based on plant extracts (see Whittle and Guy 2004 for an analysis of issues related to cannabis). Nevertheless, pharmaceutical extracts and formulations based on extracts represent a particularly promising sector of the medical marijuana industry. Indeed, for decades, governments

and the medical profession have been sympathetic to the idea of completely replacing herbal marijuana with pure chemicals for medical purposes (a movement termed "pharmaceuticalization of
marijuana"). While some proprietary products seem promising, none has reached the status of a
breakthrough drug, and like all ventures, the possibility exists of market failure. Moreover, non-
patented extracts can be prepared relatively easily, leading to the possibility that products intended
for the medical market could face cheap competition. Candidly, extracts for use in vaporizers for
the recreational market may represent the most profitable new market niche, but like tobacco-based
vaporization, social pressure may limit usage (at least in public).

Why Not Harvest CBD from Hemp?

CBD represents a particularly promising investment opportunity. As detailed in Chapter 13, it has
more promise for treating illnesses than any other cannabinoid, including THC. Since it is nonintoxicant, it has little of the abuse potential of THC (although it can be converted to THC, as noted
in Chapter 11, there is little practical danger of this being done illicitly). CBD is the chief cannabinoid of hemp, which can now be cultivated very cheaply outdoors in most countries. Indeed, CBD
can be harvested as a "waste product" from the remains of harvesting hempseed. There is a high
demand for CBD, which currently is being obtained from expensive indoor cultivation, because of
legal requirements. Common sense dictates that current legal constraints to harvesting CBD from
hemp should be removed, and the hemp industry permitted to harvest CBD and indeed other nonintoxicant cannabinoids. Cheaply produced CBD would, however, be disastrous for those who have
invested in its production from expensive indoor plants.

SYNTHETICS

One of the very disturbing aspects of illicit drugs of abuse is synthetic designer drugs, which periodically are introduced, usually with very harmful results (see Chapter 12). Whether harmful or
benign, the possibility exists that a new drug will capture the public's attention, diverting interest
from marijuana, and disappointing those who have invested heavily in it. Synthetic preparations of
THC, while expensive and often considered to be less effective than conventional marijuana, may
one day represent competition for *C. sativa* for medicinal purposes, much like synthetic (artificial)
vanilla has largely replaced vanilla from the vanilla plant.

GENETICALLY ENGINEERED *CANNABIS SATIVA*

Many of the world's major crops have been genetically engineered, and transformed cultivars are
becoming increasingly dominant. Genetic engineering is controversial and is a roadblock to marketing in some countries, especially in Europe. Both low-THC (industrial hemp) and high-THC
(marijuana) forms of *C. sativa* are symbolically viewed by many as representative of freedom from
the unreasonable constraints of the power brokers of society, including the international corporations that now dominate modern agriculture. As noted in Chapter 11, a gene controlling THC synthesis has been transferred to a tissue culture of tobacco, so that if precursors are fed to the culture,
it will transform them to the acidic form of THC. This suggests that tissue or cell cultures could
be employed for the biotechnological production of cannabinoids in the future (Sirikantaramas et
al. 2007; Zirpel et al. 2015). The possibilities of genetically engineering *C. sativa* go far beyond
simply producing "Frankenpot"—"super plants" (facetiously illustrated in Figure 15.8) could be
produced with extraordinary abilities with respect to size, longevity, growth rate, tolerance to
diseases, pests, climate, soils, odor, taste, and novelty (for example, plants that glow in the dark).
There have been fears that some of the very high-THC strains now in circulation were generated
by genetic engineering (Cascini 2012). However, the possibilities of creating novel life forms go
beyond plants—as noted previously, tissue cultures can be a basis for producing a desired cannabinoid, and valuable chemicals can also be biotically generated by transformed microorganisms

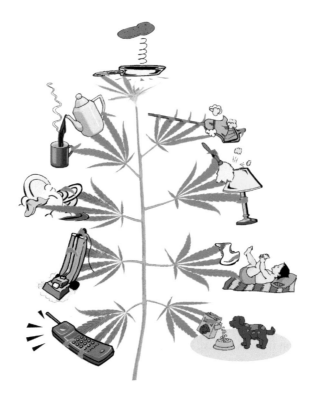

FIGURE 15.8   A hypothetical genetically engineered cannabis plant. Prepared by B. Brookes.

such as yeasts in giant vats. The opposition to genetic engineering has been mostly unsuccessful in preventing its advance, and it is certain that *C. sativa* will be transformed, if not already achieved. Some crops are now dominated by transformed cultivars, and this has generated winners and losers in the marketplace. There would seem to be a strong possibility that this pattern will also develop for marijuana.

MARKETPLACE CONSTRAINTS RELATED TO "CHEAP HIGHS"

COMPARATIVE COSTS OF POT AND BOOZE

While it may appear indelicate, cost is a major constraint to intoxication. Most people cannot afford expensive wines and liquors, and many find that their love (or addiction) for the intoxicant of their choice is a major financial drain. The price of cannabis is very low in most countries, even by comparison with other illegal drugs (Stockwell et al. 2010). "In producer countries in the developing world, it is sometimes cheaper to get 'high' on cannabis than it is to get drunk on beer" (Leggett 2006). Prices in the United States are comparatively expensive, but nevertheless, a casual user can become intoxicated for less than $5.00, which is remarkable when one considers that black market prices for illegal drugs are highly inflated. Moreover, cannabis is widely produced by individuals for their own personal use, cutting down costs. Additionally, marijuana is a remarkably social drug, frequently shared, so that some contribute very little to its purchase. Because marijuana is so cheap to produce and buy, it suggests that large amounts would have to be sold to be profitable, which is probably not a constraint for recreational marijuana (already very profitable as a black market commodity). On the other hand, those faced with buying expensive medical marijuana or pharmaceuticals may consider purchasing much less expensive street-available material.

COEXISTENCE OF BLACK AND WHITE MARKETS

Expensive products sold in the legitimate marketplace are frequently also available illicitly at much cheaper prices, as a way of avoiding taxes or license fees, because they are cheaply produced counterfeits or because they are stolen. Governments view legalized marijuana as an opportunity to collect large taxes on all aspects of the industry, and the resulting inflated prices are likely to incentivize continued existence of the underground economy for cannabis. In the case of marijuana, illicit material is supplied both by organized crime on an industrial scale and by small entrepreneurs on a cottage industry basis. The illicit trade in alcohol involving moonshine (illegal production of high-proof distilled spirits) and bootlegging (illegal transportation or smuggling of alcoholic beverages) provides a parallel example of an intoxicant that is much more expensive legally than illegally. Black markets do provide employment and generate economic activity, but the costs in law enforcement and health costs associated with inferior products can be significant. Until cures for dishonesty and greed are discovered, the illicit market in marijuana is likely to continue.

ABSENCE OF FOREIGN COMPETITION FOR AUTHORIZED MARIJUANA

Marijuana, as well as many other inebriating drugs, is illicitly transported across borders. Most marijuana is in fact a black market commodity, and its geographical production is governed by interregional competition. Because herbaceous forms of marijuana are very bulky, odorous, and easily detected, most illicit marijuana is locally produced today or at least rarely crosses more than one border. In the past, however, the United States was mostly supplied by foreign sources—Mexico from the 1930s to the mid-1970s, Colombia during the 1980s, and appreciable importation from British Columbia subsequently.

Medical marijuana could in theory be authorized for export and import, but at this time, it rarely crosses country borders, except in the European Union; Italy, Finland, Germany, and Switzerland import products from the Dutch program. Most states prefer to maintain close control over production, quality, and distribution, so medical marijuana is almost entirely produced domestically. The possibility exists, however, that in the future, much cheaper foreign production costs could greatly alter the advantage to producers of having an exclusively domestic market. Presumably Morocco, which has been a significant supplier for Europe, could develop a large legal market for the continent, and Mexico and the Caribbean could similarly be major suppliers for the United States and Canada. At present, there are very large investments in cannabis production underway in developing nations, where costs are low, on the speculation that importation into rich Western countries will be possible. Transportation costs for bulk herbal materials are appreciable, so imports of high-THC preparations, particularly extracts, are likely to be the leading forms of imported marijuana.

CONSUMER TRENDS

At present, vaporizers (allowing smoke-free consumption), concentrates (decreasing the amount of material required), and edibles are in high demand in U.S. medical marijuana dispensaries. There is developing interest in pharmaceuticals based on particular cannabinoids, particularly the extraordinarily versatile CBD.

MARKET PRODUCTION MODELS: BIG VS. SMALL-SCALE BUSINESSES

A "cottage industry" is a small business managed from a residence, typically a part-time operation run by one or two individuals with incomes of less than $100,000. "Small-scale industry" refers to operations with limited physical facilities centered in one or a few cities, typically with dozens of employees and generating millions of dollars in income. "Big business" operations typically have hundreds of employees, often are international, and generate much larger incomes. Illicit marijuana

"The debate over how to legalise cannabis tends to assume that for-profit commercial enterprise is the default option. Legalising cannabis on the alcohol model may, however, be the second-worst option (behind only continued prohibition); commercialisation creates an industry with a strong incentive to promote heavy use and appeal to minors through aggressive marketing. No system of legal availability is likely to entirely prevent an increase in problem use. But pioneering jurisdictions should consider alternative approaches including non-profit regimes and state monopoly. Both sides of the legalisation debate should acknowledge that the question is complex and the range of uncertainties wide. Such modesty, alas, is in short supply."

Kleiman and Ziskind (2014)

is currently generated in huge quantities at all three scales: at the cottage industry level, by small-scale gangs, and by large cartels. What levels of production are appropriately authorized in a legal market? Large-scale operations of most commodities have advantages of scale, and the capital and resources to ensure quality and safety, but habitually restrict their product range to the most profitable items (as do the major beer and liquor companies). Small-scale businesses are well known to provide the bulk of business employment and often provide specialty products (as do microbreweries). Cottage industry production of marijuana is very unlikely to be conducted as safely as regulated industries, but is so easy and common that whatever regulations and laws are enacted, it will continue (legally or not). Governments tend to prefer restricting production to large operations, which are much easier to monitor. Over time, one can expect that smaller operations will be authorized, especially if costs of authorized products are so high that the black market industry remains large.

MARKET OUTLET FORMATS

There are too few models of legal marijuana distribution to evaluate the comparative impacts, both positively and negatively. There is considerable concern that "wild west legalization" and accompanying rampant commercialization may be damaging, and so there is often a belief that public health requires tight control of production, distribution sales, and marketing. The following discussion examines basic alternatives. Extremely restrictive distribution systems could involve monopolistic "nanny state" or designated corporate control, but various degrees of marketing freedom are conceivable.

STATE MONOPOLY VS. PRIVATE SECTOR

Regulated control of production and/or delivery of marijuana could be based primarily or exclusively on (1) government monopoly, either as a designated function of government employees or assigned to one private sector company (the jurisdiction could be an entire country or a political subdivision); (2) an oligopoly (a market structure in which a few firms dominate), but unlike some oligopolies, which also permit many small firms, just a few large firms meeting stringent requirements are usually authorized (the current model in Canada); or (3) independent growers and dealers on a free market basis, with minimal requisites.

Generally, two basic models are common for herbal medical marijuana dispensation (assuming both source and patient are licensed). On the one hand, the firm that produces the material from plant to package directly provides it to the patient, either by courier or at a central dispensing clinic. On the other, dispensing is by a retailer, which could be as large as a drug store chain or as small as an individual physician. Private sector distribution could be restricted to very large suppliers (this has been referred to as the "Marlboro-ization of marijuana") or left to develop in the free market, with the result that there will be many small sources. Both the monopolistic and free market alternatives have developed in the Netherlands: pharmaceutical-grade herbal marijuana is available from a single state-authorized supplier on a prescription basis, but the widespread acceptance of street-available marijuana and its sale in private retail outlets represents a free-market distribution system (that is rarely accepted elsewhere in the Western world). Despite the existence of only one authorized national medical marijuana supplier in both the Netherlands and the United States, most medical marijuana is purchased from other sources in these countries. Until recently, only one national medical marijuana supplier was authorized in Canada, and once again, most authorized patients purchased their supplies from other sources (mostly small legal growers designated by the patients). One cannot avoid noticing that governmental attempts to supply commodities on a monopoly basis are rarely competitive with the private sector.

Gettman and Kennedy (2014) argue that production, sale, and marketing of cannabis should be left to the free market, with minimal governmental control, because "Marijuana, as a commodity for production, has unique attributes that distinguish it from alcohol and tobacco. It is relatively easy to grow and does not require industrial processing. Marijuana can be produced anywhere by just about anyone. It is grown throughout the country, in backyards, closets, attics, basements, and warehouses. While little technology is needed to grow marijuana, ample technology to maximize production and yield are widely, legally, available. This is a considerable factor in why prohibition has failed to control the production of marijuana. This will also be a considerable factor in the success or failure of any alternative regulatory regime."

"Weed Bars"

Once a euphoric substance becomes legalized, public consumption in dedicated dispensing establishments becomes a possibility. In the Netherlands, perhaps the world's most permissive country with respect to recreational marijuana, supplies can be purchased and consumed in so-called "coffee shops" (more appropriately termed pot cafés or hash bars; Figure 15.9). The prospect that marijuana could be similarly consumed in Starbucks-like outlets seems remote at present in most nations

FIGURE 15.9   "Coffee shop" (a euphemism for marijuana bar) in Amsterdam, illustrative of the permissiveness of retailing marijuana in the Netherlands. Photo by Bachrach44 (released into the public domain).

but is within the realm of possibility. In most cities, purveyors of cannabis can be expected to be as welcome as porn shops, strip joints, and even houses of ill repute. The likelihood is that frequently zoning will be similar to that afforded bars, with the possibility of being forbidden near schools, churches, and perhaps city centers and tourist areas. Indeed, a study of the placement of dispensaries in California showed that they were located in areas with higher rates of poverty, just outside city boundaries, often near alcohol outlets (Morrison et al. 2014). Unfortunately, marijuana-based dispensaries may often be confined to the seediest part of town.

Pot Restaurants

Weed bars, discussed in the previous paragraph, are rather parallel in nature to conventional bars, where food offerings are rarely substantial. Weed bars in the Netherlands are mostly devoted to smoking and vaping, not onsite consumption of edibles. However, it may be anticipated that the recent growing interest in marijuana edibles may eventually spawn their consumption in dedicated premises. Cannabis-infused food, like alcoholic beverages, has a special attraction because it satisfies not just a desire for euphoria but also provides the pleasures of taste, odor, and hunger satiation. Coincidentally, both alcohol and cannabis are powerful social lubricants, and so cannabis has the potential of displacing some of the usage of alcohol in the hospitality industries. Legalization of recreational marijuana, where it has occurred, has been a strong stimulus to an associated pot edibles industry, involving mostly common desert items like brownies, sweet beverages, and candies. At present, edibles are being marketed in dispensaries and boutiques, but one can anticipate that they may also be offered in restaurants, which could offer THC-laced haute cuisine. Of course, just as with alcohol, excessive consumption of cannabis could generate hazards and liabilities when customers become excessively "baked." A difficulty with the idea of consuming cannabis edibles in commercial establishments is that the effects from digestion are substantially slower, less predictable, and longer-lasting than pulmonary absorption. Eating increases stomach contents, affecting the rate of THC assimilation, contributing to the unpredictability of how long and how intensely the effects of the THC consumed will persist. Nevertheless, given that restaurant operators and patrons have learned how to safely accommodate onsite alcohol consumption, it is conceivable that the same may be true for cannabis edibles.

Home Delivery ("Weed on Wheels")

While the concept may seem odd given the persisting illegality of purchasing marijuana from street vendors, there are advantages to services that transport cannabis products to residences on a demand basis. In some cities, home delivery of alcohol is available, and apart from the possible issue of encouraging alcoholism, this can reduce the incidence of driving under the influence, at least for those who remain at the place where the beverages are received. A marijuana edibles delivery service, modeled on pizza delivery (Figure 15.10), could be particularly attractive for patients who are too debilitated to cook for themselves, combining provision of healthy food (like "meals on wheels" programs) in combination with therapeutic cannabis. Indeed, home delivery of pharmaceuticals to those who are house-bound is an established practice, and this model can be followed for medical cannabis. In several U.S. states, dispensaries frequently offer a delivery service for medical marijuana, occasionally under names like "Speed Weed," "Foggy Daze Delivery Service," and "Your Friendly Neighborhood Drug Dealer." Indeed, many cannabis-based retail businesses prefer to operate exclusively as a mobile service, as this avoids the necessity of establishing a storefront, which currently usually is an expensive and cumbersome exercise. Given the stigmatized nature of marijuana and its attraction to criminals, home delivery vehicles presumably would best be inconspicuous, although some enterprising entrepreneurs may prefer garishly decorated vans. There are areas of the United States that have enacted "dry laws" to prevent medical marijuana dispensaries from being established, and such "pot deserts" deny ready access to medical cannabis to residents.

FIGURE 15.10   Cannabis home delivery. Prepared by B. Brookes.

Cannabis delivery services would seem to represent a reasonable compromise, satisfying the wishes of local areas to prevent the presence of marijuana retail outlets, while still allowing convenient availability. In many areas of the United States, there is considerable legal conflict about what is and isn't permissible with regard to cannabis delivery, and investing in this field will remain somewhat hazardous until regulatory policies become established.

ADVERTISING

Advertisers are well aware that promoting their wares in the mass media is not merely a way of educating customers, but a determinant of profitability. They are of course not primarily interested in public health; rather, the motivation is to maximize profits, build market share, and induce as many consumers as possible to purchase the products. Because of the inevitable likelihood that unfettered advertising of medical (and even recreational) marijuana will persuade many who should not consume the substance, there is strong pressure to prohibit all marijuana advertising. However, the Internet is impossible to control completely, and market representatives are highly skilled at bringing their company's products to public attention. There is particular concern regarding how marijuana advertising affects adolescents (D'Amico et al. 2015). Limitations on advertising for alcohol, tobacco, gambling, and indeed on pharmaceuticals provide guidelines on how marijuana advertising may be constrained in the future. Advertising on medical marijuana dispensation by private companies is currently banned in Canada. Advertising for the government-regulated pharmacy sales in Uruguay is also banned. Advertising of marijuana in the United States is covered to an uncertain extent by federal regulations covering television, the mail, and interstate commerce, and this has raised uncertainties about what is occurring in states that have permissive rules. In Colorado and Washington, advertising has been relatively unfettered. At the minimum, glossy enlarged photos of glistening buds are featured; more provocatively, scantily clad models recline amidst encircling plumes of colorful smoke. This is reminiscent of how the tobacco industry "targeted women, children, and vulnerable groups by associating smoking with images of freedom, sex appeal, cartoon characters, and—in the early days—health benefits" (Richter and Levy 2014).

POTENCY AS A DETERMINANT OF SALES RESTRICTION

Control of production and sales of alcoholic beverages provides an interesting parallel example of how an inebriant substance is marketed in relation to concentration. In some jurisdictions, sales of high-alcohol beverages are entirely government controlled, while in others, distribution is in the hands of the private sector. Of course, there are restrictions based on age and customer sobriety. Low-alcohol content (beer, cider, and wine) are sometimes permitted in certain stores (e.g., super-markets), while high-proof beverages (liquors) are occasionally confined to designated outlets. It seems uncertain whether a similar system based on strength of the intoxicating ingredient could be devised to separate sales of low-THC herbal marijuana and high-THC hashish. Most stores are free to sell dealcoholized or very low-alcohol beverages; conceivably low-THC/high-CBD medical marijuana is conceptually similar. (At present, high-CBD hemp, despite lacking genuine intoxicant ability, is treated in much of the world as falling under the same laws that govern high-THC marijuana.) The most important restriction should deal with purified extracts or synthetic cannabinoids (and their precursor chemicals), since concentrated preparations can be especially dangerous. In a similar vein, pure alcohol (widely employed for industrial and scientific research) is usually subject to rigid licensing and monitoring. There has been consideration given in several countries (notably the United Kingdom, Uruguay, and the Netherlands) to having a regulatory framework that considers the percentage of THC in cannabis (Freeman and Swift 2016), but this has not yet resulted in a system of distribution on the basis of potency.

THE AGE OF MAJORITY—A PROBLEMATICAL DETERMINANT OF SALES AND CONSUMPTION RESTRICTIONS

Cannabis consumption in numerous places is rampant among children (Chapter 12), and indeed, an alarming number of adolescents have become Illicit dealers. In the light of this, it is somewhat academic to address the issue of attempting to limit availability to the very young. Nevertheless, the issue is of considerable importance.

The "age of majority," indicative of transition from legal minor status to adulthood, generally varies from 18 to 21. As noted in Chapters 12 and 13, the greatest harm potential of cannabis appears to be consumption by the young, defined as ranging up to 25 years of age. Because a proportion of young people between the conventional age of majority and 25 years seem significantly susceptible to brain impairment from cannabis usage, the minimum age for legal consumption is problematical. As with other privileges accompanying the age of majority, young people are eager to take full advantage of their legal status, and it is very difficult to conceive that they would accept restrictions to the ripe old age of 25 insofar as recreational cannabis is concerned. However, the medical profession needs to take account of the relative inadvisability of utilization until this age is reached.

LEGAL CONSIDERATIONS

CAUTIONS

Production, distribution, and usage of marijuana are governed by different legislations according to region, and indeed, policies are evolving rapidly. For the possible development of medical and/or recreational marijuana, it is important to appreciate the legal status of marijuana locally. In some countries (notably Indonesia, Malaysia, Iran, Singapore, and United Arab Emirates), penalties are severe, especially for trafficking, and may include death. Many countries (including Canada, Denmark, Germany, Spain, the Netherlands, New Zealand, the United Kingdom, and the United States) allow marijuana to be used for medical purposes. The precise regulations governing cannabis in areas where it is currently authorized are often complex and often periodically undergo changes. Authorization to produce, prescribe, sell, and employ medical marijuana (occasionally also recreational cannabis) differ considerably depending on political region. As with other complex laws and regulations governing substances and activities that are subject to extreme penalties, even when transgressions occur innocently, it is sometimes too easy to run afoul of requirements or to be accused of such. Accordingly, anyone entering into business aspects related to marijuana needs to be thoroughly aware of the rules and to understand potential pitfalls. It is probably also true that given the possible danger of accidental exposure to legal prosecution (to say nothing of the hazards from the criminal element), engaging in the developing marijuana commerce requires courage.

"Compassion clubs" providing marijuana to sick people have often run afoul of the law, and those selling marijuana in for-profit establishments need to be particularly cautious because they will not be viewed as sympathetically. "Pot speakeasies" are regularly established by brave (or foolish) entrepreneurs, testing the tolerance of the police and local government. Frequent tactics of such ventures include claims that the establishment is a "club" or "clinic" not subject to local regulations. Often, customers are required to sign long waivers absolving the business of health and legal consequences. Such testing of the boundaries of current laws governing cannabis reflects not only uncertainties about the tolerances of society and the political system but also a conviction that the future will be considerably more permissive.

INTERNATIONAL TREATIES

At the beginning of the twentieth century, international conferences concerned with the damage associated with opium use laid the groundwork for later international sanctions against harmful plant-based drugs (Mead 2014). More than a dozen international agreements dealing with psychoactive substances were signed. These were combined in the United Nations Single Convention Treaty on Narcotic Drugs of 1961. This obligated signing countries to enact controls on cannabis, while allowing certain activities to proceed with restrictions. Cannabis and its products were defined as "narcotics with a high potential for abuse and no accepted medicinal value." THC was not characterized until 1964, so early legislation did not use it as a legal criterion. The laws that countries agreed to establish were subsequently modified by two international treaties: the 1971 U.N. Convention on Psychotropic Substances and the 1988 U.N. Convention against Illicit Traffic in Narcotic Drugs and Psychotropic Substances. The agreements require countries that are signatories to prohibit numerous activities concerned with marijuana and its constituents but allow exceptions, particularly for medical purposes. Just how much discretion individual countries may exercise in these respects has been argued (Mead 2014). Room (2014) states that, while medical use is allowed under the conventions, legalization of recreational marijuana (as in Uruguay and the states of Alaska, Colorado, Oregon, and Washington) contravenes the 1961 Single Convention on Narcotic Drugs, as well as the 1988 Convention. Room concluded that "The world is now saddled with drug treaties which are not fit for purpose." Leggett (2006) noted that "Signatories to a range of international drug control treaties have agreed that cannabis should be deemed an illicit drug. Despite these agreements, many States have, in various ways, relaxed their controls over cannabis. Even where these changes do not amount to a breach of the treaties, there appears to be a divergence in spirit between international agreements and individual State action. This discontinuity has not been addressed at an international level and thus international efforts to address cannabis have also fallen by the wayside." Rodman (2015) discusses the need for reform of the United Nations treaties dealing with cannabis.

LEGAL MODELS

THE NETHERLANDS

The Netherlands, for several decades the center of economic development of high-THC cannabis, is uniquely permissive with respect to marijuana. Beginning in 1976, penalties for cannabis were softened, and eventually retail outlets euphemistically termed "coffee shops" were allowed to sell small amounts of marijuana. Although cultivation and supply are technically limited to a national supplier and usage to prescriptions, enforcement is nominal or weak, and the situation has been described as "de facto legalization."

UNITED STATES

The United States is not only the world's largest consumer of herbal marijuana, it is also highly influential in establishing public policies in the Western world. "Since the 1970s, the national trend has moved toward decriminalization, increased social acceptance, and legalization for medical use. Today, more than half the states in the United States have decriminalized the possession of small amounts of marijuana, approved it for medical use, or legalized it completely. Numerous other states—both liberal and conservative—are considering legalization, indicating that the recreational use of marijuana is no longer a partisan issue" (Hickenlooper 2014). Although the federal U.S. government still prohibits marijuana, it has not blocked states from legalizing it for medical or recreational purposes, provided that a number of conditions are met (such as preventing distribution to minors and not allowing diversion to states where marijuana is still illegal). The states of Washington and Colorado legalized the recreational use of marijuana in 2012, authorizing commercial cultivation, processing, and sales to adults, who are allowed to possess up to 1 ounce (28 g) for personal use. In Colorado, up to six plants can be cultivated by individuals. California's lack of appreciable restrictions on medical marijuana has been said to represent de facto legalization for recreational purposes.

URUGUAY

Uruguay became the first country to legalize the recreational use of marijuana in 2014. Individuals can purchase up to 40 g/month from registered pharmacies, cultivate up to six plants producing up to 480 g annually, or join a club that has the same personal limits.

OTHER COUNTRIES

Recreational usage is prohibited by law in almost all Western countries. Medical usage is permitted in many countries, generally with strict conditions. In some cases, there is just one authorized supplier, the government itself or a designated private company. In other cases, there are private sector suppliers. Some countries like Spain allow organizations ("clubs") to dispense marijuana to its membership. Sometimes, individuals are authorized to cultivate a small personal supply.

PATENTS

An "invention" is "a new product or process that solves a technical problem" (WIPO 2010). Some inventions are improvements on previous solutions. Patents are exclusive rights granted for inventions. Patents are provided by governments, giving inventors rights over copying, using, distributing, or selling their inventions, for a specific period (typically 20 years). Before a patent is issued, an inventor must provide convincing evidence that the new invention is (1) useful, (2) novel, and (3) not so obvious that others could have easily deduced it. Inventions are a major part of "intellectual property," but to be profitably commercial, other aspects of intellectual property such as copyright, trademarks, and industrial design are often required. Because marijuana has been used for thousands of years, it is often difficult to discern when a given form or process employed to produce it is patentable.

There have been several patents relating to the biochemistry of the cannabinoids. Especially interesting is Patent 6630507: Cannabinoids as antioxidants and neuroprotectants, filed by the U.S. government in 1999, rather ironically since at the time, it vehemently opposed cannabis and viewed it as a substance utterly without medical value. The patent was based on findings of Aiden Hampson and Julius Axelrod (a Nobel Prize-winning neuroscientist), who discovered that CBD showed promise in limiting neurological damage in patients with Alzheimer's disease and Parkinson's disease and in those who have suffered a stroke or head trauma (Barcott and Scherer 2015).

In Chapter 12, considerable information was provided on the production of resin powder, a high-THC form of cannabis. Equipment for production of cannabis resin powder traces to (1) ancient techniques and apparatus that evolved in Asia during past centuries, (2) techniques and apparatus invented during the past half century in the drug subculture of Western countries, and (3) techniques and apparatus invented during recent decades by merchants servicing the drug subculture of Western countries. Because the technology and apparatus developed for the preparation of cannabis resin powder were invented largely in the counterculture and indeed often constitutes "drug paraphernalia," intellectual claims are problematical. There are conflicting patent claims as to who invented and/or adapted apparatus and techniques for wet and dry sieving production of purified, high-THC preparations of the secretory glands (these two basic classes were described in Chapter 12). The techniques and apparatus (or at least similar apparatus) may have been used for decades, perhaps even for centuries. Moreover, the intellectual property issue is uncertain because the equipment and techniques have been largely utilized in an illegal or at least legally problematical setting. Mila Jansen of the Netherlands and the company she established have been particularly significant sources of equipment for production of cannabis resin powder (Jansen and Teris 2002).

A variety of modalities for administering cannabinoids are available or may be produced in the future. Commercial utilization of equipment and techniques to produce authorized cannabis preparations needs to consider the potentially problematical nature of intellectual property claims.

The added-value potential of proprietary drug derivatives of the cannabinoids and drug-delivery systems is huge. The British firm GW Pharmaceuticals has been very active in the development of cannabis technologies, both extracts and in herbal forms, which are advertised as providing the consumer with a more efficient delivery system (more THC, less tar, etc.) and therefore greater safety. Other firms as well are researching cannabinoid delivery systems. The long-term significance of such patented technological developments, which could make obsolete the use of medicinal marijuana as currently supplied, needs to be examined.

As described in Chapter 13, noncannabinoid compounds that are naturally in herbal marijuana, particularly terpenes and flavonoids, may augment the therapeutic effects of cannabinoids. In principal, there is no reason to restrict the composition of a marijuana-based formulation to compounds that are naturally in the plant, and one can foresee the marketing of imaginative combinations of ingredients. Moreover, just as the tobacco industry has marketed cigarettes with flavorants, humectants, and other materials to improve organoleptic qualities, one can foresee the same for marijuana. The extent to which given formulations constitute intellectual property may be open to debate.

There are numerous cultivars or biotypes of crops that have been created through genetic engineering, and the documentation of the genome of *C. sativa* (Van Bakel et al. 2011) stimulated interest in exploiting its genes. Russo (2011b) commented: "It is certain that the production of genetically modified organism (GMO) cannabis plants would provoke tremendous controversy among consumers, and that battles over patents and breeding rights would be obvious sequelae of such a development. Any individual or corporation anticipating dipping their toes into such an endeavor may expect to encounter a veritable regulatory minefield while attempting to license such a product."

OPIUM POPPY AS A BUSINESS MODEL FOR MARIJUANA PRODUCTION

Opium poppy (*Papaver somniferum* L.) provides an instructive example of how the world has made compromises in order to grow a plant for both drug and nondrug products (Small 2010). Reminiscent of how *C. sativa* has been altered by humans into three different categories of plant (fiber, oilseed and drugs), so *P. somniferum* has been domesticated into three different groups of plants: drug cultivars, oilseed/condiment cultivars, and ornamentals. Most people are unaware that the source of poppy seeds commonly used on bakery products is the opium poppy, exactly the same species that produces opiate drugs such as heroin. Indeed, the very same plants are often the source of both condiments and drugs. As noted in the following, most people are also unaware that ornamental poppies, often cultivated on their property, are frequently opium poppies, and that opium poppy grows wild in numerous areas.

Like *C. sativa*, the opium poppy has become established as a weed throughout much of the world. The species is believed to grow wild in the Mediterranean region, from the Canary Isles eastward, but is found as an escape from cultivation in fields, roadsides, and waste places in scattered localities almost everywhere, including North America. However, unlike governmental attempts to eradicate wild "ditchweed," there have not been any significant efforts to exterminate the wild-growing poppies.

While millions have been prosecuted for cultivation or possession of marijuana forms of *C. sativa*, almost no one has run afoul of the law because they grow ornamental opium poppies. Opium poppy seeds for growing the plants are widely available in stores and from seed supply firms, but dealers are virtually never charged with traffic in narcotics. As noted in the next paragraph, ornamental opium poppies do not have as high a potential for harm as do selected drug strains, but they still do possess opiate constituents. The tolerance for opium poppy cultivation exhibited by society in general and law enforcement in particular stands in stark contradiction to the intolerance of cannabis cultivation. It demonstrates that the mere cultivation of potentially dangerous drug plants is not necessarily dangerous.

However, it cannot be disputed that some plant-derived drug substances are dangerous, and in the history of human civilization, opium has been the most evil (Figure 15.11). Crude opium is typically just the hardened latex (milky sap) of the unripe fruit (capsule) of the opium poppy. Opium contains a mixture of many constituents, including the alkaloids morphine and codeine. Morphine is normally the most abundant alkaloid present in opium. Opium has traditionally been obtained by making incisions into the nearly ripe poppy capsules 10–20 days after flowering. In cooler climates, incisions do not seem to result in good exudation of latex, and mature capsules are simply collected for chemical extraction. Morphine is often extracted from the capsules of oilseed cultivars after the seeds have been harvested, although drug cultivars are more productive. The capsules of some ornamental forms of opium poppy have less than 1% opiate alkaloids, while those of drug cultivars

(a)   (b)

FIGURE 15.11   The evil side of opium poppies. (a) An opium poppy field in Afghanistan. Notice that the capsules have been slit longitudinally, and the exuded latex has turned brown. Photo (public domain) by Davric. (b) An artist's conception of an opium den in France. Source (public domain): Cover of *Le Petit Journal*, July 5, 1903.

(a)   (b)

FIGURE 15.12   The good side of opium poppies. (a) A field of opium poppies cultivated for pharmaceuticals in North Dorset, England. Photo by Marilyn Peddle (CC BY 2.0). (b) A vial of medicinal morphine. Photo (public domain) by Stickpen.

can have more than 20%. The opium of some selected pharmacological varieties is more than 25% by dry weight.

Opiate alkaloids are harvested from cultivated opium poppies for medicinal use (Figure 15.12) as well as for illicit narcotics. As a source of drugs of abuse, the opium poppy has caused more human pain than any other plant, but as a source of medicines, it has relieved more human pain than any other plant. Morphine is considered to be the most important analgesic used for severe pain. While it is the premier medication for agonizing pain and suffering, it has high potential for addiction, and so it is strongly regulated. Heroin, which is produced by chemical conversion of morphine, acts relatively rapidly to produce euphoria and is a chief illegal drug of abuse. The widely used pain reliever codeine is also produced by chemical conversion from morphine.

Just as it is possible to cultivate low-THC "industrial hemp" cultivars that have virtually no abuse potential, it is possible to grow poppy cultivars for certain medicinally important opiate alkaloids that have very limited potential of abuse. Some selected varieties of the opium poppy produce opium with virtually no morphine. Opium poppies that do not produce morphine are considered to be much less dangerous and in some jurisdictions are allowed to be cultivated with relatively limited oversight. Some varieties of the thebaine poppy (also known as Iranian poppy and scarlet poppy), *Papaver bracteatum* Lindl., can produce appreciable morphine, but this poppy has generally been cultivated for the production of two other alkaloids, codeine and thebaine. Some varieties of this species have low amounts of morphine and almost the only opiate alkaloid present is the nonnarcotic thebaine. As with low-morphine opium poppy, thebaine poppies are allowed to be grown in some jurisdictions where the cultivation of normal opium poppy is forbidden or allowed only under stringent conditions. Because the chief commercial product of opium poppies is codeine (as noted previously, produced by chemical conversion from morphine), the thebaine poppy has been considered to be a desirable substitute for the much more dangerous opium poppy (thebaine can also be easily converted to codeine). Thebaine poppy is increasingly being grown in Western countries.

Medicinal poppies are grown outdoors, not in glass houses. There are several advantages to growing plants indoors, especially medicinal plants; nevertheless, not only poppies but also almost all other cultivated medicinal plants (indeed, almost all crops) are in fact grown outside because it is very much cheaper to do so. Indeed, the only critical reason to produce cannabis indoors is for security. In the long-term, the fear of cannabis that still dictates most policies concerning its production is likely to lessen to the point that medicinal *C. sativa* will be grown mostly outdoors, as nature intended.

Viewed in the context of developing *C. sativa*, the opium poppy industry demonstrates that even dangerously addictive species can be managed for the good of society by enactment of sensible policies.

LAUGHING GAS AS A BUSINESS MODEL FOR MARIJUANA PRODUCTION

Laughing gas (nitrous oxide, N_2O) is a well-known, respected anesthetic and analgesic in surgery and dentistry. It is often employed as an oxidizer in rocket and racing car fuels. It is also used as an aerosol spray propellant, such as in whipped cream canisters and cooking sprays. Beginning in the nineteenth century in Britain, laughing gas became a party drug (Figure 15.13), and it is still employed recreationally to produce a pleasant giddiness and orgasm-like euphoria (Lynn et al. 1972), effects related to its opiate-like abilities (Gillman and Lichtigfeld 1994). Depending on jurisdiction, laughing gas is unregulated or somewhat regulated (for example, availability may be restricted to adults or to licensed professionals). Nonmedical use is occasionally subject to legal

FIGURE 15.13 Prints (public domain) showing the recreational usage of laughing gas. Top: A satirical picture (published in 1830) of chemist Sir Humphrey Davy administering laughing gas to a woman. Davy popularized the use of laughing gas as a party drug. He also invented the miner's lamp and became president of the Royal Society of Great Britain. The other man pictured is Sir Benjamin Thompson ("Count Rumford"), American-born British physicist and inventor. Bottom: A view of a laughing gas party in a doctor's office (published in 1820). Notice the large flask at left containing nitrous oxide gas, which was traditionally produced by heating the (dangerously explosive) mineral ammonium nitrate.

penalties but is rarely treated seriously. There is a general consensus that the harm potential of nitrous oxide is limited (at least directly to people; this greenhouse gas is thought to be a significant contributor to atmospheric pollution).

Comparing laughing gas to marijuana may appear to be a stretch, since nitrous oxide is accepted as a valuable medical material with limited recreational harm potential. However, Myles et al. (2007) indicate that it has been found to have some deleterious effects after surgery, and Cousaert et al. (2013) point out that laughing gas is one of the commonest inhalant drugs of abuse among the young. Breathing in pure nitrous oxide has led to death by oxygen deprivation (Zuck et al. 2012). In any event, laughing gas has been tested as a freely available substance in the public marketplace for over two centuries, while marijuana has almost universally been subject to extremely severe restrictions. It is at least conceivable that in the long-term, marijuana will be as normalized a business commodity as laughing gas.

THE GREAT IRONY: FROM COUNTERCULTURE DRUG TO MASS-MARKETED COMMODITY

The Hippie (hippy) subculture, which championed the huge expansion of marijuana during the mid-1960s and 1970s in the United States and eventually in much of the world, was a countercultural movement that defiantly rejected the mores of mainstream American life. In essence, smoking marijuana symbolized repudiation of capitalism, conventional values, and state-mandated regulation of pleasurable drugs. Many of the same idealistic rebellious youth who once espoused a life free of greed and hard work have been transformed into money-grubbing cannabis entrepreneurs with MBAs, firmly rooted in the profit-based establishment they once despised (Figure 15.14). Today, marijuana is becoming one of capitalism's most promisingly lucrative developments, and marijuana is increasingly being defended (albeit with reservations) by physicians, politicians, and even police—the most conservative professions of society.

THE CANNABIS ROADS TO HEAVEN AND HELL

THE PRECAUTIONARY PRINCIPLE IN RELATION TO COMMERCIAL DEVELOPMENT OF CANNABIS

As discussed in Chapter 13, the "precautionary principle" is a respected way of managing the suspected risk(s) of causing harm to the public or to the environment, in the absence of scientific consensus that the action or policy is not harmful. This approach places the burden of proof that the issue under consideration is not harmful on those advocating or tolerating the activity in question.

FIGURE 15.14   Transformation of the idealistic hippie generation to capitalism. Prepared by B. Flahey.

The precautionary principle has been cited as a justification for the prohibition of recreational drugs, including marijuana. In its extreme application, the precautionary principle leads to reluctance to alter the regulatory status quo. However, societal demand for change in the status of cannabis has developed such momentum that inaction is no longer a choice. Prohibition of alcohol consumption in the United States, followed by widespread scofflaw contravention, a tidal wave of protest, and eventual political reacceptance of booze, is eerily analogous to what is happening with respect to marijuana.

Historically, commerce has been conducted with considerable freedom, with limitations imposed only *after* significantly deleterious results become evident from a product or practice. In effect, business activities have been considered innocent until proven guilty, a situation that can be catastrophic when a product becomes extremely popular but proves to be dangerously defective (as evidenced by recent massive recalls of automobiles). It seems certain that legalized marijuana will become extremely popular (indeed, exceeding the popularity of illegal marijuana), and given the consensus that marijuana is not risk-free, there is evident need for caution in formulating new regulations.

Reasonable Regulatory Risk Management

Fundamental innovations and changes that affect many people almost invariably have unintended results, and there is always concern about negative spinoff. Governments are responsible for enacting wise regulatory landscapes maximizing the availability of safe, effective products while controlling abuse potential with constraints that are acceptable to the majority of society. Income from taxes, stimulation of employment, and control of health costs are chief governance considerations. The models of control of production, sale, and consumption of cannabis products outlined in this chapter are currently being tested in various jurisdictions, and the comparative risks and benefits of these frameworks are not yet clear. The natural tendency of governments is to retain centralized control of production and distribution, but governmental constraints inevitably seem to inflate costs, and this is certain to encourage a substantial black market distributing cheaper but unsafe material. It is essential that, over time, regulatory frameworks evolve and adapt, presumably becoming more permissive, as experience dictates.

Private sector business has thrived throughout history, and where there has been an unquenchable demand, it has always been met, legally or not. Capitalism best serves society when restraints are minimal, but sufficient, to minimize abuse and harm. Unfortunately, free-market capitalism, by its nature, cannot be relied on, either to proactively address health and safety issues or to redress negative developments. Nevertheless, it is to be hoped that leadership within the business communities will exercise the caution and responsibility necessary to minimize the problems that will occur.

It is impossible to predict with certainly whether the present momentum popularizing both medical and recreational marijuana will continue. The degree and manner of the future availability of cannabis ultimately are matters for society to constrain through its democratic institutions. Nevertheless, marijuana seems destined to become an important medication, for the betterment of mankind, and a legal social inebriant for the masses, for better or worse. The commercial prospects of both medical and recreational cannabis appear to be extraordinarily profitable both for the private and public sectors.

CURIOSITIES OF SCIENCE, TECHNOLOGY, AND HUMAN BEHAVIOR

- In 1996, the Adidas shoe company introduced the "hemp shoe," a model with the upper portion made of hemp (Figure 15.15). The U.S. White House Office of National Drug Control Policy severely criticized the company and asked it to withdraw the shoe from the marketplace. In denying the request, Adidas president Steve Wynne replied, "I don't believe you will encounter anyone smoking our shoes any time soon."

FIGURE 15.15   A pair of classic Adidas hemp shoes. Photo by Janne Toivoniemi (CC BY 2.0).

- During Japan's feudal era, merchants carried gold, silver, copper, and iron coins with square holes in the centre so that they could be carried on strings of hemp.
- In colonial America, citizens of several colonies were required by law to grow hemp. In 1682, the Virginia legislature made hemp fiber legal tender for up to one-quarter of all debts. Similar laws were enacted in Maryland in 1683 and Pennsylvania in 1706. By 1810, hemp was Kentucky's major crop and was also used as money.
- American presidents George Washington (1732–1799) and Thomas Jefferson (1743–1826) encouraged the growing of hemp. However, both lost money trying to grow the crop.
- The "Goldilocks price" for marijuana is a compromise that governments legislating cannabis sales need to consider. (The "Goldilocks principle" states that something should fall within certain margins, neither too large nor too small.) The price should not be too low (to avoid spurring consumption), nor too high (to undercut the black market).
- In Morocco, "the amount of land area dedicated to producing one gram of hashish is at least 25 times greater than that needed to produce one gram of cannabis herb outdoors" (Leggett 2006). While there is value added in such production, because of the extensive labor and processing involved, the profit margin is much too small to justify producing hashish rather than herbal marijuana. It appears that tradition and a reliable demand for Moroccan hashish in Europe is responsible for continuation of the hashish trade, despite this not making economic sense.
- In 2014, what has alleged to be the first medicinal marijuana-related TV ad (from MarijuanaDoctors.com) aired on Comcast-owned channels such as Fox, CNN, ESPN, Comedy Central, AMC, and Discovery in Chicago and New Jersey (viewable at http://www.brandingmagazine.com/2014/03/05/worlds-first-major-medicinal-marijuana-tv-ad/).
- One of the most effective of modern marketing techniques is to associate a product with a famous personality known to be associated with it. Legendary singer-songwriter Willie Nelson is a longtime proponent and enthusiast of marijuana, perhaps only secondary to comedy duo Richard "Cheech" Marin and Tommy Chong. In 2014, Nelson announced plans to produce his own brand of marijuana ("Willie's Reserve") and to establish a chain of dispensaries.

FIGURE 15.16   Statue (by Alvin Marriott) of Bob Marley in Kingston, Jamaica. The names of celebrities like Marley are being associated commercially with cannabis products. Photo by Avda (CC BY SA 3.0).

- Another example of the brand power of a famous name is provided by Jamaican reggae icon Bob Marley (Figure 15.16), who was a vocal advocate of marijuana before his death in 1981. In 2014, Privateer Holdings, a New York equity firm investing in the legal cannabis industry, acquired the rights from the Marley family to use his name to label a variety of products, including "heirloom Jamaican cannabis strains." According to Forbes' annual list of dead celebrities' earnings, in 2014, Marley was the ninth highest paid dead celebrity, making $20 million, more than Marilyn Monroe or John Lennon. Since cannabis is illegal federally in the United States, there could be problems in registering a brand name associated with prohibited usages.

16 Sustainability

"Sustainable" has become the byword for human long-term utilization of physical and energy resources in ways that do not degrade living things, habitats, ecosystems, and planetary resources such as the atmosphere, water, soil, and landscapes. Unsustainable agriculture, mostly related to crops, is the principal way that people harm the world. The cannabis plant has quite extraordinary significance in mankind's current attempts to create a more sustainable world, and this chapter deals with the considerations that make it useful in many respects for increasing sustainability, although less so in other ways.

THE REPUTATION OF THE CANNABIS PLANT FOR SUSTAINABILITY

For several decades, industrial varieties of *Cannabis sativa* have been touted as phenomenally beneficial for the environment and biodiversity, admittedly particularly by individuals sometimes considered to be "Fringe Greenies" by those with limited sympathy for environmental issues. Nevertheless, hemp has become the world's leading crop symbol of sustainable agriculture. Although the benefits of growing hemp have been greatly exaggerated in the popular press and by hemp entrepreneurs, *C. sativa* is nevertheless exceptionally suitable for organic agriculture and is remarkably less "ecotoxic" in comparison to most other crops. There are considerable ecological and sustainable advantages, as presented in this chapter, but there are also some disadvantages, particularly with respect to the illicit cultivation of marijuana.

HOW MAJOR CROPS HARM THE WORLD AND WHY *CANNABIS SATIVA* CAN BE BENEFICIAL

Humans have modified plants since the beginnings of agriculture about 13,000 years ago, so that they will be more useful and productive. Such modification has almost invariably weakened the resistance of the plants against stresses, necessitating protective measures. Crops are particularly affected by biotic stresses from "pests" in the broad sense, including animals (especially insects, slugs and snails, mites, nematodes, rodents, and birds), plant pathogens (viruses, bacteria, and fungi), and weeds. These diminish crop growth and deteriorate stored harvests, reducing productivity by as much as 40% (Flood 2010). Measures to compete against harmful organisms have resulted in the widespread use of pesticides, which have been very harmful to the natural world. In parallel, measures to compensate for such abiotic stresses as soil infertility and lack of moisture have resulted in the massive use of fertilizers and the profligate consumption of water. Invariably, mechanized or factory agriculture today uses huge amounts of fossil fuels. All of these strategies have drastically deteriorated the landscape and atmosphere of the world. Crops that minimize "agricultural inputs" (pesticides, irrigation, energy) can contribute substantially to reducing ecological damage.

Agriculture today is dominated by about a dozen major crops grown as huge monocultures, i.e., as continuous plantations with just a single species. The huge expenditures of water, herbicides, pesticides, fungicides, bactericides, fertilizer, and fossil fuels result in chemical pollution of soil, water, and the atmosphere and displacement or destruction of native animals, plants, and soil organisms. Fortunately, it is possible to find less ecologically damaging crops, and *C. sativa* is extremely promising in this regard, as discussed in this chapter.

Simply adding a new crop, like *C. sativa*, to those being grown can be beneficial. The topic of "crop diversification" deals with the addition of crops to those currently cultivated in a region. There may be economic benefits (such as increased profitability, new markets, and cushioning the effects of crop and market failures of the crops that have been grown), but there are also possible ecological benefits. Growing many crops (although in some respects less efficient than dependence on a small number of major monocultures) can lessen the attractiveness of farms to pests and diseases and may even provide niches for some wildlife. However, some crops are more beneficial than others to the natural world, and, as noted later in the chapter, hemp is remarkably advantageous.

COMPARATIVE ENVIRONMENTAL FRIENDLINESS OF *CANNABIS SATIVA* AND OTHER CROPS

Figure 16.1 compares the environmental compatibility of *Cannabis* crops (fiber, oilseed, and marijuana) and 21 of the world's major crops, based on more than two dozen criteria measuring the ecological friendliness of crops (see Montford and Small 1999a,b for details). Oilseed and fiber forms

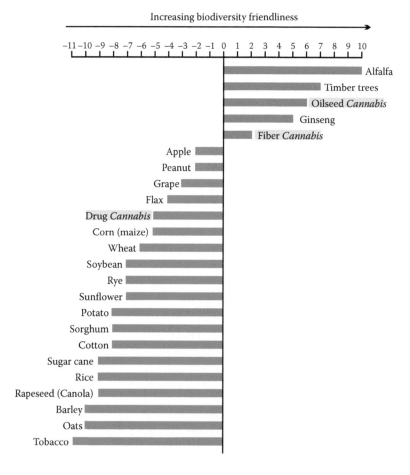

FIGURE 16.1 A comparison of the biodiversity friendliness of selected major crops and three *C. sativa* crops (fiber, oilseed, drug) based on 26 criteria (after Montford, S., Small, E., *Global Biodivers.*, 8, 2–13, 1999; and Montford, S., Small, E., *J. Int. Hemp Assoc.*, 6: 53–63, 1999).

of *C. sativa* were found to be exceptionally compatible with the environment. Illicit drug forms were less friendly to nature, but as described later, this is because of the practices of irresponsible growers.

BIOCIDE REDUCTION

Agriculture makes heavy use of biocides (pesticides, fungicides, and herbicides), which can be extremely detrimental to biodiversity. Although conscientious attempts are generally made to localize application, accidental drift of biocides can be harmful to nontargeted organisms (for example, fish in nearby streams) and soil organisms (such as earthworms) near the area of application. Innocent resident animals and plants often cannot escape exposure, nor can visiting pollinators, birds, and other foragers. Toxic biocides threaten the survival of some species and sometimes even poison humans. Although various techniques (most importantly organic agriculture, integrated pest management [IPM], and genetic engineering) offer means of decreasing biocide use, most crops are dependent on heavy application of biocides, particularly pesticides and herbicides. Crops that are naturally resistant to pests and weeds are therefore very important in reducing damage to biodiversity. *Cannabis sativa* is known to be significantly resistant to most harmful organisms and rarely requires protective treatment. Indeed, the most valid claims for the environmental friendliness of hemp relate to its very limited need for agricultural biocides.

Fields intended for hemp cultivation are still conventionally cleared of weeds using herbicides, but so long as hemp is thickly seeded (as is always done when it is grown for fiber), the rapidly developing young plants normally shade out competing weeds that appear subsequently. Young, growing plants that are more widely spaced (as is typically done for production of hempseed and marijuana) are less able to smother out weeds and may require mechanized weeding, which is wasteful of fuel.

Cannabis sativa is remarkably resistant to insects (Figure 16.2). However, the degree of immunity to attacking organisms has been greatly exaggerated, with several insects and fungi specializing on hemp. In very damp (e.g., maritime) environments, fungi can cause severe damage (Van der Werf et al. 1996), and should the plant become much more widely cultivated than at present, it is likely that significant pest problems will develop. Nevertheless, the use of pesticides and fungicides on hemp is usually unnecessary, although introduction of *C. sativa* to regions where it has not been grown for many years should be expected to attract pests.

FIGURE 16.2 Grasshopper on hemp. Grasshoppers and their orthopteran relatives (crickets and locusts) rarely damage the foliage of *C. sativa*. Indeed, few insects significantly harm the species, so the use of insecticides is very rarely required.

USING UP SURPLUS MANURE

As discussed in Chapter 3, *C. sativa* is naturally adapted to soils where animal excrement has accumulated—i.e., sites with considerable nitrogen. Although the weedy form of the species is known to be capable of surviving on poor soils with limited nitrogen, to maximize production, input of fertilizer high in the element is needed. Hemp is extraordinarily well adapted to the use of manure as a nitrogen source. Unfortunately, most hemp is grown using conventional synthetic fertilizers, which are environmentally damaging. Livestock production inevitably produces large stores of manure, and *C. sativa* has the potential of using up these surpluses. Properly managed, manure can be far friendlier to the environment than synthetic fertilizers, which (1) require the expenditure of fossil fuels to manufacture, (2) reduce soil organism biodiversity, (3) tend to be overused, resulting in eutrophication (overfertilization) of waterways, and (4) produce significant amounts of atmospheric pollutants and greenhouse gases. (Runoff from manure can be as damaging to biodiversity as runoff from inorganic fertilizers, so that the use of manure is not problem-free.)

MISCELLANEOUS ENVIRONMENTAL ADVANTAGES

There is widespread concern over the depleting supply of fossil fuels (coal, petroleum, peat), as well as the environmental degradation associated with transporting them (such as caused by oil tanker spills), and the atmospheric pollution generated from burning them. As discussed in Chapter 10, *C. sativa* is a candidate for biomass production. Plants that generate biomass can be used as ethanol fuel sources, and in some cases, this may alleviate problems associated with the use of fossil fuels. However, at this time, the prospects for *C. sativa* in this regard seem limited, mostly because the technology for transforming cellulosic plant tissues to fuel alcohol is expensive. As discussed in Chapter 8, the species is also a candidate for biodiesel production. Casas et al. (2005) analyzed the life cycle value and carbon economy of hemp biodiesel but did not find clear environmental benefits. Similarly, Van der Werf (2004) concluded that hemp was comparable to wheat and sugar beet with respect to relative contributions of polluting substances emitted to the atmosphere and resources employed. Although the price of hemp oil is too expensive at present for it to be used as biodiesel, this is a possibility in the future (Figure 16.3).

Drug and oilseed types of *C. sativa* (to a much lesser extent, fiber cultivars) are very high value crops in terms of productivity per unit area. Such crops tend to increase local wealth and indirectly

FIGURE 16.3 Biodiesel produced from hempseed oil as a possible aid to reducing environmental problems associated with burning fossil fuels. Prepared by B. Brookes.

decrease pressure on the environment and habitats since a smaller acreage needs to be farmed to produce a reasonable income.

Like most crops, at least while they are growing, *Cannabis* can reduce the supply of weed seeds in the soil (since an established crop outcompetes weeds), control soil erosion, and aerate the soil.

The deep roots of *Cannabis* are efficient at water uptake from lower soil levels, which may be ecologically desirable in some circumstances. For example, crops with shallow roots can exacerbate salinization of soils by bringing salt from shallow layers to the surface, but deeper-rooted plants can sometimes avoid this difficulty. Also, where water is only available at depth, its utilization from deeper parts of the soil can prevent excessive drying of the surface of the soil.

ENVIRONMENTAL DISADVANTAGES

As with almost all crops, there are significant environmental impacts associated with growing *C. sativa* (Van der Werf and Turunen 2008).

Where natural rainfall is limited, *C. sativa* must have moderate irrigation to be productive, especially during early growth. Irrigation can greatly alter ecosystems and is especially detrimental to indigenous plant species that are adapted to dry areas. Accordingly, irrigation is a key consideration of the extent to which a crop can be considered environmentally friendly. Agriculture is responsible for 70% of the world's fresh water consumption (Heywood 1996), and chronic shortages are developing in many countries, so the need for irrigation is a significant limitation. Indirect harmful effects of irrigation include possible soil salinization and pollution from run-off.

Current market forces tend to pressure farmers to narrow the diversity of crops and livestock produced. Modern agriculture and plantation forestry are highly productive but constitute artificial, low-diversity ecosystems. Today, much of the world is occupied by vast monocultures, particularly cereals (notably wheat, barley, oats, rice, corn, millets, and sorghum), which are so very highly domesticated that they require heavy inputs of energy and agrochemicals and are therefore rather environmentally unfriendly. Also, such systems are very susceptible to climate variations and outbreaks of pests and pathogens. Large monocultures, by their nature, exterminate local habitats and their constituent wild species. Industrial hemp (for fiber and oilseed) is most efficiently grown as a large-scale monocrop, and so it tends to add to the environmental burden in the same way as other large field crops.

Agriculture is a heavy user of fossil fuels for tilling, planting, harvesting, and processing. Such consumption generates atmospheric gases contributing to climate change. *Cannabis sativa* does not have advantages in most of these respects by comparison with other crops. Textile crops (as exemplified by fiber hemp) and oilseed crops (such as oilseed hemp) are associated with extensive extraction and processing machinery, which in turn necessitate mined resources, including fuels. Hemp oil can be prone to rancidity, requiring refrigeration to prolong quality, which requires energy. However, much of the energy expenditure of agriculture is concerned with the manufacture and application of herbicides, pesticides, fungicides, and chemical fertilizers. As noted previously, *Cannabis* crops require lower use of these agrochemicals than most other crops, and this factor tends to lower overall energy consumption costs. As noted later, several products made with hemp (such as insulation and hemp-lime concrete) contribute to conservation of energy, and some hemp products simply require less energy to make, or to transport (because they are light).

Minimum tillage, the practice of planting seeds or seedlings in fields that have received little or no plowing (indeed, sometimes directly into stubble or sod), is a promising agricultural technique that minimizes soil disturbance and reduces expenditure of fossil fuels. For the relatively few crops that can be raised with minimum tillage, it is an admirable practice for making agriculture more sustainable and less damaging to the environment. Unfortunately *C. sativa* requires a well-prepared seedbed, and minimum tillage is inappropriate.

Cannabis sativa is rich in bioactive chemicals (of course, this is why marijuana is of interest), and decaying plants can produce toxic residues in the soil. Traditional water retting of hemp to

FIGURE 16.4   Painting titled "Hanfeinlegen" by Theodor von Hörmann (1840–1895) showing traditional water retting in Europe (public domain—http://www.imkinsky.com/).

extract the fiber (described in Chapter 7; Figure 16.4) is notorious for polluting waterways, killing fish, and producing intolerable odors. Environmental regulations in Western nations prevent such obsolete technology, which unfortunately continues to be practiced in Eastern Europe and Asia.

Saving Trees

Wood fiber (from trees) and synthetic fiber (from petroleum) dominate the fiber market (Chapter 7), but ecological and economic concerns about depleting forests and petroleum have increased interest in using natural fibers as primary or at least supplemental raw material. In recent decades, the pulp and paper industry has been criticized for negatively affecting the environment by deforestation, replacing old-growth forests with tree plantations, pollution of air and land (including the production of toxic and mutagenic wastes by chlorine bleaching), and high energy use. To an extent, these problems have been alleviated by paper recycling, sustainable management of natural and planted forests, and adopting less harmful processing technologies. Interestingly, Silva Viera et al. (2010) found that for pulp and paper usage, wood (from eucalyptus trees) had less overall negative environmental impacts than hemp (hardly surprising, since trees can grow with virtually no agricultural inputs and minimum care). Wood remains the cheapest source of fiber and wood is likely to remain dominant, but as discussed in the following, the use of *C. sativa* as a source of fiber may reduce the damage associated with forestry.

The most widespread claim for environmental friendliness of *C. sativa* is that it has the potential to save trees that otherwise would be harvested for production of lumber and pulp. Several factors appear to favor increased use of wood substitutes, especially agricultural fibers such as provided by hemp (Figure 16.5). Deforestation, particularly the destruction of old growth forests, and the world's decreasing supply of wild timber resources are today major ecological concerns. Agroforestry using tree species is one useful response but nevertheless sacrifices wild lands and biodiversity and is less preferable than sustainable wildland forestry. The use of agricultural residues (e.g., straw bale construction) is an especially environmentally friendly solution to sparing trees, but material limitations restrict use. Nevertheless, agricultural residues often provide a cheap material that otherwise might simply be burned, contributing to air pollution. Another chief advantage of several annual

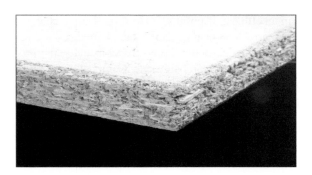

FIGURE 16.5   Hemp fiberboard. Photo by Elke Wetzig (Elya) (CC BY SA 3.0).

fiber crops over forestry crops is relative productivity, annual fiber crops sometimes producing of the order of four times as much material per unit of land. Still another important advantage is the precise control over production quantities and schedule that is possible with annual crops. In many parts of the world, tree crops are simply not a viable alternative. "Three billion people…live in areas where wood is cut faster than it grows or where fuelwood is extremely scarce" (World Commission on Environment and Development 1987). "Since mid-century, lumber use has tripled, paper use has increased six-fold, and firewood use has soared as Third World populations have multiplied" (Brown et al. 1998). Insofar as hemp reduces the need to harvest trees for building materials or other products, its use as a wood substitute tends to contribute to preserving forests. Hemp may also enhance forestry management by responding to short-term fiber demand while trees need to grow for many years to reach their ideal maturation. In areas depleted of natural stands of wood, annual crops such as *Cannabis* can be efficient sources of agricultural fibers to replace forestry products and therefore preserve trees (Montford and Small 1999a,b; Small 2012). In developing countries where fuelwood is becoming increasingly scarce and food security is a concern, the introduction of a dual-purpose crop such as hemp to meet food, shelter, and fuel needs may contribute significantly to preserving biodiversity.

BIOREMEDIATION

Some crops are notable for their tolerance of and ability to absorb substantial amounts of heavy metals and so serve for "phytoremediation." Preliminary work in Germany (noted in Karus and Leson 1994) suggested that hemp could be grown on soils contaminated with heavy metals, while the fiber remained virtually free of the metals, but as noted in the following, there is now considerable evidence that *C. sativa* can absorb heavy metals and thereby decontaminate soil and water. Baraniecki and Mankowski (1995) observed that hemp can reduce the soil content of copper, zinc, and cadmium; Loeser et al. (2002) recorded accumulation of zinc, cadmium, and nickel; Piotrowska-Cyplik and Czarnecki (2003) observed accumulation of zinc, copper, and nickel; and Angelova et al. (2004) noted the same for lead, copper, zinc, and cadmium. Koznlowski et al. (1995) observed that hemp grew very well on copper-contaminated soil in Poland (although seeds absorbed high levels of copper). Baraniecki (1997) found similar results. Mölleken et al. (1997) studied effects of high concentration of salts of copper, chromium, and zinc on hemp and demonstrated that some hemp cultivars have potential application for growth in contaminated soils. Shi and Cai (2009) recorded hemp absorbing substantial cadmium. Pejic et al. (2009) found that waste hemp fiber could be employed to remove lead, cadmium, and zinc ions from contaminated water. Petrová et al. (2012) demonstrated that chelating agents could be used to improve extraction of metals by hemp. Mihoc et al. (2012) found that hempseed seems susceptible to excessive accumulation of metals. Accordingly, it would seem unwise to grow hemp as an oilseed on contaminated soils. However, polluted habitats might be suitable for a fiber or biomass crop. In fact, Eerens (2003) found that hemp could be grown economically for fiber in an effluent disposal site. However, Linger et al. (2002) warned that edible and clothing uses of hemp grown on heavy metal contaminated soil are inadvisable. Nevertheless, for many nonedible applications, particularly construction materials, products made with hemp that has accumulated heavy metals are probably safe. Campbell et al. (2002) found that hemp was useful for treating soil contaminated with two polycyclic aromatic hydrocarbons, chrysene and benzo(*a*)pyrene. The possibility of using hemp for bioremediation deserves additional study (Griga and Bjelková 2013).

WILDLIFE SUPPORT

Cannabis sativa is plagued by bird predation, which takes a heavy toll on seed production. The seeds are well known to provide extremely nutritious food for both wild birds (see Chapter 8 for examples of wild bird species on wild hemp) and domestic fowl. Hunters and birdwatchers who discover wild patches of *Cannabis* often keep this information secret, knowing that the area will be a magnet for birds in the fall when seed maturation occurs. Upland game birds in the Midwest of the United States have been observed to utilize wild hemp as cover for nesting and foraging (Vance 1971).

Increasingly in North America, plants of various species are being employed to revegetate and maintain landscapes, especially to provide habitat and food for wildlife, most notably for wild birds. The seeds of *C. sativa* are nutritious and extremely attractive to birds, and were it not for its reputation, wild hemp could be planted for the benefit of wildlife. *Cannabis sativa* has not yet been seriously considered for this purpose. However, the species is not an aggressive weed and certainly has great potential for being used as a wildlife plant. Of course, such usage is forbidden in North America, current policies requiring the eradication of wild hemp wherever encountered.

ENVIRONMENTALLY FRIENDLY COMMERCIAL PRODUCTS

Today, there is a search for novel materials that address societal concerns and regulations regarding the environmental costs associated with products. A "green image" has economic value, and conversely, a negative environmental reputation is a liability (of course, it is often difficult to determine which information presented in advertising is factual and which is merely promotional). As stated by Fletcher et al. (1995), industrial hemp has a large devotional following because of its green image and its products seem to be almost self-promoting. Market segmentation for ethically produced goods and growing support for biodegradable and natural products have led to a wide range of new industrial hemp products being developed.

HEMP VS. COTTON

In the last several decades, synthetic fibers (particularly polyester) have come to dominate the textile fiber market (as illustrated in Figure 7.1). Synthetic fibers are largely made from petroleum and so contribute to the depletion of fossil energy resources. Accordingly, many environmentally conscious consumers favor textiles made from natural fibers. The natural fiber market is dominated by cotton, considered by many to be the bête noire of all environmentally damaging crops because of its huge consumption of pesticides, fertilizers, and water (Soth et al. 1999; Small 2013b; Figure 16.6). In the European Union (EU), where legislation has strongly favored both natural fibers and sustainable crops, hemp has been significantly subsidized. Another consideration is the "locavore" ideal of producing crops close to home. By this credo, hemp is preferable in temperate regions to the use of tropical and semitropical fibers like cotton, which need to be imported. It must be conceded, however, that synthetic fibers and cotton are dominant because of cost and quality considerations, and despite its environmental and ethical advantages, hemp is unlikely to develop more than a niche fabric market.

FIGURE 16.6 Former floor of the Aral Sea in Uzbekistan, central Asia, showing abandoned ships in a toxically polluted wasteland, caused by draining rivers feeding the sea in order to irrigate cotton and overusing pesticides and fertilizers. The Aral Sea was once the world's fourth largest inland body of water but is in a desert area and exists only because it is fed by rivers. This is one of the world's most tragic examples of the destructive effects of unsustainable agricultural practices and one of the greatest environmental catastrophes ever recorded. Photo taken in 2011 by S. Kluger (CC BY 3.0).

RECYCLABILITY

As noted in the following, in contrast to many products manufactured from traditional plastics and metal, hemp biocomposites can be much more recyclable. Plastics made with hemp fiber (described in Chapter 7) are one of the products touted as having great potential to increase sustainability while reducing environmental damages. Singha et al. (2011) summed up the advantages of fiber-hemp-based composites as follows: "Sustainability, 'cradle to grave' design, industrial ecology, eco-efficiency, and biocompatibility are the guiding principles of development of new generations of materials. Lignocellulosic reinforced composites are the materials of the new paradigm. The use of biodegradable and environment friendly plant-based fibers in the composites reduces waste disposal problems, environment pollution, and ecological concerns. Light weight, decreased wearing of machines, low abrasiveness, as well as an absence of health hazards during processing, application, and upon disposal are added advantages of these composites. Also these fibers can be incinerated, are CO_2 neutral (when they are burned), and because of their hollow and cellular structure, perform well as acoustical and thermal insulators."

CARBON SEQUESTERING

The ecology of carbon cycling is of great concern today because combustion and other activities contribute to carbon-based greenhouse gases and climate change. The ecological value of manufactured products is determined in part by the extent to which they sequester carbon (i.e., capture and store carbon compounds, thereby preventing release of carbon-based greenhouse gases to the atmosphere). Since plastics have long life cycles, bioplastics sequester carbon for long periods, which is environmentally beneficial. "Cradle-to-grave" analyses of energy expenditure and carbon sequestration suggest that hemp (indeed many natural fibers) is valuable in these respects (Haufe and Carus 2011a,b; Piotrowski and Carus 2011; Ip and Miller 2012). However, as noted later, the indoor production of marijuana is associated with astronomical production of carbon dioxide that contributes to atmospheric pollution.

THERMAL INSULATION PRODUCTS

Insulation to regulate temperature is an extremely important way of reducing energy consumption and thereby limit damage to the terrestrial and atmospheric environments and biodiversity. As noted in Chapter 7, hemp-straw-reinforced materials have excellent insulating properties and can replace conventional brick, cement and wood, vapor barriers, insulation, and plaster board in buildings. Hemp is also useful for producing insulation (e.g., Figure 16.7). Thermal insulation products are the second most important sector of the hemp industry of the EU (Carus et al. 2013). These are in demand because of the alarmingly high costs of heating fuels, ecological concerns about conservation of nonrenewable resources, and political-strategic priorities about the dependence on current sources of oil. Thermal insulation is a market segment that is growing, and hemp insulation products are increasing in popularity. Although hemp fiber is suitable for insulation (Kymäläinen and Sjöberg 2005), in the EU, glass and mineral wool (rock wool) insulation cost one-quarter to one-half as much as hemp insulation. The attraction of hemp insulation products is that they are nonirritating to installers, have notable moisture flow and heat-retaining characteristics, and are appealing to customers who view hemp as an ecologically superior choice, regardless of cost.

EROSION CONTROL BLANKETS

Sometimes referred to as "environmental blankets," these ground-covering, biodegradable mats are designed to prevent soil erosion along steep highway banks and prevent soil slippage. They stabilize new plantings and natural vegetation, which grow through the mats, developing root systems that retain the soil. Hemp erosion control mats have been manufactured in Europe and Canada.

FIGURE 16.7   Hemp fiber batt insulation. Photo by Christian Gahle, nova-Institut GmbH (CC BY 3.0).

STATE SUBSIDIZATION OF INDUSTRIAL HEMP—GOOD OR BAD?

The major crops of most rich countries are usually subsidized by direct grants or indirectly through supports for research, needed materials, or transportation. Additional support is often available in the form of protectionist policies. While such state sponsorship protects domestic producers against international competition and encourages local business and employment, it promotes inefficiency. By its nature, agriculture is the world's leading source of environmental damage, and therefore, agricultural subsidies can be especially environmentally damaging. Especially galling is subsidization of corn (maize) ethanol production in the name of environmental sustainability, despite doing more harm than good (see, for example, Conca 2014). (*Cannabis* has limited prospects for use as an ethanol source; Barta et al. 2010.)

As a comparatively minor crop, *C. sativa* has not received a great amount of subsidization, and indeed, expensive security requirements have put it at an economic disadvantage. Nevertheless, the EU has been a leading promoter of agricultural subsidies for crops that are thought to benefit the environment, particularly those that can produce biofuel and bio-energy, and for decades, *C. sativa* has been a major beneficiary of this policy. Without substantial subsidization in Europe, the start-up capital that was necessary to establish the hemp industry in the 1990s would not have been available. In recent years, however, subsidization has decreased substantially, particularly because the European industrial hemp industry has been based mostly on fiber, which has not proven to have notable growth potential and is significantly limited competitively without financial support. As of 2012, the EU eliminated most of the subsidies for hemp.

In Canada, limited subsidies (both national and provincial) have been provided to the industrial hemp industry (legalized in 1998) to encourage the development of novel hemp enterprises, improve processing technologies, and develop new hemp cultivars for the Canadian environments (Salentijn et al. 2015). In contrast to Europe, the Canadian hemp industry has been mostly centered on oilseed, a natural "fit" since Canadian agriculture is based on cereals and oilseeds (i.e., "grains"), and no crop (including hemp) is grown significantly for fiber. Indeed, as made clear in several chapters of this book, the future of industrial hemp lies much more in oilseed usage rather than in fiber applications. Nevertheless, the Canadian government subsidized the establishment of a large decortication plant (Parkland Industrial Hemp Processing) for hemp stems, which was established in Gilbert Plains, Manitoba, in 2013.

As reviewed in this chapter, industrial hemp has many genuinely admirable features contributing to sustainability, as well as a few attributes that are somewhat troubling. Compared to most crops currently being subsidized, *C. sativa* is relatively deserving of support.

ENVIRONMENTAL COST OF INDOOR PRODUCTION OF MARIJUANA

Because of security concerns, marijuana is often grown indoors. There are a few environmental advantages of growing crops indoors. Such cultivation is invariably extremely efficient, with a much higher yield per unit area occupied compared to field crops, so this reduces pressure to find agricultural lands in a world that has almost run out of unused arable land. As with all crop cultivation, water (which is scarce) and fertilizers (which tend to pollute) are consumed, and waste materials are generated, but these are relatively easy to control in the confined space of a greenhouse. Pests and diseases always accompany crops, but at least in a greenhouse, they are easy to locate and control, especially using nonchemical techniques.

Indoor cultivation does have several significant environmental costs. Wearing a T-shirt emblazoned with a slogan like "Save the world: smoke dope" ignores the reality that marijuana is not entirely benign to the planet. Glass houses or at least transparent roofs take advantage of natural sunlight, but frequently, marijuana "grow rooms" are completely artificially illuminated. Extremely high energy expenditure is required to produce marijuana indoors, primarily because of the need for lighting, but also to provide ventilation (to assist temperature control for living plants and for drying harvested marijuana), heating/cooling for climate control, and cool storage (to prevent deterioration of the product). Mills (2012) reported that 1% of the entire energy consumption of the United States is dedicated to the production of indoor marijuana, equivalent to $6 billion annually. Building materials to house marijuana production facilities are expensive to purchase but are also costly in that energy was required for their construction. All factors considered, a very large expenditure of energy and consequent "environmental imprint" is associated with the indoor cultivation of marijuana.

Indoor production of marijuana is also associated with the production of carbon dioxide, which acts as a greenhouse gas contributing to climate change. Much of the electrical energy utilized results in CO_2 production, and often, fuels are burned directly in support of greenhouse operations, releasing CO_2. Mills (2012) calculated that 1 kg of marijuana produced indoors is associated with the release of 4600 kg of CO_2 emission to the atmosphere, equivalent to operating 3 million cars for a year. Occasionally, CO_2 is injected into grow rooms to increase photosynthesis and yields, and this also contributes to atmospheric pollution. However, since productivity is increased, the overall carbon footprint of introducing CO_2 may actually decrease negative environmental impacts (BOTEC Analysis Corporation 2013).

BOTEC Analysis Corporation (2013) is a report of the environmental costs of producing marijuana for a legal market in Washington State. Several observations in the report are worth noting. It was pointed out that indoor lighting carried out during the night period is relatively efficient and would have a smaller deleterious effect on climate than lighting during daylight hours, when there is high demand for electricity. It was noted that although the environmental costs of cannabis production are substantial, they are significantly less than associated with other activities such as large-scale agriculture, mining, metallurgy, and other industries. As in all indoor plant production requiring lighting with high-intensity discharge bulbs, there is an environmental cost associated with the nonrecyclable bulbs containing mercury and other toxins.

ECOLOGICAL DAMAGE FROM IRRESPONSIBLE ILLEGAL CULTIVATION OF MARIJUANA

"Grow-ops" (grow operations, known as "cannabis factories" in the United Kingdom) are often located in suburban houses modified with stolen electricity to power lighting, ventilation, and irrigation systems (note Figure 16.8). Irresponsible and incompetent installations have resulted in heat,

FIGURE 16.8   Police in England raiding a grow-op. Photo by West Midlands Police (CC BY SA 2.0).

moisture, and electrical shorts ruining or burning down houses. Preparation of hashish by butane extraction has produced explosions and fires.

Illicit outdoor operations are frequently carried out in plots hidden in forested areas (Figures 16.9 and 16.10). Using public lands is motivated in part by the threat of forfeiting assets that are present when using personal residences for production. Those who establish plots and visit them only to maintain and harvest the plants have been called "guerrilla growers." The illegal cultivation of cannabis in preserved wildlands is extremely deleterious to biodiversity and its supporting habitats (U.S. Senate Committee on Agriculture, Nutrition, and Forestry 1988; Montford and Small 1999a, 1999b; Mallery 2011). Regrettably, illicit marijuana producers usually have little respect for delicate ecosystems. Garbage is dumped in national parks (Figure 16.11), and groundwater and creeks are contaminated with pesticides, herbicides, and spilled fuel carried to the sites to run diesel generators. Diesel production of electricity produces considerably more greenhouse gases than the relatively low-carbon electricity used

FIGURE 16.9 Illegal marijuana grow site in the White River National Forest near Redstone, Colorado (plants being removed for destruction in the top photo). Discovered in 2013, the plantation contained 3375 plants, with an estimated value of $8.4 million (based on a value of $5500 per kilogram and an estimated yield of 0.45 kg of processed material per plant). Photos (public domain) by the U.S. Forest Service.

FIGURE 16.10   Left: Aerial view of illegal marijuana grow site in 2008 in the Cleveland National Forest, California. Photo (public domain) by U.S. Drug Enforcement Agency. Right: U.S. Forest Service using a helicopter in 2010 to remove marijuana plants from a grow site in the Arapaho–Roosevelt National Forest, Colorado (public domain photo).

FIGURE 16.11   Garbage and debris left at a marijuana grow site in the Shasta-Trinity National Forest in California. Photo (public domain) by U.S. Forest Service.

conventionally ("diesel dope" is a pejorative phrase descriptive of marijuana produced using diesel energy). In California, rodenticides are often used to prevent small mammals from destroying illegal marijuana plants. Rats consume the poison, and then northern spotted owls, fishers, foxes, and bobcats eat the rats and become sick. Thompson et al. (2014) documented the considerable deleterious effects of rodenticides on fishers in California as a result of illicit marijuana plantations.

An additional negative result of the widespread clandestine cultivation of marijuana is that it stimulates law enforcement personnel to use chemical eradication at extremely toxic levels so as to ensure that there are no surviving plants (although herbicides are often of limited effectiveness for plants taller than about 60 cm). Paraquat has been widely applied to illicit *Cannabis* plantations in Mexico (Figure 12.13b).

A recent study of illicit marijuana cultivation in California (Bauer et al. 2015) observed that outdoor plants at a density of about one per square meter were consuming 22.7 L of water per plant per day. So extensive was water withdrawal from rivers in northern California that marijuana cultivation was using up 50% more water than all residents combined in San Francisco. This threatened fish and other aquatic species, particularly federally listed salmon and steelhead trout, as well as sensitive amphibian species, in the drought-prone state.

Because of constant searching for illicit plantings of marijuana, plantations tend to be small, growers often establishing several gardens in separate areas. Large monocultures of agricultural crops as grown in most of the temperate world are very productive but are environmentally harmful because they exclude almost all plants and animals from a region. Small "cottage gardens," as widely grown in tropical areas, are much friendlier to the environment because they permit survival in the natural areas between the crops. Curiously, the small, well-separated marijuana gardens that are frequent in some locations are not entirely without redeeming value, because they avoid the negative effects of large monocultures.

THE DANGEROUS CONCEPT OF EMPLOYING BIOWEAPONS TO ELIMINATE MARIJUANA PLANTS

In the last several decades, the idea of using natural pests, parasites, and diseases to eliminate illegally cultivated plants has surfaced. Biocontrol agents for destroying or controlling a given plant species can be identified by systematically searching for the natural parasites of the species as well as its closest relatives (McPartland and Nicholson 2003). The U.S. Department of Agriculture (2004) conducted unsuccessful studies from 1999 to 2004 to find insects that could be employed to destroy marijuana. The use of fungi as "mycoherbicides" to control marijuana plants has also been proposed (Hildebrand and McCain 1978; McCain and Noviello 1985; Tiourebaev et al. 2001). In fact, there have been attempts to genetically engineer fungi to destroy drug plants, including marijuana, as well as ruderal hemp (Baloch et al. 1974; McCain and Noviello 1985; McPartland 1997a; Committee on Mycoherbicides for Eradicating Illicit Drug Crops 2011). *Fusarium oxysporum* is a fungal pathogen that produces fusarium wilt disease in over a hundred plant species. It includes several physiological forms known as formae speciales (singular: forma specialis, abbreviated f. sp.), one of which is a specialist on *Cannabis*: *F. oxysporum* f. sp. *cannabis*. When employed to kill weeds, it is sometimes called "Agent Green." McPartland and West (1999) reviewed problems associated with use of this, noting that such new pathogens would inevitably endanger hemp cultivation, to say nothing of wild hemp, which constitutes a reservoir of genes for the improvement of hemp. These authors also raised the prospect that once pathogens are released into the environment, mutation and recombination with native fungi could produce new strains of virulent transgenic pathogens that could endanger crop plants. The possibility of employing microorganisms as weapons is particularly controversial. Natural pests have sometimes proven to be useful biocontrol agents for various pest organisms, and with appropriate research prior to release, the risk of unforeseen consequences can be minimized. Unfortunately, microbes, including fungi, are exceptionally difficult to control. They can evolve much faster than higher organisms like insects (which are the main biocontrol agents), and once released, they cannot be recaptured. Eliminating an entire species (and perhaps its relatives) from a region can require very widespread distribution of the control agent and represents an experiment with natural ecosystems that is dangerous.

SUSTAINABILITY DEPENDS ON USING SUSTAINABLE METHODS

In conclusion, it should be stressed that irrespective of the inherent advantages and disadvantages of *C. sativa* from the perspectives of sustainability and environmental friendliness, the impacts of cultivation of *Cannabis* depend on the agricultural, processing, and manufacturing practices employed. While it is possible to produce and process plants employing sustainable practices, not all growers do so. Processing and manufacturing of both industrial hemp and marijuana also have environmental costs, which can be controlled to at least some extent. So while *C. sativa* can be employed to reduce environmental damage, this can only be achieved by conscientious human effort.

CURIOSITIES OF SCIENCE, TECHNOLOGY, AND HUMAN BEHAVIOR

- The energy needed to produce indoor marijuana for a single joint could power a 100-W light bulb for 25 hours and would generate 1.5 kg of polluting CO_2 emissions (Mills 2012).
- The energy required to produce one marijuana joint is about equal to the requirement to manufacture 18 pints of beer (Mills 2012).
- Hawken (2007) noted "…we cannot save our planet unless human kind undergoes a widespread spiritual and religious awakening." Fortunately, religions are playing roles in advocating for ecological sustainability. Polytheistic and naturalistic spiritual traditions have long exhibited profound respect for nature and are natural supporters of environmentalism. Buddhism's tenets of reincarnation and karma require respect for plants, animals, and their habitats. Hinduism too has profound respect for life and the need to cherish the earth. Indigenous people everywhere are deeply tied to the welfare of their native lands. By contrast, monotheistic religions have been less obvious supporters of issues concerned with ecological conservation of land and living things, but in recent times, this is changing.

The Vatican has recently set an excellent example, installing solar panels on its 10,000-seat main auditorium building, arranging to reforest land in Hungary to offset Vatican City's carbon emissions and urging protection of the environment in a number of major addresses by Pope Benedict.

- As pointed out in Chapter 7, hemp fiber has been employed to produce strong, lightweight plastic used in car bodies, contributing to sustainability by virtue of reducing the use of fossil fuels that contribute to atmospheric pollution. In addition to hemp, fiber from other crops is being investigated as a possible component of plastic car bodies. One of the most interesting projects is being conducted by ketchup maker H.J. Heinz in conjunction with the Ford motor company, other manufacturers, and the World Wildlife Fund—members of the Bioplastic Feedstock Alliance, whose goal is to use waste agricultural materials to produce useful plant products. Heinz harvests over two million tons of tomatoes annually, and while some of the residue is fed to livestock, most of the refuse is wasted. To date, the use of tomato waste, left after the pulp has been extracted to produce ketchup, hasn't resulted in a plastic strong enough for car bodies.

17 Germplasm Resources

DEFINITION OF GERMPLASM

"Germplasm" is material that can be used to reproduce or propagate organisms of any kind. This can include living creatures, their reproductive sex cells, or nonsexual ("somatic") parts or structures that can be employed to reconstitute the original organisms. Germplasm of economically significant plants mostly includes either or both of (1) living plants in nature or in special cultivated collections and (2) viable seeds preserved in climate-controlled conditions. Infrequently (because it is expensive and technologically difficult), viable plant material of special significance is maintained as living or frozen tissue or cell cultures.

IN SITU VS. *EX SITU* CONSERVATION OF WILD GERMPLASM

The diversity of wild plants is ideally preserved by allowing them to grow in their natural undisturbed habitats (Figure 17.1), where the often considerable range of genic variation can continue to exist. Unfortunately, human domination of the planet is degrading or exterminating habitats that support wild species. In some cases, wild areas are reserved to allow the organisms to continue to survive. Alternatively, selected individuals of species that are highly threatened with extinction are sometimes cared for by people in nonnatural circumstances (parks, arboreta, institutional gardens, zoos, and the like). Preservation in natural habitats has come to be categorized as *in situ* conservation, preservation outside of such habitats as *ex situ* conservation. *In situ* conservation is usually far less costly and is capable of maintaining much more genic diversity than is possible with *ex situ* conservation, which necessarily is based only on selected samples. The normally small samples of material kept in collections are subject to mutations and accidental hybridization during periodic replication and loss of alleles of genes that occurs naturally when reproduction occurs in small populations, so over time *ex situ* collections tend to become less representative of the original wild population. For flowering plants such as *Cannabis sativa*, both *in situ* and *ex situ* conservations are important, as discussed further in this chapter.

PRIVATE OWNERSHIP OF GERMPLASM VS. LONG-TERM CONSERVATION IN PUBLIC COLLECTIONS

Domesticated crops (land races or cultivars) have a kind of "natural" existence in agro-ecosystems, mostly in cultivated fields and in the hands of farmers and the agricultural industry, and a kind of "unnatural" existence in special long-term biorepositories such as gene banks and botanical gardens. However, distinctive crop variants usually have a much shorter life expectancy than distinctive wild plant variants. Because new cultivars regularly replace old cultivars and land races, the *only* way to maintain the range of germplasm in land races and old cultivars, over the long-term, is to conserve them in special *ex situ* collections. Experience has shown that governmental and nonprofit institutions are far more likely to conserve germplasm over many years, by comparison with the private sector. Most collections maintained by governments will not be used for many years. The point of preserving them is to allow research to explore their usefulness over decades, indeed over centuries. By contrast, in business, the "shelf life" of materials that have lost their current utility is necessarily limited. The ethical way for the private sector to rid itself of germplasm that is not of further financial interest is to deposit it in long-term public germplasm banks (gene banks). Unfortunately, the private sector is motivated primarily by the profit potential of its property and accordingly wishes

FIGURE 17.1   Wild *Cannabis* growing as a weed at the foot of Dhaulagiri massif, Nepal. Such free-living plants are invaluable sources of genetic variability that can be employed to breed improved cultivars. Photo by Arne Hückelheim (CC BY SA 3.0).

to keep some seed stocks out of the hands of competitors, so without regulations requiring contributions to public repositories, considerable germplasm is destined for extinction. Just as patents and other forms of intellectual property are time-limited (Chapter 15), it should be possible for governments to insist that valuable germplasm in the hands of the private sector should be made available to society after a reasonable period of exclusive ownership. As noted in this chapter, there is a great need for germplasm of *C. sativa* to be preserved, and it would be tragic if the large collections held by business interests were not maintained for the long-term welfare of everyone.

WORLD GENE BANK SITUATION

At present, there are approximately 1750 plant gene banks in over 100 countries, storing more than 7.4 million germplasm accessions (Food and Agriculture Organization [FAO] 2010). Major gene banks include those in Canada, China, Germany, India, Japan, Russia, South Korea, the United Kingdom, and the United States. Although all of the world's plant species, especially those that are rare or endangered, are eligible for deposit in gene banks, in practice, most gene banks are dedicated to crop germplasm. Modern gene banks conduct several key activities: they ensure that acquired material is authoritatively identified and record collection data; they examine seeds for purity and viability; and they preserve material under appropriate conditions, regenerating seeds every few years to ensure continuing viability. Frequently, there are associated research programs, often involving characterization of the accessions. Normally, gene banks distribute material to each other and to researchers.

England's Millennium Seed Bank (Figure 17.2), appropriately labeled a "seed bank" rather than a "gene bank," is dedicated to collecting all plant species, and in this respect is unlike most long-term seed collections, which are usually dedicated to crops and their wild relatives. In 2008, one accession of *C. sativa* was listed (FAO 2016; Table 17.1).

Plant breeders and researchers employ the accessions in gene banks as raw material for plant breeding and for basic biological research, and naturally, the importance of a particular crop determines how many accessions are deposited. The most important crops, especially the cereals, have the largest numbers of accessions. Seed collections of the major crops are often duplicated in different gene banks as a security precaution against the destruction of a particular gene bank.

FIGURE 17.2 The Millennium Seed Bank building in Wakehurst Place Garden, West Sussex, England. The associated Millennium Seed Bank project is an international effort that has collected seeds of over 33,000 plant species, representing about 11% of the world's plants. A particular effort is made to collect seeds of species in danger of extinction. Seeds are sometimes available by negotiation. Photo by Patche99z (CC BY 3.0).

FIGURE 17.3   The Svalbard Global Seed Vault, a secure seed collection in an abandoned coal mine on the Norwegian island of Spitsbergen in the remote Arctic, providing a continuously frozen environment. Seeds of crops are stored at −18°C, and if the electrical supply fails, the permafrost keeps temperatures at no higher than −3.5°C. Almost 900,000 seed samples of more than 5000 different species have been deposited. The facility is expected to grow to become the world's largest collection of seeds of crops, with a capacity to store about five million samples. Known as "the doomsday vault," the purpose is to back up important crop germplasm in other seedbanks that could be destroyed by disaster. Unlike conventional seed banks that exchange seeds with numerous individuals and other institutions, only the depositor seed banks have access to their materials. Left: Containers of seeds stored on shelves. Public domain photo by NordGen/Dag Terje Filip Endresen. Right: Entrance. Photo by Bjoertvedt (CC BY 3.0).

The Svalbard Global Seed Vault (Figure 17.3) is particularly concerned with crops and their wild relatives but is a backup facility for other institutions. In 2009, it contained three accessions of *C. sativa* (FAO 2016; Table 17.1).

SEED STORAGE IN GENE BANKS

Ninety percent of material stored in plant germplasm banks is in the form of seeds (Pritchard and Nadarajan 2008). However, some tropical species with so-called "recalcitrant" seeds, which do not tolerate freezing and/or drying, require more complicated methods for conservation. The seeds of some species can remain viable under ambient conditions for decades, rarely for centuries, but those of most plants will not germinate after a few years, unless preserved under conditions of low temperature and/or humidity. Under cold storage, the otherwise short-lived seeds of some species will last decades, others for hundreds of years, lower temperatures tending to prolong viability. Many seeds can be maintained in liquid nitrogen at its boiling point of −196°C (as in Figure 17.4).

CLONAL STORAGE IN GENE BANKS

Some very valuable root or tuber crops (e.g., cassava, potato, sweet potato, taro, and yam), fruit crops (e.g., apple, banana, cranberry, date, hop, orange, pear, and strawberry), nut crops (e.g., hazelnut, hickory nut, macadamia, and walnut), and indeed many others are mostly propagated vegetatively, and living clones are maintained in special gardens. All of the preceding species mentioned are perennials (although yams are grown as annuals), and long-lived plants are naturally adapted to vegetative reproduction, hence to being propagated as clones. *Cannabis sativa* is an annual,

FIGURE 17.4 Ultracold cryopreservation laboratory of the Agricultural Research Service, USDA. Materials are stored long-term in "cryovats" of liquid nitrogen. (Public domain photo.)

normally reproduced by seed, but many annuals can easily be maintained as clones by the use of hormones that stimulate cuttings of the plant to root (in a sense, cloning an annual plant turns it into a perennial). Plants that humans have judged merit propagation as clones are generally outstanding in some desired respects and indeed are often hybrids exhibiting heterosis (hybrid vigor). Since all plants established from a clone share the same genetic makeup, they are extremely uniform in performance. The main disadvantage is that genetic uniformity makes clones very susceptible to the possibility that a mutant disease can become disastrously effective. While the agricultural departments of many nations maintain numerous clonal cultivars, clones of drug strains of *Cannabis* are being maintained almost exclusively by the private sector and by research institutions. *Cannabis* clones are rarely provided in commerce or shared in research circles in the same way as propagating material of potatoes, apples, and many other crops.

CELL AND TISSUE STORAGE IN GENE BANKS

Living materials are sometimes preserved as continuously propagated cell cultures (i.e., they are maintained as single cells or proliferating cells not organized into tissues) or as tissue cultures, which can be employed to grow innumerable identical plantlets (Figure 17.5). Animal tissues (principally semen, ova [unfertilized eggs], and embryos) of very valuable livestock are now often stored cryogenically in liquid nitrogen, and some gene banks today similarly conserve apical meristems (growing points or "buds") of some species. Pollen grains also can be maintained

FIGURE 17.5 Plants raised from tissue cultures being grown by the USDA. Photo by Lance Cheung. (Public domain photo.)

long-term under cold storage. These advanced biotechnological methods are in use for *C. sativa* by some private firms assembling genetic collections, and usually, the information is guarded as intellectual property. The U.S. Department of Agriculture (USDA) has supported research into the conservation of marijuana clones as shoot cultures and "synthetic seeds" (Lata et al. 2012; see Chapter 4).

CORE COLLECTIONS

Long-term storage of seeds (for decades or more) is expensive because it requires stringent control of temperature and, often, humidity. Seeds lose viability over time even under controlled storage, and periodic regeneration (typically once every one to three decades) to grow a new batch of seeds is also expensive. "Genetic erosion" is the loss of alleles (variations of genes) or of allele combinations, and this commonly occurs because some new cultivars become so popular that old cultivars (along with their unique genes) are lost. To combat this loss of potentially useful breeding material, gene bank managers have tried to maintain many different and large samples of the varieties and landraces of a crop. However, this is costly. In recent years, attempts have been made to identify "core collections"—made up of selected key accessions that represent the bulk of the genetic variation of a species present in a germplasm collection—so that demands by researchers for seeds will be more limited, reducing the need to generate seeds of many of the accessions. Core collections have not been prepared to date for *C. sativa*.

THE IMPORTANCE OF N.I. VAVILOV

More than any other individual, the Russian geneticist and agronomist Nikolai Ivanovich Vavilov (1887–1943; Figure 17.6) was responsible for persuading the world about the importance of collecting and preserving germplasm that can be used by breeders to create new crops and improve old crops. He is widely considered to have been the foremost plant geographer. He pioneered theories concerning centers of origin of cultivated plants and developed concepts regarding the origin of crops (Vavilov 1926a, 1992) that are still widely respected. In the 1930s, the bureaucrat T.D. Lysenko (1898–1976), a pseudo-scientist whose ideology suited the totalitarian, communist Soviet Union and its dictator Joseph Stalin (1879–1953), challenged Vavilov's ideas. At great personal risk and demonstrating extraordinary intellectual honesty, Vavilov tried to defend his scientific conclusions, but this resulted in his imprisonment in 1940 (Figure 1.14) and death in 1943. Lysenko's ideas led to the discredit of Soviet genetics, the failure of Soviet agriculture, and in no small way to the end of the Soviet Union. Vavilov's studies considerably clarified the nature of wild *Cannabis*, and it is fitting that the largest and most important germplasm collection of *C. sativa* now resides in the N.I. Vavilov Institute of Plant Industry, which commemorates his achievements.

FIGURE 17.6   N.I. Vavilov (1887–1943), illustrious Russian/Soviet geneticist and germplasm collector. He made several significant reports on variation of *Cannabis*. He also deposited many seed collections in the crop germplasm gene bank of the N.I. Vavilov Institute of Plant Genetic Resources in St. Petersburg, which was named in his honor and contains the world's largest collection of preserved seedstocks of *Cannabis*. (a) 1977 U.S.S.R. postage stamp showing Vavilov (public domain photo). (b) Monument honoring Vavilov at the Poltava Agricultural Experiment Station, Ukraine (photo by Batsv, CC BY SA 3.0). (c) Enormous ceramic tile mural honoring Vavilov, at the Kuban seed bank, Russia (photo by E. Small in 1982).

SEED STORAGE CONDITIONS FOR *CANNABIS SATIVA*

Almost all material of *C. sativa* conserved long-term is in the form of seeds. The longevity of *Cannabis* seeds is known to decrease fairly rapidly, to about 75% after two years of storage in a sheltered but otherwise uncontrolled climate, a level that is generally considered too low for commercial planting. The literature regarding factors influencing seed longevity was reviewed by Small and Brookes (2012). Small and Brookes (2012) also experimentally examined the interactions of temperature, humidity, and an oxygen-free environment as they affected seed longevity of industrial cultivars, drug strains, and ruderal plants. Progressive lowering of the temperature (from 20°C to −80°C) increased seed longevity, so did progressive lowering of moisture content (from 11% to 4%). A high moisture content (11%) at room temperature was fatal to all of the seeds examined within 18 months (fungi generally attack the seeds when they are so moist). Either reducing the temperature to at least 5°C or reducing the seed moisture content to at least 6% had a huge beneficial effect on maintaining seed viability. Additional reduction of temperature, but not additional reduction of moisture content, had a small supplementary beneficial effect.

Small and Brookes (2012) found that storage under nitrogen gas (i.e., in an oxygen-free environment) had no effect on longevity. The seeds of *C. sativa* are known to be appreciably impermeable to air, as evidenced by the observation that the seed oil oxidizes (becomes rancid) with distressing rapidity once extracted (so preservation in cold, dark conditions is required for commercial purposes). The presence of the antioxidant vitamin E in the seeds also provides protection against the deleterious effects of oxygen. These considerations likely explain why exclusion from oxygen did not improve seed germination.

Small and Brookes (2012) made the following practical recommendations regarding storage of *Cannabis* seed (for more detailed information, see Chapter 7):

- For long-term germplasm banking of *C. sativa* seed in a viable state for up to a decade, a moisture content of 6% coupled with a storage temperature of −20°C, is sufficient.
- For periods of more than a decade of storage before seed is regenerated, lower moisture content and temperature are appropriate.

THE CHALLENGE OF PREVENTING POLLEN CONTAMINATION OF *CANNABIS SATIVA* GERMPLASM

As detailed in Chapter 4, *C. sativa* pollen is produced in prodigious quantities and is carried by the wind for very long distances. In Canada and Europe, an isolation distance of 5 km is required for the production of seed that is to be used to produce industrial hemp crops (although Small and Antle 2003 found that the amount of pollen distributed downwind was about six times the amount distributed upwind). Ensuring that weedy or clandestine plants are not present for a distance of 5 km is very challenging outdoors. Accordingly, renewing seeds of large collections of *C. sativa* in gene banks, outdoors or even in greenhouses not isolated from outdoor pollen, is very difficult. For this reason, it is wise to preserve *C. sativa* seeds in gene banks for very long terms before renewal is required.

In the Gatersleben (Germany) gene bank, one of the world's finest, a distance of 80 to 100 m between different populations has been deemed sufficient for reproducing seed of small plots (100 plants/accession), provided that no large hemp fields are in proximity, wind direction is favorable, and hedges provide some protection (personal communication, Axel Diederichsen).

THE SHAMEFULLY INADEQUATE STATE OF GERMPLASM PRESERVATION OF *CANNABIS SATIVA*

As discussed in the following, germplasm resources for *C. sativa* are quite unsatisfactory. As expressed by Watson and Clarke (1997): "The last 60–70 years have been disastrous for the *Cannabis* gene pool, and many local landraces, the result of hundreds of years of selection for local use, have been lost because of *Cannabis* eradication, neglect on the part of agricultural officials and industry, anti-*Cannabis* propaganda and the general trend (until recently) to reduce industrial hemp breeding and research." At present in North America, there are no conventional public gene banks from which one can obtain material for scientific study and technological development or in which one can deposit valuable germplasm for potential long-term exploitation. As detailed in the following, gene bank resources for *C. sativa* are largely in a small number of European institutions, and they are limited in extent and availability.

THE GERMPLASM IMPORTANCE OF WILD-GROWING *CANNABIS SATIVA*

All cultivated plants originated from wild plants, and for numerous of these, including most major crops, either the wild ancestors or close relatives still exist in nature. As discussed in Chapter 18, wild-growing plants of *C. sativa* are likely mostly or entirely escapes from cultivation that have re-evolved adaptations to wild existence. The wild plants represent a natural reservoir of genic variation that serves for the improvement of cultivars. Cultivars have always been selected for characteristics desired by people, and this narrows their range of genetic variation and makes them more susceptible to environmental and biotic stresses. Genes from the wild plants can be bred into cultivars to toughen them against stresses, as well as improve agronomic and product characteristics.

Cannabis sativa is of Old World origin and has been present in Eurasia for many thousands of years. As noted in Chapter 3 (dealing with the ecology of wild-growing plants), uncultivated *C. sativa* is found growing in a very wide range of habitats in Eurasia. Such plants have evolved adaptations to the very different circumstances of these habitats and consequently harbor an enormous range of genic variation. In response to the worldwide condemnation of marijuana in the twentieth century, there have been some efforts to eliminate wild-growing plants in certain areas of Eurasia, but the attempts have been insignificant. Accordingly, Old World wild plants of *C. sativa* represent an extremely valuable genetic resource.

By contrast with the general tolerance of wild-growing *C. sativa* in most of Eurasia, there have been concerted efforts to eliminate the wild plants of North America. Wild North American hemp is derived mostly from escaped European cultivated hemp imported in past centuries. Hemp was introduced to North America in Port Royal, Acadia (Nova Scotia), in 1606. It was a popular crop in Eastern and Central Canada during the eighteenth and nineteenth centuries, but by the mid-1930s, production had ceased, except for a brief revival during World War II. Wild Canadian hemp is concentrated along the St. Lawrence and lower Great Lakes (Small 1972b), where considerable cultivation occurred in the 1800s. In the United States, wild hemp is best established in the American Midwest and Northeast, where hemp was grown historically in large amounts. Decades of eradication have exterminated many of the naturalized populations in North America. In the United States, wild plants are rather contemptuously called "ditch weed" by law enforcement personnel. However, the attempts to destroy the wild populations are short-sighted because they are mostly low in tetrahydrocannabinol (THC) and are not employed as a source of marijuana. Mehmedic et al. (2010) analyzed 1371 confiscated U.S. samples that they termed "ditch weed" collected from 1993 to 2008, and the mean THC concentration was only 0.4%. However, no information was provided that the samples actually were ruderal, and they employed the term ditch weed simply to classify low-THC collections. Wild North American plants have undergone many generations of natural adaptation to local conditions of climate, soil, and pests, and accordingly, it is safe to conclude that they have genes that are invaluable for the improvement of hemp cultivars. Nevertheless, present policies in North America still require the eradication of wild hemp wherever encountered.

THE GERMPLASM IMPORTANCE OF LANDRACES OF *CANNABIS SATIVA*

The term landrace (land race) refers to populations of domesticated plants that were selected over many generations by farmers in a region. Landraces stand in contrast to cultivars (although occasionally landraces are treated as cultivars). Both landraces and cultivars have some genetic and phenotypic (morphological and/or physiological) distinctiveness. Cultivars are usually the product of concerted (often consciously directed) selection, in recent times mostly by plant breeders; landraces are usually the product of relatively unsophisticated (but often remarkably effective) selection by farmers. Cultivars usually have a quite narrow genetic base, corresponding with a narrow range of variability and a narrow range of adaptation to stresses. Landraces usually have a much wider genetic base, corresponding with a broader range of variability and a broader range of adaptation to stresses, especially to the local environmental conditions and biotic agents where they were selected. Cultivars are currently named according to a formal code of nomenclature (Brickell et al. 2009), while landraces are often given local names or may simply not have been named. Prior to the twentieth century, farmers almost exclusively grew landraces; by the end of the twentieth century, advanced ("Westernized") farming of major crops has become substantially based on the use of crop cultivars, most of which are replaced in a decade or two by more advanced cultivars. However, third world/developing nations often continue to grow landraces. In many cases, cultivars for tropical and subtropical conditions have simply not been bred, and only landraces are available. Landraces and cultivars are both important as material from which new cultivars are created, but since landraces have a much broader range of genetic variation corresponding with adaptation for desirably agronomic characteristics, they are of much greater value in the long run. The loss of a cultivar can mean that a decade of effort by a plant breeder has been lost; the loss of a landrace can mean that centuries of effort and perhaps irreplaceable genes have been lost. Because agriculture is increasingly abandoning landraces (and indeed older cultivars) of most crops, critically important germplasm is facing extinction unless preservation efforts are made. As noted in this chapter, conservation of industrial hemp landraces is very inadequate.

CHINESE GERMPLASM OF *CANNABIS SATIVA*

Land races of fiber hemp from China represent the oldest cultivated forms of *C. sativa*, perhaps tracing back for thousands of years. Genetic studies of these are limited. By contrast, land races from Europe are more recent, many probably dating back at least hundreds of years.

Because of the genetic distinctiveness of *C. sativa* in China and its cultivation there for millennia, Chinese domesticated variants of the species are of special importance. Wang and Wei (2012) surveyed the availability of *C. sativa* cultivars and land races in China, noting that a general decline in growing hemp in recent times was resulting in an alarming reduction in germplasm. Most Chinese cultivars are dedicated to fiber production, but some are used for oilseed and others are dual purpose. In regard to China, Salentijn et al. (2015) wrote: "Hundreds of hemp landraces have been established. Examples are Liuan HuoMa and Liuan HangMa from Anhuiprovince, Laiwu DaMa and Laiyang DaMa from Shan Dong province, Gushi KuiMa in Henan province, Wenxian DaBaiPi in Hebei province, Liuzhi DaMa in Guizhou province, and DayaoDaMa and Weishan DaMa in Yunnan province. Industrial hemp cultivars in China include YunMa 1, YunMa 2, YunMa 3, YunMa 4, YunMa 5 (all of which are widely cultivated in China) and the less frequently cultivated LongDaMa 1, JinMa 1, WangDaMa 1, and WangDaMa 2." According to Salentijn et al. (2015), "Large collections of germplasm resources have been collected and maintained in the Yunnan Academy of Agricultural Sciences, which comprise approximately 350 accessions with a good representation of fiber/seed hemp groups."

WORLD GENE BANK COLLECTIONS OF *CANNABIS SATIVA*

The FAO compiles gene banks holdings. Based on FAO (2016), there were a total of 1530 accessions (including duplicates) of seeds of *C. sativa*, almost all from Europe (Table 17.1).

EUROPEAN "NATIONAL COLLECTIONS" OF *CANNABIS SATIVA*

While hemp has been cultivated in Asia and South America for centuries, it is basically in Europe that germplasm banks have made efforts to preserve hemp seeds for the long-term.

Germplasm collections of *C. sativa* have been assembled in the principal European nations in which industrial hemp was cultivated in the late twentieth or early twenty-first century (Tables 17.1 and 17.2). These are a mix of public and private collections, and the availability of seeds for use by those unaffiliated with the institutions varies, as indicated in Table 17.2. Compared to the very extensive preserved collections of most major crops, there are disturbingly few collections of seeds of *C. sativa* (Van Soest et al. 1993).

THE VAVILOV INSTITUTE COLLECTION

The N.I. Vavilov Institute of Plant Genetic Resources in St. Petersburg (formerly Leningrad), Russia, has by far the largest germplasm collection of hemp of any public gene bank, with about 500 collections, although in the past, it had accumulated 1400 accessions (Grigoryev, undated). Detailed information on the majority of hemp accessions of the Vavilov Institute can be found in Anonymous (1975;

TABLE 17.1
FAO Compilation of World Gene Bank Collections of *Cannabis sativa*

Country	Gene Bank (Location)	Number of Accessions in Gene Bank	Number of Accessions in Country
Austria	AGES Linz—Austrian Agency for Health and Food Safety (Linz)	3	4
	Office of the Styrian Regional Government, Department for Plant Health and Special Crops (Wies)	1	
Bulgaria	Institute for Plant Genetic Resources "K. Malkov" (Sadova)	35	35
Czech Republic	Agritec Research, Breeding and Services Ltd. (Sumperk)	14	14
Ecuador	Departamento Nacional de Recursos Fitogenéticos y Biotecnologiad (Quito)	3	3
France	Collection Nationale Céréales à Paille, Unité expérimentale du Magneraud, Groupe d'Étude et de contrôle des Variétés et des Semences (Sainte Pierre-d'Amilly)	18	18
Germany	Genebank, Leibniz Institute of Plant Genetics and Crop Plant Research (Gatersleben)	51	51
Hungary	Fleischmann Rudolph Agricultural Research Institute, University of Agricultural Sciences (Kompolti)	25	142
	Institute for Agrobotany (Tápiószele)	117	
India	National Bureau of Plant Genetic Resources (New Delhi)	19	19
Italy	CRA-Centro di Ricerca per le Colture Industriali (Bologna)	26	61
	CRA-Centro di Ricerca per le Colture Industriali (Rovigo)	35	
Japan	Department of Genetic Resources I, National Institute of Agrobiological Sciences (Tsukuba)	10	10
Norway	Safety Base Collection of NORDGEN (Svalbard)	3	3
Poland	Plant Breeding and Acclimatization Institute (Blonie, Radzikow)	6	6
Romania	Agricultural Research Station Secuieni-Neamt (Secuieni)	46	141
	Suceava Genebank (Suceava)	92	
	University of Agricultural Sciences and Veterinary Medicine Timisoara (Timisoara)	3	
Russian Federation	N.I. Vavilov All-Russian Scientific Research Institute of Plant Industry (St. Petersburg)	491	491
Slovakia	Plant Production Research Center Piestany (Piestany)	27	27
Slovenia	Crops and Seed Production Department, Agricultural Institute of Slovenia (Ljubljana)	3	3
Spain	Comunidad de Madrid. Universidad Politécnica de Madrid. Escuela Técnica Superior de Ingenieros Agrónomos. Banco de Germoplasma (Madrid)	3	11
	Gobierno de Aragón. Centro de Investigación y Tecnología Agroalimentaria. Banco de Germoplasma de Hortícolas (Montañana)	1	
	Instituto Nacional de Investigación y Tecnología Agraria y Alimentaria. Centro Nacional de Recursos Fitogenéticos (Madrid)	7	
Sweden	Nordic Genetic Resource Center (Alnarp)	3	3

(Continued)

TABLE 17.1 (CONTINUED)
FAO Compilation of World Gene Bank Collections of *Cannabis sativa*

Country	Gene Bank (Location)	Number of Accessions in Gene Bank	Number of Accessions in Country
Turkey	Plant Genetic Resources Department (Izmir)	52	52
Ukraine	Institute of Bast Crops (Hlukhiv)	373	435
	Ustymivka Experimental Station of Plant Production (S. Ustymivka)	62	
United Kingdom	Millennium Seed Bank Project, Seed Conservation Department, Royal Botanic Gardens, Kew (Wakehurst Place)	1	1
Total			1530

Source: Based on FAO, The World Information and Early Warning System on Plant Genetic Resources for Food and Agriculture. http://www.fao.org/wiews-archive/wiews.jsp, 2016. At the FAO website (http://www.fao.org/wiews -archive/wiews.jsp), click on the extreme left box on the top, "PGR." In the drop-down menu, click on "Germplasm," enter "Cannabis," and click on all the search parameters offered. Click on "Search" or "Report." Depending on institution, details regarding accessions may be available.

Note: Some accessions duplicate each other, notably between the Russian Federation and Ukraine, which in many cases originated from the same collections.

TABLE 17.2
Summary of European "National Collections" by Bas et al. (2010), Based on a 2006 Report

Country	Institution	Number of collections	Availability
Czech Republic	AGRITEC Ltd.	13 (including 11 modern cultivars)	"Available for users free of charge."
Germany	Leibniz Institute of Plant Genetics and Crop Plant Research, Gatersleben	38	"Available for distribution."
Hungary	Research Centre for Agrobotany, National Institute for Agricultural Quality Control	114 (including 77 landraces from Hungary)	"Available for distribution."
Italy	Istituto Sperimentale per le Colture Industriali	98 (mainly of Italian origin, included breeding lines and research crosses)	Material transfer agreements required for seed distribution.
Netherlands	Plant Research International, B.V., Wageningen	200 accessions	"The collection is available for users, with a charge of £450 per accession."
Poland	Plant Institute of Natural fibres	131 collections	No information
Romania	Agricultural research stations (ARS)	ARS-Lovrin: 43 cultivars ARS-Secuieni: 33 lines	No information

Source: Bas, N., Pavelek, M., Maggioni, L., Lipman, E., *Report of a Working Group on Fibre Crops (Flax and Hemp)*. First meeting, June 14–16, 2006, Wageningen, the Netherlands. Bioversity International, Rome, Italy, 2010.

Note: These collections are a mix of public and private (privatized) institutions. The most important European collection, at the Vavilov Institute, was not included, but is discussed later in this chapter. Compare Table 17.1 (which has more recent information for some institutions, but does not report some of the information provided here).

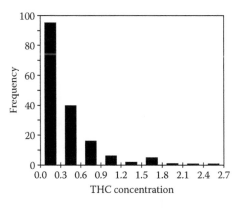 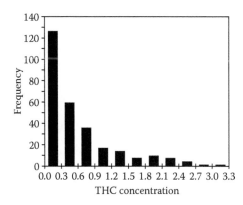

THC concentration   THC concentration

FIGURE 17.7 Frequency histograms of THC concentration in germplasm collections of the Vavilov Institute, St. Petersburg. Left: 167 accessions examined in Small, E., Marcus, D., *Econ. Bot.*, 57, 545–558, 2003; and Small, E., Marcus, D., *Econ. Bot.*, 58, 329, 2004, with 43% having THC levels >0.3%. Right: 278 accessions reported in Anonymous, *Catalogue of the Global Collection of VIR. Issue 162, Fiber Crops*, Vavilov Institute, Leningrad, USSR, 1975, with about 55% having THC levels >0.3%.

also see Table 17.3). Budgetary problems in Russia have endangered the survival of this invaluable collection. Maintenance and seed generation issues for the Vavilov hemp germplasm collection are discussed in a number of articles in the *Journal of the International Hemp Association* (e.g., Clarke 1998b; Lemeshev et al. 1994, 1995; Kutuzova et al. 1996, 1997).

It is particularly disappointing that more than half of the *Cannabis* accessions of the Vavilov Institute develop THC levels that exceed 0.3% (Small and Marcus 2003; Figure 17.7) and that there is evidence that the genetic purity of accessions has been compromised by hybridization (Hillig 2004b). Nevertheless, the value of this collection for future breeding remains outstanding, and every effort needs to be made to find new funding to preserve it.

The Gatersleben Collection

The Gatersleben gene bank of Germany, the second largest public gene bank in Europe, has a much smaller *Cannabis* collection compared to the Vavilov collection, with less than 60 accessions. Information on the Gatersleben gene bank is available at http://www.ipk-gatersleben.de/en/genebank/.

NORTH AMERICAN PUBLIC GERMPLASM COLLECTIONS OF *CANNABIS SATIVA*

The acquisition of germplasm of *C. sativa* in public institutions of North America has been rigorously discouraged to date. My own collections of close to 1000 accessions, prepared for the Canadian Department of Agriculture, were necessarily destroyed following completion of the studies in which they were employed, and there are no accessions currently maintained for public access in Canada as of the writing of this book. The situation in the United States is comparable. In 1971, for my cultivation experiments in Ottawa, I received a collection of 57 accessions of *C. sativa* from the USDA (analyses on these are in Table 4 in Small and Beckstead 1973a). These had been conserved from a USDA fiber breeding program that was, perhaps surprisingly, still in progress in the early 1950s (Feaster 1956a,b). In the late 1990s, requests for seeds resulted in the response that no seeds of *C. sativa* were maintained by USDA, and indeed, all such seeds had to be destroyed. Whether low-THC ("hemp") material still exists in the hands of the USDA is an open question. In 1971, I received several high-THC accessions from the U.S. National Institute of Mental Health program of marijuana investigation centered at the University of Mississippi (Oxford) (see Table 1 in Small and Beckstead 1973a). Official intergovernmental requests for seeds in 2005 went unanswered. Based on publications from the Mississippi group (see cited publications of M.A. ElSohly and colleagues), a range of high-THC strains are in the possession of the U.S. National Institute on Drug Abuse.

GERMPLASM COLLECTIONS AND THE FUTURE OF INDUSTRIAL HEMP

As pointed out in this chapter, there are relatively limited germplasm collections of *C. sativa*, and those pertaining to industrial hemp are mostly in Europe. There are also collections in China, which are much less available to other countries. Because industrial hemp is regaining its ancient status as an important crop, a number of private germplasm collections have been assembled in recent decades for the breeding of low-THC cultivars as commercial ventures (for examples, see De Meijer 1998; De Meijer and Van Soest 1992; Man'kowska and Grabowska 2009). Commercially produced breeding lines are treated as intellectual property, and of course, these are available only on a restricted basis, if at all, while cultivars associated with these programs are marketed.

GERMPLASM RESOURCES FOR FIBER

Almost all of the publically available germplasm resources for industrial hemp, discussed in this chapter, relate to cultivars and land races that have been used entirely or primarily for fiber, and these are in European germplasm banks. Unfortunately, European cultivars seem to have a relatively narrow genetic base. According to an analysis by De Meijer (1995a), the four dozen or so hemp cultivars of Europe, registered at that time, trace their heritage mostly to just a few local landraces. The Italian cultivar Carmagnola, the oldest landrace of Italy, has contributed parentage to many European cultivars. Chinese germplasm is not well represented in Europe, and as noted previously, interest in fiber hemp in China may be waning.

GERMPLASM RESOURCES FOR OILSEED

As discussed in Chapter 8, *C. sativa* has been grown historically mostly for fiber, with the result that there are almost no landraces dedicated to oilseed production available, and only in very recent times have cultivars dedicated to oilseed been bred. There may not even be extant land races of the kind of hemp oilseed strains that were once grown in Russia. The most pressing need of the hempseed industry is for the breeding of more productive oilseed cultivars. Most fiber strains (cultivars and land races) have relatively low seed production, and most hemp germplasm has certainly not been selected for oilseed characteristics. At present, most available registered cultivars are unsuitable for specialized oilseed production. To be competitive with the major oilseeds, hemp should produce approximately 2 tonnes/ha; at present, 1 tonne/ha is considered average to good production. Doubling the productive capacity of a conventional crop would normally be considered impossible, but it needs to be understood just how little hemp has been developed as an oilseed. There is therefore a desperate need to examine the germplasm of *C. sativa* in order to find genes to improve oilseed hemp. Wild plants of *C. sativa* have naturally undergone selection for high seed productivity and are a particularly important potential source of breeding germplasm.

Curiously, marijuana strains have potential for providing genes to breed oilseed cultivars. Drug varieties have been selected for very high yield of flowers and accordingly produce very high yield of seeds. Drug strains have been observed to produce more than a kilogram of seed per plant, so that a target yield of several tonnes per hectare is conceivable (Watson and Clarke 1997). Of course, the high THC in drug strains makes these a problematical source of germplasm.

GERMPLASM RESOURCES FOR ESSENTIAL OIL

As discussed in Chapter 9, essential oil is a very minor economic product of *C. sativa*, and indeed, the essential oil components (terpenes) are available far more cheaply from other crops. Nevertheless, biotypes grown for other purposes (fiber, oilseed, or marijuana) sometimes have desirable terpene profiles and have occasionally been used for the purpose of harvesting essential oil. It is likely that future research will characterize some populations as possessing outstanding essential oil traits (indeed, private pharmaceutical firms likely already possess such strains), and hopefully, these will be deposited eventually in a public germplasm repository.

GERMPLASM COLLECTIONS AND THE FUTURE OF MEDICINAL AND RECREATIONAL MARIJUANA

In the distant past, *C. sativa* was unregulated, and numerous marijuana landraces were selected. Many of these are still being grown in Asia, although law enforcement has reduced cultivation in most areas. Some Central Asian marijuana landraces were distributed to Africa, Southeast Asia, and the Americas, where local landraces were selected. With the explosion of interest in marijuana that began in the 1960s, landraces were employed in Western nations (notably in the Netherlands and the United States) by clandestine illicit breeders. During the last several decades, illicit breeders utilized most of the techniques available to scientific breeders and created a wide range of strains. Hundreds of named strains are currently distributed in the illicit and medical marijuana trades. Thus, an impressive range of germplasm variability exists for marijuana forms of *C. sativa*, albeit mostly in a state of illegality. Not surprisingly, the world's public gene banks currently do not (or at least do not make it public that they) possess marijuana germplasm. Rarely, collections of marijuana strains have been made by governmental organizations for research and for law enforcement purposes, most notably by the National Institutes of Health at the University of Mississippi (Oxford).

Marijuana germplasm is almost entirely in the possession of the private sector. Over 100 commercial firms, especially in the Netherlands (Figure 17.8), created or assembled collections of strains, and these have been marketed internationally (substantially illicitly to date) through the Web. A guide to many of these is at http://marijuanaseedbanks.com/ (also see "High-THC Strains" in Chapter 11 for a list of books describing strains). Especially in North America, marijuana strains are now offered by purveyors of medicinal marijuana, the legality of such transactions differing depending on jurisdiction. In Canada and in some U.S. states, numerous strains have been allowed to acquire legal status for medicinal purposes. The scientific status of marijuana strains available either in the illicit trade or from licensed medicinal sources is very uncertain. (An exception is Medisins, a medicinal cultivar registered in 1998.) The extent to which named medicinal strains are genuinely different, and just what their differences are, is unclear. Documentation concerning these strains is mostly unreliable and suspect. Sawler et al. (2015) found that strain names accompanying materials that they had acquired "often do not reflect a meaningful genetic identity." Literature produced by those in the illicit trade is commonly available but is unreliable on aspects dealing with breeding and genetics. Commercial companies involved with medicinal marijuana are obviously retaining information for their own commercial motives. Some authorized sellers simply provide new names to materials that they have acquired under older names. Companies engage in exaggerated claims in order to inflate the value of the material they market, so determination of the characteristics of strains is problematical.

Notwithstanding the uncertain scientific status of most marijuana strains, both those in the illicit trade and those available legally from authorized sources constitute a vast reservoir of material that is potentially useful, either directly for experimental and medicinal applications or for starting material for selecting or breeding material suitable for medicinal applications. The principal concern is that there is currently no attempt to collect and preserve germplasm of marijuana forms of

FIGURE 17.8   Storefronts of commercial "seed banks" in Amsterdam. These so-called seed banks mostly furnish marijuana seeds. Top: A store of the Sensi Seed Bank organization, which has been said to be "perhaps the largest supplier of cannabis seeds in the world" (Hazekamp and Fischedick 2012). Photo by Sergio Calleja (CC BY SA 2.0). Bottom: Photo by Eric Borda (CC BY ND 2.0).

C. sativa in public institutions for long-term research and development, in the manner that all other legitimate economically significant crops are treated.

DATABASES

Databases providing information on germplasm holdings in gene banks are critical to locating valuable breeding material.

INDUSTRIAL HEMP

Several of the largest public industrial hemp germplasm collections are associated with basic databases (providing information such as site of origin, collector, year of collection, and whether ruderal or a cultivar). Most collections today are backed up by information databases, the private collections generally restricting access. Bas et al. (2010, based on a 2006 meeting) presented a proposal to prepare a comprehensive database for *C. sativa* accessions, but subsequently, little had been initiated (Pavelek and Lipman 2011, based on a 2010 meeting).

MEDICINAL AND RECREATIONAL CANNABIS

With the proliferation of private firms offering medicinal strains, online information for many of these is presented at their websites, although a genuine database does not seem to be available. Mighell et al. (2013) provide details of the potential usefulness and nature of a strain registry. "Leafly," a notable online guide to strains, is at https://www.leafly.com/. Another guide to marijuana strains is GM. 2015. A-Z Strain Reviews. Grow Marijuana. http://grow-marijuana.com /strain-reviews.

GUIDE TO THE MOST COMMONLY GROWN INDUSTRIAL HEMP CULTIVARS

Plant breeders frequently prefer to employ crop cultivars rather than wild plants or even land races as starting material because cultivars are already highly refined. This can be short-sighted, as wild plants and primitive land races often contain useful genes that have not been preserved in cultivars. Nevertheless, currently available industrial hemp cultivars that have been licensed for cultivation in the European Union and Canada have all been demonstrated to have limited THC (see Chapter 11), and so breeders have little concern that their use in creating new cultivars will produce forms that are too high in THC to be grown legally. Table 17.3 provides basic information on most of the cultivars recently and currently grown in the Western World. The majority of these are currently available and their identification is highly reliable.

There are many additional cultivars or land races (often considered to be cultivars) that are grown, depending on whether permitted in given countries. Sometimes, these are known to exceed the THC limits that are commonly accepted in the European Union and Canada. However, often, particular old cultivars or land races are available only in germplasm collections or from private individuals, and the nature and identification of the material are frequently unreliable. I have often grown plants from seeds from different sources, identified as the same land race or cultivar, and discovered that they have produced quite different plants. Although such material may be misidentified and/or hybridized, it can still be valuable as sources of breeding germplasm. The same is true for so-called "common seed," which refers to relatively undefined seeds from a region that have been generated in an uncontrolled fashion (with respect to parentage) from local plants (for example, "Swissmix," a dioecious seed stock of Swiss origin grown for fiber). DNA-based genetic analysis has the potential of clarifying the identity and history of *C. sativa*, so the lack of reliable characterization should not eliminate given collections from consideration as germplasm sources. As noted previously, identification of a core collection is a way of narrowing down large collections to a much smaller sample representing the range of genetic variation.

CURIOSITIES OF SCIENCE, TECHNOLOGY, AND HUMAN BEHAVIOR

- During the Second World War, the population of the Soviet Union often had to tolerate extreme shortages of food. This was particularly true during the Siege of Leningrad (now St. Petersburg). Curators of N.I. Vavilov's seed bank in Leningrad could have pillaged the edible seeds in the collection to survive, but some heroically endured starvation rather than do so.
- The largest and most important germplasm bank in the world is the U.S. National Plant Germplasm System, under the control of the USDA. This has more than 500,000 collections numbered consecutively starting with P.I.1 ("Plant Introduction 1"), which was catalogued in 1898. The collection P.I.1 is a cabbage variety introduced from Russia.

TABLE 17.3

Guide to Registered Industrial Hemp Cultivars in the European Union, Approved Cultivars in Canada, and OECD Cultivars Certified as Low in THC, All for 2015

Name	Country of Origin or Association	Sexual Type[a]	Purpose[b]	Registered in European Union[c]	Approved in Canada[d]	OECD Certified[e]
Alyssa	Canada	Female predominant	Dual		×	×
Anka	Canada	Monoecious	Dual		×	×
Antal	Czech Republic/Hungary			×		
Armanca (=Bialobrzeskie?)	Romania	Dioecious	Seed	×		×
Asso (=Férimon)	Italy	Dioecious				×
Beniko	Poland	Monoecious	Dual	×		×
Bialobrzeskie (Bialobrzeskie)	Poland	Monoecious	Dual	×		×
Canda	Canada	Monoecious	Seed		×	×
CanMa	Canada	Dioecious	Seed		×	
Cannacomp (Kannakomp)	Hungary	Dioecious		×		×
Carma	Italy	Monoecious		×		×
Carmagnola (landrace)	Italy	Dioecious		×	×	×
Carmen	Canada	Dioecious	Dual	×	×	×
Chamaeleon	Netherlands	Female predominant		×		
CFX-1	Canada	Dioecious	Seed		×	
CFX-2	Canada	Dioecious	Seed		×	
Codimono	Italy	Monoecious		×		×
Crag	Canada	Dioecious	Dual		×	
CRS-1	Canada	Dioecious	Seed		×	
CS (Carmagnola Selezionata)	Italy	Dioecious		×	×	×
Carma	Italy	Monoecious		×		
Dacia Secuieni	Romania	Monoecious		×		×
Debbie	Canada	Monoecious	Seed			×
Delores	Canada	Monoecious	Dual		×	×
Delta 405	Spain	Monoecious		×		×

(*Continued*)

TABLE 17.3 (CONTINUED)

Guide to Registered Industrial Hemp Cultivars in the European Union, Approved Cultivars in Canada, and OECD Cultivars Certified as Low in THC, All for 2015

Name	Country of Origin or Association	Sexual Type[a]	Purpose[b]	Registered in European Union[c]	Approved in Canada[d]	OECD Certified[e]
Delta-llosa	Spain	Monoecious		×		×
Deni	Canada	Monoecious	Dual		×	×
Denise	Romania	Monoecious	Dual	×		×
Diana	Romania	Monoecious	Dual	×		×
Dioica 88	France	Dioecious		×		×
Epsilon 68	France	Monoecious		×		×
ESTA-1	Canada	Dioecious	Seed		×	
Fasamo	Germany	Monoecious	Dual		×	×
Fedora 17	France	50% female, 50% monoecious		×	×	
Fedrina 74	France	Monoecious			×	
Felina 32	France	Monoecious		×		×
Felina 34	France	Monoecious	Dual		×	×
Ferimon (Férimon)	Germany	Monoecious		×	×	×
Fibranova	Italy	Dioecious		×	×	×
Fibriko	Hungary	Female predominant?			×	
Fibrimon 24	France	Monoecious			×	×
Fibrimon 56	France	Monoecious			×	
Fibrimor	Italy	Dioecious			×	
Fibrol	Hungary	Monoecious	Dual	×		×
FINOLA (=FIN 314)	Finland	Dioecious	Seed	×	×	×
Futura 75	France	Monoecious		×	×	×
Georgina	Canada	Dioecious			×	×
Grandi	Canada		Seed		×	×
GranMa	Canada				×	
Helena	Serbia	Dioecious	Dual?			×

(Continued)

TABLE 17.3 (CONTINUED)

Guide to Registered Industrial Hemp Cultivars in the European Union, Approved Cultivars in Canada, and OECD Cultivars Certified as Low in THC, All for 2015

Name	Country of Origin or Association	Sexual Type[a]	Purpose[b]	Registered in European Union[c]	Approved in Canada[d]	OECD Certified[e]
Hempnut (=X59, which see)						
IDA	Canada	Monoecious				x
Ivory	Netherlands	Monoecious		x		x
Joey	Canada	Monoecious			x	x
Judy	Canada	Monoecious				x
Jutta	Canada	Monoecious	Dual		x	x
Katani	Canada	Monoecious	Seed		x	
KC Dora (=KC Dóra)	Hungary	Monoecious		x		x
KC Virtus	Hungary	Monoecious		x		x
KC Zuzana	Hungary	Monoecious		x		x
Kompolti	Hungary	Dioecious		x	x	x
Kompolti Hibrid TC (Kompolti Hybrid TC)	Hungary	Dioecious		x	x	x
Kompolti Sargaszaru[f]	Hungary	Dioecious			x	
Lipko	Hungary	Monoecious		x		x
Lovrin 110	Romania	Dioecious		x	x	x
Marcello	Netherlands	Monoecious		x		
Markant	Netherlands	Monoecious		x		
Monoica	Hungary	Monoecious		x		x
Novosadska	Serbia	Dioecious				x
Petera	Canada	Dioecious			x	x
Picolo	Canada		Seed		x	
Rajan	Poland			x		
Santhica 23	France	Monoecious		x		x
Santhica 27	France	Monoecious		x		x
Santhica 70	France	Monoecious		x		x

(Continued)

TABLE 17.3 (CONTINUED)

Guide to Registered Industrial Hemp Cultivars in the European Union, Approved Cultivars in Canada, and OECD Cultivars Certified as Low in THC, All for 2015

Name	Country of Origin or Association	Sexual Type[a]	Purpose[b]	Registered in European Union[c]	Approved in Canada[d]	OECD Certified[e]
Secuieni jubilee	Romania	Monoecious		×		×
Silesia	Poland	Monoecious			×	×
Silistrenski	Bulgaria	Dioecious				×
Silvana	Romania	Dioecious	Dual	×		×
Szarvasi	Hungary	Monoecious		×		×
Tiborszállási (Tiborszallasi)	Hungary	Dioecious	Dual	×		×
Tisza	Hungary	Monoecious		×		×
Tygra	Poland	Monoecious	Dual	×		×
UC-RGM	Canada	Monoecious			×	×
Uniko B[g]	Hungary	Unisexual female	Dual	×	×	×
USO 11 (=Yuso 11 = Zolotonoshskaja 11 = Zolotonosha 11)	Ukraine	Monoecious			×	
USO 14 (= Yuzhnosozrevayushchaya Odnodomnaya 14 = JSO-14 = Yuso 14)	Ukraine	Monoecious	Dual		×	
USO 15 (=Zolotonosha 15)	Ukraine	Monoecious			×	

(Continued)

TABLE 17.3 (CONTINUED)
Guide to Registered Industrial Hemp Cultivars in the European Union, Approved Cultivars in Canada, and OECD Cultivars Certified as Low in THC, All for 2015

Name	Country of Origin or Association	Sexual Type[a]	Purpose[b]	Registered in European Union[c]	Approved in Canada[d]	OECD Certified[e]
USO 31 (=Juso 31 = JSO-31 = Yuso 31)	Ukraine	Monoecious	Dual	×	×	
Victoria	Canada	Dioecious			×	×
Wielkopolskie	Poland	Monoecious		×		
Wojko	Poland	Monoecious		×		×
X59 (Hemp Nut)	Canada	Dioecious	Seed		×	
Yvonne	Canada	Monoecious	Dual		×	
Zenit	Romania	Monoecious		×		×

[a] "Monoecious" means at least a substantial proportion of the plants are monoecious; many may also be female predominant, but male plants are absent or rare in the generation (F_1 or close to it) that is commercially marketed as the pure variety or hybrid variety.

[b] All cultivars are grown for fiber, unless otherwise stated. Where clear information is available, "seed" indicates usage primarily for seed, and "dual" indicates substantial use for both oil-seed and fiber.

[c] European Commission Plant Variety Database—hemp (2015): http://ec.europa.eu/food/plant/plant_propagation_material/plant_variety_catalogues_databases/search/public/index .cfm?event=SearchVariety&ctl_type=A&species_id=240&variety_name=&listed_in=0&show_current=on&show_deleted=.

[d] Approved Canadian hemp cultivars for 2015: http://www.hanfplantage.de/wp-content/uploads/2015/04/LOAC_2015_EN_-_Health_Canada_-_List_of_approved_Cultivars_Cannabis _Sativa.pdf.

[e] Organization for Economic Co-operation and Development (OECD) List of Varieties eligible for seed certification: http://www.oecd.org/tad/code/Crucifers-and-other-oil-or-fibre-species .pdf (hemp cultivars certified as being low in THC).

[f] Chlorophyll-deficient mutant of Kompolti, employed in research.

[g] A hybrid cultivar: F_1 is unisexual-female; F_2 segregates 30% male.

- "Biopiracy" refers to unilaterally adopting for profit, without permission, recognition, or compensation, the materials or knowledge of the (usually indigenous) people of a region. Historically, germplasm of and traditional knowledge about most crops (including *Cannabis*) have simply been "stolen" and transferred from their areas of origin to much richer countries where the plants have been cultivated profitably. The 1993 Convention on Biological Diversity provided recommendations intended to benefit financially poor regions possessing valuable germplasm and traditional knowledge. Over 90 countries are signatories to the anti-biopiracy Nagoya Protocol, a part of the UN Convention on Biological Diversity. While noble in intent, identifying and penalizing examples of biopiracy have been controversial. Unique biotypes of industrial and medicinal *Cannabis*, as well as clever techniques of cultivation, processing, and product development and usage, originate from indigenous people of several poor regions of the world. Should users of cannabis be required to provide compensation?

18 Botanical Classification and Nomenclatural Issues

Given that *Cannabis sativa* is the world's most controversial plant from the perspectives of the law and medicine, it should not be surprising that there have also been profound disagreements with respect to its taxonomy (scientific classification). This chapter examines *C. sativa* in the light of the criteria that botanists employ to classify plants like it, in which variation deserving to be categorized has been brought into existence by both nature and humans. As has been documented in this book, *C. sativa* occurs widely in nature as free-living populations adapted to local climates, as well as domesticated kinds differentially selected for fiber in the stem, a multipurpose oil in the "seeds" (achenes), or an intoxicating resin secreted by pin-sized epidermal glands. The variation pattern of *C. sativa* is complex, but the causes of variation are clear and provide guidance for an appropriate interpretive classification scheme. The following relatively extensive presentation of classification theory and practice is required because, with the exception of how living populations of the human species *Homo sapiens* should be classified (note Figure 18.1) and how extinct relatives in the genus *Homo* should be interpreted (Figure 18.2), no other species has generated so much misunderstanding, argument, and contradictory literature.

THEORETICAL CLASSIFICATION ISSUES

Biological classification (taxonomy or systematics) is based on scientific evaluation of characters and genes of organisms, which are employed to assess their similarities or evolutionary relationships, to construct a sort of (usually hierarchical) organization chart that efficiently reflects relationships, and provides unequivocal names for all of the groups within the system. At least, this is the sort of technical definition that one would find in a modern textbook. In fact, people have been classifying and naming plants and animals as long as there have been people because it's important (often a matter of life or death) to be able to recognize distinctive creatures, what group they belong to (because other members of the group may also have useful or dangerous features), and to have unambiguous names (to help identify the organisms and recall information about them). Much biological classification is intuitive, and often, a young child can classify some groups as well as a modern taxonomist using sophisticated modern tools. However, there are subtleties and complicated issues that make some classification issues very difficult. Almost all of the time, the world is content to leave such classification problems to the academics, since it doesn't seem to matter to the daily lives of most people. However, the classification of *Cannabis* is an exception—indeed, the issue has been debated more in the public sphere than the classification of any other plant, and understanding the conflicting views is important to the welfare of society.

SCIENTIFIC CLASSIFICATIONS OFTEN DIFFER

Classification of organisms is often controversial because nature presents an extraordinary range of variation patterns, so that a "one-size-fits-all" or "cookie-cutter" approach is unwarranted. Also contributing to disagreement, there are several dogmatic schools of thought regarding assessment procedures and usage of various kinds of genetic information as bases for taxonomic systems. As a result, there are often competing classification and naming systems for the same set of living creatures. Harlan and de Wet (1971) remarked, "The inconsistencies and lack of agreement among taxonomists dealing with the same materials are remarkable, to say the least, and are even more

FIGURE 18.1 "Races of mankind," illustrating that despite the extensive geographically based biological variation among humans, we are all just one species. From Roe, E.T., Leonard-Stuart, C., *Webster's New Illustrated Dictionary*, Syndicate Publishing Company, New York, 1911; photo by Sue Clark (CC BY 2.0).

FIGURE 18.2 Reconstructions of extinct forms of the genus *Homo*, whose classification has been disputed. For comparison, males are shown (facial hair and pigmentation are often arbitrarily interpreted). (a) *Homo habilis*. Photo by Lillyundfreya (CC BY 3.0). (b) *Homo erectus*. Photo by Lillyundfreya (CC BY 3.0). (c) *Homo floresiensis*. Photo by Cicero Moraes et al. (CC BY 4.0). (d) *Homo heidelbergenis*. Photo by Tim Evanson (CC BY 2.0). (e) *Homo neanderthalensis*. Photo credit: Stefanie Krull, Neanderthal Museum, Picture Library, Talstr. 300, 40822 Mettmann, Germany (CC BY 3.0). (f) Cro-Magnon man (early *H. sapiens*). Photo by Cicero Moraes (CC BY 3.0). *Homo heidelbergensis* and *H. neanderthalensis* are closely related to each other and have been considered to be subspecies of *H. sapiens*.

striking when the treatments of different crops are compared." The following information is not intended to be a complete primer on classification theory but to highlight aspects that have troubled recent attempts to classify *Cannabis*. Additional considerations are discussed in Small (1979a, 1979b, 2015a).

SCIENTIFIC NAMES ARE OFTEN AMBIGUOUS

American literary figure James Whitcomb Riley (1849–1916) famously wrote, "When I see a bird that walks like a duck and swims like a duck and quacks like a duck, I call that bird a duck." However, defining (and consequently recognizing) a duck, or indeed most groups of living creatures that seem to merit a unique name, is frequently not as obvious as it seemed to Riley. Had Riley been an ornithological specialist on waterfowl, he would have learned that the swimming behaviors of birds called ducks differ greatly among species, some ducks do not walk like ducks (even if extant ducks do have webbed feet), and most ducks do not quack. Among many duck specialists, the inclusiveness of the word "duck" depends on recent evaluations of avian phylogenetic relationships (e.g., Johnson and Sorenson 1999). For example, whistling ducks (tree ducks; subfamily, *Dendrocygninae* of the duck, goose, and swan family of birds, *Anatidae*) are often considered to belong to tribe *Dendrocygnini* of the goose subfamily *Anserinae*.

It may seem disturbing that one person's duck may be another person's goose or swan (note Figure 18.3), but as long as what is meant by the user of a word or phrase is understood, the terminology is useful for purposes of communicating information. Conversely, an ambiguous word or phrase hinders understanding when it is not clear what meaning is meant. The public and, indeed, most scientists have little appreciation of how ambiguous biological "scientific names" can be. As discussed in this chapter, names applied to *Cannabis* have been plagued with ambiguity.

FIGURE 18.3 Subtleties of identification, exemplified by Hans Christian Andersen's *The Ugly Duckling*. As related in the familiar fairy tale, the ugly duckling (the small dark bird at bottom center), in fact a beautiful swan, is being persecuted by ducks, among which the egg from which it emerged was accidentally placed. As noted in the text, even when the aberrant duckling is identified as a swan, it may still be classified as a "duck." Illustration (public domain) by T. van Hoytema, published in 1893 by C.M. van Gogh, Amsterdam.

GEOGRAPHICAL UNCERTAINTIES COMPLICATE INTERPRETATION OF GROUPS

All domesticated plants arose ultimately from wild ancestors, which may no longer be extant. Plants growing outside of cultivation are commonly said to be "wild," but (as noted in Chapter 3) the term is ambiguous. Basically, a species is "indigenous" (or "native") to a given geographical area if it reproduces there and is present in that location as the result of natural processes, without the influence of humans. (For rigorous analyses of the concept of indigenous status, see Ratcliffe 1977 and Peterken 1981.) Contrarily, if a species has been transported (deliberately or not) to a location because of human activity and reproduces there without the assistance of humans, it is "introduced" (or "naturalized"). A nonindigenous species that occurs with some frequency in an area because it is often released or escapes, but does not persist indefinitely because of a lack of adaptation to that area, is said to be "spontaneous," "adventive," or "casual." The chief difficulty with determining whether a species is indigenous or introduced is the time dimension. Of course, because of geological and climate changes during the billions of years of Earth's history, most species migrated extensively. In many circumstances, indigenous status should be assessed starting with the end of the last ice age. However, determining the pre-recorded history location of some plants is very difficult and uncertain with respect to the possible influence of humans. Because the precise native homeland of *C. sativa* is unknown with any degree of confidence, and indeed the existence of truly wild forms of the species that have never been altered by contact with humans is also not known with confidence, traditional treatment in the manner that taxonomists categorize exclusively wild species with known primeval distribution ranges is doubtfully warranted.

POSSIBLE RELATIONSHIPS AMONG WILD ANCESTORS, DOMESTICATES, AND ESCAPES

Plants closely related to domesticated plants and growing outside of cultivation may be (1) ancestors of the domesticates; (2) escapes from cultivation, either identical to the domesticates or altered by generations of selection for existence in nature; or (3) hybrids or introgressants (discussed later) between a wild relative and the domesticate. Often, a domesticate arises from a weedy wild species, and conversely, often, a weed arises from domesticated plants. When one can distinguish three phases: (a) domesticated crop(s), (b) ancestral or closely related (at least somewhat interfertile) wild plants that still have natural distribution ranges, and (c) weedy or ruderal relatives of the crop that interbreed with it, the assemblage is referred to as a "wild-weed-crop complex." When only (a) and (c) can be distinguished, it is simply a "crop-weed complex." Many crops like *Cannabis* exist in crop-weed complexes (Andersson and de Vicente 2010), with domesticated forms in cultivation, and related ruderal (weedy) forms growing outside of cultivation. The issue of whether all *Cannabis* plants growing outside of cultivation are derived from escapes from cultivation, or whether some of these are free of genes altered by humans, cannot be conclusively settled with available information (in some respects, it's like trying to prove a negative). Some botanists have recognized wild-growing *Cannabis* as constituting taxonomic groups at one or more ranks (the most widely used nomenclatural epithets for these are *kafiristanica*, *ruderalis*, and *spontanea*), which is also contentious and is examined later.

The point of view taken in this book is that no persuasive evidence has been documented that there are truly wild populations of *C. sativa* (pristine genetically, never having been altered by human selection, and having natural distributions). By no means are cultivated plants lacking wild ancestors unusual: there are hundreds of domesticated plants known only in cultivation. Like *Cannabis*, many other ancient important crops are also thought to lack extant living relatives from which they originated *directly* (which is not to say that they lack living relatives). Examples of familiar crops for which direct living ancestors are believed (sometimes debatably) to be extinct include avocado (*Persea americana*), cassava (*Manihot esculenta*), corn (maize; *Zea mays*), eggplant (*Solanum melongena*), European plum (*Prunus domestica*), lemongrass (*Cymbopogon citratus*), onion (*Allium cepa*), peanut (*Arachis hypogaea*), rice (*Oryza sativa*, *O. glaberrima*), and safflower (*Carthamus tinctorius*).

CLASSIFICATION DIFFICULTIES DUE TO HYBRIDIZATION

Hybridization is a genetic combining of representatives of (at least) two different groups. Consider the entertaining quotation "if my grandmother had wheels, she'd be a bus" (or bicycle, wagon, or tractor), or the Italian counterpart, "If my grandmother had wheels, she would be a wheelbarrow," expressions indicating frustration with someone's excessive assumptions. Such extreme hybrids are difficult to conceptualize (Figure 18.4) but serve to point out that intermediacy between concepts challenges their separateness. Hybridization especially complicates classification of crop-weed complexes. In biological taxonomy, the term "hybrid" often covers more than simply entities that combine two entire genomes (F_1 hybrids, i.e., the first-generation progeny generated between the pure parental kinds). The term "hybrid" is also frequently applied to a range of backcrosses (crosses between the original hybrid and a parent) and segregants (forms with assorted gene combinations based on additional crossing). In addition, introgression (gene flow from one population to another), a special form of hybridization, often occurs. Frequent hybridization and introgression between the cultivated and ruderal phases of crop-weed complexes, and sometimes also between these and related wild species, can make classification so difficult that the exercise becomes pointless or arbitrary.

In a limited sense, every individual resulting from sexual union is a hybrid between its parents. However, biological classification is concerned not with individuals but with recognizing distinctive groups of individuals. No one has succeeded in hybridizing *C. sativa* with a species of any other genus in the Cannabaceae. However, more or less distinctive populations of *C. sativa* do hybridize readily with each other, obscuring differences, and so making the delimitation and identification of such populations as distinctive groups (whether labeled as strains, varieties, or even species) problematical. The issue of hybridization in *C. sativa* is examined additionally later.

FIGURE 18.4 Conceptual hybrid between a grandmother and a bus. Prepared by B. Brookes.

STEREOTYPICAL THINKING—A ROADBLOCK TO BIOLOGICAL CLASSIFICATION

The classification problems discussed previously regarding hybridization are allied to the issue of stereotypical thinking. Stereotypical thinking (better known in philosophical analysis as "typological thinking") is a mental set, or way of thinking about things, whereby objects are viewed as belonging to perfectly distinctive classes or categories (things are necessarily either this or that, but neither both nor something in-between). This is the way most people think most of the time and represents an efficient means of understanding the universe. Stereotypical thinking is acceptable so long as one has either fish or fowl, but when one is confronted with something which is neither but manifests attributes of both, a more sophisticated kind of conceptualization is necessary. Unless the reader can evade the mental straitjacket of stereotypical thinking, the true nature of biological classification in general, and the classification of *Cannabis* in particular, cannot be accurately understood.

However, it is very difficult for many unfamiliar with the subtleties of biological classification to escape stereotypical thinking because conceptualization in terms of discrete entities is embedded in normal human psychology. We normally assign individuals to different classes, with no middle ground (philosophers refer such thinking to the "law of the excluded middle"). Children viewing animals in a farmyard readily perceive different classes of creatures but find a continuous variation pattern, such as that presented by the racially intermixed population of people in Hawaii, much more difficult to categorize. Like children, many individuals appear unable to conceptualize things except in separate classes, and unfortunately, such a rigid mental set precludes appreciation of biological classification at the species level—the critical classification problem posed by *Cannabis*. Surprisingly perhaps, stereotypical thinking is common among scientists and not uncommon among professional taxonomists, although it is almost unknown in theoretically or experimentally oriented classification experts. The relationship of stereotypical thinking and biological classification is a complex topic and is dealt with in detail in Small (1979a, Chapter 1).

CLASSIFICATION DIFFICULTIES DUE TO OBLITERATION OF POPULATIONS BY HUMANS

People often distribute crops to areas where they previously did not exist, providing opportunities for genetic exchange with related species and creating habitats (frequently weedy) where hybrids will survive. On occasion, the result is the extermination of the genetic differences between once distinct groups and their natural distribution ranges. For example, this has happened to alfalfa, a complex species in which the two major wild parents were once the distinct species *Medicago sativa* L. and *M. falcata* L. Over the last six millennia, both in cultivation and in nature, these parental lineages have hybridized so extensively that most plants everywhere are of hybrid origin, one can no longer identify the overwhelming majority of plants as belonging to the original species, and so it is preferable to reduce the original rank of the parents to subspecies of one species (Small 2011b). The carrot species (*Daucus carota* L.) also illustrates how once distinct classes of domesticated plants can be homogenized. More than a century ago, there was a major class of domesticated carrot with purplish roots (dominated by anthocyanins) centered in Afghanistan; however, hybridization and preference for the familiar European orange carrot (the root pigments dominated by carotenes) have virtually eliminated the pure form of purple carrot, except in gene banks (Small 1978b). Cultivated assemblages are especially prone to losing their distinctness or simply becoming extinct (Jeffrey 1968), as their human masters' needs and tastes change. In *Cannabis*, hybridization between the most distinctive variations has largely obliterated populational differences, especially between the two kinds of fiber forms and between the two kinds of marijuana forms. As noted later, the two kinds of fiber plants that have been recognized taxonomically have been widely hybridized, by legal breeders, because of the resulting heterosis (hybrid vigor), and the two kinds of marijuana plants that have been recognized have been widely hybridized (mostly illicitly) to provide for the different psychological states that many have come to appreciate and also to generate plants with desired photoperiodic and size characteristics to meet local needs. Indeed, according to Clarke and Merlin (2013), "hybrids have become the predominant form of drug *Cannabis* grown throughout Europe and the New World." Hillig (2004b) concluded that most *Cannabis* accessions in the Vavilov Research Institute (St. Petersburg) germplasm bank (most of these are fiber land races), by far the world's largest such collection, are of hybrid origin. Taxonomy is a practical activity, and when most individuals encountered are hybrids, this needs to be considered for classification purposes. This means that since the fiber (low-THC) populations of the world are being homogenized by hybridization, they are doubtfully split formally into separate taxonomic groups. (The difference between formal and nonformal classification is examined later; formal classification is indicated by exclusively Latin names, and nonformal, by entirely or partly non-Latin names.) Similarly, since the marijuana (moderate- to high-THC) populations of the world are also being homogenized by hybridization, it also is inadvisable to split them into formal taxonomic groups. Just how to treat the fiber plants collectively and the marijuana plants collectively is examined later in this chapter.

TAXONOMIC SPLITTING AND RANK INFLATION

Biological classification frequently involves some degree of subjective assessment and arbitrary decision, and this is particularly evident at the species level. Darwin (1859) wrote, "I was much struck by how entirely vague and arbitrary is the distinction between species and varieties... I look at the term species as one arbitrarily given for the sake of convenience to a set of individuals closely resembling each other, and that it does not essentially differ from the term variety, which is given to less distinct and more fluctuating forms. The term variety, again, in comparison with mere individual differences, is also applied arbitrarily, and for mere convenience sake." However, this should not be interpreted to mean that biological taxonomists lack standards and consistency with respect to what constitutes a "species." As Darwin (1859) also commented, "various definitions...have been given of the term species. No one definition has satisfied all naturalists; yet every naturalist knows vaguely what he means when he speaks of a species." Since Darwin's time, it has become apparent that in practice, human psychology and motivations are important in determining how species are recognized, and these factors are discussed in this section.

Even when they agree that a set of organisms is distinctive by virtue of shared traits, taxonomists often differ with respect to (1) whether formal nomenclatural recognition is even appropriate and (2) if appropriate, the rank that should be assigned (e.g. species or subspecies). Historically and to this day, some taxonomists (facetiously referred to as "splitters") have a "liberal" approach, formally recognizing more groupings than would be accepted by most of their professional peers, and conversely, some "lumpers" have a "conservative" approach, recognizing fewer groupings than most taxonomists consider appropriate (Figure 18.5). Taxonomic splitting is one cause of "taxonomic inflation," the generation of more scientific names than justified.

Splitting is often accompanied by "rank inflation"—the elevation of groupings to a higher rank (especially to the species level) than justified (Figure 18.6). Taxonomic splitting and rank elevation are attractive to some scientists because these practices amplify the quantity and ranking of taxonomic groups for which they receive credit. However, overrecognition of some groups has resulted in distortion of the nature and significance of studies of biodiversity, ecology, and conservation (Chaitra et al. 2004; Padial and de la Riva 2006).

Isaac et al. (2004) noted that populations assigned species rather than a lower rank are often regarded as more important and that "This encourages elevation to species rank of populations that need protection, regardless of whether there is scientific support for this status... Such inflation will be biased towards charismatic, large-bodied, rare and endangered forms...that attract high public, scientific and conservation interest." Consistent with this motivation and the fact that *Cannabis* is one of the most charismatic of plants, Hillig (2004b) argued that formal recognition of Chinese fiber hemp as a separate taxon "may foster genetic conservation of this agronomically important group."

FIGURE 18.5 "Splitters" (top row) tend to place variants into more groups; "lumpers" (bottom row) tend to assign the same material to fewer groups. Prepared by B. Brookes.

FIGURE 18.6 "Rank inflation" is the tendency to enlarge (inflate) membership in the higher, more important categories of a hierarchy by excessive promotion of those who properly should be assigned membership in lower categories. In this figurative representation (drawn by B. Brookes), an army has too many generals in relation to the number of foot soldiers. As noted in the text, the rank of "species" is widely regarded as having much more importance than lower ranks, and some taxonomists deliberately elevate variations to species rank to emphasize the importance of their study material and consequently their own importance. While permissible, this distorts the relative status of such "species" and misleads science and society in regard to their nature and status.

FIGURE 18.7 Scale comparison of Pluto and the eight planets of the Solar System. Arguing that varieties within *C. sativa* deserve species status is reminiscent of the viewpoint that the "dwarf planet" Pluto deserves planetary status. To be consistent, the Earth's moon, which is five times as large as Pluto, would also have to be classified as a planet. Photos by IStoleThePies (CC BY SA 4.0), Pluto added by B. Brookes.

The tendency for humans to elevate some high-profile minor entities to the same status as major ones is illustrated by the debate over the former planet Pluto, which astronomers voted in 2006 to demote to the status of a "dwarf planet" (note Figure 18.7). Although much smaller than the eight planets currently recognized (Pluto's mass is less than a fifth of the Earth's moon and less than one four hundredth of the Earth), there are many who argue that not only should Pluto regain its planetary distinction but also that some even smaller neighbors of Pluto should be recognized as planets (Stern et al. 2015). What really matters is consistency—whether the discipline is biology or astronomy, the standards of the majority of professional scientists should be respected.

THE SEMANTIC "LEGAL SPECIES" ISSUE (OR DRESSING UP A WOLF IN SHEEP'S CLOTHING)

In the 1970s, a curious forensic debate was founded on splitting what had been widely understood up to that time as the species *C. sativa* into three species (called *C. sativa* in a narrow nonconventional sense, *C. ruderalis* Janischevsky, and *C. indica* Lamarck). In many Western countries, legislation governing illicit cannabis preparations defines the material as originating from "*Cannabis sativa* L." Court cases prior to 1970 witnessed some defenses of individuals accused of marijuana offences on the argument that the material in question came from one or more "legal species" of *Cannabis* (i.e., species in addition to *C. sativa*). This claim failed until 1971 because of the prevailing opinion (at least in the Western world) that there is only one species of *Cannabis*, *C. sativa*. However, in 1971, a court challenge was successful, based on the testimony of several botanists that there is more than one species of *Cannabis*. Subsequently, for a decade, the legal issue was raised in hundreds of courtrooms, especially in the United States and Canada. The ploy was successful because talented lawyers represented taxonomy as simply a factual assessment of existential groups called species (hence expert witnesses were sufficient to decide the "facts"), whereas in reality, one taxonomist's species is another's variety. The issue eventually became moot as judges came to realize that recognition of more than one species of *Cannabis* is based merely on splitting of *C. sativa* into several species and that taxonomic opinion on whether splitting is scientifically correct is irrelevant because the *intent* of legislation using the name "*Cannabis sativa* L." was clearly to designate all forms of *Cannabis* (and certainly the marijuana forms, which many lawyers had speciously argued were exempt from prosecution because they belonged to the "legal species" *C. indica*). In essence, the clever tactic employed was to dress up a wolf (high-THC *C. sativa*) in the guise of a sheep (low-THC *C. sativa*; see Figure 18.8). The history of the legal-taxonomic debate is detailed in Small (1974, 1975b,c,d, 1976, 1977, 1979a,b).

FIGURE 18.8 A wolf in sheep's clothing. As discussed in the text, a widespread legal ploy in the 1970s was based on the proposition that some highly intoxicating kinds of marijuana were actually "legal species" not subject to the law. Prepared by B. Brookes.

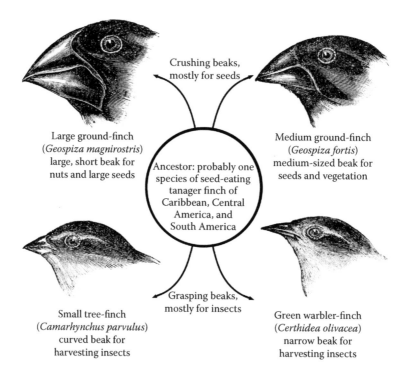

Crushing beaks, mostly for seeds

Large ground-finch (*Geospiza magnirostris*) large, short beak for nuts and large seeds

Medium ground-finch (*Geospiza fortis*) medium-sized beak for seeds and vegetation

Ancestor: probably one species of seed-eating tanager finch of Caribbean, Central America, and South America

Small tree-finch (*Camarhynchus parvulus*) curved beak for harvesting insects

Grasping beaks, mostly for insects

Green warbler-finch (*Certhidea olivacea*) narrow beak for harvesting insects

FIGURE 18.9 Four of the 14 "Darwin finches," exemplifying the natural evolution of species. The endemic Galapagos Island species studied by Charles Darwin are now appreciated to belong to four genera of tanagers (family Thraupidae), not to the true finch family (Fringillidae). Feeding behavior, enabling the birds to acquire different food resources, reflected particularly by beak characteristics, was critical to their adaptive radiation from a common ancestor into the different species. Bird drawings from Darwin, C., *Journal of Researches into the Natural History and Geology of the Countries Visited during the Voyage of H.M.S. Beagle Round the World, under the Command of Capt. Fitz Roy, R.N.* 2nd ed., John Murray, London, 1845.

DOMESTICATION COMPLICATES CLASSIFICATION

Prior to 1970, there was essential unanimity that only one species of *Cannabis* merited recognition. Since then, virtually without exception, those who have espoused the recognition of more than one species of *Cannabis* have done so without addressing the theory and practices of classification of domesticates and their closely related wild populations. Without this background, it is not possible to understand clearly the merits of competing systems of classification of *Cannabis*.

Charles Darwin (1809–1882), the father of evolution, coined the phrase "artificial selection" in the first edition of his work *On the Origin of Species* (Darwin 1859). He concluded that starting from a wild species, human selection could produce divergent breeds so spectacularly different that they mimicked related species produced by natural selection (compare Darwin's analysis of wild birds of different species, Figure 18.9, and his analysis of domesticated birds of a single species, Figure 18.10). Darwin (1859) wrote: "There are hardly any domestic races, either amongst animals or plants, which have not been ranked by some competent judges as…distinct species." Although he more clearly appreciated than anyone previously that classifications of domesticated and wild organisms are debatably comparable, Darwin did not explore the issue of appropriate scientific cataloguing of organisms originated by humans. As detailed in the following discussion, the so-called "species" of *Cannabis* that have been recognized are in fact domesticates (i.e., selections made by humans) or their related escapes, and accordingly, their recognition as conventional species, while permissible, is misleading.

In common language, "domestication" often refers to taming of wild animals, i.e., habituating them to humans so that they are relatively manageable. In biology, domestication is the process of choosing individuals of a species that have characteristics making them useful to people, the

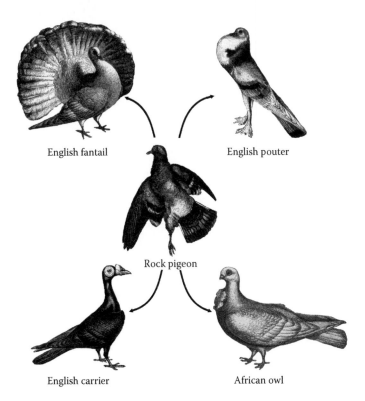

FIGURE 18.10 Four of "Darwin's pigeons," exemplifying the artificial selection of variations desired by humans. The four breeds shown here originated from the wild rock dove or rock pigeon (*Columba livia*, center), the ancestor of all fancy and racing pigeons. Bird drawings from Darwin, C., *The Variation of Animals and Plants under Domestication, Vol. 1.*, John Murray, London, 1868.

selection usually occurring over generations, so that the desired traits become genetically fixed. Almost all important species currently employed in agriculture or for other human purposes are domesticated (for examples, see Figure 18.11). Although the phrase "cultivated plant" is widespread and is often used to refer to domesticated plants, many cultivated plants are simply wild plants that are cultivated, and the different concepts should not be confused. The term "cultigen" has been used to refer to domesticates in a broad sense but has been employed in such different ways (Spencer 1999; Spencer and Cross 2007a,b) that its use can be confusing. *Cultigen* can be used to refer to all or individually recognizable classes of cultivated plants of a given species that have been genetically altered by human selection. As discussed in this chapter, since the cultigens of *Cannabis* intergrade with each other and with widespread weedy forms, all classifications of *C. sativa* are necessarily inexact. Within a cultigen, *landraces* are (typically) geographical groups that have been unconsciously selected over long periods by traditional farmers, and *cultivars* are (typically) named selections produced by breeders or at least deliberately preserved by horticulturalists.

FIGURE 18.11 Examples of selection of extraordinarily diverse variations from a wild species. (a) Gray wolf (*Canis lupus*), the basic ancestor of dogs. Photo by Alois Staudacher (CC BY 2.0). (b) A selection of dog breeds (*Canis lupus familiaris*). From Roe, E.T., Leonard-Stuart, C., *Webster's New Illustrated Dictionary*. Syndicate Publishing Company, New York, 1911. (c) Common pigeon (rock dove, *Columbia livia*), the basic ancestor of fancy pigeons. Photo by Sean MacEntee (CC BY 2.0). (d) Fancy pigeons. From an English poster showing Victorian breeds, published in 1891.

FIGURE 18.12 A "farmer" (ant) tending her "cows" (aphids). Such evolved caretaker-slave symbiotic relationships between nonhumans shows that human domestication of *C. sativa* and other species is fundamentally like natural evolution. Photo by Stuart Williams (CC BY 2.0).

Most domestication has been more or less unconscious, occurring over millennia (Zohary 2004). By contrast, deliberate breeding for desired characteristics has become important mainly during the last 100 years. Whether domestication is in some fundamental way different from natural selection has been the subject of debate (reviewed in Ross-Ibara et al. 2007). Domestication is usually conceived of as a form of "artificial" selection, which is true if one defines artificial selection as the result of human influence that alters the genetics of other species in ways that make them more useful to people. However, some (e.g., Darwin 1859; Darlington 1973) have argued that unconscious, i.e., nondeliberate, selective breeding by humans is as "natural" as the selection that occurs in nature. Domestication is, in fact, a form of evolution (which can be simply defined as the alteration of gene frequencies over time). Rindos (1984) stated, "Domestication clearly cannot be held to be an exclusively human-mediated phenomenon." This is because man is not the only animal that has usurped the freedom of other species, caring for them but at the same time altering their genome so that they can be more efficiently exploited as a source of food. For example, wood wasps and over 40 species of ambrosia beetles cultivate fungi as food sources (they inoculate wood with a fungus which they consume after it has multiplied). Some ants and termites also cultivated fungi (see Rindos 1984 and Schultz et al. 2005 for references and additional examples), and there are also ant species that herd, protect, and breed mutualistic aphids and other homopterans (Hölldolber and Wilson 1990; Schultz and McGlynn 2000; Figure 18.12). In emphasizing that the contrast of "artificial selection" and "natural selection" is in fact an artificial distinction, McNeill (1998) stated "It is not good evolutionary thinking to suppose that man is not inescapably a part of the ecosystem."

A COMPARISON OF COMPETING CLASSIFICATION SYSTEMS FOR *CANNABIS*

Several botanists have contributed to clarification of the taxonomy of *Cannabis* in recent decades, notably, Small and Cronquist (1976), Small (1979a,b, 2015a), Hillig (2004a,b, 2005a,b), Hillig and Mahlberg (2004), McPartland and Guy (2004a), Clarke and Merlin (2013). Based on these studies collectively, the following groups of domesticated plants have been recognized as warranting particular taxonomic attention (compare the postulated ancient Eurasian distribution ranges shown by the same numbers in Figure 18.13 and the key information given in Table 18.1A):

1. Hemp plants domesticated for stem fiber (and to a minor extent for oilseed) in western Asia and Europe; cannabinoids low in THC and high in cannabidiol (CBD) (part of Small's *C. sativa* subsp. *sativa* var. *sativa*, Hillig's *C. sativa* "hemp biotype")

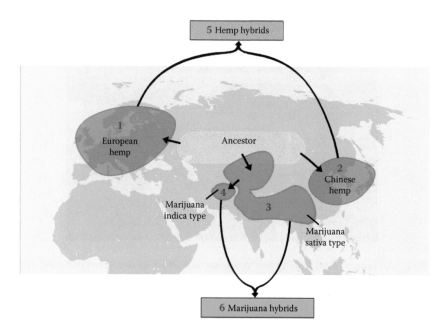

FIGURE 18.13 Approximate postulated geographical locations of ancestral, predomesticated *C. sativa* and the four principal groups (1–4) domesticated more than a millennium ago and subsequently transported to other parts of the world. Table 18.1 provides summary information on the four domesticated groups and Table 12.1 provides additional information on the two marijuana groups, 3 and 4. Hybridization, mostly during the last century, has obscured differences between the two fiber groups, 1 and 2 (generating hybrid group 5) and between the two marijuana groups, 3 and 4 (generating hybrid group 6). Detailed information concerning the evolution, classification, and nomenclature of these groups is presented in this chapter. Prepared by B. Brookes.

2. Hemp plants domesticated for stem fiber (and to a minor extent for oilseed) in East Asia, especially China; cannabinoids low to moderate in THC and high in CBD (part of Small's *C. sativa* subsp. *sativa* var. *sativa*, Hillig's *C. indica* "hemp biotype," Clarke and Merlin's *C. indica* subsp. *chinensis*)

3. Marijuana plants domesticated in a wide area of south-central Asia for very high THC content; cannabinoids mostly or almost completely THC (part of Small's *C. sativa* subsp. *indica* var. *indica*, Hillig's *C. indica* "narrow-leaflet drug biotype," the marijuana trade's "sativa type")

4. Marijuana plants domesticated in southern Asia, particularly in Afghanistan and neighboring countries, for substantial amounts of both THC and CBD (part of Small's *C. sativa* subsp. *indica* var. *indica*, Hillig's *C. indica* "wide-leaflet drug biotype," the marijuana trade's "indica type")

In addition, two hybrid classes of cultivated plants have been widely generated: (5) between the two hemp groups (1 and 2) and (6) between the two marijuana groups (3 and 4). It should be understood that the hybrid cultivars or strains are not simply first-generation hybrids but represent various degrees of stabilized intermediacy, essentially representing all degrees of variation between the parental groups, so that there is continuous variation among hemp biotypes and, similarly, continuous variation among marijuana biotypes.

TABLE 18.1

A Comparison of Taxonomic Concepts and Terminology for *Cannabis* Groupings

A. Domesticated Groupings (Excluding Hybrid Groups)

Classification System				Drug Trade Terminology	THC Content	CBD Content	Principal Early Eurasian Cultivation Area (See Figure 18.13)	Use of Landraces, Cultivars or Strains
Small and Cronquist (1976)	Hillig (2004a, 2005a)	McPartland and Guy (2004a)	Clarke and Merlin (2013)					
C. sativa subsp. *sativa* var. *sativa*	*C. sativa* "hemp biotype"	*C. sativa* subsp. *sativa*	*C. sativa* subsp. *sativa* ("narrow leaf hemp")	–	Low	High	1	Fiber and oilseed
	C. indica "hemp biotype"	*C. indica* subsp. *chinensis*	*C. indica* subsp. *chinensis* ("broad leaf hemp")	–	Low to moderate	High	2	Fiber and oilseed
C. sativa subsp. *indica* var. *indica*	*C. indica* "narrow-leaflet drug biotype"	*C. indica* subsp. *indica*	*C. indica* subsp. *indica* ("narrow leaf drug")	Sativa type	High	Low or absent	3	Marijuana
	C. indica "wide-leaflet drug biotype"	*C. indica* subsp. *afghanica*	*C. indica* subsp. *afghanica* ("broad leaf drug")	Indica type	Moderate to high	Moderate to high	4	Marijuana

B. Uncultivated Groupings (Ruderal, Possibly Including Some Truly Wild Populations)

Classification System				THC Content	CBD Content	Principal Early Eurasian Area
Small and Cronquist (1976)	Hillig (2004a, 2005a)	McPartland and Guy (2004a)	Clarke and Merlin (2013)			
C. sativa subsp. *sativa* var. *spontanea*	*C. sativa* "feral biotype"	*C. sativa* subsp. *spontanea* + *C. ruderalis*	*C. sativa* subsp. *spontanea* ("narrow leaf hemp ancestor")	Low (occasionally moderate)	High	Europe; western to north-central Asia (Small and Cronquist include ruderal low-THC plants of eastern Asia)
C. sativa subsp. *indica* var. *kafiristanica*	*C. ruderalis* + *C. indica* "feral biotype"	*C. indica* subsp. *kafiristanica*	*C. indica* subsp. *kafiristanica* ("narrow leaf drug ancestor")	Low to moderate	Low to moderate (occasionally absent)	Asia

SEMANTIC DIFFICULTIES CONCERNING SATIVA TYPE AND INDICA TYPE *CANNABIS SATIVA*

Sativa type and indica type marijuana strains are contrasted in detail in Chapter 12. Beginning with the rise of marijuana as the leading illicit counterculture drug in the 1960s and persisting to the present day with marijuana strains being marketed in the quasi-legal and legal medicinal markets, there has been a fundamental confusion in much of the popular literature over what the terms "sativa" and "indica" designate. Taxonomists have utilized the epithets *sativa* and *indica* to distinguish two taxa (taxonomic groups), the term "sativa" traditionally designating nonintoxicating hemp plants in contrast to the term "indica," which has been used to designate marijuana plants. The marijuana trade, however, uses both "sativa" and "indica" as labels for different classes of marijuana plants and (contradictory to taxonomic tradition) uses the term "sativa" to designate plants with *more* intoxicating potential (i.e., very high THC content, low or no CBD content) and the term "indica" to designate plants with *less* but still substantial intoxicating potential (i.e., moderate THC content and moderate CBD content). Without appreciation of these contradictory usages, it is often impossible for botanists familiar with taxonomic terminology to understand the information in popular articles that use the terms "indica" and "sativa." Indeed, the authors of some recent scientific publications clearly were confused about what the terms do and do not designate, and if professional scientists are confused, it is understandable that the general public is also uncertain.

HOMOGENIZATION DUE TO GENE FLOW

The domesticated groups of *Cannabis* mentioned previously are of Eurasian origin but, especially in the last several hundred years, have been transported to and cultivated in much of the world. In many regions, they have escaped, reevolved characteristics suited to wild existence, and established as self-perpetuating populations outside of cultivation. Because both domesticated and wild *Cannabis* populations are extremely widespread, interbreed spontaneously over vast distances, have a common diploid chromosome number ($2n = 20$), and possess no biological barriers to interbreeding (Small 1972a), wild-growing and domesticated plants exchange genes easily and extensively. In nature, one finds a complete spectrum of intermediate forms, demonstrating continuity of variation between wild and domesticated forms (Small 1975a). Sawler et al. (2015) found genetic evidence of intergradation between the indica type and sativa type forms of high-THC marijuana. Because domesticated selections are highly susceptible to gene influx from other domesticated selections and from wild-growing forms, to maintain their characteristics, they must be protected from "genetic contamination." Genetic infiltration into *Cannabis* from wild populations has not actually been demonstrated but has been confirmed in *Humulus* (Small 1980, 1981), the very close relative of *Cannabis*, and there is no reason why the two should be different in this respect. Moreover, as with many other crops (and domesticated animals), the mutations selected by humans are usually advantageous to humans but disadvantageous to the plants, and unless stabilizing selection is practiced, natural selection can result in degeneration or reversion (sometimes termed "atavism") of the genome, with wild characteristics appearing in cultivated plants. Patterns of gene change from various factors are summarized in Figure 18.14. The extensive intergradation that has resulted from interbreeding is the chief cause of classification difficulties. The following presentation is concerned primarily with the arrangement of the domesticated groups and wild populations into a classification and naming system.

ALTERNATIVE TAXONOMIC AND NOMENCLATURAL TREATMENTS

The professional taxonomic treatment of plants (indeed of all living things) is largely a standardized activity, involving three phases. The first phase is grouping (recognition of assemblages). The second phase is ordering of these assemblages, conventionally in a hierarchical system (like a series of smaller boxes within progressively larger boxes), involving fixed ranks (e.g., subspecies, species,

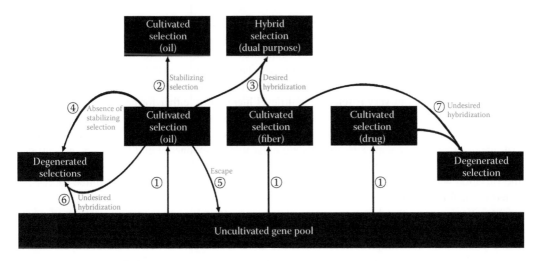

FIGURE 18.14 Patterns of gene flow, genetic stabilization, and genetic destabilization among wild and domesticated races of *C. sativa*. (1) Humans cultivate selections, principally for stem fiber, oilseed, and intoxicating resin. (2) Such selections retain their desirable characteristics only if maintained by stabilizing selection (shown here for simplicity only for the oilseed form). (3) In recent times, deliberate hybridization among oilseed and fiber kinds has generated valuable new selections. (4) In the absence of stabilizing selection, cultivated plants are likely to undergo populational genetic changes over several generations that are undesirable agriculturally (degenerative) since the highly selected characters of interest to humans are usually deleterious to the plants (for simplicity, such degeneration is shown only for the oilseed form). (5) Genes from cultivated plants may be released to the uncultivated gene pool. Selections may escape directly from cultivation and reestablish populations outside of cultivation, or pollen from cultivated selections may fertilize wild plants (for simplicity, such gene escape is shown only for the oilseed form). (6) Pollen from uncultivated plants may fertilize a cultivated selection, reducing the desired characteristics of the latter (for simplicity, this is shown only for the oilseed form). (7) Pollen from cultivated plants with undesirable characteristics (e.g., from clandestine marijuana plants) may pollinate a cultivated selection (e.g., grown for fiber or oilseed), reducing the desired characteristics of the latter.

genus, and family), although as noted later, there are other kinds of arrangements. The third and final phase is naming: the provision of appropriate nomenclature in an unambiguous manner that reflects the nature of the classification system. The possibilities differ somewhat according to the rules of current nomenclatural codes (for general information on nomenclatural codes for the principal kinds of organisms, see David et al. 2012). In the case of *Cannabis*, two botanical nomenclatural codes are particularly relevant, as well as noncodified classification systems, as discussed in the following.

CANNABIS ASSEMBLAGES AS CONVENTIONAL TAXA

Beginning with a code governing botanical nomenclature prepared in 1867, improved internationally accepted versions have been published periodically. The latest is *The International Code of Nomenclature for Algae, Fungi, and Plants* (ICNAFP; McNeill et al. 2012). This is the most respected and universally applied way of determining plant names (the third phase of taxonomic procedures mentioned in the previous paragraph). There is no impediment to treating groups that are completely or partly domesticated under this code. All groups that are recognized are assigned a particular rank, and the Latin name (if not newly coined) is determined by examining all eligible names that correspond with the group that have previously been accepted and by reference to the rules of the code to determine the single, correct name. The code specifies the conventions that must be followed for naming taxonomic groups, but different taxonomists can disagree about which individuals fall within given groups (i.e., the circumscription of groups) and about the hierarchical organization (i.e., ranks assigned to groups), and these disagreements can mean that a given plant may be identified "correctly" but differently by different taxonomists and that a given plant name can be interpreted differently by different taxonomists. When a name has been used in different senses so extensively that it is a source of confusion, Article 57 of the ICNAFP provides for stabilizing usage of or simply abandoning that name. Certainly, there has been extensive confusion over how to use some of the species names associated with *Cannabis*, but no one has yet suggested that Article 57 be applied.

Some traditional taxonomists (especially in Europe in the twentieth century) subcategorized important crop plants in very extensive, multilevel hierarchies, either formally (i.e., in strict conformity with the botanical code) or quasi-formally. Sometimes, hundreds of groups were recognized. Examples of categories that have been used are presented in Jirásek (1961); examples and a critique of excessively complex treatments are presented in Spooner et al. (2003). The eccentricity and unworkability of this approach led to efforts to find a standardized, simple way of classifying the variation within cultivated plants in relation to their wild relatives (but with limited success, as noted in the following discussion).

A particular issue that has troubled plant taxonomists is how to categorize groups in which there are both wild and domesticated kinds using traditional formal categories. There have been many proposals. For example, Harlan and de Wet (1971) suggested that where both ruderal and domesticated races exist within one species, all of the ruderal races should be recognized as a collective subspecies, and similarly, all of the domesticated forms should be placed in a collective cultivated subspecies. Similarly, Nesom (2011) treated apparent wild progenitors and their domesticated derivatives in the family Cucurbitaceae as separate subspecies of a given species. However, there is no agreed way of taxonomically separating domesticated plants and their close wild relatives and indeed very limited prospects for the adoption of a universal solution to this issue.

CANNABIS CULTIVAR ASSEMBLAGES AS "GROUPS" UNDER THE CULTIVATED PLANT CODE

Carl Linnaeus (1707–1778), the father of modern taxonomy, was aware that some species included domesticated forms differing from those found in nature. He was disinterested, indeed hostile, to the expansion of his method of designating species by binomial names to domesticated plants (Hetterscheid et al. 1996). Notably, Linnaeus used Latin phrases (mostly with more than the two terms he standardly employed in binary species names) to describe 12 kinds of *Brassica oleracea* (Linnaeus 1753), which include wild plants as well as distinctive domesticated crops known as coles, cabbages, and kohlrabis (Oost 1989). (Today, Linnaeus' cabbage-type groups have been assigned to the formal category *varietas*, a rank translated as variety [var.], although widely confused with the vernacular nonformal term "variety," which is equivalent to the term "cultivar".) As noted previously, plant taxonomists have not reached a consensus on how to classify plants in which there are both domesticated and wild representatives.

Confronted by a growing body of plant names applied to cultivated plants, taxonomists created a special code using non-Latin or "fancy" names (Stearn 1952). Since the middle of the twentieth century, domesticated selections of plants termed "cultivars," which satisfy certain descriptive and publication requirements, have been the subject of a special, at least partly non-Latinized, code of nomenclature (International Code of Nomenclature for Cultivated Plants; ICNCP; latest edition: Brickell et al. 2009). The ICNCP provides the following definition: "A cultivar is an assemblage of plants that (a) has been selected for a particular character or combination of characters, (b) is distinct, uniform, and stable in these characters, and (c) when propagated by appropriate means, retains those characters." Article 9.1, Note 1, restricts the meaning of cultivar as follows: "No assemblage of plants can be regarded as a cultivar...until its category, name, and circumscription has [sic] been published." (*Webster's Third New International Dictionary* [Gove 1981] provides a more general definition of a cultivar: "an organism of a kind [as a variety, strain, or race] that has originated and persisted under cultivation.") Cultivars as defined by the ICNCP can be of quite different nature (e.g., they may be hybrids, clones, grafts [i.e., combinations of species], chimeras [with genetically different tissues], and even plants that are distinct simply because they are infected by a microorganism), but frequently, many of the cultivars within a given species differ very little genetically from each other. There are more than a hundred recognized cultivars of nonintoxicating forms of *Cannabis*, currently grown for fiber and/or oilseed (many are listed in Table 17.3). Only a handful of forms bred for authorized medicinal usage at present are regarded as cultivars under the ICNCP (there are also numerous breeding lines that are not afforded cultivar recognition). There are also over a thousand illicit or quasi-licit marijuana "strains" (or at least allegedly different strains) that are currently circulated in the black, gray, and medicinal marijuana trades (as noted earlier, *Cannabis* strains are biologically equivalent to cultivars, although not nomenclaturally). Many cultivated plants of *Cannabis* are "land races"—populations domesticated in a locale, typically selected over long periods by unconscious (nonplanned, undeliberate) selection by traditional farmers, usually adapted to local stresses, and often much more variable than modern cultivars. (In numerous crops, land races have provided the raw materials from which cultivars have been selected.) The ICNCP does not adequately address nomenclature for land races (unless they have been recognized as cultivars, which is quite infrequent) but does provide a context for classifying and naming cultivars. There is no provision under the cultivated plant code for special recognition of uncultivated, wild (ruderal) plants, but it is understood that nomenclature for the wild phases of a species normally falls under the comprehensive plant code (ICNAFP). As noted later, the ICNCP is mainly concerned with names of plant groups that differ mostly in minor ways (terms such as "biotype" or "strain" are usually applicable). Except for the "group" category discussed next, the ICNCP has not served to address the issue of names for major divisions of domesticated plants within species or species groups, nor how to distinguish such major divisions from related wild plants.

The cultivated plant code (ICNCP) has been the subject of debate, particularly as it relates to the plant code applying to all plants (ICNAFP). There have been attempts to introduce a parallel term, "culton," for the term "taxon" (see McNeill 1998 for a critique). Mostly in the past, cultivars were sometimes grouped in "convarieties," a troublesome category because it has been used to indicate rank according to the comprehensive nomenclatural code for plants. A peculiarity of the ICNCP, pointed out by McNeill (2004), is that it does "not presume that desirable groupings are necessarily non-overlapping" (i.e., according to Article 3.4, a given cultivar can simultaneously belong to more than one group).

A key feature of the ICNCP provides for recognition of "groups" of cultivars, allowing considerable flexibility in their formation ("Criteria for forming and maintaining a group vary according to the required purposes of particular users") but insisting that "All members of a Group must share the character(s) by which that Group is defined." (A special group category, "grex," applies only to horticultural hybrids of orchids.) The group concept is flexible in choice of characters serving to define membership (of course, there may be disagreements among specialists about which characters should be the basis for group recognition). Because the group concept of the cultivated plant code has only a single rank (really no rank), it does not provide for using taxonomic rankings as an indication of phylogenetic history.

The group concept provides a simple, sound alternative way of labeling variation of domesticated forms in the genus *Cannabis*. It eliminates the need to consider rank; what various authors may have treated as species, subspecies, or varieties can be reduced to the same level. The four domesticated assemblages noted in Table 18.1A can simply be recognized as groups. There is considerable hybridization in *Cannabis*, which often makes identification problematical, but the same is true of most important domesticated plants. Groups that are hybrids between other groups can simply be recognized as separate groups.

The following classification accounts for variations of domesticated forms of *C. sativa*, under the cultivated plant code, with synonymous terminology shown in parenthesis (the designation of groups by number is consistent in this chapter; cf. Figure 18.13 and Table 18.1). The same classification was presented in Small (2015a) but with different terminology.

1. *Cannabis* European Hemp Group: Plants tracing to European and western Asian fiber and oilseed races, cannabinoids low in THC and high in CBD (part of Small and Cronquist's *C. sativa* subsp. *sativa* var. *sativa* and Hillig's *C. sativa* "hemp biotype").

2. *Cannabis* Chinese Hemp Group: Plants tracing to East Asian fiber and oilseed races; cannabinoids low to moderate in THC and high in CBD (part of Small and Cronquist's *C. sativa* subsp. *sativa* var. *sativa*, Hillig's *C. indica* "hemp biotype," and Clarke and Merlin's *C. indica* subsp. *chinensis*).

3. *Cannabis* High-THC Marijuana Group: Marijuana strains in which the cannabinoids are mostly or almost completely THC (part of Small and Cronquist's *C. sativa* subsp. *indica* var. *indica*, Hillig's *C. indica* "narrow-leaflet drug biotype," and the marijuana trade's "sativa type").

4. *Cannabis* THC/CBD Balanced Marijuana Group: Marijuana strains in which populations have substantial amounts of both THC and CBD (part of Small and Cronquist's *C. sativa* subsp. *indica* var. *indica*, Hillig's *C. indica* "wide-leaflet drug biotype," the marijuana trade's "indica type," and *C. indica* of Schultes et al. 1974).

5. *Cannabis* Hemp hybrids: A group of hybrids between groups 1 and 2.

6. *Cannabis* Marijuana Hybrids: A group of hybrids between groups 5 and 6.

CROP-WILD ASSEMBLAGES AS NONFORMAL GROUPS

"Formal" taxonomic treatment refers to the strict use of the categories and nomenclatural conventions for designating groups of organisms specified in at least one of the codes of nomenclature governing plants. "Informal" classification refers to organizational and naming systems that do not conform to one of the codes.

A number of theorists of plant classification have espoused the view that classification of crop-wild complexes, in which there is at least some interbreeding, is preferably carried out informally (also note the discussion later of natural and artificial classification). There are endless definitions of "species," no universally accepted criterion or criteria for this fundamental grouping, and considerable heterogeneity in the nature of groups that are called species. Nevertheless, the ability to interbreed and the actual degree to which interbreeding occurs are critical considerations in recognizing species of plants because gene exchange among populations tends to eliminate the differences that are employed to define species. The so-called "biological species concept" defines species on the basis of actual or potential breeding separateness (and clearly, on this basis, there is only one species of *Cannabis*). Above the biological species level, evolution is largely bifurcating (although there is debate about the degree to which hybridization among groups at the genus level and above has occurred), a pattern that is compatible with the hierarchical structure of conventional plant taxonomy. However, some systematists (e.g., Minelli 1993; Pickersgill et al. 2003) have concluded that variants below the biological species level (often classified as subspecies and varieties) are usually not generated in a hierarchical fashion, either in nature or in cultivation, and so using more than one infraspecific rank for crop-weed complexes, as has been commonly done in an attempt to reflect evolutionary patterns, is usually unjustified.

Harlan and de Wet (1971), frustrated with the inconsistent treatment of crops and their closely related wild relatives, proposed a nonformal system of classification, which is in fact an elaboration of the biological species concept (Spooner et al. 2003). Their so-called "gene pool classification" recognizes (a) a "primary genepool," based on the crop and wild populations (whether or not recognized as different species) that interbreed readily with it (Harlan and de Wet characterized their primary gene pool as equivalent to the traditional biological species concept); (b) a "secondary genepool," made up of populations that can interbreed with the crop but only with some difficulty; and (c) a "tertiary genepool," made up of populations that can interbreed with the crop but only with considerable difficulty (this group is the equivalent of a "coenospecies" in the terminology of Clausen et al. 1948). Harlan and de Wet further proposed a scheme of hierarchical subpartitioning using nonformal categories (i.e., independent of the codes of nomenclature). No one has succeeded in hybridizing *C. sativa* with any other species in the Cannabaceae, and all plants of *Cannabis* interbreed freely, so classification of *Cannabis* according to Harlan and de Wet's concept is simple: all plants belong to the primary genepool of the one biological species, *C. sativa*.

Jeffrey (1968), consistent with his view that "cultivated plants differ from one another so greatly in their variation patterns that a formal system applicable to all is not only impossible but undesirable," recommended a nonformal system of classification with a maximum of two hierarchical categories to classify cultivars and a new term ("subspecioid") to separate the domesticated from the related wild-growing plants. Other schemes have been proposed to treat crop classification in ways that are distinctive from the conventional way of classifying wild plants (for examples, see Styles 1986; for reviews, see Hetterscheid et al. 1996 and Hammer and Morimoto 2012). A comprehensive nonformal classification system for *Cannabis* has not yet been proposed.

OCCAM'S RAZOR IN RELATION TO THE EVOLUTION AND CLASSIFICATION OF *CANNABIS*

Conventional biological classifications are, at least to some degree, scientific hypotheses theorizing that certain individuals deserve to be grouped together based on consideration of all or some of their characteristics. In scientific theory, Occam's (Ockham's) razor is a recommendation that explanations be as simple as possible, limiting unproven assumptions (Figure 18.15). (Einstein's razor, variously phrased, holds that scientific explanations should be as simple as consistent with facts.) Stephen Hawking, in his classic *A Brief History of Time*, wrote "It seems better to employ the principle known as Occam's razor and cut out all the features of the theory that cannot be observed." The following are chief, unnecessary presumptions or assumptions that have contributed to confusion concerning the evolution and classification of *Cannabis*.

1. Assertion: There are wild plants growing outside of cultivation that coincide with predomestication populations, and so these can be recognized as conventional taxa (species, subspecies, or varieties).

   Observations: There might be genuinely wild *Cannabis* plants that are completely or substantially unaffected by domestication, but no one has demonstrated their existence. It is commonplace for crops that have been domesticated for very long periods to lack any evidence of genuinely wild (not merely escaped-ruderal) extant ancestral populations. Given the long history, extensive distribution of *Cannabis* by humans and the ease of genetic exchange between cultivated and uncultivated populations, it is unlikely that unaltered wild forms still exist.

2. Assertion: The four basic domesticated groups of *Cannabis* were generated over past millennia from different genuinely wild ancestral populations.

FIGURE 18.15 William of Ockham (ca. 1287–1347), Franciscan philosopher. Left: Razor photo by Fred the Oyster (CC BY 3.0). Right: Photo of a stained glass church window in Surrey, Ockham, England. Photo by Moscarlop (CC BY SA 3.0).

Observations: This viewpoint is adopted extensively by Clarke and Merlin (2013), who assign formal scientific names to seven putatively wild ancestors (including the putative ancestor of all forms of *Cannabis*) of the four domesticated groups recognized in this review, while conceding that these are "either extant and unrecognized or extinct." The far simpler and more likely explanation is that humans generated the domesticated groups from a single original wild species, as indeed is the case for innumerable domesticated plants.

HOW MANY SPECIES OF *CANNABIS* MERIT RECOGNITION?

In much of the literature debating the issue of how many species of *Cannabis* deserve recognition, the viewpoint that there are several species has been termed the "polytypic species" concept, and the view that there is just one has been called the "monotypic" view. This is simply a misinterpretation of the term "polytypic," which in taxonomy simply means that a group is composed of several elements (taxa or races). A polytypic genus has more than one species; a polytypic species has more than one infraspecific taxon, or is simply variable, containing more than one kind. A genus with more than one species is correctly described as polyspecific; a genus with just one species is monospecific.

Much of the preceding discussion explains that the contention that there are several species of *Cannabis* is simply a semantic preference, not dictated just by scientific considerations, and that taxonomists are familiar with such competing taxonomic interpretations. However, most taxonomists are suspicious of alleged species that are 100% interfertile, as are the putative species of *Cannabis*. More critically, when no one has provided a reliable means of morphologically distinguishing the proposed species, few taxonomists would accept their recognition. There is no supreme organization or authority that judges the comparative merit of given taxonomic treatments. However, competing taxonomies are judged by users, the most knowledgeable of which are those who prepare guides to the flora of regions. Today, virtually all authoritative floras recognize only one species of *Cannabis*, *C. sativa* (see, for example, Qaiser 1973; Tutin and Edmonson 1993; Small 1997; Wu et al. 2003), indicating that the designation of more than one species is inappropriate by contemporary standards. Moreover, as stated by De Meijer (2014): "A monospecific concept…has implicitly been adopted in virtually all, nontaxonomic, publications on *Cannabis*… The current pattern of *Cannabis* diversity is primarily due to intentional actions of humans and reflects a long, intense, and divergent process of domestication which has blurred any natural evolutionary pattern of diversity. It is even questionable if truly wild *Cannabis* still exists."

As discussed previously, the recognition of more than one species of *Cannabis* is typical of the overclassification of domesticated crops. Harlan and de Wet (1971) wrote about this problem: "Man has been very active in manipulating the gene pools through repeated introductions or migrations, followed by natural or artificial hybridization. The germ plasm of domesticated plants has been repeatedly and periodically stirred. The environment provided has been artificial, unstable and often very extensive geographically. Selection pressures have been very strong, but biologically capricious and often in diverse directions. The end result is an enormous amount of conspicuous variation among very closely related forms. Faced with this situation, the traditional taxonomist tends to overclasssify. He finds conspicuous either-or characters, often without intermediates, and frequently bases 'species' on them. The characters may be controlled by one or a few genes and have little biological significance. Too many species and too many genera are named."

Based on multivariate statistical similarities of allozyme frequency, Hillig (2005a,b) separated European fiber plants from the three more easterly domesticated groups: the two marijuana groups and Chinese fiber plants. Additional but less clear support for this separation was found by examination of terpene chemistry (Hillig 2004a) and cannabinoid chemistry (Hillig and Mahlberg 2004), and the evidence was clearer for cultivated accessions than for ruderal ones. In these studies, Hillig assigned the European fiber plants to "*C. sativa*" and the three eastern groups to "*C. indica*," noting that this had the unexpected effect of combining within *C. indica* the two marijuana groups and

Chinese hemp. Hillig's data are valuable in indicating that there was probably in ancient times a genetic differentiation trend between the plants of western Eurasia (and consequently Europe) and those of eastern Eurasia. Likely, European hemp went through a genetic bottleneck as it was being selected from the more eastern plants. However, by evolutionary standards, this trend seems very minor, since not a single reliable character has been found to distinguish the western (European) and eastern kinds collectively, nor has a combination of morphological characters been suggested that could serve to separate them reliably, as is necessary in conventional plant taxonomic identification keys. Recent DNA evidence does indicate that at the molecular level, combined genetic loci may be usable to discriminate European hemp strains, indica type plants, and sativa type plants (Lynch et al. 2015; Sawler et al. 2015). The situation is perhaps analogous to human blood group geography, thought to have resulted from a combination of random drift and selection for disease resistance (Anstee 2010), and certainly not warranting formal taxonomic recognition. The information is, however, useful for tracing genetic relationships and identifying strains and cultivars.

A RATIONALE FOR EMPHASIZING THE PRINCIPAL SELECTED CHARACTER COMPLEXES IN CLASSIFICATION

Aside from groups resulting from hybrid origin or lateral gene transfer, it is usually assumed that organisms sharing a unique set of characteristics arose from a single ancestor. Indeed, the cladistics school of classification insists that recognized taxonomic groups must have a single origin and uses a phyletic pattern of bifurcating groups as the theoretical justification for hierarchical classification. However, adaptive gene complexes within taxonomic groups frequently appear to have arisen recurrently, i.e., repeatedly, independently, and in parallel (e.g., Levin 2001; Arendt and Reznick 2007). Many crops appear to have arisen repeatedly and independently within the same species (Diamond 2002). In the long course of history, fiber strains of *Cannabis* were probably selected independently in different geographic regions, and the same is likely true for marijuana strains, a phyletic pattern that is not hierarchical in organization and reflects the difficulty of classifying variation within many species. In arguing against the application of hierarchical classification below the species level, Jeffrey (1968) pointed out: "Similar selection pressures, operating on genetically similar but distinct lines, may evoke similar responses in those lines, giving rise to parallel variation, the homologous series of Vavilov, a phenomenon by no means confined to cultivated plants, but often exhibited by them to a marked degree." This consideration complicates the classification of crop complexes because it means that critical aspects of the genome may be arrayed in complex ways within a group, and taxonomic recognition of this partitioning may be a debatable issue.

In biological taxonomy, "natural classifications" (sometimes termed general classifications) are based on overall genetic similarities and/or phylogeny, while so-called "artificial" or "special-purpose" classifications are based on selected similarities of particular (practical) interest to people. Artificial classification is unrelated to the concept of artificial selection and is a phrase sometimes used pejoratively to indicate that the merit of such classifications is limited. It is often claimed that restricting the character base to only certain economic considerations means that the resulting classification is not based on evolution and so not an acceptable basis for biological taxonomy. However, characteristics of domesticated organisms *are* the result of evolution, and when they are produced by strong selective pressures, they may merit special taxonomic consideration. This is important for classifying domesticated plants, particularly for *Cannabis*, because biological taxonomy is, above all, intended to convey information, and for useful plants like crops, the most useful information often resides in a particular aspect of the genome, not necessarily the entire genome. Characters or character complexes that are selected by humans are adaptive for domesticated plants, at least in the context of cultivation, and using such characters in recognizing taxa does constitute evolutionary classification. For *Cannabis*, my own classification summarized in the following is based on the recurrent selective pressures (and associated gene selection) for stem fiber or THC content (between groups of domesticated plants) and for achene retention or shattering (between wild and cultivated

plants). These principal selective evolutionary pressures on *Cannabis* are responsible for the generation of the most obvious and important variation within the genus and are accordingly appropriate
bases for taxonomic delimitation.

A PRACTICAL AND NATURAL TAXONOMY FOR *CANNABIS*

The following four-group taxonomic subdivision of *Cannabis* under the ICNAFP code (based on
Small and Cronquist 1976) is an alternative to the six-group classification under the ICNCP code
presented in the preceding. The key presented first divides the one species recognized into two
groups on the basis of THC and CBD content. As noted in Chapter 11, the genetic determination of
these compounds is probably under the partial genetic control of codominant alleles, and this may
provoke the criticism that the division on the basis of predominant cannabinoid is essentially a
"one-character taxonomy" (a rather pejorative phrase in classification science). Keys are, by their
nature, simplifications of available knowledge and necessarily limit characters used for identification. As discussed in this book, there are in fact numerous trends that differ between plants of the
hemp class and those of the marijuana class.

As shown in Figure 18.16, divergent selection for high THC content and high stem fiber content
represents a principal dimension of disruptive evolutionary forces that are responsible for differences
in *Cannabis*. All plants domesticated for fiber tend to share a common set of selected characters (e.g.,
primary fiber constitutes a large percentage of the stem, CBD makes up a large percentage of the
cannabinoids, THC rarely is present in large amounts, and the plants are photoperiodically adapted to
flower in relatively high latitudes of the Northern Hemisphere), and all plants domesticated for intoxicating effect tend to share a different set of contrasting characters (e.g., secondary, not primary fiber
constitutes a large percentage of the stem, THC makes up a large percentage of the cannabinoids,
and photoperiodic adaptation is usually for relatively lower latitudes of the Northern Hemisphere).

As shown in Figure 18.17, divergent selection for "seed" (achene) shattering (separation from the
maternal plant) in ruderal plants and achene retention in domesticated plants is a second principal
dimension of disruptive selection in *Cannabis* (reflective of a more general disruptive selection for
existence in cultivation or existence in nature).

The two kinds of disruptive selection described in the preceding paragraphs are combined in the
classification shown in Figure 18.18.

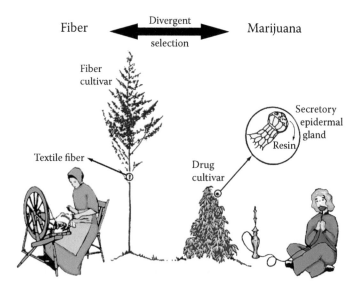

FIGURE 18.16   Divergent selection for fiber and intoxicating drug content.

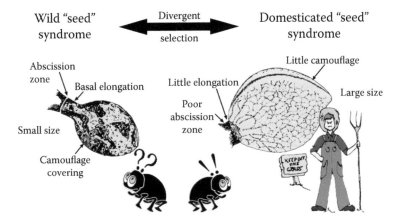

FIGURE 18.17 Divergent selection for adaptive achene ("seed") characteristics between domesticated and wild plants.

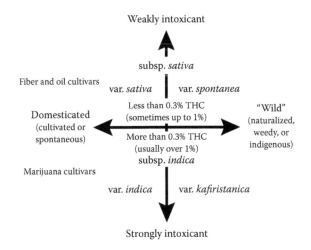

FIGURE 18.18 Classification of *C. sativa* by Small and Cronquist (1976), illustrating conceptual bases of delimitation.

Regarding THC concentration, diagnostic for subspecies: as discussed in Chapter 11, THC concentration in *C. sativa* is known to vary somewhat with environment, maturity, and other factors, and often there are differences among plants of a population. A minimum level of 1% is indicative of plants that can be used to prepare marijuana, and frequently, it is known whether material available is used for marijuana usage and is therefore assignable to subsp. *indica*. Most fiber and oilseed cultivars (with the exception of some East Asian cultivars), by contrast, have less than 1% and are assignable to subsp. *sativa*.

Regarding achenes ("seeds"), diagnostic for varieties: Only substantially mature achenes exhibit the identification characteristics clearly. In the North Temperate region of the world, geography alone frequently serves to distinguish the cultivated from the wild varieties of a given subspecies: in North America, plants growing in uncultivated situations north of 30° latitude are almost always var. *spontanea*, and in Eurasia, the same is true for plants growing in uncultivated situations north of 35° latitude. Wild-growing plants in southern Asia and northern Africa are frequently var. *kafiristanica*. In many other areas of the world, wild populations are derived from escapes either from cultivated high-THC or low-THC strains, and an analysis of THC levels is required for identification.

Since there is extensive intergradation among the taxa, the classification is necessarily inexact (some plants or populations will be found to be intermediate and not easily assigned to one of the groups, but this is a well-known limitation of classifying groups within a species).

IDENTIFICATION KEY TO SUBSPECIES AND VARIETIES OF *C. SATIVA* L.

1. Plants of limited intoxicant ability, Δ^9-THC usually comprising less than 0.3% (dry weight) of upper third of flowering plants (sometimes up to 1%) and usually less than half of cannabinoids of resin. Plants cultivated for fiber or oil or growing wild in regions where such cultivation has occurred .... *C. sativa* subsp. *sativa*

    2. Mature achenes relatively large, seldom less than 3.8 mm long, tending to be persistent, without a basal constricted zone, not mottled or marbled, the perianth poorly adherent to the pericarp and frequently more or less sloughed off .... *C. sativa* subsp. *sativa* var. *sativa*

    2. Mature achenes relatively small, commonly less than 3.8 mm long, readily disarticulating from the pedicel, with a more or less definite, short, constricted zone toward the base, tending to be mottled or marbled in appearance because of irregular pigmented areas of the largely persistent and adnate perianth .... *C. sativa* subsp. *sativa* var. *spontanea* Vavilov

1. Plants of considerable intoxicant ability, Δ^9-THC usually comprising more than 1% (dry weight) of upper third of flowering plants and frequently more than half of cannabinoids of resin. Plants cultivated for intoxicant properties or growing wild in regions where such cultivation has occurred .... *C. sativa* subsp. *indica* (Lam.) E. Small & Cronquist

    3. Mature achenes relatively large, seldom less than 3.8 mm long, tending to be persistent, without a basal constricted zone, not mottled or marbled, the perianth poorly adherent to the pericarp and frequently more or less sloughed off .... *C. sativa* subsp. *indica* var. *indica* (Lam.) Wehmer

    3. Mature achenes relatively small, usually less than 3.8 mm long, readily disarticulating from the pedicel, with a more or less definite, short, constricted zone toward the base, tending to be mottled or marbled in appearance because of irregular pigmented areas of the largely persistent and adnate perianth .... *C. sativa* subsp. *indica* var. *kafiristanica* (Vavilov) E. Small & Cronquist

CURIOSITIES OF SCIENCE, TECHNOLOGY, AND HUMAN BEHAVIOR

- Bacteria sometimes colonize the interiors of plant roots, reminiscent of endomycorrhizal fungi (endomycorrhizae). Such bacteria have been shown to be capable of taxonomic discrimination among marijuana strains of *C. sativa*, showing preferences for some over others (Winston et al. 2014).
- The butterfly *Pieris brassicae* has also been shown to have some ability to recognize different kinds of *C. sativa*. It has been observed to prefer to oviposit its eggs into a fiber Turkish strain by comparison with an intoxicating Mexican strain (Rothschild and Fairbairn 1980).
- Tiger moths (*Arctia caja*) have proven to have poor ability to discriminate kinds of *C. sativa*. Experimentally given a choice of consuming a high-THC strain of *C. sativa*, which proved fatal, and a low-THC strain that they were able to tolerate, they committed suicide by choosing the former (Rothschild et al. 1977).

FIGURE 18.19 Reconstruction of Nessie, the Loch Ness Monster, as a plesiosaur (a long-necked aquatic reptile that went extinct 66 million years ago) outside the Museum of Nessie, Loch Ness, Scotland. As noted in the text, hypothetical taxonomic groups require verifiable evidence before they are accepted. Photo by StaraBlazkova (CC BY SA 3.0).

- Plant taxonomists often deposit reference specimens in herbaria to document their observations. Such specimens often serve as vouchers, which can be studied to determine the accuracy of conclusions. A good example of this is what was alleged to be a peculiar variant of "*Cannabis*" found on a herbarium sheet housed in the Field Museum of Natural history in Chicago. The plant was determined to actually be a male *Datisca cannabina*, which is a remarkable mimic of *C. sativa* (Small 1975e; Figure 1.19b).
- About 1.4 million species have been named to date, most of them with unmemorable names. However, biological taxonomists sometimes compose humorous names, for example, *Abra cadabra* (a clam), *Ba humbugi* (a snail), *Oedipus complex* (a salamander), *Pieza pi* (a fly), and *Pison eu* (a wasp). For additional examples, see Isaak, M. Curiosities of biological nomenclature: http://www.curioustaxonomy.net/index.html.
- What appear to be legitimate scientific names for species are sometimes just not accepted by science because of a lack of evidence. In most cases, the most tangible evidence required is a "type specimen," a physical sample or, at least in some cases, an illustration. The Loch Ness monster (note Figure 18.19) was described as *Nessiteras rhombopteryx* in 1975 based on an alleged underwater photo, and if one day this is proven to have been a truthful observation, the name would be valid.

Literature Cited

Cannabis sativa. Female plant at left, male at right. (Courtesy of Baillon, M.H., *Dictionnaire de Botanique, Vol. 1.* Librairie Hachette, Paris, France, 1876.)

Abel, E.L. 1980. *Marihuana—The first twelve thousand years*. New York: Plenum Press. 289 pp.

Abel, E.L. 1982. *A marijuana dictionary. Words, terms, events and persons relating to cannabis*. Westport, CT: Greenwood Press. 136 pp.

Aboulaich, N., Mar Trigo, M., Bouziane, H., Cabezudo, B., Recio, M., El Kadiri, M. et al. 2013. Variations and origin of the atmospheric pollen of *Cannabis* detected in the province of Tetouan (NW Morocco): 2008–2010. *Science of the Total Environment* 443: 413–419.

Abrams, D.I., and Guzman, M. 2014. Cannabinoids and cancer. In: Abrams, D.I., and Weil, A.T. eds. *Integrative oncology*. Oxford, U.K., Oxford University Press, 246–284.

Abrams, D.I., Couey, P., Shade, S.B., Kelly, M.E., and Benowitz, N.L. 2011. Cannabinoid-opioid interaction in chronic pain. *Clinical Pharmacology and Therapeutics* 90: 844–851.

Abrams, D.I., Vizoso, H.P., Shade, S.B., Jay, C., Kelly, M.E., and Benowitz, N.L. 2007. Vaporization as a smokeless cannabis delivery system: A pilot study. *Clinical Pharmacology & Therapeutics* 82: 572–578.

Acharya, S.L., Howard, J., Panta, S.B., Mahatma, S.S., and Copeland, J. 2014. Cannabis, Lord Shiva and holy men: Cannabis use among Sadhus in Nepal. *Journal of Psychiatrists' Association of Nepal* 3: 9–14.

Adams, R., Pease, D.C., Cain, C.K., and Clark, J.H. 1940. Structure of cannabidiol. VI. Isomerization of cannabidiol to tetrahydrocannabinol, a physiologically active product. Conversion of cannabidiol to cannabinol. *Journal of the American Chemical Society* 62: 2402–2405.

Adams, R., Cain, C.K., McPhee, W.D., and Wearn, R.B. 1941. Structure of cannabidiol. XII. Isomerization to tetrahydrocannabinols. *Journal of the American Chemical Society* 63: 2209–2213.

Adler, J.N., and Colbert, J.A. 2013. Medicinal use of marijuana—Polling results. *New England Journal of Medicine* 368: e30. doi:10.1056/NEJMclde1305159.

Afzal, I., Shinwari, Z.K., and Iqrar, I. 2015. Selective isolation and characterization of agriculturally beneficial endophytic bacteria from wild hemp using canola. *Pakistan Journal of Botany* 47: 1999–2008.

Aggarwal, S.K. 2013. Cannabinergic pain medicine: A concise clinical primer and survey of randomized-controlled trial results. *Clinical Journal of Pain* 29: 162–171.

Aggarwal, S.K., Carter, G.T., Sullivan, M.D., ZumBrunnen, C., Morrill, R., and Mayer, J.D. 2009. Medicinal use of cannabis in the United States: Historical perspectives, current trends, and future directions. *Journal of Opioid Management* 5: 153–168.

Agurell, S., Halldin, M., Lindgren, J.E., Ohlsson, A., Widman, M., Gillespie et al. 1986. Pharmacokinetics and metabolism of delta 1-tetrahydrocannabinol and other cannabinoids with emphasis on man. *Pharmacological Reviews* 38: 21–43.

Ahmad, M., Ullah, K., Khan, M.A., Zafar, M., Tariq, M., Ali, S. et al. 2011. Physicochemical analysis of hemp oil biodiesel: A promising non edible new source for bioenergy. *Energy Sources, Part A: Recovery, Utilization, and Environmental Effects* 33: 1365–1374.

Ainsworth, C. 2000. Boys and girls come out to play: The molecular biology of dioecious plants. *Annals of Botany* 86: 211–221.

Aldrich, M.R. 1977. Tantric cannabis use in India. *Journal of Psychedelic Drugs* 9: 227–233.

Aldrich, M. 1997. History of therapeutic cannabis. In: Mathre, M.L. ed. *Cannabis in medical practice. A legal, historical and pharmacological overview of the therapeutic use of marijuana.* Jefferson, NC: McFarland & Company, 35–55.

Al-Khalifa, A., Maddaford, T.G., and Chahine, M.N. 2007. Effect of dietary hempseed intake on cardiac ischemia-reperfusion injury. *American Journal of physiology—Regulatory Integrative and Comparative Physiology* 292: 1198–1203.

Allegret, S. 2013. The history of hemp. In: Bouloc, P., Allegret, S., and Arnaud, L. eds. *Hemp: Industrial production and uses.* Wallingford, U.K.: CABI, 4–26.

Allen, J.L. 1980. *The reign of law, a tale of the Kentucky hemp fields.* New York: Macmillan. 290 pp.

Allen, J.H., de Moore, G.M., Heddle, R., and Twartz, J.C. 2004. Cannabinoid hyperemesis: Cyclical hyperemesis in association with chronic cannabis abuse. *Gut* 53: 1566–1570.

Allin, S. 2012. *Building with hemp.* 2nd edition. Rusheens, Kerry, Ireland: Seed Press. 191 pp.

Allsop, D.J., Copeland, J., Norberg, M.M., Fu, S., Molnar, A., Lewis, J. et al. 2012. Quantifying the clinical significance of cannabis withdrawal. *PLoS ONE* 7(9): e44864. doi:10.1371/journal.pone.0044864.

Alshaarawy, O., and Anthony, J.C. 2015. Brief report: Cannabis smoking and diabetes mellitus: Results from meta-analysis with eight independent replication samples. *Epidemiology* 26: 597–600.

Aluru, M., Kukk, L., Astover, A., Shanskiy, M., and Loit, E. 2013. An agro-economic analysis of briquette production from fibre hemp and energy sunflower. *Industrial Crops and Products* 51: 186–193.

Amaducci, S. 2005. Hemp production in Italy. *Journal of Industrial Hemp* 10(1): 109–115.

Amaducci, S., Errani, M., and Venturi, G. 2002a. Response of hemp to plant population and nitrogen fertilisation. *Italian Journal of Agronomy* 6(2): 103–111.

Amaducci, S., Errani, M., and Venturi, G. 2002b. Plant population effects on fibre hemp morphology and production. *Journal of Industrial Hemp* 7(2): 33–60.

Amaducci, S., Pelatti, F., and Bonatti, P.M. 2005. Fibre development in hemp (*Cannabis sativa* L.) as affected by agrotechnique: Preliminary results of a microscopic study. *Journal of Industrial Hemp* 10(1): 31–48.

Amaducci, S., Colauzzi, M., Bellocchi, G., and Venturi, G. 2008a. Modelling post-emergent hemp phenology (*Cannabis sativa* L.): Theory and evaluation. *European Journal of Agronomy* 28: 90–102.

Amaducci, S., Colauzzi, M., Zatta, A., and Venturi, G. 2008b. Flowering dynamics in monoecious and dioecious hemp genotypes. *Journal of Industrial Hemp* 13(1): 5–19.

Amaducci, S., Zatta, A., Pelatti, F., and Venturi, G. 2008c. Influence of agronomic factors on yield and quality of hemp (*Cannabis sativa* L.) fibre and implication for an innovative production system. *Field Crops Research* 107: 161–169.

Amaducci, S., Zatta, A., Raffanini, M., and Venturi, G. 2008d. Characterisation of hemp (*Cannabis sativa* L.) roots under different growing conditions. *Plant Soil* 313: 227–235.

Amaducci, S., Scordia, D., Liu, F.H., Zhang, Q., Guo, H., Testa, G. et al. 2015. Key cultivation techniques for hemp in Europe and China. *Industrial Crops and Products* 68: 2–16.

American Herbal Products Association. 2014a. *Cannabis Committee recommendations for regulators— Cannabis operations: Laboratory operations.* Silver Spring, MD: American Herbal Products Association. 12 pp. http://www.ahpa.org/Default.aspx?tabid=267.

American Herbal Products Association. 2014b. *Cannabis Committee recommendations for regulators— Cannabis operations: Dispensing operations.* Silver Spring, MD: American Herbal Products Association. 12 pp. http://www.ahpa.org/Default.aspx?tabid=267.

American Herbal Products Association. 2014c. *Cannabis Committee recommendations for regulators— Cannabis operations: Manufacturing, packaging, labeling, and holding operations.* Silver Spring, MD: American Herbal Products Association. 48 pp. http://www.ahpa.org/Default.aspx?tabid=267.

American Herbal Products Association. 2014d. *Cannabis Committee recommendations for regulators— Cannabis operations: Cultivation and processing operations.* Silver Spring, MD: American Herbal Products Association. 15 pp.

American Lung Association. 2015. *Marijuana and lung health.* http://www.lung.org/stop-smoking/about-smoking /health-effects/marijuana-lung-health.html.

American Psychiatric Association. 2013. *Diagnostic and statistical manual of mental disorders.* 5th edition. Arlington, VA: American Psychiatric Publishing. 947 pp.

Anderson, E. 1954. *Plants, life, and man.* London: Melrose. 208 pp.

Anderson, J.T., Saldaña Rojas, J., and Flecker, A.S. 2009. High-quality seed dispersal by fruit-eating fishes in Amazonian floodplain habitats. *Oecologia* 161: 279–290.

Andersson, M.S., and de Vicente, M.C. 2010. *Gene flow between crops and their wild relatives.* Baltimore, MD: Johns Hopkins University Press. 564 pp.

Andries, A., Frystyk, J., Flyvbjerg, A., and Støving, R.K. 2014. Dronabinol in severe, enduring anorexia nervosa: A randomized controlled trial. *International Journal of Eating Disorders* 47: 18–23.

Angelova, V., Ivanova, R., Delibaltova, V., and Ivanov, K. 2004. Bio-accumulation and distribution of heavy metals in fibre crops (flax, cotton and hemp). *Industrial Crops and Products* 19: 197–205.

Annas, G.J. 1997. Reefer madness—The federal response to California's medical-marijuana law. *New England Journal of Medicine* 337: 435–439.

Anonymous. 1822. *Galerie industrielle.* Paris, France: Eymery. 165 pp. + 30 plates. (In French.)

Anonymous. 1933. Discovery of sexuality in plants. *Nature* 131: 392–392.

Anonymous. 1975. *Catalogue of the global collection of VIR.* Issue 162, Fiber crops. Vavilov Institute, Leningrad, USSR. 29 pp. + 12 page unpaginated table. (See Clarke 1998b, which stated that an English version was obtainable for $50 from the International Hemp Association, Amsterdam.)

Anonymous. 2006. Medical marijuana: Reefer madness, marijuana is medically useful, whether politicians like it or not. *The Economist* 379(8475): 83–84.

Anstee, D.J. 2010. The relationship between blood groups and disease. *Blood* 115: 4635–4643.

Anwar, F., Latif, S., and Ashraf, M. 2006. Analytical characterization of hemp (*Cannabis sativa*) seed oil from different agro-ecological zones of Pakistan. *Journal of the American Oil Chemist's Society* 84: 323–329.

Appendino, G., Chianese, G., and Taglialatela-Scafati, O. 2009. Cannabinoids: Occurrence and medicinal chemistry. *Current Medicinal Chemistry* 18: 1085–1099.

Appendino, G., Gibbons, S., Giana, A., Pagani, A., Grassi, G., Stavri, M. et al. 2008. Antibacterial cannabinoids from *Cannabis sativa*: A structure-activity study. *Journal of Natural Products* 71: 1427–1430.

Arain, M., Khan, M., Craig, L., and Nakanishi, S.T. 2015. Cannabinoid agonist rescues learning and memory after a traumatic brain injury. *Annals of Clinical and Translational Neurology* 2: 289–294.

Arendt, J., and Reznick, D. 2007. Convergence and parallelism reconsidered: What have we learned about the genetics of adaptation. *Trends in Ecology and Evolution* 23: 26–32.

Armentano, P. 2013. Should per se limits be imposed for cannabis? Equating cannabinoid blood concentrations with actual driver impairment: Practical limitations and concerns. *Humboldt Journal of Social Relations* 35: 45–55.

Armentia, A., Castrodeza, J., Ruiz-Muñoz, P., Martínez-Quesada, J., Postigo, I., Herrero, M. et al. 2011. Allergic hypersensitivity to cannabis in patients with allergy and illicit drug users. *Allergologia et Immunopathologia* 39: 271–279.

Armstrong, M.J., and Miyasaki, J.M. 2012. Evidence-based guideline: Pharmacologic treatment of chorea in Huntington disease. Report of the Guideline Development Subcommittee of the American Academy of Neurology. *Neurology* 79: 597–603.

Arnoux, M. 1963. Influence des facteurs du milieu sur l'expression de la sexualité du chanvre monoïque (*Cannabis sativa* L.). I. Action du cycle photopériodique. *Annales de l'amélioration des plantes* 13: 27–49.

Arnoux, M. 1966a. Influence des facteurs du milieu sur l'expression de la sexualité du chanvre monoïque (*Cannabis sativa* L.). II. Action de la nutrition azotée. *Annales de l'amélioration des plantes* 16: 123–134.

Arnoux, M. 1966b. Influence des facteurs du milieu sur l'expression de la sexualité du chanvre monoïque (*Cannabis sativa* L.). III. Note sur l'interaction entre le cycle photopériodique et la nutrition azotée. *Annales de l'amélioration des plantes* 16: 259–262.

Asbridge, M., Hayden, J.A., and Cartwright, J.L. 2012. Acute cannabis consumption and motor vehicle collision risk: Systematic review of observational studies and meta-analysis. *British Medical Journal* 344: e536. doi:10.1136/bmj.e536.

Ascherson, P., and Graebner, P. 1908–1913. *Cannabis*. In: Ascherson, P., and Graebner, P. eds. *Synopsis der Mitteleuropäischen flora [Synopsis of the Central European Flora], Volume 4*. Leipzig, Germany: Wilhelm Engelmann Verlag, 598–601. (In German.)

Ash, A.L. 1948. Hemp: Production and utilization. *Economic Botany* 2: 158–169.

Ashton, C.H., Moore, P.B., Gallagher, P., and Young, A.H. 2005. Cannabinoids in bipolar affective disorder: A review and discussion of their therapeutic potential. *Journal of Psychopharmacology* 19: 293–300.

Ashton, J.C. 2012. Synthetic cannabinoids as drugs of abuse. *Current Drug Abuse Reviews* 5: 158–168.

ASTHO. 2008. *Prescription drug overdose: State health agencies respond*. Arlington, VA: Association of State and Territorial Health Officials. 21 pp. http://www.cdc.gov/HomeandRecreationalSafety/pubs/RXReport_web-a.pdf.

Atal, C.K. 1959. Sex reversal in hemp by application of gibberellin. *Current Science* 28: 408–409.

Avico, U., Pacifici, R., and Zuccaro, P. 1985. Variations of tetrahydrocannabiol content in cannabis plants to distinguish the fibre-type from drug-type plants. *Bulletin on Narcotics* 37(4): 61–65.

Azorlosa, J.L., Heishman, S.J., Stitzer, M.L., and Mahaffey, J.M. 1992. Marijuana smoking: Effect of varying delta 9-tetrahydrocannabinol content and number of puffs. *Journal of Pharmacology and Experimental Therapeutics* 261: 114–122.

Azorlosa, J.L., Greenwald, M.K., and Stitzer, M.L. 1995. Marijuana smoking: Effects of varying puff volume and breathhold duration. *Journal of Pharmacology and Experimental Therapeutics* 272: 560–569.

Bab, I. 2011. Editorial: Themed issue on cannabinoids in biology and medicine. *British Journal of Pharmacology* 163: 1327–1328.

Bab, I.A. 2012. Cannabinoids in bone repair. In: Sela, J.J., and Bab, I.A. eds. *Principles of bone regeneration*. New York: Springer, 67–78.

Bab, I., and Ofek, O. 2011. Targeting the CB_2 cannabinoid receptor in osteoporosis. *Expert Review of Endocrinology & Metabolism* 6: 135–138.

Bab, I., Zimmer, A., and Melamed, E. 2009. Cannabinoids and the skeleton: From marijuana to reversal of bone loss. *Annals of Medicine* 41: 560–567.

Backes, M. 2014. *Cannabis pharmacy: The practical guide to medical marijuana*. New York: Black Dog & Leventhal. 272 pp.

Bagheri, M., and Mansouri, H. 2015. Effect of induced polyploidy on some biochemical parameters in *Cannabis sativa* L. *Applied Biochemistry and Biotechnology* 175: 2366–2375.

Baillon, M.H. 1876. *Dictionnaire de Botanique, Vol. 1*. Paris, France: Librairie Hachette. 759 pp.

Baker, D., Pryce, G., Jackson, S.J., Bolton, C., and Giovannoni, G. 2012. The biology that underpins the therapeutic potential of cannabis-based medicines for the control of spasticity in multiple sclerosis. *Multiple Sclerosis and Related Disorders* 1: 64–75.

Baker, H.G. 1974. The evolution of weeds. *Annual Review of Ecology and Systematics* 5: 1–34.

Balatinecz, J., and Sain, M. 2007. Cars made of wood and hemp fibres? Why not? *Forestry Chronicle* 83: 482–484.

Balistreri, E., Garcia-Gonzalez, E., Selvi, E., Akhmetshina, A., Palumbo, K., Lorenzini, S. et al. 2011. The cannabinoid WIN55, 212-2 abrogates dermal fibrosis in scleroderma bleomycin model. *Annals of the Rheumatic Diseases* 70: 695–699.

Baloch, G.M., Mushtaque, M., and Ghani, M.A. 1974. *Natural enemies of* Papaver *spp. and* Cannabis sativa. Annual report, Pakistan Station. Pakistan: Commonwealth Institute of Biological Control, 56–57.

Baraniecki, P. 1997. Industrial plants in clean-up of heavy metal polluted soils. In: Nova Institute, corporate ed. *Bioresource hemp—Proceedings of the symposium* (Frankfurt am Main, Germany, February 27–March 2, 1997). Hürth, Germany: Nova Institute, 277–283.

Baraniecki, P., and Mankowski, J. 1995. Hemp fibre as a raw material for paper production in the aspect of natural environmental protection. *Zemedelska Technikia UZPI (Czech Republic)* 41(3): 85–88.

Barceloux, D.G. 2012. Marijuana (*Cannabis sativa* L.) and synthetic cannabinoids. In: Barceloux, D.G., and Palmer, R.B. eds. *Medical toxicology of drug abuse: Synthesized chemicals and psychoactive plants.* Hoboken, NJ: Wiley, 886–931.

Barcott, B., and Scherer, M. 2015. The great pot experiment. *Time* 185(19): 38–45.

Baron, E.P. 2015. Comprehensive review of medicinal marijuana, cannabinoids, and therapeutic implications in medicine and headache: What a long strange trip it's been… *Headache: The Journal of Head and Face Pain* 55: 885–916.

Barrett, S.C.H. 1982. Genetic variation in weeds. In: Charudattan, R., and Walker, H. eds. *Biological control of weeds with plant pathogens.* New York: John Wiley & Sons, 73–98.

Barry, R.A., Hilamo, H., and Glantz, S.A. 2014. Waiting for the opportune moment: The tobacco industry and marijuana legalization. *The Milbank Quarterly* 92: 202–247.

Barta, Z., Oliva, J.M., Ballesteros, I., Dienes, D., Ballesteros, M., and Réczey, K. 2010. Refining hemp hurds into fermentable sugars or ethanol. *Chemical and Biochemical Engineering Quarterly* 24: 331–339.

Bas, N., Pavelek, M., Maggioni, L., and Lipman, E. 2010. *Report of a working group on fibre crops (flax and hemp)* (First meeting, June 14–16, 2006, Wageningen, the Netherlands). Rome, Italy: Bioversity International. 22 pp.

Baser, K.H.C., and Buchbauer, G. eds. 2010. *Handbook of essential oils: Science, technology, and applications.* Boca Raton, FL: CRC Press. 975 pp.

Baskett, M. 1999. *Making beautiful hemp and bead jewelry.* New York: Sterling. 96 pp.

Basu, S., and Dittel, B.N. 2011. Unraveling the complexities of cannabinoid receptor 2 (CB_2) immune regulation in health and disease. *Immunologic Research* 51: 26–38.

Bauer, S., Olson, J., Cockrill, A., van Hattem, M., Miller, L., Tauzer, M. et al. 2015. Impacts of surface water diversions for marijuana cultivation on aquatic habitat in four northwestern California watersheds. *PLoS ONE* 10(3): e0120016. doi:10.1371/journal.pone.0120016.

Bawa, K.S. 1980. Evolution of dioecy in flowering plants. *Annual Review of Ecology and Systematics* 11: 15–39.

Baxter, W.J., and Scheifele, G. 2009. *Growing industrial hemp in Ontario.* Toronto, ON: Ontario Ministry of Agriculture and food. http://www.omafra.gov.on.ca/english/crops/facts/00-067.htm.

Bazzaz, F.A., Dusek, D., Seigler, D.S., and Haney, A.W. 1975. Photosynthesis and cannabinoid content of temperate and tropical populations of *Cannabis sativa. Biochemical Systematics and Ecology* 3: 15–18.

Beaubrun, M.I. 1983. Jamaica: Contrasting patterns of cannabis use. In: Edwards, G., Arif, A., and Jaffe, J. eds. *Drug use and misuse—Cultural perspectives.* London: Croom Helm, 70–73.

Bedse, G., Romano, A., Lavecchia, A.M., Cassano, T., and Gaetani, S. 2015. The role of endocannabinoid signaling in the molecular mechanisms of neurodegeneration in Alzheimer's disease. *Journal of Alzheimer's Disease* 43: 1115–1136.

Belendiuk, K.A., Babson, K.A., Vandrey, R., and Bonn-Miller, M.O. 2015. *Cannabis* species and cannabinoid concentration preference among sleep-disturbed medicinal cannabis users. *Addictive Behaviors* 50: 178–181.

Ben Amar, M. 2006. Cannabinoids in medicine: A review of their therapeutic potential. *Journal of Ethnopharmacology* 105: 1–25.

Benet, S. 1975. Early diffusion and folk uses of hemp. In: Rubin, V. ed. *Cannabis and culture.* The Hague, the Netherlands: Mouton, 39–49.

Benhaim, B. 2000. H.E.M.P. *Healthy eating made possible.* London: Fusion Press. 332 pp.

Benhaim, P., Marosszeky, K., and Marosszeky, M. 2011. *How to build a hemp house.* Create Space (Amazon self-publisher). 98 pp.

Bennett, C. 2010. Early/ancient history. In: Holland, J. ed. *The pot book: A complete guide to cannabis: Its role in medicine, politics, science and culture.* Rochester, VT: Park Street Press, 17–26.

Ben-Shabat, S., Fride, E., Sheskin, T., Tamiri, T., Rhee, M.H., Vogel, Z. et al. 1998. An entourage effect: Inactive endogenous fatty acid glycerol esters enhance 2-arachidonoyl-glycerol cannabinoid activity. *European Journal of Pharmacology* 353: 23–31.

Berenji, J., Dimić, E, and Romanić, R. 2005. Hemp—Potential raw material for cold pressed oil. In: *Proceedings of the 46th oil industry conference: Production and processing of oilseeds.* Serbia and Montenegro: Petrovac na moru, 127–136. (In Serbian.)

Bergamaschi, M.M., Queiroz, R.H., Chagas, M.H., de Oliveira, D.C., De Martinis, B.S., Kapczinski, F. et al. 2011. Cannabidiol reduces the anxiety induced by simulated public speaking in treatment-naïve social phobia patients. *Neuropsychopharmacology* 36: 1219–1226.

Berger, J. 1969. *The world's major fibre crops, their cultivation and manuring*. Zurich, Switzerland: Centre d'Étude de l'Azote. 294 pp.

Bergfjord, C., and Holst, B. 2010. A procedure for identifying textile bast fibres using microscopy: Flax, nettle/ramie, hemp and jute. *Ultramicroscopy* 110: 1192–1197.

Berkman, A.H. 1939. *Seedling anatomy of* Cannabis sativa *L*. Chicago, IL: The University of Chicago. 21 pp. (Published doctoral dissertation.)

Bermudez-Silva, F.J., Viveros, M.P., McPartland, J.M., and Rodriguez de Fonseca, F. 2010. The endocannabinoid system, eating behavior and energy homeostasis: The end or a new beginning? *Pharmacology, Biochemistry and Behavior* 95: 375–382.

Bertoli, A., Tozzi, S., Pistelli, L., and Angelini, L.G. 2010. Fibre hemp inflorescences: From crop-residues to essential oil production. *Industrial Crops and Products* 32: 329–337.

Beutler, J.A., and Der Marderosian, A.H. 1978. Chemotaxonomy of *Cannabis* I. Crossbreeding between *Cannabis sativa* and *C. ruderalis*, with analysis of cannabinoid content. *Economic Botany* 32: 387–394.

Bevan, R., and Woolley, T. 2008. *Hemp lime construction—A guide to building with hemp lime composites*. Bracknell, U.K.: IHS BRE Press. 120 pp.

Bey, H., and Zug, A. eds. 2004. *Orgies of the hemp eaters. Cuisine, slang, literature and ritual of cannabis culture*. Brooklyn, NY: Autonomedia. 694 pp.

Bhattacharyya, S., Morrison, P.D., Fusar-Poli, P., Martin-Santos, R., Borgwardt, S., Winton-Brown, T. et al. 2010. Opposite effects of Δ^9-tetrahydrocannabinol and cannabidiol on human brain function and psychopathology. *Neuropsychopharmacology* 35: 764–774.

Biewinga, E.E., and van der Bijl, G. 1996. *Sustainability of energy crops in Europe. A methodology developed and applied*. Report No. 234. Utrecht, the Netherlands: Centre for Agriculture and Environment. 209 pp.

Bifulco, M., Laezza, C., Pisanti, S., and Gazzerro, P. 2006. Cannabinoids and cancer: Pros and cons of an antitumour strategy. *British Journal of Pharmacology* 148: 123–135.

Bin Saif, G.A., Ericson, M.E., and Yosipovitch, G. 2011. The itchy scalp—Scratching for an explanation. *Experimental Dermatology* 20: 959–968.

Bíró, T., Tóth, B.I., Haskó, G., Paus, R., and Pacher, P. 2009. The endocannabinoid system of the skin in health and disease: Novel perspectives and therapeutic opportunities. *Trends in Pharmacological Sciences* 30: 411–420.

Bishop, L.R. 1966. Process for segregating lupulen from dried hops. U.S. Patent 3,271,162.

Blade, S. ed. 1998. *Alberta Hemp Symposia proceedings* (Red Deer, Alberta, March 10, 1998, and Edmonton, Alberta, April 8, 1998). Edmonton, AB: Alberta Agriculture, Food and Rural Development. 85 pp.

Blair, R.E., Deshpande, L.S., and DeLorenzo, R.J. 2015. Endocannabinoids and epilepsy. In: Fattore, L. ed. *Cannabinoids in neurologic and mental disease*. New York: Academic Press, 125–172.

Blake, D.R., Robson, P., Ho, M., Jubb, R.W., and McCabe, C.S. 2006. Preliminary assessment of the efficacy, tolerability and safety of a cannabis-based medicine (Sativex) in the treatment of pain caused by rheumatoid arthritis. *Rheumatology* 45: 50–52.

Bloor, R.N., Wang, T.S., Španěl, P., and Smith, D. 2008. Ammonia release from heated 'street' cannabis leaf and its potential toxic effects on cannabis users. *Addiction Biology* 103: 1671–1677.

Bócsa, I. 1998. Genetic improvement: Conventional approaches. In: Ranalli, P. ed. *Advances in hemp research*. London: Food Products Press (of Haworth Press), 153–184.

Bócsa, I., and Karus, M. 1998. *The cultivation of hemp: Botany, varieties, cultivation and harvesting*. Sebastopol, CA: Hemptech. 184 pp.

Bócsa, I., Máthé, M., and Hangyel, L. 1997. Effect of nitrogen on tetrahydrocannabinol (THC) content in hemp (*Cannabis sativa* L.) leaves at different positions. *Journal of the International Hemp Association* 4(2): 80–81.

Bócsa, I., Finta-Korpelová, Z., and Máthé, P. 2005. Preliminary results of selection for seed oil content in hemp (*Cannabis sativa* L.). *Journal of Industrial Hemp* 10(1): 5–15.

Bohlmann, F., and Hoffmann, E. 1979. Cannabigerol-ähnliche Verbindungen aus *Helichrysum braculigerum*. *Phytochemistry* 18: 1371–1374.

Bolognini, D., Costa, B., Maione, S., Comelli, F., Marini, P., Di Marzo, V. et al. 2010. The plant cannabinoid Δ^9-tetrahydrocannabivarin can decrease signs of inflammation and inflammatory pain in mice. *British Journal of Pharmacology* 160: 677–687.

Bolton, J. 1995. The potential of plant fibres as crops for industrial use. *Outlook on Agriculture* 24: 85–89.

Bonnet, A.E., and Marchalant, Y. 2015. Potential therapeutical contributions of the endocannabinoid system towards aging and Alzheimer's disease. *Aging and Disease* 6: 400–405.

Booth, M. 2004. *Cannabis: A history*. New York: St Martin's Press. 429 pp.

Borgelt, L.M., Franson, K.L., Nusbaum, A.M., and Wang, G.S. 2013. The pharmacologic and clinical effects of medical cannabis. *Pharmacotherapy* 33: 195–209.

Borlaug, N.E. 2000. *The green revolution revisited and the road ahead.* Oslo, Norway: Norwegian Nobel Institute. http://www.nufs.sjsu.edu/clariebh/borlaug-lecture.pdf.

Borrelli, F., Fasolino, I., Romano, B., Capasso, R., Maiello, F., Coppola, D. et al. 2013. Beneficial effect of the non-psychotropic plant cannabinoid cannabigerol on experimental inflammatory bowel disease. *Biochemical Pharmacology* 85: 1306–1316.

Borrelli, F., Pagano, E., Romano, B., Panzera, S., Maiello, F., Coppola, D. et al. 2014. Colon carcinogenesis is inhibited by the TRPM8 antagonist cannabigerol, a *Cannabis*-derived non-psychotropic cannabinoid. *Carcinogenesis* 35: 2787–2797.

Borthwick, H.A., and Scully, N.J. 1954. Photoperiodic responses in hemp. *Botanical Gazette* 116: 14–29.

Bossong, M.G., Jansma, J.M., Bhattacharyya, S., and Ramsey, N.F. 2014. Role of the endocannabinoid system in brain functions relevant for schizophrenia: An overview of human challenge studies with cannabis or Δ^9-tetrahydrocannabinol (THC). *Progress in Neuro-Psychopharmacology and Biological Psychiatry* 52: 53–69.

Bostwick, J.M. 2012. Blurred boundaries: The therapeutics and politics of medical marijuana. *Mayo Clinic Proceedings* 87: 172–186.

Bosy, T.Z., and Cole, K.A. 2000. Consumption and quantitation of delta-9 tetrahydrocannabinol in commercially available hemp seed oil products. *Journal of Analytical Toxicology* 24: 562–566.

BOTEC Analysis Corporation. 2013. *Environmental risks and opportunities in cannabis cultivation.* 31 pp. http://lcb.wa.gov/publications/Marijuana/SEPA/BOTEC_Whitepaper_Final.pdf.

Botta, B., Gacs-Baitz, E., Vinciguerra, V., and Delle Monache, G. 2003. Three isoflavanones with cannabinoid-like moieties from *Desmodium canum*. *Phytochemistry* 64: 599–602.

Bouloc, P., Allegret, S., and Arnaud, L. eds. 2013. *Hemp. Industrial production and uses.* Wallingford, Oxfordshire, U.K.: CABI. 313 pp. [Translation, and updating of some of the chapters, of Bouloc P. ed. 2006. *Le chanvre industriel: Production et utilisations.* Paris, France: Groupe France Agricole. 432 pp.]

Bouquet, R.J. 1950. *Cannabis. Bulletin on Narcotics* 2(4): 14–30.

Bourmaud, A., and Baley, C. 2007. Investigations on the recycling of hemp and sisal fibre reinforced polypropylene composites. *Polymer Degradation and Stability* 92: 1034–1045.

Boutain, J.R. 2014. *On the origin of hops: Genetic variability, phylogenetic relationships, and ecological plasticity of* Humulus *(Cannabaceae).* Dissertation. Manoa: University of Hawaii. 211 pp. https://scholar space.manoa.hawaii.edu/handle/10125/100298.

Bowles, D.W., O'Bryant, C.L., Camidge, D.R., and Jimeno, A. 2012. The intersection between cannabis and cancer in the United States. *Critical Reviews in Oncology Hematology* 83: 1–10.

Boyce, S.S. 1900. *Hemp* (Cannabis sativa). *A practical treatise on the culture of hemp for seed and fiber with a sketch of the history and nature of the hemp plant.* New York: Orange Judd Company. 112 pp.

Bradshaw, A.D. 1965. Evolutionary significance of phenotypic plasticity in plants. *Advances in Genetics* 13: 115–156.

Bradshaw, W.E., and Holzapfel, C.M. 2007. Evolution of animal photoperiodism. *Annual Review of Ecology, Evolution, and Systematics* 38: 1–25.

Braemer, R., and Paris, M. 1987. Biotransformation of cannabinoids by a cell suspension culture of *Cannabis sativa* L. *Plant Cell Reports* 6: 150–152.

Braga, R.J., Abdelmessih, S., Tseng, J., and Malhotra, A. 2015. Cannabinoids and bipolar disorder. In: Fattore, L. ed. *Cannabinoids in neurologic and mental disease.* New York: Academic Press, 205–225.

Braut-Boucher, F. 1980. Effet des conditions ecophysiologiques sur la croissance, le developpement et le contenu en cannabinoides de clones correspondant aux deux types chimiques du *Cannabis sativa* L. originaire d'Afrique du Sud. *Physiologie Vegetale* 18: 207–221.

Bredemann, G., Schwanitz, F., and von Sengbusch, R. 1956. Problems of modern hemp breeding, with particular reference to the breeding of varieties of hemp containing little or no hashish. *Bulletin on Narcotics* 8(3): 31–35.

Brickell, C.D., Alexander, C., David, J.C., Hetterscheid, W.L.A., Leslie, A.C., Malecot, V. et al. 2009. *International code of nomenclature for cultivated plants.* Leuven, Belgium: International Society for Horticultural Science. 184 pp. http://www.actahort.org/chronica/pdf/sh_10.pdf.

Briosi, G., and Tognini, F. 1894. *Anatomia della canapa* (Cannabis sativa *L.) Parte prima: Organi sessuali.* Milan, Italy: Istituto Botanico Della R. Universita de Pavia. Series 2, 3: 91–209.

Bronson, F.H. 2004. Are humans seasonally photoperiodic? *Journal of Biological Rhythms* 19: 180–192.

Broséus, J., Anglada, F., and Esseiva, P. 2010. The differentiation of fibre- and drug type *Cannabis* seedlings by gas chromatography/mass spectrometry and chemometric tools. *Forensic Science International* 200: 87–92.

Brown, L.R., Abramovitz, J., Bright, C., Flavin, C., French, H., Gardner, G. et al. 1998. State of the world, 1998, Worldwatch Institute report on progress toward a sustainable society. 15th ed. New York: W.W. Norton & Co., 251 pp.

Brown, V.K., Lawton, J.H., and Grubb, P.J. 1991. Herbivory and the evolution of leaf size and shape. *Philosophical Transactions of the Royal Society, B, Biological Sciences* 333: 265–272.

Buchbauer, G. 2010. Biological activities of essential oils. In: Baser, K.H.C., and Buchbauer, G. eds. *Handbook of essential oils: Science, technology, and applications.* Boca Raton, FL: CRC Press, 235–280.

Budney, A.J., and Hughes, J.R. 2006. The cannabis withdrawal syndrome. *Current Opinion in Psychiatry* 19: 233–238.

Budney, A.J., Roffman, R., Stephens, R.S., and Walker, D. 2007. Marijuana dependence and its treatment. *Addiction Science & Clinical Practice* 4: 4–16.

Bureau, L. 2010. Hemp oil. *Phytothérapie* 8: 109–112. (In French.)

Burns, T.L., and Ineck, J.R. 2006. Cannabinoid analgesia as a potential new therapeutic option in the treatment of chronic pain. *Annals of Pharmacotherapy* 40: 251–260.

Burstein, S. 2015. Cannabidiol (CBD) and its analogs: A review of their effects on inflammation. *Bioorganic & Medicinal Chemistry* 23: 1377–1385.

Bushdid, C., Magnasco, M.O., Vosshall, L.B., and Keller, A. 2014. Humans can discriminate more than 1 trillion olfactory stimuli. *Science* 343: 1370–1372.

Busquets-Garcia, A., Gomis-González, M., Guegan, T., Agustín-Pavón, C., Pastor, A., Mato, S. et al. 2013. Targeting the endocannabinoid system in the treatment of fragile X syndrome. *Nature Medicine* 19: 603–607.

Busse, F.P., Fiedler, G.M., Leichtle, A., Hentschel, H., and Stumvoll, M. 2008a. Lead poisoning due to adulterated marijuana in Leipzig. *Deutsches Ärzteblatt International* 105: 757–762.

Busse, F., Omidi, L., Leichtle, A., Windgassen, M., Kluge, E., and Stumvoll, M. 2008b. Lead poisoning due to adulterated marijuana. *New England Journal of Medicine* 358: 1641–1642.

Buys, Y.M., and Rafuse, P.E. 2010. Canadian Ophthalmological Society policy statement on the medical use of marijuana for glaucoma. *Canadian Journal of Ophthalmology* 45: 324–326.

Cabezudo, B., Recio, M., Sánchez-Laulhé, J.M., Del Mar Trigo, M., Toro, F.J., and Polvorinos, F. 1997. Atmospheric transportation of marihuana pollen from North Africa to the southwest of Europe. *Atmospheric Environment* 31: 3323–3328.

Cabral, G.A., Rogers, T.J., and Lichtman, A.H. 2015. Turning over a new leaf: Cannabinoid and endocannabinoid modulation of immune function. *Journal of Neuroimmune Pharmacology* 10: 193–203.

Caffarel, M.M., Andradas, C., Mira, E., Pérez-Gómez, E., Cerutti, C., Moreno-Bueno, G. et al. 2010. Cannabinoids reduce ErbB2-driven breast cancer progression through Akt inhibition. *Molecular Cancer* 9: 196. http://www.molecular-cancer.com/content/9/1/196.

Calabria, B., Degenhardt, L., Hall, W., and Lynskey, M. 2010. Does cannabis use increase the risk of death? Systematic review of epidemiological evidence on adverse effects of cannabis use. *Drug and Alcohol Review* 29: 318–330.

Callaghan, R.C., Allebeck, P., and Sidorchuk, A. 2013. Marijuana use and risk of lung cancer: A 40-year cohort study. *Cancer Causes Control* 24: 1811–1820.

Callaway, J.C. 1998. Formation of *trans*-fatty acids in heated hempseed oil: A rebuttal. *Journal of the International Hemp Association* 5(2): 106–108.

Callaway, J.C. 2002. Hemp as food at high latitudes. *Journal of Industrial Hemp* 7(1): 105–117.

Callaway, J.C. 2004. Hempseed as a nutritional resource: An overview. *Euphytica* 140: 65–72.

Callaway, J.C. 2008. A more reliable evaluation of hemp THC levels is necessary and possible. *Journal of Industrial Hemp* 13(2): 117–144.

Callaway, J.C., and Laakkonen, T.T. 1996. Cultivation of *Cannabis* oil seed varieties in Finland. *Journal of the International Hemp Association* 3(1): 32–34.

Callaway, J.C., and Pate, D.W. 2009. Hempseed oil. In: Moreau, R.A., and Kamal-Eldin, A. eds. *Gourmet and health-promoting specialty oils.* Urbana IL: American Oil Chemists Society Press, 185–213.

Callaway, J.C., Tennilä, T., and Pate, D.W. 1996. Occurrence of "omega-3" stearidonic acid (cis-6,9,12,15-octadecatetraenoic acid) in hemp (*Cannabis sativa* L.) seed. *Journal of the International Hemp Association* 3(2): 61–63.

Callaway, J.C., Weeks, R.A., Raymon, L.P., Walls, H.C., and Hearn, W.L. 1997. A positive urinalysis from hemp (*Cannabis*) seed oil. *Journal of Analytical Toxicology* 21: 319–320.

Callaway, J., Schwab, U., Harvima, I., Halonen, P., Mykkänen, O., Hyvönen, P. et al. 2005. Efficacy of dietary hempseed oil in patients with atopic dermatitis. *Journal of Dermatological Treatment* 16: 87–94.

Campbell, V.A., and Gowran, A. 2007. Alzheimer's disease; taking the edge off with cannabinoids? *British Journal of Pharmacology* 152: 655–662.

Campbell, S., Paquin, D., Awaya, J.D., and Li, Q.X. 2002. Remediation of benzo(*a*)pyrene and chrysene-contaminated soil with industrial hemp (*Cannabis sativa*). *International Journal of Phytoremediation* 4: 157–168.

Campora, L., Miragliotta, V., Ricci, E., Cristino, L., Di Marzo, V., Albanese, F. et al. 2012. Cannabinoid receptor type 1 and 2 expression in the skin of healthy dogs and dogs with atopic dermatitis. *American Journal of Veterinary Research* 73: 988–995.

Campos, A.C., Moreira, F.A., Gomes, F.V., Del Bel, E.A., and Guimarães, F.S. 2012. Multiple mechanisms involved in the large-spectrum therapeutic potential of cannabidiol in psychiatric disorders. *Philosophical Transactions of the Royal Society of London, Series B: Biological Sciences* 367: 3364–3378.

Capelle, A. 1996. Hemp: Specialty crop for the paper industry. In: Janick, J. ed. *Progress in new crops.* Arlington, VA: ASHS Press, 384–388.

Caraceni, P., Borrelli, F., Giannone, F.A., and Izzo, A.A. 2014. Potential therapeutic applications of cannabinoids in gastrointestinal and liver diseases: Focus on Δ^9-tetrahydrocannabinonl pharmacology. In: Di Marzo, V. ed. *Cannabinoids.* Chichester, U.K.: John Wiley, 219–260.

Carlini, E.A., and Cunha, J.M. 1981. Hypnotic and antiepileptic effects of cannabidiol. *Journal of Clinical Pharmacology* 21: 417–427.

Carroll, C.B., Bain, P.G., Teare, L., Liu, X., Joint, C., Wroath, C. et al. 2004. Cannabis for dyskinesia in Parkinson disease: A randomized double-blind crossover study. *Neurology* 63: 1245–1250.

Carroll, L. 1865. *Alice's adventures in wonderland.* London: Macmillan. 192 pp.

Carter, G.T., and Rosen, B.S. 2001. Marijuana in the management of amyotrophic lateral sclerosis. *American Journal of Hospice & Palliative Care* 18: 264–270.

Carter, G.T., Abood, M.E., Aggarwal, S.K., and Weiss, M.D. 2010. Cannabis and amyotrophic lateral sclerosis: Hypothetical and practical applications, and a call for clinical trials. *American Journal of Hospice & Palliative Care* 27: 347–356.

Carter, G.T., Flanagan, A.M., Earleywine, M., Abrams, D.I., Aggarwal, S.K., and Grinspoon, L. 2011. Cannabis in palliative medicine: Improving care and reducing opioid-related morbidity. *American Journal of Hospice & Palliative Care* 28: 297–303.

Carter, G.T., Javaher, S.P., Nguyen, M.H.V., Garret, S., and Carlini, B.H. 2015. Re-branding cannabis: The next generation of chronic pain medicine? *Pain Management* 5: 13–21.

Carus, M., Karst, S., Kauffmann, A., Hobson, J., and Bertucelli, A. 2013. *The European hemp industry: Cultivation, processing and applications for fibres, shivs and seeds.* Huerth, Germany: European Industrial Hemp Association. 9 pp. http://www.eiha.org/attach/8/13-03%20European%20Hemp%20Industry.pdf.

Carver, G.W. 1997. *How to grow marijuana indoors for medicinal use.* Seattle, WA: Sun Magic Publications. 127 pp.

Casano, S., Grassi, G., Martini, V., and Michelozzi, M. 2011. Variation in terpene profiles of different strains of *Cannabis sativa* L. *Acta Horticulturae* 925: 115–121.

Casas, X.A., and Rieradevall i Pons, J. 2005. Environmental analysis of the energy use of hemp—Analysis of the comparative life cycle: Diesel oil vs. hemp–diesel. *International Journal of Agricultural Resources, Governance and Ecology* 4: 133–139.

Cascini, F. 2012. Investigations into the hypothesis of transgenic *Cannabis. Journal of Forensic Sciences* 57: 718–721.

Cascini, F., Aiello, C., and Di Tanna, G. 2012. Increasing delta-9-tetrahydrocannabinol (Δ^9-THC) content in herbal cannabis over time: Systematic review and meta-analysis. *Current Drug Abuse Reviews* 5: 32–40.

Cascio, M.G., and Pertwee, R.G. 2014. Known pharmacological actions of nine nonpsychotropic phytocannabinoids. In: Pertwee, R.G. ed. *Handbook of cannabis.* Oxford, U.K.: Oxford University Press, 137–156.

Cascio, M.G., Zamberletti, E., Marini, P., Parolaro, D., and Pertwee, R.G. 2015. The phytocannabinoid, Δ^9-tetrahydrocannabivarin, can act through 5-HT$_{1A}$ receptors to produce antipsychotic effects. *British Journal of Pharmacology* 172: 1305–1318.

Casteels, C., Ahmad, R., Vandenbulcke, M., Vandenberghe, W., and Van Laere, K. 2015. Cannabinoids and Huntington's disease. In: Fattore, L. ed. *Cannabinoids in neurologic and mental disease.* New York: Academic Press, 61–97.

Caulkins, J.P. 2010. *Estimated cost of production for legalized cannabis.* Washington, DC: Rand Drug Policy Research Center. 28 pp. http://130.154.3.8/content/dam/rand/pubs/working_papers/2010/RAND_WR764 .pdf.

Caulkins, J.P., Hawken, A., Kilmer, B., and Kleiman, M.A.R. 2012. *Marijuana legalization: What everyone needs to know.* Oxford, U.K.: Oxford University Press. 266 pp.

Cawich, S.O., Downes, R., Martin, A.C., Evans, N.R., Mitchell, D.I.G., and Williams, E. 2010. Colonic perforation: A lethal consequence of cannabis body packing. *Journal of Forensic and Legal Medicine* 17: 269–271.

Ceapoiu, N. 1958. *Hemp, monographic study.* Bucharest, Romania: Editura Academiei Republicii Populare Rominae. 734 pp. (In Romanian.)

Centonze, D., Mori, F., Koch, G., Buttari, F., Codecà, C., Rossi, S. et al. 2009. Lack of effect of cannabis-based treatment on clinical and laboratory measures in multiple sclerosis. *Neurological Science* 30: 531–534.

Centre for Reviews and Dissemination. 2009. *Systematic Reviews: CRD's guidance for undertaking reviews in health care*. York, England: University of York. https://www.york.ac.uk/media/crd/Systematic_Reviews.pdf.

Cerdá, M., Wall, M., Keyes, K.M., Galea, S., and Hasin, D. 2012. Medical marijuana laws in 50 states: Investigating the relationship between state legalization of medical marijuana and marijuana use, abuse and dependence. *Drug and Alcohol Dependence* 120: 22–27.

Cervantes, J. 2006. *Marijuana horticulture: The indoor/outdoor medical grower's bible*. Vancouver, WA: Van Patten Publishing. 464 pp.

Cervantes, J. 2015. *The cannabis encyclopedia: The definitive guide to cultivation and consumption of medical marijuana*. Vancouver, WA: Van Patten Publishing. 596 pp.

Cescon, D.W., Page, A.V., Richardson, S., Moore, M.J., Boerner, S., and Gold, W.L. 2008. Invasive pulmonary aspergillosis associated with marijuana use in a man with colorectal cancer. *Journal of Clinical Oncology* 26: 2214–2215.

Chailakhan, M.K. 1979. Genetic and hormonal regulation of growth, flowering and sex expression in plants. *American Journal of Botany* 66: 717–736.

Chaitra, M.S., Vasudevan, K., and Shanker, K. 2004. The biodiversity bandwagon: The splitters have it. *Current Science* 86: 897–899.

Chandra, S., Lata, H., Khan, I.A., and ElSohly, M.A. 2008. Photosynthetic response of *Cannabis sativa* L. to variations in photosynthetic photon flux densities, temperature and CO_2 conditions. *Physiology and Molecular Biology of Plants* 14: 299–306.

Chandra, S., Lata, H., Mehmedic, Z., Khan, I.A., and ElSohly, M.A. 2010a. Assessment of cannabinoids content in micropropagated plants of *Cannabis sativa* and their comparison with conventionally propagated plants and mother plant during developmental stages of growth. *Planta Medica* 76: 743–750.

Chandra, S., Lata, H., Mehmedic, Z., Khan, I.A., and ElSohly, M.A. 2010b. Propagation of elite *Cannabis sativa* L. for the production of Δ^9-tetrahydrocannabinol (THC) using biotechnological tools. In: Arora, R. ed. *Medicinal plant biotechnology*. Wallingford, U.K.: CAB International, 98–114.

Chandra, S., Lata, H., Khan, I.A., and ElSohly, M.A. 2011a. Photosynthetic response of *Cannabis sativa* L., an important medicinal plant, to elevated levels of CO_2. *Physiology and Molecular Biology of Plants* 17: 291–295.

Chandra, S., Lata, H., Khan, I.A., and ElSohly, M.A. 2011b. Temperature response of photosynthesis in different drug and fiber varieties of *Cannabis sativa* L. *Physiology and Molecular Biology of Plants* 17: 297–303.

Chandra, S., Lata, H., Khan, I.A., and ElSohly, M.A. 2013. The role of biotechnology in *Cannabis sativa* propagation for the production of phytocannabinoids. In: Chandra, S., Lata, H., and Varma, A. eds. *Biotechnology for medicinal plants—Micropropagation and improvement*. Berlin, Germany: Springer Verlag, 123–148.

Chang, K. 1968. *The archeology of ancient China*. New Haven, CT: Yale University Press. 483 pp.

Cheatham, S., Johnston, M.C., and Marshall, L. 2009. *The useful wild plants of Texas, the Southeastern and Southwestern United States, the Southern Plains, and Northern Mexico*. Volume 3. Austin, TX: Useful Wild Plants, Inc. 617 pp. (Treatment of *Cannabis*: pp. 13–126.)

Chen, M., Chory, J., and Fankhauser, C. 2004. Light signal transduction in higher plants. *Annual Review of Genetics* 38: 87–117.

Chen, T., He, J., Zhang, J., Zhang, H., Qian, P., Hao, J., and Li, L. 2010. Analytical characterization of hempseed (seed of *Cannabis sativa* L.) oil from eight regions in China. *Journal of Dietary Supplements* 7: 117–129.

Cheng, C.-W., Bian, Z.-X., Zhu, L.-X., Wu, J.C.Y., and Sung, J.J.Y. 2011. Efficacy of a Chinese herbal proprietary medicine (hemp seed pill) for functional constipation. *American Journal of Gastroenterology* 106: 120–129.

Cheuvart, C. 1954. Studies on the development of *Cannabis sativa* L. at constant temperatures and under different photoperiods (sexuality and foliage pigments). *Bulletin de l'Académie Royale de Médecine de Belgique* 40: 1152–1168.

Chiurchiù, V., Leuti, A., and Maccarrone, M. 2015. Cannabinoid signaling and neuroinflammatory diseases: A melting pot for the regulation of brain immune responses. *Journal of Neuroimmune Pharmacology* 10: 268–280.

Choudhary, N., Siddiqui, M.B., Bi, S., and Khatoon, S. 2014. Effect of seasonality and time after anthesis on the viability and longevity of *Cannabis sativa* pollen. *Palynology* 38: 235–241.

Chowdhury, R., Warnakula, S., Kunutsor, S., Crowe, F., Ward, H.A., Johnson, L. et al. 2014. Association of dietary, circulating, and supplement fatty acids with coronary risk: A systematic review and meta-analysis. *Annals of Internal Medicine* 160: 398–406.

Chu, Y.W. 2014. The effects of medical marijuana laws on illegal marijuana use. *Journal of Health Economics* 38: 43–61.

Chusid, M.J., Gelfand, J.A., Nutter, C., and Fauci, A.S. 1975. Letter: Pulmonary aspergillosis, inhalation of contaminated marijuana smoke, chronic granulomatous disease. *Annals of Internal Medicine* 82: 682–683.

Cicero, D. 2001. *The galaxy global eatery hemp cookbook: More than 200 recipes using hemp oil, seeds, nuts, and flour.* Berkeley, CA: North Atlantic Books. 302 pp.

Cilio, M.R., Thiele, E.A., and Devinsky, O. 2014. The case for assessing cannabidiol in epilepsy. *Epilepsia* 55: 787–790.

Clapper, J.R., Mangieri, R.A., and Piomelli, D. 2009. The endocannabinoid system as a target for the treatment of cannabis dependence. *Neuropharmacology* 56(Suppl. 1): 235–243.

Clark, P.A., Capuzzi, K., and Fick, C. 2011. Medical marijuana: Medical necessity versus political agenda. *Medical Science Monitor* 17(12): RA249–261.

Clarke, R.C. 1977. *The botany and ecology of* Cannabis. Ben Lomond, CA: Pods. 57 pp.

Clarke, R.C. 1981. *Marijuana botany: An advanced study: The propagation and breeding of distinctive Cannabis.* Berkeley, CA: And/Or Press. 197 pp.

Clarke, R. 1996. The future of *Cannabis* as a source of nutraceuticals and pharmaceuticals. Zbornik radova Naučnog instituta za ratarstvo i povrtarstvo [Scientific meeting: Renaissance of hemp]. *A Periodical of Scientific Research on Field and Vegetable Crops (Novi Sad)* 27: 121–130.

Clarke, R.C. 1998a. *Hashish!* Los Angeles, CA: Red Eye Press. 387 pp.

Clarke, R.C. 1998b. Maintenance of *Cannabis* germplasm in the Vavilov Research Institute gene bank—Five year report. *Journal of the International Hemp Association* 5(1): 75–79.

Clarke, R.C. 1998c. Botany of the genus *Cannabis*. In: Ranalli, P. ed. *Advances in hemp research*. New York: Food Products Press, 1–19.

Clarke, R.C., and Merlin, M.D. 2013. Cannabis: *Evolution and ethnobotany*. Los Angeles, CA: University of California Press. 434 pp.

Clarke, R.C., and Merlin, M.D. 2015. Letter to the editor: Small, Ernest. 2015. Evolution and classification of *Cannabis sativa* (marijuana, hemp) in relation to human utilization. *Botanical Review* 81: 295–305.

Clarke, R.C., and Pate, D.W. 1994. Medical marijuana. *Journal of the Industrial Hemp Association* 1: 9–12.

Clarke, R.C., and Watson, D.P. 2002. Botany of natural *Cannabis* medicines. In: Grotenhermen, F., and Russo, E. eds. Cannabis *and cannabinoids: Pharmacology, toxicology, and therapeutic potential*. New York: Haworth Integrative Healing Press, 1–14.

Clarke, R.C., and Watson, D.P. 2006. *Cannabis* and natural *Cannabis* medicines. In: ElSohly, M.A. ed. *Marijuana and the cannabinoids*. Totowa, NJ: Humana Press, 1–17.

Clausen, J., Keck, D., and Hiesey, W. 1948. *Experimental studies on the nature of species. 3. Environmental responses of climatic races of Achillea.* Publication No. 581: 1–129. Stanford, CA: Carnegie Institute of Washington.

Cleveland, H.H., and Wiebe, R.P. 2008. Understanding the association between adolescent marijuana use and later serious drug use: Ggateway effect or developmental trajectory? *Development and Psychopathology* 20: 615–632.

Coffman, C.B., and Gentner, W.A. 1975. Cannabinoid profile and elemental uptake of *Cannabis sativa* L. as influenced by soil characteristics. *Agronomy Journal* 67: 491–497.

Coffman, C.B., and Gentner, W.A. 1977. Responses of greenhouse-grown *Cannabis sativa* L. to nitrogen, phosphorus, and potassium. *Agronomy Journal* 69: 832–836.

Collins, J. ed. 2014. *Ending the drug wars. Report of the LSE expert group on the economics of drug policy.* London: The London School of Economics and Political Science. 82 pp. https://www.lse.ac.uk/IDEAS /publications/reports/pdf/LSE-IDEAS-DRUGS-REPORT-FINAL-WEB.pdf.

Committee on Mycoherbicides for Eradicating Illicit Drug Crops. 2011. *Feasibility of using mycoherbicides for controlling illicit drug crops.* Washington, DC: National Research Council of the National Academies Press. 186 pp.

Conca, J. 2014. It's final—Corn ethanol is of no use. *Forbes* (magazine) April 20. http://www.forbes.com/sites /jamesconca/2014/04/20/its-final-corn-ethanol-is-of-no-use/.

Concannon, R., Finn, D.P., and Dowd, E. 2015. Cannabinoids in Parkinson's disease. In: Fattore, L. ed. *Cannabinoids in neurologic and mental disease*. New York: Academic Press, 35–59.

Connor, W.E., DeFrancesco, C.A., and Connor, S.L. 1993. N-3 fatty acids from fish oil. Effects on plasma lipoproteins and hypertriglyceridemic patients. *Annals of the New York Academy of Sciences* 683: 16–34.

Conrad, C. 1997. *Hemp for health: The medicinal and nutritional uses of* Cannabis sativa. Rochester, VT: Healing Arts Press. 264 pp.

Consroe, P., Laguna, J., Allender, J., Snider, S., Stern, L., Sandyk, R. et al. 1991. Controlled clinical trial of cannabidiol in Huntington's Disease. *Pharmacology, Biochemistry, and Behavior* 40: 701–708.

Corey-Bloom, J., Wolfson, T., Gamst, A., Jin, S., Marcotte, T.D., Bentley, H. et al. 2012. Smoked cannabis for spasticity in multiple sclerosis: A randomized, placebo-controlled trial. *CMAJ (Canadian Medical Association Journal)* 184: 1143–1150.

Correia, F, Roy, D.N., and Chute, W. 2003. Hemp chemical pulp: A reinforcing fibre for hardwood kraft pulps: A look at the merits of hemp as a fibre additive. *Pulp & Paper Canada* 104(5): 51–54.

Costa, B., and Comelli, F. 2014. Pain. In: Pertwee, R.G. ed. *Handbook of cannabis.* Oxford, U.K.: Oxford University Press, 473–486.

Costa, B., Trovato, A.E., Comelli, F., Giagnoni, G., and Colleoni, M. 2007. The non-psychoactive cannabis constituent cannabidiol is an orally effective therapeutic agent in rat chronic inflammatory and neuropathic pain. *European Journal of Pharmacology* 556: 75–83.

Cottone, E., Pomatto, V., Cerri, F., Campantico, E., Mackie, K., Delpero, M. et al. 2013. Cannabinoid receptors are widely expressed in goldfish: Molecular cloning of a CB2-like receptor and evaluation of CB1 and CB2 mRNA expression profiles in different organs. *Fish Physiology and Biochemistry* 39: 1287–1296.

Cougle, J.R. Bonn-Miller, M.O. Vujanovic, A.A. Zvolensky, M.J., and Hawkins, K.A. 2011. Posttraumatic stress disorder and cannabis use in a nationally representative sample. *Psychology of Addictive Behaviors* 25: 554–558.

Cousaert, C., Heylens, G., and Audenaert, K. 2013. Laughing gas abuse is no joke. An overview of the implications for psychiatric practice. *Clinical Neurology and Neurosurgery* 115: 859–862.

Crean, R.D., Crane, N.A., and Mason, B.J. 2011. An evidence-based review of acute and long-term effects of cannabis use on executive cognitive functions. *Journal of Addiction Medicine* 5: 1–8.

Cridge, B.J., and Rosengren, R.J. 2013. Critical appraisal of the potential use of cannabinoids in cancer management. *Cancer Management and Research* 5: 301–313.

Crippa, J.A., Zuardi, A.W., Martín-Santos, R, Bhattacharyya, S., Atakan, Z., McGuire, P. et al. 2009. Cannabis and anxiety: A critical review of the evidence. *Human Psychopharmacology* 24: 515–523.

Crippa, J.A., Zuardi, A.W., and Hallak, J.E. 2010. Therapeutic use of the cannabinoids in psychiatry. *Revista Brasileira de Psiquiatria* 32(Suppl. 1): 556–566.

Cristiana Moliterni, V.M., Cattivelli, L., Ranalli, P., and Mandolino, G. 2004. The sexual differentiation of *Cannabis sativa* L.: A morphological and molecular study. *Euphytica* 140: 95–106.

Cristino, L., and Di Marzo, V. 2014. Established and emerging concepts of cannabinoid action on food intake and their potential application to the treatment of anorexia and cachexia. In: Pertwee, R.G. ed. *Handbook of cannabis.* Oxford, U.K.: Oxford University Press, 455–472.

Crocioni, A. 1950. *Duration of the germinating capacity in hemp seed in relation to storage conditions. Bologna Univ. Cent. Di Studi per le Richerche Sulla Lavorazione Coltivaz. Ed Econ. Della Canapa Quaderni 9.* Bologna, Italy: Bologna University. 30 pp. (In Italian.)

Crombie, L., and Crombie, W.M.L. 1975. Cannabinoid formation in *Cannabis sativa* grafted inter-racially, and with two *Humulus* species. *Phytochemistry* 14: 409–412.

Crônier, D., Monties, B., and Chabbert, B. 2005. Structure and chemical composition of bast fibers isolated from developing hemp stem. *Journal of Agricultural and Food Chemistry* 53: 8279–8289.

Croxford, J.L., and Yamamura, T. 2005. Cannabinoids and the immune system: Potential for the treatment of inflammatory diseases? *Journal of Neuroimmunology* 166: 3–18.

Cunha, P., Romão, A.M., Mascarenhas-Melo, F., Teixeira, H.M., and Reis, F. 2011. Endocannabinoid system in cardiovascular disorders—New pharmacotherapeutic opportunities. *Journal of Pharmacy & Bioallied Sciences* 3: 350–360.

Curran, H.V., and Morgan, C.J.A. 2014. Desired and undesired effects of cannabis on the human mind and psychological well-being. In: Pertwee, R.G. ed. *Handbook of cannabis.* Oxford, U.K.: Oxford University Press, 647–660.

Curtis, A., Clarke, C.E., and Rickards, H.E. 2009. Cannabinoids for Tourette's syndrome. *Cochrane Database Systematic Review.* doi:10.1002/14651858.CD006565.pub2.

Dahl, H.V., and Frank, V.A. 2011. Medicinal marijuana—Exploring the concept in relation to small scale cannabis growers in Denmark. In: Decorte, T., Potter, G., and Bouchard, M. eds. *World wide weed: Global trends in cannabis cultivation and its control.* Burlington, VT: Ashgate, 57–73.

Dalotto, T. 1999. *The hemp cookbook: From seed to shining seed.* Rochester, VT: Healing Arts Press. 184 pp.

D'Amico, E.J., Miles, J.N.V., and Tucker, J.S. 2015. Gateway to curiosity: Medical marijuana ads and intention and use during middle school. *Psychology of Addictive Behaviors* 29: 613–619.

Dang, V., and Nguyen, K.L. 2006. Characterisation of the heterogeneous alkaline pulping kinetics of hemp woody core. *Bioresource Technology* 97: 1353–1359.

Danko, D. 2010. *The official High Times field guide to marijuana strains.* New York: High Times Corporation. 180 pp.

Da Porto, C., Decorti, D., and Tubaro, F. 2012a. Fatty acid composition and oxidation stability of hemp (*Cannabis sativa* L.) seed oil extracted by supercritical carbon dioxide. *Industrial Crops and Products* 36: 401–404.

Da Porto, C., Voinovich, D., Decorti, D., and Natolino, A. 2012b. Response surface optimization of hemp seed (*Cannabis sativa* L.) oil yield and oxidation stability by supercritical carbon dioxide extraction. *Journal of Supercritical Fluids* 68: 45–51.

Da Porto, C., Decorti, D., and Natolino, A. 2015. Potential oil yield, fatty acid composition, and oxidation stability of the hempseed oil from four *Cannabis sativa* L. cultivars. *Journal of Dietary Supplements* 12: 1–10.

Darlington, C.D. 1973. *Chromosome botany and the origins of cultivated plants.* 3rd edition. London: Allen & Unwin. 237 pp.

Darwin, C. 1845. *Journal of researches into the natural history and geology of the countries visited during the voyage of H.M.S. Beagle round the world, under the command of Capt. Fitz Roy, R.N.* 2nd edition. London: John Murray. 519 pp.

Darwin, C. 1859. *On the origin of species by means of natural selection, or the preservation of favoured races in the struggle for life.* London: John Murray. 502 pp.

Darwin, C. 1868. *The variation of animals and plants under domestication. Vol. 1.* London: John Murray. 411 pp.

David, J., Garrity, G.M., Greuter, W., Hawksworth, D.L., Jahn, R., Kirk, P.M. et al. 2012. Biological nomenclature terms for facilitating communication in the naming of organisms. *ZooKeys* 192: 67–72.

Davidyan, G.G. 1972. Hemp: Biology and initial material for breeding. *Trudy po Prikladnoi Botanike, Genetikei Selektsii [Bulletin of Applied Botany, of Genetics, and Plant-Breeding]* 48(3): 3–160. (In Russian.)

Dayanandan, P., and Kaufman, P.B. 1976. Trichomes of *Cannabis sativa* L. (Cannabacaeae). *American Journal of Botany* 63: 578–591.

De Backer, B., Maebe, K., Verstraete, A.G., and Charlier, C. 2012. Evolution of the content of THC and other major cannabinoids in drug-type cannabis cuttings and seedlings during growth of plants. *Journal of Forensic Sciences* 57: 918–922.

De Bruijn, P.B., Jeppsson, K.-H., Sandin, K., and Nilsson, C. 2009. Mechanical properties of lime-hemp concrete containing shives and fibres. *Biosystems Engineering* 103: 474–479.

De Candolle, A. 1885. *Origin of cultivated plants.* New York: D. Appleton & Co. 468 pp.

De Ceballos, M.L. 2015. Cannabinoids for the treatment of neuroinflammation. In: Fattore, L. ed. *Cannabinoids in neurologic and mental disease.* New York: Academic Press, 3–14.

De Faubert Maunder, M.J. 1976. The forensic significance of the age and origin of *Cannabis. Medicine, Science, and the Law* 16: 78–89.

Deferne, J.-L., and Pate, D.W. 1996. Hemp seed oil: A source of valuable essential fatty acids. *Journal of the International Hemp Association* 3(1): 1, 4–7.

Degenhardt, L., Ferrari, A.J., Calabria, B., Hall, W.D., Norman, R.E., McGrath, J. et al. 2013. The global epidemiology and contribution of cannabis use and dependence to the global burden of disease: Results from the GBD 2010 study. *PLoS ONE* 8(10): e76635. doi:10.1371/journal.pone.0076635.

De Groot, B., van Dam., J.E.G., van der Zwan, R.P., and van't Riet, K. 1994. Simplified kinetic modelling of alkaline delignification of hemp woody core. *Holzforschung* 18: 207–214.

De Groot, B., van Roekel, G.J. Jr., and van Dam, J.E.G. 1998. Alkaline pulping of fiber hemp. In: Ranalli, P. ed. *Advances in hemp research.* London: Food Products Press (of Haworth Press), 213–242.

De Guzman, D. 2001. Hemp oil shows huge gains in food and personal care. *Chemical Market Reporter* 259(10): 7.

Deiana, S. 2013. Medical use of cannabis. Cannabidiol: A new light for schizophrenia? *Drug Testing and Analysis* 5: 46–51.

De Lago, E., Moreno-Martet, M., Espejo-Porras, F., and Fernández-Ruiz, J. 2015. Endocannabinoids and amyotrophic lateral sclerosis. In: Fattore, L. ed. *Cannabinoids in neurologic and mental disease.* New York: Academic Press, 99–123.

De Larramendi, C.H., Carnés, J., García-Abujeta, J.L., García-Endrino, A., Muñoz-Palomino, E., Huertas et al. 2008. Sensitization and allergy to *Cannabis sativa* leaves in a population of tomato (*Lycopersicon esculentum*)-sensitized patients. *International Archives of Allergy and Immunology* 146: 195–202.

Dell, B., and McComb, A.J. 1978. *Plant resins—Their formation, secretion and possible functions.* San Diego, CA: Academic Press. 316 pp.

De Luca, M.A., and Fattore, L. 2015. Cannabinoids and drug addiction. In: Fattore, L. ed. *Cannabinoids in neurologic and mental disease.* New York: Academic Press, 289–313.

De Luca, P., Rothman, D., and Zurier, R.B. 1995. Marine and botanical lipids as immunomodulatory and therapeutic agents in the treatment of rheumatoid arthritis. *Rheumatic Diseases in the Clinics of North America* 21: 759–777.

DeLyser, D.Y, and Kasper, W.J. 1994. Hopped beer: The case for cultivation. *Economic Botany* 48: 166–170.

De Meijer, E.P.M. 1993. Hemp variations as pulp source. Research in the Netherlands. *Pulp & Paper* 67(7): 41–43.

De Meijer, E. 1994a. *Diversity in* Cannabis. Wageningen, the Netherlands: Wageningen Agricultural University. 131 pp. (Published doctoral thesis.)

De Meijer, E. 1994b. Variation of *Cannabis* with reference to stem quality for paper pulp production. *Industrial Crops and Products* 3: 201–211.

De Meijer, E.P.M. 1995a. Fibre hemp cultivars: A survey of origin, ancestry, availability and brief agronomic characteristics. *Journal of the International Hemp Association* 2(2): 66–73.

De Meijer, E.P.M. 1995b. Diversity in *Cannabis.* In: Nova Institute, corporate ed. *Bioresource hemp— Proceedings of the symposium* (Frankfurt am Main, Germany, March 2–5, 1995). 2nd edition. Ojai, California: Hemptech, 143–151.

De Meijer, E.P.M. 1998. *Cannabis* germplasm resources. In: Ranalli, P. ed. *Advances in hemp research.* New York: Food Products Press (of Haworth Press), 133–151.

De Meijer, E.P.M. 2014. The chemical phenotypes (chemotypes) of *Cannabis.* In: Pertwee, R.G. ed. *Handbook of cannabis.* Oxford, U.K.: Oxford University Press, 89–110.

De Meijer, E.P.M., and Hammond, K.M. 2005. The inheritance of chemical phenotype in *Cannabis sativa* L. (II): Canabigerol predominant plants. *Euphytica* 145: 189–198.

De Meijer, E.P.M., and Keizer, L.C.P. 1996a. Variation of *Cannabis* for phenological development and stem elongation in relation to stem production. *Field Crops Research* 38: 37–46.

De Meijer, E.P.M., and Keizer, L.C.P. 1996b. Patterns of diversity in *Cannabis. Genetic Resources and Crop Evolution* 43: 41–52.

De Meijer, E.P.M., and Van Soest, L.J.M. 1992. The CPRO *Cannabis* germplasm collection. *Euphytica* 62: 201–211.

De Meijer, E.P.M., Kamp, H.J. van der, and Eeuwijk, F.A. van. 1992. Characterisation of *Cannabis* accessions with regard to cannabinoid content in relation to other plant characters. *Euphytica* 62: 187–200.

De Meijer, E.P.M., Bagatta, M., Carboni, A., Crucitti, P., Moliterni, V.M.C., Ranalli, P. et al. 2003. The inheritance of chemical phenotype in *Cannabis sativa* L. *Genetics* 163: 335–346.

De Meijer, E.P.M., Hammond, K.M., and Micheler, M. 2009a. The inheritance of chemical phenotype in *Cannabis sativa* L. (III): Variation in cannabichrome proportion. *Euphytica* 165: 293–311.

De Meijer, E.P.M., Hammond, K.M., and Sutton, A. 2009b. The inheritance of chemical phenotype in *Cannabis sativa* L. (IV): Cannabinoid-free plants. *Euphytica* 168: 95–112.

De Meijer, W.J.M., van der Werf, H.M.G., van Roekel, G.J., de Meijer, E.P.M., and Huisman, W. 1995. Fibre hemp: Potentials and constraints. In: Blade, S. ed. *Opportunities and profits, proceedings special crops conference* (Calgary, July, 25–27, 1995). Edmonton, AB: Alberta Agricultural Research Food and Rural Development, 67–79.

De Mello Schier, A.R., de Oliveira Ribeiro, N.P., Coutinho, D.S., Machado, S., Arias-Carrion, O., A. Crippa et al. 2014. Antidepressant-like and anxiolytic-like effects of cannabidiol: A chemical compound of *Cannabis sativa. CNS & Neurological Disorders—Drug Targets* 13: 953–960.

Demkin, A.P., and Romanenko, V.I. 1978. How to store hemp seed of carryover insurance funds. *Len Konoplia (Moskva, "Kolos")* 1978(8): 29–30. (In Russian.)

Dempsey, J.M. 1975. *Fiber crops.* Gainesville, FL: University of Florida. 457 pp.

Denning, D.W., Follansbee, S.E., Scolaro, M., Norris, S., Edelstein, H., and Stevens, D.A. 1991. Pulmonary aspergillosis in the acquired immunodeficiency syndrome. *New England Journal of Medicine* 324: 654–662.

Denson, T.F., and Earleywine, M. 2006. Decreased depression in marijuana users. *Addictive Behaviors* 31: 738–742.

De Pasquale, A., Tumino, G., Ragusa, S., and Moschonas, D. 1979. Influenza del trattamento con colchicina sulla produzione di cannabinoidi delle infiorescenze femminili della *Cannabis sativa* L. [Influence of treatment with colchicine on the production of cannabinoids in the female inflorescence of *Cannabis sativa* L.]. *Farmaco* 34: 841–853.

De Petrocellis, L., Vellani, V., Schiano-Moriello, A., Marini, P., Magherini, P.C., Orlando, P. et al. 2008. Plant-derived cannabinoids modulate the activity of transient receptor potential channels of ankyrin type-1 and melastatin type-8. *Journal of Pharmacology and Experimental Therapeutics* 325: 1007–1015.

De Petrocellis, L., Ligresti, A., Schiano Moriello, A., Iappelli, M., Verde, R., Stott, C.G. et al. 2013. Non-THC cannabinoids inhibit prostate carcinoma growth in vitro and in vivo: Pro-apoptotic effects and underlying mechanisms. *British Journal of Pharmacology* 168: 79–102.

Devane, W.A., Hanus, L., Breuer, A., Pertwee, R.G., Stevenson, L.A., Griffin, G. et al. 1992. Isolation and structure of a brain constituent that binds to the cannabinoid receptor. *Science* 258(5090): 1946–1949.

Devinsky, O., Cilio, M.R., Cross, H., Fernandez-Ruiz, J., French, J., Hill, C. et al. 2014. Cannabidiol: Pharmacology and potential therapeutic role in epilepsy and other neuropsychiatric disorders. *Epilepsia* 55: 791–802.

Dewey, L.H. 1902. Hemp. In: U.S. Department of Agriculture, corporate ed. *1901 Yearbook of the United States Department of Agriculture*. Washington, DC: Government Printing Office, 250–251.

Dewey, L.H. 1913. A purple-leaved mutation in hemp. *USDA Plant Industry Circular* 113: 23–24.

Dewey, L.H. 1914. Hemp. In: U.S. Department of Agriculture, corporate ed. *Yearbook of the United States Department of Agriculture 1913*. Washington, DC: U.S. Department of Agriculture, 283–146.

Dewey, L.H. 1916. *The production and handling of hemp hurds*. Washington, DC: U.S. Department of Agriculture. 29 pp.

Dewey, L.H. 1927. Hemp varieties of improved type are result of selection. In: U.S. Department of Agriculture, corporate ed. *USDA Yearbook*. Washington, DC: U.S. Department of Agriculture, 358–361.

Dewey, L.H., and Merrill, J.L. 1916. *Hemp hurds as paper-making material*. Washington, DC: U.S. Department of Agriculture. 25 pp.

Dewick, P.M. 2002. The biosynthesis of C_5–C_{25} terpenoid compounds. *Natural Products Reports* 19: 181–222.

Diamond, J. 2002. Evolution, consequences and future of plant and animal domestication. *Nature* 418: 700–707.

Di Candilo, M., Bonatti, P.M., Guidetti, C., Focher, B., Grippo, C., Tamburini, E. et al. 2010. Effects of selected pectinolytic bacterial strains on water-retting of hemp and fibre properties. *Journal of Applied Microbiology* 108: 194–203.

Di Carlo, G., and Izzo, A.A. 2003. Cannabinoids for gastrointestinal diseases: Potential therapeutic applications. *Expert Opinion on Investigational Drugs* 12: 39–49.

Diederichsen, A., and Raney, J.P. 2006. Seed colour, seed weight and seed oil content in *Linum usitatissimum* accessions held by Plant Gene Resources of Canada. *Plant Breeding* 125: 372–377.

Di Forti, M, Morgan, C., Dazzan, P., Pariante, C., Mondelli, V., Marques, T.R. et al. 2009. High-potency cannabis and the risk of psychosis. *British Journal of Psychiatry* 195: 488–491.

Di Maio, R. 2013. Cannabinoid 1 receptor as therapeutic target in preventing chronic epilepsy. *FASEB Journal* 27(1_MeetingAbstracts): 660.2.

Di Marzo, V. 2008a. Targeting the endocannabinoid system: To enhance or reduce? *Nature Reviews Drug Discovery* 7: 438–455.

Di Marzo, V. 2008b. The endocannabinoid system in obesity and type 2 diabetes. *Diabetologia* 51: 1356–1367.

Di Marzo, V. 2009. The endocannabinoid system: Its general strategy of action, tools for its pharmacological manipulation and potential therapeutic exploitation. *Pharmacological Research* 60: 77–84.

Di Marzo, V., and De Petrocellis, L. 2014. Fifty years of 'cannabinoid research' and the need for a new nomenclature. In: Di Marzo, V. ed. *Cannabinoids*. Chichester, U.K.: John Wiley, 261–289.

Di Marzo, V., and Piscitelli, F. 2011. Gut feelings about the endocannabinoid system. *Neurogastroenterology & Motility* 23: 391–398.

Di Marzo, V., Sepe, N., De Petrocellis, L., Berger, A., Crozier, G., Fride, E. et al. 1998. Trick or treat from food endocannabinoids? *Nature* 396: 636.

Di Marzo, V., Piscitelli, F., and Mechoulam, R. 2011. Cannabinoids and endocannabinoids in metabolic disorders with focus on diabetes. *Handbook of Experimental Pharmacology* 203(75): 75–104.

Dimić, E. 2005. *Cold pressed oils*. Novi Sad, Serbia and Montenegro: University of Novi Sad, Faculty of Technology. 249 pp. (In Serbian.)

Dimić, E., Romanić, R., and Vujasinović, V. 2009. Essential fatty acids, nutritive value and oxidative stability of cold pressed hempseed (*Cannabis sativa* L.) oil from different varieties. *Acta Alimentaria* 38: 229–236.

DiPatrizio, N.V., and Piomelli, D. 2012. The thrifty lipids; endocannabinoids and the neural control of energy conservation. *Trends in Neurosciences* 35: 403–411.

Dippenaar, M.C., du Toit, C.L.N., and Botha-Greeff, M.S. 1996. Response of hemp (*Cannabis sativa* L.) varieties to conditions in Northwest Province, South Africa. *Journal of the International Hemp Association* 3(2): 63–66.

Ditchfield, C., Bredt, J., and Warner, P. 1997. Whither Australian hemp? In: Nova Institute. Corporate ed. *Bioresource hemp—Proceedings of the symposium* (Frankfurt am Main, Germany, February 27–March 2, 1997). Hürth, Germany: Nova Institute, 35–45.

Divashuk, M.G., Alexandrov, O.S., Razumova, O.V., Kirov, I.V., and Karlov, G.I. 2014. Molecular cytogenetic characterization of the dioecious *Cannabis sativa* with an XY chromosome sex determination system. *PLoS ONE* 9(1): e85118. doi:10.1371/journal.pone.0085118.

Donaldson, C.W. 2002. Marijuana exposure in animals. *Veterinary Medicine* 97: 437–439.

Doty, R.L., Wudarski, T., Marshall, D.A., and Hastings, L. 2004. Marijuana odor perception: Studies modeled from probable cause cases. *Law and Human Behavior* 28: 232–233.

Drake, B. 1979. *Marijuana, the cultivator's handbook.* Berkeley, CA: Ronin Publishing. 223 pp.

Dresen, S., Ferreirós, N., Pütz, M., Westphal, F., Zimmermann, R., and Auwärter, V. 2010. Monitoring of herbal mixtures potentially containing synthetic cannabinoids as psychoactive compounds. *Journal of Mass Spectrometry* 45: 1186–1194.

Driemeier, D. 1997. Marijuana (*Cannabis sativa*) toxicosis in cattle. *Veterinary and Human Toxicology* 39: 351–352.

Drug Enforcement Administration. 2013. *The DEA position on marijuana.* 62 pp. http://www.justice.gov/dea/docs/marijuana_position_2011.pdf.

D'Souza, D.C., and Ranganathan, M. 2015. Medical marijuana: Is the cart before the horse? *JAMA* 313: 2431–2432.

Duffy, E., Lawrence, M., and Walker, P. 2013. Hemp-lime: Highlighting room for improvement. *Civil and Environmental Research* 4: 16–21.

Duke, S.B. 1995. Drug prohibition: An unnatural disaster. *Yale Law School Faculty Scholarship Series.* Paper 812. http://digitalcommons.law.yale.edu/cgi/viewcontent.cgi?article=1824&context=fss_papers.

Duke, S.O., Canel, C., Rimando, A.M., Tellez, M.R., Duke, M.V., and Paul, R.N. 2000. Current and potential exploitation of plant glandular trichome productivity. *Advances in Botanical Research* 31: 121–151.

Duncan, M., and Izzo, A.A. 2014. Phytocannabinoids and the gastrointestinal system. In: Pertwee, R.G. ed. *Handbook of cannabis.* Oxford, U.K.: Oxford University Press, 227–244.

Dunn, S.L., Wilkinson, J.M., Crawford, A., Le Maitre, C.L., and Bunning, R.A.D. 2012. Cannabinoids: Novel therapies for arthritis? *Future Medicinal Chemistry* 4: 713–725.

DuPont, R.L. 2000. *The selfish brain: Learning from addiction.* Center City, MN: Hazelden. 505 pp.

DuPont, R.L. 2010. *Why we should not legalize marijuana.* 4 pp. Consumer News and Business Channel. http://www.cnbc.com/id/36267223.

Duran, M., Pérez, E., Abanades, S., Vidal, X., Saura, C., Majem, M. et al. 2010. Preliminary efficacy and safety of an oromucosal standardized cannabis extract in chemotherapy-induced nausea and vomiting. *British Journal of Clinical Pharmacology* 70: 656–663.

Dvorak, J.E. 2004. Hip hemp happenings. *Journal of Industrial Hemp* 9(1): 83–88.

Earleywine, M. 2010. Pulmonary harm and vaporizers. In: Holland, J. ed. *The pot book: A complete guide to cannabis: Its role in medicine, politics, science and culture.* Rochester, VT: Park Street Press, 153–160.

Earleywine, M., and Barnwell, S.S. 2007. Decreased respiratory symptoms in cannabis users who vaporize. *Harm Reduction Journal* 4: 11.

Eaton, B.J. 1972. *Identifying and controlling wild hemp.* Manhattan, KS: Agricultural Experiment Station, Kansas State University of Agriculture and Applied Science, Bulletin 555. 12 pp.

Eaton, B.J., Hartowicz, L.E., Latta, R.P., Knutson, H., Paulsen, A., and Esbaugh, E. 1972. *Controlling wild hemp.* Report of Progress 188. Manhattan, KS: Agricultural Experiment Station, Kansas State University of Agriculture and Applied Science. 10 pp.

Ebers, G. 1878. *Egypt: Descriptive, historical, and picturesque. Vol. 1.* New York: Cassell & Company.

Ebo, D.G., Swerts, S., Sabato, V., Hagendorens, M.M., Bridts, C.H., Jorens, P.G. et al. 2013. New food allergies in a European non-Mediterranean region: Is *Cannabis sativa* to blame? *International Archives of Allergy and Immunology* 161: 220–228.

Eerens, J.P.J. 2003. Potential economic viability of growing industrial hemp (*Cannabis sativa*) at the Taupo, New Zealand, effluent disposal site. *New Zealand Journal of Crop and Horticultural Science* 31: 203–208.

European Food Safety Authority (EFSA) Panel on Additives and Products or Substances Used in Animal Feed. 2011. Scientific opinion on the safety of hemp (*Cannabis* genus) for use as animal feed. *European Food Safety Authority Journal* 9(3). 41 pp. www.efsa.europa.eu/efsajournal.

Ehrensing, D.T. 1998. *Feasibility of industrial hemp production in the United States Pacific Northwest.* Department of Crop and Soil Science, Oregon State University Experiment Station Bulletin 681. Corvallis, OR: Oregon State University. http://extension.oregonstate.edu/catalog/html/sb/sb681/.

Eisner, T. 1970. Chemical defense against predation in arthropods. In: Sondheimer, E., and Simeone, J.B. eds. *Chemical ecology.* New York: Academic Press, 157–217.

Elfordy, S., Lucas, F., Tancret, F., Scudeller, Y., and Goudet, L. 2008. Mechanical and thermal properties of lime and hemp concrete (hempcrete) manufactured by a projection process. *Construction and Building Materials* 22: 2116–2123.

Ellis, R.J., Toperoff, W., Vaida, F., van den Brande, G., Gonzales, J., Gouaux, B. et al. 2009. Smoked medicinal cannabis for neuropathic pain in HIV: A randomized, crossover clinical trial. *Neuropsychopharmacology* 34: 672–680.

Elphick, M.R. 2012. The evolution and comparative neurobiology of endcannabinoid signalling. *Philosophical Transactions of the Royal Society of London, Series B: Biological Sciences* 367: 3201–3215.

Elphick, M.R., and Egertová, M. 2001. The neurobiology and evolution of cannabinoid signalling. *Philosophical Transactions of the Royal Society, B, Biological Sciences* 356: 381–408.

Elphick, M.R., and Egertová, M. 2015. The phylogenetic distribution and evolutionary origins of endocannabinoid signalling. *Handbook of Experimental Pharmacology* 168: 283–297.

ElSohly, M.A. 1995. Stable suppository formulations effecting [sic] bioavailability of Δ^9-THC. US Patent 5389375 A. http://www.google.com/patents/US5389375.

ElSohly, M.A. ed. 2006. *Marijuana and the cannabinoids*. Totowa, NJ: Humana Press. 322 pp.

ElSohly, M.A., and Gul, W. 2014. Constituents of *Cannabis sativa*. In: Pertwee, R.G. ed. *Handbook of cannabis*. Oxford, U.K.: Oxford University Press, 3–22.

ElSohly, M.A., and Slade, D. 2005. Chemical constituents of marijuana: The complex mixture of natural cannabinoids. *Life Sciences* 78: 539–548.

ElSohly, H., Turner, C.E., Clark, A.M., and ElSohly, M.A. 1982. Synthesis and antimicrobial properties of certain cannabichromene and cannabigerol related compounds. *Journal of the Pharmaceutical Sciences* 71: 1319–1323.

ElSohly, M.A., Mehmedic, Z., Foster, S., Gon, C., Chandra, S., and Church, J.C. 2016. Changes in cannabis potency over the last two decades (1995–2014)—Analysis of current data in the United States. *Biological Psychiatry* 79: 613–619.

Elzinga, S., Fischedick, J., Podkolinski, R., and Raber, J.C. 2015. Cannabinoids and terpenes as chemotaxonomic markers in *Cannabis*. *Natural Products Chemistry & Research* 3: 181. doi:10.4172/2329-6836.1000181.

EMCDDA. 2008. *A cannabis reader: Global issues and local experiences*. Lisbon, Portugal: European Monitoring Centre for Drugs and Drug Addiction. 2 volumes. http://www.emcdda.europa.eu/publica tions/monographs/cannabis.

EMCDDA. 2012. *Cannabis production and markets in Europe*. Luxembourg: Publications Office of the European Union. 268 pp.

Emerich, D.W., and Krishnan, H.B. eds. 2009. *Nitrogen fixation in crop production*. Madison, WI: American Society of Agronomy. 421 pp.

Erasmus, U. 1993. *Fats that heal, fats that kill*. 2nd edition. Burnaby, BC: Alive Books. 456 pp.

Eriksson, M., and Wall, H. 2012. Hemp seed cake in organic broiler diets. *Animal Feed Science and Technology* 171: 205–213.

Erkelens, J.L., and Hazekamp, A. 2014. That which we call *Indica*, by any other name would smell as sweet. *Cannabinoids* 9: 9–15.

Esposito, G., De Filippis, D., Cirillo, C., Iuvone, T., Capoccia, E., Scuderi et al. 2013. Cannabidiol in inflammatory bowel diseases: A brief overview. *Phytotherapy Research* 27: 633–636.

Etter, J.-F. 2015. Electronic cigarettes and cannabis: An exploratory study. *European Addiction Research* 21: 124–130.

Ettorre, E., and Riska, E. 1995. *Gendered moods: Psychotropics and society*. London: Psychology Press (Taylor & Francis Group). 177 pp.

Eubanks, L.M., Rogers, C.J., Beuscher, A.E. IV, Koob, G.F., Olson, A.J., Dickerson, T.J. et al. 2006. A molecular link between the active component of marijuana and Alzheimer's disease pathology. *Molecular Pharmaceutics* 3: 773–777.

Evans, A.G. 1989. Allergic inhalant dermatitis attributable to marijuana exposure in a dog. *Journal of the American Veterinary Medical Association* 195: 1588–1590.

Faegri, K., Iverson, J., Kaland, P.E., and Krzywinski, K. 1989. *Textbook of pollen analysis*. 4th edition. New York: John Wiley & Sons. 340 pp.

Faeti, V., Mandolino, G., and Ranalli, P. 1996. Genetic diversity of *Cannabis sativa* germplasm based on RAPD markers. *Plant Breeding* 115: 367–370.

Fagan, S.G., and Campbell, V.A. 2015. Endocannabinoids and Alzheimer's disease. In: Fattore, L. ed. *Cannabinoids in neurologic and mental disease*. New York: Academic Press, 15–33.

Fairbairn, J.W. 1972. The trichomes and glands of *Cannabis sativa* L. *Bulletin on Narcotics* 24: 29–33.

Fairbairn, J.W., and Pickens, J.T. 1981. Activity of cannabis in relation to its Δ^1-trans-tetrahydrocannabinol content. *British Journal of Pharmacology* 72: 401–409.

Fairbairn, J.W., Liebmann, J.A., and Rowan, M.G. 1976. The stability of cannabis and its preparations on storage. *Journal of Pharmacy and Pharmacology* 28: 1–7.

Fairholt, F.W. 1859. *Tobacco, its history and associations*. London: Chapman and Hall. 332 pp.

Fan, M. 2010. Characterization and performance of elementary hemp fibres: Factors influencing tensile strength. *BioResources* 5: 2307–2322.

Fan, X., and Gates, R.A. 2001. Degradation of monoterpenes in orange juice by gamma radiation. *Journal of Agricultural and Food Chemistry* 49: 2422–2426.

Fan, Y.-Y., and Chapkin, R.S. 1998. Importance of dietary γ-linolenic acid in human health and nutrition. *Journal of Nutrition* 128: 1411–1414.

Fankhauser, M. 2002. History of cannabis in Western medicine. In: Grotenhermen, F., and Russo, E. eds. Cannabis *and cannabinoids. Pharmacology, toxicology, and therapeutic potential.* Binghamton, NY: Haworth Integrative Healing Press, 37–51.

FAO. 1995. *Flavours and fragrances of plant origin.* Rome, Italy: Food and Agricultural Organization of the United Nations. 101 pp.

FAO. 2010. *The second report on the state of the world's plant genetic resources.* Rome, Italy: Food and Agricultural Organization of the United Nations. http://www.fao.org/agriculture/crops/core-themes /theme/seeds-pgr/sow/sow2/en/.

FAO. 2016. The World Information and Early Warning System on Plant Genetic Resources for Food and Agriculture. http://www.fao.org/wiews-archive/wiews.jsp.

Farkas, J., and Andrassy, E. 1976. The sporostatic effect of cannabidiolic acid. *Acta Alimentaria* 5: 57–67.

Farrimond, J.A., Mercier, M.S., Whalley, B.J., and Williams, C.M. 2011. *Cannabis sativa* and the endogenous cannabinoid system: Therapeutic potential for appetite regulation. *Phytotherapy Research* 25: 170–188.

Faux, A.M., and Bertin, P. 2014. Modelling approach for the quantitative variation of sex expression in monoecious hemp (*Cannabis sativa* L.). *Plant Breeding* 133: 782–787.

Faux, A.-M., Draye, X., Lambert, R., d'Andrimont, R., Raulier, P., and Bertin, P. 2013. The relationship of stem and seed yields to flowering phenology and sex expression in monoecious hemp (*Cannabis sativa* L.). *European Journal of Agronomy* 47: 11–22.

Faux, A.M., Berhin, A., Dauguet, N., and Bertin, P. 2014. Sex chromosomes and quantitative sex expression in monoecious hemp (*Cannabis sativa* L.). *Euphytica* 196: 183–197.

Feaster, C.V. 1956a. Monoecious hemp breeding in the United States. *Fibres (Engineering and Chemistry)* 17: 339–340.

Feaster, C.V. 1956b. Genetic and environmental variability of percent fibre and other characters in monoecious hemp, *Cannabis sativa* L. *Textile Quarterly* 6: 43–47.

Fedor, D., and Kelley, D.S. 2009. Prevention of insulin resistance by n-3 polyunsaturated fatty acids. *Current Opinion in Clinical Nutrition and Metabolic Care* 12: 138–146.

Feeney, M., and Punja, Z.K. 2003. Tissue culture and Agrobacterium-mediated transformation of hemp (*Cannabis sativa* L.). *In Vitro Cellular and Developmental Biology* 39: 578–585.

Feldman, H.W., and Mandel, J. 1998. Providing medical marijuana: The importance of cannabis clubs. *Journal of Psychoactive Drugs* 30: 179–186.

Fellermeier, M., and Zenk, M.H. 1998. Prenylation of olvetolate by a hemp transferase yields cannabigerolic acid, the precursor of tetrahydrocannabinol. *FEBS Letters* 427: 283–285.

Ferenczy, L. 1956. Antibacterial substances in seeds of *Cannabis. Nature* 178: 639.

Ferenczy, L., Grazca, L., and Jakobey, I. 1958. An antibacterial preparation from hemp (*Cannabis sativa* L.) *Naturwissenschaften* 45: 188.

Fergusson, D.M., Boden, J.M., and Horwood, L.J. 2006. Testing the cannabis gateway hypothesis: Replies to Hall, Kandel et al. and Maccoun (2006). *Addiction* 101: 474–476.

Fernández-Artamendi, S., Fernández-Hermida, J.R., Secades-Villa, R., and García-Portilla, P. 2011. Cannabis and mental health. *Actas Españolas Psiquiatría* 39: 180–190.

Fernández-López, D., Lizasoain, I., Moro, M.Á., and Martínez-Orgado, J. 2013. Cannabinoids: Well-suited candidates for the treatment of perinatal brain injury. *Brain Sciences* 3: 1043–1059.

Fernández-Ruiz, J., Sagredo, O., Pazos, M.R., García, C., Pertwee, R., Mechoulam, R. et al. 2013. Cannabidiol for neurodegenerative disorders: Important new clinical applications for this phytocannabinoid? *British Journal of Clinical Pharmacology* 75: 323–333.

Fernández-Ruiz, J., Hernández, J., and García-Movellán, Y. 2014a. Cannabinoids and the brain: New hopes for new therapies. In: Di Marzo, V. ed. *Cannabinoids.* Chichester, U.K.: John Wiley, 175–218.

Fernández-Ruiz, J., de Lago, E., Gómez-Ruiz, M., García, C., Sagredo, O., and García-Arencibia, M. 2014b. Neurodegenerative disorders other than multiple sclerosis. In: Pertwee, R.G. ed. *Handbook of cannabis.* Oxford, U.K.: Oxford University Press, 505–525.

Fertig, M. 1996. Analysis of the profitability of hemp cultivation for paper. *Journal of the International Hemp Association* 3(1): 42–43.

Fetterman, P.S., Keith, E.S., Waller, C.W., Guerrero, O., Doorenbos, N.J., and Quimby, M.W. 1971. Mississippi-grown *Cannabis sativa* L. *Journal of Pharmaceutical Sciences* 60: 1246–1477.

Finnan, J., and Burke, B. 2013. Potassium fertilization of hemp (*Cannabis sativa*). *Industrial Crops and Products* 41: 419–422.

Finnan, J., and Styles, D. 2013. Hemp: A more sustainable annual energy crop for climate and energy policy. *Energy Policy* 58: 152–162.

Finta-Korpel'ová, Z., and Berenji, J. 2007. Trends and achievements in industrial hemp (*Cannabis sativa* L.) breeding. *Bulletin for Hops, Sorghum & Medicinal Plants* 39(80): 63–75.

Fišar, Z. 2009a. Cannabinoids and atherosclerosis. *Prague Medical Report* 110: 5–12.

Fišar, Z. 2009b. Phytocannabinoids and endocannabinoids. *Current Drug Abuse Reviews* 2: 51–75.

Fischedick, J.T., Hazekamp, A., Erkelens, T., Choi, Y.H., and Verpoorte, R. 2010. Metabolic fingerprinting of *Cannabis sativa* L., cannabinoids and terpenoids for chemotaxonomic and drug standardization purposes. *Phytochemistry* 71: 2058–2073.

Fisher, B., and Johnston, D. 2002. *Marijuana for medicinal purposes: An evidence-based assessment*. Alberta, Canada: Medical Services Workers' Board Compensation Board. 72 pp. http://www.worksafebc .com/health_care_providers/Assets/PDF/marijuana_medicinal_purposes.pdf.

Fitzcharles, M.-N. 2015. Expanding medical marijuana access in Canada: Considerations for the rheumatologist. *Journal of Rheumatology* 42: 143–145.

Fitzcharles, M.-N., Clauw, D.J., Peter, A., Ste-Marie, P.A., and Shir, Y. 2014. The dilemma of medical marijuana use by rheumatology patients. *Arthritis Care & Research* 66: 797–801.

Fitzgerald, K.T., Bronstein, A.C., and Newquist, K.L. 2013. Marijuana poisoning. *Topics in Companion Animal Medicine* 28: 8–12.

Fiz, J., Durán, M., Capellà, D., Carbonell, J, and Farré, M. 2011. Cannabis use in patients with fibromyalgia: Effect on symptoms relief and health-related quality of life. *PLoS ONE* 6(4): e18440. doi:10.1371/journal.pone.0018440.

Flachenecker, P., Henze, T., and Zettl, U.K. 2014. Nabiximols (THC/CBD oromucosal spray, Sativex®) in clinical practice—Results of a multicenter, non-interventional study (MOVE 2) in patients with multiple sclerosis spasticity. *European Journal of Neurology* 71: 271–279.

Flachowsky, H., Schumann, E., Weber, W.E., and Peil, A. 2001. Application of AFLP for the detection of sex-specific markers in hemp. *Plant Breeding* 120: 305–309.

Fleming, M.P., and Clarke, R.C. 1998. Physical evidence for the antiquity of *Cannabis sativa* L. *Journal of the International Hemp Association* 5(2): 80–92.

Fletcher, R., Steggles, G., and Kregor, G. 1995. Current international market for industrial hemp. In: Field, S. ed. *Industrial hemp—The potential for an industrial hemp industry in Australia*. The Australian Institute of Agricultural Science and The Rural Industries Research and Development Corporation Joint Conference. 13–14 December 1995, Hilton Hotel, Melbourne. Carlton, Victoria, Australia: Australian Institute of Agricultural Science, 1–25.

Flood, J. 2010. The importance of plant health to food security. *Food Security* 2: 215–231.

Flores-Sanchez, I.J., and Verpoorte, R. 2008. Secondary metabolism in cannabis. *Phytochemistry Reviews* 7: 615–639.

Flores-Sanchez, I.J., Peč, J., Fei, J., Choi, Y.H., Dušek, J., and Verpoorte, R. 2009. Elicitation studies in cell suspension cultures of *Cannabis sativa* L. *Journal of Biotechnology* 143: 157–168.

Földy, C., Malenka, R.C., and Südhof, T.C. 2013. Autism-associated neuroligin-3 mutations commonly disrupt tonic endocannabinoid signaling. *Neuron* 78: 498–509.

Foroughi, M., Hendson, G., Sargent, M.A., and Steinbok, P. 2011. Spontaneous regression of septum pellucidum/forniceal pilocytic astrocytomas—Possible role of cannabis inhalation. *Child's Nervous System* 27: 671–679.

Forrest, C., and Young, J.P. 2006. The effects of organic and inorganic nitrogen fertilizer on the morphology and anatomy of *Cannabis sativa* "Fédrina" (industrial fibre hemp) grown in northern British Columbia, Canada. *Journal of Industrial Hemp* 1(2): 3–24.

Forster, E. 1996. History of hemp in Chile. *Journal of the International Hemp Association* 3(2): 72–77.

Fortenbery, T.R., and Bennett, M. 2004. Opportunities for commercial hemp production. *Applied Economic Perspectives and Policy* [Appeared originally under the former title *Review of Agricultural Economics*] 26: 97–117.

Fournier, G. 2000. The selection of hemp (*Cannabis sativa* L.) in France. Hemp and THC. *Comptes Rendus de l'Academie d'Agriculture de France (France)* 86(7): 209–217. (In French.)

Fournier, G., and Paris, M.R. 1978. Variability of chemical composition of the essential oil of hemp (*Cannabis sativa* Linnaeus). *Rivista Italiana E.P.P.O.S.* 60: 504–510. (In French.)

Fournier, G., and Paris, M. 1980. Determination of chemotypes from cannabinoids in fibre monoecious hemp (*Cannabis sativa* L.). Possibility of selection. *Physiologie Vegetale (France)* 18(2): 349–356. (In French.)

Fournier, G., Paris, M.R., Fourniat, M.C., and Quero, A.M. 1978. Bacteriostatic activity of the essential oils of *Cannabis sativa* L. *Annales Pharmaceutiques Françaises* 36: 603–606. (In French.)

Fournier, G., Richez-Dumanois, C., Duvezin, J., Mathieu, J.-P., and Paris, M. 1987. Identification of a new chemotype in *Cannabis sativa*: Cannabigerol-dominant plants, biogenetic and agronomic prospects. *Planta Medica* 53: 277–280.

Fowler, C.J. 2015. Delta(9)-tetrahydrocannabinol and cannabidiol as potential curative agents for cancer: A critical examination of the preclinical literature *Clinical Pharmacology & Therapeutics* 97: 587–596.

Frank, M. 1997. *Marijuana grower's guide*. Revised edition. Los Angeles, CA: Red Eye Press. 330 pp.

Fratta, W., and Fattore, L. 2013. Molecular mechanisms of cannabinoid addiction. *Current Opinion in Neurobiology* 23: 487–492.

Freeman, G.L. 1983. Allergic skin test reactivity to marijuana in the Southwest. *Western Journal of Medicine* 138: 829–831.

Freeman, T.P., and Swift, W. 2016. Cannabis potency: The need for global monitoring. *Addiction* 111: 376–377.

Fride, E. 2004. The endocannabinoid-CB(1) receptor system in pre- and post-natal life. *European Journal of Pharmacology* 500: 289–297.

Fride, E., and Russo, E. 2005 [imprinted 2006]. Neuropsychiatry: Schizophrenia, depression, and anxiety. In: Onaivi, E.S., Sugiura, T., and di Marzo, V. eds. *Endocannabinoids: The brain and body's marijuana and beyond*. Boca Raton, FL: CRC Press, 371–382.

Friedman, D., and Devinsky, O. 2015. Cannabinoids in the treatment of epilepsy. *New England Journal of Medicine* 373: 1048–1058.

Friedman, D.A. 2014. Public health regulation and the limits of paternalism. *Connecticut Law Review* 46: 1687–770.

Frisher, M., White, S., Varbiro, G. Voisey, C., Perumal, D., Crome, I. et al. 2010. The role of cannabis and cannabinoids in diabetes. *The British Journal of Diabetes & Vascular Disease* 10: 267–273.

Fuller, D.Q., and Allaby, R. 2009. Seed dispersal and crop domestication: Shattering, germination and seasonality in evolution under cultivation. *Annual Plant Reviews* 38: 238–295.

Furr, M., and Mahlberg, P.G. 1981. Histochemical analyses of laticifers and glandular trichomes in *Cannabis sativa*. *Journal of Natural Products* 44: 153–159.

Gabbay, F.H., Choi, K.H., Wynn, G.H., and Ursano, R.J. 2015. The role of endocannabinoid function in post-traumatic stress disorder: Modulating the risk phenotype and rendering effects of trauma. In: Fattore, L. ed. *Cannabinoids in neurologic and mental disease*. New York: Academic Press, 247–288.

Gable, R.S. 2006. The toxicity of recreational drugs. *American Scientist* 94: 206–208.

Gagne, S.J., Stout, J.M., Liu, E., Boubakir, Z., Clark, S.M., and Page, J.E. 2012. Identification of olivetolic acic cyclase from *Cannabis sativa* reveals a unique catalytic route to plant polyketides. *PNAS* 109: 12811–12816.

Gakhar, N., Goldberg, E., Jing, M., Gibson, R., and House, J.D. 2012. Effect of feeding hemp seed and hemp seed oil on laying hen performance and egg yolk fatty acid content: Evidence of their safety and efficacy for laying hen diets. *Poultry Science* 91: 707–711.

Gal, I.E., and Vajda, O. 1970. Influence of cannabidiolic acid on microorganisms. *Elelmezesi Ipar* 23: 336–339.

Galosh, E. 1978. The hormonal control of sex differentiation in dioecious plants of hemp (*Cannabis sativa*). The influence of plant growth regulators on sex expression in male and female plants. *Acta Societatis Botanicorum Poloniae* 47: 153–162.

Galston, W.A., and Dionne, E.J., Jr. 2013. *The new politics of marijuana legalization*. Washington, DC: The Brookings Institution. 17 pp.

Gamboa, P., Sanchez-Monge, R., Sanz, M.L., Palacín, A., Salcedo, G., and Diaz-Perales, A. 2007. Sensitization to *Cannabis sativa* caused by a novel allergenic lipid transfer protein, Can s 3. *Journal of Allergy and Clinical Immunology* 120: 1459–1460.

Gao, J., León, F., Radwan, M.M., Dale, O.R., Husni, A.S., Manley, Shari Lupien, S.P. et al. 2011. Benzyl derivatives with in vitro binding affinity for human opioid and cannabinoid receptors from the fungus *Eurotium repens*. *Journal of Natural Products* 74: 1636–1639.

Gaoni, Y., and Mechoulam, R. 1964. Isolation, structure and partial synthesis of an active constituent of hashish. *Journal of the American Chemical Society* 86: 1646.

Gaoni, Y., and Mechoulam, R. 1966. Hashish—VII, the isomerization of cannabidiol to tetrahydrocannabinols. *Tetrahedron* 22: 1481–1488.

Gaoni, Y., and Mechoulam, R. 1968. The iso-tetrahydrocannabinols. *Israel Journal of Chemistry* 6: 679–690.

García, C., Palomo-Garo, C., García-Arencibia, M., Ramos, J.A., Pertwee, R.G., and Fernández-Ruiz, J. 2011. Symptom-relieving and neuroprotective effects of the phytocannabinoid Δ^9-THCV in animal models of Parkinson's disease. *British Journal of Pharmacology* 163: 1495–1506.

Garcia-Jaldon, C., Dupreyre, D., and Vignon, M.R. 1998. Fibres from semi-retted hemp bundles by steam explosion treatment. *Biomass & Energy* 14: 251–260.

Gardner, E.L. 2014. Cannabinoids and addiction. In: Pertwee, R.G. ed. *Handbook of cannabis*. Oxford, U.K.: Oxford University Press, 173–188.

Gardner, F. 2010. Cannabidiol as a treatment for acne? Interview with Tamas Biro. *O'Shaughnessy's—The Journal of Cannabis in Clinical Practice* 2010(Summer): 31.

Gardner, F. 2011. Terpenoids, 'minor' cannabinoids contribute to 'entourage effect' of *Cannabis*-based medicines. *O'Shaughnessy's—The Journal of Cannabis in Clinical Practice* 2011(Autumn): 18–21.

Gargani, Y., Bishop, P., and Denning, D.W. 2011. Too many mouldy joints—Marijuana and chronic pulmonary aspergillosis. *Mediterranean Journal of Hematology and Infectious Diseases* 3: e2011005. doi:10.4084 /MJHID.2011.005.

Garner, W.W., and Allard, H.A. 1920. Effect of the relative length of day and night and other factors of the environment on growth and reproduction of plants. *Journal of Agricultural Research* 18: 553–606.

Gehl, D. 1995. *A summary of hemp research in Canada conducted by the fibre division of Agriculture Canada, 1923–1942*. Indian Head, SK: Agriculture and Agri-Food Canada. 10 pp.

Geiwitz, J. (and the Ad Hoc Committee on Hemp Risks). 2001. THC in hemp foods and cosmetics: The appropriate risk assessment. http://www.hempreport.com/response/response_january_2001.html.

Gertsch, J., Pertwee, R.G., and Di Marzo, V. 2010. Phytocannabinoids beyond the *Cannabis* plant—Do they exist? *British Journal of Pharmacology* 160: 523–529.

Gettman, J. 2006. Marijuana production in the United States. *Bulletin of Cannabis Reform* 2006(Dec, issue 2): 1–28. http://www.drugscience.org/Archive/bcr2/MJCropReport_2006.pdf.

Gettman, J., and Kennedy, M. 2014. Let it grow—The open market solution to marijuana control. *Harm Reduction Journal* 11: 32. doi:10.1186/1477-7517-11-32.

Gibb, D.J., Shah, M.A., Mir, P.S., and McAllister, T.A. 2005. Effect of full-fat hemp seed on performance and tissue fatty acids of feedlot cattle. *Canadian Journal of Animal Science* 85: 223–230.

Gieringer, D. 1994. *Economics of cannabis legalization: Detailed analysis of the benefits of ending cannabis prohibition*. Washington, DC: NORML (National Organization for the Reform of Marijuana Laws). 12 pp. http://pcosts.mpp.netdna-cdn.com/wp-content/resources/NORML_Economics_Cannabis_Legalization .pdf.

Gieringer, D.H. 2001. Cannabis "vaporization": A promising strategy for harm reduction. *Journal of Cannabis Therapeutics* 1: 153–170.

Gieringer, D. 2015. A warning re dabs. O'Shaughnessy's Online. http://www.beyondthc.com/a-warning-re-dabs/.

Gieringer, D., St. Laurent, J., and Goodrich, S. 2004. Cannabis vaporizer combines efficient delivery of THC with effective suppression of pyrolytic compounds. *Journal of Cannabis Therapeutics* 4(1): 7–27.

Giese, M.W., Lewis, M.A., Giese, L., and Smith, K.M. 2015. Method for the analysis of cannabinoids and terpenes in *Cannabis*. *Journal of AOAC International* 98: 1503–1522.

Gill, A. 2005. Bong lung: Regular smokers of cannabis show relatively distinctive histologic changes that predispose to pneumothorax. *American Journal of Surgical Pathology* 29: 980–982.

Gillman, M.A., and Lichtigfeld, F.J. 1994. Opioid properties of psychotropic analgesic nitrous oxide (laughing gas). *Perspectives in Biology and Medicine* 38: 125–138.

Gilman, J.M., Kuster, J.K., Lee, S., Lee, M.J., Kim, B.W., Makris, N. et al. 2014. Cannabis use is quantitatively associated with nucleus accumbens and amygdala abnormalities in young adult recreational users. *Journal of Neuroscience* 34: 5529–5538.

Girgih, A.T., Udenigwe, C.C., Li, H., Adebiyi, A.P., and Aluko, R.E. 2011. Kinetics of enzyme inhibition and antihypertensive effects of hemp seed (*Cannabis sativa* L.) protein hydrolysates. *Journal of the American Oil Chemists' Society* 88: 1767–1774.

Givnish, T.J. 1988. Adaptation to sun and shade: A whole-plant perspective. *Australian Journal of Plant Physiology* 15: 63–92.

Glas, J.J., Schimmel, B.C.J., Alba, J.M., Escobar-Bravo, R., Schuurink, R.C., and Kant, M.R. 2012. Plant glandular trichomes as targets for breeding or engineering of resistance to herbivores. *International Journal of Molecular Science* 13: 17077–17103.

Gloss, D., and Vickrey, B. 2012. Cannabinoids for epilepsy. *Cochrane Database Systematic Review*. doi:10.1002/14651858.CD009270.pub2.

Gold, D. 1973 (1990 reprint). *Cannabis alchemy: The art of modern hashmaking. Methods for preparation of extremely potent cannabis products*. Oakland, CA: Ronin Publishing. 96 pp.

Goldberg, E.M., Gakhar, N., Ryland, D., Aliani, M., Gibson, R.A., and House, J.D. 2012. Fatty acid profile and sensory characteristics of table eggs from laying hens fed hempseed and hempseed oil. *Journal of Food Science* 77: 153–160.

Gómez-Gálvez, Y., Palomo-Garo, C., Fernández-Ruiz, J., and García, C. 2015. Potential of the cannabinoid CB$_2$ receptor as a pharmacological target against inflammation in Parkinson's disease. *Progress in Neuro-Psychopharmacology and Biological Psychiatry* 64: 200–208.

González-García, S., Luo, L., Moreira, M.T., Feijoo, G., and Huppes, G. 2012. Life cycle assessment of hemp hurds use in second generation ethanol production. *Biomass & Bioenergy* 36: 268–279.

González-Naranjo, P., Campillo, N.E., Pérez, C., and Páez, J.A. 2013. Multitarget cannabinoids as novel strategy for Alzheimer disease. *Current Alzheimer Research* 10: 229–239.

Gordon, A.J., Conley, J.W., and Gordon, J.M. 2013. Medical consequences of marijuana use: A review of current literature. *Current Psychiatry Reports* 15(12): 419. doi:10.1007/s11920-013-0419-7.

Gorelick, D.A., Levin, K.H., Copersino, M.L., Heishman, S.J., Liu, F., Boggs, D.L. et al. 2012. Diagnostic criteria for cannabis withdrawal syndrome. *Drug and Alcohol Dependence* 123: 141–147.

Gorman, D.M., and Huber, J.C. Jr. 2007. Do medical cannabis laws encourage cannabis use? *International Journal on Drug Policy* 18: 160–167.

Gorshkova, L.M., Senchenko, G.I., and Virovets, V.G. 1988. Method of evaluating hemp plants for content of cannabinoid compounds. *Referativnyi Zhurnal* 12.65.322. (Abstract, in Russian.)

Gorski, R., Szklarz, M., and Kaniewski, R. 2009. Efficacy of hemp essential oil in the control of rosy apple aphid (*Dysaphis plantaginea* Pass.) occurring on apple tree. *Progress in Plant Protection* 49: 2013–2016.

Goss, W.L. 1924. The viability of buried seeds. *Journal of Agricultural Research* 29: 349–362.

Gove, P.B. ed. 1981. *Webster's third new international dictionary of the English language unabridged.* Springfield, MA: Merriam-Webster Inc. 2764 pp.

Graham, R. 2014. What if the rise in marijuana smoking prompted by legalization brings more than just tolerable negative side effects? What if it is actually good for public health. *JSTOR Daily.* http://daily.jstor.org/marijuana-and-public-health/.

Grant, I. 2013. Medicinal cannabis and painful sensory neuropathy. *American Medical Association Journal of Ethics* 15: 466–469.

Grant, I., Atkinson, J.H., Gouaux, B., and Wilsey, B. 2012. Medical marijuana: Clearing away the smoke. *Open Neurology Journal* 6: 18–25.

Greco, R., and Tassorelli, C. 2015. Endocannabinoids and migraine. In: Fattore, L. ed. *Cannabinoids in neurologic and mental disease.* New York: Academic Press, 173–189.

Green, G. 2003. *The cannabis grow bible. The definitive guide to growing marijuana for recreational and medical use.* San Francisco, CA: Green Candy Press. 305 pp.

Greer, G.R., Grob, C.S., and Halberstadt, A.L. 2014. PTSD symptom reports of patients evaluated for the New Mexico medical cannabis program. *Journal of Psychoactive Drugs* 46: 73–77.

Gregg, J.M., Small, E.W., Moore, R., Raft, D., and Toomey, T.C. 1976. Emotional response to intravenous delta9tetrahydrocannabinol during oral surgery. *Journal of Oral Surgery* 34: 301–313.

Greydanus, D.E., Hawver, E.K., and Merrick, J. 2013. Marijuana: Current concepts. *Frontiers in Public Health* 1(article 42): 1–17.

Griga, M., and Bjelková, M. 2013. Flax (*Linum usitissimum* L.) and hemp (*Cannabis sativa* L.) as fibre crops for phytoextraction of heavy metals: Biological, agro-technological and economical point of view. In: Gupta, D.K. ed. *Plant-based remediation processes.* Heidelberg, Germany: Springer, 199–237.

Grigoryev, S.V. 1988. Cold-resistance of hemp (*Cannabis sativa* L.). *Natural Fibres* 42: 98–102.

Grigoryev, S.V. Undated. Hemp (*Cannabis sativa* L.) genetic resources at the VIR: From the collection of seeds, through the collection of sources, towards the collection of donors of traits. http://www.vir.nw.ru/hemp/hemp1.htm.

Grigoryev [Grigor'ev], S.V., Shelenga, T.V., Baturin, V.S., and Sarana, Y.V. 2010. Biochemical characteristics of hemp seeds from various regions of Russia. *Russian Agricultural Sciences* 36: 262–264.

Grinspoon, L. 2010. A novel approach to the symptomatic treatment of Autism. *O'Shaughnessy's—The Journal of Cannabis in Clinical Practice* 2010(Spring): 3–4.

Grisswell, J., and Young, V. 2011. *Professor Grow's book of strains. The 50 Cannabis strains most commonly found at dispensaries.* Firestone, CO: Professor Grow, LLC. 160 pp.

Grotenhermen, F. 2003. Pharmacokinetics and pharmacodynamics of cannabinoids. *Clinical Pharmacokinetics* 4: 327–360.

Grotenhermen, F. 2004a. Clinical pharmacodynamics of cannabinoids. *Journal of Cannabis Therapeutics* 4(1): 29–78.

Grotenhermen, F. 2004b. The cannabinoid system—A brief review. *Journal of Industrial Hemp* 9(2): 87–92.

Grotenhermen, F. 2007. The toxicology of cannabis and cannabis prohibition. *Chemistry & Biodiversity* 4: 1744–1769.

Grotenhermen, F., and Karus, M. 1998. Industrial hemp is not marijuana: Comments on the drug potential of fiber *Cannabis*. *Journal of the International Hemp Association* 5(2): 96–101.

Grotenhermen, F., and Müller-Vahl, K. 2012. The therapeutic potential of *Cannabis* and cannabinoids. *Deutsches Ärzteblatt International* 109: 495–501.

Grotenhermen F, and Russo, E. eds. 2002. *Cannabis and cannabinoids: Pharmacology, toxicology, and therapeutic potential*. New York: Haworth Integrative Healing Press (of Haworth Press). 439 pp.

Grotenhermen, F., Karus, M., and Lohmeyer, D. 1998. THC-limits for food: A scientific study. *Journal of the International Hemp Association* 5(2): 101–105.

Grotenhermen, F., Leson, G., and Pless, P. 2003. Evaluating the impact of THC in hemp foods and cosmetics on human health and workplace drug tests: An overview. *Journal of Industrial Hemp* 8(2): 5–36.

Grotenhermen, F., Leson, G., Berghaus, G., Drummer, O.H., Krüger, H.-P., Longo, M. et al. 2005. *Developing science-based per se limits for driving under the influence of cannabis: Findings and recommendations by an expert panel*. Hürth, Germany: Nova Institute. 49 pp. http://www.canorml.org/healthfacts /DUICreport.2005.pdf.

Grotenhermen, F., Leson, G., Berghaus, G., Drummer, O.H., Krüger, H.-P., Longo, M. et al. 2007. Developing limits for driving under cannabis. *Addiction* 102: 1910–1917.

Grudzinskaya, I.A. 1988. The taxonomy of the family Cannabaceae. *Botanicheskiy Zhurnal (Leningrad)* 73: 589–593. (In Russian.)

Guenther, E. ed. 1972. *The essential oils, Vol. 1. Individual essential oils of the plant families*. Malabar, FL: Krieger Publishing Company. 444 pp.

Gui, H., Tong, Q., Qu, W., Mao, C.-M., and Dai, S.-M. 2015. The endocannabinoid system and its therapeutic implications in rheumatoid arthritis. *International Immunopharmacology* 26: 86–91.

Guindon, J., and Hohmann, A.G. 2011. The endocannabinoid system and cancer: Therapeutic implication. *British Journal of Pharmacology* 163: 1447–1463.

Gümüşkaya, E., M. Usta, and Balaban, M. 2007. Carbohydrate components and crystalline structure of organosolv hemp (*Cannabis sativa* L.) bast fibers pulp. *Bioresource Technology* 98: 491–497.

Gurley, R.J., Aranow, R., and Katz, M. 1998. Medicinal marijuana: A comprehensive review. *Journal of Psychoactive Drugs* 30: 137–147.

Gutierrez, T., and Hohmann, A.G. 2011. Cannabinoids for the treatment of neuropathic pain: Are they safe and effective? *Future Neurology* 6: 129–133.

Guy, G.W., and Stott, C.G. 2005. The development of Sativex—A natural cannabis-based medicine. In: Mechoulam, R. ed. *Cannabinoids as therapeutics*. Basel, Switzerland: Birkhäuser Verlag, 231–263.

Guzmán, M. 2003. Cannabinoids: Potential anticancer agents. *Nature Reviews Cancer* 3: 745–755.

Hackleman, J.C., and Domingo, W.E. 1943. *Hemp, an Illinois war crop. Extension Circular* 547. Urbana, IL: University of Illinois. 8 pp. https://www.ideals.illinois.edu/bitstream/handle/2142/33294/1104715 .pdf?sequence=2.

Hakim, H.A., El Kheir, Y.A., and Mohamed, M.I. 1986. Effect of climate on the content of a CBD-rich variant of *Cannabis*. *Fitoterpia* 57: 239–241.

Hale, W.J. 1934. *The farm chemurgic: Farmward the star of destiny lights our way*. Boston: Stratford Company. 201 pp.

Hall, J., Bhattarai, S.P., and Midmore, D.J. 2012. Review of flowering control in industrial hemp. *Journal of Natural Fibers* 9: 23–36.

Hall, J., Bhattarai, S.P., and Midmore, D.J. 2014. The effects of photoperiod on phenological development and yields of industrial hemp. *Journal of Natural Fibers* 11(1): 87–106.

Hall, W. 2014. What has research over the past two decades revealed about the adverse health effects of recreational cannabis use? *Addiction* 110: 19–35.

Hall, W., and Degenhardt, L. 2009. Adverse health effect of non-medical cannabis use. *Lancet* 374: 1383–1391.

Hall, W.D., and Lynskey, M. 2005. Is cannabis a gateway drug? Testing hypotheses about the relationship between cannabis use and the use of other illicit drugs. *Drug and Alcohol Review* 24: 39–48.

Hall, W., and Pacula, R. 2003. *Cannabis use and dependence. Public health and public policy*. Cambridge, U.K.: Cambridge University Press. 289 pp.

Hall, W., and Solowij, N. 1998. Adverse effects of cannabis. *Lancet* 352: 1611–1616.

Hamadeh, R., Ardehali, A., Locksley, R.M., and York, M.K. 1988. Fatal aspergillosis associated with smoking contaminated marijuana, in a marrow transplant recipient. *Chest* 94: 432–433.

Hamayun, M., and Shinwari, Z.K.S. 2004. Folk methodology of charas (hashish) production and its marketing at Afridi Tirah, Federally Administered Tribal Areas (FATA), Pakistan. *Journal of Industrial Hemp* 9(2): 41–50.

Hammer, K., and Morimoto, Y. 2012. Chapter 7: Classifications of infraspecific variation in crop plants. In: Guarino, L., and Ramanatha Rao, V. eds. *Collecting plant genetic diversity: Technical guidelines.* 2011. Rome, Italy. 15 pp. http://biodiversity-l.iisd.org/news/cgiar-releases-updated-guidelines-for -collecting-plant-genetic-diversity/.

Hammersvik, E., Sandberg, S., and Pedersen, W. 2012. Why small-scale cannabis growers stay small: Five mechanisms that prevent small-scale growers from going large scale. *International Journal of Drug Policy* 23: 458–464.

Hammond, C.T., and Mahlberg, P.G. 1977. Morphogenesis of capitate glandular hairs of *Cannabis sativa* (Cannabaceae). *American Journal of Botany* 64: 1023–1031.

Hammond, C.T., and Mahlberg, P.G. 1978. Ultrastructural development of capitate glandular hairs of *Cannabis sativa* L. (Cannabaceae) *American Journal of Botany* 65: 140–151.

Hancock, J.F. 2012. *Plant evolution and the origin of crops.* Wallingford, U.K.: CABI. 245 pp.

Haney, A., and Bazzaz, F.A. 1970. Some ecological implications of the distribution of hemp (*Cannabis sativa* L.) in the United States of America. In: Joyce, C.R.B, and Curry, S.H. eds. *The botany and chemistry of* Cannabis. London: J. & A. Churchill, 39–48.

Haney, A., and Kutscheid, B.B. 1973. Quantitative variation in the chemical constituents of marijuana from stands of naturalized *Cannabis sativa* L. in east-central Illinois. *Economic Botany* 27: 193–203.

Haney, A., and Kutscheid, B.B. 1975. An ecological study of naturalized hemp (*Cannabis sativa* L.) in east-central Illinois. *American Midland Naturalist* 93: 1–24.

Happyana, N., Agnoletc, S., Muntendamd, R., Van Dame, A., Schneiderc, B., and Kaysera, O. 2013. Analysis of cannabinoids in laser-microdissected trichomes of medicinal *Cannabis sativa* using LCMS and cryogenic NMR. *Phytochemistry* 87: 51–59.

Hari, P., and Kulmala, L. eds. 2008. *Boreal forest and climate change.* Dordrecht, the Netherlands: Springer. 582 pp.

Harlan, J.R. 1951. Anatomy of gene centers. *The American Naturalist* 85: 97–103.

Harlan, J.R. 1995. *The living fields, our agricultural heritage.* Cambridge, U.K.: Cambridge University Press. 271 pp.

Harlan, J.R., and de Wet, J.M.J. 1971. Toward a rational classification of cultivated plants. *Taxon* 20: 509–517.

Harper, S., Strumpf, E.C., and Kaufman, J.S. 2012. Do medical marijuana laws increase marijuana use? *Annals of Epidemiology* 22: 207–212.

Harris, W.S. 2006. The omega-6/omega-3 ratio and cardiovascular disease risk: Uses and abuses. *Current Atherosclerosis Reports* 8: 453–459.

Hart, C.L., van Gorp, G.W., Haney, M., Foltin, R.W., and Fischman, M.W. 2001. Effects of acute smoked marijuana on complex cognitive performance. *Neuropsychopharmacology* 25: 757–765.

Hartman, R.L., and Huestis, M.A. 2013. Cannabis effects on driving skills. *Clinical Chemistry* 59: 478–492.

Hartsel, S.C., Loh, W.H.Y., and Robertson, L.W. 1983. Biotransformation of cannabidiol to cannabielsoin by suspension cultures of *Cannabis sativa* L. and *Saccharum officinalis* L. *Planta Medica* 48: 17–19.

Harvey, D.J. 1990. Stability of cannabinoids in dried samples of cannabis dating from around 1896–1905. *Journal of Ethnopharmacology* 28: 117–128.

Hasan, K.A. 1975. Social aspects of the use of cannabis in India. In: Rubin, V. ed. *Cannabis and culture.* The Hague, the Netherlands: Mouton, 235–246.

Hasin, D.S., Saha, T.D., Kerridge, B.T., Goldstein, R.B., Chou, S.P., Zhang, H. et al. 2015. Prevalence of marijuana use disorders in the United States between 2001–2002 and 2012–2013. *JAMA Psychiatry* 72: 1235–1242.

Haufe, J., and Carus, M. 2011a. Assessment of life cycle studies on hemp fibre composites. Biowerkstoff-Report 8/2011. *Bioplastics Magazine* 6: 26–27.

Haufe, J., and Carus, M. 2011b. *Hemp fibres for green products—An assessment of life cycle studies on hemp fibre applications.* Huerth, Germany: Nova Institute. 20 pp. http://eiha.org/media/2014/10/Hemp-Fibres -for-Green-Products-----An-assessment-of-life-cycle-studies-on-hemp-fibre-applications-2011.pdf.

Hautala, M., Pasila, A., and Pirilä, J. 2004. Use of hemp and flax in composite manufacture: A search for new production methods. *Composites. Part A: Applied Science and Manufacturing* 35: 11–16.

Hawken, A., and Prieger, J. 2013. *Economies of scale in the production of cannabis.* Cambridge, MA: Biotech Analysis Corporation. 44 pp.

Hawken, P. 2007. *Blessed unrest: How the largest movement in the world came into being, and why no one saw it coming.* New York: Viking. 342 pp.

Hayward, H.E. 1938. *The structure of economic plants.* New York: MacMillan. 674 pp.

Hazekamp, A., and Fischedick, J.T. 2012. *Cannabis—From cultivar to chemovar. Drug Testing and Analysis* 4: 660–667.

Hazekamp, A., and Grotenhermen, F. 2010. Review on clinical studies with cannabis and cannabinoids 2005–2009. *Cannabinoids* 2010(5, special issue): 1–21.

Hazekamp, A, and Pappus, G. 2014. Self-medication with cannabis. In: Pertwee, R.G. ed. *Handbook of cannabis*. Oxford, U.K.: Oxford University Press, 319–338.

Hazekamp, A., Bastola, K., Rashidi, H., Bender, J., and Verpoorte, R. 2007. Cannabis tea revisited: A systematic evaluation of the cannabinoid composition of cannabis tea. *Journal of Ethnopharmacology* 113: 85–90.

Hazekamp, A., Ware, M.A., Muller-Vahl, K.R., Abrams, D., and Grotenhermen, F. 2013. The medicinal use of cannabis and cannabinoids—An international cross-sectional survey on administration forms. *Journal of Psychoactive Drugs* 45: 199–210.

Health Canada. 2013. Information for health care professionals. Cannabis (marihuana, marijuana) and the cannabinoids. http://www.hc-sc.gc.ca/dhp-mps/marihuana/med/infoprof-eng.php.

Heiser, C.B. 1988. Aspects of unconscious selection and the evolution of domesticated plants. *Euphytica* 37: 77–81.

Heitrich, A., and Binder, M. 1982. Identification of (3R,4R)-$\Delta^{1(6)}$-tetrahydrocannabinol as an isolation artifact of cannabinoid acids formed by callus cultures. *Experientia* 38: 898–899.

Hemphill, J.K., Turner, J.C., and Mahlberg, P.G. 1978. Studies on growth and cannabinoid composition of callus derived from different strains of *Cannabis sativa*. *Lloydia* 41: 453–462.

Hemphill, J.K., Turner, J.C., and Mahlberg, P.G. 1980. Cannabinoid content of individual plant organs from different geographical strains of *Cannabis sativa* L. *Journal of Natural Products* 43: 112–122.

Hemptech. 1995. *Industrial hemp: Practical products—Paper to fabric to cosmetics*. Ojai, CA: Hemptech. 48 pp.

Hendrickson, B., and Ferraro III, F.M. 2013. Assessing the clinical value of cannabis-based treatments on fibromyalgia pain. *Journal of Psychological Inquiry* 18: 6–14.

Hendriks, H., Malingré, T., Battermann, S., and Boss, R. 1975. Mono- and sesquiterpene hydrocarbons of the essential oil of *Cannabis sativa*. *Phytochemistry* 14: 814–815.

Hennink, S. 1994. Optimisation of breeding for agronomic traits in fibre hemp (*Cannabis sativa* L.) by study of parent-offspring relationships. *Euphytica* 78(1–2): 69–76.

Herrmann, E.S., Cone, E.J., Mitchell, J.M., Bigelow, G.E., LoDico, C., Flegel, R. et al. 2015. Non-smoker exposure to secondhand cannabis smoke. II: Effect of room ventilation on the physiological, subjective, and behavioral/cognitive effects. *Drug and Alcohol Dependence* 151: 194–202.

Heslop-Harrison, J. 1956. Auxin and sexuality in *Cannabis sativa*. *Physiologia Plantarum* 4: 588–597.

Heslop-Harrison, J., and Heslop-Harrison, Y. 1957. Studies on flowering-plant growth and organogenesis. II. The modification of sex expression in *Cannabis sativa* by carbon monoxide. *Proceedings of the Royal Society of Edinburgh. Section B. Biology* 66: 424–434.

Heslop-Harrison, J., and Heslop-Harrison, Y. 1969. *Cannabis sativa* L. In: Evans LT. ed. *The induction of flowering. Some case histories*. Ithaca, NY: Cornell University Press, 205–226.

Hessle, A., Eriksson, M., Nadeau, E., Turner, T., and Johansson, B. 2008. Cold-pressed hempseed cake as a protein feed for growing cattle. *Acta Agriculturae Scandinavica Section A, Animal Science* 58: 136–145.

Hetterscheid, W.L.A, van Den Berg, R.G., and Brandeburg, W.A. 1996. An annotated history of the principles of cultivated plant classification. *Acta Botanica Neerlandica* 45: 123–134.

Heywood, V.H. ed. 1996. *Global biodiversity assessment*. Cambridge, U.S.: Cambridge University Press. 1152 pp.

Hickenlooper, J.W. 2014. Experimenting with pot: The state of Colorado's legalization of marijuana. *The Milbank Quarterly* 92: 243–249.

Hiener, R., and Mack, B. 1999. *The hemp cookbook*. Berkeley, CA: Ten Speed Press. 143 pp.

Higgins, J.P.T., and Green, S. eds. 2011. *Cochrane handbook for systematic reviews of interventions. Version 5.1.0 (updated March 2011)*. The Cochrane Collaboration website. http://handbook.cochrane.org/.

Higgins, J.P.T., Altman, D.G., Gøtzsche, P.C., Jüni, P., Moher, D., Oxman, A.D. et al. 2011. The Cochrane Collaboration's tool for assessing risk of bias in randomised trials. *BMJ* 343: d5928. http://www.bmj.com/content/343/bmj.d5928.

Hii, S.W., Tam, J.D., Thompson, B.R., and Naughton, M.T. 2008. Bullous lung disease due to marijuana. *Respirology* 13: 122–127.

Hildebrand, D.C., and McCain, A.H. 1978. The use of various substrates for large-scale production of *Fusarium oxysporum* f. sp. *cannabis*. *Phytopathology* 68: 1099–1101.

Hill, A.J., Weston, S.E., Jones, N.A., Smith, I., Bevan, S.A., Williamson, E.M. et al. 2010. Δ^9-tetrahydrocannabivarin suppresses in vitro epileptiform and in vivo seizure activity in adult rats. *Epilepsia* 51: 1522–1532.

Hill, A.J., Williams, C.M., Whalley, B.J., and Stephens, G.J. 2012. Phytocannabinoids as novel therapeutic agents in CNS disorders. *Pharmacology & Therapeutics* 133: 79–97.

Hill, K.P. 2015. Medical marijuana for treatment of chronic pain and other medical and psychiatric problems. A clinical review. *JAMA* 313(24): 2474–2483.

Hill, M., and Reed, K. 2013. Pregnancy, breast-feeding, and marijuana: A review article. *Obstetrics and Gynecological Survey* 68: 710–718.

Hillard, C.J. 2008. Role of cannabinoids and endocannabinoids in cerebral ischemia. *Current Pharmaceutical Design* 14: 2347–2361.

Hillig, K. 2002. Letter to the editor. *Journal of Industrial Hemp* 7(1): 5–6.

Hillig, K.W. 2004a. A chemotaxonomic analysis of terpenoid variation in *Cannabis*. *Biochemical Systematics and Ecology* 32: 875–891.

Hillig, K.W. 2004b. A multivariate analysis of allozyme variation in 93 *Cannabis* accessions from the VIR germplasm collection. *Journal of Industrial Hemp* 9(2): 5–22.

Hillig, K.W. 2005a. Genetic evidence for speciation in *Cannabis* (Cannabaceae). *Genetic Research and Crop Evolution* 52(2): 161–180.

Hillig, K.W. 2005b. A combined analysis of agronomic trains and allozyme allele frequencies for 69 *Cannabis* accessions. *Journal of Industrial Hemp* 10(1): 17–30.

Hillig, K.W., and Mahlberg, P.G. 2004. A chemotaxonomic analysis of cannabinoid variation in *Cannabis* (Cannabaceae). *American Journal of Botany* 91: 966–975.

Hilmer, S.N., and Gnjidic, D. 2013. Rethinking psychotropics in nursing homes. *Medical Journal of Australia* 198(2): 77.

Hirata, K. 1924. Sex reversal in hemp (preliminary report). *Journal of the Society of Agriculture and Forestry, Sapporo* 16: 145–168 (English summary 166–167.)

Hirata, K. 1927. Sex determination in hemp (*Cannabis sativa* L.). *Journal of Genetics* 19: 65–79.

Hirata, K. 1929. Cytological basis of the sex determination in *Cannabis sativa* L. *Idengaku Zasshi [Japanese Journal of Genetics]* 4: 198–201.

Hirst, R.A., Lambert, D.G., and Notcutt, W.G. 1998. Pharmacology and potential therapeutic uses of cannabis. *British Journal of Anaesthesia* 81: 77–84.

Hirvonen, J., Goodwin, R.S., Li, C.T., Terry, G.E., Zoghbi, S.S., Morse, C. et al. 2012. Reversible and regionally selective downregulation of brain cannabinoid CB1 receptors in chronic daily cannabis smokers. *Molecular Psychiatry* 17: 642–649.

Hoch, E., Bonnet, U., Thomasius, R., Ganzer, F., Havemann-Reinecke, U., and Preuss, U.W. 2015. Risks associated with the non-medicinal use of cannabis. *Deutsches Ärzteblatt International* 112: 271–278.

Hoffmann, W. 1970. Hemp (*Cannabis sativa* L.). In: Hoffmann, W., Mudra, A., and Plarre, W. eds. *Textbook of breeding agricultural cultivated plants. Vol. 2*. Berlin, Germany: P. Parey, 415–430. (In German.)

Hofmann, M.E., and Frazier, C.J. 2013. Marijuana, endocannabinoids and epilepsy: Potential and challenges for improved therapeutic intervention. *Experimental Neurology* 244: 43–50.

Hohmann, A.G., and Suplita, R.L. 2006. Endocannabinoid mechanisms of pain modulation. *AAPS Journal* 8: E693–E708.

Hölldolber, B., and Wilson, E.O. 1990. *The ants*. Cambridge, MA: Belknap Press. 732 pp.

Holoborodko, P., Virovets, V., Laiko, I., Bertucelli, S., Beherec, O., and Fournier, G. 2014. Results of efforts by French and Ukrainian breeders to reduce cannabinoid levels in industrial hemp (*Cannabis sativa* L.). www.interchanvre.com/docs/article-Laiko.pdf.

Holub, D.J., and Holub, B.J. 2004. Omega-3 fatty acids from fish oils and cardiovascular disease. *Molecular and Cellular Biochemistry* 263: 217–225.

Holubek, W. 2010. Medical risks and toxicology. In: Holland, J. ed. *The pot book: A complete guide to cannabis: Its role in medicine, politics, science and culture*. Rochester, VT: Park Street Press, 141–152.

Hong, S., Song, S.-J., and Clarke, R.D. 2003. Female-associated DNA polymorphisms of hemp (*Cannabis sativa* L.). *Journal of Industrial Hemp* 8(1): 5–9.

Hood, L.V.S., Dames, M.E., and Barry, G.T. 1973. Headspace volatiles of marijuana. *Nature* 242: 402–403.

Hopkins, J.F.A. 1951. *A history of the hemp industry in Kentucky*. Lexington, KY: University Press of Kentucky. 239 pp.

Horkay, E. 1986. Establishing the proportions of self and cross fertilization in a monoecious hemp stand by means of population genetics. *Novenytermeles* 35(3): 177–182. (In Hungarian.)

Horn, M.R.L. 2010. Short-fibre hemp for high value textiles. *Aspects of Applied Biology* 101: 127–132.

Horváth, B., Mukhopadhyay, P., Haskó, G., and Pacher, P. 2012. The endocannabinoid system and plant-derived cannabinoids in diabetes and diabetic complications. *American Journal of Pathology* 180: 432–442.

House, J.D., Neufeld, J., and Leson, G. 2010. Evaluating the quality of protein from hemp seed (*Cannabis sativa* L.) products through the use of the protein digestibility-corrected amino acid score method. *Journal of Agricultural and Food Chemistry* 58: 11801–11807.

House of Lords Select Committee on Science and Technology. 1998. *Cannabis: The scientific and medical evidence*. London: The Stationery Office, Parliament. 2 volumes.

Howard, P., Twycross, R., Shuster, J., Mihalyo, M., and Wilcock, A. 2013. Cannabinoids. *Journal of Pain and Symptom Management* 46: 142–149.

Howlett, A.C., Reggio, P.H., Childers, S.R., Hampson, R.E., Ulloa, N.M., and Deutsch, D.G. 2011. Endocannabinoid tone versus constitutive activity of cannabinoid receptors. *British Journal of Pharmacology* 163: 1329–1343.

Huang, H.T. 2000. *Science and civilization in China, Vol. 6: Biology and biological technology, Part V: Fermentations and food science*. Cambridge, U.K.: Cambridge University. 769 pp.

Huang, Y.-H.J., Zhang, Z.-F., Tashkin, D.P., Feng, B., Straif, K., and Hashibe, M. 2015. An epidemiologic review of marijuana and cancer: An update. *Cancer Epidemiology, Biomarkers & Prevention* 24: 51–31.

Huestis, M.A. 2005. Pharmacokinetics and metabolism of the plant cannabinoids, delta9-tetrahydrocannabinol, cannabidiol and cannabinol. *Handbook of Experimental Pharmacology* 168: 657–690.

Hullar, I., Meleg, I., Fekete, S., and Romvari, R. 1999. Studies on the energy content of pigeon feeds. I. Determination of digestibility and metabolizable energy content. *Poultry Science* 78: 1757–1762.

Idler, C., and Pecenka, R. 2007. Ensiling—A new practice for preservation of hemp. *LŽŪU ŽŪI Instituto ir LŽŪ Universiteto mokslo darbai, 2007 [Research papers of IAg Eng LUA & LU of Ag]* 39(1): 40–48.

Idler, C., Pecenka, R., Fürll, C., and Gusovius, H.-J. 2011. Wet processing of hemp: An overview. *Journal of Natural Fibers* 8: 59–90.

Idris, A.I. 2010. Cannabinoid receptors as target for treatment of osteoporosis: A tale of two therapies. *Current Neuropharmacology* 8: 243–253.

Idris, A.I., and Ralston, S.H. 2010. Cannabinoids and bone: Friend or foe? *Calcified Tissue International* 87: 285–297.

Idris, A.I., Sophocleous, A., Landao-Bassonga, E., Canals, M., Milligan, G., Baker, D. et al. 2009. Cannabinoid receptor type 1 protects against age-related osteoporosis by regulating osteoblast and adipocyte differentiation in marrow stromal cells. *Cell Metabolism* 10: 139–147.

Illinois Bureau of Investigation. Undated. *Eradicating marihuana plants*. Illinois: Illinois Bureau of Investigation, Department of Law Enforcement; Division of Plant Industry, Illinois Department of Agriculture; and College of Agriculture of the University of Illinois at Urbana-Champaign. 12 pp.

Inam, B., Hussain, F., and Bano, F. 1989. *Cannabis sativa* L. is allelopathic. *Pakistan Journal of Scientific and Industrial Research* 32: 617–620.

Ip, K., and Miller, A. 2012. Life cycle greenhouse gas emissions of hemp-lime wall constructions in the UK. *Resources, Conservation and Recycling* 69: 1–9.

Isaac, N.J.B., Mallet, J., and Mace, G.M. 2004. Taxonomic inflation: Its influence on macroecology and conservation. *Trends in Ecology and Evolution* 19: 464–469.

Iskedjian, M., Bereza, B., Gordon, A., Piwko, C., and Einarson, T.R. 2007. Meta-analysis of cannabis based treatments for neuropathic and multiple sclerosis-related pain. *Current Medical Research and Opinion* 23: 17–24.

Iuvone, T., Esposito, G., De Filippis, D., Caterina Scuderi, C., and Steardo, L. 2009. Cannabidiol: A promising drug for neurodegenerative disorders? *CNS Neuroscience & Therapeutics* 15: 65–75.

Iványi, I. 2011. Relationship between leaf nutrient concentration and the yield of fibre hemp (*Cannabis sativa* L.). *Research Journal of Agricultural Science* 43(3): 70–76.

Iványi, I., and Izsáki, Z. 2009. Effect of nitrogen, phosphorus, and potassium fertilization on nutrional [sic] status of fiber hemp. *Communications in Soil Science and Plant Analysis* 40: 974–986.

Iványi, I., and Izsáki, Z. 2010. Effect of nutrient supply on quantity and quality of hempseed. *Research Journal of Agricultural Science* 42: 187–191.

Iversen, L.L. 2000. *The science of marijuana*. Oxford, U.K.: Oxford University Press. 283 pp.

Iversen, L.L. 2002. Foreword. In: Groterhermen, F., and Russo, E. eds. Cannabis *and cannabinoids: Pharmacology, toxicology, and therapeutic potential*. New York: Haworth Integrative Healing Press (of Haworth Press), xxi–xxii.

Izzo, A.A., and Camilleri, M. 2008. Emerging role of cannabinoids in gastrointestinal and liver diseases: Basic and clinical aspects. *Gut* 57: 1140–1155.

Izzo, A.A., Borrelli, F., Capasso, R., Di Marzo, V., and Mechoulam, R. 2009. Non-psychotropic plant cannabinoids: New therapeutic opportunities from an ancient herb. *Trends in Pharmacological Sciences* 30: 515–527.

Izzo, A.A., Capasso, R., Aviello, G., Borrelli, F., Romano, B., Piscitelli, F. et al. 2012. Inhibitory effect of cannabichromene, a major non-psychotropic cannabinoid extracted from *Cannabis sativa*, on inflammation-induced hypermotility in mice. *British Journal of Pharmacology* 166: 1444–1460.

Jager, G. 2012. Cannabis. In: Verster, J., Brady, K., Galanter, M., and Conrod, P. eds. *Drug abuse and addiction in medical illness: Causes, consequences and treatment.* New York: Springer-Verlag, 151–162.

Jain, M.C., and Arora, N. 1988. Ganja (*Cannabis sativa*) refuse as cattle feed. *Indian Journal of Animal Science* 58: 865–867.

Jakobey, I. 1965. Experiments to produce hemp with fine fiber. *Novenytermeles* 14: 45–54.

Jampel, H. 2010. American Glaucoma Society position statement: Marijuana and the treatment of glaucoma. *Journal of Glaucoma* 19: 75–76.

Jancin, B. 2010. Marijuana use may protect against diabetes. *Clinical Endocrinology News* December: 14.

Janischevsky, D.E. 1924. A form of hemp in wild areas of southeastern Russia. *Učenye zapiski Saratovskogo Gosudarstvennogo imeni N.G. Černyševskogo Universiteta* 2(2): 3–17. (In Russian.)

Jankauskienė, Z., Gruzdevienė, E., and Lazauskas, S. 2014. Potential of industrial hemp (*Cannabis sativa* L.) genotypes to suppress weeds. *Zemdirbyste-Agriculture* 101: 265–270.

Jansen, M., and Teris, R. 2002. One woman's work in the use of hashish in a medical context. *Journal of Cannabis Therapeutics* 2(3–4): 133–141.

Jarillo, J.A., del Olmo, I., Gómez-Zambrano, A., Lázaro, A., López-González, L., Miguel, E. et al. 2008. Review. Photoperiodic control of flowering time. *Spanish Journal of Agricultural Research* 6(Special issue): 221–244.

Jeffrey, C. 1968. Systematic categories for cultivated plants. *Taxon* 17: 109–114.

Jeffries, K.A., Dempsey, D.R., Behari, A.L., Anderson, R.L., and Merkler, D.J. 2014. *Drosophila melanogaster* as a model system to study long-chain fatty acid amide metabolism. *FEBS Letters* 588: 1596–1602.

Jensen, R.P., Luo, W., Pankow, J.F., Strongin, R.M., and Peyton, D.H. 2015. Hidden formaldehyde in E-cigarette aerosols. *New England Journal of Medicine* 372: 392–394.

Jiang, H.-E., Li, X., Zhao, Y.-X., Ferguson, D.K., Hueber, F., Bera, S. et al. 2006. A new insight into *Cannabis sativa* (Cannabaceae) utilization from 2500-year-old Yanghai Tombs, Xinjiang, China. *Journal of Ethnopharmacology* 108: 414–422.

Jirásek, V. 1961. Evolution of the proposals of taxonomical categories for the classification of cultivated plants. *Taxon* 10(2): 34–45.

John, R., and Ross, H. 2010. *The global economic cost of cancer.* Atlanta, GA: American Cancer Society and the Livestrong Organization. 10 pp.

Johnson, J.M., Lemberger, L., Novotny, M., Forney, R.B., Dalton, W.S., and Maskarinec, M.P. 1984. Pharmacological activity of the basic fraction of marihuana whole smoke condensate alone and in combination with delta-9-tetrahydrocannabinol in mice. *Toxicology and Applied Pharmacology* 72: 440–448.

Johnson, J.R., Burnell-Nugent, M., Lossignol, D., Ganae-Motan, E.D., Potts, R., and Fallon, M.T. 2010. Multicenter, double-blind, randomized, placebo-controlled, parallel-group study of the efficacy, safety, and tolerability of THC:CBD extract and THC extract in patients with intractable cancer-related pain. *Journal of Pain and Symptom Management* 39: 167–179.

Johnson, J.R., Lossignol, D., Burnell-Nugent, M., and Fallon, M.T. 2013. An open-label extension study to investigate the long-term safety and tolerability of THC/CBD oromucosal spray and oromucosal THC spray in patients with terminal cancer-related pain refractory to strong opioid analgesics. *Journal of Pain and Symptom Management* 46: 207–218.

Johnson, K.P., and Sorenson, M.D. 1999. Phylogeny and biogeography of the dabbling ducks (Genus: *Anas*): A comparison of molecular and morphological evidence. *The Auk* 116: 792–805.

Johnson, P. 1999. Industrial hemp: A critical review of claimed potential for *Cannabis sativa. Tappi Journal* 82: 77, 81, 85, 113–123.

Johnson, R. 2015. *Hemp as an agricultural commodity.* Washington, DC: Congressional Research Service. http://nationalaglawcenter.org/wp-content/uploads/assets/crs/RL32725.pdf.

Jones, K. 1995. *Nutritional and medicinal guide to hemp seed.* Gibsons, BC: Rainforest Botanical Laboratory. 60 pp.

Jones, N.A., Hill, A.J., Smith, I., Bevan, S.A., Williams, C.M., Whalley, B.J. et al. 2010. Cannabidiol displays antiepileptiform and antiseizure properties in vitro and in vivo. *Journal of Pharmacology and Experimental Therapeutics* 332: 569–577.

Joshi, M., Joshi, A., and Bartter, T. 2014. Marijuana and lung diseases. *Current Opinion in Pulmonary Medicine* 20: 173–179.

Journal of the IHA. 1994. Interview: Professor Dr. Iván Bócsa, the breeder of Kompolti hemp. *Journal of the International Hemp Association* 1(2): 61–62.

Joy, J.E., Stanley, J., Watson, S.J. Jr., and Benson, J.A. Jr. eds. 1999. *Marijuana and medicine: Assessing the science base.* Washington, DC: National Academy Press. 288 pp. http://www.nap.edu/books/0309071550/html/.

Kabelik, J., Krejci, Z., and Santavy, F. 1960. Cannabis as a medicament. *Bulletin on Narcotics* 12(3): 5–23.

Kagen, S.L. 1981. Aspergillus: An inhalable contaminant of marihuana. *New England Journal of Medicine* 304: 483–484.

Kagen, S.L., Kurup, V.P., Sohnle, P.G., and Fink, J.N. 1983. Marijuana smoking and fungal sensitization. *Journal of Allergy and Clinical Immunology* 71: 389–393.

Kalant, H. 2001. Medicinal use of cannabis: History and current status. *Pain Research & Management* 6: 80–91.

Kalant, H. 2008. Smoked marijuana as medicine: Not much future. *Clinical Pharmacology & Therapeutics* 83: 517–519.

Kalant, H. 2015. Cannabis in the treatment of rheumatic diseases: Suggestions for a reasoned approach. *Journal of Rheumatology* 42: 146–148.

Kalant, H., Corrigal, W.A., Hall, W., and Smart, R.G. eds. 1999. *The health effects of cannabis.* Toronto, ON: Centre for Addiction and Mental Health. 526 pp.

Kamal-Eldin, A., and Appelqvist, L.A. 1996. The chemistry and antioxidant properties of tocopherols and tocotrienols. *Lipids* 31: 671–701.

Kang, H.K., and Park, B.S. 2007. Effects of dietary γ-fatty acids on the fatty acid composition of pork and plasma lipids in swine. *Journal of the Korean Society of Food Science and Nutrition* 36: 563–568.

Karila, L., Roux, P., Rolland, B., Benyamina, A., Reynaud, M., Aubin, H.J. et al. 2014. Acute and long-term effects of cannabis use: A review. *Current Pharmaceutical Design* 20: 4112–4118.

Karimi, I., and Hayatghaibi, H. 2006. Effect of *Cannabis sativa* L. seed (hempseed) on lipid and protein profiles of rat. *Pakistan Journal of Nutrition* 5: 585–588.

Karlsson, L., Finell, M., and Martinsson, K. 2010. Effects of increasing amounts of hempseed cake in the diet of dairy cows on the production and composition of milk. *Animal* 4: 1854–1860.

Karsak, M., Gaffal, E., Date, R., Wang-Eckhardt, L., Rehnelt, J., Petrosino, S. et al. 2007. Attenuation of allergic contact dermatitis through the endocannabinoid system. *Science* 316: 1494–1497.

Karup, V.P., Resnick, A., Kagen, S.L., Cohen, S.H., and Fink, J.N. 1983. Allergenic fungi and actinomycetes in smoking materials and their health implications. *Mycopathologia* 82: 61–64.

Karus, M., and Leson, G. 1994. Hemp research and market development in Germany. *Journal of the International Hemp Association* 1(2): 52–56.

Karus, M., and Leson, G. 1996. Opportunities for German hemp. Results of the 'hemp product line project.' *Journal of the International Hemp Association* 4(1): 26–31.

Karus, M., and Vogt, D. 2004. European hemp industry: Cultivation, processing and product lines. *Euphytica* 140: 7–12.

Keller, A., Leupin, M., Mediavilla, V., and Wintermantel, E. 2001. Influence of the growth stage of industrial hemp on chemical and physical properties of the fibres. *Industrial Crops and Products* 13: 35–48.

Kepp, K.K., and Raich, A.L. 2014. *Marijuana and health: A comprehensive review of 20 years of research.* Washington County, Hilsboro, OR: Department of Health and Human Services. 58 pp. http://learnabout marijuanawa.org/Reports/Marijuana_review_ReppRaich_Oct2014.pdf.

Kepple, N.J., and Freisthler, B. 2012. Exploring the ecological association between crime and medical marijuana dispensaries. *Journal of Studies on Alcohol and Drugs* 73: 523–530.

Kemal, M., Wahba Khalil, S.K., Rao, N.G., and Woolsey, N.F. 1979. Isolation and identification of a cannabinoid-like compound from *Amorpha* species. *Journal of Natural Products* 42: 463–468.

Khalil, R.B. 2012. Would some cannabinoids ameliorate symptoms of autism? *European Child & Adolescent Psychiatry* 21: 237–238.

Khan, J.I., Kennedy, T.J., and Christian, D.R. Jr. 2012. *Basic principles of forensic chemistry.* New York: Springer. 354 pp.

Khan, M.M.R., Chen, Y., Belsham, T., Laguë, C., Landry, H., Peng, Q. et al. 2011. Fineness and tensile properties of hemp (*Cannabis sativa* L.) fibres. *Biosystems Engineering* 108: 9–17.

Khan, R.U., Durrani, F.R., Chand, N., and Anwar, H. 2010. Influence of feed supplementation with *Cannabis sativa* on quality of broilers carcass. *Pakistan Veterinary Journal* 30: 34–38.

Kim, E.S., and Mahlberg, P.G. 1995. Glandular cuticle formation in *Cannabis* (Cannabaceae). *American Journal of Botany* 82: 1207–1214.

Kim, E.S., and Mahlberg PG. 1997. Immunochemical localization of tetrahydrocannabinol (THC) in cryofixed glandular trichomes of *Cannabis* (Cannabaceae). *American Journal of Botany* 84: 336–342.

Kim, E.S., and Mahlberg, P.G. 2003. Secretory vesicle formation in the secretory cavity of glandular trichomes of *Cannabis sativa* L. (Cannabaceae). *Molecules & Cells* 15: 387–395.

King, L.A., Carpentier, C., and Griffiths, P. 2005. Cannabis potency in Europe. *Addiction* 100: 884–886.

Kirby, R.H. 1963. *Vegetables fibres: Botany, cultivation, and utilization.* London: Leonard Hill. 464 pp.

Kirk, J.M., Doty, P., and de Wit, H. 1998. Effects of expectancies on subjective responses to oral delta-9-tetra hydrocannabinol. *Pharmacology, Biochemistry, and Behavior* 59: 287–293.

Kleiman, M.A.R. 2015. *Legal commercial cannabis sales in Colorado and Washington: What can we learn?* Washington, DC: Brookings Institute. 13 pp. http://www.globalinitiative.net/download/drugs/north -america/Kleiman%20%20Wash%20and%20Co%20final.pdf.

Kleiman, M.A.R., and Ziskind, J. 2014. Lawful access to cannabis: Gains, losses and design criteria. In: Collins, J. ed. 2014. *Ending the drug wars. Report of the LSE expert group on the economics of drug policy.* London: The London School of Economics and Political Science. 82 pp. https://www.lse.ac.uk /IDEAS/publications/reports/pdf/LSE-IDEAS-DRUGS-REPORT-FINAL-WEB.pdf. 77–82.

Kluger, B. Triolo, P., Jones, W., and Jankovic, J. 2015. The therapeutic potential of cannabinoids for movement disorders. *Movement Disorders* 30: 313–327.

Kluyver, T.A., Charles, M., Jones, G., Rees, M., and Osborne, C.P. 2013. Did greater burial depth increase the seed size of domesticated legumes? *Journal of Experimental Botany* 64: 4101–4108.

Knight, G., Hansen, S., Connor, M., Poulsen, H., McGovern, C., and Stacey, J. 2010. The results of an experimental indoor hydroponic *Cannabis* growing study using the "Screen of Green" (ScrOG) method— Yield, tetrahydrocannabinol (THC) and DNA analysis. *Forensic Science International* 202: 36–44.

Kogan, N.M., and Mechoulam, R. 2007. Cannabinoids in health and disease. *Dialogues in Clinical Neuroscience* 9: 413–443.

Kogan, N.M., Melamed, E., Wasserman, E., Raphael, B., Breuer, A., Stok, K.S. et al. 2015. Cannabidiol, a major non-psychotropic cannabis constituent enhances fracture healing and stimulates lysyl hydroxylase activity in osteoblasts. *Journal of Bone and Mineral Research* 30: 1905–1913.

Köhler, F.E. 1887. *Medizinal-Pflanzen, Volume 2.* Berlin, Germany: Gera-Untermhaus. (Mostly illustration plates.)

Kok, C.J., Coenen, G.C.M., and de Heij, A. 1994. The effect of fiber hemp (*Cannabis sativa* L.) on selected soil-borne pathogens. *Journal of the International Hemp Association* 1: 6–9.

Kolarikova, M., Ivanova, T., and Havrland, B. 2013. Energy balance of briquettes made of hemp (*Cannabis sativa* L.) cultivars (Ferimon, Bialobrzeskie) from autumn harvest to produce heat for household use. In: Malinovska, L., and Osadcuks, V. eds. *Conference proceedings "Engineering for rural development"* (Latvia University of Agriculture, May 23–24, 2013). Jelgava, Latvia: Latvia University of Agriculture, 504–508.

Kolarikova, M., Ivanova, T., Hutla, P., and Havrland, B. 2015. Economic evaluation of hemp (*Cannabis sativa*) grown for energy purposes (briquettes) in the Czech Republic. *Agronomy Research* 13: 328–336.

Konca, Y., and Beyzi, S.B. 2012. Effect of hempseed (*Cannabis sativa* sp.) supplementation to diet on performance, carcass and intestinal organ traits in Japanese quail (*Coturnix coturnix japonica*). In: Levic, J., Sredanovic, S., and Đuragic, O. eds. *XV International Feed Technology Symposium. COST—"Feed for Health" Joint Workshop, Proceedings* (Novi Sad, Serbia, October 3–5, 2012). Novi Sad, Serbia: University of Novi Sad, 340–345.

Kondo, M., Kasahara, Y., and Akita, S. 1950. Germination of hemp seeds stored for 19 years and their growth. *Nogaku Kenkyu* 39: 37–39.

Koppel, B.S., Brust, J.C., Fife, T., Bronstein, J., Youssof, S., Gronseth, G. et al. 2014. Systematic review: Efficacy and safety of medical marijuana in selected neurologic disorders: Report of the Guideline Development Subcommittee of the American Academy of Neurology. *Neurology* 82: 1556–1563.

Kostić, M.D., Joković, N.M., Stamenković, O.S., Rajković, K.M., Milić, P.S., and Veljković, V.B. 2013. Optimization of hempseed oil extraction by n-hexane. *Industrial Crops and Products* 48: 133–143.

Kostić, M.D., Joković, N.M., Stamenković, O.S., Rajković, K.M., Milić, P.S., and Veljković, V.B. 2014. The kinetics and thermodynamics of hempseed oil extraction by *n*-hexane by n-hexane. *Industrial Crops and Products* 52: 679–686.

Koven, P. 2015. Making inexpensive pot. *Financial Post Ottawa/Ottawa Citizen* Nov. 7: B6.

Koven, P., and Pett, D. 2015. Medical marijuana companies Tweed, Bedrocan to merge. *Financial Post Ottawa/ Ottawa Citizen* June 24: C8.

Koznlowski, R., Baraniecki, P., Grabowska, L., and Mankowski, J. 1995. Recultivation of degraded areas through cultivation of hemp. In: Nova Institute, corporate ed. *Bioresource hemp proceedings of the symposium* (Frankfurt am Main, Germany, March 2–5, 1995). 2nd edition. Ojai, CA: Hemptech, 259–267.

Kraenzel, D.G., Petry, T., Nelson, B., Anderson, M.J., Mathern, D., and Todd, R. 1998. *Industrial hemp as an alternative crop in North Dakota. Agricultural Economics Report 402.* Fargo, ND: North Dakota State University. 22 pp. http://www.industrialhemp.net/pdf/aer402.pdf.

Kramer, J.L. 2015. Medical marijuana for cancer. *CA: A Cancer Journal for Clinicians* 65: 109–122.

Kreuger, E. Sipos, B., Zacchi, G., Svensson, S.-E., and Björnsson, L. 2011a. Bioconversion of industrial hemp to ethanol and methane: The benefits of steam pretreatment and co-production. *Bioresource Technology* 102: 3457–3465.

Kreuger, E., Prade, T., Escobar, F., Svensson, S.E., Englund, J.E., and Björnsson, L. 2011b. Anaerobic digestion of industrial hemp—Effect of harvest time on methane energy yield per hectare. *Biomass & Bioenergy* 35: 893–900.

Krieger, M.C.R., and Krieger, G.W. 2000. *Cooking for life: Recipes with cannabis butter: Research in search of wellness*. Calgary, AB: Lark. 136 pp.

Kriese, U., Schumann, E., Weber, W.E., Beyer, M., Brühl, L., and Matthäus, B. 2004. Oil content, tocopherol composition and fatty acid patterns of the seeds of 51 *Cannabis sativa* L. genotypes. *Euphytica* 137: 339–351.

Krings, M., Taylor, T.N., and Kellogg, D.W. 2002. Touch-sensitive glandular trichomes: A mode of defence against herbivorous arthropods in the carboniferous. *Evolutionary Ecology Research* 4: 779–786.

Krishnan, S., Cairns, R., and Howard, R. 2009. Cannabinoids for the treatment of dementia. *Cochrane Database Systematic Review*. doi:10.1002/14651858.CD007204.pub2.

Kristenová, M., Exnerová, A., and Štys, P. 2011. Seed preferences of *Pyrrhocoris apterus* (Heteroptera: Pyrrhocoridae): Are there specialized trophic populations? *European Journal of Entomology* 108: 581–586.

Ksir, C., and Hart, C.L. 2016. Cannabis and psychosis: A critical overview of the relationship. *Current Psychiatry Reports* 18: 12. doi:10.1007/s11920-015-0657-y.

Kuchera, L. 1960. *Root atlas of central European arable weeds and crops*. Frankfurt am Main, Germany: DLG Verl.-Ges. 574 pp. (In German.)

Kuglarz, M., Alvarado-Morales, M., Karakashev, D., and Angelidaki, I. 2016. Integrated production of cellulosic bioethanol and succinic acid from industrial hemp in a biorefinery concept. *Bioresource Technology* 200: 639–647.

Kung, C.T. 1959. *Archaeology in China*. Vol. 1. Toronto, ON: University of Toronto Press.

Kupzow, A.J. 1975. Vavilov's law of homologous series at the fiftieth anniversary of its formulation. *Economic Botany* 29: 372–379.

Kurt, S. 2014. AAN systematic review: How effective and safe is medical marijuana for different neurologic disorders? *Neurology Today* 14(9): 1, 12–13.

Kurz, R., and Blaas, K. 2010. Use of dronabinol (delta-9-THC) in autism: A prospective single-case-study with an early infantile autistic child. *Cannabinoids* 5(4): 4–6.

Kushima, H., Shoyama, Y., and Nishioka, I. 1980. *Cannabis*. XII. Variations of cannabinoid contents in several strains of *Cannabis sativa* L. with leaf-age, season and sex. *Chemical and Pharmaceutical Bulletin* 28: 594–598.

Kutschera, L. 1960. *Wurzelatlas mitteleuropäischer Ackerunkräuter und Kulturpflanzen [Root atlas of central European arable weeds and crops]*. Frankfurt am Main, Germany: DLG Verlag. 574 pp. (In German.)

Kutuzova, S., Rumyantseva, L., Grigoryev, S., and Clarke, R.C. 1996. Maintenance of *Cannabis* germplasm in the Vavilov Research Institute gene bank—1995. *Journal of the International Hemp Association* 3(1): 10–12.

Kutuzova, S., Rumyantseva, L., Grigoryev, S., and Clarke, R.C. 1997. Maintenance of *Cannabis* germplasm in the Vavilov Research Institute gene bank—1996. *Journal of the International Hemp Association* 4(1): 17–21.

Kymäläinen, H.R., and Sjöberg, A.M. 2005. Cellulosic bast fibres of flax and hemp as alternative raw materials for thermal insulations. In: *Indoor Air 2005: Proceedings of the 10th international conference on indoor air quality and climate* (Beijing, China, September 4–9, 2005). Beijing, China: Tsinghua University Press, 1038–1042.

Lachenmeier, D.W., and Walch, S.G. 2005. Analysis and toxicological evaluation of cannabinoids in hemp food products—A review. *Electronic Journal of Environmental, Agricultural and Food Chemistry* 4(1): 812–826.

Lago, P.K., and Stanford, D.F. 1989. Phytophagous insects associated with cultivated marijuana, *Cannabis sativa*, in northern Mississippi. *Journal of Entomological Science* 24: 437–445.

Lakhan, S.E., and Rowland, M. 2009. Whole plant cannabis extracts in the treatment of spasticity in multiple sclerosis: A systematic review. *BMC Neurology* 9: 59. doi:10.1186/1471-2377-9-59.

Lallemand, M.G., and Levy, M. 1860. *L'illustration Journal Universel* 926. (Weekly newspaper, Nov. 24, Paris, France; accompanying an article by M. Leon Loiseau.)

Lambert, D.M. 2007. Allergic contact dermatitis and the endocannabinoid system: From mechanisms to skin care. *ChemMedChem* 2: 1701–1702.

Lambert, D.M., and Fowler, C.J. 2005. The endocannabinoid system: Drug targets, lead compounds, and potential therapeutic applications. *Journal of Medicinal Chemistry* 48: 5059–5087.

Landrigan, P.J., Powell, K.E., James, L.M., and Taylor, P.R. 1983. Paraquat and marijuana: Epidemiologic risk assessment. *American Journal of Public Health* 73: 784–788.

Langenheim, J.H. 1994. Higher plant terpenoids: A phytocentric overview of their ecological roles. *Journal of Chemical Ecology* 20: 1223–1279.

Lanyon, V.S., Turner, J.C., and Mahlberg, P.G. 1981. Quantitative analysis of cannabinoids in the secretory product from capitate-stalked glands of *Cannabis sativa* L. (Cannabaceae). *Botanical Gazette* 142: 316–319.

Lanz, C, Mattsson, J., Soydaner, U., and Brenneisen, R. 2016. Medicinal cannabis: In vitro validation of vaporizers for the smoke-free inhalation of cannabis. *PLoS ONE* 11(1): e0147286. doi:10.1371/journal.pone.0147286.

La Porta, C., Bura, S.A., Negrete, R., and Maldonado, R. 2014. Involvement of the endocannabinoid system in osteoarthritis pain. *European Journal of Neuroscience* 39: 485–500.

Laprairie, R.P., Bagher, A.M., Kelly, M.E.M., and Denovan-Wright, E.M. 2015. Cannabidiol is a negative allosteric modulator of the type 1 cannabinoid receptor. *British Journal of Pharmacology* 172: 4790–4805.

Larramendi, C.H., López-Matas, M.Á., Ferrer, A., Huertas, A.J., Pagán, J.A., Navarro, L.Á. et al. 2013. Prevalence of sensitization to *Cannabis sativa*. Lipid-transfer and thaumatin-like proteins are relevant allergens. *International Archives of Allergy and Immunology* 162: 115–122.

Laskos, J. 1970. The influence of temperature on the energy of germination and the germination of hemp seeds. *Len a Konopi* 8: 29–35. (In Slovak.)

Lata, H., Chandra, S., Khan, I.A., and Elsohly, M.A. 2009. Propagation through alginate encapsulation of axillary buds of *Cannabis sativa* L.—An important medicinal plant. *Physiology and Molecular Biology of Plants* 15: 79–86.

Lata, H., Chandra, S., Khan, I.A., and Elsohly, M.A. 2010a. High frequency plant regeneration from leaf derived callus of high Δ^9-tetrahydrocannabinol yielding *Cannabis sativa* L. *Planta Medica* 76: 1629–1633.

Lata, H., Chandra, S., Khan, I.A., Techen, N., Khan, I.A., and Elsohly, M.A. 2010b. Assessment of the genetic stability of micropropagated plants of *Cannabis sativa* by ISSR markers. *Planta Medica* 76: 97–100.

Lata, H., Chandra, S., Techen, N., Khan, I.A., and ElSohly, M.A. 2011. Molecular analysis of genetic fidelity in *Cannabis sativa* L. plants grown from synthetic (encapsulated) seeds following in vitro storage. *Biotechnology Letters* 33: 2503–2508.

Lata, H., Chandra, S., Mehmedic, Z., Khan, I.A., and ElSohly, M.A. 2012. In vitro germplasm conservation of high Δ^9-tetrahydrocannabinol yielding elite clones of *Cannabis sativa* L. under slow growth conditions. *Acta Physiologiae Plantarum* 34: 743–750.

Latta, R., and Eaton, B. 1975. Seasonal fluctuations in cannabinoid content of Kansas marijuana. *Economic Botany* 29: 153–163.

Lavins, E.S., Lavins, B.D., and Jenkins, A.J. 2004. Cannabis (marijuana) contamination of United States and foreign paper currency. *Journal of Analytical Toxicology* 28: 439–442.

Le Dain, G. (Chair). 1972. *Cannabis. A report of the commission of inquiry into the non-medical use of drugs.* Ottawa, ON: Information Canada. 426 pp.

Ledbetter, M.C., and Krikorian, A.D. 1975. Trichomes of *Cannabis sativa* as viewed with scanning electron microscope. *Phytomorphology* 25: 166–176.

Lee, M.H., and Hancox, R.J. 2011. Effects of smoking cannabis on lung function. *Expert Review of Respiratory Medicine* 5: 537–546.

Lee, M.J., Park, M.S., Hwang, S., Hong, Y.K., Choi, G., Suh, Y.S. et al. 2010. Dietary hempseed meal intake increases body growth and shortens the larval stage via the upregulation of cell growth and sterol levels in *Drosophila melanogaster. Molecules and Cells* 30: 29–36.

Le Foll, B., Trigo, J.M., Sharkey, K.A., and Le Strat, Y. 2013. Cannabis and Δ^9-tetrahydrocannabinol (THC) for weight loss? *Medical Hypotheses* 80: 564–567.

Leggett, T. 2006. A review of the world cannabis situation. *Bulletin on Narcotics* 58: 1–155.

Lehmann, T., Sager, F., and Brenneisen, R. 1997. Excretion of cannabinoids in urine after ingestion of cannabis seed oil. *Journal of Analytical Toxicology* 21: 373–375.

Lehtonen, M., Storvik, M., Tupala, E., Hyytiä, P., Tiihonen, J., and Callaway, J.C. 2010. Endogenous cannabinoids in post-mortem brains of Cloninger type 1 and 2 alcoholics. *European Neuropsychopharmacology* 20: 245–252.

Leizer, C., Ribnicky, D., Poulev, A., Dushenkov, S., and Raskin, I. 2000. The composition of hemp seed oil and its potential as an important source of nutrition. *Journal of Nutraceuticals Functional and Medical Foods* 2: 35–53.

Lemberkovics, E., Veszki, P., Verzar-Petri, G., and Trka, A. 1981. Study on sesquiterpenes of the essential oil in the inflorescence and leaves of *Cannabis sativa* L. var. Mexico. *Scientia Pharmaceutica* 49: 401–408.

Lemeshev, N., Rumyantseva, L., and Clarke, R.C. 1994. Maintenance of *Cannabis* germplasm in the Vavilov Research Institute gene bank—1993. *Journal of the International Hemp Association* 1(1): 1, 3–5.

Lemeshev, N., Rumyantseva, L., and Clarke, R.C. 1995. Maintenance of hemp (*Cannabis sativa*) germplasm accessioned in the Vavilov Research Institute gene bank—1994. *Journal of the International Hemp Association* 2(1): 10–13.

Leson, G. 2001. Evaluating interference of THC in hemp food produces with employee drug testing. In: Nova Institute, corporate ed. *Bioresource hemp and other fibre crops: Proceedings of the symposium* (Wolfsburg, Germany, September 13–16, 2000). Hürth, Germany: Nova Institute. Irregularly paginated.

Leson, G. 2013. Hemp seeds for nutrition. In: Bouloc, P., Allegret, S., and Arnaud, L. eds. *Hemp: Industrial production and uses*. Wallingford, U.K.: CABI, 229–238.

Leson, G., and Pless, P. 2002. Hemp seed and hemp oil. In: Grotenhermen F, and Russo, E. eds. *Cannabis and cannabinoids. Pharmacology, toxicology, and therapeutic potential*. Binghamton, NY: Haworth Integrative Healing Press, 411–425.

Leson, G., Pless, P., and Roulac, J.W. 1999. *Hemp foods and oils for health*. Sebastopol, CA: Hemptech. 62 pp.

Leson, G., Pless, P., Grotenhermen, F., Kalant, H., and Elsohly, M.A. 2001. Evaluating the impact of hemp food consumption on workplace drug tests. *Journal of Analytical Toxicology* 25: 691–698.

Le Strat, Y., and Le Foll, B. 2011. Obesity and cannabis use: Results from 2 representative surveys. *American Journal of Epidemiology* 174: 929–933.

Letniak, R., Weeks, C., Blade, S., and Whiting, A. 2000. *Low THC hemp (Cannabis sativa L.) Research Report 99-10028-R11999*. Hemaruka, AB: Alberta Agriculture and Forestry.

Levin, D.A. 1973. The role of trichomes in plant defence. *Quarterly Review of Biology* 48: 3–15.

Levin, D.A. 2001. The recurrent origin of plant races and species. *Systematic Botany* 26: 197–204.

Levitz, S.M., and Diamond, R.D. 1991. Aspergillosis and marijuana. *Annals of Internal Medicine* 115: 578–579.

Lev-Ran, S., Roerecke, M., Le Foll, B., George, T.P., McKenzie, K., and Rehma, J. 2014. The association between cannabis use and depression: A systematic review and meta-analysis of longitudinal studies. *Psychological Medicine* 44: 797–810.

Leweke, F.M., and Koethe, D. 2008. Cannabis and psychiatric disorders: It is not only addiction. *Addiction Biology* 13: 264–275.

Leweke, F.M., Piomelli, D., Pahlisch, F., Muhl, D., Gerth, C.W., Hoyer, C. et al. 2012. Cannabidiol enhances anandamide signaling and alleviates psychotic symptoms of schizophrenia. *Translational Psychiatry* 2: e94. doi:10.1038/tp.2012.15.

Lewis, D. 1942. The evolution of sex in flowering plants. *Biological Reviews* 17: 46–67.

Lewis, E.B., Card, D.G., and McHargue, J.S. 1948. Tobacco stalks, hemp hurds, and sorghum bagasse as source of cellulose for making high-quality paper. *Kentucky Agricultural Experimental Station Bulletin* 515. Lexington, KY: University of Kentucky. 14 pp.

Lewis, G.S., and Turner, C.E. 1978. Constituents of *Cannabis sativa* L. XIII: Stability of dosage form prepared by impregnating synthetic (–)-delta 9-trans-tetrahydrocannabinol on placebo *Cannabis* plant material. *Journal of Pharmaceutical Sciences* 67: 876–878.

Lewis, W.H., Dixit, A.B., and Wedner, H.J. 1991. Aeropollen of weeds of the western United States Gulf Coast. *Annals of Allergy* 67: 47–52.

Lewis, W.H., and Elvin-Lewis, M.P.F. 2003. *Medical botany: Plants affecting human health*. 2nd edition. Hoboken, N.J.: John Wiley. 832 pp.

Li, H.-L. 1973. An archaeological and historical account of cannabis in China. *Economic Botany* 28: 437–448.

Li, H.-L. 1974. The origin and use of cannabis in eastern Asia: Linguistic-cultural implications. *Economic Botany* 28: 293–301.

Li, M.-C., Brady, J.E., DiMaggio, C.J., Lusardi, A.R., Tzong, K.Y., and Li, G. 2012. Marijuana use and motor vehicle crashes. *Epidemiological Reviews* 34: 65–72.

Li, S.-Y., Stuart, J.D., Li, Y., and Parnas, R.S. 2010. The feasibility of converting *Cannabis sativa* L. oil into biodiesel. *Bioresource Technology* 101: 8457–8460.

Li, X., Wang, S., Du, G., Wu, Z., and Meng, Y. 2013. Variation in physical and mechanical properties of hemp stalk fibers along height of stem. *Industrial Crops and Products* 42: 344–348.

Li, Y., Pickering, K.L., and Farrell, R.L. 2009. Analysis of green hemp fibre reinforced composites using bag retting and white rot fungal treatments. *Industrial Crops and Products* 29: 420–426.

Licata, M., Verrri, P., and Beduschi, G. 2005. Δ⁹-THC content in illicit cannabis products over the period 1997–2004 (first four months). *Annali dell'Istituto Superiore di Sanità* 241: 483–485.

Lin, J.-Y., Zeng, Q.-X., An, Q., Zeng, Q.-Z., Jian, L.-X., and Zhu, Z.-W. 2012. Ultrasonic extraction of hempseed oil. *Journal of Food Process Engineering* 35: 76–90.

Lindemayr, H., and Jager, S. 1980. Occupational immediate type allergy to hemp pollen and hashish. *Dermatosen in Beruf und Umwelt* 28(1): 17–19. (In German.)

Lindholst, C. 2010. Long term stability of cannabis resin and cannabis extracts. *Australian Journal of Forensic Sciences* 42: 181–190.

Linger, P., Müssig, J., Fischer, H., and Kobert, J. 2002. Industrial hemp (*Cannabis sativa* L.) growing on heavy metal contaminated soil: Fibre quality and phytoremediation potential. *Industrial Crops and Products* 16: 33–42.

Linnaeus, C. 1753. *Species plantarum*. Stockholm, Sweden: Laurentius Salvius. 2 volumes.

Lisson, S.N., and Mendham, N.J. 1998. Response of fiber hemp (*Cannabis sativa* L.) to varying irrigation regimes. *Journal of the International Hemp Association* 5(1): 9–15.

Lisson, S.N., and Mendham, N.J. 2000a. The potential for using hemp (*Cannabis sativa*) and flax (*Linum usitatissimum*) fibre as a reinforcing agent in newsprint manufacture. In: Michalk, D.L., and Pratley, J.E. eds. *Agronomy, growing a greener future? Proceedings of the 9th Australian Agronomy Conference* (Wagga, July 20–23, 1998). Wagga, NSW, Australia: Charles Sturt University. http://www.regional.org .au/au/asa/1998/5/203lisson.htm.

Lisson, S.N., and Mendham, N.J. 2000b. Cultivar, sowing date and plant density studies of fiber hemp (*Cannabis sativa* L.) in Tasmania. *Australian Journal of Experimental Agriculture* 40: 975–986.

Lisson, S.N., Mendham, N.J., and Carberry, P.S. 2000a. Development of a hemp (*Cannabis sativa* L.) simulation model. 1. General introduction and the effect of temperature on the pre-emergent development of hemp. *Australian Journal of Experimental Agriculture* 40: 405–411.

Lisson, S.N., Mendham, N.J., and Carberry, P.S. 2000b. Development of a hemp (*Cannabis sativa* L.) simulation model. 2. The flowering response of two hemp cultivars to photoperiod. *Australian Journal of Experimental Agriculture* 40: 413–417.

Lisson, S.N., Mendham, N.J., and Carberry, P.S. 2000c. Development of a hemp (*Cannabis sativa* L.) simulation model. 3. The effect of plant density on leaf appearance, expansion and senescence. *Australian Journal of Experimental Agriculture* 40: 419–423.

Lit, L., Schweitzer, J.B., and Oberbauer, A.M. 2011. Handler beliefs affect scent detection dog outcomes. *Animal Cognition* 14: 387–394.

Liu, M., Fernando, D., Daniel, G., Madsen, B., Meyer, A.S., Ale, M.T. et al. 2015. Effect of harvest time and field retting duration on the chemical composition, morphology and mechanical properties of hemp fibers. *Industrial Crops and Products* 69: 29–39.

Llamas, R., Hart, D.R., and Schneider, N.S. 1978. Allergic bronchopulmonary aspergillosis associated with smoking moldy marihuana. *Chest* 73: 871–872.

Lloyd, E.H., and Seber, D.H. 1996. Bast fiber applications for composites. In: *Thirtieth International particle-board/composite materials symposium*. Pullman, WA: Washington State University, 215–236.

Loeser, C., Zehnsdorf, A., Fussy, M., and Staerk, H.-J. 2002. Conditioning of heavy metal-polluted river sediment by *Cannabis sativa* L. *International Journal of Phytoremediation* 4: 27–45.

Loewe, S. 1946. Studies on the pharmacology and acute toxicity of compounds with marihuana activity. *Journal of Pharmacology and Experimental Therapeutics* 88: 154–161.

Loflin, M., and Earleywine, M. 2014. A new method of cannabis ingestion: The dangers of dabs? *Addictive Behaviors* 39: 1430–1433.

Loh, W.H.T., Hartsel, S.C., and Robertson, L.W. 1983. Tissue culture of *Cannabis sativa* L. and in vitro biotransformation of phenolics. *Zeitschrift für Pflanzen physiologie* 111: 395–400.

Lopez, H.H. 2010. Cannabinoid-hormone interactions in the regulation of motivational processes. *Hormones and Behavior* 58: 100–110.

Lorentzos, M.S., and Webster, R. 2015. Cannabinoids for paediatric epilepsy: Weeding out the issues. *Journal of Paediatrics and Child Health* 51: 476–477.

Lotan, I., Treves, T.A., Roditi, Y., and Djaldetti, R. 2014. Cannabis (medical marijuana) treatment for motor and non-motor symptoms of Parkinson disease: An open-label observational study. *Clinical Neuropharmacology* 37: 41–44.

Lourens, A.C., Viljoen, A.M., and Van Heerden, F.R. 2008. South African *Helichrysum* species: A review of the traditional uses, biological activity and phytochemistry. *Journal of Ethnopharmacology* 119: 630–652.

Lu, N., and Korman, T. 2012. Engineering sustainable construction material: Hemp fiber reinforced composite with recycled high density polyethylene matrix. In: Lu, N., and Korman, T. eds. *ICSDC 2011: Integrating sustainability practices in the construction industry*. Reston, VA: American Society of Civil Engineers, 569–576.

Lucas, P. 2012. Cannabis as an adjunct to or substitute for opiates in the treatment of chronic pain. *Journal of Psychoactive Drugs* 44: 125–133.

Lucchese, C., Venturi, G., Amaducci, M.T., and Lovato, A. 2001. Electrophoretic polymorphism of *Cannabis sativa* L. cultivars: Characterisation and geographical classification. *Seed Science and Technology* 29: 239–248.

Lunger, M. 1999. *Hemp masters: Ancient hippie secrets for knotting hip hemp jewelry*. Liberty, UT: Eagle's View. 100 pp.

Lutge, E.E., Gray, A., and Siegfried, N. 2013. The medical use of cannabis for reducing morbidity and mortality in patients with HIV/AIDS. *Cochrane Database Systematic Review*. doi:10.1002/14651858.CD005175 .pub3. http://onlinelibrary.wiley.com/doi/10.1002/14651858.CD005175.pub3/abstract;jsessionid=3D2E 34926392468EE758BAA97F96ADA8.f03t01.

Lydon, J., Teramura, A.H., and Coffman, C.B. 1987. UV-B radiation effects on photosynthesis, growth and cannabinoid production of two *Cannabis sativa* chemotypes. *Photochemistry and Photobiology* 46: 201–206.

Lynch, M.E., and Campbell, F. 2011. Cannabinoids for treatment of chronic non-cancer pain; a systematic review of randomized trials. *British Journal of Clinical Pharmacology* 72: 735–744.

Lynch, M.E., and Ware, M.A. 2015. Cannabinoids for the treatment of chronic non-cancer pain: An updated systematic review of randomized controlled trials. *Journal of Neuroimmune Pharmacology*. doi:10.1007 /s11481-0159600-6.

Lynch, R.C., Vergara, D., Tittes, S., White, K., Schwartz, C.J., Gibbs, M.J. et al. 2015. Genomic and chemical diversity in *Cannabis*. *BioRxiv*. http://dx.doi.org/10.1101/034314.

Lynn, E.J., Walter, R.G., Harris, L.A., Dendy, R., and James, M. 1972. Nitrous oxide: It's a gas. *Journal of Psychoactive Drugs (Journal of Psychedelic Drugs)* 5: 1–7.

Lynne-Landsman, S.D., Livingston, M.D., and Wagenaar, A.C. 2013. Effects of state medical marijuana laws on adolescent marijuana use. *American Journal of Public Health* 103: 1500–1506.

Maa, E., and Figi, P. 2014. The case for medical marijuana in epilepsy. *Epilepsia* 55: 783–786.

Machado Rocha, F.C., Stéfano, S.C., De Cássia Haiek, R., Rosa Oliveira, L.M., and Da Silveira, D.X. 2008. Therapeutic use of *Cannabis sativa* on chemotherapy-induced nausea and vomiting among cancer patients: Systematic review and meta-analysis. *European Journal of Cancer Care* 17: 431–443.

Macleod, J., and Hickman, M. 2010. How ideology shapes the evidence and the policy: What do we know about cannabis use and what should we do? *Addiction* 105: 1326–1330.

Mahew, H., and Binny, J. 1864. *The criminal prisons of London, and scenes of prison life*. London: Griffin, Bohn, and Company. 634 pp.

Mahlberg, P.G., and Hemphill, J.K. 1983. Effect of light quality on cannabinoid content of *Cannabis sativa* L. (Cannabaceae). *Botanical Gazette* 144: 43–48.

Mahlberg, P.G., and Kim, E.S. 1991. Cuticle development on glandular trichomes of *Cannabis sativa* (Cannabaceae). *American Journal of Botany* 78: 1113–1122.

Mahlberg, P.G., and Kim, E.S. 1992. Secretory vesicle formation in glandular trichomes of *Cannabis sativa* (Cannabaceae). *American Journal of Botany* 79: 166–173.

Mahlberg, P.G., and Kim, E.S. 2004. Accumulation of cannabinoids in glandular trichomes of *Cannabis* (Cannabaceae). *Journal of Industrial Hemp* 9(1): 15–36.

Mahlberg, P.G., Hammond, C.T., Turner, J.C., and Hemphill, J.K. 1984. Structure, development and composition of glandular trichomes of *Cannabis sativa* L. In: Rodriguez, E., Healey, P.L., and Mehta, I. eds. *Biology and chemistry of plant trichomes*. New York: Plenum Press, 23–51.

Maiden, J.H. 1894. Hemp. *Agricultural Gazette of New South Wales* 4: 899–908.

Majmudar, V., Azam, N.A.M., and Finch, T. 2006. Contact urticaria to *Cannabis sativa*. *Contact Dermatitis* 54: 127.

Malfait, A.M., Gallily, R., Sumariwalla, P.F., Malik, A.S., Andreakos, E., Mechoulam, R. et al. 2000. The nonpsychoactive cannabis constituent cannabidiol is an oral anti-arthritic therapeutic in murine collagen-induced arthritis. *Proceedings of the National Academy of Sciences of the United States of America* 97: 9561–9566.

Malingré, T., Hendricks, H., Batterman, S., Bos, R., and Visser, J. 1975. The essential oil of *Cannabis sativa*. *Planta Medica* 28: 56–61.

Malini, T., and Vanithakumari, G. 1990. Rat toxicity studies with β-sitosterol. *Journal of Ethnopharmacology* 28: 221–234.

Mallat, A., Teixeira-Clerc, F., Deveaux, V., Manin, S., and Lotersztajn, S. 2011. The endocannabinoid system as a key mediator during liver diseases: New insights and therapeutic openings. *British Journal of Pharmacology* 163: 1432–1440.

Mallery, M. 2011. Marijuana national forest: Encroachment on California public lands for cannabis cultivation. *Berkeley Undergraduate Journal* 23(2). 51 pp. https://escholarship.org/uc/item/7r10t66s.

Mallet, C., Daulhac, L., Bonnefont, J., Ledent, C., Etienne, M., Chapuy, E. et al. 2008. Endocannabinoid and serotonergic systems are needed for acetaminophen-induced analgesia. *Pain* 139: 190–200.

Mallik, M.K., Singh, U.K., and Ahmad, N. 1990. Batch digester studies on biogas production from *Cannabis sativa*, water hyacinth and crop wastes mixed with dung and poultry litter. *Biological Wastes* 31: 315–319.

Maloney, E., and Brodkey, M. 1940. Hemp pollen sensitivity in Omaha. *Nebraska State Medical Journal* 5: 190–191.

Mandolino, G. 2004. Again on the nature of inheritance of chemotype. Letter to the editor. *Journal of Industrial Hemp* 9(1): 5–7.

Mandolino, G., and Carboni, A. 2004. Potential of marker-assisted selection in hemp genetic improvement. *Euphytica* 140: 107–120.

Mandolino, G., and Ranalli, P. 1998. Advances in biotechnological approaches for hemp breeding and industry. In: Ranalli, P. ed. *Advances in hemp research.* New York: Food Products Press (of Haworth Press), 185–212.

Mandolino, G., and Ranalli, P. 2002. The applications of molecular markers in genetics and breeding of hemp. *Journal of Industrial Hemp* 7(1): 7–23.

Mandolino, G., Faeti, V., Zottini, M., Moschella, A., and Ranalli, P. 1996. Hemp breeding: Biotechnological aspects. *Sementi-Elette* 42(2): 57–60. (In Italian.)

Mandolino, G., Carboni, A., Forapani, S., Faeti, V., and Ranalli, P. 1999. Identification of DNA markers linked to the male sex in dioecious hemp (*Cannabis sativa* L.). *Theoretical and Applied Genetics* 98: 86–92.

Mandolino, G., Carboni, A., Bagatta, M., Cristiana Moliterni, V.M., and Ranalli, P. 2002. Occurrence and frequency of putatively Y chromosome linked DNA markers in *Cannabis sativa* L. *Euphytica* 126: 211–218.

Mandolino, G., Bagatta, M., Carboni, A., Ranalli, P., and de Meijer, E.P.M. 2003. Qualitative and quantitative aspects of the inheritance of chemical phenotype in *Cannabis*. *Journal of Industrial Hemp* 8(2): 52–72.

Man'kowska, G., and Grabowska, L. 2009. Genetic resources of *Cannabis sativa* L. at the Institute of Natural Fibres and Medicinal Plants in Poznan. *Herba Polonica* 55: 178–184.

Marcus, D. 1998. Commercial hemp cultivation in Canada: An economic justification. http://naihc.org/hemp_information/content/dmarcustx.html.

Marcus, D., and Small, E. 2002. Breeding hemp for North America. *Ontario Hemp Report* 2002(5): 4.

Marks, M.A., Chaturvedi, A.K., Kelsey, K., Straif, K., Berthiller, J., Schwartz, S.M. et al. 2014. Association of marijuana smoking with oropharyngeal and oral tongue cancers: Pooled analysis from the INHANCE consortium. *Cancer Epidemiology, Biomarkers & Prevention* 23: 160–171.

Marks, W.H., Florence, L., Lieberman, J., Chapman, P., Howard, D., Roberts, P. et al. 1996. Successfully treated invasive pulmonary aspergillosis associated with smoking marijuana in a renal transplant recipient. *Transplantation* 61: 1771–1774.

Martin, L., Smith, D., and Farmilo, C.G. 1961. Essential oil from fresh *Cannabis sativa* and its use in identification. *Nature* 191(4790): 774–776.

Martín-Sánchez, E., Furukcamplawa, T.A., Taylor, J., and Martin, J.L. 2009. Systematic review and meta-analysis of cannabis treatment for chronic pain. *Pain Medicine* 10: 1353–1368.

Martin-Santos, R., Crippa, J.A., Batalla, A., Bhattacharyya, S., Atakan, Z., Borgwardt, S. et al. 2012. Acute effects of a single, oral dose of d9-tetrahydrocannabinol (THC) and cannabidiol (CBD) administration in healthy volunteers. *Current Pharmacological Design* 18: 4966–4979.

Martyny, J.W., Serrano, K.A., Schaeffer, J.W., and Van Dyke, M.V. 2013. Potential exposures associated with indoor marijuana growing operations. *Journal of Occupational and Environmental Hygiene* 10: 622–639.

Massi, P., Solinas, M., Cinquina, V., and Parolaro, D. 2013. Cannabidiol as potential anticancer drug. *British Journal of Clinical Pharmacology* 75: 303–312.

Matos, I., Bento, A.F., Marcon, R., Claudino, R.F., and Calixto, J.B. 2013. Preventive and therapeutic oral administration of the pentacyclic triterpene α,β-amyrin ameliorates dextran sulfate sodium-induced colitis in mice: The relevance of cannabinoid system. *Molecular Immunology* 54: 482–492.

Matsunaga, T., Watanabe, K., Yoshimura, H., and Yamamoto, I. 1998. Quantitative analysis and pharmacotoxicity of cannabinoids in commercially available cannabis seed. *Yakugaku Zasshi* 118: 408–414. [In Japanese.]

Matthäus, B., and Brühl, L. 2008. Virgin hemp seed oil: An interesting niche product. *European Journal of Lipid Science and Technology* 110: 655–661.

Mayoral, M., Calderón, H., Cano, R., and Lombardero, M. 2008. Allergic rhinoconjunctivitis caused by *Cannabis sativa* pollen. *Journal of Investigational Allergology & Clinical Immunology* 18: 73–74.

McCain, A.H., and Noviello, C. 1985. Biological control of *Cannabis sativa*. In: Delfosse, E.S. ed. *Proceedings VI International Symposium, Biological Control of Weeds* (Vancouver, Canada, August 19–25, 1984). Ottawa, ON: Agriculture Canada, 635–642.

McClure, H.E. 1943. Ecology and management of the mourning dove in Iowa. *Iowa Agricultural Experiment Station Research Bulletin* 310: 353–415.

McGeeney, B.E. 2013. Cannabinoids and hallucinogens for headache. *Headache: The Journal of Head and Face Pain* 53: 447–458.

McLaren, J., Swift, W., Dillon, P., and Allsop, S. 2008. Cannabis potency and contamination: A review of the literature. *Addiction* 103: 1100–1109.

McNaughton, S.J. 1983. Compensatory plant growth as a response to defoliation by gypsy moth larvae. *Oikos* 49: 329–336.

McNeill, J. 1998. Culton: A useful term, questionably argued. *Hortax News* 1(4): 15–22.

McNeill, J. 2004. Nomenclature of cultivated plants: A historical botanical standpoint. In: Davidson, C.G., and Trehane, P. eds. *Fourth international symposium on taxonomy of cultivated plants*. Leuven, Belgium: International Society for Horticultural Science, 29–36.

McNeill, J., Barrie, F.R., Buck, W.R., Demoulin, V., Greuter, W., Hawksworth, D.L. et al. eds. 2012. *International code of nomenclature for algae, fungi, and plants (Melbourne Code)*. Koenigstein, Germany: Koelz Scientific Books. 492 pp. (Regnum Vegetabile 154.) http://www.iapt-taxon.org/nomen/main.php?page=title.

McNulty, S. ed. 1995. *Report to the Governor's hemp and related fiber crops task force*. Frankfort, KE: Commonwealth of Kentucky. Irregularly paginated. Ca. 150 pp.

McPartland, J.M. 1984. Pathogenicity of *Phomopsis ganjae* on *Cannabis sativa* and the fungistatic effect of cannabinoids produced by the host. *Mycopathologia* 87: 149–153.

McPartland, J.M. 1994. Microbiological contaminants of marijuana. *Journal of the International Hemp Association* 1(1): 41–44.

McPartland, J.M. 1996a. A review of *Cannabis* diseases. *Journal of the International Hemp Association* 3(1): 19–23.

McPartland, J.M. 1996b. *Cannabis* pests. *Journal of the International Hemp Association* 3(2): 49, 52–55.

McPartland, J.M. 1997a. *Cannabis* as repellent and pesticide. *Journal of the International Hemp Association* 4(2): 87–92.

McPartland, J.M. 1997b. Diseases and pests of *Cannabis*. In: *Nova Institute, Bioresource hemp—Proceedings of the symposium* (Frankfurt am Main, Germany, February 27–March 2, 1997). Hürth, Germany: Nova Institute, 284–290.

McPartland, J.M. 1998a. A survey of hemp diseases and pests. In: Ranalli, P. ed. *Advances in hemp research*. London: Food Products Press (of Haworth Press), 109–131.

McPartland, J.M. 1998b. Diseases and pests of hemp in Canada. *Commercial Hemp Magazine* 2(5): 33–34.

McPartland, J.M. 2001. Advantages of polypharmaceutical herbal *Cannabis* compared to single ingredient, synthetic tetrahydrocannabinoid. In: Nova Institute, corporate ed. *Bioresource hemp and other fibre crops: Proceedings of the symposium* (Wolfsburg, Germany, September 13–16, 2000). Hürth, Germany: Nova Institute. Irregularly paginated.

McPartland, J.M. 2002. Contaminants and adulterants in herbal cannabis. In: Grotenhermen, F., and Russo, E. eds. Cannabis *and cannabinoids. Pharmacology, toxicology, and therapeutic potential*. Binghamton, NY: Haworth Integrative Healing Press, 337–343.

McPartland, J.M. 2004. Phylogenomic and chemotaxonomic analysis of the endocannabinoid system. *Brain Research Reviews* 45: 18–29.

McPartland, J.M. 2005 [imprinted 2006]. Distribution of endocannabinoids and their enzymes and receptors on the tree of life. In: Onaivi, E.S., Sugiura, T., and di Marzo, V. eds. *Endocannabinoids: The brain and body's marijuana and beyond*. Boca Raton, FL: CRC Press, 517–534.

McPartland, J.M. 2006. Cannabinoid receptors in invertebrates. *Journal of Evolutionary Biology* 19: 366–373.

McPartland, J.M. 2008a. The endocannabinoid system: An osteopathic perspective. *Journal of the American Osteopathic Association* 108: 586–600.

McPartland, J.M. 2008b. Adulteration of cannabis with tobacco, calamus, and other cholinergic compounds. *Cannabinoids* 3: 16–20.

McPartland, J.M. 2009. Obesity, the endocannabinoid system, and bias arising from pharmaceutical sponsorship. *PLOS ONE*. doi:10.1371/journal.pone.0005092.

McPartland, J.M., and Cubeta, M.A. 1997. New species, combinations, host associations and location records of fungi associated with hemp (*Cannabis sativa*). *Mycological Research* 101: 853–857.

McPartland, J.M., and Guy, G.W. 2004a. The evolution of *Cannabis* and coevolution with the cannabinoid receptor—A hypothesis. In: Guy, G.W., Whittle, B.A., and Robson, P.J. eds. *The medicinal uses of* Cannabis *and cannabinoids*. London: Pharmaceutical Press, 71–101.

McPartland, J.M., and Guy, G.W. 2004b. Phylogenetics of the endocannabinoid system. In: Wenger, T. ed. *Recent advances in pharmacology and physiology of cannabinoids*. Trivandrum, Kerala, India: Research Signpost, 21–35.

McPartland, J.M., and Mediavilla, V. 2002. Noncannabinoid components. In: Grotenhermen, F., and Russo, E. eds. Cannabis *and cannabinoids. Pharmacology, toxicology, and therapeutic potential*. Binghamton, NY: Haworth Integrative Healing Press, 401–409.

McPartland, J.M., and Nicholson, J. 2003. Using parasite databases to identify potential nontarget hosts of biological control organisms. *New Zealand Journal of Botany* 41: 699–706.

McPartland, J.M., and Pruitt, P.L. 1997. Medical marijuana and its use by the immunocompromised. *Alternative Therapies in Health and Medicine* 3(3): 39–45.

McPartland, J.M., and Pruitt, P.L. 1999. Side effects of pharmaceuticals not elicited by comparable herbal medicines: The case of tetrahydrocannabinol and marijuana. *Alternative Therapies in Health and Medicine* 5(4): 57–62.

McPartland, J.M., and Russo, E.B. 2001. *Cannabis* and cannabis extracts: Greater than the sum of their parts? *Journal of Cannabis Therapeutics* 1(3–4): 103–132.

McPartland, J.M., and Russo, E.B. 2014. Non-phytocannabinoid constituents of cannabis and herbal synergy. In: Pertwee, R.G. ed. *Handbook of cannabis*. Oxford, U.K.: Oxford University Press, 280–295.

McPartland, J.M., and West, D. 1999. Killing *Cannabis* with mycoherbicides. *Journal of the International Hemp Association* 6(1): 1, 4–8.

McPartland, J.M., Clarke, R.C., and Watson, D.P. 2000. *Hemp diseases and pests: Management and biological control*. Wallingford, Oxon, U.K.: CABI. 251 pp.

McPartland, J.M., Di Marzo, V., De Petrocellis, L., Mercer, A., and Glass, M. 2001. Cannabinoid receptors are absent in insects. *Journal of Comparative Neurology* 436: 423–429.

McPartland, J.M., Agraval, J., Gleeson, D., Heasman, K., and Glass, M. 2006. Cannabinoid receptors in invertebrates. *Journal of Evolutionary Biology* 19: 366–373.

McPartland, J.M., Norris, R.W., and Kilpatrick, C.W. 2007a. Coevolution between cannabinoid receptors and endocannabinoid ligands. *Gene* 397: 126–135.

McPartland, J.M., Norris, R.W., and Kilpatrick, C.W. 2007b. Tempo and mode in the endocannaboinoid system. *Journal of Molecular Evolution* 65: 267–276.

McPartland, J.M., Guy, G.W., and Di Marzo, V. 2014. Care and feeding of the endocannabinoid system: A systematic review of potential clinical interventions that upregulate the endocannabinoid system. *PLoS ONE* 9(3): e89566. doi:10.1371/journal.pone.0089566.

McPartland, J.M., Duncan, M., Di Marzo, V., and Pertwee, R.G. 2015. Are cannabidiol and Δ^9-tetrahydrocannabivarin negative modulators of the endocannabinoid system? A systematic review. *British Journal of Pharmacology* 172: 737–753.

McPhee, H.C. 1925 [published 1926]. The genetics of sex in hemp. *Journal of Agricultural Research* 31: 935–943.

McRae-Clark, A.L., Carter, R.E., Killeen, T.K., Carpenter, M.J., White, K.G., and Brady, K.T. 2010. A placebo-controlled trial of atomoxetine in marijuana-dependent individuals with attention deficit hyperactivity disorder. *American Journal on Addictions* 19: 481–489.

Mead, A.P. 2014. International control of cannabis. In: Pertwee, R.G. ed. *Handbook of cannabis*. Oxford, U.K.: Oxford University Press, 44–64.

Mechoulam, R. 1986. The pharmacohistory of *Cannabis sativa*. In: Mechoulam, R. ed. *Cannabinoids as therapeutic agents*. Boca Raton, FL: CRC Press, 1–19.

Mechoulam, R. 2002. Discovery of endocannabinoids and some random thoughts on their possible roles in neuroprotection and aggression. *Prostaglandins, Leukotrienes, and Essential Fatty Acids* 66: 93–99.

Mechoulam, R. 2004. Preface. In: Wenger, T. ed. *Recent advances in pharmacology and physiology of cannabinoids*. Trivandrum, Kerala, India: Research Signpost. (Unpaginated.)

Mechoulam, R. 2012. Cannabis—A valuable drug that deserves better treatment. *Mayo Clinic Proceedings* 87(2): 107–109.

Mechoulam, R., and Ben-Shabat, S. 1999. From gan-zi-gun-nu to anandamide and 2arachidonoylglycerol: The ongoing story of cannabis. *Natural Products Report* 16: 131–143.

Mechoulam, R., and Gaoni, Y. 1967. Recent advances in the chemistry of hashish. *Fortschritte der Chemie Organischer Naturstoffe* 25: 175–213.

Mechoulam, R., and Parker, L.A. 2013a. The endocannabinoid system and the brain. *Annual Review of Psychology* 64: 21–47.

Mechoulam, R., and Parker, L.A. 2013b. Towards a better cannabis drug. *British Journal of Pharmacology* 170: 1363–1364.

Mechoulam, R., Devane, W.A., Breuer, A., and Zahalka, J. 1991. A random walk through a cannabis field. *Pharmacology Biochemistry & Behavior* 40: 461–464.

Mechoulam, R., Peters, M., Murillo-Rodriguez, E., and Hanus, L.O. 2007. Cannabidiol—Recent advances. *Chemistry and Biodiversity* 4: 1678–1692.

Mechoulam, R., Hanuš, L.O., Pertwee, R., and Howlett, A.C. 2014. Early phytocannabinoid chemistry to endocannabinoids and beyond. *Nature Reviews Neuroscience* 15: 757–764.

Mechtler, K., Bailer, J., and de Hueber, K. 2004. Variations of Δ^9-THC content in single plants of hemp varieties. *Industrial Crops and Products* 19: 19–24.

Mediavilla, V. 1998. The production of essential hemp oil in Switzerland. In: *Proceedings of the hemp, flax and other bast fibrous plants production, technology and ecology symposium* (Poznan, Poland, September 24–25, 1998). Europe: FAO European Cooperative Research Network on Flax and Other Bast Plants, 117–118.

Mediavilla, V., and Steinemann, S. 1997. Essential oil of *Cannabis sativa* L. strains. *Journal of the International Hemp Association* 4(2): 80–82.

Mediavilla, V., Jonquera, M., Schmid-Slembrouck, I., and Soldati, A. 1998. Decimal code for growth stages of hemp (*Cannabis sativa* L.). *Journal of the International Hemp Association* 5(2): 65, 68–72.

Mediavilla, V., Bassetti, P., and Leupin, M. 1999. Agronomic characteristics of some hemp genotypes. *Journal of the International Hemp Association* 6(2): 45, 48–53.

Mediavilla, V., Leupin, M., and Keller, A. 2001. Influence of the growth stage of industrial hemp on the yield formation in relation to certain fibre quality traits. *Industrial Crops and Products* 13: 49–56.

Mehmedic, Z., Chandra, S., Slade, D., Denham, H., Foster, S., Patel, A.S. et al. 2010. Potency trends of Δ^9-THC and other cannabinoids in confiscated cannabis preparations from 1993 to 2008. *Journal of Forensic Sciences* 55: 1209–1217.

Meier, C., and Mediavilla, V. 1998. Factors influencing the yield and the quality of hemp (*Cannabis sativa* L.) essential oil. *Journal of the International Hemp Association* 5(1): 16–20.

Meier, M.H., Hill, M.L., Small, P.J., and Luthar, S.S. 2015. Associations of adolescent cannabis use with academic performance and mental health: A longitudinal study of upper middle class youth. *Drug and Alcohol Dependence* 156: 207–212.

Meijer, J.G. 1904. Structuur van de Inflorescentieschubben van *Cannabis sativa*. http://www.geheugenvanned erland.nl/?/en/items/SAE01:2074.

Meijer, W.J.M., Vanderwerf, H.M.G., Mathijssen, E.W.J.M., and Vandenbrink, P.W.M. 1995. Constraints to dry-matter production in fiber hemp. *European Journal of Agronomy* 4: 109–117.

Melberg, H.O., Jones, A.M., and Bretteville-Jensen, A.L. 2010. Is cannabis a gateway to hard drugs? *Empirical Economics* 38: 583–603.

Mendizábal, V.E., and Adler-Graschinsky, E. 2007. Cannabinoids as therapeutic agents in cardiovascular disease: A tale of passions and illusions. *British Journal of Pharmacology* 151: 427–440.

Menzel, M.Y. 1964. Meiotic chromosomes of monoecious Kentucky hemp (*Cannabis sativa*). *Bulletin of the Torrey Botanical Club* 91: 193–205.

Meola, S.D., Tearney, C.C., Haas, S.A., Hackett, T.B., and Mazzaferro, E.M. 2012. Evaluation of trends in marijuana toxicosis in dogs living in a state with legalized medical marijuana: 125 dogs (2005–2010). *Journal of Veterinary Emergency and Critical Care* 22: 690–696.

Merlin, M.D. 1972. *Man and marijuana: Some aspects of their ancient relationship.* Cranbury, NJ: Associated University Presses. 120 pp.

Metz, T.D., and Stickrath, E.H. 2015. Marijuana use in pregnancy and lactation: A review of the evidence. *American Journal of Obstetrics and Gynecology* 213: 761–778.

Micale, V., Tabiova, K., Kucerova, J., and Drago, F. 2015. Role of the endocannabinoid system in depression: From preclinical to clinical evidence. In: Campolongo, P., and Fattore, L. eds. *Cannabinoid modulation of emotion, memory, and motivation.* New York: Springer, 97–129.

Miettinen, T.A., and Gylling, H. 2004. Plant stanol and sterol esters in prevention of cardiovascular diseases. *Annals of Medicine* 36: 126–134.

Migalj, N.D. 1969. Morphology of hemp (*Cannabis* L.) pollen. *Botanicheskiy Zhurnal (Leningrad)* 54: 274–276. (In Russian.)

Mighell, D.J., Corval, D., and Sexton, M. 2013. A proposal to establish a comprehensive cannabis varietal registry in Washington State. http://visionwashington.files.wordpress.com/2013/06/public_proposal_cvr_06162013.pdf.

Mignoni, G. 1997–1998 [Double issue.]. *Cannabis* as a licit crop: Recent developments in Europe. *Bulletin on Narcotics* 49(1–2) [1997], 50(1–2) [1998]: 23–44.

Mihoc, M., Pop, G., Alexa, E., and Radulov, I. 2012. Nutritive quality of Romanian hemp varieties (*Cannabis sativa* L.) with special focus on oil and metal contents of seeds. *Chemistry Central Journal* 6(1): 122. doi:10.1186/1752-153X-6-122.

Mikuriya, T.H. 1969. Marijuana in medicine: Past, present and future. *California Medicine* 110(1): 34–40.

Miles, J.H. 2011. Autism spectrum disorders—A genetics review. *Genetics in Medicine* 13: 278–294.

Miller, N.G. 1970. The genera of the Cannabaceae in the Southeastern United States. *Journal of the Arnold Arboretum* 51: 185–203.

Mills, E. 2012. The carbon footprint of indoor *Cannabis* production. *Energy Policy* 46: 58–67.

Minelli, A. 1993. *Biological systematics: The state of the art*. London: Chapman & Hall. 387 pp.

Minozzi, S., Davoli, M., Bargagli, A.M., Amato, L., Vecchi, S., and Perucci, C.A. 2010. An overview of systematic reviews on cannabis and psychosis: Discussing apparently conflicting results. *Drug and Alcohol Review* 29: 304–317.

Miron, M., and Waldock, K. 2010. *The budgetary impact of ending drug prohibition*. Washington, DC: Cato Institute. 54 pp. http://www.cato.org/publications/white-paper/budgetary-impact-ending-drug-prohibition.

Misra, S.B., and Dixit, S.N. 1979. Antifungal activity of leaf extracts of some higher plants. *Acta Botanica Indica* 7: 147–150.

Mittleman, M.A., Lewis, R.A., Maclure, M., Sherwood, J.B., and Muller, J.E. 2001. Triggering myocardial infarction by marijuana. *Circulation* 103: 2805–2809.

Moes, J. 1998. Hemp research in Manitoba—1995–1997. In: Blade, S. ed. *Alberta hemp symposia proceedings* (Red Deer, Alberta, March 10, 1998, and Edmonton, Alberta, April 8, 1998). Edmonton, AB: Food and Rural Development, 43–48.

Mohan Ram, H.Y., and Jaiswal, V.S. 1970. Induction of female flowers on male plants of *Cannabis sativa* by 2-chlorothane phosphonic acid. *Experimentia* 26: 214–216.

Mohan Ram, H.Y., and Jaiswal, V.S. 1972. Induction of male flowers on female plants of *Cannabis sativa* by gibberellins and its inhibition by abscisic acid. *Planta (Berlin)* 105: 263–266.

Mohan Ram, H.Y., and Sett, R. 1982a. Modification of growth and sex expression in *Cannabis sativa* by aminoethoxyvinylglycine and ethephon. *Zeitschrift für Pflanzenphysiologie* 105: 165–172.

Mohan Ram, H.Y., and Sett, R. 1982b. Induction of fertile male flowers in genetically female *Cannabis sativa* plants by silver nitrate and silver thiosulphate anionic complex. *Theoretical and Applied Genetics* 62: 369–375.

Mohanty, A.K., Misra, M., and Hinrichsen, G. 2000. Biofibres, biodegradable polymers and biocomposites: An overview. *Macromolecular Materials and Engineering* 276–277: 1–24.

Moir, D., Rickert, W.S., Levasseur, G., Larose, Y., Maertens, R., White, P. et al. 2008. A comparison of mainstream and sidestream marijuana and tobacco cigarette smoke produced under two machine smoking conditions. *Chemical Research in Toxicology* 21: 494–502.

Mölleken, H., and Husmann, H. 1997. Cannabinoids in seed extracts of *Cannabis sativa* cultivars. *Journal of the International Hemp Association* 4(2): 73, 76–79.

Mölleken, H., and Theimer, R.R. 1997. Survey of minor fatty acids in *Cannabis sativa* L. fruits of various origins. *Journal of the International Hemp Association* 4(1): 13–17.

Mölleken, H., Oswald, A., and Theimer, R.R. 1997. In vitro studies of the cultivation of various hemp varieties on heavy metal polluted soils and the evidence for phytochelatins in hemp. In: Nova Institute, corporate ed. *Bioresource hemp—Proceedings of the symposium* (Frankfurt am Main, Germany, February 27–March 2, 1997). Hürth, Germany: Nova Institute. 265–276. (In German.)

Mölleken, H., Mothes, R., and Dudek, S. 2001. Quality of hemp fruits and hemp oil in relation to the maturity of the fruits. In: Nova Institute, corporate ed. *Bioresource hemp and other fibre crops: Proceedings of the symposium* (Wolfsburg, Germany, September 13–16, 2000). Hürth, Germany: Nova Institute. Irregularly paginated.

Moninckx, J. 1682–1709. *Moninckx atlas, Vol. 2*. Amsterdam, the Netherlands.

Moore, T.H., Zammit, S., Lingford-Hughes, A., Barnes, T.R., Jones, P.B., Burke, M. et al. 2007. Cannabis use and risk of psychotic or affective mental health outcomes: A systematic review. *Lancet* 370: 319–328.

Monteleone, A.M., Di Marzo, V., Aveta, T., Piscitelli, F., Grave, R.D., Scognamiglio, P. et al. 2015. Deranged endocannabinoid responses to hedonic eating in underweight and recently weight-restored patients with anorexia nervosa. *American Journal of Clinical Nutrition* 101: 262–269.

Montford, S., and Small, E. 1999a. Measuring harm and benefit: The biodiversity friendliness of *Cannabis sativa*. *Global Biodiversity* 8(4): 2–13.

Montford, S., and Small, E. 1999b. A comparison of the biodiversity friendliness of crops with special reference to hemp (*Cannabis sativa*). *Journal of the International Hemp Association* 6(2): 53–63.

Montgomery, B. 1954. The bast fibers. In: Mauersberger, H.R. ed. *Matthews' textile fibers*. 6th edition. New York: Wiley, 257–359.

Morgan, C.J., Freeman, T.P., Schafer, G.L., and Curran, H.V. 2010. Cannabidiol attenuates the appetitive effects of Δ^9-tetrahydrocannabinol in humans smoking their chosen cannabis. *Neuropsychopharmacology* 35: 1879–1885.

Morral, A.R., McCafrey, D., and Paddock, S.M. 2002. Reassessing the marijuana gateway effect. *Addiction* 97: 1493–1504.

Morris, R.G., TenEyck, M., Barnes, J.C., and Kovandzic, T.V. 2014. The effect of medical marijuana laws on crime: Evidence from state panel data, 1990–2006. *PLoS ONE* 9(3): e92816. doi:10.1371/journal.pone.0092816.

Morrison, C., Gruenewald, P.J., Freisthler, B., Ponicki, W.R., and Remer, L.G. 2014. The economic geography of medical cannabis dispensaries in California. *International Journal of Drug Policy* 25: 508–515.

Morrow, R.C. 2008. LED lighting in horticulture. *HortScience* 43: 1947–1950.

Muhammad, I., Li, X.C., Dunbar, D.C., ElSohly, M.A., and Khan, I.A. 2001. Antimalarial (+)-trans-hexahydrodibenzopyran derivatives from *Machaerium multiflorum*. *Journal of Natural Products* 64: 1322–1325.

Mukamal, K.J., Maclure, M., Muller, J.E., and Mittleman, M.A. 2008. An exploratory prospective study of marijuana use and mortality following acute myocardial infarction. *American Heart Journal* 155: 465–470.

Mukherjee, A., Roy, S.C., De Bera, S., Jiang, H.E., Li, X., Li, C.S. et al. 2008. Results of molecular analysis of an archaeological hemp (*Cannabis sativa* L.) DNA sample from North West China. *Genetic Resources and Crop Evolution* 55: 481–485.

Mukhtar, T., Kayani, M.Z., and Hussain, M.A. 2013. Nematicidal activities of *Cannabis sativa* L. and *Zanthoxylum alatum* Roxb. against *Meloidogyne incognita*. *Industrial Crops and Products* 42: 447–453.

Müller-Vahl, K.R. 2013. Treatment of Tourette syndrome with cannabinoids. *Behavioural Neurology* 27: 119–124.

Müller-Vahl, K.R., and Emrich, H.M. 2008. Cannabis and schizophrenia: Towards a cannabinoid hypothesis of schizophrenia. *Expert Review of Neurotherapeutics* 8: 1037–1048.

Murari, G., Puccini, A.M., Sanctis, R. de, and Lombardi, S. 1983. Influence of environmental conditions on tetrahydrocannabinol (delta-9-THC) in different cultivars on *Cannabis sativa* L. *Fitoterapia* 54: 195–202.

Murillo-Rodríguez, E., Aguilar-Torton, L., Mijangos-Moreno, S., Sarro-Ramírez, A., and Arias-Carrión, Ó. 2014. Phytocannabinoids as novel therapeutic agents for sleep disorders. In: Pertwee, R.G. ed. *Handbook of cannabis*. Oxford, U.K.: Oxford University Press, 538–546.

Müssig, J. 2003. Quality aspects in hemp fiber production—Influence of cultivation, harvesting and retting. *Journal of Industrial Hemp* 8(1): 11–32.

Mustafa, A.F., McKinnon, J.J., and Christensen, D.A. 1999. The nutritive value of hemp meal for ruminants. *Canadian Journal of Animal Science* 79: 91–95.

Myles, P.S., Leslie, K., Chan, M.T., Forbes, A., Paech, M.J., Peyton, P. et al. 2007. Avoidance of nitrous oxide for patients undergoing major surgery: A randomized controlled trial. *Anesthesiology* 107: 221–231.

Myrstol, B.A., and Brandeis, J.D. 2012. *The predictive validity of marijuana odor detection: An examination of Alaska state trooper case reports 2006–2010*. Anchorage, AK: Justice Center, University of Alaska. 53 pp.

Naftali, T., Mechulam, R., Lev, L.B., and Konikoff, F.M. 2014. Cannabis for inflammatory bowel disease. *Digestive Diseases* 32: 468–474.

Naftali, T., Lev, L.B., Yablecovitch, D., Half, E., and Konikoff, F.M. 2011. Treatment of Crohn's disease with cannabis: An observational study. *Israel Medical Association Journal* 13: 455–458.

Naftali, T., Bar-Lev Schleider, L., Dotan, I., Lansky, E.P., Benjaminov, F.S., and Konikoff, F.M. 2013. Cannabis induces a clinical response in patients with Crohn's disease: A prospective placebo-controlled study. *Clinical Gastroenterology and Hepatology* 11: 1276–1280.

Nahas, G.G. 1982. Hashish in Islam 9th to 18th century. *Bulletin of the New York Academy of Medicine* 58: 814–831.

Narayanaswami, K., Golani, H.C., Bami, H.L., and Dau, R.D. 1978. Stability of *Cannabis sativa* L. samples and their extracts, on prolonged storage in Delhi. *Bulletin on Narcotics* 30(4): 57–69.

Nash, J., Haghe, L., and Roberts, D. 1852 ["1854"]. *Dickinson's comprehensive pictures of the Great Exhibition of 1851*. London: Dickinson Brothers. 230 pp.

Nathan, D. 2013. Why marijuana should be legal for adults. CNN. http://www.cnn.com/2013/01/09/opinion/nathan-legal-marijuana/.

National Cancer Institute. 2015. Cannabis and cannabinoids–for health professionals (PDQ®). http://www.cancer.gov/about-cancer/treatment/cam/hp/cannabis-pdq.

Nature. 1933. Discovery of sexuality in plants. *Nature* 131: 392.

Nature. 2001. Gathering the evidence on medical marijuana. *Nature* 410: 613.

Nayak, A.P., Green, B.J., Sussman, G., Berlin, N., Lata, H., Chandra, S. et al. 2013. Characterization of *Cannabis sativa* allergens. *Annals of Allergy, Asthma & Immunology* 111: 32–37.

Neavyn, M.J., Blohm, E., Babu, K.M., and Bird, S.B. 2014. Medical marijuana and driving: A review. *Journal of Medical Toxicology* 10: 269–279.

Neff, G.W., O'Brien, C.B., Reddy, K.R., Bergasa, N.V., Regev, A., Molina, E. et al. 2002. Preliminary observation with dronabinol in patients with intractable pruritus secondary to cholestatic liver disease. *American Journal of Gastroenterology* 97: 2117–2119.

Nerio, L.S., Olivero-Verbel, J., and Stashenko, E. 2010. Repellent activity of essential oils: A review. *Bioresource Technology* 101: 372–378.

Nesom, G.L. 2011. Toward consistency of taxonomic rank in wild/domesticated Cucurbitaceae. *Phytoneuron* 13: 1–33.

Neumeister, A., Normandin, M.D., Pietrzak, R.H., Piomelli, D., Zheng, M.Q., Gujarro-Anton, A. et al. 2013. Elevated brain cannabinoid CB1 receptor availability in post-traumatic stress disorder: A positron emission tomography study. *Molecular Psychiatry* 18: 1034–1040.

New England Journal of Medicine. 2013. Medicinal use of marijuana. 368: 866–868. doi:10.1056/NEJMclde1300970.

New York Times Editorial Board. 2014. Repeal prohibition, again. http://www.nytimes.com/interactive/2014/07/27/opinion/sunday/high-time-marijuana-legalization.html?_r=0.

Nguyen, B.M., Kim, D., Bricker, S., Bongard, F., Neville, A., Putnam, B. et al. 2014. Effect of marijuana use on outcomes in traumatic brain injury. *American Surgeon* 80: 979–983.

Nicholson, A.N., Turner, C., Stone, B.M., and Robson, P.J. 2004. Effect of Δ^9-tetrahydrocannabinol and cannabidiol on nocturnal sleep and early-morning behavior in young adults. *Journal of Clinical Psychopharmacology* 24: 305–313.

Nicolson, S.E., Denysenko, L., Mulcare, J.L., Vito, J.P., and Chabon, B. 2012. Cannabinoid hyperemesis syndrome: A case series and review of previous reports. *Psychosomatics* 53: 212–329.

Nigam, M.C., Handa, K.L., Nigam, J.C., and Levi, L. 1965. Essential oils and their constituents. XXIX. The essential oil of marihuana: Composition of genuine Indian *Cannabis sativa* L. *Canadian Journal of Chemistry* 43: 3372–3376.

Nigam, R.K., Varkey, M., and Reuben, D.E. 1981. Irradiation induced changes in flower formation in *Cannabis sativa* L. *Biologia Plantarum (Praha)* 23: 389–391.

Nikvash, N, Kraft, R., Kharazipour, A., and Euring, M. 2010. Comparative properties of bagasse, canola, and hemp particle boards. *European Journal of Wood and Wood Products* 68: 323–327.

Nissen, L., Zatta, A., Stefanini, I., Grandi, S., Sgorbati, B., Biavati, B. et al. 2009. Characterization and antimicrobial activity of essential oils of industrial hemp varieties (*Cannabis sativa* L.). *Fitoterapia* 81: 413–419.

Nobel, P.S. 1976. Photosynthetic rates of sun versus shade leaves of *Hyptis emoryi* Torr. *Plant Physiology* 58: 218–223.

Norberg, M.M., Battisti, R.A., Copeland, J., Hermens, D.F., and Hickie, I.B. 2012. Two sides of the same coin: Cannabis dependence and mental health problems in help-seeking adolescent and young adult outpatients. *International Journal of Mental Health and Addiction* 10: 818–828.

Notcutt, W. 2004. Cannabis in the treatment of chronic pain. In: Guy, G.W., Whittle, B.A., and Robson, P.J. eds. *The medicinal uses of* Cannabis *and cannabinoids.* London: Pharmaceutical Press, 271–299.

Novak, J., and Franz, C. 2003. Composition of the essential oils and extracts of two populations of *Cannabis sativa* L. ssp. *spontanea* from Austria. *Journal of Essential Oil Research* 15: 158–160.

Novak, J., Zitterl-Eglseer, K., Deans, S.G., and Franz, C.M. 2001. Essential oils of different cultivars of *Cannabis sativa* L. and their antimicrobial activity. *Flavour and Fragrance Journal* 16: 259–262.

Novotna, A., Mares, J., Ratcliffe, S,, Novakova, I., Vachova, M., Zapletalova, O. et al. 2011. A randomized, double-blind, placebo-controlled, parallel-group, enriched-design study of nabiximols (Sativex(®)), as add-on therapy, in subjects with refractory spasticity caused by multiple sclerosis. *European Journal of Neurology* 18: 1122–1131.

Nucci, C., Bari, M., Spanò, M., Corasaniti, M.T., Bagetta, G., Maccarrone, M. et al. 2008. Potential roles of (endo)cannabinoids in the treatment of glaucoma: From intraocular pressure control to neuroprotection. *Progress in Brain Research* 173: 451–464.

Nutt, D.J., King, L.A., and Phillips, L.D. 2010. Drug harms in the UK: A multi-criteria decision analysis. *Lancet* 376: 1558–1565.

Ocampo, T.L., and Rans, T.S. 2015. *Cannabis sativa*: The unconventional "weed" allergen. *Annals of Allergy, Asthma & Immunology* 14: 187–192.

Oerke, E.C. 2006. Crop losses due to pests. *Journal of Agricultural Science* 144: 31–43.

Okosun, R.E., Osadolor, H.B., Uso, O., and Adu, E.M. 2014. Serum testosterone levels in Nigerian male marijuana and cigarette smokers. *Journal of Medicine and Biomedical Research* 13: 93–98.

Oláh, A., Tóth, B.I., Borbíró, I., Sugawara, K., Szöllősi, A.G., Czifra, G. et al. 2014. Cannabidiol exerts sebostatic and anti-inflammatory effects on human sebocytes. *Journal of Clinical Investigation* 124: 3713–3724.

Olsen, J.K. 2004. *An information paper on industrial hemp (industrial cannabis).* Australia: Department of Primary Industries and Fisheries, Queensland Government. 24 pp. http://www.420magazine.com/forums/hemp-facts-information/80195-information-paper-industrial-hemp-industrial-cannabis.html.

Olson, D. 1997. Hemp culture in Japan. *Journal of the International Hemp Association* 4(2): 40–50.

Onaivi, E.S., Sugiura, T., and di Marzo, V. eds. 2005a [imprinted 2006]. *Endocannabinoids: The brain and body's marijuana and beyond.* Boca Raton, FL: CRC Press. 563 pp.

Onaivi, E.S., Ishiguro, H., Zhang, P.W., Lin, Z., Akinshola, B.E., Leonard, C.M. et al. 2005b [imprinted 2006]. Endocannabinoid receptor genetics and marijuana use. In: Onaivi, E.S., Sugiura, T., and di Marzo, V. eds. *Endocannabinoids: The brain and body's marijuana and beyond.* Boca Raton, FL: CRC Press, 57–118.

Onaivi, E.S., Benno, R., Halpern, T., Mehanovic, M., Schanz, N., Sanders, C. et al. 2011. Consequences of cannabinoid and monoaminergic system disruption in a mouse model of autism spectrum disorders. *Current Neuropharmacology* 9: 209–214.

Oner, S.T. 2011. Cannabis indica: *The essential guide to the world's finest marijuana strains.* Volume 1. San Francisco: Green Candy Press. 280 pp.

Oner, S.T. 2012a. Cannabis indica: *The essential guide to the world's finest marijuana strains.* Volume 2. San Francisco: Green Candy Press. 208 pp.

Oner, S.T. 2012b. Cannabis sativa: *The essential guide to the world's finest marijuana strains.* San Francisco: Green Candy Press. 228 pp.

Oner, S.T. 2013a. Cannabis indica: *The essential guide to the world's finest marijuana strains.* Volume 3. San Francisco: Green Candy Press. 208 pp.

Oner, S.T. 2013b. Cannabis sativa: *The essential guide to the world's finest marijuana strains.* Volume 3. San Francisco: Green Candy Press. 228 pp.

Oner, S.T. 2014. Cannabis sativa: *The essential guide to the world's finest marijuana strains.* Volume 3. San Francisco: Green Candy Press. 208 pp.

Onofri, C., de Meijer, E.P.M., and Mandolino, G. 2015. Sequence heterogeneity of cannabidiolic- and tetrahydrocannabinolic acid-synthase in *Cannabis sativa* L. and its relationship with chemical phenotype. *Phytochemistry* 116: 57–68.

Oomah, B.D., Busson, M., Godfrey, D.V., and Drover, J.C.G. 2002. Characteristics of hemp (*Cannabis sativa* L.) seed oil. *Food Chemistry* 76: 33–43.

Oost, E.H. 1989. Typification of *Brassica oleracea* L. (Cruciferae) and its Linnaean varieties. *Botanical Journal of the Linnean Society* 101: 329–345.

Oreja-Guevara, C. 2012. Treatment of spasticity in multiple sclerosis: New perspectives regarding the use of cannabinoids. *Revista de Neurologia* 55: 421–30. (In Spanish.)

Orr, J., and Starodub, M.E. 1999. *Industrial hemp risk assessment.* Ottawa, ON: Product Safety Bureau, Health Canada. 199 pp.

Osler, W., and McCrae, T. 1915. *The principles and practice of medicine: Designed for the use of practitioners and students of medicine.* New York: Appleton and Company. 1266 pp.

Osman, A., Thorpe, J.W., and Caddy, B. 1985. Comparison of *Cannabis* samples from different origins by the headspace technique and an assessment of chromatographic traces using the *r*-matrix. *Journal of the Forensic Society* 25: 427–433.

Owen, K.P., Sutter, M.E., and Albertson, T.E. 2014. Marijuana: Respiratory tract effects. *Clinical Reviews in Allergy & Immunology* 46: 65–81.

Pacher, P. 2013. Towards the use of non-psychoactive cannabinoids for prostate cancer. *British Journal of Pharmacology* 168: 76–78.

Pacher, P., and Kunos, G. 2013. Modulating the endocannabinoid system in human health and disease—successes and failures. *FEBS Journal* 280: 1918–1943.

Pacher, P., and Mechoulam, R. 2011. Is lipid signaling through cannabinoid 2 receptors part of a protective system? *Progress in Lipid Research* 50: 193–211.

Pacher, P., Bátkai, S., and Kunos, G. 2006. The endocannabinoid system as an emerging target of pharmacotherapy. *Pharmacological Reviews* 58: 389–462.

Pacifico, D., Miselli, F., Micheler, M., Carboni, A., Ranalli, P., and Mandolino, G. 2006. Genetics and marker-assisted selection of the chemotype in *Cannabis sativa* L. *Molecular Breeding* 17: 257–268.

Pacifico, D., Miselli, F., Carboni, A., Moschella, A., and Mandolino, G. 2008. Time course of cannabinoid accumulation and chemotype development during the growth of *Cannabis sativa* L. *Euphytica* 160: 231–240.

Pacioni, G., Rapino, C., Zarivi, O., Falconi, A., Leonardi, M., Battista, N. et al. 2015. Truffles contain endocannabinoid metabolic enzymes and anandamide. *Phytochemistry* 110: 104–110.

Padial, J.M., and de la Riva, I. 2006. Taxonomic inflation and the stability of species lists: The perils of ostrich's behavior. *Systematic Biology* 55: 859–867.

Paduch, R., Kandefer-Szerszeń, M., Trytek, M., and Fiedurek, J. 2007. Terpenes: Substances useful in human healthcare. *Archivum Immunologiae et Therapiae Experimentalis* 55: 315–327.

Page, S.A., and Verhoef, M.J. 2006. Medicinal marijuana use: Experiences of people with multiple sclerosis. *Canadian Family Physician* 52: 64–65.

Pakarinen, A., Zhang, J., Brock, T., Maijala, P., and Viikari, L. 2012. Enzymatic accessibility of fiber hemp is enhanced by enzymatic or chemical removal of pectin. *Bioresource Technology* 107: 275–281.

Pamplona, F.A., and Takahashi, R.N. 2012. Psychopharmacology of the endocannabinoids: Far beyond anandamide. *Journal of Psychopharmacology* 26: 7–22.

Pandey, K.N. 1982. Antifungal activity of some medicinal plants on stored seeds of *Eleusine coracana*. *Journal of Indian Phytopathology* 35: 499–501.

Panlilio, L.V., Goldberg, S.R., and Justinova, Z. 2015. Cannabinoid abuse and addiction: Clinical and preclinical findings. *Clinical Pharmacology and Therapeutics* 97: 616–627.

Panza, F., Frisardi, V., Capurso, C., D'Introno, A., Colacicco, A.M., Di Palo, A. et al. 2009. Polyunsaturated fatty acid and *S*-adenosylmethionine supplementation in predementia syndromes and Alzheimer's disease: A review. *Scientific World Journal* 9: 373–389.

Parfieniuk, A., and Flisiak, R. 2008. Role of cannabinoids in chronic liver diseases. *World Journal of Gastroenterology* 14: 6109–6114.

Parihar, S.S., Dadlani, M., Lal, S.K., Tonapi, VA., Nautiyal, P.C., and Basu, S. 2014. Effect of seed moisture content and storage temperature on seed longevity of hemp (*Cannabis sativa*). *Indian Journal of Agricultural Sciences*, 84: 1303–1309.

Paris, M., Boucher, F., and Cosson, L. 1975. The constituents of *Cannabis sativa* pollen. *Economic Botany* 29: 245–253.

Parker, L.A. Rock, E.M., and Limebeer, C.L. 2011. Regulation of nausea and vomiting by cannabinoids. *British Journal of Pharmacology* 163: 1411–1422.

Parolaro, D., Zamberletti, E., and Rubino, T. 2014. Cannabidiol/phytocannabinoids: A new opportunity for schizophrenia treatment? In: Pertwee, R.G. ed. *Handbook of cannabis*. Oxford, U.K.: Oxford University Press, 526–537.

Passie, T., Emrich, H.M., Karst, M., Brandt, S.D., and Halpern, J.H. 2012. Mitigation of post-traumatic stress symptoms by cannabis resin: A review of the clinical and neurobiological evidence. *Drug Testing and Analysis* 4: 649–659.

Patch, C.S., Tapsell, L.C., Williams, P.G., and Gordon, M. 2006. Plant sterols as dietary adjuvants in the reduction of cardiovascular risk: Theory and evidence. *Vascular Health and Risk Management* 2: 157–162.

Pate, D.W. 1983. Possible role of ultraviolet radiation in evolution of *Cannabis* chemotypes. *Economic Botany* 37: 396–405.

Pate, D.W. 1994. Chemical ecology of *Cannabis*. *Journal of the International Hemp Association* 1(2): 29, 32–37.

Pate, D.W. 1998a. The phytochemistry of *Cannabis*: Its ecological and evolutionary implications. In: Ranalli, P. ed. *Advances in hemp research*. London: Food Products Press (of Haworth Press), 21–42.

Pate, D.W. 1998b. Hemp seed: A valuable food source. In: Ranalli, P. ed. *Advances in hemp research*. London: Food Products Press (of Haworth Press), 243–255.

Pate, D.W. 1999. *Anandamide structure-activity relationships and mechanisms of action on intraocular pressure in the normotensive rabbit model*. Dissertation, University of Kuopio, Finland. 99 pp.

Patel, S., and Hillard, C.J. 2009. Role of endocannabinoid signaling in anxiety and depression. *Current Topics in Behavioral Neurosciences* 1: 347–371.

Patel, S., Hill, M.N., and Hillard, C.J. 2014. Effects of phytocannabinoids on anxiety, mood, and the endocrine system. In: Pertwee, R.G. ed. *Handbook of cannabis*. Oxford, U.K.: Oxford University Press, 189–207.

Pavelek, M., and Lipman, E. 2011. *Report of a working group on fibre crops (flax and hemp)* (Second meeting, July 7–9, 2010, Šumperk—Velké Losiny, Czech Republic). Rome, Italy: Bioversity International. 12 pp.

Peck, C., and Coleman, G. 1991. Implications of placebo theory for clinical research and practice in pain management. *Theoretical Medicine* 12: 247–270.

Pedersen, W. 2014. From badness to illness: Medical cannabis and self-diagnosed attention deficit hyperactivity disorder. *Addiction Research & Theory* 23: 177–186.

Pedrosa, K.P. 2008. *Dietary hemp by-products containing low levels of endocannabinoids stimulate growth in juvenile turbot (Scophthalmus maximus)*. Master's thesis, Vila Real, Portugal: Universidade de Trás-os-Montes e Alto Douro. 76 pp. http://hdl.handle.net/10348/429.

Peil, A., Flachowsky, H., Schumann, E., and Weber, W.E. 2003. Sex-linked AFLP markers indicate a pseudoautosomal region in hemp (*Cannabis sativa* L.). *Theoretical and Applied Genetics* 107: 102–109.

Peiretti, P.G. 2009. Influence of the growth stage of hemp (*Cannabis sativa* L.) on fatty acid content, chemical composition and gross energy. *Agricultural Journal* 4: 27–31.

Pejic, B., Vukcevic, M., Kostic, M., and Skundric, P. 2009. Biosorption of heavy metal ions from aqueous solutions by short hemp fibers: Effect of chemical composition. *Journal of Hazardous Materials* 164: 146–153.

Penner, E.A., Buettner, H., and Mittleman, M.A. 2013. The impact of marijuana use on glucose, insulin, and insulin resistance among US adults. *American Journal of Medicine* 126: 583–589.

Pertwee, R.G. 2006. Cannabinoid pharmacology: The first 66 years. *British Journal of Pharmacology* 147(Suppl. 1): S163–S171.

Pertwee, R.G. 2008. The diverse CB1 and CB2 receptor pharmacology of three plant cannabinoids: Delta9-tetrahydrocannabinol, cannabidiol and delta9-tetrahydrocannabivarin. *British Journal of Pharmacology* 153: 199–215.

Pertwee, R.G., and Cascio, M.G. 2014. Known pharmacological actions of delta-9-tetrahydrocannabinol and of four other chemical constituents of cannabis that activate cannabinoid receptors. In: Pertwee, R.G. ed. *Handbook of cannabis*. Oxford, U.K.: Oxford University Press, 115–136.

Pertwee, R.G., Howlett, A.C., Abood, M.E., Alexander, S.P.H., Di Marzo, V., Elphick, M.R. et al. 2010. International Union of Basic and Clinical Pharmacology. LXXIX. Cannabinoid receptors and their ligands: Beyond CB_1 and CB_2. *Pharmacological Reviews* 62: 588–631.

Peterken, G.F. 1981. *Woodland conservation and management*. London: Chapman and Hall. 374 pp.

Petrová, Š., Benešová, D., Soudek, P., and Vaněk, T. 2012. Enhancement of metal(loid)s phytoextraction by *Cannabis sativa* L. *Journal of Food, Agriculture & Environment* 10: 631–641.

Petrović, M., Debeljak, Ž., Kezić, N., and Džidara, P. 2015. Relationships between cannabinoids content and composition of fatty acids in hempseed oils. *Food Chemistry* 170: 218–225.

Phatak, H.C., Lundsgaard, T., Verma, V.S., and Singh, S. 1975. Mycoplasma-like bodies associated with *Cannabis* phyllody. *Phytopathologische Zeitschrift* 83: 281–294.

Phillips, L.J., Curry, C., Yung, A.R., Yuen, H.P., Adlard, S., and McGorry, P.D. 2002. Cannabis use is not associated with the development of psychosis in an 'ultra' high-risk group. *Australian and New Zealand Journal of Psychiatry* 36: 800–806.

Phillips, R., Turk, R., Manno, J., Crim, D., and Forney, R. 1970. Seasonal variation in cannabinolic content of Indiana marihuana. *Journal of Forensic Sciences* 15: 191–200.

Pickens, J.T. 1981. Sedative activity of cannabis in relation to its Δ^1-trans-tetrahydrocannabinol and cannabidiol content. *British Journal of Pharmacology* 72: 649–656.

Pickersgill, B., Chacón Sánchez, M.I., and Debouck, D.G. 2003. Multiple domestications and their taxonomic consequences: The example of *Phaseolus vulgaris*. *Schriften zu Genetischen Ressourcen* 22: 71–83.

Pietropaolo, S., Bellocchio, L., Ruiz-Calvo, A., Cabanas, M., Du, Z., Guzmán, M. et al. 2015. Chronic cannabinoid receptor stimulation selectively prevents motor impairments in a mouse model of Huntington's disease. *Neuropharmacology* 89: 368–374.

Piluzza, G., Delogu, G., Cabras, A., Marceddu, S., and Bullitta, S. 2013. Differentiation between fiber and drug types of hemp (*Cannabis sativa* L.) from a collection of wild and domesticated accessions. *Genetic Resources and Crop Evolution* 60: 2331–2342.

Piñeiro, R., and Falasca, M. 2012. Lysophosphatidylinositol signalling: New wine from an old bottle. *Biochimica et Biophysica Acta* 1821: 694–705.

Pinfold Consulting. 1998. *A maritime industrial hemp product marketing study*. Prepared for Nova Scotia Agriculture and Marketing (Marketing and Food Industry Development), and New Brunswick Agriculture & Rural Development (Marketing and Business Development). G. Pinfold Consulting Economists Ltd. and J. White, InfoResults Ltd. Irregularly paginated. http://www.novascotia.ca/agri/marketing/research/hempms02.shtml.

Pinto, A.C., Guarieiro, L.L.N., Rezende, M.J.C., Ribeiro, N.M., Torres, E.A., Lopes, W. et al. 2005. Biodiesel: An overview. *Journal of the Brazilian Chemical Society* 16: 1313–1330.

Piomelli, D. 2003. The molecular logic of endocannabinoid signalling. *Nature Reviews Neuroscience* 4: 873–884.

Piomelli, D., and Russo, E.B. 2016. The *Cannabis sativa* versus *Cannabis indica* debate: An interview with Ethan Russo, MD. *Cannabis and Cannabinoid Research* 1(1): 44–46.

Piontek, D., Kraus, L., Legleye, S., and Bühringer, G. 2011. The validity of *DSM-IV* cannabis abuse and dependence criteria in adolescents and the value of additional cannabis use indicators. *Addiction* 106: 1137–1145.

Piotrowska-Cyplik, A., and Czarnecki, Z. 2003. Phytoextraction of heavy metals by hemp during anaerobic sewage sludge management in the nonindustrial sites. *Polish Journal of Environmental Studies* 12: 779–784.

Piotrowski, S., and Carus, M. 2011. *Ecological benefits of hemp and flax cultivation and products*. Huerth, Germany: Nova Institute. 6 pp. http://eiha.org/media/2014/10/Ecological-benefits-of-hemp-and-flax-cultivation-and-products-2011.pdf.

Piper, A. 2005. *The mysterious origins of the word 'marihuana.'* Sino-Platonic Papers 153. Philadelphia, PA: University of Pennsylvania. 17 pp.

Pletcher, M.J., Vittinghoff, E., Kalhan, R., Richman, J., Safford, M., Sidney, S., Lin, F., and Kertesz, S. 2012. Association between marijuana exposure and pulmonary function over 20 years. *JAMA* 307: 173–181.

Poiša, L., Adamovicšs, A., Jankauskiene, Z., and Gruzdeviene, E. 2010. Industrial hemp (*Cannabis sativa* L.) as a biomass crop. In: Marques dos Santos Cordovil, C.S.C., and Ferreira, L. eds. *Treatment and use of organic residues in agriculture: Challenges and opportunities towards sustainable management: Proceedings of the 14th Ramiran International Conference of the FAO ESCORENA Network on the Recycling in Agricultural, Municipal and Industrial Residues in Agriculture* (Lisboa, Portugal). Lisbon, Portugal: Instituto Superior de Agronomia. Universidade Técnica de Lisboa, 326–330.

Popular Mechanics. 1938. New billion-dollar crop. *Popular Mechanics* 144A: 238–239.

Portenoy, R.K., Ganae-Motan, E.D., Allende, S., Yanagihara, R., Shaiova, L., Weinstein, S. et al. 2012. Nabiximols for opioid-treated cancer patients with poorly-controlled chronic pain: A randomized, placebo-controlled, graded-dose trial. *Journal of Pain* 13: 438–449.

Porter, B.E., and Jacobson, C. 2013. Report of a parent survey of cannabidiol-enriched cannabis use in pediatric treatment-resistant epilepsy. *Epilepsy & Behavior* 29: 574–577.

Potter, D. 2004. Growth and morphology of medicinal cannabis. In: Guy, G.W., Whittle, B.A., and Robson, P.J. eds. *The medicinal uses of* Cannabis *and cannabinoids*. London: Pharmaceutical Press, 17–54.

Potter, D. 2009. *The propagation, characterisation and optimisation of* Cannabis sativa *L. as a phytopharmaceutical*. Dissertation, London: King's College. 224 pp. http://www.gwpharm.com/publications-1.aspx.

Potter, D.J. 2014. A review of the cultivation and processing of cannabis (*Cannabis sativa* L.) for production of prescription medicines in the UK. *Drug Testing and Analysis* 6: 31–38.

Potter, D.J., and Duncombe, P. 2012. The effect of electrical lighting power and irradiance on indoor-grown *Cannabis* potency and yield. *Journal of Forensic Sciences* 57: 618–622.

Prade, T., Svensson, S.E., and Mattson, J.E. 2012. Energy balances for biogas and solid biofuel production from industrial hemp. *Biomass & Bioenergy* 40: 36–52.

Prade, T., Svensson, S.E., Andersson, A., and Mattson, J.E. 2011. Biomass and energy yield of industrial hemp grown for biogas and solid fuel. *Biomass & Bioenergy* 35: 3040–3049.

Pritchard, H.W., and Nadarajan, J. 2008. Cryopreservation of orthodox (desiccation tolerant) seeds. In: Reed, B.M. ed. *Plant cryopreservation: A practical guide*. Berlin, Germany: Springer, 485–501.

Prociuk, M., Edel, A., Gavel, N., Deniset, J., Ganguly, R., Austria, J. et al. 2006. The effects of dietary hempseed on cardiac ischemia/reperfusion injury in hypercholesterolemic rabbits. *Experimental and Clinical Cardiology* 11: 198–205.

Prociuk, M.A., Edel, A.L., Richard, M.N., Gavel, N.T., Ander, B.P., Dupasquier, C.M. et al. 2008. Cholesterol-induced stimulation of platelet aggregation is prevented by a hempseed enriched diet. *Canadian Journal of Physiology and Pharmacology* 86: 153–159.

Proctor, M., Yeo, P., and Lack, A. 1996. *The natural history of pollination*. Portland, OR: Timber Press. 479 pp.

Prud'homme, M., Cata, R., and Jutras-Aswad, D. 2015. Cannabidiol as an intervention for addictive behaviors: A systematic review of the evidence. *Substance Abuse: Research and Treatment* 9: 33–38.

Pryce, G., and Baker, D. 2012. Potential control of multiple sclerosis by cannabis and the endocannabinoid system. *CNS & Neurological Disorders—Drug Targets* 11: 624–641.

Pryce, G., and Baker, D. 2014. Cannabis and multiple sclerosis. In: Pertwee, R.G. ed. *Handbook of cannabis*. Oxford, U.K.: Oxford University Press, 487–501.

Przybylski, R., Moes, J., and Sturko, A. 1997. Effect of growing conditions on composition of hemp oils. In: Nova Institute. corporate ed. *Bioresource hemp: Proceedings of the symposium* (Frankfurt am Main, Germany, February 27–March 2, 1997). Hürth, Germany: Nova Institute, 505–514.

Pudełko, K., Majchrzak, L., and Narozna, D. 2014. Allelopathic effects of fibre hemp (*Cannabis sativa* L.) on monocot and dicot plant species. *Industrial Crops and Products* 56: 191–199.

Purohit, V., Ahluwahlia, B.S., and Vigersky, R.A. 1980. Marihuana inhibits dihydrotestosterone binding to the androgen receptor. *Endocrinology* 107: 848–850.

Qaiser, M. 1973. Cannabaceae. In: Nasir, E., and Ali, S.I. eds. *Flora of West Pakistan, issue 44*. Karachi, Pakistan: University of Karachi, 1–3.

Quaghebeur, K., Coosemans, J., Toppet, S., and Compernolle, F. 1994. Cannabiorci- and 8-chlorocannabiorci-chromenic acid as fungal antagonists from *Cylindrocarpon olidum*. *Phytochemistry* 37: 159–161.

Quitkin, F.M. 1999. Placebos, drug effects and study design: A clinician's guide. *American Journal of Psychiatry* 156: 829–836.

Radhakrishnan, R., Wilkinson, S.T., and D'Souza, D.C. 2014. Gone to pot—A review of the association between cannabis and psychosis. *Frontiers in Psychiatry* 5, 54. doi:10.3389/fpsyt.2014.00054.

Radosevic, A., Kupinic, M., and Grlic, L. 1962. Antibiotic activity of various types of *Cannabis* resin. *Nature* 195: 1007–1009.

Radwan, M.M., ElSohly, M.A., Slade, D., Ahmed, S.A., Khan, I.A., and Ross, S.A. 2009. Biologically active cannabinoids from high-potency *Cannabis sativa*. *Journal of Natural Products* 72: 906–911.

Radwan, M.M., Elsohly, M.A., Slade, D., Ahmed, S.A., Wilson, L., El-Alfy, A.T. et al. 2008. Non-cannabinoid constituents from a high potency *Cannabis sativa* variety. *Phytochemistry* 69: 2627–2633.

Raedestorff, D., Schwager, J., and Schueler, G. 2012. *Nutraceutical and pharmaceutical compositions and use thereof for the treatment, co-treatment or prevention of inflammatory disorders*. U.S. patent 8158681. https://www.google.com/patents/US8158681.

Ragit, S.S., Mohapatra, S.K., Gill, P., and Kundu, K. 2012. Brown hemp methyl ester: Trans-esterification process and evaluation of fuel properties. *Biomass & Bioenergy* 41: 14–20.

Raharjo, T.J., Chang, W.T., Verberne, M.C., Peltenburg-Looman, A.M., Linthorst, H.J., and Verpoorte, R. 2004. Cloning and over-expression of a cDNA encoding a polyketide synthase from *Cannabis sativa*. *Plant Physiology and Biochemistry* 42: 291–297.

Rahn, E.J., and Hohmann, A.G. 2009. Cannabinoids as pharmacotherapies for neuropathic pain: From the bench to the bedside. *Neurotherapeutics* 6: 713–737.

Raie, M.Y., Ahmad, A., Ashraf, M., and Hussain, S. 1995. Studies of *Cannabis sativa* and *Sorghum bicolor* oils. *Fat Science Technology* 97: 428–429.

Rajesh, M., Mukhopadhyay, P., Bátkai, S., Patel, V., Saito, K., Matsumoto, S. et al. 2010. Cannabidiol attenuates cardiac dysfunction, oxidative stress, fibrosis, and inflammatory and cell death signaling pathways in diabetic cardiomyopathy. *Journal of the American College of Cardiology* 14: 2115–2125.

Ram, M. 1960. Occurrence of endosperm haustorium in *Cannabis sativa* L. *Annals of Botany* (New Series) 24: 79–82.

Ramírez, B.G., Blázquez, C., Gómez del Pulgar, T., Guzmán, M., and de Ceballos, M.L. 2005. Prevention of Alzheimer's disease pathology by cannabinoids: Neuroprotection mediated by blockade of microglial activation. *Journal of Neuroscience* 25: 1904–1913.

Ranalli, P. 1998. Agronomical and physiological advances in hemp crops. In: Ranalli, P. ed. *Advances in hemp research*. London: Food Products Press (of Haworth Press), 61–84.

Ranalli, P. 2004. Current status and future scenarios of hemp breeding. *Euphytica* 140: 121–131.

Ranalli, P., and Venturi, G. 2004. Hemp as a raw material for industrial applications. *Euphytica* 140: 1–6.

Ratcliffe, D.A. ed. 1977. *A nature conservation review*. Cambridge, U.K.: Cambridge University Press. 2 volumes.

Ravi, D., and Anand, P. 2012. Production and applications of artificial seed: A review. *International Research Journal of Biological Sciences* 1: 74–78.

Reardon, S. 2015. Marijuana gears up for production high in US labs. *Nature* 519: 269–270.

Reddy, D.S., and Golub, V. 2016. The pharmacological basis of cannabis therapy for epilepsy. *Journal of Pharmacology* 357: 45–55.

Reece, A.S. 2009. Chronic toxicology of cannabis. *Clinical Toxicology (Philadelphia)* 47: 517–524.

Regan, T. 2011. *Joint ventures. Inside America's almost legal marijuana industry*. Hoboken, NJ: Wiley. 265 pp.

Reggio, P.H. 2005 [imprinted 2006]. The relationship between endocannabinoid conformation and endocannabinoid interaction at the cannabinoid receptors. In: Onaivi, E.S., Sugiura, T., and di Marzo, V. eds. *Endocannabinoids: The brain and body's marijuana and beyond*. Boca Raton, FL: CRC Press, 11–53.

Rehman, M.S.U., Rashid, N., Saif, A., Mahmood, T., and Han, J.-I. 2013. Potential of bioenergy production from industrial hemp (*Cannabis sativa*): Pakistan perspective. *Renewable and Sustainable Energy Reviews* 18: 154–164.

Reid, P.T., Macleod, J., and Robertson, J.R. 2010. Cannabis and the lung. *Journal of the Royal College of Physicians of Edinburgh* 40: 328–334.

Reinarman, C., Nunberg, H., Lanthier, F., and Heddleston, T. 2011. Who are medical marijuana patients? Population characteristics from nine California assessment clinics. *Journal of Psychoactive Drugs* 43: 128–135.

Rema, P., Pedrosa, K., Grassi, G., and Dias, J. 2010. Dietary hempseed meal and hempseed oil enhanced growth performance in juvenile turbot (*Scophthalmus maximus*). In: *Program and Abstracts of the 14th International Symposium on Fish Nutrition and Feeding*. http://cpfd.cnki.com.cn/Article/CPFDTOTAL -HDSC201005001150.htm.

Renard, D., Taieb, G., Gras-Combe, G., and Labauge, P. 2012. Cannabis-related myocardial infarction and cardioembolic stroke. *Journal of Stroke and Cerebrovascular Diseases* 21: 82–83.

Renner, S.S., and Ricklefs, R.E. 1995. Dioecy and its correlates in the flowering plants. *American Journal of Botany* 82: 596–606.

Rey, A.A., Purrio, M., Viveros, M.-P., and Lutz, B. 2012. Biphasic effects of cannabinoids in anxiety responses: CB1 and GABA$_B$ receptors in the balance of GABAergic and glutamatergic neurotransmission. *Neuropsychopharmacology* 37: 2624–2634.

Rhyne, D.N., Anderson, S.L., Gedde, M., and Borgelt, L.M. 2016. Effects of medical marijuana on migraine headache frequency in an adult population. *Pharmacotherapy* 36: 505–510.

Rice, B. 2008. Hemp as a feedstock for biomass-to-energy conversion. *Journal of Industrial Hemp* 13(2): 145–156.

Rice, S., and Koziel, J.A. 2015. Characterizing the smell of marijuana by odor impact of volatile compounds: An application of simultaneous chemical and sensory analysis. *PLoS ONE* 10(12): e0144160. doi:10.1371 /journal.pone.0144160.

Richter, K.P., and Levy, S. 2014. Big Marijuana—Lessons from Big Tobacco. *New England Journal of Medicine* 371: 399–401.

Riddlestone, S., Desai, P., Evans, M., and Skyring, A. 1994. *Bioregional fibers, the potential for a sustainable regional paper and textile industry based on flax and hemp*. Carshalton, Surrey, U.K.: Bioregional Development Group, Sutton Ecology Centre. 130 pp.

Rigby, F.L. 2000. *Process and apparatus for obtaining lupulin from hops*. World Intellectual Property Organization Patent No. W0/2000/006691.

Rindos, D. 1984. *The origins of agriculture: An evolutionary perspective*. New York: Academic Press. 325 pp.

Roach, B., Eisner, T., and Meinwald, J. 1990. The defense mechanisms of arthropods. 83. Alpha-and beta-necrodol, novel terpens from a carrion beetle (*Necrodes surinamensis*, Silphidae, Coleoptera). *Journal of Organic Chemistry* 55: 4047–4051.

Robbins, L., Snell, W., Halich, G., Maynard, L., Dillon, C., Spalding, D. 2013. *Economic considerations for growing industrial hemp: Implications for Kentucky's farmers and Agricultural Economy Department of Agricultural Economics (AEC)*. Lexington, KY: University of Kentucky, 25 pp. http://www2.ca.uky.edu /cmspubsclass/files/EconomicConsiderationsforGrowingIndustrialHemp.pdf.

Robel, R.J. 1969. Food habits, weight dynamics, and fat content of bobwhites in relation to food planting in Kansas. *Journal of Wildlife Management* 33: 237–294.

Robinson, B.B. 1935. Hemp. *USDA farmers' bulletin no. 1935*. Washington, DC: U.S. Department of Agriculture. 16 pp. (Revised 1952.)

Robson, P. 2005. Human studies of cannabinoids and medicinal cannabis. *Handbook of Experimental Pharmacology* 168: 719–756.

Robson, P. 2014. Therapeutic potential of cannabinoid medicines. 2014. *Drug Testing and Analysis* 6: 24–30.

Robson, P.J., Guy, G.W., and Di Marzo, V. 2014. Cannabinoids and schizophrenia: Therapeutic prospects. *Current Pharmaceutical Design* 20: 2194–2204.

Rock, E.M., Bolognini, D., Limebeer, C.L., Cascio, M.G., Anavi-Goffer, S., Fletcher, P.J. et al. 2012. Cannabidiol, a non-psychotropic component of cannabis, attenuates vomiting and nausea-like behaviour via indirect agonism of 5-HT(1A) somatodendritic autoreceptors in the dorsal raphe nucleus. *British Journal of Pharmacology* 165: 2620–2634.

Rock, E.M., Sticht, M.A., and Parker, L.A. 2014. Effect of phytocannabinoids on nausea and vomiting. In: Pertwee, R.G. ed. *Handbook of cannabis*. Oxford, U.K.: Oxford University Press, 435–454.

Rode, J., In-Chol, K., Saal, B., Flachowsky, H., Kriese, U., and Weber, W.E. 2005. Sex-linked SSR markers in hemp. *Plant Breeding* 124: 167–170.

Rodman, E. 2015. From criminalization to regulation: New classifications of cannabis necessitate reform of united national drug treaties. *Brooklyn Journal of International Law* 40: 647–683.

Rodriguez-Leyva, D., and Pierce, G.N. 2010. The cardiac and haemostatic effects of dietary hempseed. *Nutrition & Metabolism* 7: 32. doi:10.1186/1743-7075-7-32.

Roe, E.T., and Leonard-Stuart, C. 1911. *Webster's new illustrated dictionary*. New York: Syndicate Publishing Company. 1046 pp.

Rohleder, C., and Leweke, F.M. 2015. Cannabinoids and schizophrenia. In: Fattore, L. ed. *Cannabinoids in neurologic and mental disease*. New York: Academic Press, 193–204.

Rojas Pérez-Ezquerra, P., Sánchez-Morillas, L., Davila-Ferandez, G., Ruiz-Hornillos, F.J., Carrasco García, I., Herranz Mañas, M. et al. 2015. Contact urticaria to *Cannabis sativa* due to a lipid transfer protein (LTP). *Allergologia et Immunopathologia* 43: 231–233.

Romano, L.I., and Hazekamp, A. 2013. Cannabis oil: Chemical evaluation of an upcoming cannabis-based medicine. *Cannabinoids* 1: 1–11.

Room, R. 2014. Legalizing a market for cannabis for pleasure: Colorado, Washington, Uruguay and beyond. *Addiction* 109: 345–351.

Rosales-Corral, S., Hernández, L., and Gallegos, M. 2015. Cannabinoids in neuroinflammation, oxidative stress and neuro excitotoxicity. *Pharmaceutica Analytica Acta* 6: 346. doi:10.4172/2153-2435.1000346.

Ros Barceló, A. 1997. Lignification in plant cell walls. *International Review of Cytology* 176: 87–132.

Rose, R., Mars, B., and Pirello, C. 2004. *The hempnut cookbook: Ancient foods for a new millennium.* Summertown, TN: Book Publishing Company. 180 pp.

Rosenbaum, C.D., Carreiro, S.P., and Babu, K.M. 2012. Here today, gone tomorrow…and back again? A review of herbal marijuana alternatives (K2, Spice), synthetic cathinones (bath salts), Kratom, *Salvia divinorum*, methoxetamine, and piperazines. *Journal of Medical Toxicology* 8: 15–32.

Rosenberg, E.C., Tsien, R.W., Whalley, B.J., and Devinsky, O. 2015. Cannabinoids and epilepsy. *Neurotherapeutics* 12: 747–768.

Rosenthal, E. 1998. *Marijuana grower's handbook. The indoor high yield guide.* 3rd edition. Oakland, CA: Quick American Archives. 261 pp.

Rosenthal, E. 2001. *The big book of buds. Marijuana varieties from the world's great seed breeders.* Oakland, CA: Quick American Archives. 214 pp.

Rosenthal, E. 2004. *The big book of buds: 2. More marijuana varieties from the world's great seed breeders.* Oakland, CA: Quick American Archives. 192 pp.

Rosenthal, E. 2007. *The big book of buds: 3. More marijuana varieties from the world's great seed breeders.* Oakland, CA: Quick American. 256 pp.

Rosenthal, E. 2010. *The big book of buds: 4. Marijuana varieties from the world's great seed breeders.* Oakland, CA: Quick American Publishing. 240 pp.

Rosenthal, E., Gieringer, D., and Mikuriya, T.H. 1997. *Marijuana medical handbook: A guide to therapeutic use.* Oakland, CA: Quick American Archives. 270 pp.

Ross, S.A., and ElSohly, M.A. 1996. The volatile oil composition of fresh and air-dried buds of *Cannabis sativa. Journal of Natural Products* 59: 49–51.

Ross, S.A., and ElSohly, M.A. 1997. CBN and Δ^9-THC concentration ratio as an indicator of the age of stored marijuana samples. *Bulletin on Narcotics* 49/50: 139–147.

Ross, S.A., ElSohly, H.N., ElKashoury, E.A., and ElSohly, M.A. 1996. Fatty acids of Cannabis seeds. *Phytochemical Analysis* 7: 279–283.

Ross, S.A., Mehmedic, Z., Murphy, T.P., and ElSohly, M.A. 2000. GC-MS analysis of the total Δ^9-THC content of both drug- and fiber-type *Cannabis* seeds. *Journal of Analytical Toxicology* 24: 715–717.

Ross, S.A., ElSohly, M.A., Sultana, G.N.N., Mehmedic, Z., Hossain, C.F., and Chandra, S. 2005. Flavonoid glycosides and cannabinoids from the pollen of *Cannabis sativa* L. *Phytochemical Analysis* 16: 45–48.

Rossi, F., Bellini, G., Luongo, L., Torella, M., Silvia Mancusi, S., De Petrocellis, L. et al. 2011. The endovanilloid/endocannabinoid system: A new potential target for osteoporosis therapy. *Bone* 48: 997–1007.

Ross-Ibara, J., Morrell, P.L., and Gaut, B.S. 2007. Plant domestication, a unique opportunity to identify the genetic basis of adaptation. In: Avise, J.C., and Ayala, F.J. eds. *In the light of evolution. Volume 1: Adaptation and complex design.* Washington, DC: National Academies Press, 205–224.

Rothschild, M., and Fairbairn, J.W. 1980. Ovipositing butterfly (*Pieris brassicae* L.) distinguishes between aqueous extracts of two strains of *Cannabis sativa* L. and THC and CBD. *Nature* 286: 56–59.

Rothschild, M., Rowen, M.G., and Fairbairn, J.W. 1977. Storage of cannabinoids by *Arctia caja* and *Zonocerus elegans* fed on chemically distinct strains of *Cannabis sativa. Nature* 266: 650–651.

Rothschild, M., Bergström, G., and Wängberg, S.-Å. 2005. *Cannabis sativa*: Volatile compounds from pollen and entire male and female plants of two variants, Northern Lights and Hawaiian Indica. *Botanical Journal of the Linnean Society* 147: 387–397.

Roulac, J.W. 1997. *Hemp horizons: The comeback of the world's most promising plant.* White River Junction, VT: Chelsea Green Publishing. 211 pp.

Rowan, M.G., and Fairbairn, J.W. 1977. Cannabinoid patterns in seedlings of *Cannabis sativa* and their use in the determination of chemical race. *Journal of Pharmacy and Pharmacology* 29: 491–494.

Rubin, V., and Comitas, L. eds. 1975. *Ganja in Jamaica: A medical anthropological study of chronic marijuana use.* The Hague, the Netherlands: Mouton. 205 pp.

Rubino, T., and Parolaro, D. 2014. Cannabis abuse in adolescence and the risk of psychosis: A brief review of the preclinical evidence. *Progress in Neuro-Psychopharmacology & Biological Psychiatry* 52: 41–44.

Ruchlemer, R., Amit-Kohn, M., Raveh, D., and Hanuš, L. 2015. Inhaled medicinal cannabis and the immunocompromised patient. *Supportive Care in Cancer* 23: 819–822.

Rudenko, S.I. 1970. *Frozen tombs of Siberia: The Pazyryk burials of Iron Age horsemen.* London: Dent. 309 pp.

Russo, E. 1998. Cannabis for migraine treatment: The once and future prescription? An historical and scientific review. *Pain* 76: 3–8.

Russo, E. 2001. Hemp for headache: An in-depth historical and scientific review of cannabis in migraine treatment. *Journal of Cannabis Therapeutics* 1(2): 21–92.

Russo, E. 2002. Cannabis treatments in obstetrics and gynecology: A historical review. *Journal of Cannabis Therapeutics* 2(3–4): 5–34.

Russo, E. 2003a. Introduction: Cannabis from pariah to prescription. *Journal of Cannabis Therapeutics* 3(1): 1–29.

Russo, E. 2003b. Future of cannabis and cannabinoids in therapeutics. *Journal of Cannabis Therapeutics* 3(4): 163–174.

Russo, E.B. 2004a. History of cannabis as a medicine. In: Guy, G.W., Whittle, B.A., and Robson, P.J. eds. *The medicinal uses of* Cannabis *and cannabinoids*. London: Pharmaceutical Press, 1–16.

Russo, E.B. 2004b. Clinical endocannabinoid deficiency (CECD): Can this concept explain therapeutic benefits of cannabis in migraine, fibromyalgia, irritable bowel syndrome and other treatment-resistant conditions? *Neuro Endocrinology Letters* 25: 31–39. [Republished as Russo, E.B. 2008. *Neuro Endocrinology Letters* 29: 192–200.]

Russo, E.B. 2007. History of cannabis and its preparations in saga, science, and sobriquet. *Chemistry & Biodiversity* 4: 1614–1648.

Russo, E.B. 2011a. Taming THC: potential cannabis synergy and phytocannabinoid-terpenoid entourage effects. *British Journal of Pharmacology* 163: 1344–1364.

Russo, E.B. 2011b. *Cannabis genome uncloaked: Commentary on the scientific implications*. The International Cannabinoid Research Society. http://www.icrs.co/content/Cannabis_Genome_Uncloaked.pdf.

Russo, E.B. 2014. The pharmacological history of cannabis. In: Pertwee, R.G. ed. *Handbook of cannabis*. Oxford, U.K.: Oxford University Press, 23–43.

Russo, E.B. 2015. Synthetic and natural cannabinoids: The cardiovascular risk. *British Journal of Cardiology* 22: 7–9.

Russo, E.B., and Grotenhermen, F. eds. 2006. *Handbook of cannabis therapeutics: From bench to bedside*. Binghamton, NY: Haworth Press. 471 pp.

Russo, E.B, and Guy, G.W. 2006. A tale of two cannabinoids: The therapeutic rationale for combining tetrahydrocannabinol and cannabidiol. *Medical Hypotheses* 66: 234–246.

Russo, E., and Hohmann, A. 2013. Role of cannabinoids in pain management. In: Deer, T., Gordin, V., Kim, P.S., Panchal, S.J., and Ray, A.L. eds. *Comprehensive treatment of chronic pain by medical, interventional and behavioral approaches*. New York: Springer, 181–197.

Russo, E.B., and McPartland, J.M. 2003. Cannabis is more than simply Δ^9-tetrahydrocannabinol. *Psychopharmacology* 165: 431–432.

Russo, E.B., Merzouki, A., Mesa, J.M., Frey, K.A., and Bach, P.J. 2004. Cannabis improves night vision: A case study of dark adaptometry and scotopic sensitivity in kif smokers of the Rif mountains of northern Morocco. *Journal of Ethnopharmacology* 93: 99–104.

Russo, E.B., Jiang, H.-E., Li, X., Sutton, A., Carboni, A., del Bianco, F. et al. 2008. Phytochemical and genetic analyses of ancient cannabis from Central Asia. *Journal of Experimental Botany* 59: 4171–4182.

Russo, E.B., Guy, G.W., and Robson, P.J. 2009. Cannabis, pain, and sleep: Lessons from therapeutic clinical trials of *Sativex®*, a cannabis-based medicine. In: Lambert, D.M. ed. *Cannabinoids in nature and medicine*. New York: Wiley-VCH, 141–155.

Russo, E.B., Mead, A.P., and Sulak, D. 2015. Current status and future of cannabis research. *Clinical Researcher* April: 58–62.

Russo, R., and Reggiani, R. 2013. Variability in antinutritional compounds in hempseed meal of Italian and French varieties. *Plant* 1(2): 25–29.

Ryan, D., Drysdale, A.J., Pertwee, R.G., and Platt, B. 2006. Differential effects of cannabis extracts and pure plant cannabinoids on hippocampal neurones and glia. *Neuroscience Letters* 408: 236–241.

Saeglitz, C., Pohl, M., and Bartsch, D. 2000. Monitoring gene flow from transgenic sugar beet using cytoplasmic male-sterile bait plants. *Molecular Ecology* 9: 2035–2040.

Sakamoto, K., Akiyama, Y., Fukui, K., Kamada, H., and Satoh, S. 1998. Characterization: Genome size and morphology of sex chromosomes in hemp (*Cannabis sativa* L.). *Cytologia* 3: 459–464.

Sakamoto, K., Shimomura, K., Komeda, Y., Kamada, H., and Satoh, S. 1995. A male-associated DNA sequence in a dioecious plant, *Cannabis sativa* L. *Plant Cell Physiology* 36: 1549–1554.

Sakamoto, K., Ohmido, N., Fukui, K., Kamada, H., and Satoh, S. 2000. Site-specific accumulation of a LINE-like retrotransposon in a sex chromosome of the dioecious plant *Cannabis sativa*. *Plant Molecular Biology* 44: 723–732.

Sakamoto, K., Abe, T., Matsuyama, T., Yoshida, S., Ohmido, N., Fukui, K. et al. 2005. RAPD markers encoding retrotransposable elements are linked to the male sex in *Cannabis sativa* L. *Genome* 48: 931–936.

Sakuma, S., Salomon, B., and Komatsuda, T. 2011. The domestication syndrome genes responsible for the major changes in plant form in the Triticeae crops. *Plant & Cell Physiology* 52: 738–749.

Salentijn, E.M.J., Zhang, Q., Amaducci, S., Yang, M., Luisa, M., and Trindade, L.M. 2015. New developments in fiber hemp (*Cannabis sativa* L.) breeding. *Industrial Crops and Products* 68: 32–41.

Samudre, S., Hosseini, A., and Lattanzio, F. 2014. Cannabinoids: A novel treatment for glaucoma. *Acta Ophthalmologica* 92(Suppl. S253). doi:10.1111/j.1755-3768.2014.T022.x.

Sapino, S., Carlotti, M.E., Peira, E., and Gallarate, M. 2005. Hemp-seed and olive oils: Their stability against oxidation and use in O/W emulsions. *Journal of Cosmetic Science* 56: 227–251.

Sarfaraz, S., Adhami, V.M., Syed, D.N., Afaq, F., and Mukhtar, H. 2008. Cannabinoids for cancer treatment: Progress and promise. *Cancer Research* 68: 339–342.

Sarlikioti, V., de Visser, P.H.B., Buck-Sorlin, G.H., and Marcelis, L.F.M. 2011. How plant architecture affects light absorption and photosynthesis in tomato: Towards an ideotype for plant architecture using a functional-structural plant model. *Annals of Botany* 108: 1065–1073.

Sawler, J., Stout, J.M., Gardner, K.M., Hudson, D., Vidmar, J., Butler, L. et al. 2015. The genetic structure of marijuana and hemp. *PLoS ONE* 10(8): e0133292. doi:10.1371/journal.pone.0133292.

Schäfer, T., and Honermeier, B. 2006. Effect of sowing date and plant density on the cell morphology of hemp (*Cannabis sativa* L.) *Industrial Crops and Products* 23: 88–98.

Schaffner, J.H. 1919. Complete reversal of sex in hemp. *Science* 50: 311–312.

Schaffner, J.H. 1921. Influence of environment on sexual expression in hemp. *Botanical Gazette* 75: 45–59.

Schaffner, J.H. 1923. The influence of relative length of daylight on the reversal of sex in hemp. *Ecology* 4: 327–334.

Schaffner, J.H. 1926. The change of opposite to alternate phyllotaxis and repeated rejuvenations in hemp by means of changed photoperiodicity. *Ecology* 7: 315–325.

Schaffner, J.H. 1931. The functional curve of sex-reversal in staminate hemp plants induced by photoperiodicity. *American Journal of Botany* 18: 424–430.

Schauer, G.L., King, B.A., Bunnell, R.E., Promoff, G., and McAfee, T.A. 2016. Toking, vaping, and eating for health or fun: Marijuana use patterns in adults, U.S., 2014. *American Journal of Preventive Medicine* 50: 1–8.

Schicho, R., and Storr, M. 2014. Cannabis finds its way into treatment of Crohn's disease. *Pharmacology* 93: 1–3.

Schier, A.R., Ribeiro, N.P., Silva, A.C., Hallak, J.E., Crippa, J.A., Nardi, A.E. et al. 2012. Cannabidiol, a *Cannabis sativa* constituent, as an anxiolytic drug. *Revista Brasileira de Psiquiatria* 34(Suppl. 1): S104–S110.

Schley, M., Legler, A., Skopp, G., Schmelz, M., Konrad, C., and Rukwied, R. 2006. Delta-9-THC based monotherapy in fibromyalgia patients on experimentally induced pain, axon reflex flare, and pain relief. *Current Medical Research and Opinion* 22: 1269–1276.

Scholz, H. 1957. Wild hemp as a ruderal plant of Central Europe. *Verhandlungen des Botanischen Vereins der Provinz Brandenburgund der angrenzenden Länder* 83(97): 61–64. (In German.)

Schubart, C.D., Sommera, I.E.C., Fusar-Polib, P., de Wittea, L., Kahnc, R.S., and Boksa, M.P.M. 2014. Cannabidiol as a potential treatment for psychosis. *European Neuropsychopharmacology* 24: 51–64.

Schultes, R.E. 1970. Random thoughts and queries on the botany of *Cannabis*. In: Joyce, R.B., and Curry, S.H. eds. *The botany and chemistry of* Cannabis. London: J. & A. Churchill, 11–38.

Schultes, R.E. 1973. Man and marijuana. *Natural History* 82: 58–63, 80, 82.

Schultes, R.E, and Hofmann, A. 1980. *The botany and chemistry of hallucinogens*. 2nd edition. Springfield, IL: Thomas. 437 pp.

Schultes, R.E., Klein, W.M., Plowman, T., and Lockwood, T.E. 1974. *Cannabis*: An example of taxonomic neglect. *Botanical Museum Leaflets, Harvard University* 23: 337–367.

Schultz, O.E., and Haffner, G. 1959. Information on a sedative and antibacterial agent from the German fiber hemp (*Cannabis sativa*). *Zeitschrift für Naturforschung (Sec. B)* 14: 98–100. (In German.)

Schultz, T.R., and McGlynn, T.P. 2000. The interactions of ants with other organisms. In: Agosti, D, Majer, J., Alonso, L., and Schultz, T.R. eds. *Ants: Standard methods for measuring and monitoring biodiversity*. Washington, DC: Smithsonian Institution Press, 35–44.

Schultz, T.R., Mueller, U.G., Currie, C.R., and Rehner, S.A. 2005. Reciprocal illumination: A comparison of agriculture in humans and in fungus-growing ants. In: Vega, F.E., and Blackwell, M. eds. *Insect–fungal associations: Ecology and evolution*. New York: Oxford University Press, 149–190.

Schwab, U.S., Callaway, J.C., Erkkilä, A.T., Gynther, J., Uusitupa, M.I., and Järvinen, T. 2006. Effects of hempseed and flaxseed oils on the profile of serum lipids, serum total and lipoprotein lipid concentrations and haemostatic factors. *European Journal of Nutrition* 45: 470–477.

Schwartz, I.S. 1985. Marijuana and fungal infection. *American Journal of Clinical Pathology* 84: 256.

Scotter, E.L., Abood, M.E., and Glass, M. 2010. The endocannabinoid system as a target for the treatment of neurodegenerative disease. *British Journal of Pharmacology* 160: 480–498.

Scuderi, C., Filippis, D.D., Iuvone, T., Blasio, A., Steardo, A., and Esposito, G. 2009. Cannabidiol in medicine: A review of its therapeutic potential in CNS disorders. *Phytotherapy Research* 23: 597–602.

Scully, C. 2007. Cannabis; adverse effects from an oromucosal spray. *British Dental Journal* 203: 334–335.

Sèbe, G., Cetin, N.S., Hill, C.A.S., and Hughes, M. 2000. RTM [resin transfer molding] hemp fibre-reinforced polyester composites. *Applied Composite Materials* 7: 341–349.

Seelly, K.A., Lapoint, J., Moran, J.H., and Fattore, L. 2012. Spice drugs are more than harmless herbal blends: A review of the pharmacology and toxicology of synthetic cannabinoids. *Progress in Neuro-Psychopharmacology & Biological Psychiatry* 39: 234–243.

Segelman, A.B., Sofia, R.D., Segelman, F.P., Harakal, J.J., and Knobloch, L.C. 1974. *Cannabis sativa* L. (marijuana) V: Pharmacological evaluation of marijuana aqueous extract and volatile oil. *Journal of Pharmaceutical Sciences* 26: 962–964.

Selkirk, S.W., and Spencer, R.D. 1999. Economics of fibre production from industrial hemp and blue gum plantations. *Australian Forestry* 63: 193–201.

Seo, J.-H., Jeong, E.-S., Lee, K.-S., Heo, S.-H., Jeong, D.-G., Lee, S.-J. et al. 2012. Hempseed water extract ameliorates atherosclerosis in apolipoprotein E knockout mice. *Food Science and Biotechnology* 21: 927–932.

Serebriakova, T.Y. 1940. Fiber plants, Volume 5, Part 1. In: Wulff, E.V. ed. *Flora of cultivated plants*. Moscow, Russia: State Printing Office. (In Russian.)

Serebriakova, T.Y., and Sizov, I.A. 1940. Cannabinaceae Lindl. In: Vavilov, N.I. ed. *Kulturnaja Flora SSSR Vol. 5*. Moscow, Russia: Kolos, 1–53. (In Russian.)

Sewell, R.A., Poling, J, and Sofuoglu, M. 2009. The effect of cannabis compared with alcohol on driving. *American Journal on Addictions* 18: 185–193.

Shahzad, A. 2012. Hemp fiber and its composites—A review. *Journal of Composite Materials* 46: 973–986.

Shao, H., Song, S.-J., and Clarke, R.C. 2003. Female-associated DNA polymorphisms of hemp (*Cannabis sativa* L.). *Journal of Industrial Hemp* 8(1): 5–9.

Sharma, G.K. 1975. Altitudinal variation in leaf epidermal patterns of *Cannabis sativa*. *Bulletin of the Torrey Botanical Club* 102: 199–200.

Sharma, G.K. 1979. Significance of eco-chemical studies of *Cannabis*. *Science Culture* 45: 303–307.

Shepard, E.M., and Blackley, P.R. 2016. Medical marijuana and crime: Further evidence from the Western States. *Journal of Drug Issues* 46: 122–134.

Sherratt, A.G. 1991. Sacred and profane substances: The ritual use of narcotics in later Neolithic Europe. In: Garwood, P., Jennings, D., Skeates, R., and Toms, J. eds. *Sacred and profane: Proceedings of a conference on archaeology, ritual and religion*. Oxford, U.K.: Oxford University Committee for Archaeology, 50–64.

Shi, G., and Cai, Q. 2009. Cadmium tolerance and accumulation in eight potential energy crops. *Biotechnology Advances* 27: 555–562.

Shohami, E., Cohen-Yeshurun, A., Magid, L., Algali, M., and Mechoulam, R. 2011. Endocannabinoids and traumatic brain injury. *British Journal of Pharmacology* 163: 1402–1410.

Shoyama, Y., Taura, F., and Morimoto, S. 2001. Expression of tetrahydrocannabinolic acid synthase in tobacco. *Proceedings, 2001 Symposium on the Cannabinoids*. Burlington, VT: International Cannabinoid Research Society, 9.

Sibthrop, J., and Smith, J.E. 1840. *Flora Graeca. Vol. 10*. London: Taylor.

Sidorenko, N.M. 1978. Anatomical and cytological characteristics of tetraploid monoecious hemp. *Tsitologiya i Genetika* 12: 115–117.

Sikora, V., Berenji, J., and Latković, D. 2011. Influence of agroclimatic conditions on content of main cannabinoids in industrial hemp (*Cannabis sativa* L.). *Genetika* 43: 449–456.

Silva Viera, R., da Canaveira, A., da Simões, A., and Domingos, T. 2010. Industrial hemp or eucalyptus paper? An environmental comparison using life cycle assessment. *International Journal of Life Cycle Assessment* 15: 368–375.

Silversides, F.G., and LeFrançois, M.R. 2005. The effect of feeding hemp seed meal to laying hens. *British Poultry Science* 46: 231–235.

Simão da Silva, K.A.B., Paszcuk, A.F., Passos, G.F., Silva, E.S., Bento, A.F., Meotti, F.C. et al. 2011. Activation of cannabinoid receptors by the pentacyclic triterpene α,β-amyrin inhibits inflammatory and neuropathic persistent pain in mice. *Pain* 152: 1872–1887.

Simmonds, N.W. 1976. Hemp. In: Simmonds, N.W. ed. *Evolution of crop plants*. New York: Longman, 203–204.

Simonetto, D.A., Oxentenko, A.S., Herman, M.L., and Szostek, J.H. 2012. Cannabinoid hyperemesis: A case series of 98 patients. *Mayo Clinic Proceedings* 87: 114–119.

Simopoulos, A.P. 2002. The importance of the ratio of omega-6/omega-3 essential fatty acids. *Biomedicine & Pharmacotherapy* 56: 365–379.

Simopoulos, A.P. 2008. The importance of the omega-6/omega-3 fatty acid ratio in cardiovascular disease and other chronic diseases. *Experimental Biology and Medicine (Maywood, NJ)* 233: 674–688.

Sinclair, L. 2012. Are psychotropics overprescribed for children in foster care? *Psychiatric News.* http://psych news.psychiatryonline.org/doi/full/10.1176/pn.47.3.psychnews_47_3_12-a.

Singh, A.B., and Kumar, P. 2003. Aeroallergens in clinical practice of allergy in India: An overview. *Annals of Agricultural and Environmental Medicine* 10: 131–136.

Singha, A.S., Kaith, B.S., and Khanna, A.J. 2011. Synthesis and characterization of *Cannabis indica* fiber reinforced composites. *BioResources* 6: 2101–2117.

Siniscalco, D., Sapone, A., Giordano, C., Cirillo, A., de Magistris, L., Rossi, F. et al. 2013. Cannabinoid receptor type 2, but not type 1, is up-regulated in peripheral blood mononuclear cells of children affected by autistic disorders. *Journal of Autism and Developmental Disorders* 43: 2686–695.

Sipos, B., Kreuger, E., Svensson, S.E., Réczey, K., Björnsson, L., and Guido, Z. 2010. Steam pretreatment of dry and ensiled industrial hemp for ethanol production. *Biomass & Bioenergy* 34: 1721–1731.

Sirikantaramas, S., Taura, F, Tanaka, Y., Ishikawa, Y., Morimoto, S., and Shoyama, Y. 2005. Tetrahydrocannabinolic acid synthase, the enzyme controlling marijuana psychoactivity, is secreted into the storage cavity of the glandular trichomes. *Plant Cell Physiology* 46: 1578–1582.

Sirikantaramas, S., Taura, F., Morimoto, S., and Shoyama, Y. 2007. Recent advances in *Cannabis sativa* research: Biosynthetic studies and its potential in biotechnology. *Current Pharmaceutical Biotechnology* 8: 237–243.

Skrabek, R.Q., Galimova, L., Ethans, K., and Perry, D. 2008. Nabilone for the treatment of pain in fibromyalgia. *Journal of Pain* 9: 164–173.

Slatkin, N.E. 2007. Cannabinoids in the treatment of chemotherapy-induced nausea and vomiting: Beyond prevention of acute emesis. *Journal of Supportive Oncology* 5(5 Suppl. 3): 1–9.

Small, E. 1971. *An agricultural perspective of marijuana.* Canadian Parliamentary Commission of Inquiry into the non-medical use of drugs. Research Report No. 104. 59 pp. [On file, Health Canada, Ottawa.]

Small, E. 1972a. Interfertility and chromosomal uniformity in *Cannabis. Canadian Journal of Botany* 50: 1947–1949.

Small, E. 1972b. The hemp problem in Canada. *Greenhouse Garden Grass* 11(3): 46–52.

Small, E. 1974. American law and the species problem in *Cannabis. Microgram* 7: 131–132.

Small, E. 1975a. Morphological variation of achenes of *Cannabis. Canadian Journal of Botany* 53: 978–987.

Small, E. 1975b. American law and the species problem in *Cannabis*: Science and semantics. *Bulletin on Narcotics* 27(3): 1–20.

Small, E. 1975c. Essential considerations of the taxonomic debate in *Cannabis. Journal of Forensic Sciences* 20: 739–741.

Small, E. 1975d. On toadstool soup and legal species of *Cannabis. Plant Science Bulletin* 21: 34–39.

Small, E. 1975e. The case of the curious "*Cannabis*". *Economic Botany* 29: 254.

Small, E. 1976. The forensic taxonomic debate on *Cannabis*: Semantic hokum. *Journal of Forensic Sciences* 21: 239–251.

Small, E. 1977. Nomenclatural nonsense and legal marijuana plants. *Bulletin of the Pacific Tropical Botanical Garden* 7(1): 1–6.

Small, E. 1978a. A numerical and nomenclatural analysis of morpho-geographic taxa of *Humulus. Systematic Botany* 3: 37–76.

Small, E. 1978b. A numerical taxonomic analysis of the *Daucus carota* complex. *Canadian Journal of Botany* 56: 248–276.

Small, E. 1979a. *The species problem in Cannabis: Science and semantics. Volume 1, Science.* Toronto, ON: Corpus. 218 pp.

Small, E. 1979b. *The species problem in Cannabis: Science and semantics. Volume 2, Semantics.* Toronto, ON: Corpus. 156 pp.

Small, E. 1980. The relationships of hop cultivars and wild variants of *Humulus lupulus. Canadian Journal of Botany* 58: 676–686.

Small, E. 1981. A numerical analysis of morpho-geographic groups of cultivars of *Humulus lupulus* L. based on samples of cones. *Canadian Journal of Botany* 59: 311–324.

Small, E. 1995. Hemp. In: Smartt, J., and Simmonds, N.W. *Evolution of crop plants.* 2nd edition. Burnt Mill, Harlow, Essex, U.K.: Longman Scientific & Technical, 28–32.

Small, E. 1997. Cannabaceae. In: Flora North America Editorial Committee. ed. *Flora of North America, north of Mexico, Vol. 3.* New York: Oxford University Press, 381–387.

Small, E. 2004. Narcotic plants as sources of medicinals, nutraceuticals, and functional foods. In: Hou, F.-F., Lin, H.-S., Chou, M.-H., and Chang, T.-W. eds. *Proceedings of the international symposium on the development of medicinal plants* (Hualien, Taiwan, August 24–25 2004). Hualien, Taiwan: Hualien District Agricultural Research and Extension Station, 11–67.

Small, E. 2006. *Culinary herbs.* 2nd edition. Ottawa, ON: NRC Research Press. 1036 pp.

Small, E. 2007. *Cannabis* as a source of medicinals, nutraceuticals, and functional foods. In: Acharya, S.N., and Thomas, J.E. eds. *Advances in medicinal plant research.* Trivandrum, Kerala, India: Research Signpost/ Transworld Research Network, 1–39.

Small, E. 2009. *Top 100 food plants: The world's most important culinary crops.* Ottawa, ON: National Research Council Press. 636 pp.

Small, E. 2010. Blossoming treasures of biodiversity 33. Non-narcotic drug poppies—Benefits for people and biodiversity. *Biodiversity* 11(3 & 4): 73–80.

Small, E. 2011a. Blossoming treasures of biodiversity. 36. Castor bean—Taming the world's most poisonous plant. *Biodiversity* 12(3): 186–195.

Small, E. 2011b. *Alfalfa and relatives: Evolution and classification of Medicago.* Ottawa, ON: NRC Research Press and Wallingford, Oxon, U.K.: CABI. 727 pp.

Small, E. 2012. Hemp. In: Fredericks, S., Shen, L., Thompson, S., and Vasey, D. eds. *The Berkshire encyclopedia of sustainability: Vol. 4. Natural resources and sustainability.* Great Barrington, MA: Berkshire Publishing, 220–222.

Small, E. 2013a. *North American cornucopia: Top 100 indigenous food plants.* Boca Raton FL: CRC Press. 743 pp.

Small, E. 2013b. Blossoming treasures of biodiversity. 43. Cotton—Redeeming the world's most vilified plant threat to biodiversity. *Biodiversity* 14: 207–222.

Small, E. 2014a. Blossoming treasures of biodiversity. 44. Saguaro—Threatened monarch of the desert. *Biodiversity* 15: 39–53.

Small, E. 2014b. Hemp fiber and composites for the 21st century. In: Thakur, V.K.T, and Njuguna, J. eds. *Natural fibers and composites.* Houston, TX: Studium Press, 29–64.

Small, E. 2015a. Evolution and classification of *Cannabis sativa* (marijuana, hemp) in relation to human utilization. *Botanical Review* 81: 189–294.

Small, E. 2015b. Response to the erroneous critique of my *Cannabis* monograph by R.C. Clarke and M.D. Merlin. *Botanical Review* 81: 306–316.

Small, E. 2016. Blossoming treatures of biodiversity. 51. Hop (*Humulus lupulus*)—A bitter crop with sweet prospects. *Biodiversity* 17(1): in press.

Small, E., and Antle, T. 2003. A preliminary study of pollen dispersal in *Cannabis sativa. Journal of Industrial Hemp* 8(2): 37–50.

Small, E., and Antle, T. 2007. A study of cotyledon asymmetry in *Cannabis sativa. Journal of Industrial Hemp* 12(1): 3–14.

Small, E., and Beckstead, H.D. 1973a. Common cannabinoid phenotypes in 350 stocks of *Cannabis. Lloydia* 35: 144–165.

Small, E., and Beckstead, H.D. 1973b. Cannabinoid phenotypes in *Cannabis. Nature* 245: 147–148.

Small, E., and Beckstead, H.D. 1979. Cannabinoid composition of F_1 hybrids between "drug" and "non-drug" strains of *Cannabis.* In: Small, E. *The species problem in Cannabis: Science and semantics. Vol. 1, Science.* Toronto, ON: Corpus, 121–127.

Small, E., and Brookes, B. 2012. Temperature and moisture content for storage maintenance of germination capacity of seeds of industrial hemp, marijuana, and ditchweed forms of *Cannabis sativa. Journal of Natural Fibers* 9(4): 240–255.

Small, E., and Catling, P.M. 1999. *Canadian medicinal crops.* Ottawa, ON: NRC Press. 240 pp.

Small, E., and Catling, P.M. 2009. Blossoming treasures of biodiversity. 27. *Cannabis*—Dr. Jekyll and Mr. Hyde. *Biodiversity* 10(1): 31–38.

Small, E., and Cronquist, A. 1976. A practical and natural taxonomy for *Cannabis. Taxon* 25: 405–435.

Small, E., and Marcus, D. 2000. Hemp germplasm trials in Canada. In: Nova Institute. corporate ed. *Proceedings third international symposium bioresource hemp.* Hürth, Germany: Nova Corporation. Irregularly paginated.

Small, E., and Marcus, D. 2002. Hemp—A new crop with new uses for North America. In: Janick, J., and Whipkey, A. eds. *Trends in new crops and new uses.* Alexandria, VA: ASHS Press, 284–326.

Small, E., and Marcus, D. 2003. Tetrahydrocannabinol levels in hemp (*Cannabis sativa*) germplasm resources. *Economic Botany* 57: 545–558.

Small, E., and Marcus, D. 2004. Comment on THC development in *Cannabis sativa. Economic Botany* 58: 329.

Small, E., and Naraine, S.G.U. 2016a. Expansion of female sex organs in response to prolonged virginity in *Cannabis sativa* (marijuana). *Genetic Resources and Crop Evolution* 63: 339–348.

Small, E., and Naraine, S.G.U. 2016b. Size matters: Evolution of large drug-secreting resin glands in elite pharmaceutical strains of *Cannabis sativa* (marijuana). *Genetic Resources and Crop Evolution* 63: 349–359.

Small, E., Beckstead, H.D., and Chan, A. 1975. The evolution of cannabinoid phenotypes in *Cannabis*. *Economic Botany* 29: 219–232.

Small, E., Jui, P., and Lefkovitch, L.P. 1976. A numerical taxonomic analysis of *Cannabis* with special reference to species delimitation. *Systematic Botany* 1: 67–84.

Small, E., Pocock, T., and Cavers, P.B. 2003. The biology of Canadian weeds. 119. *Cannabis sativa* L. *Canadian Journal of Plant Science* 83: 217–237.

Small, E., Marcus, D., McElroy, A., and Butler, G. 2007. Apparent increase in biomass and seed productivity in hemp (*Cannabis sativa*) resulting from branch proliferation caused by the European corn borer (*Ostrinia nubilalis*). *Journal of Industrial Hemp* 12(1): 15–26.

Smiley, A. 1999. Marijuana: On-road and driving simulator studies. In: Kalant, H., Corrigal, W.A., Hall, W., and Smart, R.G. eds. *The health effects of cannabis*. Toronto, ON: Centre for Addiction and Mental Health, 171–191.

Smith, D., Bloor, R., George, C., Pysanenkod, A., and Španělad, P. 2015. Release of toxic ammonia and volatile organic compounds by heated cannabis and their relation to tetrahydrocannabinol content. *Analytical Methods* 7: 4104–4110.

Smith, M.H. 2012. *Heart of dankness. Underground botanists, outlaw farmers, and the race for the Cannabis Cup*. Toronto, ON: McClelland & Stewart. 256 pp.

Smith, P.H., Homish, G.G., Collins, R.L., Giovino, G.A., White, H.R., and Leonard, K.E. 2014. Couples' marijuana use is inversely related to their intimate partner violence over the first 9 years of marriage. *Psychology of Addictive Behaviors* 28: 734–742.

Smith, S.C., and Wagner, M.S. 2014. Clinical endocannabinoid deficiency (CECD) revisited: Can this concept explain the therapeutic benefits of cannabis in migraine, fibromyalgia, irritable bowel syndrome and other treatment-resistant conditions? *Neuro Endocrinology Letters* 35: 198–201.

Snoeijer, W. 2002. *A checklist of some Cannabaceae cultivars. Part A:* Cannabis. Leiden, the Netherlands: Division of Pharmacology, Leiden/Amsterdam Center for Drug Research. 60 pp.

Sokora, V.P. 1979. Breeding of monoecious hemp with minimal THC-content. *Szelek Szemen* 4: 11–12.

Solinas, M., Cinquina, V., and Parolaro, D. 2015. Cannabidiol and cancer—An overview of the preclinical data. In: Lichtor, T. ed. *Molecular considerations and evolving surgical management issues in the treatment of patients with a brain tumor*. Rijeka, Croatia: Intech, 399–421. (Open access e-book.)

Soros, G. 2010. Why I support legal marijuana. *Wall Street Journal* October 26, p. A.17.

Soth, J., Grasser, C., and Salerno, R. 1999. *The impact of cotton on fresh water resources and ecosystems*. Zurich, Switzerland: World Wildlife Fund International. 47 pp.

Spelman, K. 2009. "Silver bullet" drugs vs. traditional herbal remedies: Perspectives on malaria. *HerbalGram* 84: 44–55.

Spencer, R.D. 1999. Cultivated plants and the codes of nomenclature—Towards the resolution of a demarcation dispute. In: Andrews, S., Leslie, A.C., and Alexander, C. eds. *Taxonomy of cultivated plants: Third international symposium*. Kew, U.K: Royal Botanic Gardens, 171–181.

Spencer, R.D., and Cross, RG. 2007a. The international code of botanical nomenclature (ICBN), the international code of nomenclature for cultivated plants (ICNCP), and the cultigen. *Taxon* 56: 938–940.

Spencer, R.D., and Cross, RG. 2007b. The cultigen. In: Berg, R.G. van den, Groendijk-Wilders, N., Alexander, C., and Hetterscheid, W.L.A. eds. *Proceedings of the fifth international symposium on the taxonomy of cultivated plants*. Leuven, Belgium: International Society for Horticultural Science, 163–167.

Spielmann, D., Bracco, U., Traitler, H., Crozier, G., Holman, R., Ward, M. et al. 1988. Alternative lipids to usual omega 6 PUFAS: Gamma-linolenic acid, alpha-linolenic acid, stearidonic acid, EPA, etc. *Journal of Parenteral and Enteral Nutrition* 12(6 Suppl.): 111S–123S.

Sponner, J., Toth, L., Cziger, S., and Franck, R.R. 2005. Hemp. In: Franck, R.R. ed. *Bast and other plant fibres*. Boca Raton, FL: CRC Press, 176–206.

Spooner, D.M., van den Berg, R.G., Hetterscheid, W.L.A., and Brandenburg, W.A. 2003. Plant nomenclature and taxonomy. An horticultural and agronomic perspective. *Horticultural Reviews* 28: 1–59.

Stadtmauer, G., Beyer, K., Bardina, L., and Sicherer, S.H. 2003. Anaphylaxis to ingestion of hempseed (*Cannabis sativa*). *Journal of Allergy and Clinical Immunology* 112: 216–217.

Ständer, S., Schmelz, M., Metze, D., Luger, T., and Rukwied, R. 2005. Distribution of cannabinoid receptor 1 (CB1) and 2 (CB2) on sensory nerve fibers and adnexal structures in human skin. *Journal of Dermatological Science* 38: 177–188.

Stanwix, W., and Sparrow, A. 2014. *The hempcrete book. Designing and building with hemp-lime.* Cambridge, U.K.: Green Books. 368 pp.

Starks, M. 1990. *Marijuana chemistry: Genetics, processing and potency.* 2nd edition. Oakland, CA: Ronin Publishing. 199 pp.

Stearn, W.T. 1952. Proposed international code of nomenclature for cultivated plants. Historical introduction. *Journal of the Royal Horticultural Society* 77: 157–173.

Stearn, W.T. 1974. Typification of *Cannabis sativa* L. *Harvard University Botanical Museum Leaflets* 23: 325–336.

Steffens, S., and Pacher, P. 2015. The activated endocannabinoid system in atherosclerosis: Driving force or protective mechanism? *Current Drug Targets* 16: 334–341.

Ste-Marie, P.A., Fitzcharles, M.A., Gamsa, A., Ware, M.A., and Shir, Y. 2012. Association of herbal cannabis use with negative psychosocial parameters in patients with fibromyalgia. *Arthritis Care & Research* 64: 1202–1208.

Stepanov, G.S. 1974. Selection of hemp varieties in breeding for heterosis. *Cytology and Genetics* 8(5): 54–56.

Stepanov, G.S. 1976. Evaluation of the hybridizability of hemp varieties in breeding for heterosis. *Cytology and Genetics* 9(2): 17–20.

Stern, S.A., Bagenal, F., Ennico, K., Gladstone, G.R., Grundy, W.M., McKinnon, W.B. et al. 2015. The Pluto system: Initial results from its exploration by New Horizons. *Science* 350(6258). doi:10.1126/science. aad1815.

Stockwell, T., Vallance, K., Martin, G., Macdonald, S., Ivsins, A., Chow, C. et al. 2010. *The price of getting high, stoned and drunk in BC: A comparison of minimum prices for alcohol and other psychoactive substances. CARBC Statistical Bulletin 7.* Victoria, British Columbia: University of Victoria. 8 pp. http://www.uvic.ca/research/centres/carbc/assets/docs/bulletin7-price-of-getting-high.pdf.

Stokes, J.R., Hartel, R., Ford, L.B., and Casale, T.B. 2000. *Cannabis* (hemp) positive skin tests and respiratory symptoms. *Annals of Allergy, Asthma and Immunology* 85: 238–240.

Stone, S. 2010. *Bongology. The art of creating 35 of the world's most bongtastic marijuana ingestion devices.* Berkeley, CA: Ten Speed Press. 128 pp.

Stout, J.M., Boubakir, Z., Ambrose, S.J., Purves, R.W., and Page, J.E. 2012. The hexanoyl-CoA precursor for cannabinoid biosynthesis is formed by an acyl-activating enzyme in *Cannabis sativa* trichomes. *The Plant Journal* 71: 353–365.

Strohbeck-Kuehner, P., Skopp, P., and Mattern, R. 2008. Cannabis improves symptoms of ADHD. *Cannabinoids* 3: 1–3.

Strougo, A., Zuurman, L., Roy, C., Pinquier, J.L., van Gerven, J.M., Cohen, A.F. et al. 2008. Modelling of the concentration–effect relationship of THC on central nervous system parameters and heart rate—Insight into its mechanisms of action and a tool for clinical research and development of cannabinoids. *Journal of Psychopharmacology* 22: 717–726.

Struik, P.C., Amaducci, S., Bullard, M.J., Stutterheim, N.C., Venturi, G., and Cromack, H.T.H. 2000. Agronomy of fibre hemp (*Cannabis sativa* L.) in Europe. *Industrial Crops and Products* 11: 107–118.

Stuart, J.M., Leishman, E., and Bradshaw, H.B. 2014. Reproduction and cannabinoids: Ups and downs, ins and outs. In: Pertwee, R.G. ed. *Handbook of cannabis.* Oxford, U.K.: Oxford University Press, 245–260.

Styles, B.T. ed. 1986. *Infraspecific classification of wild and cultivated plants.* Oxford, U.K.: Clarendon Press. 435 pp.

Su, M., Yang, R., and Li, M. 2013. Biodiesel production from hempseed oil using alkaline earth metal oxides supporting copper oxide as bi-functional catalysts for transesterification and selective hydrogenation. *Fuel* 103: 398–407.

Sugiura, T, Oka, S., Ikeda, S., and Waku, K. 2005 [imprinted 2006]. Occurrence, biosynthesis, and metabolism of endocannabinoids. In: Onaivi, E.S., Sugiura, T., and di Marzo, V. eds. *Endocannabinoids: The brain and body's marijuana and beyond.* Boca Raton, FL: CRC Press, 177–214.

Sullivan, N., Elzinga, S., and Raber, J.C. 2013. Determination of pesticide residues in cannabis smoke. *Journal of Toxicology*, Article ID 378168. http://dx.doi.org/10.1155/2013/378168.

Sun, X., Xu, C.S., Chadha, N., Chen, A., and Liu, J. 2015. Marijuana for glaucoma: A recipe for disaster or treatment? *Yale Journal of Biology and Medicine* 88: 265–269.

Sutton, S., Lum, B.L., and Torti, F.M. 1986. Possible risk of invasive pulmonary aspergillosis with marijuana use during chemotherapy for small cell lung cancer. *Drug Intelligence and Clinical Pharmacy* 20: 289–291.

Suzanne, K. 2009. *Krysten Suzanne's ultimate raw vegan hemp recipes: Fast and easy raw food hemp recipes for delicious soups, salads, dressings, bread, crackers, butter, spreads, dips, breakfast, lunch, dinner and desserts.* Scottsdale, AZ: Green Butterfly Press. 116 pp.

Svrakic, D.M., Lustman, P.J., Mallya, A., Lynn, T.A., Finney, R., and Svrakic, N.M. 2012. Legalization, decriminalization and medicinal use of cannabis: A scientific and public health perspective. *Missouri Medicine* 109(2): 90–98.

Swerts, S., Van Gasse, A., Leysen, J., Faber, M., Sabato, V., Bridts, C.H. et al. 2014. Allergy to illicit drugs and narcotics. *Clinical & Experimental Allergy* 44: 307–318.

Sytnik, V.P., and Stelmah, A.F. 1999. The character of inheritance of differences in cannabinoid content in hemp (*Cannabis sativa* L.). *Journal of the International Hemp Association* 6(1): 8–9.

Sytsma, K.J., Morawetz, J., Pires, J.C., Nepokroeff, M., Conti, E., Zjhra, M. et al. 2002. Urticalean rosids: Circumscription, rosid ancestry, and phylogenetics based on rbcL, trnL-trnF, and ndhF sequences. *American Journal of Botany* 89: 1531–1546.

Szyper-Kravitz, M., Lang, R., Manor, Y., and Lahav, M. 2001. Early invasive pulmonary aspergillosis in a leukemia patient linked to aspergillus contaminated marijuana smoking. *Leukemia & Lymphoma* 42: 1433–1437.

Tamburini, E., León, A.G., Perito, B., Di Candilo, M., and Mastromei, G. 2004. Exploitation of bacterial pectinolytic strains for improvement of hemp water retting. *Euphytica* 140: 47–54.

Tanaka, H., Degawa, M., Kawata, E., Hayashi, J., and Shoyama, Y. 1998. Identification of *Cannabis* pollens using an allergic patient's immunoglobulin E and purification and characterization of allergens in *Cannabis* pollens. *Forensic Science International* 97: 139–153.

Tang, C.-H., Ten, Z., Wang, X.-S., and Yang, X.-Q. 2006. Physicochemical and functional properties of hemp (*Cannabis sativa* L.) protein isolate. *Journal of Agricultural and Food Chemistry* 54: 8945–8950.

Tang, C.-H., Wang, X.-S., and Yang, X.-Q. 2009. Enzymatic hydrolysis of hemp (*Cannabis sativa* L.) protein isolate by various proteases and antioxidant properties of the resulting hydrolysates. *Food Chemistry* 114: 1484–1490.

Tarter, R.E., Kirisci, L., Mezzich, A., Ridenour, T., Fishbein, D., Horner, M. et al. 2012. Does the "gateway" sequence increase prediction of cannabis use disorder development beyond deviant socialization? Implications for prevention practice and policy. *Drug and Alcohol Dependence* 123(Suppl. 1): S72–S78.

Tashkin, D.P. 2002. Respiratory risks from marijuana smoking. In: Grotenhermen, F., and Russo, E. eds. *Cannabis and cannabinoids: Pharmacology, toxicology, and therapeutic potential.* New York: Haworth Integrative Healing Press, 325–335.

Tashkin, D.P. 2005. Smoked marijuana as a cause of lung injury. *Monaldi Archives for Chest Disease* 63: 93–100.

Tashkin, D.P. 2013. Effects of marijuana smoking on the lung. *Annals of the American Thoracic Society* 10: 239–247.

Taura, F., Morimoto, S., Shoyama, Y., and Mechoulam, R. 1995. First direct evidence for the mechanism of delta-1-tetrahydrocannabinoloic acid biosynthesis. *Journal of the American Chemical Society* 38: 9766–9767.

Taura, F., Morimoto, S., and Shoyama, Y. 1996. Purification and characterization of cannabidiolic-acid synthase from *Cannabis sativa* L. *Journal of Biological Chemistry* 271: 17411–17416.

Taura, F., Sirikantaramas, S., Shoyama, Y., Yoshikai, K., Shoyama, Y., and Morimoto, S. 1997. Cannabidiolic-acid synthase, the chemotype-determining enzyme in the fiber-type *Cannabis sativa*. *FEBS Letters* 581: 2929–2934.

Taura, F. Dono, E., Sirikantaramas, S., Yoshimura, K., Shoyama, Y., and Morimoto, S. 2007. Production of Δ_1-tetrahydrocannabinolic acid by the biosynthetic enzyme secreted from transgenic *Pichia pastoris*. *Biochemical and Biophysical Research Communications* 361: 675–680.

Taura, F., Sirikantaramas, S., Shoyama, Y., Shoyama, Y, and Morimoto, S. 2009. Phytocannabinoids in *Cannabis sativa*: Recent studies on biosynthetic enzymes. In: Lambert, D.M. ed. *Cannabinoids in nature and medicine*. New York: Wiley-VCH, 51–65.

Taylor, F.M. 1988. Marijuana as a potential respiratory tract carcinogen: A retrospective analysis of a community hospital population. *Singapore Medical Journal* 81: 1213–1216.

Techen, N., Chandra, S., Lata, H., Elsohly, M.A., and Khan, I.A. 2010. Genetic identification of female *Cannabis sativa* plants at early developmental stage. *Planta Medica* 76: 1938–1939.

Teh, S.-S., and Birch, J. 2013. Physicochemical and quality characteristics of cold-pressed hemp, flax and canola seed oils. *Journal of Food Composition and Analysis* 30: 26–31.

Telek, A., Bíró, T., Bodó, E., Tóth, B.I., Borbíró, I., Kunos, G. et al. 2007. Inhibition of human hair follicle growth by endo- and exocannabinoids. *FASEB Journal* 21: 3534–3541.

Tennstedt, T., and Saint-Remy, A. 2011. Cannabis and skin diseases. *European Journal of Dermatology* 21: 5–11.

Tessmer, A., Berlin, N., Sussman, G., Leader, N., Chung, E.C., and Beehzhold, D. 2012. Hypersensitivity reactions to marijuana. *Annals of Allergy Asthma & Immunology* 108: 282–284.

The Furrow. 1990. Growing borage for the health of it. *The Furrow* 95(6): 19.

Theimer, R.R., and Mölleken, H. 1995. Analysis of the oils from different hemp (*Cannabis sativa* L.) cultivars—perspectives for economic utilization. In: Nova Institute. corporate ed. *Bioresource hemp—proceedings of the symposium* (Frankfurt am Main, Germany, March 2–5, 1995). 2nd edition. Ojai, CA: Hemptech, 536–545.

Theis, N., and Lerdau, M. 2003. The evolution of function in plant secondary metabolites. *International Journal of Plant Science* 164(3): S93–S102.

Thomas, B.F., Wiley, J.L., Pollard, G.T., and Grabenauer, M. 2014a. Cannabinoid designer drugs: Effects and forensics. In: Pertwee, R.G. ed. *Handbook of cannabis*. Oxford, U.K.: Oxford University Press, 710–729.

Thomas, G., Kloner, R.A., and Rezkalla, S. 2014b. Adverse cardiovascular, cerebrovascular, and peripheral vascular effects of marijuana inhalation: What cardiologists need to know. *American Journal of Cardiology* 113: 187–190.

Thomas, T.G., Sharma, S.K., Prakash, A., and Sharma, B.R. 2000. Insecticidal properties of essential oil of *Cannabis sativa* Linn. against mosquito larvae. *Entomon* 25: 21–24.

Thompson, C., Sweitzer, R., Gabriel, M., Purcell, K., Barrett, R., and Poppenga, R. 2014. Impacts of rodenticide and insecticide toxicants from marijuana cultivation sites on fisher survival rates in the Sierra National Forest, California. *Conservation Letters* 7: 91–102.

Thompson, E.C., Berger, M.C., and Allen, S. 1998. *Economic impact of industrial hemp in Kentucky*. Lexington, KY: University of Kentucky, Center for Business and Economic Research. 66 pp.

Thompson, J.W. Jr., and Koenen, M.A. 2011. Physicians as gatekeepers in the use of medical marijuana. *Journal of the American Academy of Psychiatry and the Law* 39: 460–464.

Thornton, J.I., and Nakamura, G.R. 1972. The identification of marijuana. *Journal of the Forensic Science Society* 12: 461–519.

Thygesen, A., Daniel, G., Lilholt, H., and Thomsen, A.B. 2005. Hemp fiber microstructure and use of fungal defibration to obtain fibers for composite materials. *Journal of Natural Fibers* 2(4): 19–37.

Tiourebaev, K.S., Semenchenko, G.V., Dolgovskaya, M., McCarthy, M.K., Anderson, T.W., Carsten, L.D. et al. 2001. Biological control of infestations of ditchweed (*Cannabis sativa*) with *Fusarium oxysporum* f. sp. *cannabis* in Kazakhstan. *Biocontrol Science and Technology* 11: 535–540.

Todaro, B. 2012. Cannabinoids in the treatment of chemotherapy-induced nausea and vomiting. *Journal of the National Comprehensive Cancer Network* 10: 487–492.

Tomida, I., Pertwee, R.G., and Azuara-Blanco, A. 2004. Cannabinoids and glaucoma. *British Journal of Ophthalmology* 88: 708–713.

Toole, E.H., Toole, V.K., and Nelson, E.G. 1960. *Preservation of hemp and kenaf seed*. Washington, DC: U.S. Department of Agriculture, Agricultural Research Service Technical Bulletin 1215. 16 pp.

Toonen, M., Ribot, S., and Thissen, J. 2006. Yield of illicit indoor cannabis cultivation in the Netherlands. *Journal of Forensic Science* 51: 1050–1054.

Törjék, O., Bucherna, N., Kiss, E., Homoki, H., Finta-Korpelová, Z., Bócsa et al. 2002. Novel male-specific molecular markers (MADC5, MADC6) in hemp. *Euphytica* 127: 209–218.

Torres, S., Lorente, M., Rodríguez-Fornés, F., Hernández-Tiedra, S., Salazar, M., García-Taboada, E. et al. 2011. A combined preclinical therapy of cannabinoids and temozolomide against glioma. *Molecular Cancer Therapeutics* 10: 90–103.

Tóth, A., Blumberg, P.M., and Boczán, J. 2009. Anandamide and the vanilloid receptor (TRPV1). *Vitamins and Hormones* 81: 389–419.

Tournois, J. 1912. Influence de la lumière sur la floraison du houblon japonais et du chanvre déterminées par des semis haitifs. *Comptes Rendus Hebdomadaires des Séances de l'Academie des Sciences, Paris* 155: 297–300.

Touw, M. 1981. The religious and medical uses of cannabis in China, India and Tibet. *Journal of Psychoactive Drugs* 13: 23–34.

Townshend, J.M., Boleyn, J.M., McGill, C.R., and Rowarth, J.S. 2012. Plant density effect on oil seed yield and quality of industrial hemp cv. Fasamo in Canterbury. In: McGill, C.R., and Rowarth, J.S. eds. *Seed symposium: Seeds for futures. Proceedings of a joint symposium between the Agronomy society of New Zealand and the New Zealand Grassland Association* (Massey University, Palmerston North, New Zealand, November 26–27, 2008). Palmerston North, New Zealand: Massey University, 85–91.

Toyota, M., Kinugawa, T., and Asakawa, Y. 1994. Bibenzyl cannabinoid and bisbibenzyl derivative from the liverwort *Radula perrottetii*. *Phytochemistry* 37: 859–862.

Toyota, M., Shimamura, T., Ishii, H., Renner, M., Braggins, J., and Asakawa, Y. 2002. New bibenzyl cannabinoid from the New Zealand liverwort *Radula marginata*. *Chemical and Pharmaceutical Bulletin* (Tokyo) 50: 1390–1392.

Trachtenberg, A.I. 1994. Opiates for pain: Patients' tolerance and society's intolerance. *JAMA* 271: 427.

Trillou, C.R., Delgorge, C., Menet, C., Arnone, M., and Soubrié, P. 2004. CB1 cannabinoid receptor knockout in mice leads to leanness, resistance to diet-induced obesity and enhanced leptin sensitivity. *International Journal of Obesity* 28: 640–648.

Tringale, R., and Jensen, C. 2011. Cannabis and insomnia. *O'Shaughnessy's—The Journal of Cannabis in Clinical Practice* 2011(Autumn): 31–32.

Troccaz, M., Borchard, G., Vuilleumier, C., Raviot-Derrien, S., Niclass, Y., Beccucci, S. et al. 2009. Gender-specific differences between the concentrations of nonvolatile (R)/(S)-3-methyl-3-sulfanylhexan-1-Ol and (R)/(S)-3-hydroxy-3-methyl-hexanoic acid odor precursors in axillary secretions. *Chemical Senses* 34: 203–210.

Trofin, I.G., Vlad, C.C., Dabija, G., and Filipescu, L. 2011. Influence of storage conditions on the chemical potency of herbal cannabis. *Revistade Chimie* 62: 639–645.

Trofin, I.G., Dabija, G., Văireanu, D.-I., and Filipescu, L. 2012. The influence of long-term storage conditions on the stability of cannabinoids derived from cannabis resin. *Revistade Chimie* 63: 422–427.

Truță, E., Olteanu, Z., Surdu, S., Zamfirache, M.-M., and Oprică, L. 2007. Some aspects of sex determinism in hemp. *Analele Științifice ale Universității, Alexandru Ioan Cuza, Secțiunea Genetică și Biologie Moleculară* 8: 31–39.

Tubaro, A., Giangaspero, A., Sosa, S., Negri, R., Grassi, G., Casano, S. et al. 2010. Comparative topical anti-inflammatory activity of cannabinoids and canabivarins. *Fitoterapia* 81: 816–819.

Turcotte, D., Le Dorze, J.A., Esfahani, F., Frost, E., Gomori, A., and Namaka, M. 2010. Examining the roles of cannabinoids in pain and other therapeutic indications: A review. *Expert Opinion on Pharmacotherapy* 11: 17–31.

Turner, C.E., Hadley, K.W., Fetterman, P.S., Doorenbos, N.J., Quimby, M.W., and Waller, C. 1973. Constituents of *Cannabis sativa* L. IV: Stability of cannabinoids in stored plant material. *Journal of Pharmaceutical Sciences* 62: 1601–1605.

Turner, C., Fetterman, P., Hadley, K., and Urbanek, J. 1975. Constituents of *Cannabis sativa* L. X. Cannabinoid profile of a Mexican variant and its possible correlation to pharmacological activity. *Acta Pharmaceutica Jugoslavica* 25: 7–15.

Turner, J.C., Hemphill, J.K., and Mahlberg, P.G. 1980. Trichomes and cannabinoid content of developing leaves and bracts of *Cannabis sativa* L. (Cannabaceae). *American Journal of Botany* 67: 1397–1406.

Turner, J.C., Hemphill, J.K., and Mahlberg, P.G. 1981a. Interrelationships of glandular trichomes and cannabinoid content. I. Developing pistillate bracts of *Cannabis sativa* L. (Cannabaceae). *Bulletin on Narcotics* 33(2): 59–69.

Turner, J.C., Hemphill, J.K., and Mahlberg, P.G. 1981b. Interrelationships of glandular trichomes and cannabinoid content. II. Developing vegetative leaves of *Cannabis sativa* L. (Cannabaceae). *Bulletin on Narcotics* 33(3): 63–71.

Tutin, T.G., and Edmonson, J.R. 1993. Cannabaceae. In: Tutin, T.G., and Edmonson, J.R. eds. *Flora Europaea, Volume 1*. 2nd edition. Cambridge, U.K.: University of Cambridge, 78.

Tyler, T.R. 1990. *Why people obey the law*. New Haven, CT: Yale University Press. 273 pp. (Re-published in 2006 with an afterword.)

Tyler, V.E. 1993a. *The honest herbal*. 3rd ed. Binghamton, NY: Pharmaceutical Press. 209 pp.

Tyler, V.E. 1993b. Phytomedicines in Western Europe—Potential impact on herbal medicine in the United States. In: Kinghorn, A.D., and Balandrin, M.F. eds. *Human medicinal agents from plants*. San Francisco: American Chemical Society, 25–37.

Tyler, V.E. 1996. What pharmacists should know about herbal remedies. *Journal of the American Pharmaceutical Association (Washington, New Series)* 36: 29–37.

Tyler, V.E., and Foster, S. 1996. Herbs and phytomedicinal products. In: Covington, T.R. ed. *Handbook of non-prescription drugs*. 11th ed. (2 volumes) Washington, DC: American Pharmaceutical Association, 695–713.

Tzadok, M., Uliel-Siboni, S., Linder, I., Kramer, U., Epstein, O., Menascu, S. et al. 2016. CBD-enriched medical cannabis for intractable pediatric epilepsy: The current Israeli experience. *Seizure* 35: 41–44.

United Nations. 1987. *Recommended methods for testing cannabis: Manual for use by national narcotics laboratories*. New York: United Nations. 38 pp.

United Nations. 2014. *United Nations Office on Drugs and Crime, world drug report 2014*. Vienna, Austria: United Nations Office on Drugs and Crime. 197 pp. + annexes. http://www.unodc.org/wdr2014.

Upton, R., Craker, L., ElSohly, M., Romm, A., Russo, E., and Sexton, M. eds. 2013. *American herbal pharmacopoeia: Cannabis inflorescence: Cannabis spp.: Standards of identity, analysis, and quality control.* Scott's Valley, CA: American Herbal Pharmacopoeia. 63 pp.

U.S. Department of Agriculture. 2004. *Classical biological control of narcotic plants.* [Project proposal, 1999–2004.] http://portal.nifa.usda.gov/web/crisprojectpages/0402468-classical-biological-control-of-narcotic-plants.html.

U.S. House of Representatives. 2005. *Marijuana and medicine: The need for a science-based approach: Hearing before the Subcommittee on Criminal Justice, Drug Policy, and Human Resources of the Committee on Government Reform, House of Representatives, One Hundred Eighth Congress, second session, April 1, 2004.* Washington, DC: U.S. Government Printing Office. 352 pp. http://purl.access.gpo.gov/GPO/LPS58580.

U.S. Senate Committee on Agriculture, Nutrition, and Forestry. 1988. *Public safety issues surrounding marijuana production in national forests. Joint hearing before the Subcommittee on Nutrition and Investigations and the Subcommittee on Conservation and Forestry of the Committee on Agriculture, Nutrition, and Forestry, United States Senate, One Hundredth Congress.* Washington, DC: U.S. Government Printing Office. 87 pp.

Valladares Juárez, A.G., Dreyer, J., Göpel, P.K., Koschke, N., Frank, D., Märkl, H. et al. 2009. Characterisation of a new thermoalkaliphilic bacterium for the production of high-quality hemp fibres, *Geobacillus thermoglucosidasius* strain PB94A. *Applied Microbiology and Biotechnology* 83: 521–527.

Valle, J.R., Vieira, J.E.V., Aucélio, J.G., and Valio, I.F.M. 1978. Influence of photoperiodism on cannabinoid content of *Cannabis sativa* L. *Bulletin on Narcotics* 30(1): 67–68.

Van Bakel, H., Stout, J.M., Cote, A.G., Tallon, C.M., Sharpe, A.G., Hughes, T.R. et al. 2011. The draft genome and transcriptome of *Cannabis sativa. Genome Biology.* doi:10.1186/gb-2011-12-10-r102. http://genomebiology.com/content/pdf/gb-2011-12-10-r102.pdf.

Vance, J.M. 1971. Marijuana is for the birds. *Outdoor Life* 147(6): 53–55, 96–100.

Van der Werf, H.M.G. 1991. *Agronomy and crop physiology of fibre hemp. A literature review. Report 142.* Wageningen, the Netherlands: Center for Agrobiological Research. 16 pp.

Van der Werf, H.M.G. 1993. *Fibre hemp in Ukraine. Report of a visit to the Institute of Bast Crops of the Ukrainian Academy of Agrarian Sciences at Glukhov, Ukraine, 14 to 22 August 1998.* Wageningen, the Netherlands: Agricultural University, Wageningen. Unpaginated (8 pp.).

Van der Werf, H.M.G. 1994a. *Crop physiology of fibre hemp (Cannabis sativa L.).* Dissertation, Wageningen, the Netherlands: Wageningen Agricultural University. 152 pp.

Van der Werf, H.M.G. 1994b. Hemp facts and hemp fiction. *Journal of the International Hemp Association* 1(1): 58–59.

Van der Werf, H.M.G. 2004. Life cycle analysis of field production of fibre hemp, the effect of production practices on environmental impacts. *Euphytica* 140: 13–23.

Van der Werf, H.M.G., and Turunen, L. 2008. The environmental impacts of the production of hemp and flax yarn. *Industrial Crops and Products* 27: 1–10.

Van der Werf, H.M.G., and van den Berg, W. 1995. Nitrogen fertilization and sex expression affect size variability of fibre hemp (*Cannabis sativa* L.). *Oecologia* 103: 462–470.

Van der Werf, H., Haasken, H., and Wijlhuizen, M. 1994a. The effect of daylength on yield and quality of fibre hemp (*Cannabis sativa* L.). *European Journal of Agronomy* 3: 117–123.

Van der Werf, H.M.G., van der Veen, J.E.H., Bouma, A.T.M., and Cate, M. 1994b. Quality of hemp (*Cannabis sativa* L.) stems as raw material for paper. *Industrial Crops and Products* 2: 219–227.

Van der Werf, H.M.G., Wijhuisen, M., and de Schutter, J.A.A. 1995a. Plant density and self-thinning affect yield and quality of fibre hemp (*Cannabis sativa* L.). *Field Crops Research* 40: 153–164.

Van der Werf, H.M.G., Brouwer, K., Wijhuisen, M., and Withagen, J.C.M. 1995b. The effect of temperature on leaf appearance and canopy establishment in fibre hemp (*Cannabis sativa* L.). *Annals of Applied Biology* 126: 551–561.

Van der Werf, H.M.G., van Geel, W.C.A., van Gils, L.J.C., and Haverkort, A.J. 1995c. Nitrogen fertilization and row width affect thinning and productivity of fibre hemp (*Cannabis sativa* L.). *Field Crops Research* 42: 27–37.

Van der Werf, H.M.G., Mathijssen, E.W.J.M., and Haverkort, A.J. 1996. The potential of hemp (*Cannabis sativa* L.) for sustainable fibre production: A crop physiological appraisal. *Annals of Applied Biology* 129: 109–123.

Van der Werf, H.M.G., Mathijssen, E.W.J.M., and Harverkort, A.J. 1998. Crop physiology of *Cannabis sativa* L.: A simulation study of potential yield of hemp in Northwest Europe. In: Ranalli, P. ed. *Advances in hemp research.* London: Food Products Press (of Haworth Press), 85–108.

Vandrey, R., Dunn, K.E., Fry, J.A., and Girling, E.R. 2012. A survey study to characterize use of Spice products (synthetic cannabinoids). *Drug and Alcohol Dependence* 120: 238–241.

Vanhove, W., Van Damme, P., and Meert, N. 2011. Factors determining yield and quality of illicit indoor cannabis (*Cannabis* spp.) production. *Forensic Science International* 212: 158–163.

Vanhove, W., Surmont, T., Van Damme, P., and De Ruyver, B. 2012. Yield and turnover of illicit indoor cannabis (*Cannabis* spp.) plantations in Belgium. *Forensic Science International* 220: 265–270.

Van Klingeren, B., and Ten Ham, M. 1976. Antibacterial activity of delta-9-tetrahydrocannabinol and cannabidiol. *Antonie van Leeuwenhoek Journal of Microbiology and Serology* 42: 9–12.

Van Lai, T. 1985. Effect of inbreeding on some major characteristics of hemp. *Acta Agronomica Academiae Scientiarum Hungaricae* 34(1/2): 77–84.

Van Ours, J.C, and Williams, J. 2012. The effects of cannabis use on physical and mental health. *Journal of Health Economics* 31: 564–577.

Van Roekel, G.J. Jr. 1994. Hemp pulp and paper production. *Journal of the International Hemp Association* 1(1): 12–14.

Van Roekel, G.J., Lips, S.J.J., Op den Kamp, R.G.M., Baron, G. 1995. Extrusion pulping of true hemp bast fibre (*Cannabis sativa* L.). In: *Proceedings of the 1995 TAPPI pulping conference* (Chicago, IL, October 2–6, 1995). Atlanta, GA: TAPPI Press, 477–485.

Van Soest, L.J.M., Mastebroeke, H.D., and de Meijer, E.P.M. 1993. Genetic resources and breeding: A necessity for the success of industrial crops. *Industrial Crops and Products* 1: 283–288.

Van Winter, J.M. 1981. The use of cannabis in two cookery books of the fifteenth century. In: Fenton, A., and Owen, T.M. eds. *Food in Perspective, Proceedings of the third international conference on ethnological food research, Cardiff, Wales, 1977*. Edinburgh, U.K.: John Donald, 401–407.

Vardakou, I., Pistos, C., and Spiliopoulou, C. 2010. Spice drugs as a new trend: Mode of action, identification and legislation. *Toxicology Letters* 197: 157–162.

Vavilov, N.I. 1926a. *Studies on the origin of cultivated plants*. Leningrad, U.S.S.R.: Institute of Applied Botany and *Plant Breeding*. 248 pp. (In Russian and English.)

Vavilov, N.I. 1926b. The origin of the cultivation of "primary" crops, in particular of cultivated hemp. *Trudy po Prikladnoj Botanike i Selekcii* 16(2): 221–233.

Vavilov, N.I. 1931. The role of Central Asia in the origin of cultivated plants. *Bulletin of Applied Botany, Genetics, and Plant Breeding* 26(3): 3–44. (In Russian and English.)

Vavilov, N.I. 1992. *Origin and geography of cultivated plants*. Cambridge, U.K.: Cambridge University Press. 498 pp. (Based on a 1987 collection of Vavilov's publications in Russian, translated into English by D. Löve.)

Velasco, G., Sánchez, C., and Guzmán, M. 2012. Towards the use of cannabinoids as antitumour agents. *National Reviews Cancer* 12: 436–444.

Velasco, G., Sánchez, C., and Guzmán, M. 2014. Cancer. In: Pertwee, R.G. ed. *Handbook of cannabis*. Oxford, U.K.: Oxford University Press, 626–643.

Veliky, I.A., and Genest, K. 1972. Growth and metabolites of *Cannabis sativa* cell suspension cultures. *Lloydia* 35: 450–456.

Venderová, K., Růzicka, E., Vorísek, V., and Visnovský, P. 2004. Survey on cannabis use in Parkinson's disease: Subjective improvement of motor symptoms. *Movement Disorders* 19: 1102–1106.

Venkatesan, T., Sengupta, J., Lodhi, A., Schroeder, A., Adams, K., Hogan, W.J. et al. 2014. An internet survey of marijuana and hot shower use in adults with cyclic vomiting syndrome (CVS). *Experimental Brain Research* 232: 2563–2570.

Vera, C.L., Malhi, S.S., Raney, J.P., and Wang, Z.H. 2004. The effect of N and P fertilization on growth, seed yield and quality of industrial hemp in the Parkland region of Saskatchewan. *Canadian Journal of Plant Science* 84: 939–947.

Vera, C.L., Woods, S.M., and Raney, J.P. 2006. Seeding rate and row spacing effect on weed competition, yield and quality of hemp in the Parkland region of Saskatchewan. *Canadian Journal of Plant Science* 86: 911–915.

Vera, C.L., Malhi, S.S., Phelps, S.M., May, W.E., and Johnson, E.N. 2010. N, P, and S fertilization effects on industrial hemp in Saskatchewan. *Canadian Journal of Plant Science* 90: 179–184.

Verkaar, H.J. 1986. When does grazing benefit plants? *Trends in Ecology and Evolution* 1: 168–169.

Verweij, P.E., Kerremans, J.J., Voss, A., and Meis, J.F. 2000. Fungal contamination of tobacco and marijuana. *JAMA* 284: 2875.

Virovets, V.G. 1996. Selection for non-psychoactive hemp varieties (*Cannabis sativa* L.) in the CIS (former USSR). *Journal of the International Hemp Association* 3(1): 13–15.

Virovets, V.G., Senchenko, G.I., Gorshkova, L.M., and Sazhko, M.M. 1991. Narcotic activity of *Cannabis sativa* L. and prospects of breeding for a reduction in content of cannabinoids. *Agricultural Biology* 1: 35–49. (In Russian.)

Vivanco, J.M., and Baluška, F. eds. 2012. *Secretions and exudates in biological systems*. Berlin, Germany: Springer. 284 pp.

Viveros, M.P., Rodriguez de Fonseca, F., Bermudez-Silva, F.J., and McPartland, J.M. 2008. Critical role of the endocannabinoid system in the regulation of food intake and energy metabolism, with phylogenetic, developmental, and pathophysiological implications. *Endocrine, Metabolic & Immune Disorders—Drug Targets* 8: 220–230.

Vogelmann, A.F., Turner, J.C., and Mahlberg, P.G. 1988. Cannabinoid composition in seedlings compared to adult plants of *Cannabis sativa*. *Journal of Natural Products* 51: 1075–1079.

Vogl, C.R., Mölleken, H., Lissek-Wolf, G., Surböck, A., and Kobert, J. 2004. Hemp (*Cannabis sativa* L.) as a resource for green cosmetics: Yield of seed and fatty acid composition of 20 varieties under the growing conditions of organic farming in Austria. *Journal of Industrial Hemp* 9(1): 51–68.

Volkow, N.D., Baler, R.D., Compton, W.M., and Weiss, S.R. 2014. Adverse health effects of marijuana use. *New England Journal of Medicine* 370: 2219–2227.

Vonapartis, E., Aubin, M.-P., Seguin, P., Mustafa, A.F., and Charron, J.-B. 2015. Seed composition of ten industrial hemp cultivars approved for production in Canada. *Journal of Food Composition and Analysis* 39: 8–12.

Von Henk, L.F.W., Nieth, E., and von Werner, A. 1895. *Zur See*. 3rd edition. Hamburg, Germany: Verlagsanstalt und Druckerei. 417 pp. (In German.)

Voth, E.A. 2001. Guidelines for prescribing medical marijuana. *Western Journal of Medicine* 175: 305–306.

Wagner, G.J. 1990. Secreting glandular trichomes: More than just hairs. *Plant Physiology* 96: 675–679.

Wagner, G.J., Wang, E., and Shepherd, W. 2004. New approaches for studying and exploiting an old protuberance, the plant trichome. *Annals of Botany* 93: 3–11.

Waldman, M., Hochhauser, E., Fishbein, M., Aravot, D., Shainberg, A., and Sarne, Y. 2013. An ultra-low dose of tetrahydrocannabinol provides cardioprotection. *Biochemical Pharmacology* 85: 1626–1633.

Walker, L.A., Harland, E.C., Best, A.M., and ElSohly, M.A. 1999. Δ⁹-THC hemisuccinate in suppository form as an alternative to oral and smoke THC. In: Nahas, G.G., Sutin, K.M., Harvey, D.J., and Agurell, S. eds. *Marihuana and medicine*. Totowa, NJ: Humana Press, 123–126.

Wallace, E.A., Andrews, S.E., Garmany, C.L., and Jelley, M.J. 2011. Cannabinoid hyperemesis syndrome: Literature review and proposed diagnosis and treatment algorithm. *Southern Medical Journal* 104: 659–664.

Walsh, Z., Callaway, R., Belle-Isle, L., Capler, R., Kay, R., Lucas, P. et al. 2013. Cannabis for therapeutic purposes: Patient characteristics, access, and reasons for use. *International Journal on Drug Policy* 24: 511–516.

Walter, H. 1938. *Cannabis*. In: Kirchner, O.V., Loew, E., Schröter, C., and Wangerin, W. eds. *Lebengeschichte der Blütenpflanzen Mitteleuropas, Vol. 2*. Stuttgart, Germany: Eugen Ulment, 875–909. (In German.)

Walther, S., and Halpern, M. 2010. Cannabinoids and dementia: A review of clinical and preclinical data. *Pharmaceuticals* 3: 2689–2708.

Wang, H.-D., and Wei, Y.-F. 2012. Survey on the germplasm resources of *Cannabis sativa* L. *Medicinal Plant* 3(7): 11–14.

Wang, H., Xu, Z., Kohandehghan, A., Li, Z., Cui, K., Tan, X. et al. 2013. Interconnected carbon nanosheets derived from hemp for ultrafast supercapacitors with high energy. *ACS Nano* 7: 5131–5141.

Wang, Q., and Shi, G. 1999. Industrial hemp: China's experience and global implications. *Review of Agricultural Economics* 21: 344–357.

Wang, R., He, L.-H., Xia, B., Tong, J.-F., Li, N., and Peng, F. 2009. A micropropagation system for cloning of hemp (*Cannabis sativa* L.) by shoot tip culture. *Pakistan Journal of Botany* 41: 603–608.

Wang, T., Collet, J.P., Shapiro, S., and Ware, M.A. 2008a. Adverse effects of medical cannabinoids: A systematic review. *CMAJ (Canadian Medical Association Journal)* 178: 1669–1678.

Wang, X.-S., Tang, C.-H., Yang, X.-Q., and Guo, W.-R. 2008b. Characterization, amino acid composition and in vitro digestibility of hemp (*Cannabis sativa* L.) proteins. *Food Chemistry* 107: 11–18.

Ware, M.A., Adams, H., and Guy, G.W. 2005. The medicinal use of cannabis in the UK: Results of a nationwide survey. *Journal of Clinical Practice* 59: 291–295.

Ware, M.A., Ducruet, T., and Robinson, A.R. 2006. Evaluation of herbal cannabis characteristics by medical users: A randomized trial. *Harm Reduction Journal* 3: 32. doi:10.1186/1477-7517-3-32.

Ware, M.A., Fitzcharles, M.-A., Joseph, L., and Shir, Y. 2010a. The effects of nabilone on sleep in fibromyalgia: Results of a randomized controlled trial. *Anesthesia & Analgesia* 110: 604–610.

Ware, M.A., Wang, T., Shapiro, S., Robinson, A., Ducruet, T., Huynh, T. et al. 2010b. Smoked cannabis for chronic neuropathic pain: A randomized controlled trial. *CMAJ (Canadian Medical Association Journal)* 182: E694–E701.

Wargent, E.T., Zaibi, M.S., Silvestri, C., Hislop, D.C., Stocker, C.J., Stott, C.G. et al. 2013. The cannabinoid Δ^9-tetrahydrocannabivarin (THCV) ameliorates insulin sensitivity in two mouse models of obesity. *Nutrition & Diabetes* (2013) 3, e68. doi:10.1038/nutd.2013.9.

Warmke, H.E., and Blakeslee, A.F. 1939. Effect of polyploidy upon the sex mechanism in dioecious plants [abstract]. *Genetics* 24: 88–89.

Watanabe, K., Itokawa, Y., Yamaori, S., Funahashi, T., Kimura, T., Kaji, T. et al. 2007. Conversion of cannabidiol to Δ^9-tetrahydrocannabinol and related cannabinoids in artificial gastric juice, and their pharmacological effects in mice. *Forensic Toxicology* 25: 16–21.

Watson, D.P., and Clarke, R.C. 1997. Genetic future of hemp. *Journal of the International Hemp Association* 4(1): 32–36.

Weber, J.B. 1978. Today's weed—Marihuana. *Weeds Today* 9(2): 21–22.

Weber, M., Goldman, B., and Truniger, S. 2010. Tetrahydrocannabinol (THC) for cramps in amyotrophic lateral sclerosis: A randomised, double-blind crossover trial. *Journal of Neurology Neurosurgery and Psychiatry* 81: 1135–1140.

Webster, C.D., Thompson, K.R., Morgan, A.M., Grisby, E.J., and Gannon, A.L. 2000. Use of hemp seed meal, poultry by-product meal, and canola meal in practical diets without fish meal for sunshine bass (*Morone chrysops × M. saxatilis*). *Aquaculture* 188: 299–309.

Webster, G.R.B., Sarna, L.P., and Mechoulam, R. 2008. *Converting cannabidiol to delta 8 or delta 9-tetrahydrocannabinol by mixing the diol with a catalyst and solvent and allowing the mixture to separate, removing the organic phase; and eluting the tetrahydrocannabinol from the organic phase.* US patent 7399872 B2. http://www.google.com/patents/US7399872.

Weiblen, G.D., Wenger, J.P., Craft, K.J., ElSohly, M.A., Mehmedic, Z., Treiber, E.L. et al. 2015. Gene duplication and divergence affecting drug content in *Cannabis sativa*. *New Phytologist* 208: 1241–1250.

Weightman, R., and Kindred, D. 2005. *Review and analysis of breeding and regulation of hemp and flax varieties available for growing in the UK: Final report for the Department for Environment Food and Rural Affairs.* 77 pp. http://www.grfa.org.uk.

Weil, A.T., Zinberg, N.E., and Nelsen, J.M. 1968. Clinical and psychological effects of marihuana in man. *Science* 162: 1234–1242.

Weir, T.A. 2001. Photoperiodism in humans and other primates: Evidence and implications. *Journal of Biological Rhythms* 16: 348–364.

Weiss, L., Zeira, M., Reich, S., Slavin, S., Raz, I., Mechoulam, R. et al. 2008. Cannabidiol arrests onset of autoimmune diabetes in NOD mice. *Neuropharmacology* 54: 244–249.

Welty, T.E., Luebke, A., and Gidal, B.E. 2014. Cannabidiol: Promise and pitfalls. *Epilepsy Currents* 14: 250–252.

Wendel, M., and Heller, A.R. 2009. Anticancer actions of omega-3 fatty acids—Current state and future perspectives. *Anticancer Agents in Medicinal Chemistry* 9: 457–470.

Werb, D., Nosyk, B., Kerr, T., Fischer, B., Montaner, J., and Wood, E. 2012. Estimating the economic value of British Columbia's domestic cannabis market: Implications for provincial cannabis policy. *International Journal of Drug Policy* 23: 436–441.

Werker, E. 2000. Trichome diversity and development. In: Hallahan, D.L., Gray, J.C., and Callow, J.A. eds. *Advances in botanical research incorporating advances in plant pathology 31: plant trichomes.* London: Academic Press, 1–35.

Werner, C.A. 2001. Medical marijuana and the AIDS crisis. *Journal of Cannabis Therapeutics* 1(3/4): 17–33.

Werner, E.T.C. 1922. *Myths and legends of China.* London: George G. Harrap. 453 pp.

West, M. 1997. The use of certain cannabis derivatives (Canasol) in glaucoma. In: Mathre, M.L. ed. *Cannabis in medical practice: A legal, historical and pharmacological overview of the therapeutic use of marijuana.* Jefferson, NC: McFarland, 103–111.

West-Eberhard, M.J. 2003. *Developmental plasticity and evolution.* New York: Oxford University Press. 816 pp.

Westerhuis, W., Struik, P.C., van Dam, J.E.G., and Stomph, T.J. 2009. Postponed sowing does not alter the fibre/wood ratio or fibre extractability of fibre hemp (*Cannabis sativa*). *Annals of Applied Biology* 155: 333–348.

Westfall, R.E., Janssen, P.A., Lucas, P., and Capler, R. 2006. Survey of medicinal cannabis use among childbearing women: Patterns of its use in pregnancy and retroactive self-assessment of its efficacy against 'morning sickness.' *Complementary Therapies in Clinical Practice* 12: 27–33.

Whalley, B.J., Wilkinson, J.D., Williamson, E.M., and Constanti, A. 2004. A novel component of cannabis extracts potentiates excitatory synaptic transmission in rat olfactory cortex in vitro. *Neuroscience Letters* 365: 58–63.

Whitehill, J.M., Rivara, F.P., and Moreno, M.A. 2014. Marijuana-using drivers, alcohol-using drivers and their passengers: Prevalence and risk factors among underage college students. *JAMA Pediatrics* 168: 618–624.

Whiting, P.F., Wolff, R.F., Deshpande, S., Di Nisio, M., Duffy, S., Hernandez, A.V. et al. 2015. Cannabinoids for medical use. A systematic review and meta-analysis. *JAMA* 313: 2456–2473.

Whitman, T., and Aarssen, L.W. 2010. The leaf size/number trade-off in herbaceous angiosperms. *Journal of Plant Ecology* 3: 49–58.

Whittle, B.A., and Guy, G.W. 2004. Development of cannabis-based medicines: Risk, benefit and serendipity. In: Guy, G.W., Whittle, B.A., and Robson, P.J. eds. *The medicinal uses of* Cannabis *and cannabinoids*. London: Pharmaceutical Press, 427–466.

Wichtl, M. ed. 2004. *Herbal drugs and phytopharmaceuticals: A handbook for practice on a scientific basis*. 3rd edition. Boca Raton, FL: CRC Press, 708 pp.

Wilkinson, J.D., and Williamson, E.N. 2007. Cannabinoids inhibit human keratinocyte proliferation through a non-CB_1/CB_2 mechanism and have a potential therapeutic value in the treatment of psoriasis. *Journal of Dermatological Science* 45: 87–92.

Wilkinson, J.D., Whalley, B.J., Baker, D., Pryce, G., Constanti, A., Gibbons, S. et al. 2003. Medicinal cannabis: Is Δ^9-tetrahydrocannabinol necessary for all its effects? *Journal of Pharmacy and Pharmacology* 55: 1687–1694.

Wilkinson, S.T., and D'Souza, D.C. 2014. Problems with the medicalization of marijuana. *JAMA* 311: 2377–2378.

Williams, C., Thompstone, J., and Wilkinson, M. 2008. Work-related contact urticaria to *Cannabis sativa*. *Contact Dermatitis* 58: 62–63.

Williams, C.M., Jones, N.A., and Whalley, B.J. 2014. Cannabis and epilepsy. In: Pertwee, R.G. ed. *Handbook of cannabis*. Oxford, U.K.: Oxford University Press, 547–563.

Williams, C.M., Whalley, B.J., and McCabe, C. 2015. Cannabinoids and appetite (dys)regulation. In: Fattore, L. ed. *Cannabinoids in neurologic and mental disease*. New York: Academic Press, 315–339.

Williams-Garcia, R. 1975. The ritual use of cannabis in Mexico. In: Rubin, V. ed. *Cannabis and culture*. The Hague, the Netherlands: Mouton, 133–145.

Wills, S. 1998. Cannabis use and abuse by man: An historical perspective. In: Brown, D.T. ed. *Cannabis. The genus* Cannabis. Australia (and other countries): Harwood Academic Publishers, 1–27.

Wilsey, B., Marcotte, T., Deutsch, R., Sakai, S., and Donaghe, H. 2013. Low-dose vaporized cannabis significantly improves neuropathic pain. *Journal of Pain* 14: 136–148.

Wilsie, C.P., Black, C.A., and Aandahl, A.R. 1944. Hemp production experiments, cultural practices and soil requirements. *Iowa Agricultural Experiment Station Bulletin P63*. Ames, IA: Iowa State College. 46 pp.

Wilsie, C.P., Dyas, E.S., and Norman, A.G. 1942. Hemp, a war crop for Iowa. *Iowa Agricultural Experiment Station Bulletin P49*. Ames, IA: Iowa State College. 13 pp.

Winston, M.E., Hampton-Marcell, J., Zarraonaindia, I., Owens, S.H., Moreau, C.S., Gilbert, J.A. et al. 2014. Understanding cultivar-specificity and soil determinants of the *Cannabis* microbiome. *PLoS ONE* 9(9): e107415. doi:10.1371/journal.pone.0107415.

WIPO. 2010. *Learn from the past, create the future: Inventions and patents*. Geneva, Switzerland: World Intellectual Property Organization. 67 pp.

Wirtshafter, D. 1995. Nutrition of hemp seeds and hemp seed oil? In: Nova Institute, corporate ed. *Bioresource hemp—Proceedings of the symposium* (Frankfurt am Main, Germany, March 2–5, 1995). 2nd edition. Ojai, CA: Hemptech, 546–555.

Wirtshafter, D., and Krawitz, M. 2005. Hemp stamps bring record prices. *Journal of Industrial Hemp* 10(2): 67–74.

Wiskerske, J., and Pattij, T. 2015. The cannabinoid system and impulsive behavior. In: Fattore, L. ed. *Cannabinoids in neurologic and mental disease*. New York: Academic Press, 343–364.

Wolff, V., Armspach, J.P., Lauer, V., Rouyer, O., Bataillard, M., Marescaux, C. et al. 2013. Cannabis-related stroke: Myth or reality? *Stroke* 44: 558–563.

Wolkowicz, A.H. 2012. *Edible marijuana: A new frontier in the culinary world*. Thesis, Providence, RI: Johnson & Wales University. 114 pp.

Wong, A. 1998. Using crop residues to save forests. *Global Biodiversity* 7(4): 7–11.

Woodhams, S.G., Sagar, D.R., Burston, J.J., and Chapman, V. 2015. The role of the endocannabinoid system in pain. In: Schaible, H.-G. ed. *Pain Control. Handbook of Experimental Pharmacology 227*. Heidelberg, Germany: Springer-Verlag, 119–143.

Woodland Publishing. 2005. *Healthy recipes: Delicious recipes for using hemp foods*. Orem, UT: Woodland Publishing. 38 pp.

World Commission on Environment and Development. 1987. *Our common future (Brundtland Report)*. Oxford, UK: Oxford University Press. 400 pp.

World Health Organization. 2014. Violence and injury prevention: Road traffic injuries. http://www.who.int /violence_injury_prevention/en/.

World Obesity Federation. 2015. Obesity the global epidemic. http://www.iaso.org/iotf/obesity/obesity theglobalepidemic.

Wright, S., Duncombe, P., and Altman, D.G. 2012. Assessment of blinding to treatment allocation in studies of a cannabis-based medicine (Sativex®) in people with multiple sclerosis: A new approach. *Trials* 13: 189. doi:10.1186/1745-6215-13-189.

Wu, Z., Zhou, Z.-K., and Bartholomew, B. 2003. Cannabaceae Endlicher. In: Zheng-yi, W., and Raven, P.H. eds. *Flora of China, Volume 5*. St. Louis, MO: Missouri Botanical Garden Press, 74–75.

Wynder, E.L., Higgins, I.T., and Harris, R.E. 1990. The wish bias. *Journal of Clinical Epidemiology* 43: 619–621.

Wyrofsky, R., McGonigle, P., and Van Bockstaele, E.J. 2015. Drug discovery strategies that focus on the endo-cannabinoid signaling system in psychiatric disease. *Expert Opinion on Drug Discovery* 10: 17–36.

Xu, H., and Azuaro-Blanco, A. 2014. Phytocannabinoids in degenerative and inflammatory retinal diseases: Glaucoma, age-related macular degeneration, diabetic retinopathy and uveoretinitis. In: Pertwee, R.G. ed. *Handbook of cannabis*. Oxford, U.K.: Oxford University Press, 601–618.

Yamada, I. 1943. The sex-chromosomes of *Cannabis sativa* L. *Seiken Ziho [Report of the Kihara Institute for Biological Research]* 2: 64–68.

Yampolsky, C., and Yampolsky, H. 1922. *Distribution of sex forms in the phanerogamic flora*. Leipzig, Germany: Gebruder Borntraeger. 62 pp. + table.

Yang, L., Rozenfeld, R., Wub, D., Devi, L.A., Zhang, Z., and Cederbaum, A. 2014. Cannabidiol protects liver from binge alcohol-induced steatosis by mechanisms including inhibition of oxidative stress and increase in autophagy. *Free Radical Biology and Medicine* 68: 260–267.

Yang, M.-Q., van Velzen, R., Bakker, F.T., Sattarian, A., Li, D.-Z., and Yi, T.-S. 2013. Molecular phylogenetics and character evolution of Cannabaceae. *Taxon* 62: 473–485.

Yazulla, S. 2008. Endocannabinoids in the retina: From marijuana to neuroprotection. *Progress in Retinal and Eye Research* 27: 501–526.

Yeh, N., and Chung, J.-P. 2009. High-brightness LEDs—Energy efficient lighting sources and their potential in indoor plant cultivation. *Renewable and Sustainable Energy Reviews* 13: 2175–2180.

Yousofi, M., Saberivand, A., Becker, L.A., and Karimi, I. 2011. The effects of *Cannabis sativa* L. seed (hemp seed) on reproductive and neurobehavioral end points in rats. *Developmental Psychobiology* 53: 402–412.

Yu, L.L., Zhou, K.K., and Parry, J. 2005. Antioxidant properties of cold pressed black caraway, carrot, cranberry and hemp seed oils. *Food Chemistry* 91: 723–729.

Zacny, J.P., and Chait, L.D. 1989. Breathhold duration and response to marijuana smoke. *Pharmacology, Biochemistry, and Behavior* 33: 481–484.

Zacny, J.P., and Chait, L.D. 1991. Response to marijuana as a function of potency and breathhold duration. *Psychopharmacology* 103: 223–226.

Zajicek, J.P., and Apostu, V.I. 2011. Role of cannabinoids in multiple sclerosis. *CNS Drugs* 25: 187–201.

Zammit, S., Moore, T.H., Lingford-Hughes, A., Barnes, T.R., Jones, P.B., Burke, M. et al. 2008. Effects of cannabis use on outcomes of psychotic disorders: Systematic review. *British Journal of Psychiatry* 193: 357–363.

Zander, A. 1928. Über Verlauf und Entstehung der Milchröhren des Hanfes (*Cannabis sativa*). *Flora* 23: 191–218.

Zatta, A., Monti, A., and Venturi, G. 2012. Eighty years of studies on industrial hemp in the Po Valley (1930–2010). *Journal of Natural Fibers* 9: 180–196.

Zawilska, J.B., and Wojcieszak, J. 2014. Spice/K2 drugs—More than innocent substitutes for marijuana. *International Journal of Neuropsychopharmacology* 17: 509–525.

Zeisser, C., Thompson, K., Stockwell, T., Duff, C., Chow, C., Vallance, K. et al. 2012. A 'standard joint'? The role of quantity in predicting cannabis-related problems. *Addiction Research and Theory* 20: 82–92.

Zhang, L.L., Zhu, R.Y., Chen, J.Y., Chen, J.Y., and Feng, X.X. 2008. Seawater-retting treatment of hemp and characterization of bacterial strains involved in the retting process. *Process Biochemistry* 43: 1195–1201.

Zhang, L.R., Morgenstern, H., Greenland, S., Chang, S.-H., Lazarus, P., Teare, M.D. et al. 2015. Cannabis smoking and lung cancer risk: Pooled analysis in the International Lung Cancer Consortium. *International Journal of Cancer* 136: 894–903.

Zhatov, A.I. 1983. Variability of pollen grains of polyploid hemp. *Tsitologiya i Genetika* 17: 47–51. (In Russian; translation available from Allerton Press Inc.)

Zhatov, A.I., Migal, N.D., and Kovalenko, V.M. 1969. Cytological study of the polyploidy of hemp. *Tsitologiya i Genetika* 3: 28–35.

Zirpel, B. Stehle, F., and Kayser, O. 2015. Production of Δ^9-tetrahydrocannabinolic acid from cannabigerolic acid by whole cells of *Pichia* (*Komagataella*) *pastoris* expressing Δ^9-tetrahydrocannabinolic acid synthase from *Cannabis sativa* L. *Biotechnology Letters* 37: 1869–1875.

Zlas, J., Stark, H., Seligman, J., Levy, R., Werker, E., Breuer, A. et al. 1993. Early medical use of cannabis. *Nature* 363: 215.

Zogopoulos, P. 2015. Cancer therapy—The role of cannabinoids and endocannabinoids. *Cancer Cell & Microenvironment* 2(1). doi:10.14800/ccm.583.

Zogopoulos, P., Vretakos, G., and Rologis, D. 2015. Therapeutic implications of the endocannabinoid system in neurodegenerative diseases. *Clinical Case Reports and Reviews* 1(4): 73–79.

Zohary, D. 2004. Unconscious selection and the evolution of domesticated plants. *Economic Botany* 58: 5–10.

Zorn, J., and Oskamp, D.L. 1796. *Afbeeldingen der artseny-gewassen met derzelver Nederduitsche en Latynsche beschryvingen. Vol. 1*. Amsterdam, the Netherlands: J.C. Seep en Zoon.

Zuardi, A.W. 2006. History of cannabis as a medicine: A review. *Revista Brasilera de Psiquiatria* 28: 153–157.

Zuardi, A.W. 2008. Cannabidiol: From an inactive cannabinoid to a drug with wide spectrum of action. *Revista Brasileira de Psiquiatria* 30: 271–280.

Zuardi, A.W., Crippa, J.A.S., Hallak, J.E.C., Moreira, F.A., and Guimarães, F.S. 2006. Cannabidiol, a *Cannabis sativa* constituent, as an antipsychotic drug. *Brazilian Journal of Medical and Biological Research* 39: 421–429.

Zuardi, A.W., Hallak, J.E.C., and Crippa, J.A.S. 2012. Interaction between cannabidiol (CBC) and Δ^9-tetrahydrocannabinol (THC): Influence of administration interval and dose ratio between the cannabinoids. *Psychopharmacology* 219: 247–249.

Zuck, D., Ellis, P., and Dronsfield, A. 2012. Nitrous oxide: Are you having a laugh? *Education in Chemistry* 2012(March): 26–29.

Zuurman, L., Ippel, A.E., Moin, E., and van Gerven, J.M. 2009. Biomarkers for the effects of cannabis and THC in healthy volunteers. *British Journal of Clinical Pharmacology* 67: 5–21.

Zvolensky, M.J., Bonn-Miller, M.O., Leyro, T.M., Johnson, K.A., and Bernstein, A. 2011. Marijuana: An overview of the empirical literature. In: Johnson, B.A. ed. *Addiction medicine*. New York: Springer, 445–461.

Zwenger, S.R. 2009. Bogarting that joint might decrease oral HPV among cannabis users. *Current Oncology* 16(6): 5–7.

Index

Page numbers followed by f and t indicate figures and tables, respectively.

541